THE RAE TABLE OF EARTH SATELLITES
1957–1982

THE RAE TABLE OF EARTH SATELLITES 1957–1982

compiled at

The Royal Aircraft Establishment, Farnborough, Hants, England

by

D.G. King-Hele, FRS, J.A. Pilkington, BSc, D.M.C. Walker, PhD,
H. Hiller, BSc and A.N. Winterbottom

The Table is a chronological list of the 2389 launches of satellites and space vehicles between 1957 and the end of 1982, giving the name and international designation of each satellite and its associated rocket(s), with the date of launch, lifetime (actual or estimated), mass, shape, dimensions and at least one set of orbital parameters. Other fragments associated with a launch, and space vehicles that escape from the Earth's influence, are given without details. Including fragments, more than 13000 satellites appear in the 718 pages of the tabulation, and there is a full Index.

M

MACMILLAN PUBLISHERS
LONDON

First published 1981, second edition 1983 by
MACMILLAN PUBLISHERS LTD (Journals Division).
Distributed by Globe Book Services Ltd
Canada Road, Byfleet, Surrey KT14 7JL, England.

British Library Cataloguing in Publication Data

The RAE table of earth satellites, 1957–1982.
– 2nd edition
1. Artificial satellites – Tables
I. King-Hele D.G.
629.47 TL796

ISBN 0-333-35374-9

Printed in Hong Kong.

INTRODUCTION

HOW THE TABLE BEGAN AND GREW

When the first satellite Sputnik 1 was launched on 4 October 1957, scientists at the Royal Aircraft Establishment, Farnborough, had already made several studies of Earth satellites and their orbits, stemming from work earlier in the 1950s on the ballistic missile Blue Streak and the Skylark research rocket. Within a few days of its launching, Sputnik 1 was being regularly tracked by a radio interferometer constructed at the RAE's outstation at Lasham, Hampshire. The satellite's orbit was determined from these observations, and the observed decay rate was used to evaluate upper-atmosphere density. The work was described in an article published in *Nature* on 9 November 1957 (Volume 180, pages 937-941).

On 3 November 1957, Sputnik 2 was launched, and the need for a regular prediction service was recognized. Initially the service was provided by the Royal Greenwich Observatory, Herstmonceux, and was taken over by the RAE in January 1958. The first US satellite Explorer 1 was launched on 1 February 1958, to be followed by Vanguard 1 and Explorer 3 during March, and Sputnik 3 in May. Soon there were numerous requests for a list of satellites, and Doreen Walker, who was responsible for providing the predictions, compiled the first RAE Table of satellites - a single sheet - in July 1958. From these small beginnings the Table has 'just growed', the original format being retained almost unchanged, apart from conversion to metric units. Very few copies of this 'first edition' still exist, so a facsimile of the original sheet, slightly reduced in size, is printed on page ii.

From the beginning it was apparent that there would be little information available on the sizes, shapes and masses of the many Russian rockets in orbit, and the decision was taken to make rough estimates of the size and shape from visual observations, and then to deduce the mass from the observed orbital decay rate and the (by then) known upper-atmosphere density. This policy has been pursued ever since, and the estimates have been improved over the years as more information became available.

Since 1957 the RAE has specialized in the analysis of satellite orbits to determine upper-atmosphere density and winds, and the Earth's gravitational field.

TABLE OF ARTIFICIAL SATELLITES

Name		Launch date and Lifetime	Shape and Weight	Size	Date	Orbital Inclination (deg.)	Orbital Period (min.)	Perigee Height (n.m.)	Apogee Height (n.m.)	Orbital Eccentricity	Angle from Apex to Perigee (deg.)
Sputnik 1 instrumented sphere	1957α2	1957 Oct. 4.90 ? 92 days	Sphere 184 lb.	23" dia.	1957 Oct. 4.90	65	96.2	122	512	0.052	- 39
					1957 Oct. 25.8	65	95.4				
Sputnik 1 rocket	1957α1	1957 Oct. 4.90 57.1 days	Cylinder? -	-	1957 Oct. 4.90	65	96.2	122	512	0.052	- 39
					1957 Nov. 19.00	65	92.0				
Sputnik 2	1957β	1957 Nov. 3.19 161.9 days	- -	-	1957 Nov. 4.00	65.33	103.760	122	902	0.0987	- 31
					1958 Jan. 4.00	65.29	100.505	119	739	0.0802	- 55
					1958 Feb. 21.00	65.26	97.105	114	570	0.0605	- 76
					1958 Mar. 25.00	65.23	93.785	107	402	0.0400	- 91
					1958 Apr. 9.00	65.21	90.780	97	253	0.0214	- 98
Explorer 1	1958α	1958 Feb. 1.16 4 years	Cylinder 30.8 lb.	80" long 6" dia.	1958 Feb. 1.16	33.2	114.8	199	1371	0.139	31
					1958 July 10.05	33.2	113.5	192	1318	0.134	33
Vanguard 1 instrumented sphere	1958β2	1958 Mar. 17.5 200 years?	Sphere $3\frac{1}{4}$ lb.	6.4" dia.	1958 Mar. 17.5	34.3	134.1	353	2140	0.191	-
					1958 June 19.52	34.3	134.1	353	2136	0.190	92
Vanguard 1 rocket	1958β1	1958 Mar. 17.5 -	Cylinder 50 lb.	4' long 20" dia.	1958 Mar. 17.5	34.3	134.1	353	2140	0.191	-
Explorer 3	1958γ	1958 Mar. 26.73 94 days	Cylinder 31 lb.	80" long 6" dia.	1958 Mar. 26.73	33.3	115.7	101	1511	0.166	-
					1958 Apr. 9.05	33.3	110.4	100	1251	0.140	-
					1958 June 14.13	33.3	96.6	93	565	0.063	-124
Sputnik 3 instrumented cone	1958δ2	1958 May 15.3 2 years	Cone 2926 lb.	12.3' long 68" dia.	1958 May 15.3	65	105.985	122	1013	0.111	- 32
					1958 June 5.7	65	105.700	122	1000	0.110	- 39
					1958 July 9.2	65	105.300	122	979	0.107	- 50
Sputnik 3 rocket	1958δ1	1958 May 15.3 7 months	Cylinder? -	- -	1958 May 15.3	65	105.985	122	1013	0.111	- 32
					1958 June 8.1	65	105.000	121	964	0.106	- 40
					1958 July 2.2	65	104.000	121	914	0.100	- 48

Notes: Oct. 4.90 means 21 hr. 36 min. G.M.T. on 4 Oct., 1 n.m. = 6080 ft. Perigee and apogee heights for Sputniks are over an earth of radius 3435 n.m. The values for the Sputniks are from observations and theory. Those for the U.S. satellites have been compiled from a variety of sources, and there may be inconsistencies.

Facsimile of the original issue of the Table

This work depends on choosing satellites for observation, determining the orbits from the observations, and then analysing the orbits. In order to choose suitable satellites, a listing like that in the Table is needed, including reasonably accurate estimates of the satellite lifetimes. These lifetime estimates are vital, because it is no good selecting a satellite for long-term studies of the gravitational field, only to find that it decays within two years. Conversely, it is no good selecting a satellite for studies of atmospheric winds if no useful results can be obtained for thirty years or more.

The estimation of lifetime has proved to be the most creative and difficult aspect of the Table. The orbits of most satellites are appreciably affected by the drag of the upper atmosphere, which makes the orbit contract and eventually brings the satellite to a fiery end in a plunge into the lower atmosphere. The lifetime is controlled by the upper-atmosphere density: if the density doubles, the lifetime will be halved. In fact, the density at a height of 500 km can be more than ten times greater at the maximum of the eleven-year sunspot cycle than at the minimum; and predictions of the intensity and timing of future sunspot maxima are notoriously unreliable. So it is very difficult to make good estimates for lifetimes greater than 5 years. On a shorter timescale, problems arise from the day-to-night variation in density, by a factor of up to 6; from the semi-annual variation, by a factor of up to 3; and from irregular day-to-day variations. Also there are some satellites in highly eccentric orbits for which the lifetime is governed by the gravitational attraction of the Sun and Moon; lengthy computations are then needed, extending over the life of the satellite, perhaps 10 or 20 years. Balloon satellites form another unusual group, with their lifetime controlled by the radiation pressure of sunlight.

Though the need for the Table of satellites arose from the work on orbit analysis, the Table was so appreciatively received, not only by individuals but also in official US and USSR publications, that we decided to continue distributing it to qualified recipients. By 1980, however, the printing and distribution had become expensive and burdensome, and the choice of recipients had become rather arbitrary (and irritating to those refused). So RAE welcomed the offer of Macmillan Press to publish the Table for the years 1957-1980. A new edition is now required, so the Table has been fully updated to the end of 1982.

One difficult decision had to be taken: should the Table be printed from the existing masters, suitably revised? Or should it be completely reset in type? Reprinting from the existing masters was preferred because printing would be very much quicker and errors would be minimized, the pages having already been open to critical scrutiny in their unamended form for years. In other words, we preferred to have a volume which is 99.9% readable, rather than to wait twice as long for a more elegant version with a greater number of errors.

Although the existing masters have been used here, they have been fully updated and corrected, and retyped whenever necessary; thousands of amendments have been made to the Tables as originally issued. The present publication supersedes all previous issues and will serve as the master copy for any future amendments.

The Table has grown to its present size only after much hard work, and the contributions of the compilers and others to the work over the years have been as follows. Desmond King-Hele has had general responsibility continuously from the outset. The detailed work was done by Doreen Walker between 1958 and 1961; from then until 1968 the main contributions to the detailed work were from Janice Rees, Alan Pilkington and Eileen Quinn. Since 1968 the Table has been issued monthly: the task of amassing data on new launches was done by Harry Hiller for 1968-80 and by Alan Winterbottom thereafter; Alan Pilkington, using this information, has produced the draft of each monthly issue; Doreen Walker has been responsible for the lifetime predictions, and has also undertaken the exacting editorial and organizational work of updating and preparing the present volume for publication.

THE NUMBER OF LAUNCHES

In 1957 there were two satellite launches, by the USSR; in 1958 there were eight, of which seven were by the USA and one by the USSR. From then until 1967 the total numbers of launchings of satellites and space vehicles increased every year except 1963, and the yearly total for 1967 was 127 launches. Between 1967 and 1982 the world has changed greatly and most activities have either strongly increased or seriously declined; but the annual space launchings have, strangely enough, remained almost constant. The maximum number of launches during this period was 128 in 1976 and the minimum was 105 in 1980. The average number of launches in the years 1967-1982 was about 117 (actually 116.7, if you can visualise .7 of a launch!). The total number of launches for 1957-1982 was 2389.

The diagram opposite shows the yearly numbers of launches of satellites and space vehicles in histogram form, with division into launches by the USSR (dark stippling), the USA (light stippling) and 'others' (large dots). The 'others' include launches by other countries and 'joint' launches, for example those for international organizations like Intelsat.

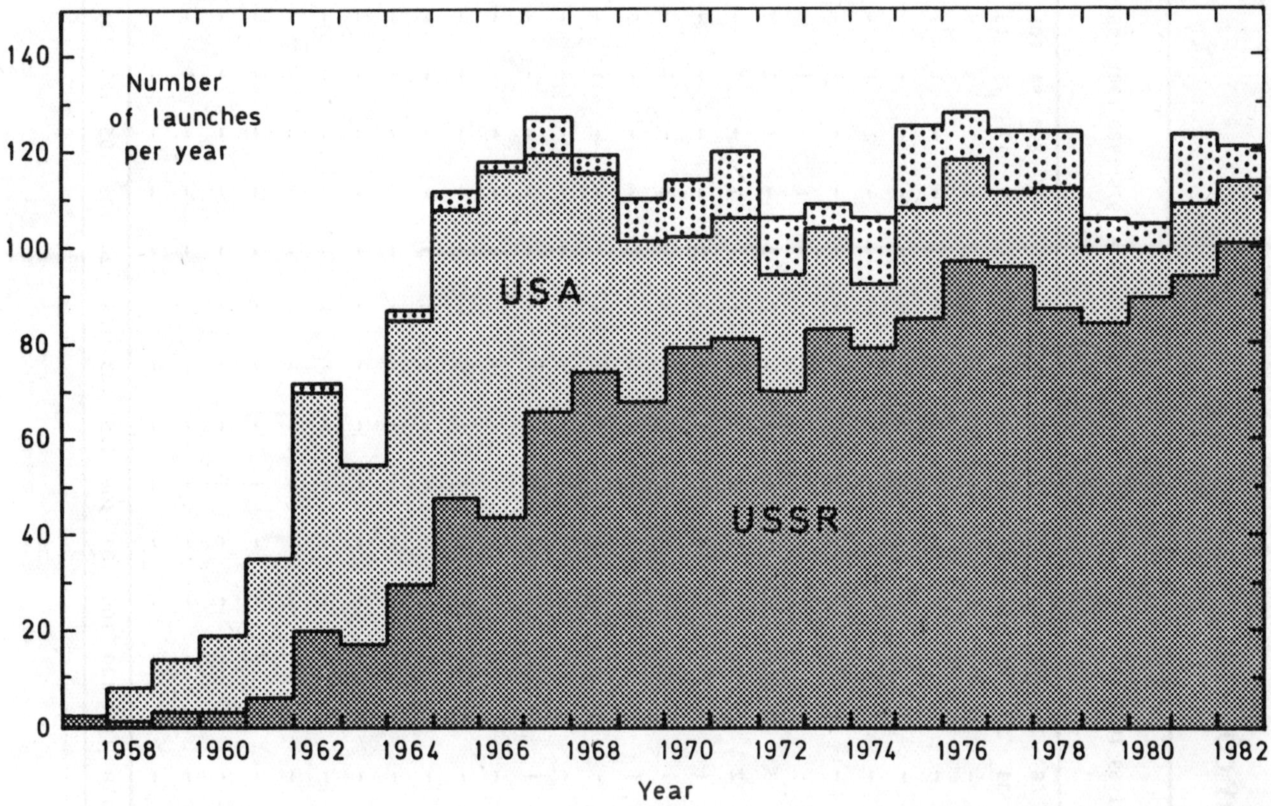

The diagram shows that the USA had more launches than the USSR in each year between 1958 and 1966, but the situation reversed in 1967. During the 1970s the preponderance of USSR launches continued, and increased, until by the early 1980s the proportions were approximately: 81% USSR; 12% USA; and 7% others.

After the USSR and USA, the next country to launch a satellite with a home-made rocket was France in 1965. Five years later, Japan became the fourth space-launching country, soon followed by China (1970) and the UK (1971). There was then another long interval before the appearance of the seventh country, India, in 1980.

The 2389 launches are tabulated year by year in 27 categories in the table occupying page vi. The first seven categories are national launches and the eighth is 'Europe' (the launchings of Ariane by the European Space Agency). Then come 19 'joint' categories, with the launching country first: the most numerous (with 29 launches) is USA/Intelsat, satellites launched by the USA for the Intelsat Corporation. Next, with 22, comes USSR/Intercosmos, satellites for the Intercosmos consortium, launched by the USSR.

NUMBERS OF LAUNCHES BY COUNTRY, 1957-1982

Country of origin \ Year of launch	1957	1958	1959	1960	1961	1962	1963	1964	1965	1966	1967	1968	1969	1970	1971	1972	1973	1974	1975	1976	1977	1978	1979	1980	1981	1982	Total
USSR	2	1	3	3	6	20	17	30	48	44	66	74	68	79	81	70	83	79	85	97	96	87	84	89	94	101	1507
USA	–	7	11	16	29	50	38	55	60	72	53	41	33	23	25	24	21	13	23	21	15	25	15	12	15	13	710
Japan	–	–	–	–	–	–	–	–	–	–	–	–	–	1	2	1	–	1	2	1	2	3	2	2	3	1	21
China	–	–	–	–	–	–	–	–	–	–	–	–	–	1	1	–	–	–	3	2	–	1	–	–	1	1	10
France	–	–	–	–	–	–	–	–	1	1	2	–	–	1	1	–	–	–	3	–	–	–	–	–	–	–	9
India	–	–	–	–	–	–	–	–	–	–	–	–	–	–	–	–	–	–	–	–	–	–	–	1	1	–	2
UK	–	–	–	–	–	–	–	–	–	–	–	–	–	–	1	–	–	–	–	–	–	–	–	–	–	–	1
Europe	–	–	–	–	–	–	–	–	–	–	–	–	–	–	–	–	–	–	–	–	–	–	1	–	1	–	2
USA/Intelsat	–	–	–	–	–	–	–	–	1	1	3	1	3	3	2	2	1	1	2	1	1	2	–	1	2	2	29
USSR/Intercosmos	–	–	–	–	–	–	–	–	–	–	–	–	2	2	1	3	2	2	2	2	1	1	2	–	2	–	22
USA/Europe	–	–	–	–	–	–	–	–	–	–	–	3	1	–	–	3	–	–	1	–	3	2	–	–	–	–	13
USA/UK	–	–	–	–	–	1	–	1	–	–	1	–	1	1	1	–	–	4	–	–	–	–	1	–	1	–	12
USA/Canada	–	–	–	–	–	1	–	–	1	–	–	–	1	–	1	1	1	–	1	1	–	1	–	–	–	2	11
USSR/France	–	–	–	–	–	–	–	–	–	–	–	–	–	–	1	1	1	–	1	–	1	–	–	–	1	–	6
USA/Italy	–	–	–	–	–	–	–	1	–	–	1	–	–	–	1	–	–	1	–	–	1	–	–	–	–	–	5
USA/FRG	–	–	–	–	–	–	–	–	–	–	–	–	1	–	–	1	–	2	–	1	–	–	–	–	–	–	5
USA/NATO	–	–	–	–	–	–	–	–	–	–	–	–	–	1	1	–	–	–	–	1	1	1	–	–	–	–	5
USA/France	–	–	–	–	–	–	–	–	1	–	–	–	–	–	1	–	–	1	1	–	–	–	–	–	–	–	4
USA/Japan	–	–	–	–	–	–	–	–	–	–	–	–	–	–	–	–	–	–	–	–	2	1	–	–	–	–	3
USSR/India	–	–	–	–	–	–	–	–	–	–	–	–	–	–	–	–	–	–	1	–	–	–	1	–	1	–	3
USA/Australia	–	–	–	–	–	–	–	–	–	–	1	–	–	1	–	–	–	–	–	–	–	–	–	–	–	–	2
USA/Indonesia	–	–	–	–	–	–	–	–	–	–	–	–	–	–	–	–	–	–	–	1	1	–	–	–	–	–	2
France/FRG	–	–	–	–	–	–	–	–	–	–	–	–	–	1	–	–	–	–	–	–	–	–	–	–	–	–	1
USA/Netherlands	–	–	–	–	–	–	–	–	–	–	–	–	–	–	–	–	–	1	–	–	–	–	–	–	–	–	1
USA/Spain	–	–	–	–	–	–	–	–	–	–	–	–	–	–	–	–	–	1	–	–	–	–	–	–	–	–	1
Europe/India	–	–	–	–	–	–	–	–	–	–	–	–	–	–	–	–	–	–	–	–	–	–	–	–	1	–	1
USA/India	–	–	–	–	–	–	–	–	–	–	–	–	–	–	–	–	–	–	–	–	–	–	–	–	–	1	1
Total	2	8	14	19	35	72	55	87	112	118	127	119	110	114	120	106	109	106	125	128	124	124	106	105	123	121	2389

THE PURPOSE OF THE LAUNCHINGS

Launching a satellite is an expensive exercise, and, on the assumption that human affairs still retain some relics of rationality, it is reasonable to ask what was the purpose of the launchings. The difficult task of assigning the launches to particular categories was tackled by the late Dr Charles Sheldon (in "United States and Soviet Progress in Space", US Congressional Research Service Report 81-27, 1981). The figures following are based on his findings.

He divides the launches into two main groups, military and civil. About 60% of the launchings have been primarily military, the proportion being greater for the USSR than for the USA. In this military group, more than half have been photographic-reconnaissance satellites, about 10% have been for communications, rather less than 10% for navigation, and the remainder for a variety of purposes, such as early warning of missile attack, ocean surveillance, electronic 'listening in', and tests of satellite interception.

In the civil group of launchings, about 40% have had scientific research as their aim: most of these satellites were designed to examine the Earth and its environment; other targets for research have been the planets, the Sun, the Moon or the stars. Nearly 20% of the civil group have been communications satellites, more than 10% weather satellites, and rather less than 10% manned satellites. Among the others have been many development satellites, for testing new instruments or engineering techniques, and a few satellites for the mapping of Earth resources.

NAMES AND DESIGNATIONS

Each launching organization likes to give each of its satellites a 'pet name' to be used by those working with the data from it. Many of these national names are familiar - Apollo, Ariel, Cosmos, Skylab, and so on - but some are weird acronyms like Spades (Solar Perturbation of Atmospheric Density Experiments Satellite); sometimes two agencies use the same name (Geos is an example); and some satellites remain nameless.

To bring all space launches into a single system, the International Committee on Space Research (known as COSPAR) has given all satellites and fragments an international designation based on the year of launch and the number of successful launches during the year. Thus the British satellite Prospero is designated

1971-93A, because it was launch 93 of the year 1971. Usually the letter A is given to the instrumented spacecraft, B is given to the rocket and C, D, E, ... to fragments, the letters I and O being omitted. Thus 1971-93B is the rocket that accompanied Prospero into orbit and 1971-93C is a fragment - an aerial that was knocked off during injection into orbit. If several spacecraft are sent into orbit in one launch, they are usually given the letters, A, B, C, ...: thus in the many eight-satellite launches by the USSR, the satellites are designated A-H and the rocket is J . When there are more than 24 pieces from one launch, as can happen after an explosion, the sequence continues after Z with AA, AB, AC, ... AZ, and then BA, BB, BC, ... BZ, and so on. The greatest number of fragments so far recorded from one launch is 462, resulting from the explosion of the satellite 1965-82A.

Besides reducing confusion, the international designation is useful because it gives not only the year but also the approximate month of launch. Since 1967 there have been approximately 10 launches per month, so the month of launch can be approximately estimated from the designation. Thus Prospero (1971-93A) would be assigned to approximately the 10th month of 1971, and it was in fact launched on 28 October 1971. Obviously there may be an error of one or two months because of the variation in the annual numbers of launches.

In the years 1957-1962 a different system was used, in terms of the 24 Greek letters. Thus the first launch of 1960 was 1960 alpha (1960 α), the second 1960 beta (1960 β), and so on. After 24 launches, double letters were used with $\alpha\alpha$, $\alpha\beta$, ... for launches 25, 26, ... The names of the Greek letters are listed on page xvi.

In the Table (pages 1-718) the national names are given first, followed by the international designations. The listing is chronological.

The index (pages 719-753) gives the national names, listed alphabetically, with the corresponding international designations and the appropriate page numbers. The full names of all satellites known by acronyms also appear in the index.

SATELLITE ORBITS

Readers who are unfamiliar with orbits may welcome an explanation of the terms used on each page of the Table to specify the size, shape and orientation of the orbits.

A satellite succeeds in entering orbit if its launching rocket can take it above the dense atmosphere and propel it nearly horizontally at a speed at least 8 km per second. In practice the satellite would be brought down quickly by air drag if its height was less than 150 km, so the 'dense atmosphere' can be regarded as extending to a height of about 150 km. If the launching rocket injects the satellite into orbit horizontally (at a high enough speed), the satellite will fly out to a greater height at the opposite side of the Earth, and will enter an elliptic orbit, as shown in the diagram. The point P on the ellipse where the satellite comes closest to the Earth is called the perigee and the point of maximum height is the apogee, A .

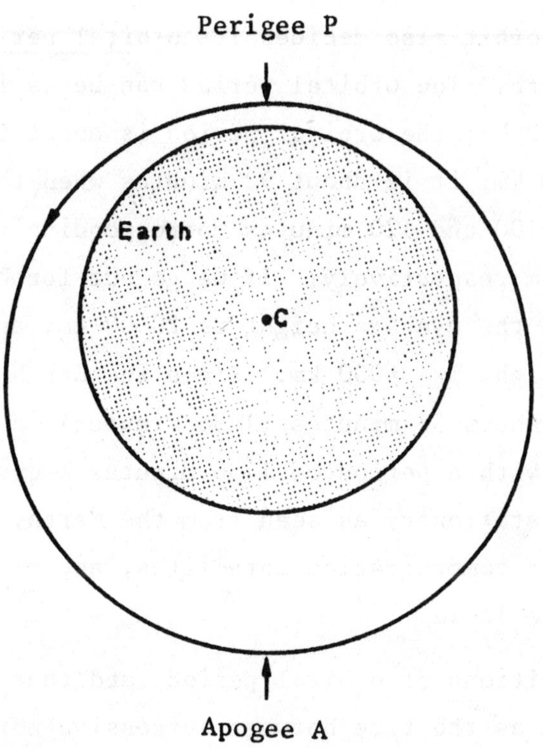

The height of a satellite's orbit is specified by its average distance from the Earth's centre, that is, half the sum of the apogee and perigee distances. This average distance, usually denoted by the symbol a , is called the semi major axis, the major axis being the distance AP between apogee and perigee, the longer diameter of the ellipse. For a circular orbit the semi major axis is of course the radius of the orbit.

The distance of perigee and apogee from the Earth's centre are not very useful quantities to tabulate: what we need to know is the height of perigee and apogee above the Earth's surface, and the Table gives perigee and apogee heights above a spherical Earth of radius 6378 km (the equatorial radius). In practice both the perigee distance and the Earth's radius vary with latitude, so the simple definition above is not exact, but it usually gives heights within 20 km of the instantaneous perigee and apogee heights and therefore provides a good general guide to the height of a satellite at perigee and apogee.

The shape of a satellite orbit is specified by its eccentricity, defined as the apogee height minus the perigee height, divided by the major axis. Thus a satellite with semi major axis 7500 km, perigee height 300 km and apogee height 1944 km, would have an eccentricity of 1644/15000 = 0.110. If the orbit is circular, the perigee and apogee heights are equal, and the eccentricity is zero. In practice, nearly circular orbits are convenient for many purposes, and a large proportion of actual orbits will be found to have eccentricities less than 0.01.

The major axis of a satellite's orbit also decides its orbital period, the time it takes to go once round the Earth. The orbital period can be as low as 88 minutes if the average height is 200 km; the orbital period is about 90 minutes when the average height is 300 km; it is about 92 minutes when the height is 400 km; and so on, with periods of 100 and 120 minutes corresponding to average heights of about 800 km and 1700 km respectively. If we go out further, an orbital period of 12 hours occurs when the average height is 20000 km; and the period is 24 hours when the average height is 36000 km. Since the Earth rotates once, relative to the stars, every 23 hours 56 minutes (1436 minutes), a satellite in an eastbound equatorial orbit with a period of 1436 minutes keeps pace with the Earth's rotation and appears stationary as seen from the Earth. This is the synchronous orbit much favoured for communication satellites, and many near-synchronous orbits will be found in the Table.

There are several possible definitions of orbital period, and that used in the Table is the nodal period, defined as the time between successive northward crossings of the Earth's equator by the satellite. (Another possible definition, not used here, is the time from one perigee to the next, the anomalistic period.)

As well as the size and shape of an orbit, we need to know whether it goes over the poles or stays near the equator. This is specified by the inclination of the orbit to the equator, that is, the angle i between the orbital plane and the plane of the Earth's equator as shown in the diagram opposite, where the track is assumed to be over a spherical 'Earth'. The point N , where the orbit crosses the equator going north, is called the ascending node, whence the name 'nodal period' previously defined.

The orbital inclination, given for each satellite in the Table, tells us the maximum latitude attained by a satellite as it travels round the world: the maximum latitude is equal to the orbital inclination, if we ignore small perturbations. Thus an inclination of 90° implies an orbit passing directly over the north and south poles on each revolution; an orbit of inclination 50° passes

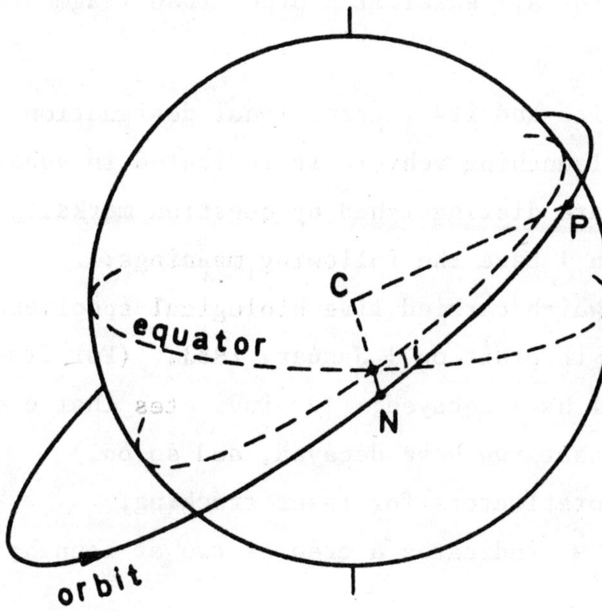

over all latitudes between 50°N and 50°S; while an orbit of inclination 10° is confined to latitudes less than 10°.

The next orbital parameter is the argument of perigee, which specifies the position of the perigee P relative to the equatorial plane. The argument of perigee is defined as the angular distance round the orbit between the ascending node N and the perigee P, and it is given by the angle $N\hat{C}P$ in the diagram, where C is the Earth's centre. The interpretation of the argument of perigee is fairly obvious from the diagram, but is listed below for reference.

Argument of perigee	Corresponding geographical position of perigee
0°	At equator going north (N in the diagram)
0°–90°	In northern hemisphere with satellite going north
90°	At maximum latitude north
90°–180°	In northern hemisphere with satellite going south
180°	At equator going south
180°–270°	In southern hemisphere with satellite going south
270°	At maximum latitude south
270°–360°	In southern hemisphere with satellite going north

One further orbital parameter, not given in the Table, is the longitude of the node; this is of less significance, though orbital specialists would have welcomed its inclusion.

GUIDE TO THE TABLE

The data given in the main Table, for all satellites other than fragments, are as follows.

Column 1 gives the name of the satellite and its international designation. If the name is unknown, the launching vehicle is indicated in square brackets. Doubtful entries are distinguished by question marks. Letters to the left of Column 1 have the following meanings:

B denotes unmanned satellites which carried live biological specimens.

D denotes satellites no longer in orbit on 1 January 1983. (For fragments, D indicates that all have decayed; 1d indicates that one has decayed; 2d indicates that two have decayed, and so on.)

L denotes satellites with retroreflectors for laser tracking.

M denotes manned satellites; 2M indicates a crew of two at launch; etc.

p indicates that pieces were picked up on Earth after re-entry.

R denotes satellites which returned to Earth and were recovered intact.

r denotes satellites carrying capsules which were successfully recovered.

T denotes satellites still transmitting radio signals on 1 January 1983.

Column 2 gives the launch date, lifetime (actual or estimated), and descent date (if appropriate). The dates are given in days and decimals of a day UT. Thus 1979 May 18.70 means "16h 48m UT (or GMT) on 18 May 1979". Actual lifetimes are given in days (and decimals of a day, if known). Estimated lifetimes are given in years, with decimals or fractions of a year as appropriate. Manoeuvrable satellites still in orbit on 1 January 1983 have 'manoeuvrable' in place of an estimated lifetime.

Column 3 gives the basic shape of the satellite and its mass in kilograms (1 kg = 2.205 lb). Sometimes the shape defies description in a few words and the description given is only approximate.

Column 4 gives the basic dimensions of the satellite in metres. Aerials, paddles carrying solar cells, and other components projecting from the main body are not normally taken into account when giving the size and shape (1 m = 3.281 ft).

Column 5 gives the date for the orbital information in Columns 6-12.

Column 6 gives the inclination of the orbit to the equator, in degrees.

Column 7 gives the nodal period of revolution - the time interval, in minutes, between successive northward equatorial crossings by the satellite.

Columns 8-11 specify the size and shape of the orbit. The quantities tabulated are the semi major axis a, in kilometres; the eccentricity e; and the perigee and apogee heights, $\{a(1 - e) - R\}$ and $\{a(1 + e) - R\}$ respectively, where R is the Earth's equatorial radius, 6378.1 km. (1 km = 0.6214 statute miles = 3281 ft = 0.5396 nautical miles.)

Column 12 gives the argument of perigee - the angle, measured round the orbit, from the northward equatorial crossing to the perigee.

The names of space vehicles (which have escaped from the dominance of the Earth's gravitational field) are given below the table, on the appropriate pages.

The index (pages 719-753) gives the names of the satellites in alphabetical order, with the international designation of each and the page on which details may be found. Satellites which are not Russian or American may be found in the index by referring to the appropriate country.

METHODS USED TO COMPILE THE TABLE

Our chief difficulty has been the lack of accurate information about the size, shape and weight of most of the satellites. The majority of launchings are military, and little information is released about these satellites or their final-stage rockets; we have to rely largely on deductions from their visual appearance in the night sky and on identifying previous launches of similar character. In contrast, we have full details of most international satellites and those launched by the US National Aeronautics and Space Administration (NASA).

Names and designations of satellites

The names given by the launching authorities are indicated when known. For unnamed United States Air Force satellites, the launch vehicle is given in square brackets: the lists issued by the United Nations have been useful in identifying the launch vehicles and orbits for these satellites.

When a name is abbreviated to initials only, the meaning of the acronym is given as a footnote for the first satellite of that name. The full names are included in the index.

The international designation of each satellite launching is allocated by the World Warning Agency on behalf of COSPAR. But the identification of particular components in a multiple launch has often depended on visual observations, since an experienced visual observer can usually distinguish between a satellite and its rocket, and may be able to recognize the species of rocket or satellite being observed. Inevitably, however, there is a possibility of confusion in identifying components in a multiple launch.

The techniques of visual observing are described in *Observing Earth Satellites* by D. King-Hele (Macmillan London, 1983).

'Fragments' may be defined loosely as 'non-functional components', or, more precisely, as any components left over after accounting for the instrumented satellite, the various rocket stages and any other major mechanical component (such as Molniya 'launchers'). Fragments thus include cast-off heat shields, de-spin weights and inter-stage structures, as well as debris from explosions - the most prolific source of fragments.

Lifetimes

The orbits of most satellites contract slowly under the action of air drag, and the severity of the drag determines their lifetimes, which can be estimated from the orbital decay rates (unless the satellites are later swept up as space rubbish, or suffer other major perturbations). The decay rate depends on air density, and the density depends critically on solar activity, which cannot be accurately predicted. So most lifetime estimates are likely to be in error by 10% or more, and lifetime estimates of over 5 years may have much greater errors: if solar activity in future cycles should decline to the low levels prevalent in the late 17th century, lifetimes of 20-50 years given here would be seriously underestimated.

For most satellites in high-eccentricity orbits, such as the Molniya satellites and rockets, the lifetimes depend primarily on lunisolar perturbations rather than air drag, and have been estimated by numerical integration of these perturbations.

The methods for predicting lifetimes are described in *Journal of the British Interplanetary Society*, Volume 31, pages 181-196 (1978).

Weights and dimensions

The weights and dimensions of the satellites come from Spacewarn launch telegrams, NASA Press Releases, and press and radio reports. Some indication of the accuracy is given by the number of significant figures. Often it is difficult to define the 'length' or 'diameter' when components of irregular size and shape are joined together, and dimensions are therefore sometimes approximate.

For satellites of unknown mass and size, the average cross-sectional area S can be approximately determined from the average brightness when observed visually; the mass/area ratio m/S can be obtained from the rate of change of orbital period and the known air density at heights near perigee, to give a value for the mass m. Many of our values for the dimensions of Russian rockets rely on the detailed studies by Dr Charles Sheldon, in *Soviet Space Programs 1971-75* (US Government Printing Office, Washington, 1976).

We hope that most of the weights and dimensions given with question marks are accurate to within a factor of 1.5, *ie* that the real values are between 2/3 and 3/2 times the value given. It seemed better to give some indication of the weights and sizes, even if approximate, rather than to leave blanks.

Orbital accuracy

Orbital information has come from many sources. Most of the orbits are based on the elements issued by the North American Air Defense Command (NORAD), and the remainder come mainly from NASA and RAE orbits.

The accuracy of the orbits varies greatly between one satellite and another, and no detailed guide can be given. Most orbits, however, are believed to have an error (sd) of about 0.02^o in orbital inclination, 0.02 min in period, 2 km in semi major axis, 5 km in perigee and apogee heights (when the apogee height is less than 2000 km), 0.001 in eccentricity e, and perhaps 3° in argument of perigee (if $e > 0.02$). Some orbits are much more accurate than this, and some, particularly those with eccentricity exceeding 0.3 or with very short lifetimes, may be much less accurate.

Radio transmissions

A satellite is given the symbol T if it transmits radio signals during its first days in orbit. The cessation of radio signals is rarely publicized, so the removal of the T is often based on the estimate that the *average* active life for radio transmission is about 2½ years for Soviet satellites and 7 to 8 years for US satellites. The most complete list of radio frequencies of satellites is in *Telecommunication Journal*, Volume 44, No.2 (1977), updated in subsequent monthly issues of *Telecommunication Journal*.

Locations of geosynchronous satellites are not given, because they are subject to drift. Current locations are available in a monthly list issued by NASA.

ACKNOWLEDGMENTS

In compiling the Table, our deepest debt is to the North American Air Defense Command: for over twenty years NORAD has (via NASA) supplied orbital elements on all satellites to the scientific community. We offer our grateful thanks to NORAD for this service, without which not only the Table but also much scientific work on orbit analysis would have been impracticable. Nearer home it is a pleasure to thank Geoffrey Perry of Kettering for supplying exact recovery times for hundreds of Cosmos satellites; and the Commanding Officer and staff of RAF, Fylingdales for their assistance in making available orbital elements. For help in identifying components in multiple launches, we are indebted to many visual observers, particularly Russell Eberst of the Royal Observatory Edinburgh, and Pierre Neirinck formerly of Appleton Laboratory, Slough. We are grateful to them and to the many people who have written with suggestions for improvements. At RAE, we thank the Deputy Director (E) and the Head of Space Department for their continued support; our warmest thanks go to Mrs Lilian Ashton, who for many years accurately filed the data on the 13000 satellites; to Peter Rolls, Head of Printing Branch; and to Mrs Margaret Davis, Superintendent of Reports Typing Pool, and the many typists over the years who have not only typed the Table skilfully and accurately but have also cheerfully tackled the far worse problem of making thousands of amendments to the masters.

GREEK LETTERS

The Greek alphabet used in the designations of the satellites of 1957-1962 is as follows:

α	alpha	η	eta	ν	nu	τ	tau
β	beta	θ	theta	ξ	xi	υ	upsilon
γ	gamma	ι	iota	o	omicron	ϕ	phi
δ	delta	κ	kappa	π	pi	χ	chi
ϵ	epsilon	λ	lambda	ρ	rho	ψ	psi
ζ	zeta	μ	mu	σ	sigma	ω	omega

Year of launch 1957

	Name		Launch date, lifetime and descent date	Shape and weight (kg)	Size (m)	Date of orbital determination	Orbital inclination (deg)	Nodal period (min)	Semi major axis (km)	Perigee height (km)	Apogee height (km)	Orbital eccentricity	Argument of perigee (deg)
D	Sputnik 1	1957α2	1957 Oct 4.81 92 days 1958 Jan 4	Sphere 83.6	0.58 dia	1957 Oct 4.8 1957 Oct 25.8 1957 Dec 25.1	65.1 65.1 65.0	96.2 95.4 91.0	6955 6916 6702	215 213 190	939 863 458	0.052 0.047 0.020	58 49 23
D	Sputnik 1 rocket	1957α1	1957 Oct 4.81 57.6 days 1957 Dec 1.4	Cylinder 4000?	28 long 2.58 to 2.95 dia	1957 Oct 4.8 1957 Nov 4.0 1957 Nov 24.3	65.1 65.1 65.0	96.2 94.0 91.0	6955 6849 6702	215 203 190	939 738 458	0.052 0.039 0.020	58 45 36
D	Fragment	1957α3											
D B	Sputnik 2*	1957β1	1957 Nov 3.1 162.0 days 1958 Apr 14.08	Cone-cylinder 4500? Payload 508.3	31.8 long 2.58 to 2.95 dia	1957 Nov 3.1 1958 Jan 4.0 1958 Feb 21.0 1958 Mar 25.0 1958 Apr 9.0	65.33 65.29 65.26 65.23 65.21	103.750 100.505 97.105 93.785 90.780	7314 7161 6999 6838 6691	212 210 200 187 166	1660 1356 1040 733 460	0.099 0.080 0.060 0.040 0.022	59 35 14 359 352
D	Fragment	1957β2											

* Decay observed over the West Indies. This satellite carried the dog Laika.

Year of launch 1958

	Name		Launch date, lifetime and descent date	Shape and weight (kg)	Size (m)	Date of orbital determination	Orbital inclination (deg)	Nodal period (min)	Semi major axis (km)	Perigee height (km)	Apogee height (km)	Orbital eccentricity	Argument of perigee (deg)
D	Explorer 1	1958 α	1958 Feb 1.16	Cylinder	2.03 long	1958 Feb 1.2	33.24	114.8	7830	356	2548	0.140	121
			4441.29 days	13.97	0.15 dia	1960 Dec 5.3	33.21	107.2	7481	347	1859	0.101	10
			1970 Mar 31.45	Payload 4.8		1967 Nov 28.7	33.18	100.90	7185	334	1281	0.066	201
	Vanguard 1	1958 β2	1958 Mar 17.51	Sphere	0.16 dia	1958 Mar 17.5	34.25	134.18	8687	650	3968	0.191	129
			300 years	1.47		1962 Nov 14.5	34.25	133.80	8671	648	3938	0.190	85
	Vanguard 1 rocket	1958 β1	1958 Mar 17.51	Cylinder	1.2 long	1958 Mar 17.5	34.25	138.50	8872	649	4340	0.208	129
			400 years	23	0.51 dia	1967 Sep 24.6	34.26	138.28	8862	658	4309	0.206	164
	Fragment	1958 β3											
D	Explorer 3	1958 γ	1958 Mar 26.73	Cylinder	2.03 long	1958 Mar 26.8	33.38	115.7	7871	186	2799	0.166	90
			93 days	14	0.15 dia	1958 May 15.0	33.35	104.8	7369	180	1802	0.110	70
			1958 Jun 28	Payload 5		1958 Jun 14.1	33.33	96.8	6990	171	1052	0.063	326
D	Sputnik 3	1958 δ2	1958 May 15.3	Cone	3.76 long	1958 May 15.3	65.18	105.97	7418	217	1864	0.111	58
			692.0 days	1327	1.73 dia	1959 Jan 1.3	65.15	102.000	7232	210	1497	0.089	331
			1960 Apr 6.3			1960 Jan 3.8	65.11	94.000	6849	190	751	0.041	182
						1960 Mar 24.5	65.06	90.000	6653	162	388	0.017	146
D	Sputnik 3 rocket	1958 δ1	1958 May 15.3	Cylinder	28 long	1958 May 15.3	65.18	105.90	7415	214	1860	0.111	58
			202.4 days	4000?	2.58 dia	1958 Aug 15.1	65.14	102.000	7232	210	1497	0.089	26
			1958 Dec 3.7		2.95 dia	1958 Oct 11.2	65.10	98.000	7042	199	1128	0.066	5
						1958 Nov 30.6	65.00	90.000	6653	162	388	0.017	339
D	Fragments	1958 δ3-5											
D	Explorer 4	1958 ε	1958 Jul 26.63	Cylinder	2.03 long	1958 Jul 26.7	50.3	110.18	7616	263	2213	0.128	50
			454 days	17.5	0.15 dia	1959 Mar 21.0	50.25	102.37	7252	257	1490	0.085	60
			1959 Oct 23	Payload 8		1959 Aug 22.0	50.25	96.05	6950	239	906	0.048	252
						1959 Oct 19.5	50.25	90.0	6656	204	351	0.011	120
D	Atlas (Score)*	1958 ζ	1958 Dec 18.96	Cylinder	25 long	1958 Dec 19.0	32.3	101.47	7213	185	1484	0.090	130
			33.6 days	3900	3.0 dia	1959 Jan 1.8	32.3	98.12	7053	181	1169	0.070	249
			1959 Jan 21.6	Payload 70		1959 Jan 17.0	32.3	92.7	6792	169	658	0.036	37

Space vehicles: Pioneer 1, 1958 η; Pioneer 3, 1958 θ

* Signal communications orbit-relay experiment

Year of launch 1959

	Name		Launch date, lifetime and descent date	Shape and weight (kg)	Size (m)	Date of orbital determination	Orbital inclination (deg)	Nodal period (min)	Semi major axis (km)	Perigee height (km)	Apogee height (km)	Orbital eccentricity	Argument of perigee (deg)
	Vanguard 2	1959 α1	1959 Feb 17.67 150 years	Sphere 9.8	0.51 dia	1959 Feb 17.7 1967 Nov 21.5	32.88 32.87	125.7 125.20	8318 8294	559 555	3320 3276	0.166 0.164	135 135
	Vanguard 2 rocket	1959 α2	1959 Feb 17.67 150 years	Cylinder 23	1.2 long 0.51 dia	1959 Feb 17.7 1967 Nov 26.4	32.92 32.91	130.0 129.42	8506 8481	563 559	3693 3646	0.184 0.182	135 -
D	Fragment	1959 α3											
D	Discoverer 1	1959 β	1959 Feb 28.91 5 days? 1959 Mar 5?	Cone-cylinder 618	6 long 1.5 dia	1959 Feb 28	89.7?	96?	6943?	163?	968?	0.058?	-
D	Discoverer 2	1959 γ	1959 Apr 13.89 12.7 days? 1959 Apr 26.6?	Cone-cylinder 1st day 743, then 650	6 long 1.5 dia	1959 Apr 13.9 1959 Apr 24.6	89.9 89.9	90.4 88.9	6671 6597	239 199	346 238	0.008 0.003	160 96
D	Explorer 6	1959 δ1	1959 Aug 7.60 23 months? 1961 Jul?	Spheroid + 4 vanes 64	Spheroid 0.74 long 0.66 dia	1959 Aug 7.6 1959 Oct 26.0 1959 Dec 19.2	47.0 47.0 47.0	765 760 754	27710 27590 27450	245 244 237	42400 42200 41900	0.761 0.760 0.759	35 53 65
D	Explorer 6 rocket	1959 δ2	1959 Aug 7.60 23 months? 1961 Jul?	Cylinder 24	1.5 long 0.46 dia	Orbit similar to 1959 δ1							
D	Discoverer 5	1959 ε1	1959 Aug 13.79 46 days 1959 Sep 28	Cone-cylinder 1st day 781, then 640	6 long 1.5 dia	1959 Aug 13.8 1959 Sep 9.1 1959 Sep 23.4	80.0 80.0 80.0	94.19 92.00 90.00	6856 6749 6651	217 209 193	739 533 353	0.038 0.024 0.012	157 70 22
D	Discoverer 5 capsule	1959 ε2	1959 Aug 13.79 547 days 1961 Feb 11	Paraboloid 140	0.6 long 0.9 dia	1960 Feb 15.1 1960 Dec 2.3 1961 Jan 31.3	78.94 78.94 78.94	104.27 94.45 90.68	7337 6869 6685	218 202 180	1700 779 434	0.101 0.042 0.019	47 320 124

Year of launch 1959, continued

	Name		Launch date, lifetime and descent date	Shape and weight (kg)	Size (m)	Date of orbital determination	Orbital inclination (deg)	Nodal period (min)	Semi major axis (km)	Perigee height (km)	Apogee height (km)	Orbital eccentricity	Argument of perigee (deg)
D	Discoverer 6	1959 ζ	1959 Aug 19.81 62.0 days 1959 Oct 20.8	Cone-cylinder 1st day 783, then 640	6 long 1.5 dia	1959 Aug 19.9 1959 Sep 28.2 1959 Oct 12.7	84.0 84.0 84.0	95.27 92.00 90.00	6908 6749 6651	212 196 186	848 547 359	0.046 0.026 0.013	143 360 297
	Vanguard 3	1959 η	1959 Sep 18.22 200 years	Rocket-sphere-rod 45 Payload 23	2.5 long 0.51 dia	1959 Sep 18.3 1965 Jul 29.6 1967 Nov 25.8	33.35 33.32 33.35	130.0 129.74 129.55	8506 8492 8485	512 514 512	3744 3714 3702	0.190 0.188 0.188	133 166 8
D	Luna 3	1959 θ1	1959 Oct 4.1 177 days? 1960 Mar 29?	Ellipsoid 278.5	1.32 long 1.19 dia	1959 Oct 18.7 1959 Dec 22.5	73.8 82.9	22700 23300	264800 269900	40300 19700	476500 507400	0.824 0.903	182 186
	Explorer 7	1959 ι1	1959 Oct 13.65 70 years	Double cone 41.5	0.76 long 0.76 dia	1959 Oct 13.7 1965 Aug 4.5 1967 Nov 28.9	50.31 50.31 50.31	101.28 101.07 100.98	7200 7190 7187	556 553 550	1088 1070 1067	0.037 0.036 0.036	55 80 86
	Explorer 7 rocket	1959 ι2	1959 Oct 13.65 40 years	Cylinder 6	1.73 long 0.15 dia	1959 Oct 13.7 1977 May 1.0	50.30 50.30	101.25 99.75	7199 7126	554 540	1087 956	0.037 0.029	56 -
D	Discoverer 7	1959 κ	1959 Nov 7.85 19.0 days 1959 Nov 26.8	Cone-cylinder 1st day 794, then 660	6 long 1.5 dia	1959 Nov 7.9 1959 Nov 15.6 1959 Nov 20.8	81.64 81.6 81.6	94.70 92.9 91.5	6881 6793 6725	159 157 152	847 673 542	0.050 0.038 0.029	165 138 120
D	Discoverer 8	1959 λ	1959 Nov 20.81 108.24 days 1960 Mar 8.05	Cone-cylinder 1st day 795, then 660	6 long 1.5 dia	1959 Nov 20.9 1960 Jan 15.5 1960 Feb 29.5	80.65 80.6 80.6	103.72 98.00 92.00	7311 7040 6749	187 176 162	1679 1147 580	0.102 0.069 0.031	156 356 206

Space vehicles: Luna 1, 1959 μ; Pioneer 4, 1959 ν; Luna 2, 1959 ξ

A rocket separated from Luna 3, but its orbit is not known

	Name	Launch date, lifetime and descent date	Shape and weight (kg)	Size (m)	Date of orbital determination	Orbital inclination (deg)	Nodal period (min)	Semi major axis (km)	Perigee height (km)	Apogee height (km)	Orbital eccentricity	Argument of perigee (deg)
	Tiros 1 * 1960 β2	1960 Apr 1.49 70 years	Cylinder 120	0.48 long 1.07 dia	1960 Apr 1.5	48.4	99.16	7100	693	750	0.004	115
					1967 Nov 12.1	48.38	99.09	7096	696	739	0.003	66
	Tiros 1 rocket 1960 β1	1960 Apr 1.49 30 years	Cylinder 24	1.5 long 0.46 dia	1960 Apr 1.5	48.41	99.15	7099	693	750	0.004	115
					1967 Oct 19.4	48.39	98.95	7090	690	733	0.003	344
	Fragments 1960 β3,4											
D	Transit 1B 1960 γ2	1960 Apr 13.50 2730.75 days 1967 Oct 5.25	Sphere 121	0.91 dia	1960 Apr 13.5	51.28	95.81	6939	373	748	0.027	261
					1963 Nov 20.6	51.25	94.10	6854	356	596	0.017	22
					1966 Sep 24.8	51.21	93.08	6806	339	516	0.013	202
					1967 Sep 28.0	51.21	89.38	6623	225	265	0.003	332
D	Transit 1B rocket 1960 γ1	1960 Apr 13.50 491.69 days 1961 Aug 18.19	Cylinder 600	4.8 long 1.4 dia	1960 Apr 13.5	51.25	95.25	6912	319	748	0.031	265
					1960 Dec 8.6	51.25	93.21	6813	285	584	0.022	64
					1961 Jun 16.8	51.25	91.05	6707	255	403	0.011	83
D	Fragments 1960 γ3,4											
D	Discoverer 11 1960 δ	1960 Apr 15.85 10.88 days 1960 Apr 26.73	Cone-cylinder 1st day 790, then 660	6 long 1.5 dia	1960 Apr 16.9	80.1	92.16	6757	170	589	0.031	150
					1960 Apr 24.7	80.1	89.75	6639	161	360	0.015	121

Space vehicle: Pioneer 5, 1960 α

* **Television and infra-red observation satellite**

	Name		Launch date, lifetime and descent date	Shape and weight (kg)	Size (m)	Date of orbital determination	Orbital inclination (deg)	Nodal period (min)	Semi major axis (km)	Perigee height (km)	Apogee height (km)	Orbital eccentricity	Argument of perigee (deg)
D	Sputnik 4*	1960 ε1	1960 May 15.00	Double	2.5 long?	1960 May 19.0	65.02	94.25	6861	290	675	0.028	87
P			844.41 days	cone 2040	2.5 dia	1961 Jun 29.4	65.02	92.53	6777	284	514	0.017	270
			1962 Sep 5.41	Payload 1477		1962 Jan 27.5	65.02	91.44	6724	272	420	0.011	183
						1962 Aug 5.8	64.95	89.50	6628	224	277	0.004	94
D	Sputnik 4 rocket	1960 ε2	1960 May 15.00	Cylinder	3.8 long	1960 May 15.0	64.89	91.25	6714	318	355	0.002	63
			63.82 days	1440	2.6 dia	1960 Jun 6.8	64.89	90.79	6692	299	329	0.002	53
			1960 Jul 17.82			1960 Jul 15.5	64.89	88.69	6588	206	215	0.0005	37
D	Sputnik 4 cabin	1960 ε3	1960 May 15.00	Sphere	2.4 dia	1960 May 19.0	65.0	94.27	6862	278	689	0.030	82
			1979.53 days	2500		1961 Jun 25.3	65.0	93.35	6817	275	602	0.024	286
			1965 Oct 15.53			1962 Nov 25.8	64.98	92.41	6771	271	515	0.018	49
						1964 Aug 20.9	64.98	91.07	6706	267	388	0.009	142
D	Fragments	1960 ε4-9											
D	Midas 2**	1960 ζ1	1960 May 24.73	Cylinder	7 long	1960 May 24.8	33.0	94.44	6876	484	511	0.002	136
			5007.10 days	2300	1.5 dia	1966 Aug 18.6	33.03	94.08	6858	469	491	0.002	232
			1974 Feb 7.83			1970 Nov 1.0	33.00	92.75	6794	409	422	0.001	-
D	Midas 2 nose-cap	1960 ζ2	1960 May 24.73	Cone?	2 long?	1960 May 24.8	33.00	94.44	6876	484	511	0.002	136
			194.5 days	20?	1.5 dia?	1960 Oct 9.5	33.00	93.02	6807	422	436	0.001	46
			1960 Dec 5.3			1960 Dec 2.6	33.00	89.79	6649	271	271	0	-

* The designation of the nine pieces of Sputnik 4 is that adopted in the United States. Russian and British prediction centres referred to Sputnik 4 as ε2 and the rocket as ε1. Between 1960 May 15.0 and May 19.0, satellite 1960 ε1 and 1960 ε3 to 9 were one piece, whose orbit was similar to that of 1960 ε2. Decay was observed over Wisconsin, USA, and pieces were picked up at Manitowoc.

** Missile defense alarm system.

Name	Launch date, lifetime and descent date	Shape and weight (kg)	Size (m)	Date of orbital determination	Orbital inclination (deg)	Nodal period (min)	Semi-major axis (km)	Perigee height (km)	Apogee height (km)	Orbital eccentricity	Argument of perigee (deg)
Transit 2A 1960η 1	1960 Jun 22.25	Sphere	0.91 dia	1960 Jun 22.3	66.69	101.66	7216	628	1047	0.029	236
	150 years	101		1965 Aug 6.1	66.71	101.59	7213	618	1050	0.030	345
SR 1* 1960η 2	1960 Jun 22.25	Sphere	0.51 dia	1960 Jun 22.3	66.69	101.66	7216	614	1061	0.031	236
	80 years	19		1967 Nov 23.2	66.70	101.54	7210	615	1048	0.030	109
Transit 2A 1960η 3	1960 Jun 22.25	Cylinder	4.8 long	1960 Jun 22.3	66.7	101.37	7202	615	1032	0.029	235
rocket	80 years	450	1.4 dia	1961 Jun 27.6	66.66	101.42	7203	615	1034	0.029	333
				1963 Mar 25.2	66.66	101.35	7201	614	1031	0.029	243
Fragments 1960η 4,5											
D Discoverer 13 1960θ	1960 Aug 10.86	Cone-cylinder	6 long	1960 Aug 10.9	82.85	94.04	6849	258	683	0.031	154
r	95.97 days	1st day 850,	1.5 dia	1960 Oct 9.9	82.85	92.00	6749	250	493	0.018	295
	1960 Nov 14.83	then 700		1960 Nov 9.4	82.85	90.00	6651	226	319	0.007	178
D Echo 1 1960ι 1	1960 Aug 12.40	Inflated sphere	30 dia	1960 Aug 12.4	47.22	118.22	7982	1524	1684	0.010	14
	2841.63 days	75.9 initially;		1960 Dec 16.0	47.27	117.28	7940	966	2157	0.075	59
	1968 May 24.03	62 after Jan 1961		1961 Jun 20.0	47.20	117.03	7929	1550	1550	0	-
				1961 Dec 6.0	47.30	116.18	7890	904	2120	0.077	78
				1962 Nov 20.0	47.20	115.35	7854	942	2010	0.068	97
				1963 Sep 4	47.29	114.82	7827	971	1926	0.061	264
				1965 Jun 24.9	47.30	113.33	7763	888	1882	0.064	238
				1967 Nov 28.5	47.26	108.50	7539	844	1477	0.042	303
Echo 1 1960ι 2	1960 Aug 12.40	Cylinder	1.5 long	1960 Aug 12.4	47.23	117.98	7972	1502	1685	0.011	12
rocket	10000 years	24	0.46 dia	1963 Dec 3.0	47.23	117.98	7971	1501	1684	0.011	8
Fragments 1960ι 3-5											

* Solar Radiation (Sunray).

	Name		Launch date, lifetime and discent date	Shape and weight (kg)	Size (m)	Date of orbital determination	Orbital inclination (deg)	Nodal period (min)	Semi major axis (km)	Perigee height (km)	Apogee height (km)	Orbital eccentricity	Argument of perigee (deg)
D r	Discoverer 14	1960 κ	1960 Aug 18.83 28.19 days 1960 Sep 16.02	Cone-cylinder 1st day 850, then 700	6 long 1.5 dia	1960 Aug 18.9 1960 Aug 30.8 1960 Sep 10.1	79.65 79.65 79.65	94.55 93.00 91.00	6874 6798 6701	186 182 175	805 658 470	0.045 0.035 0.022	168 129 94
D R B	Sputnik 5	1960 λ1	1960 Aug 19.36 1.1 days 1960 Aug 20.5	Sphere-cylinder 4600	4.3 long 2.4 dia	1960 Aug 19.4	64.95	90.72	6688	297	324	0.002	60
D	Sputnik 5 rocket	1960 λ2	1960 Aug 19.36 35.21 days 1960 Sep 23.57	Cylinder 1440	3.8 long 2.6 dia	1960 Aug 19.4 1960 Sep 10.0 1960 Sep 19.9	64.9 64.9 64.9	90.70 90.0 89.0	6687 6653 6604	296 268 225	323 282 226	0.002 0.001 0	81 72 67
D	Discoverer 15	1960 μ	1960 Sep 13.93 34.2 days 1960 Oct 18.1	Cone-cylinder 1st day 863, then 710	6 long 1.5 dia	1960 Sep 14.0 1960 Oct 3.0 1960 Oct 14.0	80.90 80.90 80.90	94.23 92.00 90.00	6858 6749 6651	199 196 180	761 547 366	0.041 0.026 0.014	162 96 55
	Courier 1B	1960 ν1	1960 Oct 4.74 1000 years	Sphere 230	1.30 dia	1960 Oct 4.8 1965 Nov 8.4	28.33 28.32	106.85 106.86	7465 7466	938 968	1237 1207	0.020 0.016	41 306
	Courier 1B rocket	1960 ν2	1960 Oct 4.74 500 years	Cylinder 450	4.8 long 1.4 dia	1960 Oct 4.8 1967 Jul 3.2	28.30 28.24	106.38 106.37	7444 7443	946 923	1184 1206	0.016 0.019	41 133
	Explorer 8	1960 ξ1	1960 Nov 3.22 50 years	Double cone 41	0.76 long 0.76 dia	1960 Nov 3.3 1967 Nov 22.0	49.95 49.94	112.69 111.82	7731 7692	417 421	2288 2206	0.121 0.116	52 110
	Explorer 8 rocket	1960 ξ2	1960 Nov 3.22 30 years	Cylinder 5	1.73 long 0.15 dia	1960 Nov 3.3 1969 Sep 16.0	49.96 49.95	112.68 109.60	7731 7590	417 415	2288 2009	0.121 0.105	52 -
D	Fragments	1960 ξ3, 4											

	Name	Launch date, lifetime and descent date	Shape and weight (kg)	Size (m)	Date of orbital determination	Orbital inclination (deg)	Nodal period (min)	Semi major axis (km)	Perigee height (km)	Apogee height (km)	Orbital eccentricity	Argument of perigee (deg)
D r	Discoverer 17 1960o	1960 Nov 12.86 46.9 days 1960 Dec 29.8	Cone-cylinder 1st 2 days 1091 then 930	8 long 1.5 dia	1960 Nov 12.9 1960 Dec 13.2 1960 Dec 25.3	81.70 81.70 81.70	96.45 93.11 90.35	6965 6804 6668	190 184 170	984 668 410	0.057 0.036 0.018	163 59 15
	Tiros 2 1960π1	1960 Nov 23.47 60 years	Cylinder 130	0.48 long 1.07 dia	1960 Nov 23.5 1967 Nov 18.9	48.5 48.52	98.20 98.11	7054 7049	619 614	732 727	0.008 0.008	334 185
D	Tiros 2 1960π2 rocket	1960 Nov 23.47 7609 days 1981 Sep 23	Cylinder 24	1.5 long 0.46 dia	1960 Nov 23.5 1977 May 1.0	48.57 48.5	98.14 97.02	7051 6997	609 579	736 659	0.009 0.006	334 -
	Fragments * 1960π3-5											
D B	Sputnik 6 1960ρ1	1960 Dec 1.31 1 day 1960 Dec 2	Sphere-cylinder 4563	4.3 long 2.4 dia	1960 Dec 1.4	64.97	88.47	6577	166	232	0.005	60?
D	Sputnik 6 1960ρ2 rocket	1960 Dec 1.31 1.6 days 1960 Dec 2.9	Cylinder 1440	3.8 long 2.6 dia	1960 Dec 2.8	65.00	87.29	6518	140	140	0	-
D r	Discoverer 18 1960σ	1960 Dec 7.85 115.9 days 1961 Apr 2.8	Cone-cylinder 1st 3 days 1240 then 950?	8 long 1.5 dia	1960 Dec 7.9 1961 Feb 5.8 1961 Mar 29.0	81.50 81.48 81.48	93.66 92.0 89.49	6830 6749 6626	243 233 205	661 510 291	0.031 0.021 0.006	164 312 121
D	Discoverer 19 1960τ	1960 Dec 20.86 33.2 days 1961 Jan 23.1	Cone-cylinder 1060	8 long 1.5 dia	1960 Dec 20.9 1961 Jan 16.6 1961 Jan 19.1	83.40 83.40 83.40	93.00 90.0 89.55	6798 6651 6629	209 186 178	631 359 324	0.031 0.013 0.011	173 75 62

* The fragment 1960π5, designated on 31 Mar 1972, probably belongs to a different launch (perhaps Echo 1).

Year of launch 1961

	Name		Launch date, lifetime and descent date	Shape and weight (kg)	Size (m)	Date of orbital determination	Orbital inclination (deg)	Nodal period (min)	Semi major axis (km)	Perigee height (km)	Apogee height (km)	Orbital eccentricity	Argument of perigee (deg)
D	Samos 2 *	1961 α1	1961 Jan 31.85 4646 days 1973 Oct 21	Cylinder 1900	7 long 1.5 dia	1961 Jan 31.9 1965 Dec 12.5 1967 Nov 28.5 1970 Nov 1.0	97.40 97.39 97.34 97.3	94.97 94.73 94.40 93.24	6894 6882 6869 6811	474 463 456 412	557 545 525 454	0.006 0.006 0.005 0.003	196 112 110 -
D	Samos 2 nose-cap	1961 α2	1961 Jan 31.85 3537.37 days 1970 Oct 9.22	Cone? 200?	2 long? 1.5 dia?	1961 Jan 31.9 1967 Aug 30.9 1969 Sep 16.0	97.40 97.36 97.30	94.97 94.25 93.02	6894 6861 6800	474 448 404	557 517 440	0.006 0.005 0.003	196 55 -
D	Sputnik 7**	1961 β1	1961 Feb 4.1 22.7 days 1961 Feb 26.8	Cylinder 6483 full	7 long? 2.0 max dia	1961 Feb 4.1	64.95	89.78	6643	212	318	0.008	59
D	Sputnik 7 rocket	1961 β2	1961 Feb 4.1 8.9 days 1961 Feb 13.0	Cylinder 2500?	7.5 long 2.6 dia	Initial orbit similar to 1961 β1							
D	Fragment	1961 β3											
	Venus 1	1961 γ1	1961 Feb 12.09 indefinite	Cylinder 643.5	2.03 long 1.05 dia	Initial earth-satellite orbit similar to 1961 γ3							
D	Sputnik 8 rocket	1961 γ2	1961 Feb 12.09 6.5 days 1961 Feb 18.6	Cylinder 2500?	7.5 long 2.6 dia	1961 Feb 13.9	65.0	89.21	6614	196	275	0.006	20
D	Sputnik 8	1961 γ3	1961 Feb 12.09 13.7 days 1961 Feb 25.8	Irregular	2 long? 2 dia?	1961 Feb 12.1	65.0	89.61	6633	229	282	0.004	23
D	Fragment	1961 γ4											

* Satellite and missile observation system.

** Sputnik 7 is believed to be a Venus Probe launcher.

	Name		Launch date, lifetime and descent date	Shape and weight (kg)	Size (m)	Date of orbital determination	Orbital inclination (deg)	Nodal period (min)	Semi major axis (km)	Perigee height (km)	Apogee height (km)	Orbital eccentricity	Argument of perigee (deg)
D	Explorer 9	1961 δ1	1961 Feb 16.55 1148.2 days 1964 Apr 9.8	Inflated sphere 6.63	3.66 dia	1961 Feb 16.6 1961 Dec 19.0 1963 Jan 1.0 1963 Dec 7.9	38.86 38.82 38.86 38.95	118.28 118.04 117.36 112.11	7986 7976 7947 7704	634 752 632 394	2583 2443 2506 2258	0.122 0.106 0.118 0.121	100 118 134 26
	Explorer 9 rocket	1961 δ2	1961 Feb 16.55 150 years	Cylinder 24	1.5 long 0.46 dia	1961 Feb 16.6	38.85	118.40	7992	639	2589	0.122	100
3d	Fragments	1961 δ3-8											
D	Discoverer 20	1961 ε1	1961 Feb 17.85 525.9 days 1962 Jul 28.7	Cone-cylinder 1st 4 days 1110 then 980	8 long 1.5 dia	1961 Feb 17.9 1962 Jan 27.2 1962 Jul 12.4	80.91 80.84 80.82	95.41 92.78 89.91	6915 6787 6641	288 267 223	786 552 303	0.036 0.021 0.006	125 36 158
D	Fragments	1961 ε2-4											
D	Discoverer 21	1961 ζ	1961 Feb 18.95 426.0 days 1962 Apr 20.9	Cone-cylinder 1100?	8 long 1.5 dia	1961 Feb 19.0 1961 Dec 17.5 1962 Apr 9.8	80.74 80.68 80.64	97.85 93.49 90.19	7033 6822 6656	240 239 212	1069 649 344	0.059 0.030 0.010	141 244 198
D	Transit 3B-Lofti 1 *	1961 η	1961 Feb 22.16 36.38 days 1961 Mar 30.54	Cylinder 600	6.5 long 1.4 dia	1961 Feb 22.2 1961 Mar 26.0	28.38 28.38	96.22 90.67	6963 6693	167 147	1002 482	0.060 0.025	29 29
D R B	Sputnik 9	1961 θ1	1961 Mar 9.27 0.07 day 1961 Mar 9.34	Sphere-cylinder 4700	4.3 long 2.4 dia	1961 Mar 9.3	64.93	88.60	6584	173	239	0.005	-
D	Sputnik 9 rocket	1961 θ2	1961 Mar 9.27 1.1 days 1961 Mar 10.4	Cylinder 1440	3.8 long 2.6 dia	1961 Mar 9.6	64.9	88.20	6564	173	199	0.002	25
D	Fragments	1961 θ3,4											

* Low-frequency trans-ionospheric satellite

	Name		Launch date, lifetime and descent date	Shape and weight (kg)	Size (m)	Date of orbital determination	Orbital inclination (deg)	Nodal period (min)	Semi major axis (km)	Perigee height (km)	Apogee height (km)	Orbital eccentricity	Argument of perigee (deg)
D R B	Sputnik 10	1961 ι1	1961 Mar 25.25 0.07 day 1961 Mar 25.32	Sphere-cylinder 4695	4.3 long 2.4 dia	1961 Mar 25.3	64.9	88.42	6575	164	230	0.005	-
D	Sputnik 10 rocket	1961 ι2	1961 Mar 25.25 1.7 days 1961 Mar 26.9	Cylinder 1440	3.8 long 2.6 dia	1961 Mar 25.9	65.0	87.80	6544	140	192	0.004	42
D	Fragment	1961 ι3											
D	Explorer 10	1961 κ	1961 Mar 25.64 87 months 1968 Jun	Sphere-cylinder 35	2.72 long 0.48 dia	1961 Mar 25.7	33	5013	97050	221	181100	0.932	-
D	Discoverer 23	1961 λ1	1961 Apr 8.75 373.1 days 1962 Apr 16.9	Cone-cylinder 1st 3 days 1150 then 950?	8 long 1.5 dia	1961 Apr 9.0 1961 Jul 2.4 1962 Jan 18.6	82.31 82.31 82.26	94.09 93.52 91.68	6851 6823 6734	295 295 268	651 595 443	0.026 0.022 0.013	168 227 224
D	Discoverer 23 capsule	1961 λ2	1961 Apr 8.75 409.4 days 1962 May 23.2	Paraboloid 150?	0.6 long 0.9 dia	1961 Apr 24.9 1962 Feb 2.5 1962 May 8.5	81.94 81.88 81.82	101.49 95.03 90.04	7206 6897 6648	208 194 180	1448 843 359	0.086 0.047 0.013	112 290 307
D	Discoverer 23 capsule rocket	1961 λ3	1961 Apr 8.75 154.8 days 1961 Sep 10.6	Frustum 50?	0.6 long? 0.9 dia ?	1961 Apr 24.9 1961 Jul 4.8 1961 Aug 29.8	81.94 81.94 81.87	101.13 97.19 91.96	7189 7001 6747	200 196 187	1422 1050 551	0.085 0.061 0.027	111 249 53

A rocket is believed to have separated from Explorer 10. From 1961 Apr 8 to Apr 11 satellites 1961 λ2 and 1961 λ3 were part of 1961 λ1.

	Name		Launch date, lifetime and descent date	Shape and weight (kg)	Size (m)	Date of orbital determination	Orbital inclination (deg)	Nodal period (min)	Semi major axis (km)	Perigee height (km)	Apogee height (km)	Orbital eccentricity	Argument of perigee (deg)
D R M	Vostok 1†	1961 μ 1	1961 Apr 12.25 108 min 1961 Apr 12.33	Sphere-cylinder 4725	4.3 long 2.4 dia	1961 Apr 12.3	64.95	89.34	6620	169	315	0.011	-
D	Vostok 1 rocket	1961 μ 2	1961 Apr 12.25 4.2 days 1961 Apr 16.5	Cylinder 1440	3.8 long 2.6 dia	1961 Apr 12.6	65.07	89.31	6618	161	320	0.012	100
	Explorer 11	1961 ν 1	1961 Apr 27.59 150 years	Cylinder 37	2.26 long 0.38 dia	1961 Apr 27.6	28.80	107.84	7512	487	1779	0.086	119
	Fragment	1961 **ν** 2				**1965 Aug 8.2**	**28.77**	**107.76**	**7508**	**484**	**1775**	**0.086**	**167**
D r	Discoverer 25	1961 ξ 1	1961 Jun 16.96 25 days 1961 Jul 12	Cone-cylinder 1st 2 days 1150 then 950?	8 long 1.5 dia	1961 Jun 17.1 1961 Jul 11.3	82.11 82.11	90.87 88.29	6694 6567	222 175	409 201	0.014 0.002	178 84
D	Fragment	1961 ξ 2											
	Transit 4A	1961 ο 1	1961 Jun 29.18 600 years	Cylinder 79	0.79 long 1.09 dia	1961 Jun 29.2	66.81	103.82	7317	881	998	0.008	319
	Injun 1 & SR 3	1961 ο 2	1961 Jun 29.18 900 years	Sphere-cylinder 25-16	0.9 long 0.51 dia	1961 Jun 29.2	66.82	103.85	7319	882	999	0.008	318
62d	Fragments*	1961 ο 3-262											
D r	Discoverer 26	1961 π	1961 Jul 7.98 150.4 days 1961 Dec 5.4	Cone-cylinder 1st 2 days 1150 then 950?	8 long 1.5 dia	1961 Jul 8.3 1961 Sep 18.4 1961 Nov 21.3	82.94 82.94 82.94	95.02 93.14 90.39	6896 6805 6670	228 223 212	808 631 372	0.042 0.030 0.012	160 260 18

† First manned space flight.

* Ablestar rocket exploded shortly after entering orbit.

	Name		Launch date, lifetime and descent date	Shape and weight (kg)	Size (m)	Date of orbital determination	Orbital inclination (deg)	Nodal period (min)	Semi major axis (km)	Perigee height (km)	Apogee height (km)	Orbital eccentricity	Argument of perigee (deg)
	Tiros 3	1961 ρ 1	1961 Jul 12.43 150 years	Cylinder 129	0.48 long 1.07 dia	1961 Jul 12.5 1965 Aug 6.4	47.90 47.90	100.33 100.32	7156 7155	735 740	820 812	0.006 0.005	42 82
	Tiros 3 rocket	1961 ρ 2	1961 Jul 12.43 75 years	Cylinder 24	1.5 long 0.46 dia	1961 Jul 12.5 1967 Jul 8.6	47.9 47.91	100.31 100.20	7154 7150	740 736	812 808	0.005 0.005	42 317
	Fragments	1961 ρ 3,4											
	Midas 3	1961 σ 1	1961 Jul 12.68 100000 years	Cylinder 1600	9 long 1.5 dia	1961 Jul 12.7 1962 Oct 12.9	91.2 91.19	161.54 161.52	9824 9820	3358 3340	3534 3544	0.009 0.010	240 95
D	Midas 3 nose-cap	1961 σ 2	1961 Jul 12.68 13.20 days 1961 Jul 25.88	Cone 20?	2 long? 1.5 dia?	1961 Jul 15.4 1961 Jul 18.5	90.80 90.80	117.25 109.7	7934 7589	138 134	2974 2289	0.179 0.142	164 -
1d	Fragments	1961 σ 3-5											
D R M	Vostok 2	1961 τ 1	1961 Aug 6.25 25.3 hours 1961 Aug 7.30	Sphere-cylinder 4730	4.3 long 2.4 dia	1961 Aug 6.3	64.93	88.46	6577	166	232	0.005	-
D	Vostok 2 rocket	1961 τ 2	1961 Aug 6.25 3 days 1961 Aug 9	Cylinder 1440	3.8 long 2.6 dia	Orbit similar to 1961 τ1							
D	Explorer 12*	1961 υ	1961 Aug 16.14 25 months 1963 Sep	Octagon + 4 vanes 38	1.29 long 0.66 dia	1961 Sep 22.5 1962 Jan 30.5	33.1 33.43	1591 1587.3	45190 45086	314 790	77310 76620	0.852 0.841	- -
D	Ranger 1	1961 φ 1	1961 Aug 23.46 6.89 days 1961 Aug 30.35	Cylinder 306	3.5 long 1.5 dia	1961 Aug 24.1 1961 Aug 29.5	32.9 32.9	90.64 88.9	6691 6605	179 174	446 280	0.020 0.008	206 -
D	Ranger 1 rocket	1961 φ 2	1961 Aug 23.46 10.68 days 1961 Sep 3.14	Cylinder 700?	6 long? 1.5 dia	1961 Aug 24.1 1961 Aug 29.5	32.93 32.93	90.71 89.7	6694 6644	175 173	456 359	0.021 0.014	206 -

* A rocket is believed to have separated from Explorer 12.

	Name	Launch date, lifetime and descent date	Shape and weight (kg)	Size (m)	Date of orbital determination	Orbital inclination (deg)	Nodal period (min)	Semi major axis (km)	Perigee height (km)	Apogee height (km)	Orbital eccentricity	Argument of perigee (deg)
D	Explorer 13 1961 χ	1961 Aug 25.81 2.3 days 1961 Aug 28.1	Cylinder 86	1.93 long 0.61 dia	1961 Aug 26.8	37.7	97.5	7023	125	1164	0.074	-
D r B	Discoverer 29 1961 ψ	1961 Aug 30.8 10.2 days 1961 Sep 9.98	Cone-cylinder 1150, then 950?	8 long 1.5 dia	1961 Aug 31.3	82.14	91.51	6725	152	542	0.029	83
D r	Discoverer 30 1961 ω 1	1961 Sep 12.83 90.1 days 1961 Dec 11.9	Cone-cylinder 1150, then 950?	8 long 1.5 dia	1961 Sep 13.6 1961 Nov 21.5 1961 Dec 5.5	82.66 82.66 82.66	92.40 90.4 89.4	6769 6671 6621	235 213 204	546 373 283	0.023 0.012 0.006	142 - -
D	Fragments 1961 ω 2,3											
D R	Mercury 4 1961αα1	1961 Sep 13.59 109 min 1961 Sep 13.66	Cone-frustum 1200	2.90 long 1.83 dia	1961 Sep 13.6	32.8	88.40	6580	156	248	0.007	-
D	Mercury 4 rocket 1961αα2	1961 Sep 13.59 5 hours 1961 Sep 13.8	Cylinder 3400	20 long 3.0 dia	1961 Sep 13.6	32.85	87.3	6526	147	147	0	-
D	Discoverer 31 1961 αβ	1961 Sep 17.88 38.57 days 1961 Oct 26.45	Cone-cylinder 1100?	8 long 1.5 dia	1961 Sep 21.0 1961 Oct 10.5	82.70 82.7	90.86 90.0	6693 6651	235 220	396 326	0.012 0.008	136 -
D r	Discoverer 32 1961αγ1	1961 Oct 13.81 30.6 days 1961 Nov 13.4	Cone-cylinder 1st day 1150 then 950?	8 long 1.5 dia	1961 Oct 14.1 1961 Nov 10.3	81.69 81.64	90.84 88.93	6692 6598	234 207	395 233	0.012 0.002	158 60
D	Fragments 1961αγ2,3											

	Name	Launch date, lifetime and descent date	Shape and weight (kg)	Size (m)	Date of orbital determination	Orbital inclination (deg)	Nodal period (min)	Semi major axis (km)	Perigee height (km)	Apogee height (km)	Orbital eccentricity	Argument of perigee (deg)
	Midas 4 1961α δ1	1961 Oct 21.58 100000 years	Cylinder 1800?	9 long? 1.5 dia	1961 Nov 2.0	95.89	166.01	10004	3496	3756	0.013	18
1d	Fragments 1961α δ 2-6											
D	Discoverer 34 1961α ε1	1961 Nov 5.83 396.4 days 1962 Dec 7.2	Cone-cylinder 1100?	8 long 1.5 dia	1961 Nov 6.1 1962 Jun 5.3 1962 Nov 24.7	82.52 82.46 82.46	97.12 94.40 89.91	6998 6863 6642	227 220 196	1011 750 332	0.056 0.039 0.010	152 149 246
D	Fragments 1961α ε 2-5											
D r	Discoverer 35 1961α ζ1	1961 Nov 15.89 17.9 days 1961 Dec 3.8	Cone-cylinder 1st day 1150 then 950?	8 long 1.5 dia	1961 Nov 21.5 1961 Dec 2.5	81.63 81.63	89.7 88.2	6636 6562	238 177	278 190	0.003 0.001	- -
D	Fragment 1961α ζ2											
	Transit 4B 1961α η1	1961 Nov 15.93 1000 years	Cylinder 86	0.79 long 1.09 dia	1961 Nov 16.6	32.43	105.63	7408	956	1104	0.010	329
	Traac * 1961α η2	1961 Nov 15.93 800 years	"Door-knob" + 32m boom 109	1.0 long 1.09 dia	1961 Nov 21.5	32.43	105.64	7409	941	1119	0.012	-
	Transit 4B rocket 1961α η3	1961 Nov 15.93 500 years	Cylinder 450?	4.8 long 1.4 dia	1961 Nov 21.5	32.41	105.49	7402	942	1105	0.011	-
	Fragment 1961α η4											

* Transit research and attitude control.

	Name		Launch date, lifetime and descent date	Shape and weight (kg)	Size (m)	Date of orbital determination	Orbital inclination (deg)	Nodal period (min)	Semi major axis (km)	Perigee height (km)	Apogee height (km)	Orbital eccentricity	Argument of perigee (deg)
D	Ranger 2	1961 αθ	1961 Nov 18.34 2 days 1961 Nov 20	Cylinder 1300?	11 long? 1.5 dia	1961 Nov 18.4 1961 Nov 19.3	33.34 33.34	88.28 87.51	6574 6536	150 145	242 171	0.007 0.002	49 59
D R B	Mercury 5	1961αι 1	1961 Nov 29.63 3.3 hours 1961 Nov 29.77	Cone-frustum 1300	2.90 long 1.83 dia	1961 Nov 29.7	32.6	88.31	6575	158	237	0.006	127
D	Mercury 5 rocket	1961αι 2	1961 Nov 29.63 1 day 1961 Nov 30	Cylinder 3400	20 long 3.0 dia	Orbit similar to 1961 αι1							
D r	Discoverer 36	1961ακ 1	1961 Dec 12.86 85.3 days 1962 Mar 8.2	Cone-cylinder 1st 4 days 1150 then 950?	8 long 1.5 dia	1961 Dec 14.7 1962 Jan 30.5 1962 Feb 27.5	81.21 81.21 81.15	91.82 90.85 89.60	6741 6691 6636	241 229 218	484 396 298	0.018 0.012 0.006	134 - -
D	Oscar 1 *	1961ακ 2	1961 Dec 12.86 49.4 days 1962 Jan 31.3	Rectangular box 5	0.30 long 0.25 wide 0.15 high	1961 Dec 14.0 1962 Jan 16.5 1962 Jan 30.5	81.21 81.21 81.21	91.76 90.4 88.2	6738 6671 6562	245 226 164	474 359 204	0.017 0.010 0.003	137 - -
D	Fragment	1961ακ 3											
D	[Atlas Agena B]	1961αλ1	1961 Dec 22.80 235 days 1962 Aug 14	Cylinder 1800?	8 long? 1.5 dia	1962 Mar 13.5 1962 Jun 5.6 1962 Aug 3.6	89.6 89.6 89.6	94.1 92.1 89.6	6851 6754 6628	244 228 201	702 524 299	0.033 0.022 0.007	- - -
D	Fragments	1961αλ2, 3											

* Orbiting satellite carrying amateur radio.

Name		Launch date, lifetime and descent date	Shape and weight (kg)	Size (m)	Date of orbital determination	Orbital inclination (deg)	Nodal period (min)	Semi major axis (km)	Perigee height (km)	Apogee height (km)	Orbital eccentricity	Argument of perigee (deg)
Tiros 4	1962 β 1	1962 Feb 8.52 150 years	Cylinder 129	0.48 long 1.07 dia	1962 Feb 13.5	48.30	100.31	7154	712	840	0.009	-
Tiros 4 rocket	1962 β 4	1962 Feb 8.52 75 years	Cylinder 24	1.5 long 0.46 dia	1965 Aug 7.8	48.24	100.21	7150	700	843	0.010	312
Fragments	1962 β 2,3											
Mercury 6 (Friendship 7)	1962 γ 1	1962 Feb 20.62 296 min 1962 Feb 20.82	Cone-frustum 1352	2.90 long 1.83 dia	1962 Feb 20.7	32.54	88.60	6590	159	265	0.008	80
Mercury 6 rocket	1962 γ 2	1962 Feb 20.62 8 hours 1962 Feb 20.9	Cylinder 3400	20 long 3.0 dia	1962 Feb 20.8	32.57	88.00	6560	156	208	0.004	84
[Thor Agena B]	1962 δ	1962 Feb 21 16 days 1962 Mar 9	Cylinder 1000?	8 long? 1.5 dia	1962 Feb 21	81.97	90.0	6649	167	374	0.016	-
Discoverer 38	1962 ε 1	1962 Feb 27.91 21 days 1962 Mar 21	Cone-cylinder 1st 4 days 1150 then 950?	8 long 1.5 dia	1962 Mar 9.8 1962 Mar 13.5	82.23 82.23	90.04 89.71	6653 6636	208 208	341 308	0.010 0.008	98 -
Fragments	1962 ε 2-4											
OSO 1 *	1962 ζ 1	1962 Mar 7.67 7155 days 1981 Oct 8	Nonagonal box 208	0.94 long 1.12 dia	1962 Jul 15.5 1971 Mar 16.5	32.83 32.83	95.99 95.28	6952 6916	553 522	595 553	0.003 0.002	77 -
OSO 1 rocket	1962 ζ 2	1962 Mar 7.67 2978.60 days 1970 May 3.27	Cylinder 24	1.5 long 0.46 dia	1962 Mar 18.3 1967 Dec 15.5 1969 Sep 16.0	32.83 32.83 32.83	95.98 95.28 93.74	6950 6917 6839	544 523 453	600 552 468	0.004 0.002 0.001	203 - -

Space vehicle: Ranger 3, 1962 α1; and rocket 1962 α2.

* Orbiting Solar Observatory.

	Name		Launch date, lifetime and descent date	Shape and weight (kg)	Size (m)	Date of orbital determination	Orbital inclination (deg)	Nodal period (min)	Semi major axis (km)	Perigee height (km)	Apogee height (km)	Orbital eccentricity	Argument of perigee (deg)
D	[Atlas Agena B]	1962 η 1	1962 Mar 7.8 457.4 days 1963 Jun 7.9	- 500?	1.5 dia?	1962 Mar 13.5 1963 Jan 5.7 1963 May 28.7	90.89 90.87 90.86	93.9 91.57 88.85	6842 6723 6600	251 223 189	676 467 255	0.031 0.018 0.005	- 2 104
D	Agena rocket	1962 η 3	1962 Mar 7.8 240.6 days 1962 Nov 3.4	Cylinder 700?	6 long? 1.5 dia	1962 May 1.6 1962 Jul 6.6 1962 Oct 24.6	90.87 90.87 90.87	93.3 92.4 89.6	6813 6769 6630	250 228 209	618 553 294	0.027 0.024 0.006	- - -
D	Fragment	1962 η 2											
D	Cosmos 1 (Sputnik 11)	1962 θ 1	1962 Mar 16.50 70 days 1962 May 25	Ellipsoid 200?	1.8 long 1.2 dia	1962 Mar 16.6 1962 May 1.6 1962 May 25	49.00 48.99 48.99	96.35 92.70 87.9	6964 6788 6552	204 194 173	967 626 175	0.055 0.032 0	104 309 -
D	Cosmos 1 rocket	1962 θ 2	1962 Mar 16.50 94 days 1962 Jun 18	Cylinder 1500?	8 long 1.65 dia	1962 Mar 19.8 1962 May 17.5 1962 Jun 5.6	49.0 49.0 49.0	96.10 92.60 91.00	6953 6783 6705	206 202 186	943 609 468	0.053 0.030 0.021	118 - 108
D	Cosmos 2 (Sputnik 12)	1962 ι 1	1962 Apr 6.72 499.3 days 1963 Aug 20.0	Ellipsoid 400?	1.8 long 1.2 dia	1962 Apr 7 1962 Dec 22.0 1963 Jul 28.3	48.97 48.94 48.90	102.25 97.17 90.61	7246 7006 6686	202 195 187	1535 1060 428	0.092 0.062 0.018	- 49 306
D	Cosmos 2 rocket	1962 ι 2	1962 Apr 6.72 182.7 days 1962 Oct 6.4	Cylinder 1500?	8 long 1.65 dia	1962 Apr 10.1 1962 Jul 16.4 1962 Oct 3.1	48.94 48.91 48.85	101.90 96.69 89.92	7230 6982 6652	215 191 169	1488 1015 379	0.088 0.059 0.016	119 147 141
	Midas 5 [Atlas Agena B]	1962 κ 1	1962 Apr 9.66 100000 years	Cylinder 2000?	9 long? 1.5 dia	1962 May 1.6	86.68	153.03	9476	2814	3382	0.030	-
3d	Fragments	1962 κ 2-6											
D	[Thor Agena B]	1962 λ 1	1962 Apr 18 40 days 1962 May 28	Cylinder 1500?	8 long? 1.5 dia	1962 May 1.6 1962 May 17.5	73.48 73.45	90.9 89.5	6699 6626	200 198	441 297	0.018 0.007	- -
D	Fragments	1962 λ 2-4											

Space vehicle: Ranger 4, 1962 μ1; and rocket 1962 μ2

Year of launch 1962, continued

	Name		Launch date, lifetime and descent date	Shape and weight (kg)	Size (m)	Date of orbital determination	Orbital inclination (deg)	Nodal period (min)	Semi major axis (km)	Perigee height (km)	Apogee height (km)	Orbital eccentricity	Argument of perigee (deg)
D	Cosmos 3	1962 ν 1	1962 Apr 24.17	Ellipsoid	1.8 long	1962 Apr 24.2	48.99	93.75	6840	216	707	0.036	-
	(Sputnik 13)		176 days	400?	1.2 dia	1962 Jul 18.3	48.97	92.43	6775	214	580	0.027	126
			1962 Oct 17			1962 Oct 10.3	48.95	89.37	6625	193	301	0.008	170
D	Cosmos 3	1962 ν 2	1962 Apr 24.17	Cylinder	8 long	1962 Apr 26.4	49.00	93.71	6837	220	699	0.035	109
	rocket		103.6 days	1500?	1.65 dia	1962 Jul 11.9	48.98	90.98	6704	212	440	0.017	101
			1962 Aug 5.8			1962 Jul 29.0	48.96	89.79	6644	209	323	0.009	183
D	Cosmos 4	1962 ξ 1	1962 Apr 26.42	Sphere-	4.3 long	1962 Apr 26.5	65.00	90.53	6679	285	317	0.002	-
R	(Sputnik 14)		3.0 days	cylinder	2.4 dia								
B			1962 Apr 29.4	4750?									
D	Cosmos 4	1962 ξ 2	1962 Apr 26.42	Cylinder	3.8 long	1962 Apr 30.4	64.95	90.52	6679	287	314	0.002	174
	rocket		52.5 days	1440	2.6 dia	1962 Jun 8.9	64.94	89.28	6615	233	240	0.001	146
			1962 Jun 17.9			1962 Jun 15.6	64.92	88.60	6584	205	207	0	-
D	Fragment	1962 ξ 3											
D	Ariel 1	1962 ο 1	1962 Apr 26.75	Cylinder +	0.53 long	1962 Apr 28.5	53.85	100.86	7180	389	1214	0.057	173
			5142.1 days	4 paddles	0.58 dia	1969 Sep 16.0	53.85	98.10	7046	377	959	0.041	-
			1976 May 24.9	60		1970 Dec 1.0	53.85	96.80	6984	368	844	0.034	-
D	Ariel 1	1962 ο 2	1962 Apr 26.75	Cylinder	1.5 long	1962 May 17.5	53.84	100.90	7182	394	1213	0.057	-
	rocket		3288.83 days	24	0.46 dia	1967 Sep 27.7	53.85	99.19	7099	384	1058	0.047	47
			1971 Apr 28.58			1969 Sep 16.0	53.84	96.71	6980	367	836	0.034	-
D	[Atlas Agena B]	1962 π	1962 Apr 26.9	Cylinder	8 long?	1962 Apr 27	74.1?	90?					
			2 days	2000?	1.5 dia								
			1962 Apr 28										
D	[Thor Agena B]	1962 ρ 1	1962 Apr 28.95	Cylinder	8 long?	1962 May 1.6	73.11	91.1	6706	180	475	0.022	-
			28 days	1500?	1.5 dia	1962 May 17.6	73.07	89.90	6647	176	362	0.014	94
			1962 May 26			1962 May 25.6	73.07	88.7	6588	166	253	0.007	-
D	Fragment	1962 ρ 2											

	Name		Launch date, lifetime and descent date	Shape and weight (kg)	Size (m)	Date of orbital determination	Orbital inclina-tion (deg)	Nodal period (min)	Semi major axis (km)	Perigee height (km)	Apogee height (km)	Orbital eccen-tricity	Argument of perigee (deg)
D	[Thor Agena B]	1962σ1	1962 May 15.82 560.0 days 1963 Nov 26.8	Cylinder 1500?	8 long? 1.5 dia	1962 May 20.6 1962 Dec 11.7 1963 Oct 17.5	82.33 82.33 82.32	94.02 93.03 90.55	6848 6795 6680	305 287 248	634 546 355	0.024 0.019 0.008	148 131 56
D	Fragments	1962σ2,3											
D R M	Mercury 7 (Aurora 7)	1962τ1	1962 May 24.53 296 min 1962 May 24.74	Cone-frustum 1349	2.90 long 1.83 dia	1962 May 24.6	32.5	88.50	6585	154	260	0.008	
D	Mercury 7 rocket	1962τ2	1962 May 24.53 1 day 1962 May 25	Cylinder 3400	20 long 3.0 dia	Orbit similar to 1962 τ1							
D	Cosmos 5 (Sputnik 15)	1962 ν1	1962 May 28.13 339.6 days 1963 May 2.7	Ellipsoid with 'hat' 400?	1.8 long 1.2 to 1.5 dia	1962 May 28.2 1962 Nov 28.8 1963 Mar 15.1	49.06 49.00 48.96	102.68 97.41 93.50	7267 7019 6826	190 187 184	1587 1095 712	0.096 0.065 0.039	112 104 205
D	Cosmos 5 rocket	1962 ν2	1962 May 28.13 201 days 1962 Dec 15	Cylinder 1500?	8 long 1.65 dia	1962 May 29.5 1962 Sep 4.3 1962 Nov 24.5	49.1 49.01 48.98	102.67 99.06 92.80	7266 7096 6792	205 194 181	1571 1242 647	0.094 0.074 0.034	116 129 113
D	[Thor Agena B]	1962φ1	1962 May 30.02 12 days 1962 Jun 11	Cylinder 1500?	8 long? 1.5 dia	1962 Jun 5.5 1962 Jun 8.3	74.10 74.10	89.70 88.96	6637 6599	199 193	319 248	0.009 0.004	 1
D	Fragment	1962φ2											

	Name		Launch date, lifetime and descent date	Shape and weight (kg)	Size (m)	Date of orbital determination	Orbital inclination (deg)	Nodal period (min)	Semi major axis (km)	Perigee height (km)	Apogee height (km)	Orbital eccentricity	Argument of perigee (deg)
D	[Thor Agena B]	1962χ1	1962 Jun 2.03 26.9 days 1962 Jun 28.9	Cylinder 1500?	8 long? 1.5 dia	1962 Jun 5.5 1962 Jun 19.5 1962 Jun 25.4	74.26 74.25 74.25	90.50 89.60 88.87	6676 6632 6596	211 193 188	385 315 247	0.013 0.009 0.004	- 91 75
D	Oscar 2	1962χ2	1962 Jun 2.03 19 days 1962 Jun 21	Rectangular box 5	0.30 long 0.25 wide 0.15 high	1962 Jun 2.9 1962 Jun 15.6	74.27 74.27	90.55 89.30	6679 6617	207 193	394 285	0.014 0.007	133 -
D	Fragment	1962χ3											
D	[Atlas Agena B]	1962ψ	1962 Jun 17 1 day 1962 Jun 18	Cylinder 2000?	8 long? 1.5 dia	Orbit unknown							
D	[Thor Agena B]	1962ω1	1962 Jun 18.85 498.1 days 1963 Oct 30.0	Cylinder 1500?	8 long? 1.5 dia	1962 Jun 22.6 1962 Dec 17.7 1963 Oct 26.8	82.14 82.12 82.10	92.49 91.86 88.69	6769 6738 6583	370 344 198	411 375 211	0.003 0.002 0.001	- 50 336
D	Fragments	1962ω2,3											
	Tiros 5	1962αα1	1962 Jun 19.51 100 years	Cylinder 129	0.56 long 1.07 dia	1962 Jul 13.4	58.08	100.44	7159	588	974	0.027	121
	Tiros 5 rocket	1962αα2	1962 Jun 19.51 50 years	Cylinder 24	1.5 long 0.46 dia	1962 Jul 17.5 1969 Sep 16.0	58.08 58.08	100.40 100.02	7157 7138	586 590	972 930	0.027 0.024	- -
1d	Fragments	1962αα3-5											
D	[Thor Agena B]	1962αβ	1962 Jun 23.02 14.7 days 1962 Jul 7.7	Cylinder 1500?	8 long? 1.5 dia	1962 Jun 27.5 1962 Jul 3.9	75.09 75.09	89.58 88.82	6631 6593	213 209	293 222	0.006 0.001	140 121

Name	Launch date, lifetime and descent date	Shape and weight (kg)	Size (m)	Date of orbital determination	Orbital inclination (deg)	Nodal period (min)	Semi major axis (km)	Perigee height (km)	Apogee height (km)	Orbital eccentricity	Argument of perigee (deg)
D [Thor Agena D] 1962αγ	1962 Jun 28.05 78 days 1962 Sep 14	Cylinder 1500?	8 long? 1.5 dia	1962 Jul 3.5 1962 Aug 20.4 1962 Sep 8.4	76.04 76.04 76.01	93.55 91.24 89.48	6828 6713 6619	211 187 176	689 482 305	0.035 0.022 0.010	- 358 297
D Cosmos 6 (Sputnik 16) 1962α δ1	1962 Jun 30.67 70 days 1962 Sep 8	Ellipsoid 400?	1.8 long 1.2 dia	1962 Jul 1.0 1962 Aug 9.5 1962 Aug 26.1	48.96 48.96 48.95	90.54 89.91 89.46	6683 6652 6628	264 247 241	344 300 258	0.006 0.004 0.001	72 258 358
D Cosmos 6 rocket 1962αδ2	1962 Jun 30.67 38.5 days 1962 Aug 8.2	Cylinder 1500?	8 long 1.65 dia	1962 Jul 1.0 1962 Jul 21.5 1962 Aug 1.6	48.97 48.97 48.95	90.49 89.86 89.22	6680 6649 6617	262 244 226	342 297 253	0.006 0.004 0.002	72 150 -
Telstar 1 1962αε1	1962 Jul 10.36 10000 years	Sphere 77	0.86 dia	1962 Jul 10.4	44.79	157.65	9670	952	5632	0.242	165
Telstar 1 rocket 1962αε2	1962 Jul 10.36 3000 years	Cylinder 24	1.5 long 0.46 dia	1962 Jul 17.5	44.78	157.53	9664	947	5625	0.242	176
D [Atlas Agena B] 1962αζ1	1962 Jul 18.87 9 days 1962 Jul 27	Cylinder 2000?	8 long? 1.5 dia	1962 Jul 22.2 1962 Jul 24.6	96.12 96.12	88.73 88.50	6588 6577	184 179	236 218	0.004 0.003	217 -
D Fragment 1962αζ2											
D [Thor Agena B] 1962αη	1962 Jul 21.04 24 days 1962 Aug 14	Cylinder 1500?	8 long? 1.5 dia	1962 Jul 21.2 1962 Aug 4.1 1962 Aug 12.8	70.29 70.29 70.29	90.42 89.69 88.42	6673 6637 6574	208 192 176	381 325 216	0.013 0.010 0.003	155 139 122
D [Thor Agena B] 1962αθ	1962 Jul 28.02 27 days 1962 Aug 24	Cylinder 1500?	8 long? 1.5 dia	1962 Jul 28.2 1962 Aug 16.6 1962 Aug 21.7	71.09 71.09 71.09	90.64 89.69 88.93	6684 6637 6599	225 192 188	386 325 254	0.012 0.010 0.005	155 119 109

	Name		Launch date, lifetime and descent date	Shape and weight (kg)	Size (m)	Date of orbital determination	Orbital inclination (deg)	Nodal period (min)	Semi major axis (km)	Perigee height (km)	Apogee height (km)	Orbital eccentricity	Argument of perigee (deg)
D R B	Cosmos 7 (Sputnik 17)	1962 αι1	1962 Jul 28.39 4.0 days 1962 Aug 1.4	Sphere-cylinder 4750?	4.3 long 2.4 dia	1962 Jul 30.3	64.95	90.04	6655	197	356	0.012	48
D	Cosmos 7 rocket	1962 αι2	1962 Jul 28.39 24 days 1962 Aug 21	Cylinder 1440	3.8 long 2.6 dia	1962 Jul 30.3 1962 Aug 12.1 1962 Aug 18.8	64.96 64.92 64.91	90.00 89.38 88.56	6653 6622 6582	208 198 171	341 291 237	0.010 0.007 0.005	67 58 52
D	Fragments	1962 αι3, 4											
D	[Thor Agena D]	1962 ακ1	1962 Aug 2.02 24 days 1962 Aug 26	Cylinder 1500?	8 long? 1.5 dia	1962 Aug 3.2 1962 Aug 17.7 1962 Aug 24.8	82.25 82.25 82.25	90.77 89.85 88.64	6689 6644 6584	204 199 179	418 332 232	0.016 0.010 0.004	149 99 71
D	Fragment	1962 ακ2											
D	[Atlas Agena B]	1962 αλ	1962 Aug 5.75 1 day 1962 Aug 6	Cylinder 2000?	8 long? 1.5 dia	1962 Aug 6.0	96.30	88.62	6583	205	205	0	-
D R M	Vostok 3	1962 αμ1	1962 Aug 11.35 3.94 days 1962 Aug 15.29	Sphere-cylinder 4722	4.3 long 2.4 dia	1962 Aug 11.5 1962 Aug 12.8 1962 Aug 13.8 1962 Aug 15.2	64.98 64.98 64.98 64.98	88.33 88.24 88.13 87.97	6570 6566 6561 6553	166 162 158 155	218 214 207 194	0.004 0.004 0.004 0.003	- - - -
D	Vostok 3 rocket	1962 αμ2	1962 Aug 11.35 2.7 days 1962 Aug 14.1	Cylinder 1440	3.8 long 2.6 dia	1962 Aug 13.0	64.82	87.50	6529	151	151	0	-
D R M	Vostok 4	1962 αν1	1962 Aug 12.33 2.96 days 1962 Aug 15.29	Sphere- cylinder 4728	4.3 long 2.4 dia	1962 Aug 12.4 1962 Aug 13.8 1962 Aug 14.8	64.95 64.95 64.95	88.39 88.26 88.18	6573 6567 6563	169 163 159	222 215 211	0.004 0.004 0.004	- - -
D	Vostok 4 rocket	1962 αν2	1962 Aug 12.33 2.4 days 1962 Aug 14.7	Cylinder 1440	3.8 long 2.6 dia	1962 Aug 13.0	64.80	88.38	6573	169	221	0.004	141

	Name		Launch date, lifetime and descent date	Shape and weight (kg)	Size (m)	Date of orbital determination	Orbital inclination (deg)	Nodal period (min)	Semi major axis (km)	Perigee height (km)	Apogee height (km)	Orbital eccentricity	Argument of perigee (deg)
D	Cosmos 8 (Sputnik 18)	1962αξ1	1962 Aug 18.21 364.7 days 1963 Aug 17.9	Ellipsoid 400?	1.8 long 1.2 dia	1962 Aug 18.3 1963 Jan 14.1 1963 Aug 14.9	48.97 48.96 48.93	92.93 91.94 88.26	6799 6751 6571	244 238 173	598 508 213	0.026 0.020 0.003	121 92 45
D	Cosmos 8 rocket	1962αξ2	1962 Aug 18.21 123.3 days 1962 Dec 19.5	Cylinder 1500?	8 long 1.65 dia	1962 Aug 19.7 1962 Oct 14.7 1962 Nov 29.0	48.98 48.98 48.96	92.92 91.95 90.50	6799 6752 6681	251 245 232	591 502 373	0.025 0.019 0.011	132 31 246
	[Blue Scout]	1962αο1	1962 Aug 23.49 80 years	- 20?	1 dia?	1962 Aug 23.5 1965 Sep 15.6	98.66 98.68	99.59 99.57	7117 7115	620 615	858 858	0.017 0.017	240 180
	Altair rocket	1962αο4	1962 Aug 23.49 40 years	Cylinder 24	1.5 long 0.46 dia	1962 Oct 10.5 1966 Oct 15.5	98.68 98.69	99.57 99.56	7115 7114	615 616	858 855	0.017 0.017	- -
1d	Fragments	1962αο 2,3											
D	Sputnik 19	1962απ1	1962 Aug 25.12 3 days 1962 Aug 28	Cylinder 6500? full	7 long? 2.0 max dia	1962 Aug 25.9 1962 Aug 27.0	64.88 64.88	88.73 88.37	6590 6572	173 168	252 221	0.006 0.004	90 -
D	Sputnik 19 rocket	1962απ2	1962 Aug 25.12 8 days 1962 Sep 2	Cylinder 2500?	7.5 long 2.6 dia	1962 Aug 25.9 1962 Aug 31.0	64.89 64.89	89.38 88.63	6623 6585	178 161	311 253	0.010 0.007	90 -
D	Fragments	1962απ3-8											
D	[Thor Agena D]	1962ασ	1962 Aug 29.05 12 days 1962 Sep 10	Cylinder 1500?	8 long? 1.5 dia	1962 Aug 30.1 1962 Sep 7.4	65.21 65.21	90.38 89.09	6672 6608	187 170	400 289	0.016 0.009	182 -
D	Sputnik 20	1962ατ1	1962 Sep 1.1 5 days 1962 Sep 6	Cylinder 6500? full	7 long? 2.0 max dia	1962 Sep 1	64.9	89.4	6623	180?	310?	0.009?	-
D	Fragments *	1962ατ2-4											

Space Vehicle: Mariner 2, 1962 αρ1; and rocket 1962 αρ2.   * Includes Sputnik 20 rocket (1962 α τ_2); life 2 days.

	Name		Launch date, lifetime and descent date	Shape and weight (kg)	Size (m)	Date of orbital determination	Orbital inclination (deg)	Nodal period (min)	Semi major axis (km)	Perigee height (km)	Apogee height (km)	Orbital eccentricity	Argument of perigee (deg)
D	[Thor Agena B]	1962 αυ	1962 Sep 1.86 785.54 days 1964 Oct 26.40	Cylinder 1500?	8 long? 1.5 dia	1962 Sep 9.2 1963 Dec 29.4 1964 Aug 22.5	82.82 82.80 82.79	94.42 92.39 90.65	6863 6764 6684	300 279 266	669 492 346	0.027 0.016 0.006	139 207 35
D	Sputnik 21*	1962 αφ1	1962 Sep 12.07 2 days 1962 Sep 14	Cylinder 6500? full	7 long? 2.0 max dia	1962 Sep 13	64.9	88.48	6578	186	213	0.002	-
D	Fragments**	1962 αφ2-7											
D	[Thor Agena B] (contained TRS 1)†	1962 αχ	1962 Sep 17.99 62.2 days 1962 Nov 19.2	Cylinder 1500?	8 long? 1.5 dia	1962 Sep 19.5 1962 Oct 17.6 1962 Nov 7.6	81.84 81.84 81.84	93.33 91.87 90.09	6814 6743 6655	204 196 191	668 533 363	0.034 0.025 0.013	154 51 -
	Tiros 6	1962 αψ1	1962 Sep 18.37 80 years	Cylinder 127	0.56 long 1.07 dia	1962 Sep 19.4 1965 Jun 21.7	58.32 58.33	98.73 98.70	7078 7076	686 685	713 710	0.002 0.002	102 62
	Tiros 6 rocket	1962 αψ2	1962 Sep 18.37 7419 days	Cylinder 24	1.5 long 0.46 dia	1962 Oct 10.6 1965 Jul 15.5	58.29 58.3	98.71 98.64	7076 7073	684 681	712 708	0.002 0.002	- -
	Fragments	1962 αψ 3, 4											
D R	Cosmos 9	1962 αω1	1962 Sep 27.40 4.0 days 1962 Oct 1.4	Sphere-cylinder 4750?	4.3 long 2.4 dia	1962 Sep 28	65.0	90.91	6697	292	346	0.004	-
D	Cosmos 9 rocket	1962 αω2	1962 Sep 27.40 86 days 1962 Dec 22	Cylinder 1440	3.8 long 2.6 dia	1962 Oct 5.8	64.94	91.00	6701	304	343	0.003	61
D	Fragments	1962 αω3-8											

*Sputniks 19, 20 and 21 are believed to have been Venus Probe launchers.

**Includes Sputnik 21 rocket (1962 α ϕ_2); life < 1 day.

†Tetrahedron Research Satellite

Year of launch 1962, continued

Name	Launch date, lifetime and descent date	Shape and weight (kg)	Size (m)	Date of orbital determination	Orbital inclination (deg)	Nodal period (min)	Semi major axis (km)	Perigee height (km)	Apogee height (km)	Orbital eccentricity	Argument of perigee (deg)
Alouette 1 1962 βα 1	1962 Sep 29.25 2000 years	Oblate spheroid 144.7	0.86 long 1.07 dia	1962 Sep 29.7	80.46	105.52	7392	996	1032	0.002	8
Alouette 1* rocket (TAVE) 1962 βα 2	1962 Sep 29.25 1500 years	Cylinder 790? (payload 91)	6 long 1.5 dia	1962 Sep 29.7	80.47	105.53	7393	1008	1023	0.001	336
Fragments 1962βα3,4											
D [Thor Agena D] 1962 ββ	1962 Sep 29.99 14 days 1962 Oct 14	Cylinder 1500?	8 long? 1.5 dia	1962 Oct 1.5 1962 Oct 10.7	65.40 65.40	90.30 89.08	6668 6607	203 196	376 262	0.013 0.005	172 166
D Explorer 14 1962 βγ 1	1962 Oct 2.92 3 years?	Octagon + 4 vanes 40	1.30 long 0.74 dia	1962 Oct 10.6 1963 Dec 27.5 1964 Feb 15.5	32.95 42.31 42.77	2185 2184.6 2184.6	55784 55772 55773	281 2558 2601	98530 96229 96189	0.881 0.840 0.839	150 191 -
D Explorer 14 rocket 1962 βγ 2	1962 Oct 2.92 3 years?	Cylinder 24	1.5 long 0.46 dia	Orbit similar to 1962 βγ1							
D R M Mercury 8 (Sigma 7) 1962 βδ 1	1962 Oct 3.51 9.22 hours 1962 Oct 3.89	Cone-frustum 1370	2.90 long 1.83 dia	1962 Oct 3.6	32.55	88.75	6597	153	285	0.010	74
D Mercury 8 rocket 1962 βδ 2	1962 Oct 3.51 1 day 1962 Oct 4	Cylinder 3400	20 long 3.0 dia	1962 Oct 3.6	32.55	88.67	6594	156	275	0.009	74
D [Thor Agena B] 1962 βε	1962 Oct 9.79 37.3 days 1962 Nov 16.1	Cylinder 1500?	8 long? 1.5 dia	1962 Oct 10.8 1962 Oct 20.7 1962 Nov 14.6	81.96 81.96 81.96	90.96 90.59 88.37	6698 6680 6569	213 209 170	427 395 212	0.016 0.014 0.003	58 20 286
D R Cosmos 10 1962 βζ 1	1962 Oct 17.39 4.0 days 1962 Oct 21.4	Sphere-cylinder 4750?	4.3 long 2.4 dia	1962 Oct 17.4	65.00	90.14	6660	197	367	0.013	-
D Cosmos 10 rocket 1962 βζ 2	1962 Oct 17.39 19 days 1962 Nov 5	Cylinder 1440	3.8 long 2.6 dia	1962 Oct 31.7	64.90	89.06	6606	196	260	0.005	61

Space vehicle: Ranger 5, 1962 βη1; and rocket 1962 βη 2.

*Carried Thor Agena Vibration Experiment.

Year of launch 1962, continued

	Name		Launch date, lifetime and descent date	Shape and weight (kg)	Size (m)	Date of orbital determination	Orbital inclination (deg)	Nodal period (min)	Semi major axis (km)	Perigee height (km)	Apogee height (km)	Orbital eccentricity	Argument of perigee (deg)
D	Cosmos 11	1962 $\beta\theta$ 1	1962 Oct 20.16	Ellipsoid	1.8 long	1962 Oct 28.2	48.97	95.95	6946	234	901	0.048	148
			575.9 days	400?	1.2 dia	1963 Dec 30.4	48.95	92.15	6762	234	533	0.022	249
			1964 May 18.1			1964 May 3.1	48.94	89.50	6631	200	306	0.008	129
D	Cosmos 11 rocket	1962 $\beta\theta$ 2	1962 Oct 20.16	Cylinder	8 long	1962 Oct 29.2	48.95	95.77	6937	233	885	0.047	154
			228.8 days	1500?	1.65 dia	1962 Dec 27.6	48.93	94.63	6883	226	784	0.041	50
			1963 Jun 6.0			1963 May 21.3	48.93	90.49	6679	221	381	0.012	351
D	Sputnik 22	1962 $\beta\iota$ 1	1962 Oct 24.75?	Cylinder	7 long?	1962 Oct 25	64.89	91.18	6711	180?	485?	0.022?	-
			5 days	6500? full	2.0 max								
			1962 Oct 29	(before explosion)	dia								
D	Fragments	1962 $\beta\iota$ 2-24											
D	Star-rad †	1962 $\beta\kappa$	1962 Oct 26.68	Cone-cylinder	9 long?	1962 Nov 10.5	71.39	147.43	9244	194	5537	0.289	147
			1805.00 days	1500?	1.5 dia	1966 Oct 18.7	71.32	114.26	7800	197	2646	0.157	166
			1967 Oct 5.68	(payload 340)		1967 Oct 2.7	71.21	90.07	6657	159	399	0.018	357
	Explorer 15	1962 $\beta\lambda$ 1	1962 Oct 27.97	Octagon + 4 vanes	1.30 long	1962 Oct 28.0	18.02	315.20	15353	313	17640	0.564	137
			100 years?	45.2	0.74 dia	1964 Aug 9.7	18.02	311.44	15247	300	17438	0.562	196
	Explorer 15 rocket	1962 $\beta\lambda$ 2	1962 Oct 27.97	Cylinder	1.5 long	Initial orbit similar to 1962 $\beta\lambda$ 1							
			100 years?	24	0.46 dia								
	Anna 1B*	1962 $\beta\mu$ 1	1962 Oct 31.34	Spheroid	0.91 long	1962 Oct 31.8	50.14	107.84	7508	1077	1182	0.007	202
			3000 years	161	1.22 dia								
	Anna 1B rocket	1962 $\beta\mu$ 2	1962 Oct 31.34	Cylinder	4.8 long	1962 Nov 7.6	50.13	107.53	7492	1069	1159	0.006	-
			1500 years	450?	1.4 dia	1966 Jun 15.5	50.13	107.56	7494	1068	1164	0.006	-
	Mars 1	1962 $\beta\nu$ 1	1962 Nov 1.68?	Cylinder	3.3 long	Initial earth-satellite orbit similar to 1962 $\beta\nu$ 3							
			Indefinite	893.5	1.1 dia								
D	Sputnik 23 rocket	1962 $\beta\nu$ 3	1962 Nov 1.68?	Cylinder	7.5 long	1962 Nov 1.7	64.9	88.65	6587	174	243	0.005	-
			0.6 day?	2500?	2.6 dia								
			1962 Nov 2.3										
D	Sputnik 23	1962 $\beta\nu$ 2	1962 Nov 1.68?	Irregular	2 long?	Initial orbit similar to 1962 $\beta\nu$ 3							
			2 days		2 dia?								
			1962 Nov 3										

In the United States, Mars 1 has been designated 1962 $\beta\nu$3 and Sputnik 23 rocket 1962 $\beta\nu$1. * Army, Navy, NASA, Air Force. † Starfish radiation.

	Name	Launch date, lifetime and descent date	Shape and weight (kg)	Size (m)	Date of orbital determination	Orbital inclination (deg)	Nodal period (min)	Semi major axis (km)	Perigee height (km)	Apogee height (km)	Orbital eccentricity	Argument of perigee (deg)
D	Sputnik 24* 1962 βξ 1	1962 Nov 4.65 1 day 1962 Nov 5	Cylinder 6500? full	7 long? 2.0 max dia	1962 Nov 4.8	64.8	88.75	6591	200?	226?	0.002?	-
D	Sputnik 24 rocket 1962 βξ 3	1962 Nov 4.65 76 days 1963 Jan 19	Cylinder 2500?	7.5 long 2.6 dia	1962 Nov 5.4 1962 Dec 1.3	64.7 64.7	92.42 91.49	6772 6726	197 193	590 503	0.029 0.023	353 355
D	Fragments 1962 βξ 2,4,5											
D	[Thor Agena B] 1962 βο	1962 Nov 5.93 27 days 1962 Dec 3	Cylinder 1500?	8 long? 1.5 dia	1962 Nov 7.7 1962 Nov 29.2	74.98 74.97	90.71 89.02	6687 6603	208 185	409 265	0.015 0.006	150 106
D	[Atlas Agena B] 1962 βπ (contained TRS)	1962 Nov 11.85 1 day 1962 Nov 12	Cylinder 2000?	8 long? 1.5 dia	1962 Nov 12.0	96.00	88.65	6584	206	206	0	-
D	[Thor Agena B] 1962 βρ	1962 Nov 24.92 18 days 1962 Dec 13	Cylinder 1500?	8 long? 1.5 dia	1962 Nov 27.0 1962 Dec 4.3	65.14 65.13	89.92 89.63	6649 6635	204 204	337 310	0.010 0.008	145 150
D	[Thor Agena D] 1962 βσ	1962 Dec 4.90 3 days 1962 Dec 8	Cylinder 1500?	8 long? 1.5 dia	1962 Dec 5.1 1962 Dec 7.0	65.1 65.1	89.16 88.40	6612 6574	194 169	273 222	0.006 0.004	154 223

*Sputniks 22 and 24 are believed to have been Mars probe launchers.

	Name		Launch date, lifetime and descent date	Shape and weight (kg)	Size (m)	Date of orbital determination	Orbital inclination (deg)	Nodal period (min)	Semi major axis (km)	Perigee height (km)	Apogee height (km)	Orbital eccentricity	Argument of perigee (deg)
D	[Thor Agena D]	1962 βτ 1	1962 Dec 13.17 1518.97 days 1967 Feb 9.14	Sphere 23	0.6 dia	1962 Dec 17.0 1964 Aug 21.3 1967 Feb 6.5	70.36 70.36 70.25	116.26 109.75 89.37	7887 7593 6621	231 228 177	2786 2202 309	0.162 0.130 0.010	145 157 343
D	Injun 3	1962 βτ 2	1962 Dec 13.17 2082.25 days 1968 Aug 25.42	Sphere 52	0.61 dia	1962 Dec 13.7 1964 Jul 3.2 1968 Jan 3.4	70.38 70.34 70.28	116.32 112.95 99.23	7888 7740 7100	235 236 225	2785 2484 1219	0.162 0.145 0.070	149 233 57
D	[Thor Agena D]	1962 βτ 3	1962 Dec 13.17 200.8 days 1963 Jul 2.0	Sphere 0.2?	0.15 dia	1962 Dec 16.5 1963 Mar 9.7 1963 Apr 21.1	70.33 70.32 70.28	115.89 108.94 101.83	7871 7564 7222	226 225 223	2763 2147 1465	0.161 0.127 0.086	146 53 353
D	Surcal 1A*	1962βτ4	1962 Dec 13.17 1132.66 days 1966 Jan 18.83	Rectangular box 3?	0.2 side	1962 Dec 16.5 1964 Aug 18.8 1965 Sep 20.0	70.34 70.31 70.31	116.24 106.25 96.56	7886 7431 6974	231 227 212	2784 1878 979	0.162 0.111 0.055	146 136 292
D	[Thor Agena D]	1962 βτ 5	1962 Dec 13.17 1515.46 days 1967 Feb 5.63	Sphere 23	0.6 dia	1962 Dec 19.5 1964 Aug 9.5 1967 Jan 29.1	70.34 70.31 70.19	116.22 109.73 90.43	7885 7592 6676	229 226 191	2785 2200 405	0.162 0.130 0.016	143 170 2
D	Injun 3 rocket	1962 βτ 6	1962 Dec 13.17 1928.04 days 1968 Mar 24.21	Cylinder 700?	6 long? 1.5 dia	1962 Dec 28.6 1964 Aug 6.9 1968 Jan 6.3	70.36 70.37 70.19	116.34 112.13 94.78	7889 7704 6888	248 238 214	2774 2412 806	0.160 0.141 0.043	- 189 358
	Relay 1	1962 βυ 1	1962 Dec 13.98 100000 years	Octagonal prism 78	0.81 long 0.74 dia	1962 Dec 14.0	47.49	185.01	10759	1322	7439	0.284	178
	Relay 1 rocket	1962 βυ 2	1962 Dec 13.98 50000 years	Cylinder 24	1.5 long 0.46 dia	1962 Dec 20.0	47.45	184.71	10750	1345	7398	0.282	184

* Surveillance calibration.

	Name		Launch date, lifetime and descent date	Shape and weight (kg)	Size (m)	Date of orbital determination	Orbital inclination (deg)	Nodal period (min)	Semi major axis (km)	Perigee height (km)	Apogee height (km)	Orbital eccentricity	Argument of perigee (deg)
D	[Thor Agena D]	1962 βφ	1962 Dec 14.89 25.0 days 1963 Jan 8.9	Cylinder 1500?	8 long? 1.5 dia	1962 Dec 15.8 1962 Dec 27.8 1963 Jan 4.5	70.97 70.95 70.95	90.46 89.85 89.08	6674 6643 6604	199 193 178	392 336 274	0.014 0.011 0.007	163 150 122
	Explorer 16	1962 βχ	1962 Dec 16.61 800 years	Cylinder 100.8	1.93 long 0.61 dia	1962 Dec 16.6	52.01	104.32	7344	750	1181	0.029	142
	Transit 5A	1962 βψ 1	1962 Dec 19.06 30 years	Octagon + 4 vanes + boom 61	0.25 long 0.46 dia	1962 Dec 20.0	90.62	99.12	7090	698	725	0.002	353
	Transit 5A rocket	1962 βψ 3	1962 Dec 19.06 40 years	Cylinder 24	1.5 long 0.46 dia	1962 Dec 20.7	90.74	99.11	7089	698	723	0.002	-
D	Fragments	1962 βψ 2,4,5											
D R	Cosmos 12	1962 βω 1	1962 Dec 22.39 7.9 days 1962 Dec 30.3	Sphere-cylinder 4750?	4.3 long 2.4 dia	1962 Dec 22.4	65.0	90.41	6673	198	392	0.015	-
D	Cosmos 12 rocket	1962 βω 2	1962 Dec 22.39 31 days 1963 Jan 22	Cylinder 1440	3.8 long 2.6 dia	1963 Jan 2.6	64.94	90.17	6662	197	370	0.013	-

	Name		Launch date, lifetime and descent date	Shape and weight (kg)	Size (m)	Date of orbital determination	Orbital inclina-tion (deg)	Nodal period (min)	Semi major axis (km)	Perigee height (km)	Apogee height (km)	Orbital eccen-tricity	Argument of perigee (deg)
D	Sputnik 25* rocket	1963-01A	1963 Jan 4.3 1 day 1963 Jan 5	Cylinder 2500?	7.5 long 2.6 dia	1963 Jan 4	64.9	87.5	6529	151?	151?	0	-
D	Fragments	1963-01B,C											
D	[Thor Agena D]	1963-02A	1963 Jan 7.88 16.3 days 1963 Jan 24.2	Cylinder 1500?	8 long? 1.5 dia	1963 Jan 7.9 1963 Jan 13.8 1963 Jan 21.4	82.23 82.23 82.19	90.54 90.01 88.75	6680 6651 6589	205 193 168	399 353 254	0.015 0.012 0.006	178 156 126
D	Fragment	1963-02B											
D	[Thor Agena D]	1963-03A	1963 Jan 16.92 2184.58 days 1969 Jan 9.50	Cylinder 1500?	8 long? 1.5 dia	1963 Jan 16.9 1968 Jan 2.2 1968 Jun 30.5	81.89 81.90 81.87	94.66 93.36 92.53	6874 6817 6775	459 418 384	533 459 410	0.005 0.003 0.002	40 216 -
D	Fragments	1963-03B,C											
	Syncom 1 **	1963-04A	1963 Feb 14.22 > million years	Cylinder 39	0.39 long 0.71 dia	1963 Feb 14.2	33.30	1425.5	41944	34392	36739	0.028	276
D	Syncom 1 rocket	1963-04B	1963 Feb 14.22 6836 days 1981 Nov 2	Cylinder 24	1.5 long 0.46 dia	1963 Apr 4.9 1969 Sep 16.0 1976 Jul 1.0	33.12 32.70 32.7	606.0 456.8 323.6	23753 19647 15615	252 276 233	34498 26262 18240	0.721 0.661 0.577	165 - -
	Altair rocket	1963-05B	1963 Feb 19.69 30 years	Cylinder 24	1.5 long 0.46 dia	1963 Mar 9.7 1970 Jul 1.0	100.48 100.48	97.79 96.68	7026 6973	505 483	791 707	0.020 0.016	289 -
D	[Blue Scout]	1963-05A	1963 Feb 19.69 6154 days 1979 Dec 26	- 40?	1 dia?	1963 Feb 19.7 1966 Oct 15.8	100.49 100.47	97.81 97.70	7028 7026	510 500	789 795	0.020 0.021	340 343
D	Fragments	1963-05C,D											
D R	Cosmos 13	1963-06A	1963 Mar 21.35 8.0 days 1963 Mar 29.3	Sphere-cylinder 4750?	4.3 long 2.4 dia	1963 Mar 21.4	64.97	89.65	6636	192	324	0.010	65?
D	Cosmos 13 rocket	1963-06B	1963 Mar 21.35 19 days 1963 Apr 9	Cylinder 1440	3.8 long 2.6 dia	Initial orbit similar to 1963-06A							

*Sputnik 25 is believed to have been a Luna launcher; 1963-01B is believed to have been a payload. (Decayed 1963 Jan 11.)
** Synchronous communication.

Year of launch 1963, continued

	Name		Launch date, lifetime and descent date	Shape and weight (kg)	Size (m)	Date of orbital determination	Orbital inclination (deg)	Nodal period (min)	Semi major axis (km)	Perigee height (km)	Apogee height (km)	Orbital eccentricity	Argument of perigee (deg)
D	[Thor Agena D]	1963-07A	1963 Apr 1.92 25.0 days 1963 Apr 26.9	Cylinder 1500?	8 long? 1.5 dia	1963 Apr 2.0 1963 Apr 10.4	75.40 75.38	90.66 90.28	6683 6661	201 198	408 367	0.015 0.013	158 134
	Luna 4*	1963-08A	1963 Apr 2.35? indefinite?	Cylinder 1422	2.5 long? 1.0 dia?	Initial earth-satellite orbit similar to 1963-08C 1963 Apr	 -	 42000	 398000	 89250	 694000	 0.760	 -
D	Luna 4 launcher rocket	1963-08C	1963 Apr 2.35? 1 day 1963 Apr 3	Cylinder 2500?	7.5 long 2.6 dia	1963 Apr 2.4	65	88	6554?	176?	176?	0?	-
D	Explorer 17	1963-09A	1963 Apr 3.08 1331.27 days 1966 Nov 24.35	Sphere 185	0.89 dia	1963 Apr 3.1 1965 Sep 15.3 1966 Oct 20.1	57.63 57.59 57.60	96.40 93.62 90.31	6964 6832 6670	255 242 218	917 666 365	0.048 0.031 0.011	49 76 68
D	Explorer 17 rocket	1963-09B	1963 Apr 3.08 235.6 days 1963 Nov 24.7	Cylinder 24	1.5 long 0.46 dia	1963 Apr 3.6 1963 May 31.6	57.59 57.59	96.32 95.12	6962 6904	247 245	920 807	0.048 0.041	51 -
D	Cosmos 14	1963-10A	1963 Apr 13.46 137.6 days 1963 Aug 29.1	Ellipsoid 400?	1.8 long 1.2 dia	1963 Apr 13.5 1963 Jun 2.4	48.95 48.88	91.99 91.29	6754 6722	252 253	499 435	0.018 0.013	- -
D	Cosmos 14 rocket	1963-10B	1963 Apr 13.46 84.2 days 1963 Jul 6.7	Cylinder 1500?	8 long 1.65 dia	1963 May 1.0 1963 Jun 9.0	48.90 48.90	91.59 90.64	6735 6689	249 237	465 384	0.016 0.011	205 42
D R	Cosmos 15	1963-11A	1963 Apr 22.35 5.0 days 1963 Apr 27.3	Sphere-cylinder 4750?	4.3 long 2.4 dia	1963 Apr 22.4	65.00	89.67	6637	160	358	0.015	-
D	Cosmos 15 rocket	1963-11B	1963 Apr 22.35 9.5 days 1963 May 1.8	Cylinder 1440	3.8 long 2.6 dia	1963 Apr 27.3	64.95	89.21	6614	170	302	0.010	58

* In the US, Luna 4 has been designated as 1963-08B and the rocket as 1963-08A. There may be a rocket in the Luna 4 orbit.

	Name		Launch date, lifetime and descent date	Shape and weight (kg)	Size (m)	Date of orbital determination	Orbital inclination (deg)	Nodal period (min)	Semi major axis (km)	Perigee height (km)	Apogee height (km)	Orbital eccentricity	Argument of perigee (deg)
D R	Cosmos 16	1963-12A	1963 Apr 28.40 9.9 days 1963 May 8.3	Sphere-cylinder 4750?	4.3 long 2.4 dia	1963 Apr 28.4	65.02	90.38	6669	194	388	0.015	57
D	Cosmos 16 rocket	1963-12B	1963 Apr 28.40 22.3 days 1963 May 20.7	Cylinder 1440	3.8 long 2.6 dia	1963 Apr 30.3	65.02	90.48	6674	196	396	0.015	73
	Telstar 2	1963-13A	1963 May 7.48 200000 years	Spheroid 79.4	0.94 long 0.86 dia	1963 May 7.5	42.73	225.05	12267	974	10803	0.401	172
	Telstar 2 rocket	1963-13B	1963 May 7.48 100000 years	Cylinder 24	1.52 long 0.45 dia	1963 May 13.3	42.76	224.81	12258	989	10770	0.399	178
	Midas [Atlas Agena B]	1963-14A	1963 May 9.84 100000 years	Cylinder 2000?	9 long? 1.5 dia	1963 May 12.2	87.42	166.48	10020	3604	3680	0.004	69
	DASH 1*	1963-14B	1963 May 9.84 50 years?	Inflated sphere 0.05?	0.31 dia	1963 May 15.2 1968 Jan 8.6	87.35 87.19	166.51 166.24	10022 10015	3604 2725	3683 4548	0.004 0.091	69 280
	TRS 2	1963-14C	1963 May 9.84 50000 years	Tetrahedron 0.7	0.17 side	1963 May 29.2	87.42	166.47	10020	3606	3678	0.004	54
	TRS 3	1963-14D	1963 May 9.84 50000 years	Tetrahedron 0.7	0.17 side	1963 May 29.2	87.41	166.47	10020	3612	3672	0.003	54
D	Westford Needles	1963-14	1963 May 9.84 2 to 4 years	Annulus 23	20000 km dia	1963 Aug 7 1964 Jan 29	87.35 87.22	166.46 166.0	10019 10007	3367 2760	3915 4497	0.027 0.087	180 349
36d	Fragments 1963-14E-DF												

* Density and scale height.

	Name		Launch date, lifetime and descent date	Shape and weight (kg)	Size (m)	Date of orbital determination	Orbital inclination (deg)	Nodal period (min)	Semi major axis (km)	Perigee height (km)	Apogee height (km)	Orbital eccentricity	Argument of perigee (deg)
D R M	Mercury 9* (Faith 7)	1963-15A	1963 May 15.54 1.44 days 1963 May 16.98	Cone frustum 1370	2.90 long 1.83 dia	1963 May 15.6	32.54	88.64	6592	161	267	0.008	-
D	Mercury 9 rocket	1963-15B	1963 May 15.54 0.6 day 1963 May 16.2	Cylinder 3400	20 long 3.0 dia	1963 May 15.7	32.54	88.22	6571	167	219	0.004	72
D	[Thor Agena D]	1963-16A	1963 May 18.94 8 days 1963 May 27	Cylinder 1500?	8 long? 1.5 dia	1963 May 20.4	74.54	91.12	6703	153	497	0.025	142
D	Cosmos 17	1963-17A	1963 May 22.13 742.79 days 1965 Jun 2.92	Ellipsoid 400?	1.8 long 1.2 dia	1963 May 22.2 1964 Jan 1.9 1965 May 28.7	49.0 49.0 48.93	94.99 93.69 88.58	6902 6839 6586	260 256 185	788 666 230	0.037 0.030 0.004	107 354 270
D	Cosmos 17 rocket	1963-17C	1963 May 22.13 316.7 days 1964 Apr 2.8	Cylinder 1500?	8 long 1.65 dia	1963 May 30.6 1964 Jan 7.4 1964 Feb 15.5	49.0 49.0 49.01	94.74 91.80 91.22	6891 6745 6716	265 238 246	761 495 430	0.036 0.019 0.014	139 66 -
D	Fragments	1963-17B-F											
D R	Cosmos 18	1963-18A	1963 May 24.45 9.0 days 1963 Jun 2.4	Sphere-cylinder 4750?	4.3 long 2.4 dia	1963 May 24.5	65.0	89.30	6620	196	288	0.007	-
D	Cosmos 18 rocket	1963-18B	1963 May 24.45 14.6 days 1963 Jun 8.0	Cylinder 1440	3.8 long 2.6 dia	1963 May 26.9 1963 Jun 3.4	65.0 64.95	89.47 88.74	6629 6593	198 195	304 235	0.008 0.003	96 92
D	[Thor Agena D]	1963-19A	1963 Jun 13.00 29.1 days 1963 Jul 12.1	Cylinder 1500?	8 long? 1.5 dia	1963 Jun 14.4 1963 Jul 10.0	81.87 81.82	90.67 88.48	6684 6577	192 173	419 225	0.017 0.004	135 36

* Mercury 9 ejected a 0.15m flashing-light capsule about 1963 May 15.72.

Year of launch 1963, continued

	Name		Launch date, lifetime and descent date	Shape and weight (kg)	Size (m)	Date of orbital determination	Orbital inclination (deg)	Nodal period (min)	Semi major axis (km)	Perigee height (km)	Apogee height (km)	Orbital eccentricity	Argument of perigee (deg)
D R M	Vostok 5	1963-20A	1963 Jun 14.50 4.96 days 1963 Jun 19.46	Sphere-cylinder 4720	4.3 long 2.4 dia	1963 Jun 14.8 1963 Jun 15.5 1963 Jun 17.5	64.97 64.97 64.97	88.27 88.23 88.00	6564 6561 6549	162 160 152	209 206 189	0.004 0.004 0.003	- - -
D	Vostok 5 rocket	1963-20B	1963 Jun 14.50 2.0 days 1963 Jun 16.5	Cylinder 1440	3.8 long 2.6 dia	1963 Jun 15.9	64.86	87.88	6547	150	187	0.003	110
D	[Thor Agena D]	1963-21A	1963 Jun 15.61 53.4 days 1963 Aug 8.0	Cylinder 1300?	8 long? 1.5 dia	1963 Jun 16.1 1963 Jul 13.2	69.87 69.87	95.65 93.34	6924 6811	172 160	919 706	0.054 0.040	152 109
D	Lofti 2A	1963-21B	1963 Jun 15.61 32.8 days 1963 Jul 18.4	Sphere 26	0.61 dia	1963 Jun 15.6 1963 Jul 11.0	69.87 69.84	95.71 91.24	6926 6709	171 161	925 501	0.054 0.026	152 110
D	SR 6A	1963-21C	1963 Jun 15.61 46.7 days 1963 Aug 1.3	Sphere 39	0.61 dia	1963 Jul 18.3 1963 Jul 31.3	69.88 69.85	92.08 88.84	6751 6590	155 146	590 278	0.032 0.010	100 74
D	Radose *	1963-21D	1963 Jun 15.61 44.9 days 1963 Jul 30.5	Sphere 25?	0.51 dia	1963 Jun 28.9 1963 Jul 26.7	69.88 69.82	94.84 89.91	6884 6644	175 152	837 379	0.048 0.017	139 85
D	[Thor Agena D]	1963-21E	1963 Jun 15.61 42.1 days 1963 Jul 27.7	Sphere 35?	0.61 dia	1963 Jun 26.4 1963 Jul 26.5	69.91 69.87	94.82 88.74	6883 6585	181 154	829 260	0.048 0.008	139 87
D	Surcal 1B	1963-21F	1963 Jun 15.61 19.7 days 1963 Jul 5.3	rectangular box 3?	0.2 side	1963 Jun 17.4 1963 Jul 2.5	69.86 69.81	95.26 90.67	6906 6680	169 168	887 435	0.052 0.020	150 123
D	Fragment	1963-21G											
	Transit 7 [Blue Scout]	1963-22A	1963 Jun 16.08 30 years	octagon + 4 vanes + boom 55	0.25 long 0.46 dia	1963 Jun 16.5	89.97	99.67	7119	724	757	0.002	64
	Altair rocket	1963-22B	1963 Jun 16.08 50 years	Cylinder 24	1.5 long 0.46 dia	1963 Jun 16.2	90.02	99.67	7119	720	759	0.002	-
1d	Fragments	1963-22C, D											

* Radiation dosimeter

	Name		Launch date, lifetime and descent date	Shape and weight (kg)	Size (m)	Date of orbital determination	Orbital inclination (deg)	Nodal period (min)	Semi major axis (km)	Perigee height (km)	Apogee height (km)	Orbital eccentricity	Argument of perigee (deg)
D R M	Vostok 6	1963-23A	1963 Jun 16.40 2.95 days 1963 Jun 19.35	Sphere-cylinder 4713	4.3 long 2.4 dia	1963 Jun 16.4 1963 Jun 17.5	65.09 65.09	88.34 88.25	6571 6566	168 164	218 212	0.004 0.004	80 -
D	Vostok 6 rocket	1963-23B	1963 Jun 16.40 1.7 days? 1963 Jun 18.1?	Cylinder 1440	3.8 long 2.6 dia	1963 Jun 16.8	65.08	88.38	6571	163	223	0.005	76
	Tiros 7	1963-24A	1963 Jun 19.41 40 years	Cylinder 135	0.56 long 1.07 dia	1963 Jun 19.5 1968 Jan 18.7	58.23 58.23	97.40 97.34	7013 7010	621 616	649 648	0.002 0.002	17 141
D	Tiros 7 rocket	1963-24B	1963 Jun 19.41 5666 days 1978 Dec 23	Cylinder 24	1.5 long 0.46 dia	1963 Jul 23.1 1970 Jun 1.0	58.21 58.22	97.32 96.27	7009 6959	612 577	650 585	0.003 0.001	56 -
1d	Fragments	1963-24C,D											
D	[Thor Agena D]	1963-25A	1963 Jun 27.03 29.7 days 1963 Jul 26.7	Cylinder 1000?	8 long? 1.5 dia	1963 Jun 29.5 1963 Jul 23.8	81.6 81.6	90.5 88.8	6674 6583	196 168	396 243	0.015 0.006	147 36
	Hitch-hiker 1	1963-25B	1963 Jun 27.03 40 years	Octagon 79.8 23 payload	0.3 long 0.9 dia	1963 Jul 20.1 1966 Oct 27.9 1968 Jan 17.9	82.11 82.11 82.13	132.55 131.81 131.14	8607 8580 8551	333 340 343	4132 4046 4002	0.221 0.217 0.214	350 54 343
	GRS*	1963-26A	1963 Jun 28.83 30 years	Cylinder 99.3	2.54 long 0.59 dia	1963 Jun 29.6 1971 Feb 15.0	49.74 49.74	102.05 100.33	7239 7156	411 409	1311 1147	0.062 0.052	136 -
D	[Thor Agena B]	1963-27A	1963 Jun 29.94 2310.72 days 1969 Oct 26.66	Cylinder 1500?	8 long? 1.5 dia	1963 Jul 10.0 1968 Jan 13.3 1968 Nov 30.5	82.3 82.33 82.33	94.84 93.92 93.11	6888 6844 6805	484 452 417	536 479 436	0.004 0.002 0.001	336 337 -
D	Fragments	1963-27B,C											

* Geophysics Research Satellite.

	Name		Launch date, lifetime and descent date	Shape and weight (kg)	Size (m)	Date of orbital determination	Orbital inclination (deg)	Nodal period (min)	Semi major axis (km)	Perigee height (km)	Apogee height (km)	Orbital eccentricity	Argument of perigee (deg)
D	[Atlas Agena D]	1963-28A	1963 Jul 12.86 5.2 days 1963 Jul 18.0	Cylinder 2000?	8 long? 1.5 dia	1963 Jul 17.1	95.37	88.2	6542	164	164	0	136
D	Fragments	1963-28B,C											
D	[Thor Agena D]	1963-29A	1963 Jul 19.00 25.8 days 1963 Aug 13.8	Cylinder 1500?	8 long? 1.5 dia	1963 Jul 20.7 1963 Jul 26.8 1963 Aug 11.6	82.86 82.86 82.86	90.44 90.37 88.65	6669 6656 6573	194 189 178	387 367 215	0.014 0.013 0.003	142 118 43
D	Fragment	1963-29B											
	Midas 7* [Atlas Agena B]	1963-30A	1963 Jul 19.16 100000 years	Cylinder 2000?	9 long? 1.5 dia	1963 Jul 23.5	88.41	167.80	10077	3670	3727	0.003	16
	TRS 4	1963-30B	1963 Jul 19.16 5000 years	Tetrahedron 0.8	0.17 side	1963 Jul 20.9 1966 Aug 24.3	88.36 88.44	167.79 167.90	10076 10081	3662 3662	3734 3743	0.003 0.004	- 334
D	DASH 2	1963-30D	1963 Jul 19.16 2824.22 days 1971 Apr 12.38	Inflated sphere 1.25?	2.4 dia?	1963 Aug 6.1 1965 Sep 22.2 1970 Aug 1.0	88.42 87.96 85.24	167.96 167.64 164.2	10083 10070 9936	3665 2906 841	3745 4477 6274	0.004 0.078 0.273	45 300 -
	Fragments	1963-30C,E-G											
	Syncom 2	1963-31A	1963 Jul 26.61 > million yr	Cylinder 39	0.39 long 0.71 dia	1963 Jul 27.8 1964 Oct 9.6	33.05 32.36	1454.0 1437.9	42512 42201	35584 35780	36693 35865	0.013 0.001	62 221
	Syncom 2 rocket	1963-31B	1963 Jul 26.61 23 years	Cylinder 24	1.5 long 0.46 dia	1963 Aug 15.0 1968 Jan 15.5 1970 Nov 1.0	33.14 32.87 32.7	636.5 393.8 248.4	24505 17796 13089	255 190 216	36029 22647 13205	0.730 0.631 0.496	- - -
D r?	[Thor Agena D]	1963-32A	1963 Jul 31.00 12.0 days 1963 Aug 12.0	Cylinder 1500?	8 long? 1.5 dia	1963 Aug 5.7	74.95	90.4	6663	157	411	0.019	110
D	Fragment	1963-32B											

*Contains a TRS.

	Name		Launch date, lifetime and descent date	Shape and weight (kg)	Size (m)	Date of orbital determination	Orbital inclina-tion (deg)	Nodal period (min)	Semi major axis (km)	Perigee height (km)	Apogee height (km)	Orbital eccen-tricity	Argument of perigee (deg)
D	Cosmos 19	1963-33A	1963 Aug 6.25 237.07 days 1964 Mar 30.32	Ellipsoid 400?	1.8 long 1.2 dia	1963 Aug 15.6 1963 Oct 10.7 1964 Jan 7.0	49.01 49.01 49.00	92.11 91.71 90.89	6760 6740 6704	267 261 252	497 463 400	0.017 0.015 0.011	153 99 98
D	Cosmos 19 rocket	1963-33B	1963 Aug 6.25 124.8 days 1963 Dec 9.0	Cylinder 1500?	8 long 1.65 dia	1963 Aug 15.5 1963 Oct 21.1 1963 Dec 3.7	49.00 49.00 48.94	92.00 91.07 89.58	6756 6712 6626	267 253 235	489 415 261	0.016 0.012 0.002	- 109 299
D	[Thor Agena D]	1963-34A	1963 Aug 25.02 18.6 days 1963 Sep 12.6	Cylinder 1500?	8 long? 1.5 dia	1963 Sep 7.3	75.01	89.4	6618	161	320	0.012	104
D	Fragment	1963-34B											
D	Lampo	1963-35A	1963 Aug 29.80 69.7 days 1963 Nov 7.5	Cylinder 1000?	8 long? 1.5 dia	1963 Sep 3.0 1963 Oct 22.1	81.89 81.86	90.80 90.00	6686 6652	292 261	324 287	0.002 0.002	151 32
D	[Thor Agena D]	1963-35B	1963 Aug 29.80 29-30 days 1963 Sep 28-29	-	-	1963 Sep 2.7	81.89	92.07	6749	310	431	0.009	261
D	Fragments	1963-35C, D											
D	[Atlas Agena D]	1963-36A	1963 Sep 6.81 7.05 days 1963 Sep 13.86	Cylinder 2000?	8 long? 1.5 dia	1963 Sep 10.8	94.37	89.06	6594	168	263	0.007	103
D	Fragments	1963-36B-F											
D	[Thor Agena D]	1963-37A	1963 Sep 23.95 18.2 days 1963 Oct 12.1	Cylinder 1500?	8 long? 1.5 dia	1963 Sep 24.1 1963 Oct 10.8	74.90 74.89	90.63 88.64	6679 6594	161 150	441 282	0.021 0.010	158 101

	Name		Launch date, lifetime and descent date	Shape and weight (kg)	Size (m)	Date of orbital determination	Orbital inclination (deg)	Nodal period (min)	Semi major axis (km)	Perigee height (km)	Apogee height (km)	Orbital eccentricity	Argument of perigee (deg)
	Ablestar rocket	1963-38A	1963 Sep 28.84 500 years	Cylinder 450?	4.8 long 1.4 dia	1963 Sep 29.7	89.80	107.13	7466	1069	1107	0.003	240
	Transit 5B-1	1963-38B	1963 Sep 28.84 1000 years	Octagon + boom 70	0.5 long 0.46 dia	1963 Oct 11.1	89.90	107.42	7479	1075	1127	0.003	232
	Radiation satellite (SN-39)	1963-38C	1963 Sep 28.84 1000 years	Octagon + 4 vanes 61	0.3 long? 0.46 dia?	1963 Oct 11.2	89.89	107.40	7479	1075	1126	0.003	234
	Fragments	1963-38D-J											
	Vela 1 [Atlas Agena D]	1963-39A	1963 Oct 17.10 >million years	Icosahedron 150	1.4 dia	1963 Oct 17.1 1968 Jan 15.5	38.3 37.1	6270 6474	113000 115128	102098 93470	111137 124031	0.040 0.133	- -
D	TRS 5	1963-39B	1963 Oct 17.10 1 year?	Tetrahedron 2.0	0.23 side	1963 Oct 17.1 1964 Apr 1.5	36.77 35.90	2329 2319.4	58240 58041	220 953	103500 102372	0.887 0.874	153 -
	Vela 2	1963-39C	1963 Oct 17.10 > million years	Icosahedron 150	1.4 dia	1963 Oct 19 1968 Jan 15.5	37.8 36.2	6370 6507	113900 115500	99300 100122	115800 118122	0.072 0.078	- -
D	Agena D rocket	1963-39D	1963 Oct 17.10 1 year?	Cylinder 700?	6 long? 1.5 dia	Orbit similar to 1963-39B							
D R	Cosmos 20	1963-40A	1963 Oct 18.40 7.9 days 1963 Oct 26.3	Sphere-cylinder 4750?	4.3 long 2.4 dia	1963 Oct 18.7 1963 Oct 20.7	64.90 64.90	89.60 89.51	6632 6628	205 204	302 296	0.007 0.007	32 32
D	Cosmos 20 rocket	1963-40B	1963 Oct 18.40 12 days 1963 Oct 30-31	Cylinder 1440	3.8 long 2.6 dia	1963 Oct 18.4 1963 Oct 28.0	64.91 64.87	89.68 88.12	6635 6586	204 185	310 231	0.008 0.004	43 91

	Name		Launch date, lifetime and descent date	Shape and weight (kg)	Size (m)	Date of orbital determination	Orbital inclination (deg)	Nodal period (min)	Semi major axis (km)	Perigee height (km)	Apogee height (km)	Orbital eccentricity	Argument of perigee (deg)
D	[Atlas Agena D]	1963-41A	1963 Oct 25.79	Cylinder	8 long?	1963 Oct 26.0	99.05	88.99	6616	144	332	0.014	97
			4.0 days	1500?	1.5 dia	1963 Oct 27.8	99.05	88.70	6584	140	272	0.010	78
			1963 Oct 29.8										
D	Capsule	1963-41B	1963 Oct 25.79	-	-	1963 Oct 27.9	99.05	88.85	6595	136	297	0.012	78
			3.2 days										
			1963 Oct 29.0										
D	Fragments	1963-41C,D											
D	[Thor Agena D]	1963-42A	1963 Oct 29.88	Cylinder	8 long?	1963 Nov 2.1	89.90	90.84	6690	279	345	0.005	-
			83.51 days	1000?	1.5 dia	1963 Nov 29.4	89.90	90.42	6670	275	308	0.002	250
			1964 Jan 21.39			1964 Jan 6.4	89.89	89.53	6623	232	258	0.002	84
D	[Thor Agena D]	1963-42B	1963 Oct 29.88	Octagon?	0.3 long?	1963 Oct 31.7	89.99	93.35	6813	285	585	0.022	32
			571.12 days	60?	0.9 dia?	1964 Nov 18.5	89.97	91.83	6741	282	444	0.012	284
			1965 May 23.00			1965 May 18.4	89.95	88.90	6596	205	231	0.002	223
D	Fragment	1963-42C											
D	Polyot 1	1963-43A	1963 Nov 1.37	Irregular	2 long?	1963 Nov 1.4	- †	94.0	6843	339	592	0.018	-
			6924 days	600?	1 dia?	1963 Nov 2.0	58.92	102.46	7268	343	1437	0.075	114
			1982 Oct 16			1968 Jan 9.5	58.90	101.55	7211	342	1323	0.068	320
						1971 Dec 1.0	58.8	99.71	7130	329	1175	0.059	-
D	Fragments	1963-43B-D											
D	Cosmos 21*	1963-44A	1963 Nov 11.27	Cylinder	7 long?	1963 Nov 11.4	64.83	88.5	6577	182	216	0.003	-
			2.86 days	6520?	2.0 max								
			1963 Nov 14.13	full	dia								
D	Cosmos 21 rocket	1963-44B	1963 Nov 11.27	Cylinder	7.5 long	Orbit similar to 1963-44A							
			1.69 days	2500?	2.6 dia								
			1963 Nov 12.96										

* Cosmos 21 may have been a Space Vehicle launcher. † Initial inclination not announced.

	Name		Launch date, lifetime and descent date	Shape and weight (kg)	Size (m)	Date of orbital determination	Orbital inclination (deg)	Nodal period (min)	Semi major axis (km)	Perigee height (km)	Apogee height (km)	Orbital eccentricity	Argument of perigee (deg)
D R	Cosmos 22	1963-45A	1963 Nov 16.45 6.0 days 1963 Nov 22.4	Sphere-cylinder 5530?	5 long? 2.4 dia	1963 Nov 16.5	64.93	90.29	6665	192	381	0.014	-
D	Cosmos 22 rocket	1963-45B	1963 Nov 16.45 16.7 days 1963 Dec 3.2	Cylinder 2500?	7.5 long 2.6 dia	1963 Nov 18.4 1963 Dec 1.1	64.86 64.84	90.14 88.06	6658 6573	189 166	369 223	0.013 0.004	33 40
D	Explorer 18 (Imp 1)*	1963-46A	1963 Nov 27.10 25 months 1965 Dec	Octagon + 4 vanes 62	0.34 long 0.74 dia	1963 Nov 28.5 1964 Sep 15.5 1965 Sep 8.8	33.34 35.29 35.20	5666 5599 5606	105,282 103,453 104,592	192 2073 4395	197,616 192,077 192,033	0.938 0.918 0.897	- - 162
D	Explorer 18 rocket	1963-46B	1963 Nov 27.10 2 years?	Cylinder 24	1.5 long 0.46 dia	Orbit similar to 1963-46A							
	Centaur AC2	1963-47A	1963 Nov 27.79 100 years	Cylinder 4620	8.6 long 3.0 dia	1963 Nov 30.8	30.34	107.46	7500	544	1699	0.077	137
2d	Fragments	1963-47B-P											
D	[Thor Agena D]	1963-48A	1963 Nov 27.88 17.3 days 1963 Dec 15.2	Cylinder 1500?	8 long? 1.5 dia	1963 Nov 30	69.99	90.2	6658	175	386	0.016	-

* Interplanetary monitoring platform.

	Name		Launch date, lifetime and descent data	Shape and weight (kg)	Size (m)	Date of orbital determination	Orbital inclination (deg)	Nodal period (min)	Semi major axis (km)	Perigee height (km)	Apogee height (km)	Orbital eccentricity	Argument of perigee (deg)
	Ablestar rocket	1963-49A	1963 Dec 5.91 500 years	Cylinder 450?	4.8 long 1.4 dia	1963 Dec 6.1	89.97	106.86	7458	1065	1095	0.002	308
	Transit 5B-2	1963-49B	1963 Dec 5.91 1000 years	Octagon + boom 75	0.5 long 0.46 dia	1963 Dec 12.4	89.98	107.18	7468	1067	1112	0.003	327
	[Thor Ablestar]	1963-49C	1963 Dec 5.91 1000 years	Octagon + 4 vanes 53	0.25 long 0.46 dia	1963 Dec 8.8	89.95	107.16	7468	1069	1111	0.003	303
4d	Fragments	1963-49D-M											
D	Cosmos 23	1963-50A	1963 Dec 13.58 104.48 days 1964 Mar 27.06	Ellipsoid + 2 panels 400?	1.8 long? 1.2 dia?	1963 Dec 13.7 1964 Jan 9.3	49.0 48.98	92.90 92.27	6805 6769	240 241	613 540	0.027 0.022	126 255
D	Cosmos 23 rocket	1963-50B	1963 Dec 13.58 84.37 days 1964 Mar 6.95	Cylinder 1500?	8 long 1.65 dia	1963 Dec 13.6 1964 Jan 12.0	49.12 48.99	92.84 92.04	6799 6757	230 230	611 527	0.028 0.022	136 268
D	Fragments	1963-50C, D											
D	[Atlas Agena D]	1963-51A	1963 Dec 18.91 1.28 days 1963 Dec 20.17	Cylinder 2000?	8 long? 1.5 dia	1963 Dec 19.1	97.89	88.48	6572	122	266	0.011	94
D R	Cosmos 24	1963-52A	1963 Dec 19.39 8.9 days 1963 Dec 28.3	Sphere-cylinder 4750?	4.3 long 2.4 dia	1963 Dec 19.8	65.03	90.51	6676	204	391	0.014	49
D	Cosmos 24 rocket	1963-52B	1963 Dec 19.39 37.1 days 1964 Jan 25.5	Cylinder 1440	3.8 long 2.6 dia	1963 Dec 21.1 1964 Jan 13.0	65.00 65.00	90.58 89.61	6679 6630	207 192	394 312	0.014 0.009	62 46

	Name		Launch date, lifetime and descent date	Shape and weight (kg)	Size (m)	Date of orbital determination	Orbital inclination (deg)	Nodal period (min)	Semi major axis (km)	Perigee height (km)	Apogee height (km)	Orbital eccentricity	Argument of perigee (deg)
D	Explorer 19	1963-53A	1963 Dec 19.78	Inflated sphere	3.65 dia	1963 Dec 19.8	78.62	115.93	7870	590	2394	0.115	-
			6352 days	7		1968 Jan 9.4	78.76	113.48	7764	749	2022	0.082	98
			1981 May 10			1972 Feb 1.0	78.90	112.02	7699	871	1770	0.058	-
	Explorer 19 rocket	1963-53B	1963 Dec 19.78	Cylinder	1.5 long	1963 Dec 23.0	78.62	115.85	7867	594	2383	0.114	154
			200 years	24	0.46 dia								
2d	Fragments	1963-53C-J											
	Tiros 8	1963-54A	1963 Dec 21.39	Cylinder	0.55 long	1963 Dec 21.5	58.48	99.33	7105	691	765	0.005	123
			80 years	119	1.05 dia								
	Tiros 8 rocket	1963-54B	1963 Dec 21.39	Cylinder	1.5 long	1963 Dec 29.9	58.47	99.27	7103	696	753	0.004	117
			40 years	24	0.46 dia	1969 Sep 16.0	58.47	99.05	7092	690	737	0.003	-
1d	Fragments	1963-54C,D											
D	[Thor Agena D]	1963-55A	1963 Dec 21.91	Cylinder	8 long?	1963 Dec 22.3	64.94	89.96	6644	176	355	0.0135	149
			18.0 days	1000?	1.5 dia								
			1964 Jan 8.9										
D	[Thor Agena D]	1963-55B	1963 Dec 21.91	Octagon?	0.3 long?	1963 Dec 23.8	64.52	91.68	6733	321	388	0.005	89
			326.89 days	60?	0.9 dia?	1964 Jul 26.1	64.52	90.73	6689	291	331	0.003	348
			1964 Nov 7.80			1964 Nov 4.1	64.52	88.72	6588	203	216	0.001	307

Name		Launch date, lifetime and descent date	Shape and weight (kg)	Size (m)	Date of orbital determination	Orbital inclination (deg)	Nodal period (min)	Semi major axis (km)	Perigee height (km)	Apogee height (km)	Orbital eccentricity	Argument of perigee (deg)
Agena D rocket	1964-01A	1964 Jan 11.84 800 years	Cylinder 700?	6 long 1.5 dia	1964 Jan 16.3	69.91	103.47	7298	905	934	0.002	84
GGSE 1*	1964-01B	1964 Jan 11.84 1000 years	Sphere 39	0.51 dia	1964 Jan 16.8	69.94	103.47	7298	898	942	0.003	92
Secor 1** (EGRS 1)†	1964-01C	1964 Jan 11.84 1500 years	Rectangular box 18	0.36 x 0.28 x 0.23	1964 Jan 16.8	69.89	103.46	7297	904	933	0.002	84
SR 7A	1964-01D	1964 Jan 11.84 1000 years	Sphere 45	0.6 dia	1964 Jan 16.8	69.90	103.47	7298	905	934	0.002	74
[Thor Agena D]	1964-01E	1964 Jan 11.84 1000 years	-	0.7 dia?	1964 Jan 21.3	69.90	103.48	7298	905	934	0.002	96
Agena D rocket	1964-02A	1964 Jan 19.45 300 years	Cylinder 700?	8 long? 1.5 dia	1964 Nov 9.5	99.07	101.33	7199	792	850	0.004	169
Capsule	1964-02B	1964 Jan 19.45 300 years	Polyhedral cylinder?	0.5 long? 1.0 dia?	1964 Jan 22.5	99.04	101.31	7194	801	830	0.002	354
[Thor Agena D]	1964-02C	1964 Jan 19.45 300 years	Polyhedral cylinder?	0.5 long? 1.0 dia?	1964 Jan 22.5	99.07	101.32	7196	811	825	0.001	349
Relay 2	1964-03A	1964 Jan 21.88 1 million years	Octagonal prism 78	0.81 long 0.74 dia	1964 Jan 22.9	46.32	194.60	11129	2091	7411	0.239	184
Relay 2 rocket	1964-03B	1964 Jan 21.88 250000 years	Cylinder 24	1.5 long 0.46 dia	1964 Jan 22.8	46.32	194.61	11132	2071	7437	0.241	186

* Gravity Gradient Stabilisation Experiment. ** Sequential collation of range. † Electronic Geodetic Ranging System.

	Name		Launch date, lifetime and descent date	Shape and weight (kg)	Size (m)	Date of orbital determination	Orbital inclination (deg)	Nodal period (min)	Semi major axis (km)	Perigee height (km)	Apogee height (km)	Orbital eccentricity	Argument of perigee (deg)
D	Echo 2	1964-04A	1964 Jan 25.58	Sphere	41 dia	1964 Jan 27.1	81.50	108.95	7551	1029	1316	0.019	104
			1960.17 days	256		1966 Sep 20.2	81.46	107.64	7496	966	1268	0.020	160
			1969 Jun 7.75			1968 Jan 15.7	81.47	105.55	7399	932	1110	0.012	348
	Echo 2 rocket	1964-04B	1964 Jan 25.58 5000 years	Cylinder 700?	6 long 1.5 dia	1964 Jan 27.1	81.50	108.96	7552	1030	1317	0.019	103
1d	Fragments	1964-04C-E											
D	Saturn SA5	1964-05A	1964 Jan 29.68	Cone-cylinder	25.6 long	1964 Jan 30.6	31.43	94.60	6890	264	760	0.036	135
			821.40 days	17100	5.5 dia	1965 Oct 12.1	31.45	91.88	6751	244	501	0.019	112
			1966 Apr 30.08			1966 Apr 16.8	31.43	89.30	6623	212	278	0.005	8
	Elektron 1	1964-06A	1964 Jan 30.40 200 years	Cylinder and 6 paddles 329	1.2 long 0.8 dia	1964 Jan 31.5	60.83	169.32	10138	394	7126	0.332	61
	Elektron 2	1964-06B	1964 Jan 30.40	Cone-cylinder	2.4 long	1964 Feb 5.0	60.87	1356.40	40593	441	67988	0.832	70
			30 years	444	1.8 dia	1971 Mar 1.0	58.62	1356.24	40588	4735	63684	0.726	-
	Elektron 2 rocket	1964-06D	1964 Jan 30.40	Cylinder	3.8 long	1964 Feb 6.1	60.87	1384.11	41145	411	69123	0.835	70
			30 years	1440	2.6 dia	1971 Mar 1.0	58.58	1383.87	41139	4974	64548	0.724	-
	Fragment	1964-06C											

Space Vehicle: Ranger 6, 1964-07A

	Name		Launch date, lifetime and descent date	Shape and weight (kg)	Size (m)	Date of orbital determination	Orbital inclination (deg)	Nodal period (min)	Semi major axis (km)	Perigee height (km)	Apogee height (km)	Orbital eccentricity	Argument of perigee (deg)
D r?	[Thor Agena D]	1964-08A	1964 Feb 15.90 23.0 days 1964 Mar 9.9	Cylinder 1st 8 days 1590 then 700	8 long? 1.5 dia	1964 Feb 17.7 1964 Mar 5.5	74.95 74.95	90.86 89.5	6690 6623	179 165	444 324	0.020 0.012	147 96
D	Fragment	1964-08B											
D	[Atlas Agena D]	1964-09A	1964 Feb 25.79 4 days 1964 Mar 1	Cylinder 2000?	8 long? 1.5 dia	1964 Feb 26.4	95.66	88.24	6560	173	190	0.001	103
D	Fragment	1964-09B											
D	Cosmos 25	1964-10A	1964 Feb 27.56 267.05 days 1964 Nov 21.61	Ellipsoid 400?	1.8 long 1.2 dia	1964 Feb 27.7 1964 Jun 10.5 1964 Nov 7.3	49.01 48.97 49.0	92.27 91.40 89.69	6769 6725 6641	255 253 225	526 441 301	0.020 0.014 0.005	121 248 236
D	Cosmos 25 rocket	1964-10B	1964 Feb 27.56 111.7 days 1964 Jun 18.3	Cylinder 1500?	8 long 1.65 dia	1964 Feb 28.4 1964 Jun 3.1	49.07 49.04	92.25 89.64	6768 6639	234 227	545 294	0.023 0.005	127 220
D	Fragments	1964-10C, D											
D	[Thor Agena D]	1964-11A	1964 Feb 28.14 1817.92 days 1969 Feb 19.06	Cylinder 1500?	8 long? 1.5 dia	1964 Feb 29.1 1966 Oct 7.3 1968 Jan 10.7	82.03 82.07 82.08	94.74 94.45 93.59	6878 6870 6828	479 485 444	520 499 456	0.003 0.001 0.001	58 347 140
D	Fragments	1964-11B-D											

	Name		Launch date, lifetime and descent date	Shape and weight (kg)	Size (m)	Date of orbital determination	Orbital inclination (deg)	Nodal period (min)	Semi major axis (km)	Perigee height (km)	Apogee height (km)	Orbital eccentricity	Argument of perigee (deg)
D	[Atlas Agena D]	1964-12A	1964 Mar 11.84 4.3 days 1964 Mar 16.1	Cylinder 2000?	8 long? 1.5 dia	1964 Mar 12.9	95.73	88.2	6561	163	203	0.003	123
D	Fragment	1964-12B											
D	Cosmos 26	1964-13A	1964 Mar 18.63 193.87 days 1964 Sep 28.50	Ellipsoid 400?	1.8 long 1.2 dia	1964 Mar 19.3 1964 Jun 20.4 1964 Sep 21.4	48.96 48.95 48.95	91.00 90.23 88.79	6705 6668 6598	266 250 207	387 330 233	0.009 0.006 0.002	122 212 304
D	Cosmos 26 rocket	1964-13B	1964 Mar 18.63 59.5 days 1964 May 17.1	Cylinder 1500?	8 long 1.65 dia	1964 Mar 22.8 1964 Apr 30.4	48.99 48.96	90.91 89.92	6702 6649	270 251	377 291	0.008 0.003	132 331
D	Fragments	1964-13C,D											
D	Cosmos 27*	1964-14A	1964 Mar 27.14 1.2 days 1964 Mar 28.3	Cylinder 6520? full	7 long? 2.0 max dia	1964 Mar 27.6	64.80	88.16	6561	167	198	0.002	323
D	Cosmos 27 rocket	1964-14B	1964 Mar 27.14 2.6 days 1964 Mar 29.7	Cylinder 2500?	7.5 long 2.6 dia	1964 Mar 28.5	64.80	88.22	6566	181	194	0.001	18
D	Fragments	1964-14C,D											

*Cosmos 27 is believed to have been a Zond probe launcher.

	Name		Launch date, lifetime and descent date	Shape and weight (kg)	Size (m)	Date of orbital determination	Orbital inclination (deg)	Nodal period (min)	Semi major axis (km)	Perigee height (km)	Apogee height (km)	Orbital eccentricity	Argument of perigee (deg)
D	Ariel 2	1964-15A	1964 Mar 27.73 1330.95 days 1967 Nov 18.68	Cylinder + 4 paddles 68	0.9 long 0.58 dia	1964 Mar 28.4 1966 Oct 12.7 1967 Nov 10.0	51.64 51.63 51.51	101.29 97.75 90.15	7201 7031 6663	285 280 211	1362 1025 358	0.075 0.053 0.011	140 175 119
D	Ariel 2 rocket	1964-15B	1964 Mar 27.73 1111 days 1967 Apr 13	Cylinder 24	1.8 long 0.46 dia	1964 Mar 29.2 1966 Oct 15.5 1967 Apr 8.9	51.67 51.65 51.57	101.27 95.95 89.81	7200 6946 6646	282 275 221	1362 861 314	0.075 0.042 0.007	142 - 132
D	Fragment	1964-15C											
D	Zond 1 launcher	1964-16A	1964 Apr 2.12 1.5 days 1964 Apr 3.6	-	-	1964 Apr 2.5	64.83	88.47	6578	187	213	0.002	345
D	Zond 1 launcher rocket	1964-16B	1964 Apr 2.12 0.6 day 1964 Apr 2.7	Cylinder 2500?	7.5 long 2.6 dia	1964 Apr 2.5	65.22	88.10	6559	122	240	0.009	139
D	Fragment	1964-16C											
D R	Cosmos 28	1964-17A	1964 Apr 4.40 7.9 days 1964 Apr 12.3	Sphere-cylinder 4750?	4.3 long 2.4 dia	1964 Apr 4.8	65.04	90.37	6671	213	373	0.012	45
D	Cosmos 28 rocket	1964-17B	1964 Apr 4.40 28.7 days 1964 May 3.1	Cylinder 1440	3.8 long 2.6 dia	1964 Apr 4.9	65.01	90.48	6676	224	371	0.011	63
D	Fragment	1964-17C											

Space Vehicle: Zond 1, 1964-16D

Year of launch 1964, continued

	Name		Launch date, lifetime and descent date	Shape and weight (kg)	Size (m)	Date of orbital determination	Orbital inclination (deg)	Nodal period (min)	Semi major axis (km)	Perigee height (km)	Apogee height (km)	Orbital eccentricity	Argument of perigee (deg)
D	Gemini 1 -Titan rocket	1964-18A	1964 Apr 8.67 4.2 days 1964 Apr 12.9	Cone-cylinder 5170 payload 3000	11.6 long 3.0 dia	1964 Apr 9.5	32.56	89.00	6605	154	299	0.011	82
D	Polyot 2	1964-19B	1964 Apr 12.39 787.49 days 1966 Jun 8.88	Irregular 600?	2 long? 1 dia?	1964 Apr 14.1 1964 May 10.8 1965 Oct 8.5	59.92 58.06 58.01	91.86 92.31 91.31	6742 6769 6718	242 303 279	485 479 400	0.018 0.013 0.009	57 98 212
D	Polyot 2 rocket	1964-19A	1964 Apr 12.39 18.4 days 1964 Apr 30.8	-	-	1964 Apr 17.2	59.92	91.54	6729	236	465	0.017	63
D	[Atlas Agena D]	1964-20A	1964 Apr 23.78 5.2 days 1964 Apr 29.0	Cylinder 2000?	8 long? 1.5 dia	1964 Apr 25.5	103.56	89.40	6621	150	336	0.014	128
D	Fragments	1964-20B-E											
D R	Cosmos 29	1964-21A	1964 Apr 25.43 7.9 days 1964 May 3.3	Sphere-cylinder 4750?	4.3 long 2.4 dia	1964 Apr 28.2	65.01	89.50	6628	203	296	0.007	22
D	Cosmos 29 rocket	1964-21B	1964 Apr 25.43 15.9 days 1964 May 11.3	Cylinder 1440	3.8 long 2.6 dia	1964 Apr 28.3	65.04	89.56	6631	220	286	0.005	69
D	Fragments	1964-21C-E											

	Name	Launch date, lifetime and descent date	Shape and weight (kg)	Size (m)	Date of orbital determination	Orbital inclination (deg)	Nodal period (min)	Semi major axis (km)	Perigee height (km)	Apogee height (km)	Orbital eccentricity	Argument of perigee (deg)
D	[Thor Agena D] 1964-22A	1964 Apr 27.98 28.19 days 1964 May 26.17	Cylinder 1500?	8 long? 1.5 dia	1964 May 1.1	79.93	90.77	6690	178	446	0.020	130
D R	Cosmos 30 1964-23A	1964 May 18.41 7.90 days 1964 May 26.31	Sphere-cylinder 5530 ?	5 long? 2.4 dia	1964 May 20.3	64.87	90.25	6664	206	366	0.012	38
D	Cosmos 30 rocket 1964-23B	1964 May 18.41 20.3 days 1964 Jun 7.7	Cylinder 2500?	7.5 long 2.6 dia	1964 May 24.2	64.84	89.94	6650	205	338	0.010	24
D	[Atlas Agena D] 1964-24A	1964 May 19.81 2.9 days 1964 May 22.7	Cylinder 2000?	8 long? 1.5 dia	1964 May 20.7	101.12	89.69	6639	141	380	0.018	120
D	Saturn SA6 (Apollo Model 1) 1964-25A	1964 May 28.71 3.31 days 1964 Jun 1.02	Cone-cylinder 16900	24.4 long 5.5 dia	1964 May 29.7	31.74	88.22	6570	179	204	0.002	122
	Transit 9 [Blue Scout] 1964-26A	1964 Jun 4.16 70 years	Octagon + 4 vanes + boom 54	0.25 long 0.46 dia	1964 Jun 5.2	90.42	103.12	7283	854	956	0.007	99
	Altair rocket 1964-26D	1964 Jun 4.16 100 years	Cylinder 24	1.5 long 0.46 dia	1964 Jun 19.2	90.45	103.13	7283	854	956	0.007	-
1d	Fragments 1964-26B,C,E-G											
D	[Thor Agena D] 1964-27A	1964 Jun 4.96 13.94 days 1964 Jun 18.90	Cylinder 1500?	8 long? 1.5 dia	1964 Jun 7.1 1964 Jun 17.0	79.96 79.95	90.27 89.15	6667 6610	149 139	429 324	0.021 0.014	107 74

Year of launch 1964, continued

	Name		Launch date, lifetime and descent date	Shape and weight (kg)	Size (m)	Date of orbital determination	Orbital inclination (deg)	Nodal period (min)	Semi major axis (km)	Perigee height (km)	Apogee height (km)	Orbital eccentricity	Argument of perigee (deg)
D	Cosmos 31	1964-28A	1964 Jun 6.25 135.97 days 1964 Oct 20.22	Ellipsoid 400?	1.8 long 1.2 dia	1964 Jun 6.7 1964 Oct 8.5	48.93 48.97	91.61 89.40	6735 6626	222 195	492 301	0.020 0.008	112 353
D	Cosmos 31 rocket	1964-28B	1964 Jun 6.25 71.35 days 1964 Aug 16.60	Cylinder 1500?	8 long 1.65 dia	1964 Jun 8.2 1964 Aug 8.3	48.97 48.94	91.60 89.51	6733 6633	220 208	490 301	0.020 0.007	118 61
D R	Cosmos 32	1964-29A	1964 Jun 10.45 7.99 days 1964 Jun 18.44	Sphere-cylinder 4750?	4.3 long 2.4 dia	1964 Jun 10.9	51.24	89.76	6644	213	319	0.008	35
D	Cosmos 32 rocket	1964-29B	1964 Jun 10.45 34.43 days 1964 Jul 14.88	Cylinder 1440	3.8 long 2.6 dia	1964 Jun 12.4	51.30	89.93	6650	232	312	0.006	82
D	Fragment	1964-29C											
D	Starflash 1A [Thor Agena D]	1964-30A	1964 Jun 13.66 354.21 days 1965 Jun 2.87	Cylinder 2500?	8 long? 1.5 dia	1964 Jun 14.6 1964 Nov 2.5 1965 May 29.8	114.98 115.00 114.98	91.67 91.34 88.75	6735 6719 6590	350 334 205	364 348 218	0.001 0.001 0.001	216 329 286
D	Fragment	1964-30B											
	Capsule [Thor Agena D]	1964-31A	1964 Jun 18.20 500 years	Cylinder?	0.5 long? 1.0 dia?	1964 Jun 18.4	99.84	101.64	7213	828	842	0.001	334
	Agena D rocket	1964-31C	1964 Jun 18.20 500 years	Cylinder 700?	8 long? 1.5 dia	1964 Jun 30.5	99.85	101.63	7212	827	840	0.002	-
	Capsule	1964-31B	1964 Jun 18.20 500 years	Cylinder?	0.5 long? 1.0 dia?	1964 Jun 24.4	99.83	101.64	7213	828	842	0.001	135

	Name		Launch date, lifetime and descent date	Shape and weight (kg)	Size (m)	Date of orbital determination	Orbital inclina-tion (deg)	Nodal period (min)	Semi major axis (km)	Perigee height (km)	Apogee height (km)	Orbital eccen-tricity	Argument of perigee (deg)
D	[Thor Agena D]	1964-32A	1964 Jun 19.97 26.81 days 1964 Jul 16.78	Cylinder 1500?	8 long? 1.5 dia	1964 Jun 21.1 1964 Jul 10.5	85.0 84.99	90.95 89.60	6697 6631	176 173	462 332	0.021 0.012	- 56
D	Fragment	1964-32B											
D R	Cosmos 33	1964-33A	1964 Jun 23.43 7.93 days 1964 Jul 1.36	Sphere-cylinder 4750?	4.3 long 2.4 dia	1964 Jun 23.6	65.0	89.50	6629	209	293	0.006	-
D	Cosmos 33 rocket	1964-33B	1964 Jun 23.43 17.37 days 1964 Jul 10.80	Cylinder 1440	3.8 long 2.6 dia	1964 Jun 23.8	65.08	89.54	6630	219	285	0.005	42
D	Fragments	1964-33C,D											
D R	Cosmos 34	1964-34A	1964 Jul 1.47 7.93 days 1964 Jul 9.40	Sphere-cylinder 5530?	5 long? 2.4 dia	1964 Jul 3.4	64.89	89.98	6653	202	348	0.011	37
D	Cosmos 34 rocket	1964-34B	1964 Jul 1.47 13.89 days 1964 Jul 15.36	Cylinder 2500?	7.5 long 2.6 dia	1964 Jul 5.4	64.89	89.80	6644	193	339	0.011	24

	Name		Launch date, lifetime and descent date	Shape and weight (kg)	Size (m)	Date of orbital determination	Orbital inclination (deg)	Nodal period (min)	Semi major axis (km)	Perigee height (km)	Apogee height (km)	Orbital eccentricity	Argument of perigee (deg)
D	[Thor Agena D]	1964-35A	1964 Jul 3.06 1861.45 days 1969 Aug 7.51	Cylinder 1500?	8 long? 1.5 dia	1964 Jul 5.5 1968 Jan 13.2 1968 Oct 31.2	82.09 82.08 82.08	94.94 94.00 93.16	6893 6848 6807	501 456 424	529 483 434	0.002 0.002 0.001	228 101 -
D	[Atlas Agena D]	1964-36A	1964 Jul 6.91 2.0 days 1964 Jul 8.9	Cylinder 1500?	8 long? 1.5 dia	1964 Jul 7.4	92.89	89.20	6612	121	346	0.017	145
D	Capsule	1964-36B	1964 Jul 6.91 180.21 days 1965 Jan 3.12	Octagon? 60?	0.3 long? 0.9 dia?	1964 Jul 8.5 1964 Nov 26.2	92.97 92.93	91.2 90.31	6715 6666	297 261	377 314	0.006 0.004	253 42
D	Fragment	1964-36C											
D	[Thor Agena D]	1964-37A	1964 Jul 10.97 26.52 days 1964 Aug 6.49	Cylinder 1500?	8 long? 1.5 dia	1964 Jul 13.4	84.98	91.00	6699	180	461	0.021	133
	Elektron 3	1964-38A	1964 Jul 10.91 200 years	Cylinder and 6 paddles 350	1.2 long 0.8 dia	1964 Jul 14.3	60.79	168.17	10093	404	7025	0.328	61
	Elektron 4	1964-38B	1964 Jul 10.91 23 years	Cone-cylinder 444	2.4 long 1.8 dia	1964 Jul 14.5 1968 Jan 8.3 1972 Dec 1.0	60.80 58.76 63.44	1313.63 1313.78 1313.58	39737 39736 39733	457 1887 413	66261 64829 66296	0.828 0.792 0.829	70 91 -
	Elektron 4 rocket	1964-38D	1964 Jul 10.91 23 years	Cylinder 1440	3.8 long 2.6 dia	1964 Jul 15.9 1967 Dec 25.9 1973 Jan 1.0	60.96 58.73 63.34	1341.37 1341.22 1336.05	40289 40287 40184	471 1921 197	67350 65900 67415	0.830 0.794 0.836	71 90 -
2d	Fragments	1964-38C,E,F											

	Name		Launch date, lifetime and descent date	Shape and weight (kg)	Size (m)	Date of orbital determination	Orbital inclination (deg)	Nodal period (min)	Semi major axis (km)	Perigee height (km)	Apogee height (km)	Orbital eccentricity	Argument of perigee (deg)
D R	Cosmos 35	1964-39A	1964 Jul 15.48 7.92 days 1964 Jul 23.40	Sphere-cylinder 4750?	4.3 long 2.4 dia	1964 Jul 16.2	51.24	89.18	6616	218	258	0.003	23
D	Cosmos 35 rocket	1964-39B	1964 Jul 15.48 17.32 days 1964 Aug 1.80	Cylinder 1440	3.8 long 2.6 dia	1964 Jul 16.3	51.32	89.40	6627	216	282	0.005	130
D	Fragments	1964-39C,D											
	Vela 3 [Atlas Agena D]	1964-40A	1964 Jul 17.35 > million years	Icosahedron 150	1.4 dia	1964 Jul 17.4 1964 Dec 15.5	39.58 39.13	6022.6 6091.5	109653 110487	101959 103048	104591 105169	0.012 0.010	149 -
	Vela 4	1964-40B	1964 Jul 17.35 > million years	Icosahedron 150	1.4 dia	1964 Jul 17.4 1964 Dec 15.5	40.88 40.90	6007.0 6070.5	109462 110233	94436 94584	111775 113125	0.079 0.084	74 -
D	TRS 6	1964-40C	1964 Jul 17.35 18 months?	Tetrahedron 2.0	0.2 side	1964 Jul 17.4 1964 Oct 24.9	36.7 38.6	2364 2350	58988 58555	220 590	105000 103764	0.888 0.881	147 159
D	Agena D rocket	1964-40D	1964 Jul 17.35 18 months?	Cylinder 700?	6 long? 1.5 dia	Orbit similar to 1964-40C							
D	Cosmos 36	1964-42A	1964 Jul 30.15 212.88 days 1965 Feb 28.03	Ellipsoid 400?	1.8 long 1.2 dia	1964 Aug 2.4 1964 Dec 8.9 1965 Feb 20.8	49.00 48.99 48.97	91.85 90.86 88.99	6747 6698 6606	261 239 203	477 400 253	0.016 0.012 0.004	127 22 28
D	Cosmos 36 rocket	1964-42B	1964 Jul 30.15 121.98 days 1964 Nov 29.13	Cylinder 1500?	8 long 1.65 dia	1964 Aug 4.7 1964 Oct 7.4 1964 Nov 24.0	49.02 49.00 48.99	91.83 91.12 89.22	6746 6710 6617	254 245 212	482 419 265	0.017 0.013 0.004	143 85 328

Space vehicle: Ranger 7, 1964-41A. Ranger 7 rocket (Agena B) is believed to be in a highly eccentric orbit.

Year of launch 1964, continued

	Name		Launch date, lifetime and descent date	Shape and weight (kg)	Size (m)	Date of orbital determination	Orbital inclination (deg)	Nodal period (min)	Semi major axis (km)	Perigee height (km)	Apogee height (km)	Orbital eccentricity	Argument of perigee (deg)
D r?	[Thor Agena D]	1964-43A	1964 Aug 5.97 26.0 days 1964 Sep 1.0	Cylinder 1st 9 days 1540, later 720	8 long? 1.5 dia	1964 Aug 6.9 1964 Aug 24.8	79.96 79.99	90.71 89.35	6687 6619	182 175	436 307	0.019 0.010	138 68
D R	Cosmos 37	1964-44A	1964 Aug 14.40 7.9 days 1964 Aug 22.3	Sphere-cylinder 4750?	4.3 long 2.4 dia	1964 Aug 15.8	64.92	89.41	6625	207	287	0.006	22
D	Cosmos 37 rocket	1964-44B	1964 Aug 14.40 19.64 days 1964 Sep 3.04	Cylinder 1440	3.8 long 2.6 dia	1964 Aug 15.8	65.01	89.54	6633	215	295	0.006	36
D	[Atlas Agena D]	1964-45A	1964 Aug 14.92 8.8 days 1964 Aug 23.7	Cylinder 1500?	8 long? 1.5 dia	1964 Aug 17.6	95.52	89.0	6606	149	307	0.012	135
D	Hitch-hiker 2 (P.11)	1964-45B	1964 Aug 14.92 5319 days 1979 Mar 8	Octagon 79	0.3 long? 0.9 dia?	1964 Aug 18.3 1969 Sep 16.0 1975 May 1.0	95.67 95.65 95.5	127.40 117.29 104.59	8390 7941 7356	275 265 254	3748 2861 1702	0.207 0.163 0.098	11 - -
D	Fragment	1964-45C											

	Name		Launch date, lifetime and descent date	Shape and weight (kg)	Size (m)	Date of orbital determination	Orbital inclination (deg)	Nodal period (min)	Semi major axis (km)	Perigee height (km)	Apogee height (km)	Orbital eccentricity	Argument of perigee (deg)
D	Cosmos 38	1964-46A	1964 Aug 18.39 82.4 days 1964 Nov 8.8	Ellipsoid? 50?	1.0 long? 0.8 dia?	1964 Aug 18.8 1964 Oct 3.5 1964 Nov 1.4	56.12 56.13 56.09	94.31 92.17 89.70	6866 6760 6641	206 193 190	769 571 336	0.041 0.028 0.011	60 157 233
D	Cosmos 39	1964-46B	1964 Aug 18.39 91.14 days 1964 Nov 17.53	Ellipsoid? 50?	1.0 long? 0.8 dia?	1964 Aug 18.8 1964 Oct 12.4 1964 Nov 11.4	56.10 56.10 56.10	94.59 91.93 89.52	6880 6751 6631	206 197 186	798 548 319	0.043 0.026 0.010	61 177 254
D	Cosmos 40	1964-46C	1964 Aug 18.39 92.50 days 1964 Nov 18.89	Ellipsoid? 50?	1.0 long? 0.8 dia?	1964 Aug 19.7 1964 Oct 5.6 1964 Nov 9.5	56.12 56.10 56.10	93.95 92.07 89.74	6851 6757 6643	206 196 185	740 561 345	0.039 0.027 0.012	61 162 247
D	Cosmos 38 rocket	1964-46D	1964 Aug 18.39 185.29 days 1965 Feb 19.68	Cylinder 2200?	7.4 long 2.4 dia	1964 Aug 19.8 1964 Dec 11.5 1965 Feb 14.8	56.12 56.15 56.11	95.13 92.51 89.09	6908 6778 6608	212 210 170	848 590 289	0.046 0.028 0.009	64 304 95
D	Fragments	1964-46E-G											
	Syncom 3	1964-47A	1964 Aug 19.51 >million years	Cylinder 39	0.39 long 0.71 dia	1964 Aug 22.2 1964 Dec 15.5 1970 Oct 1.0	0.10 0.07 4.9	1407.8 1436.5 1437.3	41609 42177 42189	34191 35790 35718	36271 35799 35903	0.025 0 0.002	117 - -
	Syncom 3 rocket	1964-47B	1964 Aug 19.51 100 000 years	Cylinder 24	1.5 long 0.46 dia	1964 Aug 20.2	16.80	694.5	25977	1113	38084	0.712	181
D	Starflash 1B [Thor Agena D]	1964-48A	1964 Aug 21.66 221.66 days 1965 Mar 31.32	Cylinder 1500?	8 long? 1.5 dia	1964 Aug 25.6 1964 Dec 5.5 1965 Mar 27.2	115.0 115.0 114.96	91.60 90.90 89.15	6734 6697 6604	349 305 219	363 332 232	0.001 0.002 0.001	108 50 336

	Name		Launch date, lifetime and descent date	Shape and weight (kg)	Size (m)	Date of orbital determination	Orbital inclination (deg)	Nodal period (min)	Semi major axis (km)	Perigee height (km)	Apogee height (km)	Orbital eccentricity	Argument of perigee (deg)
	Cosmos 41	1964-49D	1964 Aug 22.30 46 years	Cylinder + 6 vanes 1000?	3.4 long 1.6 dia	1964 Sep 10.7 1965 Aug 14.7	64.88 66.06	714.58 714.30	26477 26473	426 796	39771 39394	0.743 0.729	326 318
D	Cosmos 41 launcher rocket	1964-49A	1964 Aug 22.30 24.24 days 1964 Sep 15.54	Cylinder 2500?	7.5 long 2.6 dia	1964 Aug 23.7	64.74	91.06	6706	200	455	0.019	66
D	Cosmos 41 launcher	1964-49B	1964 Aug 22.30 37 days 1964 Sep 28	Irregular	-	1964 Aug 31.5	64.73	91.19	6712	197	471	0.020	-
	Cosmos 41 rocket	1964-49E	1964 Aug 22.30 46 years	Cylinder 440	2.0 long 2.0 dia	1966 Aug 27	67.52	718.93	26564	998	39373	0.722	-
1d	Fragment	1964-49C,F											
D	Cosmos 42	1964-50A	1964 Aug 22.46 484.14 days 1965 Dec 19.60	Ellipsoid? 100?	1.0 long? 0.8 dia?	1964 Aug 23.2 1964 Dec 8.9 1965 Oct 21.3	48.96 48.96 48.93	98.05 96.84 92.07	7047 6989 6757	230 226 210	1113 995 548	0.062 0.055 0.025	110 196 153
D	Cosmos 42 rocket	1964-50B	1964 Aug 22.46 345.21 days 1965 Aug 2.67	Cylinder 1500?	8 long 1.65 dia	1964 Aug 23.4 1964 Dec 12.2 1965 Jul 22.8	48.97 48.97 48.97	97.91 96.14 89.87	7040 6955 6648	224 222 203	1098 931 336	0.062 0.051 0.010	112 215 316
D	Cosmos 43	1964-50C	1964 Aug 22.46 492.37 days 1965 Dec 27.83	Ellipsoid? 100?	1.0 long? 0.8 dia?	1964 Aug 23.3 1964 Dec 6.3 1965 Oct 20.7	48.96 48.97 48.92	98.00 96.82 92.21	7042 6988 6764	227 225 216	1100 994 555	0.062 0.055 0.025	111 186 150
	Explorer 20	1964-51A	1964 Aug 25.57 800 years	Double cone 44	0.83 long 0.66 dia	1964 Aug 31.0 1965 Dec 12.3	79.87 79.91	103.97 103.97	7323 7323	871 864	1018 1025	0.010 0.011	289 146
	Explorer 20 rocket	1964-51B	1964 Aug 25.57 500 years	Cylinder 24	1.5 long 0.46 dia	1964 Aug 26.6	79.93	103.92	7321	877	1009	0.009	300
D	Fragments	1964-51C-E											

	Name		Launch date, lifetime and descent date	Shape and weight (kg)	Size (m)	Date of orbital determination	Orbital inclination (deg)	Nodal period (min)	Semi major axis (km)	Perigee height (km)	Apogee height (km)	Orbital eccentricity	Argument of perigee (deg)
D	Nimbus 1	1964-52A	1964 Aug 28.33 3548.55 days 1974 May 16.88	Conical skeleton + 2 paddles 376	3.00 long 1.45 dia	1964 Aug 28.8 1969 Sep 16.0 1971 Dec 1.0	98.66 98.66 98.66	98.42 96.48 94.56	7061 6968 6876	429 412 385	937 768 611	0.036 0.025 0.016	158 - -
D	Nimbus 1 rocket	1964-52B	1964 Aug 28.33 3637.22 days 1974 Aug 13.55	Cylinder 700?	6 long 1.5 dia	1964 Sep 15.5 1968 Jan 15.5 1972 Mar 1.0	98.68 98.68 98.68	98.40 97.65 94.57	7060 7024 6877	429 425 388	934 867 609	0.036 0.031 0.016	- - -
	Cosmos 44	1964-53A	1964 Aug 28.68 50 years	Cylinder + 2 vanes	3 long? 1 dia?	1964 Aug 29.3	65.04	99.48	7114	615	857	0.017	23
	Cosmos 44 rocket	1964-53B	1964 Aug 28.68 50 years	Cylinder 1440	3.8 long 2.6 dia	1964 Aug 30.1	65.05	99.54	7117	682	796	0.008	45
	OGO 1*	1964-54A	1964 Sep 5.90 32 years	Box + booms 487	1.73 long 0.84 wide 0.84 high	1964 Sep 7.8 1969 Sep 16.0 1970 Jun 1.0	31.15 57.50 58.8	3838.8 3842.8 3840.1	81211 81270 81232	281 35743 45880	149385 114040 103827	0.918 0.482 0.357	313 - -
	OGO 1 rocket	1964-54B	1964 Sep 5.90 32 years	Cylinder 700?	6 long 1.5 dia	Orbit similar to 1964-54A							
D R	Cosmos 45	1964-55A	1964 Sep 13.41 4.9 days 1964 Sep 18.3	Sphere-cylinder 5530?	5 long? 2.4 dia	1964 Sep 14.6	64.89	89.68	6638	207	313	0.008	36
D	Cosmos 45 rocket	1964-55B	1964 Sep 13.41 14.45 days 1964 Sep 27.86	Cylinder 2500?	7.5 long 2.6 dia	1964 Sep 14.5	64.88	89.60	6634	203	309	0.008	36
D r?	[Thor Agena D]	1964-56A	1964 Sep 14.95 21.7 days 1964 Oct 6.7	Cylinder 1st 11 days 1330, later 580	8 long? 1.5 dia	1964 Sep 16.1	84.96	90.88	6697	172	466	0.022	135
D	Fragment	1964-56B											

* Orbiting Geophysical Observatory. Numerical integration indicates that the perigee height will decrease to 390 km in August 1980 and then to 290 km in January 1981, before starting another 16-year cycle. The same applies to 1964-54B.

Year of launch 1964, continued

	Name		Launch date, lifetime and descent date	Shape and weight (kg)	Size (m)	Date of orbital determination	Orbital inclina-tion (deg)	Nodal period (min)	Semi major axis (km)	Perigee height (km)	Apogee height (km)	Orbital eccen-tricity	Argument of perigee (deg)
D	Saturn SA7 (Apollo Model 2)	1964-57A	1964 Sep 18.68 3.86 days 1964 Sep 22.54	Cylinder 16700	24.4 long 5.5 dia	1964 Sep 20.3	31.72	88.30	6568	178	203	0.002	142
D	[Atlas Agena D]	1964-58A	1964 Sep 23.84 4.78 days 1964 Sep 28.62	Cylinder 2000?	8 long? 1.5 dia	1964 Sep 25.2	92.91	89.00	6602	145	303	0.012	173
D	Fragment	1964-58B											
D R	Cosmos 46	1964-59A	1964 Sep 24.50 8.02 days 1964 Oct 2.52	Sphere-cylinder 4750?	4.3 long 2.4 dia	1964 Sep 25.5	51.25	89.22	6616	211	264	0.004	16
D	Cosmos 46 rocket	1964-59B	1964 Sep 24.50 13.22 days 1964 Oct 7.72	Cylinder 1440	3.8 long 2.6 dia	1964 Sep 25.5	51.27	89.40	6624	234	259	0.002	125
D	Explorer 21 (Imp 2)	1964-60A	1964 Oct 4.16 15 months 1966 Jan	Octagon + 4 vanes 62	0.20 long 0.71 dia	1964 Oct 4.2 1964 Dec 15.5 1965 Oct 15.5	33.53 33.77 33.72	2097 2080 2080	54271 53971 53981	190 362 917	95595 94825 94288	0.879 0.875 0.865	133 - -
D	Explorer 21 rocket	1964-60B	1964 Oct 4.16 1½ years?	Cylinder 24	1.5 long 0.46 dia	Orbit similar to 1964-60A							
D r?	[Thor Agena D]	1964-61A	1964 Oct 5.91 20.50 days 1964 Oct 26.41	Cylinder 1500?	8 long? 1.5 dia	1964 Oct 7.3	79.97	90.75	6689	182	440	0.019	158

Year of launch 1964, continued

	Name		Launch date, lifetime and descent date	Shape and weight (kg)	Size (m)	Date of orbital determination	Orbital inclination (deg)	Nodal period (min)	Semi major axis (km)	Perigee height (km)	Apogee height (km)	Orbital eccentricity	Argument of perigee (deg)
D R	Cosmos 47*	1964-62A	1964 Oct 6.30 1.0 days 1964 Oct 7.3	Sphere-cylinder 5000?	6 long? 2.4 dia	1964 Oct 6.7	64.62	90.07	6657	174	383	0.016	72
D	Cosmos 47 rocket	1964-62B	1964 Oct 6.30 7.9 days 1964 Oct 14.2	Cylinder 2500?	7.5 long 2.6 dia	1964 Oct 7.2	64.71	89.92	6649	168	373	0.015	67
D	Fragments	1964-62C-E											
	Ablestar rocket	1964-63A	1964 Oct 6.71 1500 years	Cylinder 450?	4.8 long 1.4 dia	1964 Oct 10.8	89.92	106.38	7436	1035	1080	0.003	57
	Transit 5B-4	1964-63B	1964 Oct 6.71 1000 years	Octagon + boom 60?	0.5 long 0.46 dia	1964 Dec 21.2	89.92	106.65	7448	1055	1085	0.002	156
	Calsphere 1 **	1964-63C	1964 Oct 6.71 200 years	Sphere 0.98	0.36 dia	1964 Oct 13.0	89.93	106.63	7447	1054	1084	0.002	29
	Calsphere 2	1964-63E	1964 Oct 6.71 2000 years	Sphere 9.8	0.36 dia	1964 Oct 14.5	89.97	106.66	7449	1056	1086	0.002	32
	Fragments	1964-63D,F											
L	Explorer 22 (Beacon B)	1964-64A	1964 Oct 10.13 250 years	Octagon + 4 vanes 52	0.30 long 0.46 dia	1964 Oct 13.3	79.69	104.82	7363	889	1081	0.013	138
	Explorer 22 rocket	1964-64B	1964 Oct 10.13 300 years	Cylinder 24	1.5 long 0.46 dia	1964 Oct 20.9	79.69	104.75	7362	888	1079	0.013	119
	Fragments	1964-64C,D											

* Voskhod test flight. ** Calibration sphere.

Year of launch 1964, continued

	Name		Launch date, lifetime and descent date	Shape and weight (kg)	Size (m)	Date of orbital determination	Orbital inclination (deg)	Nodal period (min)	Semi major axis (km)	Perigee height (km)	Apogee height (km)	Orbital eccentricity	Argument of perigee (deg)
D 3M R	Voskhod 1	1964-65A	1964 Oct 12.31 1.01 days 1964 Oct 13.32	Sphere-cylinder 5320	6 long? 2.4 dia	1964 Oct 12.7	64.90	90.04	6655	177	377	0.015	65
D	Voskhod 1 rocket	1964-65B	1964 Oct 12.31 8.1 days 1964 Oct 20.4	Cylinder 2500?	7.5 long 2.6 dia	1964 Oct 15.7	64.77	89.39	6627	169	328	0.012	57
D	Fragment	1964-65C											
D R	Cosmos 48	1964-66A	1964 Oct 14.41 5.96 days 1964 Oct 20.37	Sphere-cylinder 4750?	4.3 long 2.4 dia	1964 Oct 15.8	65.08	89.32	6622	204	284	0.006	21
D	Cosmos 48 rocket	1964-66B	1964 Oct 14.41 13.84 days 1964 Oct 28.25	Cylinder 1440	3.8 long 2.6 dia	1964 Oct 15.5	65.06	89.53	6629	211	291	0.006	23
D	Fragment	1964-66C											
D	[Thor Agena D]	1964-67A	1964 Oct 17.92 17.27 days 1964 Nov 4.19	Cylinder 1500?	8 long? 1.5 dia	1964 Oct 19.1	74.99	90.59	6681	189	416	0.017	147
D	[Atlas Agena D]	1964-68A	1964 Oct 23.77 5.06 days 1964 Oct 28.83	Cylinder 1500?	8 long? 1.5 dia	1964 Oct 24.6	95.55	88.62	6583	139	271	0.010	169
D	Capsule	1964-68B	1964 Oct 23.77 123.22 days 1965 Feb 23.99	Octagon? 60?	0.3 long? 0.9 dia?	1964 Oct 27.9 1964 Dec 22.8 1965 Feb 22.0	95.50 95.49 95.45	91.14 90.59 88.53	6708 6680 6579	323 288 198	336 315 203	0.001 0.002 0.0004	330 78 247
D	Fragments	1964-68C,D											

Year of launch 1964, continued

	Name		Launch date, lifetime and descent date	Shape and weight (kg)	Size (m)	Date of orbital determination	Orbital inclination (deg)	Nodal period (min)	Semi major axis (km)	Perigee height (km)	Apogee height (km)	Orbital eccentricity	Argument of perigee (deg)
D	Cosmos 49	1964-69A	1964 Oct 24.22 301.77 days 1965 Aug 21.99	Ellipsoid 400?	1.8 long 1.2 dia	1964 Oct 24.7 1964 Dec 16.4 1965 Aug 16.8	48.99 48.94 48.93	91.78 91.60 88.63	6743 6734 6588	264 255 197	466 457 223	0.015 0.015 0.002	117 7 119
D	Cosmos 49 rocket	1964-69B	1964 Oct 24.22 109.46 days 1965 Feb 10.68	Cylinder 1500?	8 long 1.65 dia	1964 Oct 24.7 1964 Dec 3.4 1964 Feb 6.4	48.94 48.93 48.92	91.85 91.26 89.22	6746 6717 6614	260 251 209	477 426 262	0.016 0.013 0.004	117 306 276
D	Fragment	1964-69C											
D P	Cosmos 50	1964-70A	1964 Oct 28.45 8.0 days 1964 Nov 5.5	Sphere-cylinder 4750?	4.3 long* 2.4 dia	1964 Oct 29.7	51.23	88.67	6588	190	230	0.003	312
D	Cosmos 50 rocket	1964-70B	1964 Oct 28.45 4.7 days 1964 Nov 2.2	Cylinder 1440	3.8 long 2.6 dia	1964 Oct 29.7	51.24	88.74	6592	187	240	0.004	192
D	Fragments	1964-70C-DA											
D	[Thor Agena D]	1964-71A	1964 Nov 2.90 25.33 days 1964 Nov 28.23	Cylinder 1500?	8 long? 1.5 dia	1964 Nov 3.6	79.95	90.70	6692	180	448	0.020	155
D	[Thor Agena D]	1964-72A	1964 Nov 4.09 1827.53 days 1969 Nov 5.62	Cylinder 1500?	8 long? 1.5 dia	1964 Nov 5.1 1968 Jan 11.5 1968 Dec 16.0	82.00 82.04 82.04	95.05 94.21 93.32	6897 6859 6815	512 477 431	526 485 442	0.001 0.001 0.001	303 179 -
D	Fragments	1964-72B-D											

*Size before explosion on Nov 5. Pieces of this satellite were picked up in Malawi.

Year of launch 1964, continued

	Name		Launch date, lifetime and descent date	Shape and weight (kg)	Size (m)	Date of orbital determination	Orbital inclination (deg)	Nodal period (min)	Semi Major axis (km)	Perigee height (km)	Apogee height (km)	Orbital eccentricity	Argument of perigee (deg)
	Explorer 23	1964-74A	1964 Nov 6.50 18.6 years	Cylinder 134	2.34 long 0.62 dia	1964 Nov 6.5 1973 Feb 1.0	51.95 51.95	99.17 97.86	7100 7036	466 451	977 865	0.036 0.029	138 -
D	Fragments*	1964-74B-E											
D r?	Orbis** [Thor Agena D]	1964-75A	1964 Nov 18.86 17.45 days 1964 Dec 6.31	Cylinder 1500?	8 long? 1.5 dia	1964 Nov 21.7	70.02	89.71	6638	180	339	0.012	100
D	Explorer 24	1964-76A	1964 Nov 21.72 1427 days 1968 Oct 18	Inflated-sphere 8.6	3.65 dia	1964 Nov 21.8 1965 Oct 7.3 1968 Jan 13.3	81.36 81.40 81.36	116.30 115.70 106.57	7889 7864 7446	525 589 539	2498 2382 1596	0.125 0.114 0.071	166 193 73
	Explorer 25 (Injun 4)	1964-76B	1964 Nov 21.72 200 years	Spheroid 40	0.76 long 0.61 dia	1964 Nov 21.8	81.36	116.27	7886	522	2494	0.125	166
	Explorer 24 rocket	1964-76C	1964 Nov 21.72 100 years	Cylinder 24	1.5 long 0.46 dia	1964 Dec 15.5	81.36	116.35	7891	531	2495	0.124	-
8d	Fragments	1964-76D-N											
D	Zond 2 launcher	1964-78A	1964 Nov 30.55 1.23 days 1964 Dec 1.78	-	-	1964 Nov 30.9	64.72	88.16	6564	153	219	0.005	313
D	Zond 2 launcher rocket	1964-78B	1964 Nov 30.55 2.10 days 1964 Dec 2.65	Cylinder 2500?	7.5 long 2.6 dia	1964 Dec 1.5	64.73	88.15	6562	177	190	0.001	317
D	[Atlas Agena D]	1964-79A	1964 Dec 4.79 1.2 days 1964 Dec 6.0	Cylinder 2000?	8 long? 1.5 dia	1964 Dec 5.2	97.02	89.69	6636	158	357	0.015	-

Space Vehicles: Mariner 3, 1964-73; Mariner 4, 1964-77A, and a rocket 1964-77B; Zond 2, 1964-78C.

* Fragments 1964-74D and 74E, designated in the USA about 17 Jul 1973, probably belong to the 1973-27 launch.

** Orbiting radio beacon ionospheric satellite.

	Name		Launch date, lifetime and descent date	Shape and weight (kg)	Size (m)	Date of orbital determination	Orbital inclination (deg)	Nodal period (min)	Semi major axis (km)	Perigee height (km)	Apogee height (km)	Orbital eccentricity	Argument of perigee (deg)
D	Cosmos 51	1964-80A	1964 Dec 9.96 339.93 days 1965 Nov 14.89	Ellipsoid 400?	1.8 long 1.2 dia	1964 Dec 11.9 1965 Jun 20.5 1965 Nov 10.2	48.78 48.73 48.73	92.44 91.28 88.68	6776 6718 6590	262 246 192	533 434 231	0.020 0.014 0.003	130 319 312
D	Cosmos 51 rocket	1964-80B	1964 Dec 9.96 152.4 days 1965 May 11.4	Cylinder 1500?	8 long 1.65 dia	1964 Dec 11.9 1965 Mar 12.3 1965 May 8.3	48.78 48.75 48.74	92.46 91.38 88.90	6778 6723 6601	257 244 203	542 446 243	0.021 0.015 0.003	129 207 138
D	Fragments	1964-80C-F											
D p	Transtage 2 [Titan 3A]	1964-81A	1964 Dec 10.70 2.95 days 1964 Dec 13.65	Cylinder 1700	6 long 3 dia	1964 Dec 12.3	32.15	87.60	6542	157	170	0.001	280
D	Centaur AC4	1964-82A	1964 Dec 11.60 0.63 day 1964 Dec 12.23	Cylinder 2950	14 long 3.0 dia	1964 Dec 11.7	30.71	87.81	6550	165	178	0.001	183
	Ablestar rocket	1964-83A	1964 Dec 13.01 500 years	Cylinder 450?	4.8 long 1.4 dia	1964 Dec 16.1 1966 Nov 30.5	89.99 89.95	106.06 106.20	7420 7427	1012 1024	1071 1073	0.004 0.003	352 -
	Radiation satellite (SN-43)	1964-83C	1964 Dec 13.01 1000 years	Octagonal prism 78	0.92 long 0.46 dia	1964 Dec 24.5	89.99	106.36	7435	1027	1086	0.004	286
	Transit 5B-5 [Thor Ablestar]	1964-83D	1964 Dec 13.01 1000 years	Octagon + boom 73?	0.5 long 0.46 dia	1964 Dec 24.8	89.86	106.33	7433	1025	1084	0.004	13
5d	Fragments	1964-83B,E-M											

	Name		Launch date, lifetime and descent date	Shape and weight (kg)	Size (m)	Date of orbital determination	Orbital inclination (deg)	Nodal period (min)	Semi major axis (km)	Perigee height (km)	Apogee height (km)	Orbital eccentricity	Argument of perigee (deg)
D	San Marco 1	1964-84A	1964 Dec 15.85	Sphere	0.66 dia	1964 Dec 16.9	37.77	94.94	6900	198	846	0.047	113
			270.12 days	115		1965 Apr 30.3	37.77	92.64	6788	193	627	0.032	158
			1965 Sep 11.97			1965 Sep 9.5	37.75	88.48	6582	164	243	0.006	242
D	San Marco 1 rocket	1964-84B	1964 Dec 15.85	Cylinder	1.5 long	1964 Dec 31.5	37.80	93.40	6824	194	697	0.037	-
			53.2 days	24	0.46 dia	1965 Jan 31.5	37.76	90.39	6678	186	411	0.017	-
			1965 Feb 7.1										
D	Fragment	1964-84C											
D	[Thor Agena D]	1964-85A	1964 Dec 19.88	Cylinder	8 long?	1964 Dec 22.9	74.97	90.46	6675	183	410	0.017	153
			26.06 days	1500?	1.5 dia	1965 Jan 12.9	74.95	88.74	6590	166	258	0.007	86
			1965 Jan 14.94										
D	Explorer 26	1964-86A	1964 Dec 21.38	Octagon +	0.43 long	1964 Dec 21.4	20.14	456.26	19632	316	26191	0.659	121
			13.3 years?	4 vanes	0.71 dia	1965 Oct 6.2	19.86	451.98	19515	276	25997	0.659	67
			1978 Apr ?	46		1973 Mar 1.0	18.0	282.6	14264	368	15404	0.527	-
						1976 Sep 1.0	18.1	184.5	10736	171	8545	0.390	-
D	Explorer 26 rocket	1964-86B	1964 Dec 21.38	Cylinder	1.5 long	orbit similar to 1964-86A							
			2598 days	24	0.46 dia								
			1972 Feb 1										
D	Fragment*	1964-86C											
D	[Thor Agena D]	1964-87A	1964 Dec 21.80	Cylinder	8 long?	1964 Dec 25.1	70.08	89.5	6629	238	264	0.002	111
			21.64 days	1500?	1.5 dia								
			1965 Jan 11.44										

* Fragment 1964-86C, designated on 1972 Apr 15, is probably the rediscovered satellite 1964-86A (lost during 1965 Oct).
Object 1964-86C (alias 86A) was lost about 1977 Oct, some 6 months before its expected decay.

Year of launch 1965

	Name		Launch date, lifetime and descent date	Shape and weight (kg)	Size (m)	Date of orbital determination	Orbital inclination (deg)	Nodal period (min)	Semi major axis (km)	Perigee height (km)	Apogee height (km)	Orbital eccentricity	Argument of perigee (deg)
D R	Cosmos 52	1965-01A	1965 Jan 11.40 7.89 days 1965 Jan 19.29	Sphere-cylinder 4750?	4.3 long 2.4 dia	1965 Jan 11.8	65.00	89.50	6628	203	298	0.007	18
D	Cosmos 52 rocket	1965-01B	1965 Jan 11.40 18.21 days 1965 Jan 29.61	Cylinder 1440	3.8 long 2.6 dia	1965 Jan 13.8	64.98	89.53	6633	215	295	0.006	75
D	Fragments	1965-01C,D											
D r?	[Thor Agena D]	1965-02A	1965 Jan 15.88 25.0 days 1965 Feb 9.9	Cylinder 1st 10 days 1290, later 580	8 long? 1.5 dia	1965 Jan 21.8	74.95	90.52	6678	180	420	0.018	128
D	Fragment	1965-02B											
D	[Thor Altair]	1965-03A	1965 Jan 19.21 5288 days 1979 Jul 13	Cylinder 150?	2 long? 0.5 dia?	1965 Jan 20.1 1972 Jan 1.0	98.78 98.77	97.68 95.80	7025 6933	471 434	822 675	0.025 0.017	272 -
D	Fragments	1965-03B,C											
	Tiros 9	1965-04A	1965 Jan 22.33 1000 years	Cylinder 138	0.56 long 1.07 dia	1965 Jan 31.2	96.40	119.23	8022	705	2582	0.117	166
	Tiros 9 rocket	1965-04B	1965 Jan 22.33 200 years	Cylinder 24	1.5 long 0.46 dia	1965 Feb 1.5	96.43	119.32	8028	710	2589	0.117	163
	Fragments	1965-04C,D											
D	[Atlas Agena D]	1965-05A	1965 Jan 23.84 5.2 days 1965 Jan 29.0	Cylinder 2000?	8 long? 1.5 dia	1965 Jan 25.1	102.5	88.85	6597	146	291	0.011	131
D	Fragment	1965-05B											

Year of launch 1965 continued

	Name		Launch date, lifetime and descent date	Shape and weight (kg)	Size (m)	Date of orbital determination	Orbital inclination (deg)	Nodal period (min)	Semi major axis (km)	Perigee height (km)	Apogee height (km)	Orbital eccentricity	Argument of perigee (deg)
D	Cosmos 53	1965-06A	1965 Jan 30.40	Ellipsoid	1.8 long	1965 Jan 30.9	48.72	98.71	7077	218	1180	0.068	107
			558.84 days	400?	1.2 dia	1965 Oct 30.0	48.70	95.94	6944	212	920	0.051	175
			1966 Aug 12.24			1966 Jul 29.1	48.67	89.98	6654	189	362	0.013	345
D	Cosmos 53 rocket	1965-06B	1965 Jan 30.40	Cylinder	8 long	1965 Jan 31.3	48.72	98.64	7075	230	1164	0.066	109
			346.10 days	1500?	1.65 dia	1965 Jun 21.0	48.73	96.46	6970	222	961	0.053	342
			1966 Jan 11.50			1965 Sep 30.5	48.70	94.53	6874	214	778	0.041	-
D	Fragments	1965-06C,D											
	OSO 2	1965-07A	1965 Feb 3.69	Nonagonal box	0.94 long	1965 Feb 3.7	32.87	96.40	6970	550	634	0.006	118
			25 years	247	1.12 dia	1968 Jan 3.6	32.86	96.18	6961	541	625	0.006	252
D	OSO 2 rocket	1965-07B	1965 Feb 3.69	Cylinder	1.5 long	1965 Feb 15.5	32.86	96.41	6971	545	640	0.007	-
			4603 days	24	0.46 dia	1969 Sep 16.0	32.84	95.59	6931	519	586	0.005	-
			1977 Sep 11			1973 Jul 1.0	32.8	93.90	6849	451	490	0.003	-
	Transtage 3 [Titan 3A]	1965-08A	1965 Feb 11.72	Cylinder	6 long	1965 Feb 14.5	32.15	145.47	9167	2776	2802	0.001	302
			50000 years	3700	3 dia								
	LES 1 *	1965-08C	1965 Feb 11.72	Polyhedron	0.6 dia	1965 Feb 19.7	32.15	145.55	9171	2774	2811	0.002	315
			50000 years	31									
	Fragment	1965-08B											
D	Pegasus 1 [Saturn SA9]	1965-09A	1965 Feb 16.61	Cylinder +	29.2 long	1965 Feb 18.8	31.73	96.80	6992	495	733	0.017	156
			4960.65 days	wings	4.3 wide	1969 Nov 16.0	31.75	95.81	6944	477	654	0.013	-
			1978 Sep 17.26	10400		1974 Aug 19.8	31.7	94.06	6860	433	531	0.007	-
	Apollo Model 3	1965-09B	1965 Feb 16.61	Cone-cylinder	9 long	1965 Feb 20.4	31.76	96.90	6995	498	736	0.017	172
			25 years	4600	3.91 dia	1968 Jan 15.5	31.74	96.72	6987	496	722	0.016	-

Space Vehicle: Ranger 8, 1965-10A. Ranger 8 rocket, 1965-10B is believed to be in a highly eccentric orbit.

* Lincoln Experimental Satellite.

	Name		Launch date, lifetime and descent date	Shape and weight (kg)	Size (m)	Date of orbital determination	Orbital inclination (deg)	Nodal period (min)	Semi Major axis (km)	Perigee height (km)	Apogee height (km)	Orbital eccentricity	Argument of perigee (deg)
D	Cosmos 54	1965-11A	1965 Feb 21.46 1302.42 days 1968 Sep 15.88	Ellipsoid? 60?	1.0 long? 0.8 dia?	1965 Feb 25.7 1966 Nov 5.5 1968 Jan 5.7	56.03 56.01 56.03	104.90 101.96 96.80	7370 7232 6985	256 254 251	1729 1454 963	0.100 0.083 0.051	106 115 212
D	Cosmos 55	1965-11B	1965 Feb 21.46 1075.97 days 1968 Feb 2.43	Ellipsoid? 50?	1.0 long? 0.8 dia?	1965 Feb 25.2 1966 Nov 7.4 1968 Jan 15.4	56.02 56.00 55.99	105.22 101.39 91.14	7385 7206 6711	261 259 219	1753 1397 447	0.101 0.079 0.017	110 123 281
D	Cosmos 56	1965-11C	1965 Feb 21.46 984.52 days 1967 Nov 2.98	Ellipsoid? 50?	1.0 long? 0.8 dia?	1965 Feb 24.5 1966 Nov 8.9 1967 Oct 30.8	56.04 56.01 55.96	104.52 99.96 89.42	7352 7137 6627	261 252 189	1687 1266 308	0.097 0.071 0.009	108 147 162
D	Cosmos 54 rocket	1965-11D	1965 Feb 21.46 1766.29 days 1969 Dec 23.75	Cylinder 2200?	7.4 long 2.4 dia	1965 Feb 24.4 1966 Nov 6.1 1968 Mar 13.7	56.04 56.04 56.06	106.21 104.45 100.62	7431 7349 7169	273 265 262	1833 1676 1319	0.105 0.096 0.074	110 78 240
D	Fragment	1965-11E											
D	Cosmos 57	1965-12A	1965 Feb 22.32 0.1 day 1965 Feb 22.4	Sphere-cylinder* 5500?	6 long?* 2.4 dia	1965 Feb 22.4	64.74	90.42	6674	165	427	0.020	64
D	Cosmos 57 rocket	1965-12B	1965 Feb 22.32 11 days 1965 Mar 5	Cylinder 2500?	7.5 long 2.6 dia	1965 Feb 25.6	64.74	90.42	6674	178	416	0.018	69
D	Fragments	1965-12C-FZ											

*Before disintegration (Voskhod test flight).

Year of launch 1965 continued

	Name		Launch date, lifetime and descent date	Shape and weight (kg)	Size (m)	Date of orbital determination	Orbital inclination (deg)	Nodal period (min)	Semi major axis (km)	Perigee height (km)	Apogee height (km)	Orbital eccentricity	Argument of perigee (deg)
D	[Thor Agena D]	1965-13A	1965 Feb 25.91 20.92 days 1965 Mar 18.83	Cylinder 1500?	8 long? 1.5 dia	1965 Feb 26.1	75.08	90.07	6655	177	377	0.015	163
D	Fragment	1965-13B											
	Cosmos 58	1965-14A	1965 Feb 26.21 50 years	Cylinder + 2 vanes 2000?	5 long? 1.5 dia?	1965 Feb 27.2	65.00	96.78	6983	563	647	0.006	258
	Cosmos 58 rocket	1965-14B	1965 Feb 26.21 25 years	Cylinder 1440	3.8 long 2.6 dia	1965 Feb 28.1	65.05	96.90	6988	512	708	0.014	205
D R	Cosmos 59	1965-15A	1965 Mar 7.38 7.92 days 1965 Mar 15.30	Sphere-cylinder 5530?	5 long? 2.4 dia	1965 Mar 7.5	64.97	89.78	6642	217	310	0.007	63
D	Cosmos 59 rocket	1965-15B	1965 Mar 7.38 12 days 1965 Mar 19	Cylinder 2500?	7.5 long 2.6 dia	1965 Mar 18.1	64.96	88.27	6568	170	210	0.003	296
D	Fragment	1965-15C											

Year of launch 1965 continued

	Name		Launch date, lifetime and descent date	Shape and weight (kg)	Size (m)	Date of orbital determination	Orbital inclination (deg)	Nodal period (min)	Semi major axis (km)	Perigee height (km)	Apogee height (km)	Orbital eccentricity	Argument of perigee (deg)
	SR 6B	1965-16A	1965 Mar 9.77 1000 years	Sphere? 40?	0.6 dia?	1965 May 3.5	70.09	103.52	7303	910	939	0.002	209
	GGSE 2	1965-16B	1965 Mar 9.77 500 years	Ellipsoid 4?	0.3 dia?	1965 Apr 29.0	70.11	103.50	7302	902	946	0.003	205
	GGSE 3	1965-16C	1965 Mar 9.77 500 years	Ellipsoid 4?	0.3 dia?	1965 May 4.1	70.09	103.52	7303	910	939	0.002	203
	SR 7B	1965-16D	1965 Mar 9.77 1000 years	Sphere 47	0.61 dia	1965 Apr 27.7	70.09	103.52	7303	910	939	0.002	212
	Secor 3 (EGRS 3)	1965-16E	1965 Mar 9.77 1000 years	Rectangular box 18	0.4 x 0.3 x 0.2	1965 May 4.4	70.08	103.51	7302	909	938	0.002	208
	Oscar 3	1965-16F	1965 Mar 9.77 1000 years	Rectangular box 13.6	0.4 x 0.3 x 0.2	1965 May 4.5	70.12	103.50	7302	902	946	0.003	211
D	Surcal 2	1965-16G	1965 Mar 9.77 5862 days 1981 Mar 27	Rectangular box? 5?	0.4 x 0.3 x 0.2?	1965 Apr 17.0 1972 Jan 1.0	70.06 70.06	103.50 102.50	7301 7252	901 856	945 892	0.003 0.002	189 -
	Dodecapole 1	1965-16H	1965 Mar 9.77 200 years	Dodecahedron 4?	0.6 dia?	1965 May 1.0	70.08	103.51	7303	910	939	0.002	194
	Agena D rocket	1965-16J	1965 Mar 9.77 500 years	Cylinder 700?	6 long? 1.5 dia	1965 Apr 24.6	70.09	103.48	7301	908	937	0.002	223
	Fragment	1965-16K											

	Name		Launch date, lifetime and descent date	Shape and weight (kg)	Size (m)	Date of orbital determination	Orbital inclination (deg)	Nodal period (min)	Semi major axis (km)	Perigee height (km)	Apogee height (km)	Orbital eccentricity	Argument of perigee (deg)
D	Capsule	1965-17A	1965 Mar 11.57 94.72 days 1965 Jun 14.29	-	-	1965 Apr 12.2 1965 Jun 8.1	89.97 89.98	95.19 90.67	6929 6683	211 191	890 418	0.049 0.017	7 118
D	Secor 2 (EGRS 2)	1965-17B	1965 Mar 11.57 1081.76 days 1968 Feb 26.33	Rectangular box 18	0.4 x 0.3 x 0.2	1965 May 3.8 1966 Nov 8.0 1968 Jan 15.2	89.98 89.98 89.94	97.85 96.20 91.40	7033 6952 6721	296 283 249	1014 866 437	0.051 0.042 0.014	269 72 200
D	Ablestar rocket	1965-17C	1965 Mar 11.57 906.89 days 1967 Sep 4.46	Cylinder 450?	4.8 long 1.4 dia	1965 Mar 19.6 1966 Sep 21.6 1967 Aug 21.7	90.00 89.98 90.10	97.96 95.95 90.19	7038 6940 6660	287 284 222	1033 839 342	0.053 0.040 0.009	93 231 8
D	[Thor Ablestar]	1965-17D	1965 Mar 11.57 870 days 1967 Jul 29	-	-	1965 Mar 19.7 1966 Nov 18.9 1967 Jul 22.9	89.99 90.00 90.02	97.97 95.20 89.30	7040 6904 6615	289 277 210	1035 774 263	0.053 0.036 0.004	92 15 119
D	Fragments	1965-17E-H											
D	Cosmos 60*	1965-18A	1965 Mar 12.40 5 days 1965 Mar 17	Cylinder 6530? full	7 long? 2.0 max dia	1965 Mar 13.6	64.72	88.93	6600	195	248	0.004	54
D	Cosmos 60 rocket	1965-18B	1965 Mar 12.40 4.5 days 1965 Mar 16.9	Cylinder 2500?	7.5 long 2.6 dia	1965 Mar 14.7	64.74	88.45	6577	192	205	0.001	67
D	Fragment	1965-18C											
D	[Atlas Agena D]	1965-19A	1965 Mar 12.81 4.98 days 1965 Mar 17.79	Cylinder 2000?	8 long? 1.5 dia	1965 Mar 15.3	107.69	88.51	6579	155	247	0.007	119
D	Fragments	1965-19B,C											

*Probably Luna Probe Launcher. (Payload about 1450 kg).

	Name		Launch date, lifetime and descent date	Shape and weight (kg)	Size (m)	Date of orbital determination	Orbital inclination (deg)	Nodal period (min)	Semi major axis (km)	Perigee height (km)	Apogee height (km)	Orbital eccentricity	Argument of perigee (deg)
D	Cosmos 61	1965-20A	1965 Mar 15.46 1036.05 days 1968 Jan 15.51	Ellipsoid? 50?	1.0 long? 0.8 dia?	1965 Mar 17.4 1966 Nov 7.7 1968 Jan 13.4	56.02 55.99 55.95	105.02 101.02 88.99	7378 7188 6604	262 256 180	1738 1363 272	0.100 0.077 0.007	107 91 263
D	Cosmos 62	1965-20B	1965 Mar 15.46 1288.69 days 1968 Sep 24.15	Ellipsoid? 60?	1.0 long? 0.8 dia?	1965 Mar 17.1 1966 Nov 3.7 1968 Jan 15.1	56.03 55.99 56.00	104.77 101.97 96.70	7364 7233 6982	257 254 248	1715 1455 960	0.099 0.083 0.051	106 76 195
D	Cosmos 63	1965-20C	1965 Mar 15.46 964.48 days 1967 Nov 4.94	Ellipsoid? 50?	1.0 long? 0.8 dia?	1965 Mar 17.6 1966 Nov 2.6 1967 Oct 28.9	56.03 55.98 55.99	104.37 100.05 90.32	7345 7142 6672	262 257 234	1672 1270 354	0.096 0.071 0.009	106 99 114
D	Cosmos 61 rocket	1965-20P	1965 Mar 15.46 1705 days 1969 Nov 14	Cylinder* 2200?	7.4 long* 2.4 dia	1965 May 23.5 1968 Jan 31.6 1968 Aug 21.5	56.12 56.05 56.05	105.99 101.33 98.95	7421 7203 7089	295 277 268	1791 1372 1154	0.101 0.076 0.062	- 66 -
122d	Fragments	1965-20D-FF											
	[Thor Altair]	1965-21A	1965 Mar 18.20 25 years	- 130?	-	1965 Mar 18.3 1968 Jan 16.4	99.12 99.04	97.68 97.45	7023 7014	525 524	764 748	0.017 0.016	301 290
D	Altair rocket	1965-21C	1965 Mar 18.20 5271 days 1979 Aug 23	Cylinder 24	1.5 long 0.46 dia	1965 Sep 1.3 1970 Aug 1.0	99.01 98.99	97.67 96.58	7021 6968	523 503	762 676	0.017 0.012	125 -
D	Fragments	1965-21B,D-F											
D 2M R	Voskhod 2	1965-22A	1965 Mar 18.29 1.09 days 1965 Mar 19.38	Sphere-cylinder 5682	6 long? 2.4 dia	1965 Mar 18.7	64.79	90.93	6699	167	475	0.023	70
D	Voskhod 2 rocket	1965-22B	1965 Mar 18.29 9.69 days 1965 Mar 27.98	Cylinder 2500?	7.5 long 2.6 dia	1965 Mar 20.2	64.77	90.57	6681	163	443	0.021	70
D	Fragments	1965-22C,D											
D 2M R	Gemini 3	1965-24A	1965 Mar 23.60 0.20 day 1965 Mar 23.80	Cone 3220 payload 2100	5.6 long 3.0 dia	1965 Mar 23.6	33.0	88.37	6578	160	240	0.006	-
D	Gemini 3 rocket	1965-24B	1965 Mar 23.60 0.9 day 1965 Mar 24.5	Cylinder 1900	6 long 3.0 dia	1965 Mar 23.7	32.62	87.9	6555	164	190	0.002	65

Space vehicle: Ranger 9, 1965-23A. Ranger 9 rocket, 1965-23B, is now in a heliocentric orbit.

* Before explosion.

Year of launch 1965 continued

	Name		Launch date, lifetime and descent date	Shape and weight (kg)	Size (m)	Date of orbital determination	Orbital inclina-tion (deg)	Nodal period (min)	Semi major axis (km)	Perigee height (km)	Apogee height (km)	Orbital eccen-tricity	Argument of perigee (deg)
D R	Cosmos 64	1965-25A	1965 Mar 25.42 7.92 days 1965 Apr 2.34	Sphere-cylinder 4750?	4.3 long 2.4 dia	1965 Mar 25.5	64.98	89.17	6612	201	267	0.005	15
D	Cosmos 64 rocket	1965-25B	1965 Mar 25.42 10.56 days 1965 Apr 4.98	Cylinder 1440	3.8 long 2.6 dia	1965 Mar 25.8	65.00	89.25	6617	219	259	0.003	46
D	Fragment	1965-25C											
D	[Thor Agena D]	1965-26A	1965 Mar 25.88 10.1 days 1965 Apr 5.0	Cylinder 1500?	8 long? 1.5 dia	1965 Mar 25.9	96.08	89.06	6604	186	265	0.006	153
D	Fragment	1965-26B											
	Snapshot* [Atlas Agena D]	1965-27A	1965 Apr 3.89 5000 years	Cylinder 2000? payload 440	11.6 long 1.5 dia	1965 Apr 4.1	89.97	111.58	7676	1282	1313	0.002	246
	Secor 4 (EGRS 4)	1965-27B	1965 Apr 3.89 5000 years	rectangular box 18	0.4 x 0.3 x 0.2	1965 Apr 4.1	90.03	111.58	7676	1282	1313	0.002	223
	Fragments	1965-27C-H											

* Satellite carrying SNAP 10A reactor (system for nuclear auxiliary power).

Year of launch 1965 continued

	Name		Launch date, lifetime and descent date	Shape and weight (kg)	Size (m)	Date of orbital determination	Orbital inclina-tion (deg)	Nodal period (min)	Semi major axis (km)	Perigee height (km)	Apogee height (km)	Orbital eccen-tricity	Argument of perigee (deg)
	Intelsat 1 F-1* (Early Bird)	1965-28A	1965 Apr 6.99 > million years	Cylinder 39	0.59 long 0.72 dia	1965 May 10.9 1977 May 13.0	0.13 10.43	1436.95 1436.57	42183 42174	35003 35777	36606 35815	0.019 0.0005	345 80
	Intelsat 1 F-1 rocket	1965-28B	1965 Apr 6.99 100000 years	Cylinder 24	1.5 long 0.46 dia	1965 Apr 15.5	18.1	672.5	25425	1454	36639	0.692	-
D R	Cosmos 65	1965-29A	1965 Apr 17.41 7.94 days 1965 Apr 25.35	Sphere-cylinder 5530?	5 long? 2.4 dia	1965 Apr 20.7	65.00	89.75	6641	207	319	0.008	54
D	Cosmos 65 rocket	1965-29B	1965 Apr 17.41 18.86 days 1965 May 6.27	Cylinder 2500?	7.5 long 2.6 dia	1965 Apr 21.6	65.04	89.87	6647	210	326	0.009	62
D	Fragment	1965-29C											
D	Molniya 1A	1965-30A	1965 Apr 23.08 5128? days 1979 May 8?	Cylinder + 6 vanes 1000?	3.4 long 1.6 dia	1965 Apr 25.1 1969 Sep 30.9	65.50 65.50	707.29 720.0	26297 26618	538 2303	39300 38177	0.737 0.674	324 -
D	Molniya 1A launcher rocket	1965-30C	1965 Apr 23.08 70.07 days 1965 Jul 2.15	Cylinder 2500?	7.5 long 2.6 dia	1965 Apr 29.1 1965 Jun 24.4	64.83 64.83	94.09 90.10	6855 6659	203 181	751 381	0.040 0.015	59 37
D	Molniya 1A launcher	1965-30B	1965 Apr 23.08 88 days 1965 Jul 20	Irregular	-	1965 Apr 25.8	64.83	94.52	6877	196	801	0.044	63
D	Molniya 1A rocket	1965-30D	1965 Apr 23.08 5165 days 1979 Jun 14	Cylinder 440	2.0 long 2.0 dia	1966 Feb 17.2 1969 Sep 16.0 1974 Jul 1.0	65.56 65.77 65.8	702.50 702.59 702.59	26178 26180 26180	768 2149 1981	38831 37455 37623	0.727 0.674 0.681	319 - -

* International telecommunications satellite.

Year of launch 1965 continued

	Name	Launch date, lifetime and descent date	Shape and weight (kg)	Size (m)	Date of orbital determination	Orbital inclination (deg)	Nodal period (min)	Semi major axis (km)	Perigee height (km)	Apogee height (km)	Orbital eccentricity	Argument of perigee (deg)
D	[Atlas Agena D] 1965-31A	1965 Apr 28.84 5.14 days 1965 May 3.98	Cylinder 1500?	8 long? 1.5 dia	1965 May 1.3	95.60	88.95	6598	180	259	0.006	149
D	Capsule 1965-31B	1965 Apr 28.84 1646.28 days 1969 Oct 31.12	Octagon? 60?	0.3 long? 0.9 dia?	1965 Apr 29.1 1968 Jan 9.7 1968 Nov 30.5	95.26 95.17 95.13	95.16 94.30 93.34	6903 6864 6817	490 472 427	559 500 450	0.005 0.002 0.002	229 248 -
D	Fragments 1965-31C-G											
T L	Explorer 27 (Beacon C) 1965-32A	1965 Apr 29.60 3000 years	Octagon + 4 Vanes 60	0.30 long 0.46 dia	1965 Apr 29.6	41.19	107.78	7507	941	1317	0.025	64
	Explorer 27 rocket 1965-32B	1965 Apr 29.60 500 years	Cylinder 24	1.5 long 0.46 dia	1965 May 1.1	41.16	107.71	7503	930	1320	0.026	72
	Fragments 1965-32C, D											
D	[Thor Agena D] 1965-33A	1965 Apr 29.90 26.5 days 1965 May 26.4	Cylinder 1000?	8 long? 1.5 dia	1965 May 3.4	85.04	91.05	6704	178	473	0.022	147
D	Capsule 1965-33B	1965 Apr 29.90 39.43 days 1965 Jun 8.33	-	-	1965 May 19.0 1965 Jun 4.6	84.88 84.88	95.95 90.97	6997 6700	145 134	1092 509	0.068 0.028	135 55

	Name		Launch date, lifetime and descent date	Shape and weight (kg)	Size (m)	Date of orbital determination	Orbital inclina-tion (deg)	Nodal period (min)	Semi major axis (km)	Perigee height (km)	Apogee height (km)	Orbital eccen-tricity	Argument of perigee (deg)
	Transtage 4 [Titan 3A]	1965-34A	1965 May 6.51 50000 years	Cylinder 3000	6 long 3 dia	1965 May 7.1	32.06	156.93	9642	2782	3746	0.050	293
	LES 2	1965-34B	1965 May 6.51 500000 years	Polyhedron 37	0.6 dia	1965 Jun 23.0 1968 Mar 9.0	32.10 32.20	309.85 309.62	15169 15166	2784 2782	14798 14794	0.396 0.396	327 114
	LCS 1*	1965-34C	1965 May 6.51 30000 years	Sphere 34	1.13 dia	1965 May 7.4	32.11	145.42	9165	2704	2869	0.009	267
	Fragment	1965-34D											
D R	Cosmos 66	1965-35A	1965 May 7.41 7.9 days 1965 May 15.3	Sphere-cylinder 4750?	4.3 long 2.4 dia	1965 May 7.8	65.01	89.33	6620	202	282	0.006	359
D	Cosmos 66 rocket	1965-35B	1965 May 7.41 16.47 days 1965 May 23.88	Cylinder 1440	3.8 long 2.6 dia	1965 May 7.8	65.02	89.50	6630	225	278	0.004	52
D	Fragments	1965-35C,D											
D	Luna 5 launcher	1965-36B	1965 May 9.33 1.2 days 1965 May 10.5	-	-	1965 May 9.7	64.78	88.25	6562	151	217	0.005	322
D	Luna 5 launcher rocket	1965-36C	1965 May 9.33 1.0 days 1965 May 10.3	Cylinder 2500?	7.5 long 2.6 dia	1965 May 9.7	64.75	88.24	6561	143	222	0.006	323

Space Vehicle: Luna 5, 1965-36A

* Lincoln Calibration Satellite.

	Name		Launch date, lifetime and descent date	Shape and weight (kg)	Size (m)	Date of orbital determination	Orbital inclination (deg)	Nodal period (min)	Semi major axis (km)	Perigee height (km)	Apogee height (km)	Orbital eccentricity	Argument of perigee (deg)
D	[Thor Agena D]	1965-37A	1965 May 18.75 28.24 days 1965 Jun 15.99	Cylinder 1500?	8 long? 1.5 dia	1965 May 25.3 1965 Jun 13.1	75.01 75.00	89.71 88.57	6643 6582	198 184	331 224	0.010 0.003	166 99
D	Fragment	1965-37B											
	[Thor Altair]	1965-38A	1965 May 20.69 40 years	- 130?	-	1965 May 23.8 1968 Jan 12.2	98.69 98.50	100.06 99.92	7138 7134	567 556	953 956	0.027 0.028	276 264
	Altair rocket	1965-38B	1965 May 20.69 25 years	Cylinder 24	1.5 long 0.46 dia	1965 Jun 12.8 1968 Jan 15.5	98.62 98.41	100.06 99.90	7139 7132	554 554	968 955	0.029 0.028	209 -
2d	Fragments	1965-38C-G											
D	Pegasus 2 [Saturn SA8]	1965-39A	1965 May 25.32 5275 days 1979 Nov 3	Cylinder + wings 10500	Wings 29.3 long 4.3 wide	1965 May 31.3 1970 Apr 16.0	31.73 31.75	96.99 96.01	6999 6952	502 488	740 660	0.017 0.012	141 -
	Apollo Model 4	1965-39B	1965 May 25.32 25 years	Cone-cylinder 4600	9 long 3.91 dia	1965 Jun 7.4 1967 Oct 5.2	31.74 31.78	97.04 96.95	7002 6997	512 507	736 731	0.016 0.016	254 336
D R	Cosmos 67	1965-40A	1965 May 25.45 7.99 days 1965 Jun 2.44	Sphere- cylinder 5530?	5 long? 2.4 dia	1965 May 27.5	51.81	89.89	6651	200	346	0.011	40
D	Cosmos 67 rocket	1965-40B	1965 May 25.45 10.0 days 1965 Jun 4.5	Cylinder 2500?	7.5 long 2.6 dia	1965 May 30.7	51.82	88.36	6621	190	296	0.008	39
D	Fragments	1965-40C, D											
D	[Atlas Agena D]	1965-41A	1965 May 27.81 5.11 days 1965 Jun 1.92	Cylinder 2000?	8 long? 1.5 dia	1965 May 29.0	95.78	88.67	6586	149	267	0.009	134

	Name		Launch date, lifetime and descent date	Shape and weight (kg)	Size (m)	Date of orbital determination	Orbital inclination (deg)	Nodal period (min)	Semi major axis (km)	Perigee height (km)	Apogee height (km)	Orbital eccentricity	Argument of perigee (deg)
D	Explorer 28 (Imp 3)	1965-42A	1965 May 29.50 1132.5 days 1968 Jul 5.0	Octagon + 4 vanes 59	0.2 long 0.71 dia	1965 May 29.52 1966 Aug 15.5 1967 Aug 3.0	34.0 50.08 53.61	8550 8423.2 8341.9	138473 137137 136251	190 26665 32290	264000 234852 227456	0.953 0.759 0.716	- - -
D	Explorer 28 rocket	1965-42B	1965 May 29.50 37 months 1968 Jul	Cylinder 24	1.5 long 0.46 dia	Orbit similar to 1965-42A							
D 2M R	Gemini 4	1965-43A	1965 Jun 3.64 4.07 days 1965 Jun 7.71	Cone 3574	5.6 long 3.0 dia	1965 Jun 4.7	32.53	88.82	6600	162	281	0.009	74
D	Gemini 4 rocket	1965-43B	1965 Jun 3.64 2.09 days 1965 Jun 5.73	Cylinder 1900	6 long 3.0 dia	1965 Jun 3.9	32.58	88.59	6588	164	256	0.007	74
D	Luna 6 launcher	1965-44B	1965 Jun 8.32 4 days 1965 Jun 12	-	-	1965 Jun 9.4	64.76	88.65	6585	167	246	0.006	47
D	Luna 6 launcher rocket	1965-44C	1965 Jun 8.32 2.2 days 1965 Jun 10.5	Cylinder 2500?	7.5 long 2.6 dia	1965 Jun 8.7	64.70	88.52	6580	189	215	0.002	9
D	[Thor Agena D]	1965-45A	1965 Jun 9.92 12.58 days 1965 Jun 22.50	Cylinder 1500?	8 long? 1.5 dia	1965 Jun 18.4	75.07	89.84	6647	176	362	0.014	149
D	Fragment	1965-45B											

Luna 6, 1965-44A, was probably in a highly eccentric earth orbit initially, later a heliocentric orbit.

	Name		Launch date, lifetime and descent date	Shape and weight (kg)	Size (m)	Date of orbital determination	Orbital inclination (deg)	Nodal period (min)	Semi major axis (km)	Perigee height (km)	Apogee height (km)	Orbital eccentricity	Argument of perigee (deg)
D R	Cosmos 68	1965-46A	1965 Jun 15.42 7.90 days 1965 Jun 23.32	Sphere-cylinder 4750?	4.3 long 2.4 dia	1965 Jun 15.8	65.02	89.82	6640	209	315	0.008	30
D	Cosmos 68 rocket	1965-46B	1965 Jun 15.42 21.79 days 1965 Jul 7.21	Cylinder 1440	3.8 long 2.6 dia	1965 Jun 17.1	65.00	89.89	6646	215	321	0.008	57
D	Fragments	1965-46C-E											
D	[Titan 3C]	1965-47A	1965 Jun 18.58 11.2 days 1965 Jun 29.8	Cone 9700 lead ballast	5 long 3 dia	1965 Jun 19.5	32.14	88.02	6558	167	193	0.002	333
D	Transtage 5	1965-47B	1965 Jun 18.58 3.33 days 1965 Jun 21.91	Cylinder 1500	6 long 3 dia	1965 Jun 20.5	32.13	87.75	6548	163	176	0.001	313
	Transit 5B-6	1965-48A	1965 Jun 24.94 1000 years	Octagon + boom 61	0.5 long 0.46 dia	1965 Jun 25.3	90.00	106.92	7462	1024	1144	0.008	311
	[Thor Ablestar]	1965-48C	1965 Jun 24.94 1000 years	Cylinder?	1.45 long? 0.47 dia?	1965 Jun 28.0	90.00	106.94	7463	1025	1144	0.008	298
	Ablestar rocket	1965-48B	1965 Jun 24.94 1000 years	Cylinder 450?	4.8 long 1.4 dia	1965 Jun 30.3	90.00	106.65	7448	1034	1107	0.005	293
2d	Fragments	1965-48D-K											

	Name		Launch date, lifetime and descent date	Shape and weight (kg)	Size (m)	Date of orbital determination	Orbital inclina-tion (deg)	Nodal period (min)	Semi major axis (km)	Perigee height (km)	Apogee height (km)	Orbital eccen-tricity	Argument of perigee (deg)
D R	Cosmos 69	1965-49A	1965 Jun 25.41 7.91 days 1965 Jul 3.32	Sphere-cylinder 5530?	5 long? 2.4 dia	1965 Jun 27.5	64.89	89.65	6637	212	305	0.007	47
D	Cosmos 69 rocket	1965-49B	1965 Jun 25.41 11.5 days 1965 Jul 6.9	Cylinder 2500?	7.5 long 2.6 dia	1965 Jun 26.4	64.90	89.56	6631	193	312	0.009	36
D	Fragment	1965-49C											
D	Capsule	1965-50A	1965 Jun 25.81 1153 days 1968 Aug 22	Octagon 60?	0.3 long 0.9 dia	1965 Jun 27.0 1968 Jan 9.7 1968 Apr 30.5	107.65 107.63 107.57	94.68 93.22 92.20	6881 6811 6762	496 429 381	510 436 386	0.001 0.0005 0.0004	259 135 -
D	[Atlas Agena D]	1965-50B	1965 Jun 25.81 4.9 days 1965 Jun 30.7	Cylinder 1500?	8 long? 1.5 dia?	1965 Jun 26.8	107.64	88.78	6595	151	283	0.010	137
D	Fragments	1965-50C,D											
	Tiros 10	1965-51A	1965 Jul 2.17 80 years	Cylinder 127	0.56 long 1.07 dia	1965 Jul 2.3	98.65	100.76	7172	751	837	0.006	248
	Tiros 10 rocket	1965-51B	1965 Jul 2.17 80 years	Cylinder 24	1.5 long 0.46 dia	1965 Jul 5.7	98.64	100.71	7169	748	834	0.006	241
	Fragments	1965-51C, D											
D	Cosmos 70	1965-52A	1965 Jul 2.27 534.72 days 1966 Dec 18.99	Ellipsoid 400?	1.8 long 1.2 dia	1965 Jul 3.1 1965 Nov 1.9 1966 Nov 9.1	48.74 48.75 48.72	98.29 97.41 91.35	7059 7016 6723	215 223 204	1147 1052 486	0.066 0.059 0.021	109 256 110
D	Cosmos 70 rocket	1965-52B	1965 Jul 2.27 339.15 days 1966 Jun 6.42	Cylinder 1500?	8 long 1.65 dia	1965 Jul 5.5 1965 Oct 30.3 1966 May 23.4	48.75 48.79 48.71	98.24 96.80 90.85	7057 6987 6697	227 238 206	1130 979 433	0.064 0.053 0.017	120 250 90
D	Fragment	1965-52C											

	Name		Launch date, lifetime and descent date	Shape and weight (kg)	Size (m)	Date of orbital determination	Orbital inclination (deg)	Nodal period (min)	Semi major axis (km)	Perigee height (km)	Apogee height (km)	Orbital eccentricity	Argument of perigee (deg)
D	Cosmos 71	1965-53A	1965 Jul 16.15 1852 days 1970 Aug 11	Ellipsoid? 60?	1.0 long? 0.8 dia?	1965 Jul 18.6 1968 Jan 11.8 1969 Sep 16.0	56.04 56.07 56.00	95.23 94.70 93.45	6910 6884 6824	522 492 441	542 520 450	0.002 0.002 0.001	357 102 -
D	Cosmos 72	1965-53B	1965 Jul 16.15 5152 days 1979 Aug 24	Ellipsoid? 60?	1.0 long? 0.8 dia?	1965 Jul 18.8 1972 Jan 1.0	56.06 56.06	95.87 94.72	6941 6885	538 494	588 520	0.004 0.002	357 -
D	Cosmos 73	1965-53C	1965 Jul 16.15 3169.36 days 1974 Mar 20.51	Ellipsoid? 60?	1.0 long? 0.8 dia?	1965 Jul 18.9 1969 Sep 16.0 1971 Nov 1.0	56.07 56.08 56.07	95.54 94.67 93.53	6925 6883 6828	538 498 446	556 511 453	0.001 0.001 0.001	356 - -
D	Cosmos 74	1965-53D	1965 Jul 16.15 5263 days 1979 Dec 13	Ellipsoid? 60?	1.0 long? 0.8 dia?	1965 Jul 18.9 1970 Dec 1.0	56.04 56.05	96.16 95.28	6955 6913	538 508	616 561	0.006 0.004	356 -
D	Cosmos 75	1965-53E	1965 Jul 16.15 5187 days 1979 Sep 28	Ellipsoid? 60?	1.0 long? 0.8 dia?	1965 Jul 28.8 1970 Sep 16.0	56.04 56.05	96.48 95.54	6970 6925	540 509	644 585	0.007 0.005	316 -
D	Cosmos 71 rocket	1965-53F	1965 Jul 16.15 5972 days 1981 Nov 21	Cylinder 2200?	7.4 long 2.4 dia	1965 Jul 28.3 1977 Sep 1.0	56.08 56.08	96.54 95.60	6973 6928	544 515	646 585	0.007 0.005	318 -
D	Fragments	1965-53G-J											
D	Proton 1	1965-54A	1965 Jul 16.47 86.86 days 1965 Oct 11.33	Cylinder 12,200	3 long? 4 dia?	1965 Jul 17.8 1965 Oct 9.2	63.44 63.44	92.25 88.30	6764 6569	183 145	589 237	0.030 0.007	131 64
D	Proton 1 rocket	1965-54B	1965 Jul 16.47 63.9 days 1965 Sep 18.4	Cylinder 4000?	12 long? 4 dia	1965 Jul 20.4 1965 Sep 14.0	63.47 63.44	92.27 89.05	6769 6608	181 157	601 302	0.031 0.011	121 64
D	Fragments	1965-54C,D											

	Name		Launch date, lifetime and descent date	Shape and weight (kg)	Size (m)	Date of orbital determination	Orbital inclination (deg)	Nodal period (min)	Semi major axis (km)	Perigee height (km)	Apogee height (km)	Orbital eccentricity	Argument of perigee (deg)
D	[Thor Agena D]	1965-55A	1965 Jul 17.25 1249.94 days 1968 Dec 18.19	Cylinder 1500?	8 long? 1.5 dia	1965 Jul 22.3 1968 Jan 14.4 1968 Jun 30.5	70.18 70.19 70.19	94.46 93.23 92.30	6870 6812 6769	471 417 380	512 451 401	0.003 0.002 0.002	85 182 -
D	Fragments	1965-55B-E											
D	Zond 3 launcher rocket	1965-56B	1965 Jul 18.61 1.4 days 1965 Jul 20.0	Cylinder 2500?	7.5 long 2.6 dia	1965 Jul 19.0	64.78	88.18	6565	164	210	0.004	189
D	Fragment *	1965-56C											
D	[Thor Agena D]	1965-57A	1965 Jul 19.92 29.25 days 1965 Aug 18.17	Cylinder 1500?	8 long? 1.5 dia	1965 Jul 21.0	85.05	91.01	6701	182	464	0.021	331
	Vela 5 [Atlas Agena D]	1965-58A	1965 Jul 20.35 > million years	Icosahedron 150	1.4 dia	1965 Jul 20.4 1969 Sep 16.0	35.27 32.3	5148.16 6709.8	98764 117843	88534 93297	96238 129632	0.039 0.154	220 -
	Vela 6	1965-58B	1965 Jul 20.35 > million years	Icosahedron 150	1.4 dia	1965 Jul 20.4 1969 Sep 16.0	34.99 31.4	6726.4 6718.3	118034 117942	101859 81949	121453 141179	0.083 0.251	218 -
D	ORS 3** (ERS 17)†	1965-58C	1965 Jul 20.35 3 years?	Octahedron 5.4	0.7 dia	1965 Jul 20.4 1965 Dec 31.5	34.39 36.88	2608.81 2595.4	62802 62558	153 566	112694 111793	0.896 0.889	219 -
D	Agena D rocket	1965-58D	1965 Jul 20.35 3 years?	Cylinder 700?	6 long? 1.5 dia	orbit similar to 1965-58C							
D	Cosmos 76	1965-59A	1965 Jul 23.19 236 days 1966 Mar 16	Ellipsoid 400?	1.8 long 1.2 dia	1965 Jul 24.1 1965 Nov 6.2 1966 Jan 14.5	48.78 48.78 48.80	92.17 91.49 90.65	6763 6730 6688	256 251 243	513 453 377	0.019 0.015 0.010	312 254 237
D	Cosmos 76 rocket	1965-59B	1965 Jul 23.19 134.3 days 1965 Dec 4.5	Cylinder 1500?	8 long 1.65 dia	1965 Jul 24.4 1965 Oct 31.3 1965 Dec 1.2	48.79 48.78 48.77	92.08 90.69 89.05	6760 6691 6606	253 246 201	510 380 254	0.019 0.010 0.004	310 231 22

* Zond 3 launcher, decayed 1965 Jul 21. ** Octahedron Research Satellite. † Environment Research Satellite.

Space vehicle: Zond 3, 1965-56A

	Name		Launch date, lifetime and descent date	Shape and weight (kg)	Size (m)	Date of orbital determination	Orbital inclination (deg)	Nodal period (min)	Semi major axis (km)	Perigee height (km)	Apogee height (km)	Orbital eccentricity	Argument of perigee (deg)
D	Pegasus 3 [Saturn SA10]	1965-60A	1965 Jul 30.54 1465.76 days 1969 Aug 4.30	Cylinder + wings 10500	Wings 29.2 long 4.3 wide	1965 Jul 30.6 1968 Jan 14.4 1968 Oct 31.2	28.80 28.88 28.88	95.52 94.19 93.35	6929 6864 6823	535 479 441	567 493 449	0.002 0.001 0.001	322 238 -
D	Apollo Model 5	1965-60B	1965 Jul 30.54 3767 days 1975 Nov 22	Cone- cylinder 2600?	9 long 3.91 dia	1965 Aug 3.1 1969 Sep 16.0 1971 Dec 1.0	28.86 28.88 28.87	95.02 94.25 93.36	6905 6867 6824	519 482 441	534 496 450	0.001 0.001 0.001	321 - -
D	Fragment	1965-60C											
D R	Cosmos 77	1965-61A	1965 Aug 3.46 7.93 days 1965 Aug 11.39	Sphere- cylinder 5530?	5 long? 2.4 dia	1965 Aug 4.3	51.79	89.29	6619	201	280	0.006	123
D	Cosmos 77 rocket	1965-61B	1965 Aug 3.46 4.68 days 1965 Aug 8.14	Cylinder 2500?	7.5 long 2.6 dia	1965 Aug 4.3	51.81	89.03	6607	189	268	0.006	123
D	[Atlas Agena D]	1965-62A	1965 Aug 3.80 4.11 days 1965 Aug 7.91	- 500?	1.5 dia?	1965 Aug 4.0	107.47	89.06	6606	149	307	0.012	288
D	Capsule	1965-62B	1965 Aug 3.80 1048.31 days 1968 Jun 17.11	Octagon 60?	0.3 long? 0.9 dia?	1965 Aug 4.0 1968 Jan 10.0 1968 Mar 31.2	107.36 107.32 107.32	94.78 93.06 92.17	6886 6802 6759	501 421 376	515 427 385	0.001 0.0004 0.0006	288 268 -
D	Agena D rocket	1965-62C	1965 Aug 3.80 3 days 1965 Aug 6	Cylinder 700?	6 long? 1.5 dia		Orbit similar to 1965-62A						

Name		Launch date, lifetime and descent date	Shape and weight (kg)	Size (m)	Date of orbital determination	Orbital inclination (deg)	Nodal period (min)	Semi major axis (km)	Perigee height (km)	Apogee height (km)	Orbital eccentricity	Argument of perigee (deg)
Secor 5 rocket	1965-63A	1965 Aug 10.75 10000 years	Cylinder 24	1.49 long 0.50 dia	1965 Aug 18.0	69.26	122.26	8160	1137	2426	0.079	155
Secor 5 (EGRS 5)	1965-63B	1965 Aug 10.75 10000 years	Sphere 20	0.61 dia	1965 Dec 15.5	69.26	122.24	8160	1140	2423	0.079	-
Surveyor model 1	1965-64A	1965 Aug 11.60 Indefinite	Irregular 950	2.8 long 1.3 dia	1965 Aug 12	28.59	44640	417524	165	822,128	0.984	-
Centaur AC6	1965-64B	1965 Aug 11.60 Indefinite	Cylinder 1815	8.6 long 3.0 dia	orbit similar to 1965-64A							
Surcal [Thor Ablestar]	1965-65A	1965 Aug 13.92 1000 years	Sphere? + aerial?	60 metre aerial?	1965 Aug 16.9	90.02	108.19	7520	1074	1209	0.009	289
Ablestar rocket	1965-65B	1965 Aug 13.92 1000 years	Cylinder 450?	4.8 long 1.4 dia	1965 Aug 17.3	90.02	107.83	7503	1087	1162	0.005	297
Dodecapole 2	1965-65C	1965 Aug 13.92 50 years	Sphere + aerials 4	Sphere 0.3 dia	1965 Aug 18.0	90.03	108.14	7517	1094	1184	0.006	290
Tempsat 1	1965-65E	1965 Aug 13.92 1000 years	Sphere (black) 9	0.36 dia	1965 Aug 26.2	90.03	108.17	7519	1096	1186	0.006	262
Transit 5B-7	1965-65F	1965 Aug 13.92 1000 years	Octagon + boom 61	0.5 long 0.46 dia	1965 Aug 22.5	90.01	108.19	7520	1089	1194	0.007	276
Surcal	1965-65H	1965 Aug 13.92 200 years	Sphere (white) 2.2	0.36 dia	1965 Nov 15.5	90.05	108.19	7520	1082	1201	0.008	-
Surcal	1965-65K	1965 Aug 13.92 1000 years	Rectangular box 6	0.4 x 0.3 x 0.2?	1965 Aug 21.6	90.02	108.14	7517	1094	1184	0.006	269
4d Fragments	1965-65 D,G,J,L-T											

	Name		Launch date, lifetime and descent date	Shape and weight (kg)	Size (m)	Date of orbital determination	Orbital inclination (deg)	Nodal period (min)	Semi major axis (km)	Perigee height (km)	Apogee height (km)	Orbital eccentricity	Argument of perigee (deg)
D R	Cosmos 78	1965-66A	1965 Aug 14.47 7.89 days 1965 Aug 22.36	Sphere-cylinder 4750?	4.3 long 2.4 dia	1965 Aug 14.8	68.92	89.75	6636	218	298	0.006	167
D	Cosmos 78 rocket	1965-66B	1965 Aug 14.47 20.3 days 1965 Sep 3.8	Cylinder 1440	3.8 long 2.6 dia	1965 Aug 16.5	68.98	89.76	6637	219	299	0.006	40
D	Fragment	1965-66C											
D	[Thor Agena D]	1965-67A	1965 Aug 17.87 54.40 days 1965 Oct 11.27	Cylinder 1500?	8 long? 1.5 dia	1965 Aug 19.9 1965 Oct 8.1	70.04 70.04	90.37 88.83	6672 6594	180 176	407 255	0.017 0.006	153 65
D 2 M R	Gemini 5	1965-68A	1965 Aug 21.58 7.96 days 1965 Aug 29.54	Cone 3605	5.6 long 3.0 dia	1965 Aug 24.5	32.61	89.38	6628	197	303	0.008	104
D	Gemini 5 rocket	1965-68B	1965 Aug 21.58 3.12 days 1965 Aug 24.70	Cylinder 1900	6 long 3.0 dia	1965 Aug 23.9	32.58	88.22	6576	152	244	0.007	103
D	REP*	1965-68C	1965 Aug 21.58 5.71 days 1965 Aug 27.29	Box 35	0.6 x 0.3 x 0.3	1965 Aug 23.2	32.58	89.18	6619	168	314	0.011	87
D	Fragment	1965-68D											

*Radar Evaluation Pod ejected from Gemini 5 about 1965 Aug 21.66.

	Name		Launch date, lifetime and descent date	Shape and weight (kg)	Size (m)	Date of orbital determination	Orbital inclination (deg)	Nodal period (min)	Semi major axis (km)	Perigee height (km)	Apogee height (km)	Orbital eccentricity	Argument of perigee (deg)
D R	Cosmos 79	1965-69A	1965 Aug 25.43 7.9 days 1965 Sep 2.3	Sphere-cylinder 5530?	5 long? 2.4 dia	1965 Aug 28.3	64.90	89.94	6650	205	338	0.010	54
D	Cosmos 79 rocket	1965-69B	1965 Aug 25.43 13.18 days 1965 Sep 7.61	Cylinder 2500?	7.5 long 2.6 dia	1965 Aug 30.8	64.90	89.64	6635	204	310	0.008	39
D	Fragments	1965-69C-E											
	Cosmos 80	1965-70A	1965 Sep 3.58 10000 years	Ellipsoid? 50?	1.0 long? 0.8 dia	1965 Sep 4.0	55.98	114.97	7834	1357	1555	0.013	107
	Cosmos 81	1965-70B	1965 Sep 3.58 10000 years	Ellipsoid? 50?	1.0 long? 0.8 dia?	1965 Sep 5.5	56.05	115.29	7849	1384	1557	0.011	116
	Cosmos 82	1965-70C	1965 Sep 3.58 10000 years	Ellipsoid? 50?	1.0 long? 0.8 dia?	1965 Sep 5.6	56.04	115.65	7865	1408	1565	0.010	126
	Cosmos 83	1965-70D	1965 Sep 3.58 10000 years	Ellipsoid? 50?	1.0 long? 0.8 dia?	1965 Sep 5.5	56.03	116.01	7882	1441	1567	0.008	138
	Cosmos 84	1965-70E	1965 Sep 3.58 10000 years	Ellipsoid? 50?	1.0 long? 0.8 dia?	1965 Sep 5.6	56.03	116.33	7899	1466	1576	0.007	208
	Cosmos 80 rocket	1965-70F	1965 Sep 3.58 10000 years	Cylinder 2200?	7.4 long 2.4 dia	1965 Sep 5.5	56.11	114.57	7815	1359	1515	0.010	208
	Fragment	1965-70G											

	Name		Launch date, lifetime and descent date	Shape and weight (kg)	Size (m)	Date of orbital determination	Orbital inclination (deg)	Nodal period (min)	Semi major axis (km)	Perigee height (km)	Apogee height (km)	Orbital eccentricity	Argument of perigee (deg)
D R	Cosmos 85	1965-71A	1965 Sep 9.40 7.89 days 1965 Sep 17.29	Sphere-cylinder 5530?	5 long? 2.4 dia	1965 Sep 9.8	64.90	89.53	6629	204	297	0.007	56
D	Cosmos 85 rocket	1965-71B	1965 Sep 9.40 9.0 days? 1965 Sep 18.4?	Cylinder 2500?	7.5 long 2.6 dia	1965 Sep 11.5	64.91	89.26	6617	199	279	0.006	42
	[Thor Altair]	1965-72A	1965 Sep 10.20 80 years	- 130?	-	1965 Sep 10.6	98.65	101.93	7230	649	1054	0.028	250
	Altair rocket	1965-72D	1965 Sep 10.20 50 years	Cylinder 24	1.49 long 0.50 dia	1966 Jan 15.5	98.64	101.94	7230	649	1055	0.028	-
2d	Fragments	1965-72B,C,E,F											
	Cosmos 86	1965-73A	1965 Sep 18.33 10000 years	Ellipsoid? 50?	1.0 long? 0.8 dia?	1965 Sep 20.3	56.06	115.02	7836	1277	1638	0.023	136
	Cosmos 87	1965-73B	1965 Sep 18.33 10000 years	Ellipsoid? 50?	1.0 long? 0.8 dia?	1965 Sep 22.6	56.06	115.42	7855	1304	1650	0.022	145
	Cosmos 88	1965-73C	1965 Sep 18.33 10000 years	Ellipsoid? 50?	1.0 long? 0.8 dia?	1965 Sep 20.3	56.05	115.81	7872	1321	1667	0.022	146
	Cosmos 89	1965-73D	1965 Sep 18.33 10000 years	Ellipsoid? 50?	1.0 long? 0.8 dia?	1965 Sep 19.8	56.09	116.17	7890	1346	1677	0.021	153
	Cosmos 90	1965-73E	1965 Sep 18.33 10000 years	Ellipsoid? 50?	1.0 long? 0.8 dia?	1965 Sep 20.0	56.06	116.60	7909	1373	1689	0.020	159
	Cosmos 86 rocket	1965-73F	1965 Sep 18.33 10000 years	Cylinder 2200?	7.4 long 2.4 dia	1965 Sep 20.3	56.03	116.75	7915	1379	1695	0.020	159
	Fragments	1965-73G-L											

	Name		Launch date, lifetime and descent date	Shape and weight (kg)	Size (m)	Date of orbital determination	Orbital inclination (deg)	Nodal period (min)	Semi major axis (km)	Perigee height (km)	Apogee height (km)	Orbital eccentricity	Argument of perigee (deg)
D	[Thor Agena D]	1965-74A	1965 Sep 22.89 18 days 1965 Oct 11	Cylinder 1500?	8 long? 1.5 dia	1965 Sep 25.0	80.01	90.04	6656	191	364	0.013	160
D R	Cosmos 91	1965-75A	1965 Sep 23.38 7.91 days 1965 Oct 1.29	Sphere-cylinder 5530?	5 long? 2.4 dia	1965 Sep 24.9	64.98	89.76	6642	204	324	0.009	49
D	Cosmos 91 rocket	1965-75B	1965 Sep 23.38 10.89 days 1965 Oct 4.27	Cylinder 2500?	7.5 long 2.6 dia	1965 Sep 25.8	64.97	89.53	6629	204	297	0.007	34
D	Fragments	1965-75C-E											
D	Capsule [Atlas Agena D]	1965-76A	1965 Sep 30.81 4.70 days 1965 Oct 5.51	- 500?	1.5 dia?	1965 Oct 1.2	95.60	88.77	6589	158	264	0.008	151
D	Agena D rocket	1965-76B	1965 Sep 30.81 2 days 1965 Oct 3	Cylinder 700?	6 long? 1.5 dia	orbit similar to 1965-76A							
D	Luna 7 launcher	1965-77B	1965 Oct 4.33 0.6 day 1965 Oct 4.9	-	-	1965 Oct 4.4	64.75	88.62	6586	129	286	0.012	138
D	Luna 7 launcher rocket	1965-77C	1965 Oct 4.33 0.8 day 1965 Oct 5.1	Cylinder 2500?	7.5 long 2.6 dia	1965 Oct 4.7	64.76	88.44	6577	124	272	0.011	134
	OV1-2*	1965-78A	1965 Oct 5.38 100 years	Cylinder + hemisphere 88	1.40 long 0.69 dia	1965 Oct 6.1 1968 Feb 11.5	144.30 144.2	125.58 125.2	8311 8290	403 411	3462 3412	0.184 0.181	53 -
	OV1-2 rocket	1965-78B	1965 Oct 5.38 50 years	Cylinder 70?	2.05 long 0.72 dia	1965 Oct 31.5 1968 Feb 11.5	144.30 144.2	125.56 125.1	8309 8284	414 412	3448 3399	0.183 0.180	- -

Space Vehicle: 1965-77A, Luna 7. * Orbiting vehicle.

Year of launch 1965 continued

	Name		Launch date, lifetime and descent date	Shape and weight (kg)	Size (m)	Date of orbital determination	Orbital inclination (deg)	Nodal period (min)	Semi major axis (km)	Perigee height (km)	Apogee height (km)	Orbital eccentricity	Argument of perigee (deg)
D	[Thor Agena D]	1965-79A	1965 Oct 5.74 24.01 days 1965 Oct 29.75	Cylinder 1500?	8 long? 1.5 dia	1965 Oct 6.7	75.05	89.75	6641	203	323	0.009	142
D	Molniya 1B	1965-80A	1965 Oct 14.82 518 days 1967 Mar 17	Cylinder + 6 vanes 1000?	3.4 long 1.6 dia	1965 Oct 18.3 1966 Mar 15.8	65.19 64.91	718.84 716.37	26586 26524	481 342	39935 39950	0.742 0.747	285 -
D	Molniya 1B launcher	1965-80B	1965 Oct 14.82 24.38 days 1965 Nov 8.20	Irregular	-	1965 Oct 17.1	64.85	91.09	6707	199	458	0.019	60
D	Molniya 1B launcher rocket	1965-80D	1965 Oct 14.82 33 days 1965 Nov 16	Cylinder 2500?	7.5 long 2.6 dia	1965 Oct	64.8	91.5	6724	197	496	0.022	-
D	Molniya 1B rocket	1965-80E	1965 Oct 14.82 460 days 1967 Jan 18	Cylinder 440	2.0 long 2.0 dia	Orbit similar to 1965-80A							
D	Fragment	1965-80C											
D	OGO 2	1965-81A	1965 Oct 14.54 5817 days 1981 Sep 17	Box + booms 507	1.73 long 0.84 wide 0.84 high	1965 Oct 14.7 1970 Sep 1.0	87.43 87.3	104.41 102.36	7344 7249	415 409	1517 1333	0.075 0.064	171 -
	OGO 2 rocket	1965-81B	1965 Oct 14.54 25 years	Cylinder 700?	6 long 1.5 dia	1965 Oct 15.6 1971 Aug 1.0	87.37 87.3	104.38 102.34	7342 7248	406 412	1522 1328	0.076 0.063	169 -
D	Transtage 6 [Titan 3C]	1965-82A	1965 Oct 15.79 2476.77 days 1972 Jul 27.56	Cylinder* 1500?	6 long* 3 dia	1965 Oct 18.2 1968 Jan 25.2 1970 May 1.0	32.61 32.27 32.3	99.70 99.34 97.80	7127 7110 7037	706 703 633	792 761 685	0.006 0.004 0.004	286 - -
369d	Fragments	1965-82B-UJ											
D R B	Cosmos 92	1965-83A	1965 Oct 16.34 7.94 days 1965 Oct 24.28	Sphere-cylinder 5530?	5 long? 2.4 dia	1965 Oct 17.4	64.97	89.85	6646	201	334	0.010	56
D	Cosmos 92 rocket	1965-83B	1965 Oct 16.34 13 days 1965 Oct 29	Cylinder 2500?	7.5 long 2.6 dia	1965 Oct 16.8	64.98	89.74	6641	203	322	0.009	53
D	Fragment	1965-83C											

* Before explosion; carried OV2-1 and LCS 2.

	Name		Launch date, lifetime and descent date	Shape and weight (kg)	Size (m)	Date of orbital determination	Orbital inclination (deg)	Nodal period (min)	Semi major axis (km)	Perigee height (km)	Apogee height (km)	Orbital eccentricity	Argument of perigee (deg)
D	Cosmos 93	1965-84A	1965 Oct 19.24 76.29 days 1966 Jan 3.53	Ellipsoid 400?	1.8 long 1.2 dia	1965 Oct 19.7	48.39	91.77	6743	216	513	0.022	97
D	Cosmos 93 rocket	1965-84B	1965 Oct 19.24 29 days 1965 Nov 17	Cylinder 1500?	8 long 1.65 dia	1965 Oct 19.8	48.35	91.60	6734	208	504	0.022	94
D	Fragments	1965-84C-E											
D R B	Cosmos 94	1965-85A	1965 Oct 28.35 7.93 days 1965 Nov 5.28	Sphere-cylinder 5530?	5 long? 2.4 dia	1965 Oct 28.8	64.96	89.23	6616	205	271	0.005	65
D	Cosmos 94 rocket	1965-85B	1965 Oct 28.35 6.57 days 1965 Nov 3.92	Cylinder 2500?	7.5 long 2.6 dia	1965 Oct 29.8	64.97	89.04	6607	202	255	0.004	41
D	[Thor Agena D]	1965-86A	1965 Oct 28.89 19.81 days 1965 Nov 17.70	Cylinder 1500?	8 long? 1.5 dia	1965 Oct 31.9	74.97	90.54	6681	176	430	0.019	167
D	Fragment	1965-86B											
D	Proton 2	1965-87A	1965 Nov 2.52 96.01 days 1966 Feb 6.53	Cylinder 12,200	3 long? 4 dia?	1965 Nov 5.9	63.45	92.52	6776	189	608	0.031	59
D	Proton 2 rocket	1965-87B	1965 Nov 2.52 62.67 days 1966 Jan 4.19	Cylinder 4000?	12 long? 4 dia	1965 Nov 5.8	63.46	92.54	6777	182	616	0.032	250
D	Fragments	1965-87C-E											

	Name		Launch date, lifetime and descent date	Shape and weight (kg)	Size (m)	Date of orbital determination	Orbital inclination (deg)	Nodal period (min)	Semi major axis (km)	Perigee height (km)	Apogee height (km)	Orbital eccentricity	Argument of perigee (deg)
D	Cosmos 95	1965-88A	1965 Nov 4.23 75.39 days 1966 Jan 18.62	Ellipsoid 400?	1.8 long 1.2 dia	1965 Nov 4.6 1966 Jan 9.5	48.40 48.39	91.78 89.39	6744 6624	211 186	521 309	0.023 0.009	106 77
D	Cosmos 95 rocket	1965-88B	1965 Nov 4.23 24.7 days 1965 Nov 28.9	Cylinder* 1500?	8 long* 1.65 dia	1965 Nov 5.4 1965 Nov 23.5	48.40 48.39	91.55 89.50	6731 6639	218 201	487 321	0.020 0.009	107 336
D	Fragments	1965-88C-Z											
L	Explorer 29 (Geos 1)**	1965-89A	1965 Nov 6.78 50000 years	Octahedron + Pyramid 175	0.81 high 1.22 wide	1965 Nov 6.9	59.38	120.30	8074	1115	2277	0.072	150
	Explorer 29 rocket	1965-89B	1965 Nov 6.78 10000 years	Cylinder 24	1.5 long 0.46 dia	1965 Nov 8.5	59.37	120.29	8073	1114	2276	0.072	151
	Fragments	1965-89C,D											
D	[Atlas Agena D]	1965-90A	1965 Nov 8.81 2.92 days 1965 Nov 11.73	Cylinder 2000?	8 long? 1.5 dia	1965 Nov 8.9	93.88	88.74	6589	145	277	0.010	135
D	Fragment	1965-90B											
D	Venus 2 launcher	1965-91B	1965 Nov 12.21 5 days 1965 Nov 17	-	-	1965 Nov 15.3	51.85	88.67	6588	203	216	0.001	48
D	Venus 2 launcher rocket	1965-91C	1965 Nov 12.21 13.0 days 1965 Nov 25.2	Cylinder 2500?	7.5 long 2.6 dia	1965 Nov 15.2	51.87	89.47	6628	217	283	0.005	95

Space Vehicle: Venus 2, 1965-91A.

* Before disintegration.

** Geodetic satellite.

	Name		Launch date, lifetime and descent date	Shape and weight (kg)	Size (m)	Date of orbital determination	Orbital inclination (deg)	Nodal period (min)	Semi major axis (km)	Perigee height (km)	Apogee height (km)	Orbital eccentricity	Argument of perigee (deg)
D	Venus 3 launcher	1965-92B	1965 Nov 16.18 10.19 days 1965 Nov 26.37	-	-	1965 Nov 17.5	51.85	89.54	6631	213	293	0.006	80
D	Venus 3 launcher rocket	1965-92C	1965 Nov 16.18 16.86 days 1965 Dec 3.04	Cylinder 2500?	7.5 long 2.6 dia	1965 Nov 24.4	51.85	89.34	6623	203	286	0.006	117
D	Fragment	1965-92E											
	Explorer 30 (IQSY)*	1965-93A	1965 Nov 19.20 200 years	Sphere 57	0.61 dia	1965 Nov 19.8	59.72	100.80	7176	704	891	0.013	153
	Explorer 30 rocket	1965-93B	1965 Nov 19.20 100 years	Cylinder 24	1.5 long 0.46 dia	1965 Nov 30.5	59.70	100.75	7173	706	884	0.012	-
	Fragments	1965-93C, D											
D	Cosmos 96**	1965-94A	1965 Nov 23.14 16.21 days 1965 Dec 9.35	Cylinder 6540? full	7 long? 2.0 max dia	1965 Dec 1.9	51.88	89.20	6614	209	262	0.004	111
D	Cosmos 96 rocket	1965-94B	1965 Nov 23.14 11.83 days 1965 Dec 4.97	Cylinder 2500?	7.5 long 2.6 dia	1965 Nov 28.7	51.85	89.24	6618	210	269	0.004	120
D	Fragments	1965-94C-H											
D	Cosmos 97	1965-95A	1965 Nov 26.51 492.1 days 1967 Apr 2.6	Polygonal ellipsoid 400?	1.8 long 1.2 to 1.5 dia	1965 Nov 27.6 1966 Nov 8.5 1967 Mar 31.9	48.42 48.41 48.34	108.83 99.92 89.00	7556 7135 6606	213 207 175	2144 1306 281	0.128 0.077 0.008	104 298 212
D	Cosmos 97 rocket	1965-95B	1965 Nov 26.51 451.54 days 1967 Feb 21.05	Cylinder 1500?	8 long 1.65 dia	1965 Nov 27.6 1966 Nov 8.7 1967 Feb 2.4	48.43 48.41 48.41	108.68 97.96 92.60	7546 7042 6784	211 206 182	2125 1121 630	0.127 0.065 0.033	104 326 354
D	Fragments	1965-95C, D											

Space Vehicle: Venus 3, 1965-92A, and rocket 1965-92D

*International Quiet Sun Year satellite.
**Cosmos 96 may have been a Space Vehicle launcher.

	Name		Launch date, lifetime and descent date	Shape and weight (kg)	Size (m)	Date of orbital determination	Orbital inclina-tion (deg)	Nodal period (min)	Semi major axis (km)	Perigee height (km)	Apogee height (km)	Orbital eccen-tricity	Argument of perigee (deg)
	Asterix 1 * [Diamant]	1965-96A	1965 Nov 26.62 200 years	Double-cone 40	0.53 long 0.55 dia	1965 Nov 27.9	34.24	108.61	7545	527	1808	0.085	82
	Asterix 1 rocket	1965-96B	1965 Nov 26.62 100 years	Cylinder 68	2.1 long 0.65 dia	1965 Nov 29.4	34.21	108.61	7546	524	1812	0.085	92
1d	Fragments	1965-96C,D											
D R	Cosmos 98	1965-97A	1965 Nov 27.35 8.0 days 1965 Dec 5.3	Sphere-cylinder 4750?	4.3 long 2.4 dia	1965 Nov 28.5	65.03	92.07	6754	205	547	0.025	85
D	Cosmos 98 rocket	1965-97B	1965 Nov 27.35 56.79 days 1966 Jan 23.14	Cylinder 1440	3.8 long 2.6 dia	1965 Nov 28.4 1966 Jan 10.7	65.04 65.04	92.17 90.07	6760 6658	198 187	566 373	0.027 0.014	92 72
D	Fragment	1965-97C											
	Alouette 2	1965-98A	1965 Nov 29.20 350 years	Oblate spheroid 145	0.86 long 1.07 dia	1965 Nov 30.5	79.82	121.43	8124	505	2987	0.153	335
	Explorer 31 (DME A)**	1965-98B	1965 Nov 29.20 500 years	Octagonal Cylinder 100	0.63 long 0.76 dia	1965 Nov 30.5	79.82	121.39	8120	505	2978	0.152	335
	Alouette 2 rocket	1965-98C	1965 Nov 29.20 400 years	Cylinder 700?	6 long? 1.5 dia	1965 Nov 30.5	79.84	121.39	8120	505	2979	0.152	335
	Fragments	1965-98D-J											
D	Luna 8 launcher	1965-99B	1965 Dec 3.45 1.63 days 1965 Dec 5.08	-	-	1965 Dec 4.1	51.82	88.20	6567	169	209	0.003	239
D	Luna 8 launcher rocket	1965-99C	1965 Dec 3.45 3 days 1965 Dec 6	Cylinder 2500?	7.5 long 2.6 dia	1965 Dec 4.3	51.87	88.48	6579	181	221	0.003	167

Space Vehicle: Luna 8, 1965-99A

*First French satellite. **Direct Measurement Explorer.

	Name		Launch date, lifetime and descent date	Shape and weight (kg)	Size (m)	Date of orbital determination	Orbital inclination (deg)	Nodal period (min)	Semi major axis (km)	Perigee height (km)	Apogee height (km)	Orbital eccentricity	Argument of perigee (deg)
D 2 M R	Gemini 7*	1965-100A	1965 Dec 4.81 13.78 days 1965 Dec 18.59	Cone 3663	5.6 long 3.0 dia	1965 Dec 5.5 1965 Dec 12.3	28.87 28.90	89.75 90.27	6646 6673	215 292	321 298	0.008 0.0005	94 177
D	Gemini 7 rocket	1965-100B	1965 Dec 4.81 2.34 days 1965 Dec 7.15	Cylinder 1900	6 long 3.0 dia	1965 Dec 5.9	28.89	89.20	6617	160	318	0.012	87
D	Gemini 7 adaptor module	1965-100C	1965 Dec 4.81 54.10 days 1966 Jan 27.91	Truncated cone 2450?	2.3 long 3.0 dia	1965 Dec 31.5	28.90	89.97	6657	262	296	0.003	-
	France 1	1965-101A	1965 Dec 6.88 50 years	Polyhedron 60	1.32 long 0.69 dia	1965 Dec 9.6	75.87	99.94	7133	746	762	0.001	45
	France 1 rocket	1965-101B	1965 Dec 6.88 50 years	Cylinder 24	1.5 long 0.46 dia	1965 Dec 8.3	75.83	100.04	7139	747	775	0.002	111
D	Fragments	1965-101C,D											
D	[Thor Agena D]	1965-102A	1965 Dec 9.88 16.78 days 1965 Dec 26.66	Cylinder 1500?	8 long? 1.5 dia	1965 Dec 10.6	80.04	90.72	6688	183	437	0.019	102
D	Fragment	1965-102B											
D R	Cosmos 99	1965-103A	1965 Dec 10.34 7.90 days 1965 Dec 18.24	Sphere-cylinder 4750?	4.3 long 2.4 dia	1965 Dec 10.8	64.99	89.61	6634	203	309	0.008	236
D	Cosmos 99 rocket	1965-103B	1965 Dec 10.34 18.65 days 1965 Dec 28.99	Cylinder 1440	3.8 long 2.6 dia	1965 Dec 10.8	65.02	89.71	6639	221	301	0.006	45
D 2M R	Gemini 6*	1965-104A	1965 Dec 15.57 1.08 days 1965 Dec 16.65	Cone 3546	5.6 long 3.0 dia	1965 Dec 15.8	28.89	89.64	6643	258	271	0.001	97
D	Gemini 6 rocket	1965-104B	1965 Dec 15.57 1.30 days 1965 Dec 16.87	Cylinder 1900	6 long 3.0 dia		initial orbit similar to 1965-104A						

*Rendezvous between Gemini 6 and Gemini 7 on 1965 Dec 15.81. 1965-100A and 100C were joined until 1965 Dec 18.58.

	Name		Launch date, lifetime and descent date	Shape and weight (kg)	Size (m)	Date of orbital determination	Orbital inclination (deg)	Nodal period (min)	Semi major axis (km)	Perigee height (km)	Apogee height (km)	Orbital eccentricity	Argument of perigee (deg)
D	Pioneer 6 second stage	1965-105B	1965 Dec 16.31 995.04 days 1968 Sep 6.35	Cylinder 400?	4.9 long 1.43 dia	1966 Jan 2.9 1967 Oct 8.1 1968 Mar 15.8	30.18 30.17 30.17	100.15 96.10 93.94	7150 6956 6853	273 265 252	1271 891 697	0.070 0.045 0.032	288 329 -
	Cosmos 100	1965-106A	1965 Dec 17.10 60 years	Cylinder + 2 vanes 2000?	5 long? 1.5 dia?	1965 Dec 17.8	65.00	97.58	7022	630	658	0.002	254
	Cosmos 100 rocket	1965-106B	1965 Dec 17.10 40 years	Cylinder 1440	3.8 long 2.6 dia	1965 Dec 19.5	64.99	97.74	7029	566	735	0.012	180
D	Cosmos 101	1965-107A	1965 Dec 21.26 203 days 1966 Jul 12	Ellipsoid 400?	1.8 long 1.2 dia	1965 Dec 21.8 1966 Jun 30.0	48.78 48.77	92.36 89.50	6775 6629	254 211	539 291	0.021 0.006	104 308
D	Cosmos 101 rocket	1965-107B	1965 Dec 21.26 116.1 days 1966 Apr 16.4	Cylinder 1500?	8 long 1.65 dia	1965 Dec 23.2 1966 Apr 7.9	48.77 48.74	92.23 89.70	6767 6642	253 217	524 310	0.020 0.007	108 263
D	Transtage 7* [Titan 3C]	1965-108A	1965 Dec 21.65 3526 days 1975 Aug 17	Cylinder 2000?	10 long? 3 dia	1965 Dec 26.1 1967 Nov 20.9 1969 Sep 16.0	26.38 26.45 26.53	589.26 441.56 289.4	23293 19245 14492	167 165 184	33662 25568 16043	0.719 0.660 0.547	3 226 -
D	LES 4	1965-108B	1965 Dec 21.65 4241 days? 1977 Aug 1?	Cylinder 52	0.91 long 0.85 dia	1965 Dec 28.6 1966 May 15.5	26.60 26.50	589.24 578.0	23289 23090	189 216	33632 33208	0.718 0.714	5 -
D	Oscar 4	1965-108C	1965 Dec 21.65 3765 days? 1976 Apr 12?	Tetrahedron 15	0.48 side	1965 Dec 27.8 1966 May 15.5	26.80 26.73	587.49 579.4	23240 22923	162 161	33561 32929	0.719 0.715	5 -
D	LES 3	1965-108D	1965 Dec 21.65 836 days 1968 Apr 6	Polyhedron 16	0.6 dia	1966 Jan 19.9 1967 Oct 21.9	26.46 26.36	581.41 151.93	23064 10083	195 155	33177 7254	0.715 0.352	20 17
D	Fragments	1965-108E-K											

Space Vehicle: Pioneer 6, 1965-105A * Carried OV2-3.

Fragment 1965-108G, designated about 1970 Jul 16, was probably satellite LES 4 rediscovered (lost 1966 May)
Fragment 1965-108K, designated about 1972 Apr 15, was probably satellite Oscar 4 rediscovered (lost 1966 May)

	Name		Launch date, lifetime and descent date	Shape and weight (kg)	Size (m)	Date of orbital determination	Orbital inclination (deg)	Nodal period (min)	Semi major axis (min)	Perigee height (km)	Apogee height (km)	Orbital eccentricity	Argument of perigee (deg)
	Transit 10 [Scout]	1965-109A	1965 Dec 22.19 1000 years	Octagon + 4 vanes 60?	0.25 long 0.46 dia	1965 Dec 27.9	89.11	105.09	7376	909	1086	0.012	190
	Altair rocket	1965-109B	1965 Dec 22.19 1000 years	Cylinder 24	1.5 long 0.46 dia	1965 Dec 26.3	89.10	105.09	7375	916	1078	0.011	196
	Fragments	1965-109C-E											
D	[Thor Agena D]	1965-110A	1965 Dec 24.88 26.59 days 1966 Jan 20.47	Cylinder 1500?	8 long? 1.5 dia	1965 Dec 26.7	80.01	90.83	6690	178	446	0.020	157
D	Fragment	1965-110B											
D	Cosmos 102	1965-111A	1965 Dec 27.93 17.00 days 1966 Jan 13.93	Cylinder	10 long? 2 dia?	1965 Dec 28.7	64.97	89.20	6614	203	269	0.005	241
	Cosmos 103	1965-112A	1965 Dec 28.52 25 years	Cylinder + paddles 850?	2 long? 1 dia?	1965 Dec 29.9	56.07	96.95	6993	594	636	0.003	114
D	Cosmos 103 rocket	1965-112B	1965 Dec 28.52 6013 days 1982 Jun 15	Cylinder 2200?	7.4 long 2.4 dia	1965 Dec 28.9	56.05	97.07	6999	593	649	0.004	136
12 d	Fragments	1965-112C-Q											

	Name		Launch date, lifetime and descent date	Shape and weight (kg)	Size (m)	Date of orbital determination	Orbital inclination (deg)	Nodal period (min)	Semi major axis (km)	Perigee height (km)	Apogee height (km)	Orbital eccentricity	Argument of perigee (deg)
D R	Cosmos 104	1966-01A	1966 Jan 7.35 7.90 days 1966 Jan 15.25	Sphere-cylinder 4750?	4.3 long 2.4 dia	1966 Jan 8.1	65.00	90.22	6665	193	380	0.014	79
D	Cosmos 104 rocket	1966-01B	1966 Jan 7.35 17.37 days 1966 Jan 24.72	Cylinder 1440	3.8 long 2.6 dia	1966 Jan 8.7	65.03	90.27	6667	189	389	0.015	91
D	Fragment	1966-01C											
D	Capsule	1966-02A	1966 Jan 19.84 6.00 days 1966 Jan 25.84	Sphere?	-	1966 Jan 22.0	93.86	88.72	6588	150	269	0.009	138
D	[Atlas Agena D]	1966-02B	1966 Jan 19.84 3.88 days 1966 Jan 23.72	Cylinder 1500?	8 long? 1.5 dia	1966 Jan 21.2	93.89	88.51	6578	154	246	0.007	140
D R	Cosmos 105	1966-03A	1966 Jan 22.36 7.90 days 1966 Jan 30.26	Sphere-cylinder 4750?	4.3 long 2.4 dia	1966 Jan 23.7	65.01	89.64	6635	204	310	0.008	26
D	Cosmos 105 rocket	1966-03B	1966 Jan 22.36 19 days 1966 Feb 10	Cylinder 1440	3.8 long 2.6 dia	1966 Jan 24.8	65.03	89.72	6640	215	308	0.007	64
D	Fragment	1966-03C											
D	Cosmos 106	1966-04A	1966 Jan 25.52 293.32 days 1966 Nov 14.84	Ellipsoid 400?	1.8 long 1.2 dia	1966 Jan 29.5 1966 Jun 15.5 1966 Oct 28.7	48.39 48.38 48.38	92.82 91.99 89.95	6795 6754 6652	281 275 241	553 476 307	0.020 0.015 0.005	115 - 32
D	Cosmos 106 rocket	1966-04B	1966 Jan 25.52 172.46 days 1966 Jul 16.98	Cylinder 1500?	8 long 1.65 dia	1966 Jan 28.9 1966 Jul 2.7	48.40 48.37	92.76 89.96	6792 6653	285 235	543 315	0.019 0.006	126 167

	Name		Launch date, lifetime and descent date	Shape and weight (kg)	Size (m)	Date of orbital determination	Orbital inclination (deg)	Nodal period (min)	Semi major axis (km)	Perigee height (km)	Apogee height (km)	Orbital eccentricity	Argument of perigee (deg)
	Transit 11 [Scout]	1966-05A	1966 Jan 28.71 1000 years	Octagon + 4 vanes 60?	0.25 long 0.46 dia	1966 Jan 31.4	89.78	105.99	7417	861	1217	0.024	198
	Altair rocket	1966-05B	1966 Jan 28.71 500 years	Cylinder 24	1.5 long 0.46 dia	1966 Jan 30.7	89.69	105.99	7417	864	1213	0.024	201
1d	Fragments	1966-05C-J											
D	Luna 9 launcher	1966-06B	1966 Jan 31.49 1.82 days 1966 Feb 2.31	-	-	1966 Jan 31.7	51.85	88.30	6571	167	219	0.004	353
D	Luna 9 launcher rocket	1966-06C	1966 Jan 31.49 0.45 day 1966 Jan 31.94	Cylinder 2500?	7.5 long 2.6 dia	Initial orbit similar to 1966-06B							
D	[Thor Agena D]	1966-07A	1966 Feb 2.90 24.67 days 1966 Feb 27.57	Cylinder 1500?	8 long? 1.5 dia	1966 Feb 3.8	75.05	90.64	6683	185	425	0.018	158
	Essa 1*	1966-08A	1966 Feb 3.32 70 years	Cylinder 138	0.56 long 1.07 dia	1966 Feb 3.9	97.91	100.35	7152	702	845	0.010	222
	Essa 1 rocket	1966-08B	1966 Feb 3.32 35 years	Cylinder 24	1.5 long 0.46 dia	1966 Feb 15.5	97.86	100.61	7165	703	870	0.012	-
	Fragments	1966-08C-E											
D	[Thor Agena D]	1966-09A	1966 Feb 9.84 1324.93 days 1969 Sep 26.77	Cylinder 1500?	8 long? 1.5 dia	1966 Feb 10.0 1968 Jan 9.0 1968 Nov 15.5	82.09 82.08 82.08	94.83 94.06 93.21	6888 6851 6809	508 471 429	512 474 433	0.0003 0.0002 0.0003	88 4 -
D	Fragments	1966-09B,C											

Space vehicle: Luna 9, 1966-06A; and a rocket, 1966-06D, in a highly eccentric orbit.
* Environmental science and services administration.

	Name		Launch date, lifetime and descent date	Shape and weight (kg)	Size (m)	Date of orbital determination	Orbital inclination (deg)	Nodal period (min)	Semi major axis (km)	Perigee height (km)	Apogee height (km)	Orbital eccentricity	Argument of perigee (deg)
D R	Cosmos 107	1966-10A	1966 Feb 10.37 7.90 days 1966 Feb 18.27	Sphere-cylinder 4750?	4.3 long 2.4 dia	1966 Feb 11.5	64.97	89.64	6635	204	310	0.008	25
D	Cosmos 107 rocket	1966-10B	1966 Feb 10.37 18.57 days 1966 Feb 28.94	Cylinder 1440	3.8 long 2.6 dia	1966 Feb 12.2	64.99	89.74	6640	222	302	0.006	60
D	Fragments	1966-10C,D											
D	Cosmos 108	1966-11A	1966 Feb 11.75 282.80 days 1966 Nov 21.55	Ellipsoid 400?	1.8 long 1.2 dia	1966 Feb 12.3 1966 Jun 30.5 1966 Nov 10.0	48.87 48.87 48.84	95.32 93.40 89.70	6915 6821 6639	219 215 188	855 670 334	0.046 0.033 0.011	109 - 279
D	Cosmos 108 rocket	1966-11B	1966 Feb 11.75 109.31 days 1966 Jun 1.06	Cylinder 1500?	8 long 1.65 dia	1966 Feb 13.0 1966 May 21.5	48.85 48.82	95.22 90.29	6911 6669	222 197	844 384	0.045 0.014	111 201
D	Fragment	1966-11C											
D	Capsule [Atlas Agena D]	1966-12A	1966 Feb 15.85 7.44 days 1966 Feb 23.29	- 540?	1.5 dia?	1966 Feb 16.0	96.54	89.00	6599	148	293	0.011	126
D	Bluebell 2 cylinder	1966-12B	1966 Feb 15.85 0.72 day 1966 Feb 16.57	Inflated cylinder 8.7	2.44 long 0.30 dia	1966 Feb 16.3	96.48	88.13	6562	115	253	0.011	135
D	Bluebell 2 sphere	1966-12C	1966 Feb 15.85 6 days 1966 Feb 22	Inflated sphere 4.1	0.30 dia	1966 Feb 16.5	96.50	88.65	6586	149	267	0.009	128
D	Agena D rocket	1966-12D	1966 Feb 15.85 3 days 1966 Feb 18	Cylinder 660	6 long? 1.5 dia	Orbit similar to 1966-12A							
D	Fragments	1966-12E-AR											

	Name		Launch date, lifetime and descent date	Shape and weight (kg)	Size (m)	Date of orbital determination	Orbital inclination (deg)	Nodal period (min)	Semi major axis (km)	Perigee height (km)	Apogee height (km)	Orbital eccentricity	Argument of perigee (deg)
	Diapason 1	1966-13A	1966 Feb 17.36 200 years	Cylinder 19	0.20 long 0.50 dia	1966 Feb 19.7	34.03	118.51	7997	499	2738	0.140	81
	Diapason 1 rocket	1966-13B	1966 Feb 17.36 100 years	Cylinder 68	2.1 long 0.65 dia	1966 Feb 18.2	34.04	118.59	8001	503	2743	0.140	73
8d	Fragments	1966-13C-M											
D R B	Cosmos 109	1966-14A	1966 Feb 19.37 7.91 days 1966 Feb 27.28	Sphere-cylinder 5530?	5 long? 2.4 dia	1966 Feb 19.8	64.94	89.48	6627	202	295	0.007	54
D	Cosmos 109 rocket	1966-14B	1966 Feb 19.37 7.70 days 1966 Feb 27.07	Cylinder 2500?	7.5 long 2.6 dia	1966 Feb 21.2	64.95	89.18	6613	202	268	0.005	31
D	Fragment	1966-14C											
D R B	Cosmos 110	1966-15A	1966 Feb 22.84 21.75 days 1966 Mar 16.59	Sphere-cylinder 5700	5.0 long 2.4 dia	1966 Feb 23.5	51.85	95.30	6914	190	882	0.050	70
D	Cosmos 110 rocket	1966-15B	1966 Feb 22.84 65.94 days 1966 Apr 29.78	Cylinder 2500?	7.5 long 2.6 dia	1966 Feb 23.3 1966 Apr 21.7	51.83 51.82	95.22 90.12	6910 6661	186 163	877 403	0.050 0.018	66 275
D	Fragments	1966-15C, D											
	Essa 2	1966-16A	1966 Feb 28.58 10000 years	Cylinder 132	0.56 long 1.07 dia	1966 Mar 3.9	101.00	113.57	7765	1356	1418	0.004	192
	Essa 2 rocket	1966-16B	1966 Feb 28.58 5000 years	Cylinder 24	1.5 long 0.46 dia	1966 Mar 15.5	100.98	113.58	7766	1356	1420	0.004	-
	Fragments	1966-16C-E											

	Name		Launch date, lifetime and descent date	Shape and weight (kg)	Size (m)	Date of orbital determination	Orbital inclination (deg)	Nodal period (min)	Semi major axis (km)	Perigee height (km)	Apogee height (km)	Orbital eccentricity	Argument of perigee (deg)
D	Cosmos 111*	1966-17A	1966 Mar 1.46 1.61 days 1966 Mar 3.07	Cylinder 6540? full	7 long? 2.0 max dia	1966 Mar 1.8	51.84	88.19	6566	182	194	0.001	25
D	Cosmos 111 rocket	1966-17B	1966 Mar 1.46 <0.5 day 1966 Mar 1	Cylinder 2500?	7.5 long 2.6 dia	1966 Mar 1.5	51.80	87.49	6531	102?	203?	0.007?	-
D	Fragment	1966-17C											
D	[Thor Agena D]	1966-18A	1966 Mar 9.92 19.83 days 1966 Mar 29.75	Cylinder 1500?	8 long? 1.5 dia	1966 Mar 11.7	75.03	90.59	6683	178	432	0.019	160
D	Fragment	1966-18B											
D	Target Agena 8**	1966-19A	1966 Mar 16.63 548.21 days 1967 Sep 15.84	Cylinder 3175	7.9 long 1.5 dia	1966 Mar 16.7 1966 Mar 22.5 1967 Sep 12.9	28.88 28.86 28.88	90.20 92.47 88.50	6670 6782 6585	285 401 200	298 407 213	0.001 0.0004 0.001	254 322 180
D 2M R	Gemini 8**	1966-20A	1966 Mar 16.70 0.44 day 1966 Mar 17.14	Cone 3789	5.6 long 3.0 dia	1966 Mar 16.8 1966 Mar 17.0	28.91 28.88	88.60 90.20	6590 6670	159 285	265 298	0.008 0.001	120 254
D	Gemini 8 rocket	1966-20B	1966 Mar 16.70 1.22 days 1966 Mar 17.92	Cylinder 1900	6 long 3.0 dia	1966 Mar 17.3	28.89	88.06	6563	145	224	0.006	110
D	Gemini 8 adapter module	1966-20C	1966 Mar 16.70 33 days 1966 Apr 18	Truncated cone 2525?	2.3 long 3.0 dia	1966 Mar 31.5 1966 Apr 15.5	28.84 28.85	89.84 88.88	6652 6604	264 226	284 226	0.0015 0	- -

*Cosmos 111 may have been a Luna Probe launcher

**Gemini 8 and Agena 8 docked Mar 16.97; 1966-20A and 20C seperated Mar 17.13

	Name		Launch date, lifetime and descent date	Shape and weight (kg)	Size (m)	Date of orbital determination	Orbital inclination (deg)	Nodal period (min)	Semi major axis (km)	Perigee height (km)	Apogee height (km)	Orbital eccentricity	Argument of perigee (deg)
D R	Cosmos 112*	1966-21A	1966 Mar 17.44 7.79 days 1966 Mar 25.23	Sphere-cylinder 4750?	4.3 long 2.4 dia	1966 Mar 17.8	72.07	92.09	6754	207	545	0.025	65
D	Cosmos 112 rocket	1966-21B	1966 Mar 17.44 61.06 days 1966 May 17.50	Cylinder 1440	3.8 long 2.6 dia	1966 Mar 18.4	72.08	92.22	6761	214	552	0.025	72
D	Fragments	1966-21C,D											
D	Capsule	1966-22A	1966 Mar 18.85 5 days 1966 Mar 24	-	-	1966 Mar 19.0	100.95	89.30	6613	162	308	0.011	141
D	[Atlas Agena D]	1966-22B	1966 Mar 18.85 4.92 days 1966 Mar 23.77	Cylinder 1500?	8 long? 1.5 dia	1966 Mar 20.2	101.01	88.87	6596	152	284	0.010	140
D R	Cosmos 113	1966-23A	1966 Mar 21.40 7.92 days 1966 Mar 29.32	Sphere-cylinder 5530?	5 long? 2.4 dia	1966 Mar 21.8	64.94	89.71	6638	207	313	0.008	57
D	Cosmos 113 rocket	1966-23B	1966 Mar 21.40 9 days 1966 Mar 30	Cylinder 2500?	7.5 long 2.6 dia	1966 Mar 21.8	64.98	89.65	6634	209	302	0.007	56
D	Fragment	1966-23C											
	Transit 12 [Scout]	1966-24A	1966 Mar 26.15 300 years	Octagon + 4 vanes 60?	0.25 long 0.46 dia	1966 Mar 29.0	89.73	105.37	7388	891	1128	0.016	152
	Altair rocket	1966-24B	1966 Mar 26.15 300 years	Cylinder 24	1.5 long 0.46 dia	1966 Apr 16.6	89.74	105.36	7387	891	1127	0.016	99
1d	Fragments	1966-24C-E											

* First satellite launch from Plesetsk.

	Name		Launch date, lifetime and descent date	Shape and weight (kg)	Size (m)	Date of orbital determination	Orbital inclination (deg)	Nodal period (min)	Semi major axis (km)	Perigee height (km)	Apogee height (km)	Orbital eccentricity	Argument of perigee (deg)
	OV1-4	1966-25A	1966 Mar 30.39 500 years	Cylinder + hemisphere 87.6	1.40 long 0.69 dia	1966 Mar 31.3	144.53	103.85	7323	879	1011	0.009	73
	OV1-5	1966-25B	1966 Mar 30.39 1000 years	Cylinder + hemisphere 114.3	1.40 long 0.69 dia	1966 Mar 30.8	144.66	105.48	7400	996	1048	0.004	33
	OV1-5 rocket	1966-25C	1966 Mar 30.39 500 years	Cylinder 70?	2.05 long 0.72 dia	1966 Apr 15.5	144.67	105.50	7401	987	1059	0.005	-
	OV1-4 rocket	1966-25D	1966 Mar 30.39 200 years	Cylinder 70?	2.05 long 0.72 dia	1966 Aug 27	144.53	103.96	7327	889	1009	0.008	-
1d	Fragments	1966-25E-H											
	[Thor Altair]	1966-26A	1966 Mar 31.19 50 years	- 130?	-	1966 Apr 2.3	98.60	100.56	7162	634	933	0.021	262
	Altair rocket	1966-26B	1966 Mar 31.19 35 years	Cylinder 24	1.5 long 0.46 dia	1966 Apr 19.3	98.63	100.56	7162	634	933	0.021	215
1d	Fragments	1966-26C-F											
D	Luna 10 launcher	1966-27B	1966 Mar 31.45 3.72 days 1966 Apr 4.17	-	-	1966 Apr 1.8	51.80	88.52	6582	195	212	0.001	55
D	Luna 10 launcher rocket	1966-27C	1966 Mar 31.45 1.55 days 1966 Apr 2.00	Cylinder 2500?	7.5 long 2.6 dia	1966 Mar 31.7	51.82	88.28	6571	186	199	0.001	85
D R	Cosmos 114	1966-28A	1966 Apr 6.49 7.81 days 1966 Apr 14.30	Sphere-cylinder 5530?	5 long? 2.4 dia	1966 Apr 6.9	72.94	90.06	6655	210	343	0.010	65
D	Cosmos 114 rocket	1966-28B	1966 Apr 6.49 12 days 1966 Apr 18	Cylinder 2500?	7.5 long 2.6 dia	1966 Apr 8.2	72.90	89.85	6645	200	333	0.010	45
D	Fragments	1966-28C-E											

Space Vehicle: Luna 10, 1966-27A, and rocket 1966-27D. Two fragments, 1966-27E and F, are in highly eccentric orbits.

	Name		Launch date, lifetime and descent date	Shape and weight (kg)	Size (m)	Date of orbital determination	Orbital inclination (deg)	Nodal period (min)	Semi major axis (km)	Perigee height (km)	Apogee height (km)	Orbital eccentricity	Argument of perigee (deg)
D	[Thor Agena D]	1966-29A	1966 Apr 7.92 18.43 days 1966 Apr 26.35	Cylinder 1500?	8 long? 1.5 long	1966 Apr 9.9	75.06	89.56	6631	193	312	0.009	166
D	Surveyor Model 2	1966-30A	1966 Apr 8.04 27.22 days 1966 May 5.26	Truncated cone 771	2.8 long 1.3 dia	1966 Apr 9.6	30.71	89.50	6633	175	334	0.012	258
D	Centaur 8	1966-30B	1966 Apr 8.04 9 days 1966 Apr 17	Cylinder 1815	8.6 long 3.0 dia	1966 Apr 10.0	30.74	89.39	6629	165	337	0.013	305
	OAO 1*	1966-31A	1966 Apr 8.82 500 years	Octagonal Cylinder 1769	3.05 long 2.15 dia	1966 Apr 10.0	35.03	100.71	7177	792	806	0.001	46
	OAO 1 rocket	1966-31B	1966 Apr 8.82 200 years	Cylinder 700?	6 long? 1.5 dia	1966 Apr 12.1	35.03	100.66	7174	789	803	0.001	131
D	Fragment	1966-31C											
D	Capsule [Atlas Agena D]	1966-32A	1966 Apr 19.80 6 days 1966 Apr 26	- 500?	- 1.5 dia?	1966 Apr 20.0	116.95	89.94	6650	145	398	0.019	127
D	Agena D rocket	1966-32B	1966 Apr 19.80 2 days 1966 Apr 22	Cylinder 700?	6 long? 1.5 dia	1966 Apr 20.0	116.96	89.02	6604	139	312	0.013	-
D R	Cosmos 115	1966-33A	1966 Apr 20.45 7.93 days 1966 Apr 28.38	Sphere-cylinder 4750?	4.3 long 2.4 dia	1966 Apr 20.5	65.00	89.44	6626	201	294	0.007	9
D	Cosmos 115 rocket	1966-33B	1966 Apr 20.45 10 days 1966 Apr 30	Cylinder 1440	3.8 long 2.6 dia	1966 Apr 20.9	65.03	89.44	6625	214	280	0.005	17

* Orbiting Astronomical Observatory.

Year of launch 1966 continued

	Name		Launch date, lifetime and descent date	Shape and weight (kg)	Size (m)	Date of orbital determination	Orbital inclination (deg)	Nodal period (min)	Semi major axis (km)	Perigee height (km)	Apogee height (km)	Orbital eccentricity	Argument of perigee (deg)
	OV3-1	1966-34A	1966 Apr 22.41 200 years	Octagon 69	0.74 long 0.74 dia	1966 Apr 24.3 1968 Feb 12.8	82.47 82.46	151.81 150.86	9424 9388	351 353	5741 5667	0.286 0.283	164 332
	OV3-1 rocket	1966-34B	1966 Apr 22.41 35 years	Cylinder 24	1.5 long 0.46 dia	1966 Oct 31.5 1976 Jul 21.7	82.46 82.44	151.39 142.62	9410 9040	356 346	5708 4978	0.284 0.256	- -
2d	Fragments	1966-34C-E											
D	Molniya 1C	1966-35A	1966 Apr 25.30 2604 days 1973 Jun 11	Cylinder + 6 vanes 1000?	3.4 long 1.6 dia	1966 Apr 25.4 1966 Apr 26.3 1969 Oct 1.0	64.89 65.04 65.09	91.18 710.41 705.5	6712 26377 26254	186 506 1742	482 39492 38010	0.022 0.739 0.691	74 284 -
D	Molniya 1C launcher rocket	1966-35C	1966 Apr 25.30 13.57 days 1966 May 8.87	Cylinder 2500?	7.5 long 2.6 dia	1966 Apr 28.4	64.90	90.85	6694	182	450	0.020	72
D	Molniya 1C launcher	1966-35B	1966 Apr 25.30 13.84 days 1966 May 9.14	-	-	1966 Apr 30.5	64.86	90.78	6691	176	449	0.020	-
D	Molniya 1C rocket	1966-35F	1966 Apr 25.30 2611.24 days	Cylinder 440	2.0 long 2.0 dia	1968 Jun 15.5 1969 Oct 16.3	65.15 65.02	702.75 702.71	26184 26183	1215 1747	38397 37863	0.710 0.690	- -
D	Fragments	1966-35D,E	1973 Jun 18.54			1971 Oct 1.0	65.02	702.77	26185	869	38744	0.723	-
D	Cosmos 116	1966-36A	1966 Apr 26.42 220.75 days 1966 Dec 3.17	Ellipsoid 400?	1.8 long 1.2 dia	1966 Apr 26.9 1966 Nov 6.9	48.42 48.35	92.01 90.12	6754 6662	288 250	464 317	0.013 0.005	111 7
D	Cosmos 116 rocket	1966-36B	1966 Apr 26.42 111.20 days 1966 Aug 15.62	Cylinder 1500?	8 long 1.65 dia	1966 Apr 28.2 1966 Aug 4.1	48.38 48.36	91.91 89.72	6748 6641	289 236	451 289	0.012 0.004	116 236
D R	Cosmos 117	1966-37A	1966 May 6.46 7.89 days 1966 May 14.35	Sphere-cylinder 4750?	4.3 long 2.4 dia	1966 May 6.9	64.93	89.55	6630	205	298	0.007	26
D	Cosmos 117 rocket	1966-37B	1966 May 6.46 14.6 days 1966 May 21.1	Cylinder 1440	3.8 long 2.6 dia	1966 May 7.4	64.96	89.69	6637	219	299	0.006	65
D	Fragments	1966-37C-E											

	Name		Launch date, lifetime and descent date	Shape and weight (kg)	Size (m)	Date of orbital determination	Orbital inclination (deg)	Nodal period (min)	Semi major axis (km)	Perigee height (km)	Apogee height (km)	Orbital eccentricity	Argument of perigee (deg)
	Cosmos 118	1966-38A	1966 May 11.59 35 years	Cylinder + 2 vanes 2000?	5 long? 1.5 dia?	1966 May 12.0	65.00	97.13	7000	587	657	0.005	248
	Cosmos 118 rocket	1966-38B	1966 May 11.59 35 years	Cylinder 1440	3.8 long 2.6 dia	1966 May 13.3	64.98	97.03	6995	560	673	0.008	328
D	Capsule [Atlas Agena D]	1966-39A	1966 May 14.77 6 days 1966 May 21	- 500?	- 1.5 dia?	1966 May 16.6	110.55	89.40	6624	133	358	0.017	120
D	Capsule	1966-39B	1966 May 14.77 1627.01 days 1970 Oct 27.78	Octagon 60?	0.3 long 0.9 dia	1966 May 16.4 1968 Jan 7.3 1969 Sep 16.0	109.94 109.94 109.94	95.39 94.87 93.73	6916 6892 6838	517 493 443	559 534 476	0.003 0.003 0.002	84 215 -
D	Agena D rocket	1966-39C	1966 May 14.77 2 days 1966 May 17	Cylinder 700?	6 long? 1.5 dia	1966 May 15.6	110.68	88.87	6594	150?	280?	0.01?	129
	Nimbus 2	1966-40A	1966 May 15.33 800 years	Conical skeleton + 2 paddles 414	3.00 long 1.45 dia	1966 May 16.0	100.35	108.15	7519	1103	1179	0.005	339
	Nimbus 2 rocket	1966-40B	1966 May 15.33 1000 years	Cylinder 700?	6 long 1.5 dia	1966 May 20.4	100.31	107.91	7508	1085	1175	0.006	323
	Transit 13 [Scout]	1966-41A	1966 May 19.10 200 years	Octagon + 4 vanes 60?	0.25 long 0.46 dia	1966 May 19.4	90.00	103.48	7300	863	980	0.008	239
	Altair rocket	1966-41B	1966 May 19.10 200 years	Cylinder 24	1.5 long 0.46 dia	1966 May 20.7	90.00	103.48	7300	863	980	0.008	234
	Fragments	1966-41C-F											

	Name		Launch date, lifetime and descent date	Shape and weight (kg)	Size (m)	Date of orbital determination	Orbital inclination (deg)	Nodal period (min)	Semi major axis (km)	Perigee height (km)	Apogee height (km)	Orbital eccentricity	Argument of perigee (deg)
D	[Thor Agena D]	1966-42A	1966 May 24.08 16 days 1966 Jun 9	Cylinder 1500?	8 long? 1.5 dia	1966 May 24.6	66.04	89.00	6603	179	271	0.007	77
D	Cosmos 119	1966-43A	1966 May 24.23 189.96 days 1966 Nov 30.19	Cone-cylinder 2000?	9 long? 1.65 dia	1966 May 25.2 1966 Nov 7.5	48.38 48.34	99.76 92.52	7128 6779	208 191	1292 611	0.076 0.031	106 115
	Explorer 32 (AE-B)*	1966-44A	1966 May 25.58 25 years	Sphere 225	0.89 dia	1966 May 27.3 1977 Jun 8.1	64.66 64.55	116.01 107.34	7881 7489	289 257	2716 1965	0.154 0.114	50 -
D	Explorer 32 rocket	1966-44B	1966 May 25.58 1120.51 days 1969 Jun 19.09	Cylinder 24	1.5 long 0.46 dia	1966 Jun 4.0 1968 Jan 15.5 1968 Sep 30.5	64.67 64.65 64.53	115.86 107.21 101.94	7873 7483 7232	281 250 247	2710 1960 1461	0.159 0.114 0.084	48 - -
D	Fragment	1966-44C											
D	ATDA**	1966-46A	1966 Jun 1.63 40.03 days 1966 Jul 11.66	Cylinder 794	3.7 long 1.5 dia	1966 Jun 2.4	28.87	90.24	6672	292	296	0.0003	104
D	Atlas rocket	1966-46B	1966 Jun 1.63 disintegrated 1966 Jun 22	Cylinder 3400	20 long 3.0 dia	1966 Jun 14.4	28.81	89.47	6633	248	261	0.001	28
D	Fragments	1966-46C-BF											
D R 2M	Gemini 9 †	1966-47A	1966 Jun 3.57 3.01 days 1966 Jun 6.58	Cone 3750	5.6 long 3.0 dia	1966 Jun 6.2	28.86	89.80	6649	270	272	0.0002	151
D	Gemini 9 rocket	1966-47B	1966 Jun 3.57 0.86 day 1966 Jun 4.43	Cylinder 1900	6 long 3.0 dia	1966 Jun 3.6	28.80	87.37	6526	133	162	0.002	-

Space Vehicles: Surveyor 1, 1966-45A; and Centaur 10, 1966-45B in highly eccentric orbit.

* Atmospheric Explorer B. ** Augmented Target Docking Adaptor. † Gemini 9 and ATDA rendezvous Jun 3.8.

	Name		Launch date, lifetime and descent date	Shape and weight (kg)	Size (m)	Date of orbital determination	Orbital inclination (deg)	Nodal period (min)	Semi major axis (km)	Perigee height (km)	Apogee height (km)	Orbital eccentricity	Argument of perigee (deg)
D	[Atlas Agena D]	1966-48A	1966 Jun 3.81 6.17 days 1966 Jun 9.98	Cylinder 1500?	8 long? 1.5 dia	1966 Jun 4.6	87.01	88.87	6594	143	288	0.011	132
D	Capsule	1966-48B	1966 Jun 3.81 5.43 days 1966 Jun 9.24	-	-	1966 Jun 5.9	86.97	88.70	6587	136	281	0.011	125
D	OGO 3	1966-49A	1966 Jun 7.12 15.8 years 1982 Apr?	Box + booms 515	1.73 long 0.84 wide 0.84 high	1966 Jun 19.1 1969 Sep 16.0 1972 Mar 1.0	31.39 64.5 77.6	2915.0 2912.6 2911.5	67624 67559 67541	319 6593 19519	122173 115769 102806	0.903 0.808 0.617	314 - -
D	OGO 3 rocket	1966-49B	1966 Jun 7.12 15.8 years 1982 Apr?	Cylinder 700?	6 long 1.5 dia	orbit similar to 1966-49A							
D R	Cosmos 120	1966-50A	1966 Jun 8.46 7.94 days 1966 Jun 16.40	Sphere-cylinder 5530?	5 long? 2.4 dia	1966 Jun 8.8	51.80	89.37	6623	205	285	0.006	30
D	Cosmos 120 rocket	1966-50B	1966 Jun 8.46 4.38 days 1966 Jun 12.84	Cylinder 2500?	7.5 long 2.6 dia	1966 Jun 9.3	51.78	89.11	6612	194	273	0.006	24
D	Fragments	1966-50C, D											
D	[Atlas Agena D]	1966-51A	1966 Jun 9.84 176.80 days 1966 Dec 3.64	Cylinder 700	8 long? 1.5 dia	1966 Jun 14.2 1966 Nov 23.4	90.05 89.94	124.89 95.84	8273 6936	174 155	3616 960	0.208 0.058	157 73
D	Secor 6 (EGRS 6)	1966-51B	1966 Jun 9.84 391.7 days 1967 Jul 6.5	Rectangular box 17	0.33 x 0.28 x 0.23	1966 Jun 13.2 1966 Nov 24.5 1967 Jul 2.5	90.05 90.02 89.90	125.13 115.15 91.21	8286 7838 6710	168 167 144	3648 2753 520	0.210 0.165 0.028	159 132 174
D	ORS 2 (ERS 16)	1966-51C	1966 Jun 9.84 275.44 days 1967 Mar 12.28	Octahedron 5	0.23 side	1966 Jun 14.2 1966 Nov 24.0 1967 Mar 6.3	90.03 90.00 89.87	125.02 109.90 93.28	8279 7598 6811	179 164 147	3623 2276 719	0.208 0.139 0.042	157 117 150

Name		Launch date, lifetime and descent date	Shape and weight (kg)	Size (m)	Date of orbital determination	Orbital inclination (deg)	Nodal period (min)	Semi major axis (km)	Perigee height (km)	Apogee height (km)	Orbital eccentricity	Argument of perigee (deg)
OV3-4	1966-52A	1966 Jun 10.18 600 years	Octagonal cylinder 78.5	0.74 long 0.74 dia	1966 Jun 10.6	40.77	142.99	9057	641	4718	0.225	140
OV3-4 rocket	1966-52B	1966 Jun 10.18 300 years	Cylinder 24	1.5 long 0.46 dia	1966 Jun 30.5	40.79	143.19	9065	643	4730	0.225	-
Fragments	1966-52C,D											
GGTS 1*	1966-53A	1966 Jun 16.58 > million yr	Polyhedron (26 faces) 47	0.8 long 0.9 dia	1966 Jun 17.0	0.10	1334.2	40147	33648	33889	0.003	202
IDCSP 1-1**	1966-53B	1966 Jun 16.58 > million yr	Polyhedron (26 faces) 45	0.8 long 0.9 dia	1966 Jun 17.0	0.09	1334.7	40155	33656	33897	0.003	144
IDCSP 1-2	1966-53C	1966 Jun 16.58 > million yr	Polyhedron (26 faces) 45	0.8 long 0.9 dia	1966 Jun 17.0	0.08	1335.3	40167	33668	33909	0.003	175
IDCSP 1-3	1966-53D	1966 Jun 16.58 > million yr	Polyhedron (26 faces) 45	0.8 long 0.9 dia	1966 Jun 17.0	0.12	1336.6	40194	33695	33936	0.003	197
IDCSP 1-4	1966-53E	1966 Jun 16.58 > million yr	Polyhedron (26 faces) 45	0.8 long 0.9 dia	1966 Jun 17.0	0.18	1338.6	40235	33696	34018	0.004	180
IDCSP 1-5	1966-53F	1966 Jun 16.58 > million yr	Polyhedron (26 faces) 45	0.8 long 0.9 dia	1966 Jun 17.0	0.04	1340.8	40279	33699	34102	0.005	126
IDCSP 1-6	1966-53G	1966 Jun 16.58 > million yr	Polyhedron (26 faces) 45	0.8 long 0.9 dia	1966 Jun 17.0	0.06	1344.0	40342	33722	34206	0.006	167
IDCSP 1-7	1966-53H	1966 Jun 16.58 > million yr	Polyhedron (26 faces) 45	0.8 long 0.9 dia	1966 Jun 17.0	0.04	1347.6	40414	33712	34359	0.008	120
Transtage 8 [Titan 3C]	1966-53J	1966 Jun 16.58 > million yr	Cylinder 1500?	6 long? 3 dia	1966 Jun 16.6 1966 Jun 16.7 1966 Jun 17.0	28.6 26.4 0.09?	87.9 591.2? 1351.3?	6553 23348 40488?	168 190 33730?	182 33750 34470?	0.001 0.751 0.009?	- - 164?

*Gravity Gradient Test Satellite. ** Initial Defense Communication Satellite Programme.

	Name		Launch date, lifetime and descent date	Shape and weight (kg)	Size (m)	Date of orbital determination	Orbital inclination (deg)	Nodal period (min)	Semi major axis (km)	Perigee height (km)	Apogee height (km)	Orbital eccentricity	Argument of perigee (deg)
D R	Cosmos 121	1966-54A	1966 Jun 17.46 7.80 days 1966 Jun 25.26	Sphere-cylinder 5530?	5 long? 2.4 dia	1966 Jun 17.7	72.83	89.86	6645	200	333	0.010	68
D	Cosmos 121 rocket	1966-54B	1966 Jun 17.46 9.05 days 1966 Jun 26.51	Cylinder 2500?	7.5 long 2.6 dia	1966 Jun 17.7	72.85	89.76	6640	202	322	0.009	60
D	Fragments	1966-54C-E											
D	[Thor Agena D]	1966-55A	1966 Jun 21.90 22 days 1966 Jul 14	Cylinder 1500?	8 long? 1.5 dia	1966 Jun 23.6	80.10	90.15	6659	194	367	0.013	169
D	Fragment	1966-55B											
	Pageos 1**	1966-56A	1966 Jun 24.01 disintegrated	Inflated sphere 55	30.48 dia	1966 Jun 24.9 1972 Feb 1.0	87.14 86.3	181.43 180.43	10617 10578	4207 3496	4271 4903	0.003 0.067	241 -
	Pageos 1 rocket	1966-56B	1966 Jun 24.01 100000 years	Cylinder 700?	6 long 1.5 dia	1966 Jun 25.5	86.99	181.23	10609	4209	4252	0.002	243
66d	Fragments*	1966-56C-CF											
	Cosmos 122	1966-57A	1966 Jun 25.43 50 years	Cylinder + 2 vanes 2000?	5 long? 1.5 dia?	1966 Jun 25.5 1968 Jan 26.9	65.14 64.98	97.12 96.99	6998 6993	583 589	657 643	0.005 0.004	352 339
	Cosmos 122 rocket	1966-57B	1966 Jun 25.43 25 years	Cylinder 1440	3.8 long 2.6 dia	1966 Jun 27.5 1968 Jan 19.5	64.98 65.00	97.11 96.97	6997 6991	549 522	689 704	0.010 0.013	218 335
	Explorer 33 (Imp 4)	1966-58A	1966 Jul 1.67 Indefinite?	Octagon + 4 vanes 57	1.12 long 0.71 dia	1966 Jul 8.0 1968 Feb 11.5 1971 Jul 1.0	29.0 40.9 24.35	23148 25863.5 38792	268759 289701 379600	30532 85228 265680	494230 481417 480763	0.863 0.684 0.283	- - -
D	Explorer 33 second stage	1966-58B	1966 Jul 1.67 138 days 1966 Nov 17	Cylinder 350?	4.9 long 1.43 dia	1966 Jul 4.6 1966 Nov 15.5	28.79 28.78	101.41 90.60	7205 6688	181 175	1473 444	0.090 0.020	126 -
	Explorer 33 third stage	1966-58C	1966 Jul 1.67 Indefinite?	Cylinder 24	1.5 long 0.46 dia	Initial orbit similar to 1966-58A							
D	Fragments	1966-58D, E											

* Pageos shed 27 fragments on 1975 Jul 12.77, and 44 more fragments on 1976 Jan 20.0. ** Passive geodetic satellite

	Name		Launch date, lifetime and descent date	Shape and weight (kg)	Size (m)	Date of orbital determination	Orbital inclination (deg)	Nodal period (min)	Semi major axis (km)	Perigee height (km)	Apogee height (km)	Orbital eccentricity	Argument of perigee (deg)
D	[Saturn 203] Apollo 3	1966-59A	1966 Jul 5.62 disintegrated 1966 Jul 5.85	Cylinder 26552	28.3 long 6.6 dia	1966 Jul 5.7	31.98	88.31	6576	183	212	0.002	6
D	Fragments	1966-59B-AL											
D	Proton 3	1966-60A	1966 Jul 6.54 72.20 days 1966 Sep 16.74	Cylinder 12200	3 long? 4 dia?	1966 Jul 14.4 1966 Sep 8.7	63.47 63.45	92.24 89.32	6763 6620	185 162	585 321	0.030 0.012	60 58
D	Proton 3 rocket	1966-60B	1966 Jul 6.54 46.33 days 1966 Aug 21.87	Cylinder 4000?	12 long? 4 dia	1966 Jul 13.8 1966 Aug 14.4	63.46 63.45	92.08 89.60	6756 6633	181 162	574 348	0.029 0.014	60 58
D	Fragments	1966-60C,D											
D	Cosmos 123	1966-61A	1966 Jul 8.23 155.68 days 1966 Dec 10.91	Ellipsoid 400?	1.8 long 1.2 dia	1966 Jul 12.3 1966 Nov 23.1	48.77 48.77	92.16 89.95	6762 6651	256 220	512 326	0.019 0.008	124 53
D	Cosmos 123 rocket	1966-61B	1966 Jul 8.23 85.83 days 1966 Oct 2.06	Cylinder 1500?	8 long 1.65 dia	1966 Jul 13.4 1966 Sep 24.8	48.82 48.81	92.02 89.74	6755 6641	255 216	499 309	0.018 0.007	127 130
D	Capsule [Atlas Agena D]	1966-62A	1966 Jul 12.75 7 days 1966 Jul 20	- 500?	- 1.5 dia?	1966 Jul 17.6	95.52	88.25	6565	137	236	0.008	122
D	Agena D rocket	1966-62B	1966 Jul 12.75 1 day 1966 Jul 14	Cylinder 700?	8 long? 1.5 dia	1966 Jul 13.0	95.54	88.00	6552	130	217	0.007	-
D	OV1-8	1966-63A	1966 Jul 14.09 4192 days 1978 Jan 4	Spherical skeleton 10.4	9.14 dia	1966 Jul 15.5 1969 Feb 22.5	144.27 144.1	105.12 102.96	7384 7284	998 861	1013 950	0.001 0.006	263 -
	OV1-8 rocket	1966-63C	1966 Jul 14.09 1000 years	Cylinder 70?	2.05 long 0.72 dia	1966 Jul 31.5	144.27	105.11	7383	995	1015	0.001	-
	Fragments	1966-63B,D,E											

	Name		Launch date, lifetime and descent date	Shape and weight (kg)	Size (m)	Date of orbital determination	Orbital inclination (deg)	Nodal period (min)	Semi major axis (km)	Perigee height (km)	Apogee height (km)	Orbital eccentricity	Argument of perigee (deg)
D R	Cosmos 124	1966-64A	1966 Jul 14.44 7.95 days 1966 Jul 22.39	Sphere-cylinder 5530?	5 long? 2.4 dia	1966 Jul 16.7	51.78	89.41	6624	205	286	0.006	59
D	Cosmos 124 rocket	1966-64B	1966 Jul 14.44 5.11 days 1966 Jul 19.55	Cylinder 2500?	7.5 long 2.6 dia	1966 Jul 16.7	51.78	88.91	6601	169	277	0.008	37
D	Fragments	1966-64C-E											
D	Target Agena 10*	1966-65A	1966 Jul 18.86 163.93 days 1966 Dec 29.79	Cylinder 3175	7.9 long 1.5 dia	1966 Jul 18.9 1966 Jul 19.2 1966 Jul 20.0 1966 Nov 20.0	28.85 28.80 28.86 28.87	90.23 95.04 92.31 90.21	6671 6905 6774 6669	290 298 391 288	296 755 400 293	0.0004 0.033 0.0004 0.0004	97 137 - 325
D 2 M R	Gemini 10*	1966-66A	1966 Jul 18.93 2.95 days 1966 Jul 21.88	Cone 3750	5.6 long 3.0 dia	1966 Jul 18.9 1966 Jul 19.2 1966 Jul 20.0	28.85 28.80 28.86	88.64 95.04 92.31	6592 6905 6774	160 298 391	268 755 400	0.008 0.033 0.0004	- 137 -
D	Gemini 10 rocket	1966-66B	1966 Jul 18.93 1.07 days 1966 Jul 20.00	Cylinder 1900	6 long 3.0 dia	1966 Jul 19.6	28.86	87.75	6547	143	195	0.004	123
D	Gemini 10 Adaptor module	1966-66J	1966 Jul 18.93 183 days 1967 Jan 17	Truncated cone 2510?	2.3 long 3.0 dia	1966 Sep 7.8 1966 Oct 31.5 1966 Dec 19.5	28.86 28.87 28.87	92.07 91.71 90.60	6762 6744 6690	381 361 311	388 370 312	0.0004 0.0006 0	- - -
D	Fragments	1966-66C-H,K											
D	Cosmos 125	1966-67A	1966 Jul 20.38 13.23 days 1966 Aug 2.61	Cone-cylinder 4000?	10 long? 2 dia?	1966 Jul 21.2	65.00	89.12	6610	205	258	0.004	277

*Gemini 10 and Agena 10 docked Jul 19.14; rendezvous with Agena 8 on Jul 19.93.
1966-66J was joined to 1966-66A until Jul 21.87.

	Name		Launch date, lifetime and descent date	Shape and weight (kg)	Size (m)	Date of orbital determination	Orbital inclination (deg)	Nodal period (min)	Semi major axis (km)	Perigee height (km)	Apogee height (km)	Orbital eccentricity	Argument of perigee (deg)
D R	Cosmos 126	1966-68A	1966 Jul 28.45 8.94 days 1966 Aug 6.39	Sphere-cylinder 5530?	5 long? 2.4 dia	1966 Jul 28.8	51.79	89.99	6655	204	350	0.011	53
D	Cosmos 126 rocket	1966-68B	1966 Jul 28.45 13.19days 1966 Aug 10.64	Cylinder 2500?	7.5 long 2.6 dia	1966 Jul 28.6	51.81	89.94	6652	201	347	0.011	49
D	Fragment	1966-68C											
D	[Titan 3B Agena D]	1966-69A	1966 Jul 29.78 7 days 1966 Aug 6	Cylinder 3000?	8 long? 1.5 dia	1966 Jul 31.1	94.12	88.58	6582	158	250	0.007	143
	OV3-3	1966-70A	1966 Aug 4.45 50 years	Octagon 75	0.74 long 0.74 dia	1966 Aug 4.6 1968 Jan 31.9	81.44 81.49	137.01 136.28	8804 8773	360 359	4492 4430	0.235 0.232	158 3
	OV3-3 rocket	1966-70B	1966 Aug 4.45 25 years	Cylinder 24	1.5 long 0.46 dia	1966 Aug 19.3 1972 Jan 1.0	81.49 81.4	136.98 130.32	8801 8509	363 357	4482 3905	0.234 0.208	141 -
1d	Fragments	1966-70C,D											
D R	Cosmos 127	1966-71A	1966 Aug 8.47 7.93 days 1966 Aug 16.40	Sphere-cylinder 5530?	5 long? 2.4 dia	1966 Aug 11.1	51.83	89.13	6612	201	267	0.005	46
D	Cosmos 127 rocket	1966-71B	1966 Aug 8.47 3.88 days 1966 Aug 12.35	Cylinder 2500?	7.5 long 2.6 dia	1966 Aug 9.8	51.82	88.90	6599	175	267	0.007	28
D	[Thorad* Agena D]	1966-72A	1966 Aug 9.88 32.20 days 1966 Sep 11.08	Cylinder 2000?	8 long? 1.5 dia	1966 Aug 15.1	100.12	89.35	6619	194	287	0.007	158

* Thorad: long-tank thrust-augmented Thor.

	Name		Launch date, lifetime and descent date	Shape and weight (kg)	Size (m)	Date of orbital determination	Orbital inclination (deg)	Nodal period (min)	Semi major axis (km)	Perigee height (km)	Apogee height (km)	Orbital eccentricity	Argument of perigee (deg)
D P	Capsule* [Atlas Agena D]	1966-74A	1966 Aug 16.77 7.5 days 1966 Aug 24.3	- 500?	1.5 dia?	1966 Aug 17.6	93.24	89.58	6630	146	358	0.016	150
D	Capsule	1966-74B	1966 Aug 16.77 1296.25 days 1970 Mar 5.02	Octagon 60?	0.3 long 0.9 dia	1966 Aug 18.3 1968 Feb 8.2 1969 Feb 15.3	93.17 93.15 93.15	94.99 94.33 93.47	6895 6864 6824	510 479 440	524 493 451	0.001 0.001 0.001	291 351 -
D	Agena D rocket	1966-74C	1966 Aug 16.77 1 day 1966 Aug 18	Cylinder 700?	6 long? 1.5 dia	1966 Aug 16.8	93.30	89.25	6613	144	325	0.014	-
D	Pioneer 7 second stage	1966-75B	1966 Aug 17.64 354.5 days 1967 Aug 7.1	Cylinder 400?	4.9 long 1.43 dia	1966 Aug 20.8 1966 Nov 15.5 1967 Aug 2.2	32.76 32.97 32.97	97.04 96.15 89.06	7001 6958 6612	252 249 194	994 911 274	0.053 0.048 0.006	185 - 307
D	Fragments	1966-75D,E											
	Transit 14 [Scout]	1966-76A	1966 Aug 18.10 1000 years	Octagon + 4 vanes 60?	0.25 long 0.46 dia	1966 Aug 19.0	88.86	106.85	7457	1056	1101	0.003	109
	Altair rocket	1966-76B	1966 Aug 18.10 500 years	Cylinder 24	1.5 long 0.46 dia	1966 Sep 8.0	88.85	106.85	7457	1049	1109	0.004	58
3d	Fragments	1966-76C-G											
	[Atlas Agena D]	1966-77A	1966 Aug 19.81 100 000 years	Cylinder 2000?	8 long? 1.5 dia	1966 Aug 20.1	90.07	167.59	10068	3680	3700	0.001	167
	Secor 7 (EGRS 7)	1966-77B	1966 Aug 19.81 100 000 years	Rectangular box 17	0.36 x 0.30 x 0.25	1966 Aug 19.9	90.11	167.59	10068	3680	3700	0.001	166
	ORS 1 (ERS 15)	1966-77C	1966 Aug 19.81 80 000 years	Octahedron 5	0.25 side	1966 Aug 19.9	90.11	167.56	10064	3670	3702	0.002	174

Space Vehicles: Lunar Orbiter 1, 1966-73A; and Agena D rocket 1966-73B in a highly eccentric orbit.
Pioneer 7, 1966-75A and Altair rocket 1966-75C.

* Pieces picked up in Malawi.

Year of launch 1966 continued

	Name		Launch date, lifetime and descent date	Shape and weight (kg)	Size (m)	Date of orbital determination	Orbital inclination (deg)	Nodal period (min)	Semi major axis (km)	Perigee height (km)	Apogee height (km)	Orbital eccentricity	Argument of perigee (deg)
D	Luna 11 launcher	1966-78B	1966 Aug 24.34 1.40 days 1966 Aug 25.74	-	-	1966 Aug 25.1	51.86	88.12	6562	177	190	0.001	312
D	Luna 11 launcher rocket	1966-78C	1966 Aug 24.34 2.49 days 1966 Aug 26.83	Cylinder 2500?	7.5 long 2.6 dia	1966 Aug 25.1	51.84	88.39	6575	193	201	0.0006	87
D R	Cosmos 128	1966-79A	1966 Aug 27.41 7.87 days 1966 Sep 4.28	Sphere-cylinder 5530?	5 long? 2.4 dia	1966 Aug 28.3	64.99	89.81	6644	213	319	0.008	41
D	Cosmos 128 rocket	1966-79B	1966 Aug 27.41 11 days 1966 Sep 7	Cylinder 2500?	7.5 long 2.6 dia	1966 Aug 28.4	64.98	90.00	6654	209	342	0.010	52
D	Target Agena 11*	1966-80A	1966 Sep 12.55 108.76 days 1966 Dec 30.31	Cylinder 3175	7.9 long 1.5 dia	1966 Sep 12.6 1966 Sep 16.6	28.83 28.88	90.35 91.20	6676 6718	298 326	298 353	0 0.002	- 308
D 2M R	Gemini 11*	1966-81A	1966 Sep 12.61 2.97 days 1966 Sep 15.58	Cone 3830?	5.6 long 3.0 dia	1966 Sep 12.6 1966 Sep 14.3 1966 Sep 14.5	28.83 28.83 28.83	88.78 101.55 90.31	6599 7211 6673	161 298 281	280 1368 308	0.009 0.074 0.002	- - -
D	Gemini 11 rocket	1966-81B	1966 Sep 12.61 0.99 day 1966 Sep 13.60	Cylinder 1900	6 long 3.0 dia	1966 Sep 12.6	28.80	87.71	6545	144	190	0.003	-
D	Gemini 11 Adaptor module	1966-81C	1966 Sep 12.61 26 days 1966 Oct 8	Truncated cone 2535?	2.3 long 3.0 dia	1966 Sep 30.5	28.83	89.47	6633	254	255	0	-
D	Fragment	1966-81D											
	[Thor Burner 2]	1966-82A	1966 Sep 16.19 50 years	12-sided frustum 195	1.64 long 1.31 to 1.10 dia	1966 Sep 17.1	98.46	100.86	7176	705	891	0.013	290
	Burner 2 rocket	1966-82B	1966 Sep 16.19 50 years	Sphere-cone 66	1.32 long 0.94 dia	1966 Sep 16.8	98.47	100.88	7178	699	900	0.014	289

Space vehicle: Luna 11 1966-78A. *Gemini and Agena 11 initially docked Sep 12.68. 1966-81C was joined to 1966-81A until Sep 15.57.

	Name		Launch date, lifetime and descent date	Shape and weight (kg)	Size (m)	Date of orbital determination	Orbital inclina-tion (deg)	Nodal period (min)	Semi major axis (km)	Perigee height (km)	Apogee height (km)	Orbital eccen-tricity	Argument of perigee (deg)
D	Capsule [Atlas Agena D]	1966-83A	1966 Sep 16.75 6 days 1966 Sep 23	- 500?	1.5 dia?	1966 Sep 17.9	93.98	89.37	6619	148	333	0.014	141
D	Capsule	1966-83B	1966 Sep 16.75 600.29 days 1968 May 9.04	Octagon? 60?	0.3 long 0.9 dia	1966 Sep 17.3 1968 Jan 31.8 1968 Mar 15.8	94.06 93.98 93.98	94.25 92.38 91.74	6859 6767 6736	460 385 354	501 393 361	0.003 0.0006 0.0005	11 227 -
D	Agena D rocket	1966-83C	1966 Sep 16.75 1.27 days 1966 Sep 18.02	Cylinder 700?	6 long? 1.5 dia	1966 Sep 16.8	93.90	89.10	6605	132	322	0.014	-
D	Fragment	1966-83D											
D	[Thor Agena D]	1966-85A	1966 Sep 20.88 21.90 days 1966 Oct 12.78	Cylinder 1500?	8 long? 1.5 dia	1966 Sep 21.0	85.13	90.87	6693	188	442	0.019	163
D	Fragment	1966-85B											
D	[Titan 3B Agena D]	1966-86A	1966 Sep 28.80 9.06 days 1966 Oct 7.86	Cylinder 3000?	8 long? 1.5 dia	1966 Sep 30.0	93.98	89.01	6602	151	296	0.011	125
	Essa 3	1966-87A	1966 Oct 2.44 10000 years	Cylinder 145	0.56 long 1.07 dia	1966 Oct 3.2	101.06	114.60	7816	1383	1493	0.007	125
	Essa 3 rocket	1966-87B	1966 Oct 2.44 5000 years	Cylinder 24	1.5 long 0.46 dia	1966 Oct 13.5	101.04	114.60	7816	1383	1493	0.007	99
	Fragments	1966-87C-F											
D	Cosmos U 1	1966-88A	1966 Sep 17.94 54.45 days 1966 Nov 11.39	Cone-cylinder?	6 long? 1.5 dia?	1966 Sep 18.0 1966 Oct 5.6	49.63 49.60	96.08 94.42	6982 6872	163 136	1046 851	0.063 0.052	83 156
D	Cosmos U 1 rocket	1966-88B	1966 Sep 17.94 168 days 1967 Mar 4	Cylinder 1500?	8 long? 2.5 dia?	1966 Oct 16.3 1967 Feb 8.0	49.27 49.23	97.58 92.40	7023 6772	280 259	1010 529	0.052 0.020	221 -
D	Fragments	1966-88C-BE											

Space Vehicle: Surveyor 2, 1966-84A, and Centaur 11, 1966-84B in highly eccentric Orbit.
Cosmos U 1 is an abbreviation for 'first unnumbered Cosmos'.

	Name		Launch date, lifetime and descent date	Shape and weight (kg)	Size (m)	Date of orbital determination	Orbital inclination (deg)	Nodal period (min)	Semi major axis (km)	Perigee height (km)	Apogee height (km)	Orbital eccentricity	Argument of perigee (deg)
	[Atlas Agena D]	1966-89A	1966 Oct 5.92 100 000 years	Cylinder 2000?	8 long? 1.5 dia	1966 Oct 6.9	90.20	167.63	10070	3682	3702	0.001	190
	Secor 8 (EGRS 8)	1966-89B	1966 Oct 5.92 100 000 years	Rectangular box 17	0.33 x 0.28 x 0.23	1966 Oct 6.0	90.19	167.63	10069	3676	3706	0.002	183
D	[Atlas Agena D]	1966-90A	1966 Oct 12.80 8.18 days 1966 Oct 20.98	-	-	1966 Oct 13.3	90.96	89.00	6599	155	287	0.010	173
D	[Atlas Agena D]	1966-90B	1966 Oct 12.80 8.46 days 1966 Oct 21.26	Cylinder 1500?	8 long? 1.5 dia	1966 Oct 17.2	90.88	88.99	6598	181	258	0.006	142
D	Fragments	1966-90C, D											
D R	Cosmos 129	1966-91A	1966 Oct 14.51 6.75 days 1966 Oct 21.26	Sphere-cylinder 4750?	4.3 long 2.4 dia	1966 Oct 14.7	64.65	89.45	6624	180	312	0.010	49
D	Cosmos 129 rocket	1966-91B	1966 Oct 14.51 9 days 1966 Oct 23	Cylinder 1440	3.8 long 2.6 dia	1966 Oct 15.2	64.58	89.50	6628	217	283	0.005	85
D	Molniya 1D	1966-92A	1966 Oct 20.33 692.02 days 1968 Sep 11.35	Windmill 1000?	3.4 long 1.6 dia	1966 Oct 21.4 1968 Jan 4.4 1968 May 31.2	65.35 64.91 64.91	714.4 717.93 715.8	26473 26564 26508	505 396 236	39685 39976 40024	0.740 0.745 0.751	283 275 -
D	Molniya 1D launcher	1966-92B	1966 Oct 20.33 21.62 days 1966 Nov 10.95	-	-	1966 Oct 20.4	64.76	90.83	6694	189	443	0.019	57
D	Molniya 1D launcher rocket	1966-92C	1966 Oct 20.33 13.62 days 1966 Nov 2.95	Cylinder 2500?	7.5 long 2.6 dia	1966 Oct 21.3	64.82	90.73	6689	200	422	0.017	62
D	Molniya 1D rocket	1966-92D	1966 Oct 20.33 959.06 days 1969 Jun 5.39	Cylinder 440	2.0 long 2.0 dia	1968 May 15.8 1969 Feb 16.0	64.8 64.9	710.8 220.5	26384 12091	236 104	39776 11321	0.749 0.464	- -

	Name		Launch date, lifetime and descent date	Shape and weight (kg)	Size (m)	Date of orbital determination	Orbital inclination (deg)	Nodal period (min)	Semi major axis (km)	Perigee height (km)	Apogee height (km)	Orbital eccentricity	Argument of perigee (deg)
D R	Cosmos 130	1966-93A	1966 Oct 20.37 7.91 days 1966 Oct 28.28	Sphere-cylinder 5530?	5 long? 2.4 dia	1966 Oct 23.2	64.95	89.71	6639	208	314	0.008	51
D	Cosmos 130 rocket	1966-93B	1966 Oct 20.37 9.27 days 1966 Oct 29.64	Cylinder 2500?	7.5 long 2.6 dia	1966 Oct 22.7	64.94	89.46	6627	206	292	0.007	42
D	Luna 12 launcher	1966-94B	1966 Oct 22.36 2.49 days 1966 Oct 24.85	-	-	1966 Oct 22.7	51.92	88.58	6584	199	212	0.001	81
D	Luna 12 launcher rocket	1966-94C	1966 Oct 22.36 1.07 days 1966 Oct 23.43	Cylinder 2500?	7.5 long 2.6 dia	1966 Oct 22.7	51.89	88.13	6563	174	195	0.002	309
D	Surveyor Model 3	1966-95A	1966 Oct 26.47 11.06 days 1966 Nov 6.53	Irregular 950	2.8 long 1.3 dia	1966 Nov 1.0	29.6	15933	209561	166	406200	0.969	-
D	Intelsat 2 F-1	1966-96A	1966 Oct 26.96 5795 days 1982 Sep 7	Cylinder 140	0.67 long 1.42 dia	*1966 Oct 27.0 1966 Nov 7.7	26.43 17.22	669.8 730.0	25351 26856	289 3424	37656 37531	0.737 0.635	178 181
D	Intelsat 2 F-1 second stage	1966-96B	1966 Oct 26.96 105 days 1967 Feb 9	Cylinder 400?	4.9 long 1.43 dia	1966 Oct 31.5	28.73	92.31	6773	244	545	0.022	-
	Fragment	1966-96C											
D	OV3-2	1966-97A	1966 Oct 28.50 1796.79 days 1971 Sep 29.29	Octagon 82	0.74 long 0.74 dia	1966 Oct 29.8 1968 Jan 31.0 1969 Sep 16.0	81.97 81.98 81.98	104.24 102.21 98.95	7337 7242 7086	320 313 306	1597 1414 1109	0.087 0.076 0.057	164 296 -
D	OV3-2 rocket	1966-97B	1966 Oct 28.50 1188.21 days 1970 Jan 28.71	Cylinder 24	1.5 long 0.46 dia	1966 Nov 7.7 1968 Jan 15.5 1969 Jan 31.7	81.97 81.97 81.93	104.23 101.42 97.82	7336 7204 7031	319 318 303	1596 1335 1003	0.087 0.071 0.050	139 - -
D	Fragments	1966-97C,D											

Space Vehicle: Luna 12, 1966-94A. Centaur 9, 1966-95B, is in a highly eccentric orbit.
* Intelsat 2 F-1 third-stage rocket, 1966-96D, had an initial orbit similar to the first orbit of 1966-96A.

Year of launch 1966 continued

	Name		Launch date, lifetime and descent date	Shape and weight (kg)	Size (m)	Date of orbital determination	Orbital inclination (deg)	Nodal period (min)	Semi major axis (km)	Perigee height (km)	Apogee height (km)	Orbital eccentricity	Argument of perigee (deg)
D	[Atlas Agena D]	1966-98A	1966 Nov 2.85 7.2 days 1966 Nov 10.0	Cylinder 1500?	8 long? 1.5 dia	1966 Nov 3.6	90.96	89.20	6610	159	305	0.011	150
D	[Atlas Agena D]	1966-98B	1966 Nov 2.85 13.71 days 1966 Nov 16.56	-	-	1966 Nov 4.1	91.00	89.86	6644	208	324	0.009	119
D	OV4-3 MOL canister* [Titan 3C]	1966-99A	1966 Nov 3.58 66.9 days 1967 Jan 9.5	Cylinder 9680	13.7 long 3.0 dia	1966 Nov 4.4	32.82	90.42	6680	298	305	0.0005	307
D	OV4-1R (receiver)	1966-99B	1966 Nov 3.58 62.56 days 1967 Jan 5.14	Domed cylinder 136	1.40 long 0.43 dia	1966 Nov 5.6	32.84	90.30	6673	291	298	0.0005	164
D	OV1-6	1966-99C	1966 Nov 3.58 57.71 days 1966 Dec 31.29	Cylinder + 2 hemispheres 202	1.73 long 0.69 dia	1966 Nov 4.6	32.84	90.27	6671	290	295	0.0004	276
D	OV4-1T (transmitter)	1966-99D	1966 Nov 3.58 68.6 days 1967 Jan 11.2	Domed cylinder 109	1.40 long 0.43 dia	1966 Nov 5.3	32.83	90.59	6686	294	321	0.002	348
D	Lunar-orbiter 2 rocket	1966-100B	1966 Nov 6.97 8.90 days 1966 Nov 15.87	Cylinder 700	8 long 1.5 dia	1966 Nov 11.4	29.6	12819	181242	128	349600	0.964	-
D	Cosmos U2	1966-101A	1966 Nov 2.03 15 days 1966 Nov 17	Cone-cylinder	6 long? 1.5 dia?	1966 Nov 7.6	49.58	94.50	6876	140	855	0.052	106
D	Capsule	1966-101G	1966 Nov 2.03 185 days 1967 May 6	- 1000?	0.8 dia?	1966 Dec 28.9 1967 May 1.1	49.61 49.47	94.28 88.29	6865 6571	144 114	830 272	0.050 0.012	321 148
D	Fragments	1966-101B-AS											

Space Vehicle: Lunar Orbiter 2, 1966-100A.
*Manned orbital laboratory.

	Name		Launch date, lifetime and descent date	Shape and weight (kg)	Size (m)	Date of orbital determination	Orbital inclination (deg)	Nodal period (min)	Semi major axis (km)	Perigee height (km)	Apogee height (km)	Orbital eccentricity	Argument of perigee (deg)
D	[Thor Agena D]	1966-102A	1966 Nov 8.83 20.6 days 1966 Nov 29.4	Cylinder 1500?	8 long? 1.5 dia	1966 Nov 9.7	100.09	89.42	6623	172	318	0.011	175
D	Target* Agena 12	1966-103A	1966 Nov 11.80 41 days 1966 Dec 23	Cylinder 3175	7.9 long 1.5 dia	1966 Nov 12.3 1966 Nov 21.6	28.78 28.85	89.93 89.70	6655 6643	243 248	310 281	0.005 0.002	45 303
D 2M R	Gemini 12*	1966-104A	1966 Nov 11.87 3.93 days 1966 Nov 15.80	Cone 3765	5.6 long 3.0 dia	1966 Nov 12.3	28.78	89.93	6655	243	310	0.005	45
D	Gemini 12 rocket	1966-104B	1966 Nov 11.87 0.93 day 1966 Nov 12.80	Cylinder 1900	6 long 3.0 dia	1966 Nov 12.0	28.91	88.74	6597	159	278	0.009	132
D	Gemini 12 adapter module	1966-104G	1966 Nov 11.87 19.74 days 1966 Dec 1.61	Truncated cone 2510?	2.3 long 3.0 dia	1966 Nov 30.5	28.85	89.13	6616	238	238	0	-
D	Fragments	1966-104C-F,H-Q											
D R	Cosmos 131	1966-105A	1966 Nov 12.41 7.81 days 1966 Nov 20.22	Sphere-cylinder 5530?	5 long? 2.4 dia	1966 Nov 12.7	72.86	89.94	6649	204	337	0.010	51
D	Cosmos 131 rocket	1966-105B	1966 Nov 12.41 10.19 days 1966 Nov 22.60	Cylinder 2500?	7.5 long 2.6 dia	1966 Nov 12.7	72.85	89.80	6644	206	326	0.009	50
D R	Cosmos 132	1966-106A	1966 Nov 19.34 8.0 days 1966 Nov 27.3	Sphere-cylinder 4750?	4.3 long 2.4 dia	1966 Nov 19.4	65.02	89.37	6621	210	276	0.005	19
D	Cosmos 132 rocket	1966-106B	1966 Nov 19.34 8.61 days 1966 Nov 27.95	Cylinder 1440	3.8 long 2.6 dia	1966 Nov 19.7	65.01	89.37	6621	223	263	0.003	66

*Gemini 12 and Agena 12 initially docked Nov 12.03. 1966-104G was joined to 1966-104A until Nov 15.79.

Year of launch 1966 continued

	Name		Launch date, lifetime and descent date	Shape and weight (kg)	Size (m)	Date of orbital determination	Orbital inclination (deg)	Nodal period (min)	Semi major axis (km)	Perigee height (km)	Apogee height (km)	Orbital eccentricity	Argument of perigee (deg)
D R	Cosmos 133	1966-107A	1966 Nov 28.46 2.0 days 1966 Nov 30.5	Sphere-cylinder + 2 wings 6450?	7.5 long 2.2 dia	1966 Nov 29.0	51.82	88.40	6575	171	223	0.004	323
D	Cosmos 133 rocket	1966-107B	1966 Nov 28.46 1.50 days 1966 Nov 29.96	Cylinder 2500?	7.5 long 2.6 dia	1966 Nov 29.0	51.82	88.22	6567	169	209	0.003	279
D R	Cosmos 134	1966-108A	1966 Dec 3.34 7.90 days 1966 Dec 11.24	Sphere-cylinder 5530?	5 long? 2.4 dia	1966 Dec 3.4	64.98	89.46	6626	201	294	0.007	52
D	Cosmos 134 rocket	1966-108B	1966 Dec 3.34 7.10 days 1966 Dec 10.44	Cylinder 2500?	7.5 long 2.6 dia	1966 Dec 4.1	64.97	89.39	6623	205	285	0.006	43
D	Fragment	1966-108C											
D	[Atlas Agena D]	1966-109A	1966 Dec 5.88 8.2 days 1966 Dec 14.1	- 500?	1.5 dia?	1966 Dec 6.9	104.63	89.77	6640	137	388	0.019	132
D	Agena D rocket	1966-109B	1966 Dec 5.88 3 days 1966 Dec 8	Cylinder 700?	8 long? 1.5 dia	1966 Dec 5.9	104.65	89.38	6621	121	364	0.018	-
T	ATS 1*	1966-110A	1966 Dec 7.09 > million years	Cylinder 352	1.45 long 1.42 dia	1966 Dec 7.8 1967 Feb 10	0.23 0.1	1465.89 1436.1	42748 42166	35852 35782	36887 35793	0.012 0.0001	129 -
D	ATS 1 rocket [Atlas Agena D]	1966-110B	1966 Dec 7.09 2869.36 days 1974 Oct 15.45	Cylinder 700?	6 long? 1.5 dia	1966 Dec 13.1 1969 Sep 16.0 1972 May 1.0	31.37 30.9 30.88	647.15 365.5 227.5	24733 16934 12346	167 170 178	36543 20941 11758	0.735 0.613 0.469	183 - -

* Applications Technology Satellite.

	Name		Launch date, lifetime and descent date	Shape and weight (kg)	Size (m)	Date of orbital determination	Orbital inclination (deg)	Nodal period (min)	Semi major axis (km)	Perigee height (km)	Apogee height (km)	Orbital eccentricity	Argument of perigee (deg)
	OV1-9	1966-111A	1966 Dec 11.88 200 years	Cylinder + hemisphere 104	1.40 long 0.69 dia	1966 Dec 12.6	99.14	142.30	9027	473	4824	0.241	147
	OV1-10	1966-111B	1966 Dec 11.88 40 years	Cylinder + hemispheres + 6 booms 130	1.40 long 0.69 dia	1966 Dec 12.5	93.43	98.87	7083	641	769	0.009	209
	OV1-10 rocket	1966-111C	1966 Dec 11.88 60 years	Cylinder 70?	2.05 long 0.72 dia	1966 Dec 30.4	93.42	98.85	7082	640	768	0.009	149
	OV1-9 rocket	1966-111D	1966 Dec 11.88 150 years	Cylinder 70?	2.05 long 0.72 dia	1966 Dec 21.3	99.13	142.30	9027	473	4824	0.241	135
D	Cosmos 135	1966-112A	1966 Dec 12.86 120.7 days 1967 Apr 12.6	Ellipsoid 400?	1.8 long 1.2 dia	1966 Dec 12.9 1967 Apr 7.8	48.44 48.41	93.58 89.39	6829 6624	253 206	649 286	0.029 0.006	112 320
D	Cosmos 135 rocket	1966-112B	1966 Dec 12.86 92.17 days 1967 Mar 15.03	Cylinder 1500?	8 long 1.65 dia	1966 Dec 13.3 1967 Mar 11.0	48.45 48.42	93.37 89.39	6822 6624	253 206	635 286	0.028 0.006	114 195
D	Fragments	1966-112C-E											
D	[Titan 3B Agena D]	1966-113A	1966 Dec 14.76 9 days 1966 Dec 24	Cylinder 3000?	8 long 1.5 dia	1966 Dec 17.1	109.56	89.58	6631	138	368	0.017	129
D B	Bios 1 Capsule*	1966-114A	1966 Dec 14.81 62.35 days 1967 Feb 15.16	Blunt cone 127	1.2 long 1.02 dia	1966 Dec 17.6 1967 Feb 13.8	33.51 33.50	90.44 87.85	6680 6553	295 168	309 181	0.001 0.001	352 153
D	Bios 1 rocket	1966-114B	1966 Dec 14.81 38.68 days 1967 Jan 22.49	Cylinder 400?	4.9 long 1.43 dia	1966 Dec 18.7	33.50	90.27	6673	281	308	0.002	9
D	Bios 1 Adapter	1966-114C	1966 Dec 14.81 26.55 days 1967 Jan 10.36	Cylinder 300	1.8 long 1.45 dia	1967 Jan 2.5	33.50	89.65	6642	250	277	0.002	164
D	Fragment	1966-114D											

*Before 1966 Dec 17.8, Bios 1 capsule was attached to Bios 1 adapter.

	Name		Launch date, lifetime and descent date	Shape and weight (kg)	Size (m)	Date of orbital determination	Orbital inclination (deg)	Nodal period (min)	Semi major axis (km)	Perigee height (km)	Apogee height (km)	Orbital eccentricity	Argument of perigee (deg)
D R	Cosmos 136	1966-115A	1966 Dec 19.50 7.75 days 1966 Dec 27.25	Sphere-cylinder 4750?	4.3 long 2.4 dia	1966 Dec 19.7	64.68	89.17	6612	188	280	0.007	33
D	Cosmos 136 rocket	1966-115B	1966 Dec 19.50 9.0 days 1966 Dec 28.5	Cylinder 1440	3.8 long 2.6 dia	1966 Dec 21.4	64.61	89.35	6620	209	275	0.005	74
D	Luna 13 launcher	1966-116B	1966 Dec 21.43 1.85 days 1966 Dec 23.28	-	-	1966 Dec 21.5	51.80	88.40	6575	171	223	0.004	340
D	Luna 13 launcher rocket	1966-116C	1966 Dec 21.43 1 day 1966 Dec 22	Cylinder 2500?	7.5 long 2.6 dia	1966 Dec 21.5	51.77	87.79	6546	135	201	0.005	276
D	Cosmos 137	1966-117A	1966 Dec 21.55 337.26 days 1967 Nov 23.81	Ellipsoid 400?	1.8 long 1.2 dia	1966 Dec 21.6 1967 Jun 30.5 1967 Nov 20.8	48.78 48.77 48.74	104.38 98.93 89.50	7347 7084 6630	219 213 172	1718 1199 331	0.102 0.070 0.012	106 - 29
D	Cosmos 137 rocket	1966-117B	1966 Dec 21.55 278.25 days 1967 Sep 25.80	Cylinder 1500?	8 long 1.65 dia	1966 Dec 29.4 1967 May 15.8 1967 Sep 22.4	48.72 48.80 48.75	104.08 98.81 89.80	7332 7078 6645	221 215 180	1687 1185 353	0.100 0.069 0.013	134 - 153
D	[Thor Agena D]	1966-118A	1966 Dec 29.50 828.02 days 1969 Apr 5.52	Cylinder 1500?	8 long? 1.5 dia	1966 Dec 29.6 1968 Jan 30.7 1968 Aug 31.2	75.03 75.03 75.03	94.41 93.66 92.91	6869 6831 6794	486 448 413	496 458 419	0.0007 0.0007 0.0004	88 157 -
D	Fragments	1966-118B,C											
D	Fragments	1966-00A-D*											

Space Vehicle: 1966-116A, Luna 13.

*These unidentified fragments were discovered in orbit, and catalogued on 1966 Sep 24 (A-C) and 1966 Dec 25 (D). Fragments 1966-00B and C probably decayed in 1971 or 1972. Fragment 1966-00A decayed 1977 Sep 18 and 1966-00D decayed 1966 Dec 30.

	Name		Launch date, lifetime and descent date	Shape and weight (kg)	Size (m)	Date of orbital determination	Orbital inclination (deg)	Nodal period (min)	Semi major axis (km)	Perigee height (km)	Apogee height (km)	Orbital eccentricity	Argument of perigee (deg)
	Intelsat 2 F-2	1967-01A	1967 Jan 11.45 > million years	Cylinder 87	0.67 long 1.42 dia	1967 Jan 22.9 1977 May 13.0	2.14 6.62	1448.5 1437.31	42408 42198	35563 35758	36496 35883	0.011 0.001	26 3
D	Intelsat 2 F-2 second stage	1967-01B	1967 Jan 11.45 92 days 1967 Apr 13	Cylinder 400?	4.9 long 1.43 dia	1967 Feb 5.4 1967 Mar 8.0	28.75 28.74	91.95 91.04	6754 6710	241 232	511 432	0.020 0.015	58 -
	Intelsat 2 F-2 third stage	1967-01D	1967 Jan 11.45 60 years	Cylinder 24	1.5 long 0.46 dia	1967 Jan 31.5	26.18	652.9	24927	288	36810	0.733	-
15d	Fragments	1967-01C,E-X											
D r?	[Thor Agena D]	1967-02A	1967 Jan 14.89 18.7 days 1967 Feb 2.6	Cylinder 1st 15 days 1500? then 700?	8 long? 1.5 dia	1967 Jan 15.6	80.07	90.13	6658	180	380	0.015	161
D	Fragment	1967-02B											
	IDCSP 2-1	1967-03A	1967 Jan 18.60 > million years	Polyhedron (26 faces) 45	0.8 long 0.9 dia	1967 Jan 28.7	0.07	1329.6	40055	33557	33800	0.003	352
	IDCSP 2-2	1967-03B	1967 Jan 18.60 > million years	Polyhedron (26 faces) 45	0.8 long 0.9 dia	1967 Jan 28.3	0.05	1330.0	40064	33526	33846	0.004	117
	IDCSP 2-3	1967-03C	1967 Jan 18.60 > million years	Polyhedron (26 faces) 45	0.8 long 0.9 dia	1967 Jan 28.3	0.06	1330.7	40077	33579	33819	0.003	95
	IDCSP 2-4	1967-03D	1967 Jan 18.60 > million years	Polyhedron (26 faces) 45	0.8 long 0.9 dia	1967 Jan 28.3	0.07	1332.1	40105	33606	33847	0.003	104
	IDCSP 2-5	1967-03E	1967 Jan 18.60 > million years	Polyhedron (26 faces) 45	0.8 long 0.9 dia	1967 Jan 28.3	0.03	1334.2	40147	33608	33929	0.004	89
	IDCSP 2-6	1967-03F	1967 Jan 18.60 > million years	Polyhedron (26 faces) 45	0.8 long 0.9 dia	1967 Jan 28.2	0.06	1336.5	40195	33656	33978	0.004	28

Continued on Page 126

	Name		Launch date, lifetime and descent date	Shape and weight (kg)	Size (m)	Date of orbital determination	Orbital inclination (deg)	Nodal period (min)	Semi major axis (km)	Perigee height (km)	Apogee height (km)	Orbital eccentricity	Argument of perigee (deg)
	IDCSP 2-7	1967-03G	1967 Jan 18.60 > million years	Polyhedron (26 faces) 45	0.8 long 0.9 dia	1967 Jan 28.2	0.03	1339.5	40254	33675	34077	0.005	16
	IDCSP 2-8	1967-03H	1967 Jan 18.60 > million years	Polyhedron (26 faces) 45	0.8 long 0.9 dia	1967 Jan 28.2	0.05	1343.0	40325	33665	34229	0.007	32
	Transtage 9 [Titan 3C]	1967-03J	1967 Jan 18.60 > million years	Cylinder 1500?	6 long? 3 dia	1967 Jan 18.6 1967 Jan 28.2	28 0.1?	87.88 1347.6?	6554 40414?	170 33712?	182 34359?	0.001 0.008	- -
D R	Cosmos 138	1967-04A	1967 Jan 19.53 7.73 days 1967 Jan 27.26	Sphere-cylinder 4750?	4.3 long 2.4 dia	1967 Jan 23.3	64.55	89.15	6610	191	273	0.006	27
D	Cosmos 138 rocket	1967-04B	1967 Jan 19.53 8.5 days 1967 Jan 28.0	Cylinder 1440	3.8 long 2.6 dia	1967 Jan 26.6	64.57	88.31	6570	183	201	0.001	74
D R?	Cosmos 139	1967-05A	1967 Jan 25.58 0.06 day 1967 Jan 25.64	Cylinder?	2 long? 1 dia?	1967 Jan 25.6	49.7	87.97	6555	144	210	0.005	-
D	Cosmos 139 rocket	1967-05B	1967 Jan 25.58 0.3 day? 1967 Jan 25.9?	Cylinder 1500?	8 long? 2.5 dia ?	orbit similar to 1967-05A							
D	Cosmos 139 launch platform	1967-05C	1967 Jan 25.58 0.4 day? 1967 Jan 25	Irregular	-	orbit similar to 1967-05A							

	Name		Launch date, lifetime and descent date	Shape and weight (kg)	Size (m)	Date of orbital determination	Orbital inclination (deg)	Nodal period (min)	Semi major axis (km)	Perigee height (km)	Apogee height (km)	Orbital eccentricity	Argument of perigee (deg)
	Essa 4	1967-06A	1967 Jan 26.73 10000 years	Cylinder 132	0.56 long 1.07 dia	1967 Jan 29.4	102.00	113.48	7764	1328	1443	0.007	55
	Essa 4 rocket	1967-06B	1967 Jan 26.73 5000 years	Cylinder 24	1.5 long 0.46 dia	1967 Jan 27.4	101.98	113.68	7773	1344	1444	0.006	69
	Fragments	1967-06C-E											
D	[Atlas Agena D]	1967-07A	1967 Feb 2.83 9 days 1967 Feb 12	- 500?	1.5 dia?	1967 Feb 6.0	102.96	89.47	6625	136	357	0.017	123
D	Agena D rocket	1967-07B	1967 Feb 2.83 1.5 days 1967 Feb 4.3	Cylinder 700?	8 long? 1.5 dia?	1967 Feb 3	103.13	89.75	6639	176	345	0.013	-
D R	Cosmos 140	1967-09A	1967 Feb 7.14 1.99 days 1967 Feb 9.13	Sphere- cylinder + 2 wings 6450?	7.5 long 2.2 dia	1967 Feb 8.4	51.66	88.27	6570	165	218	0.004	335
D P	Cosmos 140 rocket	1967-09B	1967 Feb 7.14 1 day 1967 Feb 8	Cylinder 2500?	7.5 long 2.6 dia	1967 Feb 8.0	51.72	87.79	6546	148	187	0.003	329
	[Thor Burner 2]	1967-10A	1967 Feb 8.33 70 years	12-sided frustum 195	1.64 long 1.31 to 1.10 dia	1967 Feb 8.5	98.84	101.55	7210	796	868	0.005	45
	Burner 2 rocket	1967-10B	1967 Feb 8.33 70 years	Sphere-cone 66	1.32 long 0.94 dia	1967 May 24.1	98.83	101.53	7209	790	871	0.006	98
L	Diademe 1	1967-11A	1967 Feb 8.40 50 years	Cylinder + 4 vanes 22.7	0.20 long 0.5 dia	1967 Feb 28.5 1973 Feb 1.0	39.98 39.98	104.17 103.7	7338 7315	569 561	1350 1312	0.053 0.051	- -
	Diademe 1 rocket	1967-11B	1967 Feb 8.40 100 years	Cylinder 68	2.1 long 0.65 dia	1967 Mar 6.0	39.92	103.97	7328	574	1325	0.051	-
11d	Fragments	1967-11C-Q											

Space Vehicle: Lunar Orbiter 3, 1967-08A

	Name		Launch date, lifetime and descent date	Shape and weight (kg)	Size (m)	Date of orbital determination	Orbital inclination (deg)	Nodal period (min)	Semi major axis (km)	Perigee height (km)	Apogee height (km)	Orbital eccentricity	Argument of perigee (deg)
D R	Cosmos 141	1967-12A	1967 Feb 8.43 7.79 days 1967 Feb 16.22	Sphere-cylinder 5530?	5 long? 2.4 dia	1967 Feb 13.4	72.85	89.74	6639	205	316	0.008	45
D	Cosmos 141 rocket	1967-12B	1967 Feb 8.43 7.74 days 1967 Feb 16.17	Cylinder 2500?	7.5 long 2.6 dia	1967 Feb 12.3	72.87	89.25	6615	197	276	0.006	35
D	Cosmos 142	1967-13A	1967 Feb 14.42 142.19 days 1967 Jul 6.61	Ellipsoid 400?	1.8 long 1.2 dia	1967 Feb 14.6 1967 May 23.5	48.3 48.39	100.25 94.61	7150 6880	207 197	1336 807	0.079 0.044	101 -
D	Cosmos 142 rocket	1967-13B	1967 Feb 14.42 121.28 days 1967 Jun 15.70	Cylinder 1500?	8 long 1.65 dia	1967 Feb 14.6 1967 May 8.0	48.4 48.38	100.35 94.44	7156 6872	210 199	1346 788	0.079 0.043	103 -
L	Diademe 2	1967-14A	1967 Feb 15.44 200 years	Cylinder + 4 vanes 22.7	0.20 long 0.5 dia	1967 Apr 15.5	39.45	110.01	7614	591	1881	0.085	-
	Diademe 2 rocket	1967-14B	1967 Feb 15.44 100 years	Cylinder 68	2.1 long 0.65 dia	1967 Apr 15.5	39.45	110.09	7618	590	1890	0.085	-
4d	Fragments	1967-14C-H											

	Name		Launch date, lifetime and descent date	Shape and weight (kg)	Size (m)	Date of orbital determination	Orbital inclination (deg)	Nodal period (min)	Semi major axis (km)	Perigee height (km)	Apogee height (km)	Orbital eccentricity	Argument of perigee (deg)
D r?	[Thor Agena D]	1967-15A	1967 Feb 22.92 17.02 days 1967 Mar 11.94	Cylinder 1st 14 days 1500? then 700?	8 long? 1.5 dia	1967 Feb 23.8	80.03	90.12	6658	180	380	0.015	159
D	[Titan 3B Agena D]	1967-16A	1967 Feb 24.83 10.15 days 1967 Mar 6.98	Cylinder 3000?	8 long? 1.5 dia	1967 Feb 25.4	106.98	90.02	6653	135	414	0.021	138
D R	Cosmos 143	1967-17A	1967 Feb 27.35 7.89 days 1967 Mar 7.24	Sphere-cylinder 4750?	4.3 long 2.4 dia	1967 Feb 27.7	64.99	89.53	6629	204	297	0.007	15
D	Cosmos 143 rocket	1967-17B	1967 Feb 27.35 9.78 days 1967 Mar 9.13	Cylinder 1440	3.8 long 2.6 dia	1967 Feb 28.5	65.02	89.65	6635	217	297	0.006	55
D	Cosmos 144	1967-18A	1967 Feb 28.61 5677 days 1982 Sep 14	Cylinder + 2 vanes 2000?	5 long? 1.5 dia?	1967 Feb 28.8	81.25	96.88	6987	574	644	0.005	274
	Cosmos 144 rocket	1967-18B	1967 Feb 28.61 20 years	Cylinder 1440	3.8 long 2.6 dia	1967 Mar 2.2	81.21	97.01	6993	521	709	0.013	190
D	Fragment	1967-18C											
D	Cosmos 145	1967-19A	1967 Mar 3.28 371.11 days 1968 Mar 8.39	Ellipsoid 400?	1.8 long 1.2 dia	1967 Mar 3.7 1967 Nov 8.0 1968 Jan 8.0	48.42 48.42 48.35	108.60 99.96 96.51	7544 7136 6972	215 204 197	2116 1312 991	0.126 0.078 0.057	101 - -
D	Cosmos 145 rocket	1967-19B	1967 Mar 3.28 272.86 days 1967 Dec 1.14	Cylinder 1500?	8 long 1.65 dia	1967 Mar 3.7 1967 Sep 8.0 1967 Oct 23.7	48.40 48.39 48.37	108.33 100.55 96.15	7533 7165 6954	213 203 194	2096 1371 958	0.125 0.081 0.055	102 - -

	Name		Launch date, lifetime and descent date	Shape and weight (kg)	Size (m)	Date of orbital determination	Orbital inclination (deg)	Nodal period (min)	Semi major axis (km)	Perigee height (km)	Apogee height (km)	Orbital eccentricity	Argument of perigee (deg)
D	OSO 3	1967-20A	1967 Mar 8.68 5506 days 1982 Apr 4	Nonagon 281	0.94 long 1.12 dia	1967 Mar 9.6 1972 Feb 1.0	32.87 32.87	95.53 94.98	6928 6900	534 510	564 533	0.002 0.002	200 -
D	OSO 3 rocket	1967-20B	1967 Mar 8.68 1137.71 days 1970 Apr 19.39	Cylinder 24	1.5 long 0.46 dia	1967 Mar 9.6 1968 Mar 14.5 1969 Sep 16.0	32.86 32.86 32.86	95.48 94.96 93.47	6925 6899 6824	540 510 441	554 531 451	0.001 0.001 0.001	206 - -
D	Fragment	1967-20C											
D	Cosmos 146*	1967-21A	1967 Mar 10.54 0.99 day 1967 Mar 11.53	Sphere-cylinder + 2 wings 5375?	5.3 long 2.2 dia	1967 Mar 10.7	51.44	89.20	6615	177	296	0.009	92
D	Service module?	1967-21B	1967 Mar 10.54 8.65 days 1967 Mar 19.19	Cylinder + 2 wings? 2615?	3.0 long? 2.3 dia?	1967 Mar 15.0	51.52	89.16	6613	175	294	0.009	107
D	Descent module?	1967-21D	1967 Mar 10.54 8.35 days 1967 Mar 18.89	Spheroid? 2760?	2.3 dia?	1967 Mar 15.0	51.51	89.15	6612	176	292	0.009	-
D	Fragment**	1967-21C											
D R	Cosmos 147	1967-22A	1967 Mar 13.51 7.76 days 1967 Mar 21.27	Sphere-cylinder 4750?	4.3 long 2.4 dia	1967 Mar 14.1	64.57	89.42	6626	195	301	0.008	39
D	Cosmos 147 rocket	1967-22B	1967 Mar 13.51 8.78 days 1967 Mar 22.29	Cylinder 1440	3.8 long 2.6 dia	1967 Mar 17.2	64.58	89.27	6617	206	272	0.005	74
D	Cosmos 148	1967-23A	1967 Mar 16.74 51.77 days 1967 May 7.51	Ellipsoid 400?	1.8 long 1.2 dia	1967 Mar 19.6	71.00	91.26	6715	270	404	0.010	68
D	Cosmos 148 rocket	1967-23B	1967 Mar 16.74 28.66 days 1967 Apr 14.40	Cylinder 1500?	8 long 1.65 dia	1967 Mar 19.4	71.02	91.08	6706	274	381	0.008	69

* Cosmos 146 apparently split into 2 modules on 1967 Mar 11.53 while over Tyuratam (1967-21D retained the 21A designation in USA)
** This was probably Cosmos 146 rocket, lifetime 1 day, similar to Cosmos 154 rocket.

	Name		Launch date, lifetime and descent date	Shape and weight (kg)	Size (m)	Date of orbital determination	Orbital inclination (deg)	Nodal period (min)	Semi major axis (km)	Perigee height (km)	Apogee height (km)	Orbital eccentricity	Argument of perigee (deg)
D	Cosmos 149	1967-24A	1967 Mar 21.42 17.28 days 1967 Apr 7.70	Ellipsoid + annular tail 300?	6.5 long 1.2 dia	1967 Mar 23.1	48.40	89.76	6643	245	285	0.003	277
D	Cosmos 149 rocket	1967-24B	1967 Mar 21.42 3.05 days 1967 Mar 24.47	Cylinder 1500?	8 long 1.65 dia	1967 Mar 23.8	48.44	88.43	6579	181	221	0.003	273
D	Fragments	1967-24C-T											
D R	Cosmos 150	1967-25A	1967 Mar 22.53 7.75 days 1967 Mar 30.28	Sphere-cylinder 5530?	5 long? 2.4 dia	1967 Mar 24.1	65.64	90.04	6655	204	350	0.011	54
D	Cosmos 150 rocket	1967-25B	1967 Mar 22.53 8.32 days 1967 Mar 30.85	Cylinder 2500?	7.5 long 2.6 dia	1967 Mar 24.1	65.67	89.65	6636	191	324	0.010	39
	Intelsat 2 F-3	1967-26A	1967 Mar 23.06 > million years	Cylinder 87	0.67 long 1.42 dia	1967 Apr 7.7 1977 May 13.0	1.37 7.30	1434 1434.95	42107 42142	35687 35687	35771 35841	0.001 0.002	81 313
D	Intelsat 2 F-3 second stage	1967-26B	1967 Mar 23.06 106.4 days 1967 Jul 7.5	Cylinder 400?	4.9 long 1.43 dia	1967 Mar 24.0 1967 May 31.5	28.80 28.76	92.98 91.16	6804 6716	242 231	610 445	0.027 0.016	145 -
D	Intelsat 2 F-3 third stage	1967-26C	1967 Mar 23.06 14¼ years 1981 Jun?	Cylinder 24	1.5 long 0.46 dia	1967 Mar 25.0 1980 Nov 1.0	26.68 26.9	669.8 232.4	25351 12521	289 158	37656 12128	0.737 0.478	178*

*Approximate orbit.

	Name		Launch date, lifetime and descent date	Shape and weight (kg)	Size (m)	Date of orbital determination	Orbital inclination (deg)	Nodal period (min)	Semi major axis (km)	Perigee height (km)	Apogee height (km)	Orbital eccentricity	Argument of perigee (deg)
	Cosmos 151	1967-27A	1967 Mar 24.49 30 years	Cylinder + paddles 850?	2 long? 1 dia?	1967 Mar 24.9	56.07	97.14	7002	596	652	0.004	125
	Cosmos 151 rocket	1967-27B	1967 Mar 24.49 25 years	Cylinder 2200?	7.4 long 2.4 dia	1967 Mar 31.5	56.06	97.13	7001	594	653	0.004	-
D	Fragments	1967-27C,D											
D	Cosmos 152	1967-28A	1967 Mar 25.29 132.88 days 1967 Aug 5.17	Ellipsoid 400?	1.8 long 1.2 dia	1967 Mar 26.6 1967 Jun 23.0	70.98 70.99	92.13 90.63	6758 6684	272 245	488 367	0.016 0.009	78 -
D	Cosmos 152 rocket	1967-28B	1967 Mar 25.29 57.78 days 1967 May 22.07	Cylinder 1500?	8 long 1.65 dia	1967 Mar 26.6 1967 May 8.0	71.00 70.97	92.00 90.27	6752 6667	273 242	475 335	0.015 0.007	79 -
D	[Thor Agena D]	1967-29A	1967 Mar 30.79 17.65 days 1967 Apr 17.44	Cylinder 1500?	8 long? 1.5 dia	1967 Mar 30.9	85.03	89.45	6625	167	326	0.012	181
D R	Cosmos 153	1967-30A	1967 Apr 4.58 7.74 days 1967 Apr 12.32	Sphere-cylinder 4750?	4.3 long 2.4 dia	1967 Apr 5.3	64.59	89.26	6617	199	279	0.006	36
D	Cosmos 153 rocket	1967-30B	1967 Apr 4.58 7.06 days 1967 Apr 11.64	Cylinder 1440	3.8 long 2.6 dia	1967 Apr 7.3	64.58	89.09	6608	210	250	0.003	-
D	ATS 2	1967-31A	1967 Apr 6.14 880.81 days 1969 Sep 2.95	Cylinder 370	1.83 long 1.42 dia	1967 Apr 9.2 1968 Mar 12.4 1969 Feb 15.3	28.40 28.42 28.27	218.9 177.5 132.3	12029 10486 8601	178 176 169	11124 8040 4276	0.455 0.375 0.239	- 15 -
D	ATS 2 rocket	1967-31B	1967 Apr 6.14 450.1 days 1968 Jun 29.2	Cylinder 700?	6 long? 1.5 dia	1967 Apr 9.8 1968 Mar 9.2 1968 May 8.0	28.51 28.39 28.30	218.2 157.9 129.4	11999 9739 8474	149 147 140	11092 6575 4052	0.456 0.330 0.231	- 43 -

	Name		Launch date, lifetime and descent date	Shape and weight (kg)	Size (m)	Date of orbital determination	Orbital inclination (deg)	Nodal period (min)	Semi major axis (km)	Perigee height (km)	Apogee height (km)	Orbital eccentricity	Argument of perigee (deg)
D	Cosmos 154*	1967-32A	1967 Apr 8.38 2.2 days 1967 Apr 10.6	Sphere-cylinder + 2 wings 5375?	5.3 long 2.2 dia	1967 Apr 8.7	51.30	88.50	6581	183	223	0.003	100
D	Cosmos 154 rocket	1967-32B	1967 Apr 8.38 2.3 days 1967 Apr 10.7	Cylinder 4000?	12 long? 4 dia	1967 Apr 10.2	51.35	87.70	6542	157	170	0.001	-
D	Fragment	1967-32C											
D R	Cosmos 155	1967-33A	1967 Apr 12.46 7.98 days 1967 Apr 20.44	Sphere-cylinder 5530?	5 long? 2.4 dia	1967 Apr 12.8	51.80	89.11	6611	193	272	0.006	17
D	Cosmos 155 rocket	1967-33B	1967 Apr 12.46 4.02 days 1967 Apr 16.48	Cylinder 2500?	7.5 long 2.6 dia	1967 Apr 13.5	51.78	89.00	6605	194	260	0.005	-
D	Fragments	1967-33C, D											
	Transit 15 [Scout]	1967-34A	1967 Apr 14.14 1000 years	Octagon + 4 vanes 60?	0.25 long 0.46 dia	1967 Apr 14.3	90.23	106.60	7446	1053	1083	0.002	101
	Altair rocket	1967-34B	1967 Apr 14.14 1000 years	Cylinder 24	1.5 long 0.46 dia	1967 May 2.4	90.25	106.59	7445	1052	1083	0.002	58
2d	Fragments	1967-34C-G											
	Essa 5	1967-36A	1967 Apr 20.47 10000 years	Cylinder 145	0.56 long 1.07 dia	1967 Apr 21.3	101.97	113.63	7770	1361	1423	0.004	353
	Essa 5 rocket	1967-36B	1967 Apr 20.47 5000 years	Cylinder 24	1.5 long 0.46 dia	1967 Apr 22.6	101.95	113.63	7770	1361	1423	0.004	349
	Fragments	1967-36C,D											

Space Vehicle: Surveyor 3, 1967-35A; and Centaur 12, 1967-35B, in highly eccentric orbit.
* Cosmos 154 and Cosmos 146 were Zond test flights.

	Name		Launch date, lifetime and descent date	Shape and weight (kg)	Size (m)	Date of orbital determination	Orbital inclination (deg)	Nodal period (min)	Semi major axis (km)	Perigee height (km)	Apogee height (km)	Orbital eccentricity	Argument of perigee (deg)
D M P	Soyuz 1 *	1967-37A	1967 Apr 23.03 1.11 days 1967 Apr 24.14	Sphere-cylinder + 2 wings 6450	7.5 long 2.2 dia	1967 Apr 23.4	51.64	88.52	6583	198	211	0.001	324
D	Soyuz 1 rocket	1967-37B	1967 Apr 23.03 1.88 days 1967 Apr 24.91	Cylinder 2500?	7.5 long 2.6 dia	1967 Apr 23.6	51.66	88.39	6576	185	211	0.002	328
D	San Marco 2 **	1967-38A	1967 Apr 26.42 171.10 days 1967 Oct 14.52	Sphere 129	0.66 dia	1967 Apr 26.6 1967 Aug 14.5	2.89 2.89	93.93 91.76	6856 6745	217 199	738 535	0.038 0.025	296 -
D	San Marco 2 rocket	1967-38B	1967 Apr 26.42 33 days 1967 May 29	Cylinder 24	1.5 long 0.46 dia	1967 Apr 30.5	3.16	93.52	6835	204	710	0.037	349
	Cosmos 156	1967-39A	1967 Apr 27.53 25 years	Cylinder + 2 vanes 2000?	5 long? 1.5 dia?	1967 Apr 28.8	81.17	96.96	6992	593	635	0.003	275
	Cosmos 156 rocket	1967-39B	1967 Apr 27.53 30 years	Cylinder 1440	3.8 long 2.6 dia	1967 Apr 29.4	81.23	97.20	7003	540	710	0.012	190
	Vela 7	1967-40A	1967 Apr 28.42 >million years	Icosahedron 231	1.17 long 1.42 dia	1967 May 1.4	33.06	6671.8	117353	107337	114612	0.031	87
	Vela 8	1967-40B	1967 Apr 28.42 >million years	Icosahedron 231	1.17 long 1.42 dia	1967 May 1.4	33.06	6671.8	117353	107337	114612	0.031	87
	ERS 18	1967-40C	1967 Apr 28.42 million years?	Octahedron 9	0.28 side	1967 May 1.4	32.8	2829.6	66295	8604	111229	0.774	173
	ERS 20 (OV5-3)	1967-40D	1967 Apr 28.42 million years?	Octahedron 8.6	0.28 side	1967 May 1.4	32.8	2829.6	66295	8604	111229	0.774	173
	ERS 27 (OV5-1)	1967-40E	1967 Apr 28.42 million years?	Octahedron 7.4	0.28 side	1967 May 1.4	32.8	2829.6	66295	8604	111229	0.774	173
	Transtage 10 [Titan 3C]	1967-40F	1967 Apr 28.42 million years?	Cylinder 1500?	6 long? 3 dia	Orbit similar to 1967-40E							

* Crashed after unstable re-entry (one solar panel may not have unfolded).   ** First satellite launch from Indian Ocean platform (Kenya).

	Name		Launch date, lifetime and descent date	Shape and weight (kg)	Size (m)	Date of orbital determination	Orbital inclination (deg)	Nodal period (min)	Semi major axis (km)	Perigee height (km)	Apogee height (km)	Orbital eccentricity	Argument of perigee (deg)
D	Ariel 3	1967-42A	1967 May 5.67 1318.37 days 1970 Dec 14.04	Cylinder + 4 paddles 89.8	0.91 long 0.76 dia	1967 May 5.8 1969 Sep 16.0 1970 May 1.0	80.17 80.18 80.1	95.69 94.33 93.40	6931 6863 6818	497 451 414	608 518 466	0.008 0.005 0.004	170 - -
D	Ariel 3 rocket	1967-42B	1967 May 5.67 1309.98 days 1970 Dec 5.65	Cylinder 24	1.5 long 0.46 dia	1967 May 9.1 1969 Sep 16.0 1970 May 1.0	80.18 80.18 80.1	95.80 94.40 93.42	6935 6866 6819	494 449 412	619 527 470	0.009 0.006 0.004	161 - -
D	Fragments	1967-42C,D											
D	[Thorad Agena D]	1967-43A	1967 May 9.91 64.62 days 1967 Jul 13.53	Cylinder 2000?	8 long? 1.5 dia?	1967 May 10.1	85.10	94.36	6867	200	777	0.042	154
	Capsule	1967-43B	1967 May 9.91 50 years	Octagon? 60?	0.3 long? 0.9 dia?	1967 May 10.0	85.10	98.38	7060	555	809	0.018	173
D	Fragment	1967-43C											
D R	Cosmos 157	1967-44A	1967 May 12.44 7.93 days 1967 May 20.37	Sphere-cylinder 4750?	4.3 long 2.4 dia	1967 May 12.8	51.26	89.60	6634	249	262	0.001	67
D	Cosmos 157 rocket	1967-44B	1967 May 12.44 11.94 days 1967 May 24.38	Cylinder 1440	3.8 long 2.6 dia	1967 May 15.7	51.28	89.40	6625	234	260	0.002	121
D	Fragments	1967-44C-E											
	Cosmos 158	1967-45A	1967 May 15.46 200 years	Cylinder? 750?	1.4 long? 2.0 dia?	1967 May 15.7	74.03	100.40	7158	738	822	0.006	148
	Cosmos 158 rocket	1967-45B	1967 May 15.46 200 years	Cylinder 2200?	7.4 long 2.4 dia	1967 May 16.1	74.01	100.59	7168	732	847	0.008	148
	Fragments	1967-45C,D											

Space Vehicle: Lunar Orbiter 4, 1967-41A; and Agena rocket, 1967-41B was initially in earth orbit.

	Name		Launch date, lifetime and descent date	Shape and weight (kg)	Size (m)	Date of orbital determination	Orbital inclination (deg)	Nodal period (min)	Semi major axis (km)	Perigee height (km)	Apogee height (km)	Orbital eccentricity	Argument of perigee (deg)
D	Cosmos 159*	1967-46A	1967 May 16.91 3832 days? 1977 Nov 11?	Cylinder 4490 full? 1465 empty?	3.0 long? 2.3 dia?	1967 May 31.5 1970 May 16.5 1974 Mar 1.0	51.60 53.38 53.6	1174.2 1174.1 1174.0	36872 36864 36865	350 2502 1124	60637 58470 59849	0.818 0.759 0.796	- - -
D	Cosmos 159 launcher rocket	1967-46B	1967 May 16.91 12.70 days 1967 May 29.61	Cylinder 2500?	7.5 long 2.6 dia	1967 May 17.3	51.84	90.80	6695	203	431	0.017	70
D	Cosmos 159 launcher	1967-46C	1967 May 16.91 17.00 days 1967 Jun 2.91	Irregular	2 long? 1 dia?	1967 May 18.2	51.74	90.50	6680	208	395	0.014	72
D	Cosmos 159 torus	1967-46F	1967 May 16.91 3986 days 1978 Apr 14	Annulus 1875 full? 215 empty?	0.62 long? 1.23 to 2.25 dia?	1967 Jun 13.0 1970 May 16.5 1974 Mar 1.0	50.4 53.37 53.6	1171.0 1171.0 1171.1	36802 36801 36804	380 2490 1129	60467 58356 59722	0.816 0.759 0.796	- - -
D	Fragments	1967-46D,E											
D	Cosmos 160**	1967-47A	1967 May 17.67 0.8 day 1967 May 18.5	Cylinder?	2 long? 1 dia?	1967 May 17.8	49.66	87.58	6535	137	177	0.003	-
D	Cosmos 160 rocket	1967-47B	1967 May 17.67 0.8 day 1967 May 18.5	Cylinder 1500?	8 long? 2.5 dia?	orbit similar to 1967-47A							
	Transit 16 [Scout]	1967-48A	1967 May 18.38 1000 years	Octagon + 4 vanes 60?	0.25 long 0.46 dia	1967 May 19.8	89.57	107.04	7467	1074	1105	0.002	187
	Altair rocket	1967-48B	1967 May 18.38 1000 years	Cylinder 24	1.5 long 0.46 dia	1967 Jun 2.7	89.58	107.04	7466	1073	1106	0.002	144
D R	Cosmos 161	1967-49A	1967 May 22.58 7.79 days 1967 May 30.37	Sphere-cylinder 5530?	5 long? 2.4 dia	1967 May 24.8	65.64	89.71	6639	201	321	0.009	51
D	Cosmos 161 rocket	1967-49B	1967 May 22.58 6.07 days 1967 May 28.65	Cylinder 2500?	7.5 long 2.6 dia	1967 May 24.3	65.64	89.45	6626	195	301	0.008	42

* Probably a test of the Soyuz propulsion module.

** Disintegrated into more than 16 pieces on first revolution

	Name		Launch date, lifetime and descent date	Shape and weight (kg)	Size (m)	Date of orbital determination	Orbital inclination (deg)	Nodal period (min)	Semi major axis (km)	Perigee height (km)	Apogee height (km)	Orbital eccentricity	Argument of perigee (deg)
D	[Atlas Agena D] *	1967-50A	1967 May 22.77 8.18 days 1967 May 30.95	Cylinder 700?	6.3 long 1.5 dia	1967 May 28.4	91.49	88.82	6592	135	293	0.012	117
D	Capsule [Atlas Agena D]	1967-50B	1967 May 22.77 4.9 days 1967 May 27.7	- 500?	1.5 dia?	1967 May 27.0	91.49	88.42	6572	148	240	0.007	125
D	Explorer 34 (Imp 5)	1967-51A	1967 May 24.59 710.47 days 1969 May 4.06	Octagon + 4 vanes 75	0.25 long 0.71 dia	1967 May 24.6 1968 Jul 26.5 1969 Feb 15.5	67.17 66.97 68.50	6346.1 6222.8 6218.3	113691 112071 112015	242 4006 2031	214383 207380 209242	0.942 0.907 0.925	180 - -
D	Explorer 34 rocket	1967-51B	1967 May 24.59 23 months? 1969 May ?	Cylinder 24	1.5 long 0.46 dia	orbit similar to 1967-51A							
D	Molniya 1E	1967-52A	1967 May 24.95 1647 days 1971 Nov 26	Windmill 1000?	3.4 long 1.6 dia	1967 May 29.9 1969 Mar 31.7 1970 Oct 31.7	64.88 64.87 64.8	715.5 710.4 711.77	26500 26376 26408	460 1188 567	39785 38807 39492	0.742 0.713 0.737	285 - -
D	Molniya 1E launcher rocket	1967-52B	1967 May 24.95 11 days 1967 Jun 5	Cylinder 2500?	7.5 long 2.6 dia	1967 May 25.4	64.87	90.88	6696	197	438	0.018	55
D	Molniya 1E launcher	1967-52C	1967 May 24.95 21.98 days 1967 Jun 15.93	Irregular	2 long? 1 dia?	1967 May 25.8	64.85	91.26	6716	203	472	0.020	62
D	Molniya 1E rocket	1967-52F	1967 May 24.95 1667 days 1971 Dec 16	Cylinder 440	2.0 long 2.0 dia	1969 May 16.3 1970 Dec 1.0	64.87 64.8	709.62 709.54	26354 26352	1194 554	38758 39394	0.713 0.737	- -
D	Fragments	1967-52D,E,G											

* Carried Logacs - low gravity accelerometer calibration system.

	Name		Launch date, lifetime and descent date	Shape and weight (kg)	Size (m)	Date of orbital determination	Orbital inclina-tion (deg)	Nodal period (min)	Semi major axis (km)	Perigee height (km)	Apogee height (km)	Orbital eccen-tricity	Argument of perigee (deg)
	Agena D rocket	1967-53A	1967 May 31.40 500 years	Cylinder 700?	6 long? 1.5 dia	1967 Jun 1.5	69.98	103.45	7299	914	928	0.001	343
	Surcal	1967-53B	1967 May 31.40 500 years	Sphere 2.48	0.51 dia	1967 Jun 1.1	69.98	103.55	7305	919	934	0.001	33
	GGSE 4	1967-53C	1967 May 31.40 500 years	Ellipsoid? 4?	0.3 dia?	1967 Jun 1.1	69.98	103.45	7300	915	929	0.001	13
	GGSE 5	1967-53D	1967 May 31.40 500 years	Ellipsoid? 4?	0.3 dia?	1967 Jun 5.7	69.98	103.45	7300	915	929	0.001	20
	[Thor Agena D]	1967-53E	1967 May 31.40 500 years	-	-	1967 Jun 15.5	69.91	103.38	7297	916	921	0.001	-
	Timation 1*	1967-53F	1967 May 31.40 500 years	Rectangular box 38	0.81 × 0.41 × 0.20	1968 Jan 15.5	69.91	103.39	7298	915	926	0.001	-
	[Thor Agena D]	1967-53G	1967 May 31.40 500 years	-	-	1968 Jan 15.5	69.91	103.40	7299	915	927	0.001	-
	[Thor Agena D]	1967-53H	1967 May 31.40 500 years	-	-	1968 Jan 15.5	69.91	103.39	7298	915	926	0.001	-
	Surcal (150B)	1967-53J	1967 May 31.40 50 years	Sphere 1.55	0.41 dia	1968 Jan 15.5	69.91	103.38	7297	917	922	0.0003	-
D R	Cosmos 162	1967-54A	1967 Jun 1.45 7.99 days 1967 Jun 9.44	Sphere-cylinder 5530?	5 long? 2.4 dia	1967 Jun 1.7	51.81	89.19	6614	196	275	0.006	19
D	Cosmos 162 rocket	1967-54B	1967 Jun 1.45 3.06 days 1967 Jun 4.51	Cylinder 2500?	7.5 long 2.6 dia	1967 Jun 3.8	51.78	88.15	6563	159	211	0.004	9
D	Fragment	1967-54C											

*Time/navigation satellite

	Name		Launch date, lifetime and descent date	Shape and weight (kg)	Size (m)	Date of orbital determination	Orbital inclination (deg)	Nodal period (min)	Semi major axis (km)	Perigee height (km)	Apogee height (km)	Orbital eccentricity	Argument of perigee (deg)
D	[Atlas Agena D]	1967-55A	1967 Jun 4.75 8.17 days 1967 Jun 12.92	- 500?	1.5 dia?	1967 Jun 6.9 1967 Jun 10.4	104.88 104.87	90.57 90.33	6681 6668	149 145	456 434	0.023 0.022	153 139
D	Agena D rocket	1967-55B	1967 Jun 4.75 1 day 1967 Jun 6	Cylinder 700?	8 long? 1.5 dia	1967 Jun 5.4	104.87	90.28	6668	143	436	0.022	150
D	Cosmos 163	1967-56A	1967 Jun 5.21 128.51 days 1967 Oct 11.72	Ellipsoid 400?	1.8 long 1.2 dia	1967 Jun 5.4 1967 Aug 15.8	48.38 48.39	93.07 92.01	6806 6754	244 247	611 504	0.027 0.019	124 -
D	Cosmos 163 rocket	1967-56B	1967 Jun 5.21 99.20 days 1967 Sep 12.41	Cylinder 1500?	8 long 1.65 dia	1967 Jun 5.9	48.43	93.05	6802	254	594	0.025	111
D R	Cosmos 164	1967-57A	1967 Jun 8.55 5.76 days 1967 Jun 14.31	Sphere-cylinder 5530?	5 long? 2.4 dia	1967 Jun 9.2	65.59	89.51	6629	185	317	0.010	44
D	Cosmos 164 rocket	1967-57B	1967 Jun 8.55 9.11 days 1967 Jun 17.66	Cylinder 2500?	7.5 long 2.6 dia	1967 Jun 9.2	65.65	89.37	6622	197	290	0.007	32
D	Fragment	1967-57C											
D	Venus 4 launcher	1967-58B	1967 Jun 12.11 1.76 days 1967 Jun 13.87	Irregular	2 long? 1 dia?	1967 Jun 12.4	51.80	87.92	6553	162	188	0.002	16
D	Venus 4 launcher rocket	1967-58C	1967 Jun 12.11 0.78 day 1967 Jun 12.89	Cylinder 2500?	7.5 long 2.6 dia	1967 Jun 12.4	51.78	88.30	6571	173	212	0.003	271
D	Fragment	1967-58D											

Space Vehicle: Venus 4, 1967-58A

	Name		Launch date, lifetime and descent date	Shape and weight (kg)	Size (m)	Date of orbital determination	Orbital inclination (deg)	Nodal period (min)	Semi major axis (km)	Perigee height (km)	Apogee height (km)	Orbital eccentricity	Argument of perigee (deg)
D	Cosmos 165	1967-59A	1967 Jun 12.76 216.90 days 1968 Jan 15.66	Ellipsoid 400?	1.8 long 1.2 dia	1967 Jun 14.0 1967 Nov 8.0	81.89 81.88	102.08 96.13	7235 6951	198 190	1515 957	0.091 0.055	71 -
D	Cosmos 165 rocket	1967-59B	1967 Jun 12.76 142.16 days 1967 Nov 1.92	Cylinder 1500?	8 long 1.65 dia	1967 Jun 15.7 1967 Sep 15.5	81.91 81.88	101.88 96.73	7226 6980	197 197	1498 1007	0.090 0.058	66 -
D	Cosmos 166	1967-61A	1967 Jun 16.20 130.89 days 1967 Oct 25.08	Ellipsoid + 8 panels 400?	1.8 long 1.2 dia	1967 Jun 24.6 1967 Sep 8.0	48.43 48.42	92.84 91.78	6795 6741	281 270	553 456	0.020 0.014	159 -
D	Cosmos 166 rocket	1967-61B	1967 Jun 16.20 116.82 days 1967 Oct 11.01	Cylinder 1500?	8 long 1.65 dia	1967 Jun 17.5 1967 Sep 8.0	48.43 48.42	92.83 91.46	6794 6726	280 266	552 430	0.020 0.012	128 -
D	[Thorad Agena D]	1967-62A	1967 Jun 16.90 33.16 days 1967 Jul 20.06	Cylinder 2000?	8 long? 1.5 dia	1967 Jun 17.7	80.02	89.97	6652	181	367	0.014	166
D	Capsule**	1967-62B	1967 Jun 16.90 493.12 days 1968 Oct 22.02	Octagon? 60?	0.3 long? 0.9 dia?	1967 Jun 30.8 1968 Mar 14.5 1968 Jun 30.5	80.20 80.19 80.13	94.81 93.76 92.89	6887 6835 6793	501 451 407	517 464 423	0.001 0.001 0.001	278 - -
D	Cosmos 167*	1967-63A	1967 Jun 17.11 7.99 days 1967 Jun 25.10	Cylinder 6560? full	7 long? 2.0 max dia	1967 Jun 17.5	51.79	89.22	6616	211	264	0.004	38
D	Cosmos 167 launcher	1967-63B	1967 Jun 17.11 4.88 days 1967 Jun 21.99	Irregular	2 long? 1 dia?	1967 Jun 18.5	51.80	89.00	6605	187	266	0.006	26
D	Cosmos 167 launcher rocket	1967-63C	1967 Jun 17.11 9.32 days 1967 Jun 26.43	Cylinder 2500?	7.5 long 2.6 dia	1967 Jun 17.7	51.87	88.10	6560	149	215	0.005	87
D	Fragments	1967-63D-F											

Space Vehicle: Mariner 5, 1967-60A; also Agena rocket, 1967-60B. *Possibly an attempted Venus Probe, (payload about 1106 kg)
**1967-62B ejected from 1967-62A on 1967 Jun 16.97.

	Name		Launch date, lifetime and descent date	Shape and weight (kg)	Size (m)	Date of orbital determination	Orbital inclination (deg)	Nodal period (min)	Semi major axis (km)	Perigee height (km)	Apogee height (km)	Orbital eccentricity	Argument of perigee (deg)
D	[Titan 3B Agena D]	1967-64A	1967 Jun 20.68 10.22 days 1967 Jun 30.90	Cylinder 3000?	8 long? 1.5 dia	1967 Jun 22.6	111.40	89.01	6604	127	325	0.015	107
	Secor 9 (EGRS 9)	1967-65A	1967 Jun 29.88 100000 years	Rectangular box 20	0.33x0.28 x 0.23	1967 Jun 30.9	89.91	172.22	10253	3803	3947	0.007	343
	Aurora 1 [Thor Burner 2]	1967-65B	1967 Jun 29.88 100000 years	Rectangular box 21.5	0.61x0.36 x 0.24	1967 Jul 19.7	89.82	172.18	10251	3800	3947	0.007	319
	Burner 2 rocket	1967-65C	1967 Jun 29.88 100000 years	Sphere-cone? 66?	1.32 long? 0.94 dia?	1967 Jul 24.5	89.83	172.18	10251	3801	3945	0.007	305
	IDCSP 3-1	1967-66A	1967 Jul 1.55 >million years	Polyhedron (26 faces) 45	0.8 long 0.9 dia	1967 Jul 24.5	7.18	1308.9	39635	32986	33528	0.007	-
	IDCSP 3-2	1967-66B	1967 Jul 1.55 >million years	Polyhedron (26 faces) 45	0.8 long 0.9 dia	1967 Jul 24.5	7.22	1309.8	39655	33006	33548	0.007	-
	IDCSP 3-3	1967-66C	1967 Jul 1.55 >million years	Polyhedron (26 faces) 45	0.8 long 0.9 dia	1967 Jul 6.2	7.2	1311.6	39695	33079	33555	0.006	306
	IDCSP 3-4 (DATS)**	1967-66D	1967 Jul 1.55 >million years	Polyhedron (26 faces) 68	0.8 long 0.9 dia	1967 Jul 6.2	7.1	1313.6	39733	33156	33553	0.005	297
	LES 5	1967-66E	1967 Jul 1.55 >million years	Cylinder 102	1.67 long 1.37 dia	1968 Jan 15.5	6.8	1316.2	39785	33178	33636	0.006	-
	DODGE 1 †	1967-66F	1967 Jul 1.55 > million years	Octagonal door-knob 195	2.41 long 1.22 dia	1968 Jan 15.5	6.2	1318.9	39843	33270	33659	0.005	-
	Transtage 11 [Titan 3C]	1967-66G	1967 Jul 1.55 > million years	Cylinder 1500?	6 long? 3 dia	1967 Sep 15.5	7.0	1320.8	39884	33327	33685	0.004	- *

* Approximate orbit

** De-spun Antenna Test Satellite.

† Department of Defense Gravity Experiment.

	Name		Launch date, lifetime and descent date	Shape and weight (kg)	Size (m)	Date of orbital determination	Orbital inclination (deg)	Nodal period (min)	Semi major axis (km)	Perigee height (km)	Apogee height (km)	Orbital eccentricity	Argument of perigee (deg)
D R	Cosmos 168	1967-67A	1967 Jul 4.25 7.98 days 1967 Jul 12.23	Sphere-cylinder 5530?	5 long? 2.4 dia	1967 Jul 4.9	51.81	89.05	6609	198	264	0.005	15
D	Cosmos 168 rocket	1967-67B	1967 Jul 4.25 3.98 days 1967 Jul 8.23	Cylinder 2500?	7.5 long 2.6 dia	1967 Jul 5.5	51.81	88.80	6592	187	240	0.004	341
D	Fragment	1967-67C											
D R?	Cosmos 169	1967-69A	1967 Jul 17.70 0.06 day 1967 Jul 17.76	Cylinder?	2 long? 1 dia?	1967 Jul 17.7	49.68	87.78	6546	135	200	0.005	69
D	Cosmos 169 rocket	1967-69C	1967 Jul 17.70 0.31 day 1967 Jul 18.01	Cylinder 1500?	8 long? 2.5 dia?	1967 Jul 17.9	49.60	86.51	6481	102?	103?	0	-
D	Cosmos 169 launch platform	1967-69B	1967 Jul 17.70 0.5 day? 1967 Jul 18	Irregular	-	1967 Jul 17.7	49.60	87.02	6507	128?	129?	0	-
D	Fragments	1967-69D,E											
D	Explorer 35 second stage	1967-70B	1967 Jul 19.60 42.92 days 1967 Aug 31.52	Cylinder 350?	4.9 long 1.43 dia	1967 Jul 20.3 1967 Aug 14.5	29.56 29.55	103.30 97.28	7298 7011	155 149	1686 1116	0.105 0.069	77 -
D	[Thor Agena D]	1967-71A	1967 Jul 25.16 681.00 days 1969 Jun 5.16	Cylinder 1500?	8 long? 1.5 dia	1967 Jul 30.4 1968 Feb 29.8 1968 Oct 15.8	75.03 75.03 75.03	94.30 93.83 93.13	6864 6841 6806	458 437 410	513 488 446	0.004 0.004 0.003	42 - -
D	Fragments	1967-71B-D											

Space Vehicles: Surveyor 4, 1967-68A; also Centaur 13, 1967-68B in an highly eccentric orbit. Explorer 35 (Imp 6), 1967-70A is orbiting the moon; also Explorer 35 third stage rocket is in an heliocentric orbit.

	Name		Launch date, lifetime and descent date	Shape and weight (kg)	Size (m)	Date of orbital determination	Orbital inclination (deg)	Nodal period (min)	Semi major axis (km)	Perigee height (km)	Apogee height (km)	Orbital eccentricity	Argument of perigee (deg)
D	OV1-86	1967-72A	1967 Jul 27.79 1670.91 days 1972 Feb 22.70	Cylinder + hemisphere 118	1.40 long 0.69 dia	1967 Aug 3.5 1969 Sep 16.0 1970 Sep 16.0	101.72 101.72 101.72	95.48 94.51 93.58	6920 6871 6826	480 446 415	604 540 481	0.009 0.007 0.005	51 - -
D	OV1-86 rocket	1967-72B	1967 Jul 27.79 1207 days 1970 Nov 15	Cylinder 70?	2.05 long 0.72 dia	1967 Aug 2.3 1969 Jan 31.7 1969 Dec 31.7	101.72 101.72 101.72	95.47 94.43 93.55	6919 6867 6825	479 426 400	603 552 493	0.009 0.009 0.007	53 - -
D	OV1-12 rocket	1967-72C	1967 Jul 27.79 3567 days 1977 May 2	Cylinder 70?	2.05 long 0.72 dia	1967 Aug 2.2 1969 Sep 16.0 1973 Feb 15.0	101.61 101.61 101.59	95.59 95.01 93.61	6925 6896 6828	540 510 445	554 526 454	0.001 0.001 0.001	287 - -
D	OV1-12	1967-72D	1967 Jul 27.79 4744 days 1980 Jul 22	Cylinder + hemisphere 141	1.40 long 0.69 dia	1967 Aug 2.2 1970 Oct 31.7	101.62 101.62	95.62 95.04	6927 6898	542 513	556 526	0.001 0.001	282 -
D	Fragment	1967-72E											
D	OGO 4	1967-73A	1967 Jul 28.60 1845.84 days 1972 Aug 16.44	Box + booms 552	1.73 long 0.84 high 0.84 wide	1967 Jul 29.2 1969 Jul 31.7 1971 Aug 1.0	86.03 85.93 85.93	97.89 96.53 93.97	7035 6967 6846	411 403 370	903 775 565	0.035 0.027 0.014	153 - -
D	OGO 4 rocket	1967-73B	1967 Jul 28.60 2483.18 days 1974 May 15.78	Cylinder 700?	8 long 1.5 dia	1967 Aug 1.3 1969 Jul 31.7 1972 Aug 1.0	86.01 85.97 85.99	97.82 96.58 93.97	7032 6970 6846	415 402 367	893 782 568	0.034 0.027 0.015	146 - -
D R?	Cosmos 170	1967-74A	1967 Jul 31.70 0.06 day 1967 Jul 31.76	Cylinder?	2 long? 1 dia?	1967 Jul 31.7	49.46	88.19	6565	121	252	0.010	35
D	Cosmos 170 rocket	1967-74B	1967 Jul 31.70 0.30 day 1967 Aug 1.00	Cylinder? 1500?	8 long? 2.5 dia?	1967 Jul 31.7	49.40	88.33	6573	126	263	0.010	-
D	Cosmos 170 launch platform	1967-74C	1967 Jul 31.70 0.5 day? 1967 Aug 1	Irregular	-	1967 Jul 31.7	49.40	87.59	6535	123	191	0.005	-
D	Fragments	1967-74D,E											

Space Vehicle: Lunar Orbiter 5, 1967-75A; and Agena rocket, 1967-75B in a highly eccentric orbit.

	Name		Launch date, lifetime and descent date	Shape and weight (kg)	Size (m)	Date of orbital determination	Orbital inclination (deg)	Nodal period (min)	Semi major axis (km)	Perigee height (km)	Apogee height (km)	Orbital eccentricity	Argument of perigee (deg)
D	[Thorad Agena D]	1967-76A	1967 Aug 7.90 24.85 days 1967 Sep 1.75	Cylinder 2000?	8 long? 1.5 dia	1967 Aug 8.3	79.94	89.72	6638	174	346	0.013	180
D	Fragment	1967-76B											
D R?	Cosmos 171	1967-77A	1967 Aug 8.67 0.06 day 1967 Aug 8.73	Cylinder?	2 long? 1 dia?	1967 Aug 8.7	49.60	87.58	6536	138	177	0.003	-
D	Cosmos 171 launch platform	1967-77B	1967 Aug 8.67 0.60 day 1967 Aug 9.27	Irregular	-	1967 Aug 9.0	49.62	87.69	6541	130	200	0.005	59
D	Cosmos 171 rocket	1967-77D	1967 Aug 8.67 0.29 day 1967 Aug 8.96	Cylinder 1500?	8 long? 2.5 dia?	1967 Aug 8.7	49.61	87.41	6528	134?	165?	0.002?	-
D	Fragments	1967-77C,E											
D R	Cosmos 172	1967-78A	1967 Aug 9.24 7.94 days 1967 Aug 17.18	Sphere-cylinder 5530?	5 long? 2.4 dia	1967 Aug 9.6	51.80	89.40	6625	200	293	0.007	33
D	Cosmos 172 rocket	1967-78B	1967 Aug 9.24 4.25 days 1967 Aug 13.49	Cylinder 2500?	7.5 long 2.6 dia	1967 Aug 9.9	51.80	89.11	6612	188	280	0.007	15
D	[Titan 3B Agena D]	1967-79A	1967 Aug 16.71 13 days 1967 Aug 29	Cylinder 3000?	8 long? 1.5 dia	1967 Aug 17.3	111.88	90.43	6674	142	449	0.023	128

	Name		Launch date, lifetime and descent date	Shape and weight (kg)	Size (m)	Date of orbital determination	Orbital inclination (deg)	Nodal period (min)	Semi major axis (km)	Perigee height (km)	Apogee height (km)	Orbital eccentricity	Argument of perigee (deg)
	[Thor Burner 2]	1967-80A	1967 Aug 23.20 100 years	12-sided frustum 195	1.64 long 1.31 to 1.10 dia	1967 Aug 23.6	98.97	102.20	7241	834	892	0.004	101
	Burner 2 rocket	1967-80B	1967 Aug 23.20 80 years	Sphere-cone 66	1.32 long 0.94 dia	1967 Oct 9.2	98.95	102.20	7241	831	895	0.004	333
D	Cosmos 173	1967-81A	1967 Aug 24.21 115.53 days 1967 Dec 17.74	Ellipsoid 400?	1.8 long 1.2 dia	1967 Aug 24.8 1967 Nov 8.0	71.03 71.02	92.10 90.85	6757 6694	277 247	480 386	0.015 0.010	81 -
D	Cosmos 173 rocket	1967-81B	1967 Aug 24.21 67.33 days 1967 Oct 30.54	Cylinder 1500?	8 long 1.65 dia	1967 Aug 26.8 1967 Oct 8.0	71.02 71.02	92.06 90.70	6754 6687	281 262	470 357	0.014 0.007	78 -
D	Cosmos 174*	1967-82A	1967 Aug 31.33 487 days 1968 Dec 30	Windmill 1000?	3.4 long 1.6 dia	1967 Sep 6.4 1968 Mar 15.8 1968 Aug 15.8	64.85 64.87 65.0	715.0 714.5 711.4	26491 26474 26399	430 347 189	39796 39845 39853	0.743 0.746 0.751	285 - -
D	Cosmos 174 launcher rocket	1967-82B	1967 Aug 31.33 15.19 days 1967 Sep 15.52	Cylinder 2500?	7.5 long 2.6 dia	1967 Sep 1.5	64.84	91.06	6705	199	454	0.019	59
D	Cosmos 174 launcher	1967-82C	1967 Aug 31.33 18.87 days 1967 Sep 19.20	Irregular	-	1967 Sep 1.2	64.79	91.33	6721	202	484	0.012	62
D	Cosmos 174 rocket	1967-82E	1967 Aug 31.33 492 days 1969 Jan 4	Cylinder 440	2.0 long 2.0 dia	1967 Dec 27	64.86	710.9	26386	381	39635	0.744	-
D	Fragment	1967-82D											
D	Bios 2** adapter	1967-83A	1967 Sep 7.92 26.18 days 1967 Oct 4.10	Cylinder 360?	1.8 long 1.45 dia	1967 Sep 16.6	33.46	90.38	6678	286	313	0.002	55
D	Bios 2 rocket	1967-83B	1967 Sep 7.92 133.84 days 1968 Jan 19.76	Cylinder 400?	4.9 long 1.43 dia	1967 Sep 8.8 1967 Dec 8.0	32.94 32.94	92.08 90.82	6761 6700	321 285	445 358	0.009 0.005	211 -

* This is probably a Molniya satellite, later replaced by Molniya 1F (1967-95A).

** Before 1967 Sep 9.8, Bios 2 adapter was attached to Bios 2 capsule, 1967-83C (see next page).

Continued on page 146

	Name		Launch date, lifetime and descent date	Shape and weight (kg)	Size (m)	Date of orbital determination	Orbital inclination (deg)	Nodal period (min)	Semi major axis (km)	Perigee height (km)	Apogee height (km)	Orbital eccentricity	Argument of perigee (deg)
D R B	Bios 2 capsule	1967-83C	1967 Sep 7.92 1.87 days 1967 Sep 9.79	Blunt cone 150?	1.2 long 1.02 dia	1967 Sep 8.4	33.48	90.53	6685	296	318	0.002	20
D R	Cosmos 175	1967-85A	1967 Sep 11.44 7.82 days 1967 Sep 19.26	Sphere-cylinder 5530?	5 long? 2.4 dia	1967 Sep 11.8	72.93	90.20	6663	211	358	0.011	56
D	Cosmos 175 rocket	1967-85B	1967 Sep 11.44 12.13 days 1967 Sep 23.57	Cylinder 2500?	7.5 long 2.6 dia	1967 Sep 12.4	72.88	90.05	6656	205	351	0.011	49
D	Fragment	1967-85C											
D	Cosmos 176	1967-86A	1967 Sep 12.71 172.91 days 1968 Mar 3.62	Ellipsoid 400?	1.8 long 1.2 dia	1967 Sep 13.0 1968 Jan 8.0	81.89 81.8	102.19 96.08	7239 6949	196 183	1525 958	0.092 0.056	74 -
D	Cosmos 176 rocket	1967-86B	1967 Sep 12.71 91.13 days 1967 Dec 12.84	Cylinder* 1500?	8 long* 1.65 dia	1967 Sep 23.0 1967 Nov 8.0	81.83 81.8	101.85 96.71	7222 6979	204 196	1484 1006	0.089 0.058	- -
D	Fragments	1967-86C-M											
D	[Thorad Agena D]	1967-87A	1967 Sep 15.82 18.69 days 1967 Oct 4.51	Cylinder 2000?	8 long? 1.5 dia	1967 Sep 16.4	80.07	89.95	6648	150	389	0.018	174
D R	Cosmos 177	1967-88A	1967 Sep 16.25 7.99 days 1967 Sep 24.24	Sphere-cylinder 5530?	5 long? 2.4 dia	1967 Sep 16.4	51.84	89.29	6618	200	280	0.006	34
D	Cosmos 177 rocket	1967-88B	1967 Sep 16.25 3.42 days 1967 Sep 19.67	Cylinder 2500?	7.5 long 2.6 dia	1967 Sep 16.7	51.81	89.04	6607	196	262	0.005	14

Space Vehicle: Surveyor 5, 1967-84A; also Centaur 14, 1967-84B is in a highly eccentric orbit.

* Before explosion.

	Name		Launch date, lifetime and descent date	Shape and weight (kg)	Size (m)	Date of orbital determination	Orbital inclination (deg)	Nodal period (min)	Semi major axis (km)	Perigee height (km)	Apogee height (km)	Orbital eccentricity	Argument of perigee (deg)
D R?	Cosmos 178	1967-89A	1967 Sep 19.62 0.06 day 1967 Sep 19.68	Cylinder	2 long? 1 dia?	1967 Sep 19.6	49.65	88.39	6576	138	258	0.009	-
D	Cosmos 178 launch platform	1967-89B	1967 Sep 19.62 0.45 day 1967 Sep 20.07	Irregular	-	1967 Sep 19.8	49.66	87.85	6548	130	209	0.006	60
D	Cosmos 178 rocket	1967-89C	1967 Sep 19.62 0.2 day 1967 Sep 19.8	Cylinder 1500?	8 long? 2.5 dia?	1967 Sep 19.7	49.60	87.41	6528	137	163	0.002	-
D	Fragment	1967-89D											
D	[Titan 3B Agena D]	1967-90A	1967 Sep 19.77 10.23 days 1967 Sep 30.00	Cylinder 3000?	8 long? 1.5 dia?	1967 Sep 21.4	106.10	89.75	6640	122	401	0.021	119
D R?	Cosmos 179	1967-91A	1967 Sep 22.59 0.06 day 1967 Sep 22.65	Cylinder	2 long? 1 dia?	1967 Sep 22.6	49.57	87.87	6551	139	207	0.005	-
D	Cosmos 179 launch platform	1967-91B	1967 Sep 22.59 0.43 day 1967 Sep 23.02	Irregular	-	1967 Sep 22.8	49.52	87.74	6544	120	212	0.007	32
D	Cosmos 179 rocket	1967-91C	1967 Sep 22.59 0.24 day 1967 Sep 22.83	Cylinder 1500?	8 long? 2.5 dia?	1967 Sep 22.7	49.40	87.39	6527	141	156	0.001	-
	Transit 17 [Scout]	1967-92A	1967 Sep 25.35 1000 years	Octagon + 4 vanes 60?	0.25 long 0.46 dia	1967 Sep 26.4	89.28	106.81	7457	1041	1116	0.005	189
	Altair rocket	1967-92B	1967 Sep 25.35 1000 years	Cylinder 24	1.5 long 0.46 dia	1967 Sep 26.4	89.29	106.80	7456	1040	1115	0.005	187
	Fragments	1967-92C-F											

	Name		Launch date, lifetime and descent date	Shape and weight (kg)	Size (m)	Date of orbital determination	Orbital inclination (deg)	Nodal period (min)	Semi major axis (km)	Perigee height (km)	Apogee height (km)	Orbital eccentricity	Argument of perigee (deg)
D R	Cosmos 180	1967-93A	1967 Sep 26.43 7.82 days 1967 Oct 4.25	Sphere-cylinder 5530?	5 long? 2.4 dia	1967 Sep 27.2	72.89	90.04	6653	208	341	0.010	56
D	Cosmos 180 rocket	1967-93B	1967 Sep 26.43 9.41 days 1967 Oct 5.84	Cylinder 2500?	7.5 long 2.6 dia	1967 Sep 26.8	72.89	90.00	6650	205	338	0.010	52
	Intelsat 2 F-4	1967-94A	1967 Sep 28.03 > million years	Cylinder 87	0.67 long 1.42 dia	1967 Sep 28.1 1967 Oct 15.5 1977 May 13.0	26.43 0.93 7.41	655.0 1438.3 1437.51	24980 42208 42192	305 35747 35805	36897 35913 35823	0.732 0.002 0.0002	179 - 63
D	Intelsat 2 F-4 second stage	1967-94B	1967 Sep 28.03 89.01 days 1967 Dec 26.04	Cylinder 400?	4.9 long 1.43 dia	1967 Oct 2.6 1967 Nov 23.0	28.74 28.74	92.65 91.33	6790 6724	242 231	582 461	0.025 0.017	182 -
	Intelsat 2 F-4 third stage	1967-94C	1967 Sep 28.03 30 years?	Cylinder 24	1.5 long 0.46 dia	1967 Sep 28.3	26.44	625.4	24235	307	35406	0.724	179
D	Fragment	1967-94D											
D	Molniya 1F	1967-95A	1967 Oct 3.21 518.76 days 1969 Mar 4.97	Windmill 1000?	3.4 long 1.6 dia	1967 Oct 9.7 1968 Mar 15.8 1968 Oct 23.5	64.96 64.93 65.2	718.03 716.23 707.35	26563 26519 26299	502 338 173	39868 39943 39668	0.741 0.747 0.751	285 - -
D	Molniya 1F launcher	1967-95B	1967 Oct 3.21 18.25 days 1967 Oct 21.46	Irregular	-	1967 Oct 9.0	64.79	90.95	6699	200	441	0.018	65
D	Molniya 1F launcher rocket	1967-95C	1967 Oct 3.21 14.16 days 1967 Oct 17.37	Cylinder 2500?	7.5 long 2.6 dia	1967 Oct 11.0	64.81	90.27	6667	195	382	0.014	64
D	Molniya 1F rocket	1967-95D	1967 Oct 3.21 496.70 days 1969 Feb 10.91	Cylinder 440	2.0 long 2.0 dia	1967 Dec 27 1968 Aug 15.8 1968 Nov 8.0	64.94 65.00 65.2	709.7 708.5 699.5	26356 26343 26104	420 227 151	39536 39703 39301	0.742 0.749 0.750	- - -

	Name		Launch date, lifetime and descent date	Shape and weight (kg)	Size (m)	Date of orbital determination	Orbital inclination (deg)	Nodal period (min)	Semi major axis (km)	Perigee height (km)	Apogee height (km)	Orbital eccentricity	Argument of perigee (deg)
	[Thor Burner 2]	1967-96A	1967 Oct 11.33 80 years	12-sided frustum 195	1.64 long 1.31 to 1.10 dia	1967 Oct 15.8	99.16	100.18	7144	667	866	0.014	355
	Burner 2 rocket	1967-96B	1967 Oct 11.33 60 years	Sphere-cone 66	1.32 long 0.94 dia	1967 Oct 13.0	99.12	100.18	7144	673	859	0.013	3
D R	Cosmos 181	1967-97A	1967 Oct 11.48 7.78 days 1967 Oct 19.26	Sphere-cylinder 5530?	5 long? 2.4 dia	1967 Oct 12.0	65.61	89.72	6639	194	327	0.010	47
D	Cosmos 181 rocket	1967-97B	1967 Oct 11.48 7.53 days 1967 Oct 19.01	Cylinder 2500?	7.5 long 2.6 dia	1967 Oct 14.1	65.62	89.15	6612	188	280	0.007	32
D R	Cosmos 182	1967-98A	1967 Oct 16.33 7.93 days 1967 Oct 24.26	Sphere-cylinder 5530?	5 long? 2.4 dia	1967 Oct 16.7	64.99	89.90	6648	210	330	0.009	54
D	Cosmos 182 rocket	1967-98B	1967 Oct 16.33 7.96 days 1967 Oct 24.29	Cylinder 2500?	7.5 long 2.6 dia	1967 Oct 17.1	64.98	89.75	6642	204	324	0.009	45
D	Fragment	1967-98C											
D R?	Cosmos 183	1967-99A	1967 Oct 18.56 0.07 day 1967 Oct 18.63	Cylinder	2 long? 1 dia?	1967 Oct 18.6	49.63	88.90	6601	130	315	0.014	63
D	Cosmos 183 launch platform	1967-99B	1967 Oct 18.56 0.48 day 1967 Oct 19.04	Irregular	-	1967 Oct 18.9	49.66	87.50	6531	127	179	0.004	41
D	Cosmos 183 rocket	1967-99C	1967 Oct 18.56 0.26 day 1967 Oct 18.82	Cylinder 1500?	8 long? 2.5 dia?	1967 Oct 18.7	49.3	88.05	6559	151	211	0.005	-

	Name		Launch date, lifetime and descent date	Shape and weight (kg)	Size (m)	Date of orbital determination	Orbital inclination (deg)	Nodal period (min)	Semi major axis (km)	Perigee height (km)	Apogee height (km)	Orbital eccentricity	Argument of perigee (deg)
D	OSO 4	1967-100A	1967 Oct 18.67 5354 days 1982 Jun 15	Nonagonal box 272	0.94 long 1.12 dia	1967 Oct 19.5	33.04	95.58	6931	546	560	0.001	140
D	OSO 4 rocket	1967-100B	1967 Oct 18.67 4580 days 1980 May 2	Cylinder 24	1.5 long 0.46 dia	1967 Oct 20.3 1972 Mar 11.3	32.99 32.9	95.58 94.78	6931 6892	539 501	567 526	0.002 0.002	261 -
D	Molniya 1G	1967-101A	1967 Oct 22.36 801 days 1969 Dec 31	Windmill 1000?	3.4 long 1.6 dia	1967 Oct 22.9 1969 Jan 15.8 1969 Sep 8.5	65.00 64.7 64.8	715.0 733.45 732.35	26487 26942 26915	508 471 197	39710 40657 40877	0.740 0.746 0.756	285 - -
D	Molniya 1G launcher	1967-101B	1967 Oct 22.36 15.78 days 1967 Nov 7.14	Irregular	-	1967 Oct 24.7	64.80	90.99	6701	202	443	0.018	63
D	Molniya 1G launcher rocket	1967-101C	1967 Oct 22.36 11.87 days 1967 Nov 3.23	Cylinder 2500?	7.5 long 2.6 dia	1967 Oct 22.7	64.82	90.95	6698	199	440	0.018	58
D	Molniya 1G rocket	1967-101F	1967 Oct 22.36 817 days 1970 Jan 16	Cylinder 440	2.0 long 2.0 dia	1967 Dec 27 1968 Aug 23.5 1969 May 8.5	64.9 64.70 64.73	711.7 711.6 711.6	26406 26405 26405	517 584 414	39539 39470 39640	0.739 0.736 0.743	- - -
D	Fragments	1967-101D,E											
	Cosmos 184	1967-102A	1967 Oct 24.96 30 years	Cylinder + 2 vanes 2000?	5 long? 1.5 dia?	1967 Oct 25.1	81.19	97.09	6997	600	638	0.003	282
	Cosmos 184 rocket	1967-102B	1967 Oct 24.96 40 years	Cylinder 1440	3.8 long 2.6 dia	1967 Oct 25.3	81.20	97.27	7006	544	711	0.012	177
D	[Titan 3B Agena D]	1967-103A	1967 Oct 25.80 9 days 1967 Nov 4	Cylinder 3000?	8 long? 1.5 dia	1967 Oct 26.1	111.57	90.15	6661	136	429	0.022	130
D	Cosmos 185	1967-104A	1967 Oct 27.10 445.61 days 1969 Jan 14.71	Spheroid + paddles	1.5 dia?	1967 Oct 28.4 1968 Mar 28.2 1968 Aug 15.8	64.09 64.0 64.0	98.67 97.46 96.52	7074 7017 6972	518 493 466	873 784 722	0.025 0.021 0.018	47 - -
	Cosmos 185 rocket	1967-104B	1967 Oct 27.10 50 years	Cylinder 1500?	6 long? 2 dia?	1967 Oct 29.5	64.09	98.60	7070	510	874	0.026	47

	Name		Launch date, lifetime and descent date	Shape and weight (kg)	Size (m)	Date of orbital determination	Orbital inclina-tion (deg)	Nodal period (min)	Semi major axis (km)	Perigee height (km)	Apogee height (km)	Orbital eccen-tricity	Argument of perigee (deg)
D R	Cosmos 186 *	1967-105A	1967 Oct 27.40 3.87 days 1967 Oct 31.27	Cylinder + 2 wings 6480?	7.5 long 2.2 dia	1967 Oct 27.5 1967 Oct 31.0	51.67 51.61	88.29 88.96	6570 6604	172 193	212 258	0.003 0.005	274 57
D	Cosmos 186 rocket	1967-105B	1967 Oct 27.40 1.91 days 1967 Oct 29.31	Cylinder 2500?	7.5 long 2.6 dia	1967 Oct 27.7	51.67	88.58	6584	199	212	0.001	41
D R?	Cosmos 187	1967-106A	1967 Oct 28.55 0.07 day 1967 Oct 28.62	Cylinder	2 long? 1 dia	1967 Oct 28.7	49.63	88.88	6600	143	301	0.012	82
D	Cosmos 187 launch platform	1967-106B	1967 Oct 28.55 0.44 day 1967 Oct 28.99	Irregular	-	1967 Oct 28.7	49.63	88.80	6597	139	298	0.012	82
D	Cosmos 187 rocket	1967-106C	1967 Oct 28.55 0.3 day 1967 Oct 28.8	Cylinder 1500?	8 long? 2.5 dia?	1967 Oct 28.7	49.6	88.23	6568	139	240	0.008	-
D R	Cosmos 188 *	1967-107A	1967 Oct 30.34 3.03 days 1967 Nov 2.37	Cylinder + 2 wings 6480?	7.5 long 2.2 dia	1967 Oct 30.5	51.65	88.70	6592	180	247	0.005	96
D	Cosmos 188 rocket	1967-107B	1967 Oct 30.34 2.74 days 1967 Nov 2.08	Cylinder 2500?	7.5 long 2.6 dia	1967 Oct 31.2	51.64	88.71	6592	185	242	0.004	81
D	Cosmos 189	1967-108A	1967 Oct 30.75 3874 days 1978 Jun 8	Cylinder + paddles 900?	2 long? 1 dia?	1967 Oct 31.1 1970 Nov 7.0 1976 Feb 16.5	74.01 73.99 73.99	95.53 94.74 93.15	6923 6883 6807	524 488 420	565 522 438	0.003 0.002 0.001	162 - -
D	Cosmos 189 rocket	1967-108B	1967 Oct 30.75 4512 days 1980 Mar 7	Cylinder 2200?	7.4 long 2.4 dia	1967 Nov 4.5 1972 May 1.0	74.01 73.99	95.77 94.94	6935 6893	531 497	582 533	0.004 0.003	115 -
D	Fragments **	1967-108C-L											

* Cosmos 186 and Cosmos 188 docked Oct 30.39 and separated Oct 30.53.

** Fragments designated 1967-108F to 108L possibly belong to the 1972-88, 1973-03 and 1973-10 launches.

	Name		Launch date, lifetime and descent date	Shape and weight (kg)	Size (m)	Date of orbital determination	Orbital inclination (deg)	Nodal period (min)	Semi major axis (km)	Perigee height (km)	Apogee height (km)	Orbital eccentricity	Argument of perigee (deg)
D	[Thorad Agena D]	1967-109A	1967 Nov 2.90 29.83 days 1967 Dec 2.73	Cylinder 2000?	8 long? 1.5 dia	1967 Nov 4.9	81.53	90.47	6675	183	410	0.017	168
D	Capsule (Module 283)	1967-109B	1967 Nov 2.90 511.93 days 1969 Mar 28.83	Octagon? 60?	0.3 long? 0.9 dia?	1967 Nov 6.9 1968 Apr 30.5 1968 Oct 15.8	81.68 81.6 81.6	94.41 93.77 93.07	6868 6836 6802	455 431 403	524 485 444	0.005 0.004 0.003	227 - -
D R	Cosmos 190	1967-110A	1967 Nov 3.47 7.80 days 1967 Nov 11.27	Sphere-cylinder 5530?	5 long? 2.4 dia	1967 Nov 3.8	65.73	89.80	6643	191	338	0.011	43
D	Cosmos 190 rocket	1967-110B	1967 Nov 3.47 6.03 days 1967 Nov 9.50	Cylinder 2500?	7.5 long 2.6 dia	1967 Nov 5.8	65.67	89.22	6614	183	289	0.008	31
T	ATS 3	1967-111A	1967 Nov 5.98 > million years	Cylinder 365	1.83 long 1.42 dia	1967 Nov 20.4 1968 Jan 8.0	0.53 0.45	1444.9 1436.8	42339 42172	35791 35776	36130 35812	0.004 0.0004	123 -
D	ATS 3 rocket [Atlas Agena D]	1967-111B	1967 Nov 5.98 276 days 1968 Aug 8	Cylinder 700?	6 long? 1.5 dia	1967 Nov 14.9 1968 Jul 15.8	28.39 28.07	624.5 205.4	24196 11489	179 123	35457 10099	0.729 0.434	186 -
D r	Apollo 4* [Saturn 501]	1967-113A	1967 Nov 9.50 0.36 day 1967 Nov 9.86	Cone-cylinder 30440	10.30 long 3.91 dia	1967 Nov 9.5 1967 Nov 9.7 1967 Nov 9.8	32.7 30.2 30.13	88.08 307.4 319.5	6564 15109 15501	183 60? 80?	188 17402 18326	0.0004 0.571 0.588	- - -
D	LEM Model 1 - Saturn IVB	1967-113B	1967 Nov 9.50 0.34 day 1967 Nov 9.84	Cone-cylinder 29300	24.2 long 6.6 dia	1967 Nov 9.5 1967 Nov 9.7	32.7 30.2	88.08 307.4	6564 15109	183 60?	188 17402	0.0004 0.571	- -
	Essa 6	1967-114A	1967 Nov 10.75 10000 years	Cylinder 132	0.56 long 1.07 dia	1967 Nov 23.0	102.12	114.82	7827	1410	1488	0.005	69
	Essa 6 rocket	1967-114B	1967 Nov 10.75 5000 years	Cylinder 24	1.50 long 0.46 dia	1967 Nov 14.8	102.12	114.83	7828	1411	1489	0.005	88
	Fragments	1967-114C-E											

Space Vehicle: Surveyor 6, 1967-112A, (landed on the Moon, Nov 10.04); also Centaur 15, 1967-112B, in highly eccentric orbit.

* Before Nov 9.63, 1967-113A and B were joined.

	Name		Launch date, lifetime and descent date	Shape and weight (kg)	Size (m)	Date of orbital determination	Orbital inclination (deg)	Nodal period (min)	Semi major axis (km)	Perigee height (km)	Apogee height (km)	Orbital eccentricity	Argument of perigee (deg)
D	Cosmos 191	1967-115A	1967 Nov 21.60 102.37 days 1968 Mar 2.97	Ellipsoid 400?	1.8 long 1.2 dia	1967 Nov 22.3 1968 Feb 7.8	70.96 70.96	92.16 90.47	6760 6676	267 235	497 360	0.017 0.009	84 -
D	Cosmos 191 rocket	1967-115B	1967 Nov 21.60 53.42 days 1968 Jan 14.02	Cylinder 1500?	8 long 1.65 dia	1967 Nov 22.3 1967 Dec 23.5	70.97 70.97	91.95 90.96	6749 6700	270 258	472 386	0.015 0.010	84 -
	Cosmos 192	1967-116A	1967 Nov 23.62 80 years	Cylinder + boom? 750?	1.4 long? 2.0 dia?	1967 Nov 23.9	73.98	99.85	7132	747	761	0.001	179
	Cosmos 192 rocket	1967-116B	1967 Nov 23.62 80 years	Cylinder 2200?	7.4 long 2.4 dia	1967 Nov 25.4	74.01	99.84	7131	746	760	0.001	260
D R	Cosmos 193	1967-117A	1967 Nov 25.48 7.74 days 1967 Dec 3.22	Sphere-cylinder 5530?	5 long? 2.4 dia	1967 Nov 25.8	65.63	89.85	6647	202	335	0.010	52
D	Cosmos 193 rocket	1967-117B	1967 Nov 25.48 6.18 days 1967 Dec 1.66	Cylinder 2500?	7.5 long 2.6 dia	1967 Nov 27.3	65.64	89.50	6628	184	316	0.010	39
D	WRESAT* [SPARTA]	1967-118A	1967 Nov 29.20 42.28 days 1968 Jan 10.48	Cone 73 (payload 45)	2.2 long 0.76 dia	1967 Nov 29.2 1967 Dec 23.5	83.35 83.2	99.27 95.24	7104 6908	193 167	1259 893	0.075 0.053	340 -
D R	Cosmos 194	1967-119A	1967 Dec 3.50 7.78 days 1967 Dec 11.28	Sphere-cylinder 5530?	5 long? 2.4 dia	1967 Dec 3.7	65.66	89.55	6632	201	307	0.008	39
D P	Cosmos 194 rocket	1967-119B	1967 Dec 3.50 6.12 days 1967 Dec 9.62	Cylinder 2500?	7.5 long 2.6 dia	1967 Dec 3.8	65.65	89.55	6632	201	307	0.008	48

*Weapons Research Establishment Satellite (Australia).

	Name		Launch date, lifetime and descent date	Shape and weight (kg)	Size (m)	Date of orbital determination	Orbital inclination (deg)	Nodal period (min)	Semi major axis (km)	Perigee height (km)	Apogee height (km)	Orbital eccentricity	Argument of perigee (deg)
D	OV3-6 (Atcos 2)*	1967-120A	1967 Dec 5.04 460.14 days 1969 Mar 9.18	Octagonal-cylinder 95	0.53 long 0.73 dia	1967 Dec 7.2 1968 Apr 30.5 1968 Sep 30.5	90.67 90.6 90.6	93.14 92.58 91.99	6804 6777 6748	412 384 358	439 413 381	0.002 0.002 0.002	242 - -
D	OV3-6 rocket	1967-120B	1967 Dec 5.04 125.40 days 1968 Apr 8.44	Cylinder 24	1.50 long 0.46 dia	1967 Dec 15.5 1968 Feb 22.5	90.6 90.6	93.02 92.07	6798 6752	404 364	436 383	0.002 0.001	- -
D	Fragments	1967-120C-E											
D	[Titan 3B Agena D]	1967-121A	1967 Dec 5.78 11.18 days 1967 Dec 16.96	Cylinder 3000?	8 long? 1.5 dia	1967 Dec 7.8	109.55	90.16	6662	137	430	0.022	138
D	[Thorad Agena D]	1967-122A	1967 Dec 9.93 15 days 1967 Dec 25	Cylinder 1350	8 long? 1.5 dia	1967 Dec 10.5 1967 Dec 22.7	81.65 81.74	88.45 88.46	6576 6577	158 146	237 251	0.006 0.008	165 132
D	TTS 1**	1967-123B	1967 Dec 13.59 137.4 days 1968 Apr 29.0	Octahedron 20	0.27 side	1967 Dec 14.9 1968 Mar 8.0	32.90 32.90	92.18 90.99	6767 6708	287 271	490 388	0.015 0.009	71 -
D	Pioneer 8 second stage	1967-123C	1967 Dec 13.59 110.0 days 1968 Apr 1.6	Cylinder 400?	4.9 long 1.43 dia	1967 Dec 19.4 1968 Feb 22.5	32.91 32.85	92.15 90.94	6765 6705	293 272	479 382	0.014 0.008	117 -
D	Fragments	1967-123D,E											
D R	Cosmos 195	1967-124A	1967 Dec 16.50 7.76 days 1967 Dec 24.26	Sphere-cylinder 5530?	5 long? 2.4 dia	1967 Dec 16.7	65.65	90.10	6658	207	353	0.011	59
D	Cosmos 195 rocket	1967-124B	1967 Dec 16.50 7.31 days 1967 Dec 23.81	Cylinder 2500?	7.5 long 2.6 dia	1967 Dec 18.1	65.64	89.76	6642	197	330	0.010	47
D	Fragment	1967-124C											

Space Vehicle: Pioneer 8, 1967-123A, and Altair rocket, 1967-123F?
* Atmospheric composition satellite.
** Test and Training Satellite.

	Name		Launch date, lifetime and descent date	Shape and weight (kg)	Size (m)	Date of orbital determination	Orbital inclination (deg)	Nodal period (min)	Semi major axis (km)	Perigee height (km)	Apogee height (km)	Orbital eccentricity	Argument of perigee (deg)
D	Cosmos 196	1967-125A	1967 Dec 19.27 202.85 days 1968 Jul 7.12	Ellipsoid 400?	1.8 long 1.2 dia	1967 Dec 21.7 1968 Feb 29.8	48.80 48.73	95.42 94.09	6920 6854	223 217	860 735	0.046 0.038	117 -
D	Cosmos 196 rocket	1967-125B	1967 Dec 19.27 63.54 days 1968 Feb 20.81	Cylinder 1500?	8 long 1.65 dia	1967 Dec 22.6 1968 Feb 7.5	48.81 48.77	95.20 91.86	6908 6749	219 200	841 542	0.045 0.025	119 -
D	Fragments	1967-125C-E											
D	Cosmos 197	1967-126A	1967 Dec 26.38 34.80 days 1968 Jan 30.18	Ellipsoid 400?	1.8 long 1.2 dia	1967 Dec 26.6	48.42	91.51	6730	217	486	0.020	95
D	Cosmos 197 rocket	1967-126B	1967 Dec 26.38 12.80 days 1968 Jan 8.18	Cylinder* 1500?	8 long* 1.65 dia	1967 Dec 26.6	48.44	91.40	6724	218	473	0.019	99
D	Fragments	1967-126C-J											
	Cosmos 198**	1967-127A	1967 Dec 27.48 500 years	Cone-cylinder	6 long? 2 dia?	1967 Dec 27.5 1967 Dec 29.7	65.14 65.15	89.70 103.43	6638 7301	249 894	270 952	0.002 0.004	275 307
D	Cosmos 198 platform	1967-127B	1967 Dec 27.48 24.74 days 1968 Jan 21.22	Irregular	-	1967 Dec 27.5	65.10	89.63	6635	241	273	0.002	305
D	Cosmos 198 rocket	1967-127C	1967 Dec 27.48 3.99 days 1967 Dec 31.47	Cylinder 1500?	8 long? 2.5 dia?	1967 Dec 30.1	65.09	89.12	6610	224	239	0.001	224

* Before explosion.

** **1967-127B and 127C attached to 1967-127A until orbit change.**

	Name		Launch date, lifetime and descent date	Shape and weight (kg)	Size (m)	Date of orbital determination	Orbital inclina-tion (deg)	Nodal period (min)	Semi major axis (km)	Perigee height (km)	Apogee height (km)	Orbital eccen-tricity	Argument of perigee (deg)
L	Explorer 36 (Geos 2)	1968-02A	1968 Jan 11.68 10000 years	Octahedron + pyramid 209	0.81 high 1.22 dia	1968 Jan 12.0	105.80	112.28	7709	1084	1577	0.032	168
	Explorer 36 rocket	1968-02B	1968 Jan 11.68 5000 years	Cylinder 24	1.5 long 0.46 dia	1968 Jan 18.8	105.79	112.22	7706	1083	1573	0.032	157
	Fragments	1968-02C,D											
D	Cosmos 199 rocket	1968-03B	1968 Jan 16.50 9.4 days 1968 Jan 25.9	Cylinder 1st 2 days 8000? then 2500?	7.5 long* 2.6 dia	1968 Jan 17.4	65.64	90.11	6660	202	362	0.012	55
D	Cosmos 199	1968-03A	1968 Jan 16.50 16.49 days 1968 Feb 1.99	Sphere-cylinder** 5530?	5 long?** 2.4 dia	1968 Jan 20.2	65.63	90.15	6662	204	364	0.012	59
D	Cosmos 199 transmitter***	1968-03C	1968 Jan 16.50 13.9 days 1968 Jan 30.4	Sphere 2500?	2.4 dia	1968 Jan 25.3	65.67	88.36	6573	149	241	0.007	354
D	Fragment	1968-03D											
D	[Thor Agena D]	1968-04A	1968 Jan 17.42 902.32 days 1970 Jul 7.74	Cylinder 1500?	8 long? 1.5 dia	1968 Jan 19.8 1969 Feb 15.3 1969 Oct 16.3	75.16 75.1 75.1	94.53 93.80 93.10	6876 6842 6804	450 436 408	546 491 444	0.007 0.004 0.003	218 - -
D	Fragments	1968-04B,C											
D	[Titan 3B Agena D]	1968-05A	1968 Jan 18.79 17.13 days 1968 Feb 4.92	Cylinder 3000?	8 long? 1.5 dia	1968 Jan 19.5 1968 Feb 4.8	111.52 111.53	89.91 89.58	6649 6632	138 130	404 377	0.020 0.019	128 113
D	Cosmos 200	1968-06A	1968 Jan 19.92 1862.56 days 1973 Feb 24.48	Cylinder + paddles? 900?	2 long? 1 dia?	1968 Jan 20.3 1969 Oct 31.7 1971 Dec 1.0	74.03 74.03 74.03	95.23 94.56 93.12	6908 6875 6806	523 490 425	537 503 430	0.001 0.001 0.0004	337 - -
D	Cosmos 200 rocket	1968-06B	1968 Jan 19.92 1476.38 days 1972 Feb 4.30	Cylinder 2200?	7.4 long 2.4 dia	1968 Jan 21.8 1969 Oct 31.7 1970 Dec 1.3	74.01 74.01 74.01	95.02 94.23 93.05	6898 6858 6802	499 465 413	540 495 435	0.003 0.002 0.002	16 - -
D	Fragments	1968-06C-H											

Space Vehicle: Surveyor 7, 1968-01A;
Centaur rocket, 1968-01B, in highly eccentric orbit.

* After Jan 19.0. Before Jan 19.0, 1968-03A, B and C were joined.
** Before explosion on Jan 24.3.
*** Part of 1968-03A before Jan 24.3.

	Name		Launch date, lifetime and descent date	Shape and weight (kg)	Size (m)	Date of orbital determination	Orbital inclination (deg)	Nodal period (min)	Semi major axis (km)	Perigee height (km)	Apogee height (km)	Orbital eccentricity	Argument of perigee (deg)
D	LEM 1 Ascent Stage*	1968-07A	1968 Jan 22.95 1 day 1968 Jan 24	Box + 2 tanks 2200 empty	2.52 high 3.76 wide 3.13 deep	Orbit similar to 1968-07C							
D	Apollo 5 LEM 1 Descent Stage*	1968-07B	1968 Jan 22.95 20.79 days 1968 Feb 12.74	Octagon + cone + 4 legs 1650 empty	1.57 high 3.13 wide	1968 Jan 27.1	31.63	89.65	6641	170	356	0.014	152
D	Saturn IV B [Saturn 204]	1968-07C	1968 Jan 22.95 0.69 day 1968 Jan 23.64	Cylinder 12500?	18.7 long 6.6 dia	1968 Jan 23.2	31.64	88.11	6566	162	214	0.004	62
D	[Thorad Agena D]	1968-08A	1968 Jan 24.93 33.54 days 1968 Feb 27.47	Cylinder 2000?	8 long? 1.5 dia	1968 Jan 26.8 1968 Feb 26.4	81.48 81.47	90.55 88.44	6681 6574	176 158	430 234	0.019 0.006	168 52
D	Capsule**	1968-08B	1968 Jan 24.93 769.76 days 1970 Mar 4.69	Octagon? 60?	0.3 long? 0.9 dia?	1968 Jan 28.5 1969 Feb 15.3 1969 Oct 31.7	81.65 81.6 81.6	94.75 93.85 92.59	6886 6841 6778	473 439 386	542 487 414	0.005 0.003 0.002	216 - -
D	Fragment	1968-08C											
D R	Cosmos 201	1968-09A	1968 Feb 6.33 7.93 days 1968 Feb 14.26	Sphere-cylinder 5530?	5 long? 2.4 dia	1968 Feb 6.4	64.91	89.91	6649	204	337	0.010	54
D	Cosmos 201 rocket	1968-09B	1968 Feb 6.33 7.53 days 1968 Feb 13.86	Cylinder 2500?	7.5 long 2.6 dia	1968 Feb 6.6	64.96	89.75	6642	204	324	0.009	52

*Before Jan 23.15, 1968-07A and B were joined. Lunar Excursion Module.
**1968-08B separated from 1968-08A on Jan 25.01.

	Name		Launch date, lifetime and descent date	Shape and weight (kg)	Size (m)	Date of orbital determination	Orbital inclination (deg)	Nodal period (min)	Semi major axis (km)	Perigee height (km)	Apogee height (km)	Orbital eccentricity	Argument of perigee (deg)
D	Cosmos 202	1968-10A	1968 Feb 20.42 32.62 days 1968 Mar 24.04	Ellipsoid 400?	1.8 long 1.2 dia	1968 Feb 21.5	48.44	91.42	6726	213	482	0.020	99
D	Cosmos 202 rocket	1968-10B	1968 Feb 20.42 12.62 days 1968 Mar 4.04	Cylinder 1500?	8 long 1.65 dia	1968 Feb 21.3	48.45	91.41	6725	212	481	0.020	102
	Cosmos 203	1968-11A	1968 Feb 20.67 3000 years	Spheroid + 2 vanes? 650?	1.6 dia?	1968 Feb 21.0	74.06	109.22	7571	1178	1208	0.002	230
	Cosmos 203 rocket	1968-11B	1968 Feb 20.67 2000 years	Cylinder 2200?	7.4 long 2.4 dia	1968 Feb 26.4	74.06	109.31	7575	1189	1204	0.001	232
D	Fragment	1968-11C											
	Transit 18 [Scout]	1968-12A	1968 Mar 2.16 1000 years	Octagon + 4 venes 60?	0.25 long 0.46 dia	1968 Mar 4.2	89.99	107.00	7465	1035	1139	0.007	230
	Altair rocket	1968-12B	1968 Mar 2.16 1000 years	Cylinder 24	1.5 long 0.46 dia	1968 Mar 6.1	90.00	107.00	7465	1029	1144	0.008	223
	Fragments	1968-12C,D											
D	Zond 4 launcher rocket	1968-13B	1968 Mar 2.77 4 days 1968 Mar 7	Cylinder 4000?	12 long? 4 dia	1968 Mar 4.8	51.53	88.41	6577	192	205	0.001	30
D	Fragments	1968-13C,D											
	OGO 5	1968-14A	1968 Mar 4.55 100 years?	Box + 6 booms 611	1.73 long 0.84 high 0.84 wide	1968 Mar 4.6 1969 Jan 31.3 1971 Jun 1.0	31.13 43.80 54.0	3795.9 3745.7 3745.7	80608 79890 79896	232 6045 27008	148228 140978 120027	0.918 0.845 0.582	314 337 -
	OGO 5 rocket	1968-14B	1968 Mar 4.55 100 years?	Cylinder 700?	8 long 1.5 dia	Orbit similar to 1968-14A							

Space Vehicle: Zond 4, 1968-13A.

	Name		Launch date, lifetime and descent date	Shape and weight (kg)	Size (m)	Date of orbital determination	Orbital inclination (deg)	Nodal period (min)	Semi major axis (km)	Perigee height (km)	Apogee height (km)	Orbital eccentricity	Argument of perigee (deg)
D	Cosmos 204	1968-15A	1968 Mar 5.47 362.45 days 1969 Mar 2.92	Ellipsoid 400?	1.8 long 1.2 dia	1968 Mar 6.1 1968 Jun 30.5 1968 Oct 31.2	70.99 70.9 70.9	95.81 94.81 93.22	6938 6886 6812	275 262 253	844 754 614	0.041 0.036 0.027	80 - -
D	Cosmos 204 rocket	1968-15B	1968 Mar 5.47 222.11 days 1968 Oct 13.58	Cylinder 1500?	8 long 1.65 dia	1968 Mar 6.3 1968 Jun 30.5	71.01 70.9	95.70 93.85	6932 6841	277 258	831 668	0.040 0.030	80 -
D R	Cosmos 205	1968-16A	1968 Mar 5.52 7.76 days 1968 Mar 13.28	Sphere-cylinder 5530?	5 long? 2.4 dia	1968 Mar 5.7	65.66	89.40	6624	199	292	0.007	45
D	Cosmos 205 rocket	1968-16B	1968 Mar 5.52 4.32 days 1968 Mar 9.84	Cylinder 2500?	7.5 long 2.6 dia	1968 Mar 6.1	65.66	89.22	6614	190	282	0.007	30
	Explorer 37 (SR 9)	1968-17A	1968 Mar 5.77 **20 years**	Dodecahedron 90	0.69 high 0.76 dia	1968 Mar 6.3	59.43	98.68	7075	513	881	0.026	274
D	Explorer 37 rocket	1968-17B	1968 Mar 5.77 4678 days 1980 Dec 25	Cylinder 24	1.5 long 0.46 dia	1968 Mar 10.3 1972 Jan 16.5	59.42 59.42	98.70 97.72	7076 7027	514 500	882 798	0.026 0.021	278 -
2d	Fragments	1968-17C-E											
D	[Titan 3B Agena D]	1968-18A	1968 Mar 13.83 11 days 1968 Mar 24	Cylinder 3000?	8 long? 1.5 dia	1968 Mar 15.7 1968 Mar 23.1	99.87 99.88	89.87 88.92	6646 6598	128 131	407 309	0.021 0.013	134 106
	Cosmos 206	1968-19A	1968 Mar 14.40 20 years	Cylinder + 2 vanes 2000?	5 long? 1.5 dia?	1968 Mar 26.1	81.23	97.08	6997	598	640	0.003	232
	Cosmos 206 rocket	1968-19B	1968 Mar 14.40 30 years	Cylinder 1440	3.8 long 2.6 dia	1968 Mar 26.3	81.23	97.28	7006	544	712	0.012	144

	Name		Launch date, lifetime and descent date	Shape and weight (kg)	Size (m)	Date of orbital determination	Orbital inclination (deg)	Nodal period (min)	Semi major axis (km)	Perigee height (km)	Apogee height (km)	Orbital eccentricity	Argument of perigee (deg)
D	[Thorad Agena D]	1968-20A	1968 Mar 14.92 26.22 days 1968 Apr 10.14	Cylinder 2000?	8 long? 1.5 dia	1968 Mar 15.6 1968 Apr 8.1	83.01 82.91	90.20 89.09	6663 6606	178 188	391 268	0.016 0.006	168 64
D	Capsule*	1968-20B	1968 Mar 14.92 659.85 days 1970 Jan 3.77	Octagon? 60?	0.3 long? 0.9 dia?	1968 Mar 17.8 1969 Feb 15.3 1969 Sep 16.0	83.09 83.0 83.0	94.66 93.81 92.58	6880 6839 6778	481 445 389	522 476 410	0.003 0.002 0.002	265 - -
D R	Cosmos 207	1968-21A	1968 Mar 16.52 7.79 days 1968 Mar 24.31	Sphere-cylinder 5530?	5 long? 2.4 dia	1968 Mar 17.5	65.64	89.71	6639	201	321	0.009	51
D	Cosmos 207 rocket	1968-21B	1968 Mar 16.52 6.22 days 1968 Mar 22.74	Cylinder 2500?	7.5 long 2.6 dia	1968 Mar 17.8	65.61	89.55	6632	207	300	0.007	42
D R	Cosmos 208	1968-22A	1968 Mar 21.41 11.85 days 1968 Apr 2.26	Sphere-cylinder 5900?	5.9 long 2.4 dia	1968 Mar 23.1	64.95	89.35	6619	208	274	0.005	32
D	Cosmos 208 rocket	1968-22B	1968 Mar 21.41 4.32 days 1968 Mar 25.73	Cylinder 2500?	7.5 long 2.6 dia	1968 Mar 22.5	64.97	89.15	6611	200	266	0.005	34
D	Gamma Flux package**	1968-22C	1968 Mar 21.41 16 days 1968 Apr 6	Ellipsoid 200?	0.9 long 1.9 dia	1968 Mar 30.6	64.95	89.09	6608	196	264	0.005	37

*1968-20B separated from 1968-20A on Mar 14.99.
**1968-22C separated from 1968-22A about Mar 29.3.

	Name		Launch date, lifetime and descent date	Shape and weight (kg)	Size (m)	Date of orbital determination	Orbital inclination (deg)	Nodal period (min)	Semi major axis (km)	Perigee height (km)	Apogee height (km)	Orbital eccentricity	Argument of perigee (deg)
	Cosmos 209*	1968-23A	1968 Mar 22.40 500 years	Cone-cylinder?	6 long? 2 dia?	1968 Mar 22.5 1968 Mar 28.6	65.04 65.33	89.74 103.13	6641 7286	183 871	343 944	0.012 0.005	317 270
D	Cosmos 209 rocket	1968-23B	1968 Mar 22.40 3.43 days 1968 Mar 25.83	Cylinder 1500?	8 long? 2.5 dia?	1968 Mar 24.9	65.09	88.95	6601	210	236	0.002	244
D	Cosmos 209 platform	1968-23C	1968 Mar 22.40 19.25 days 1968 Apr 10.65	Irregular	-	1968 Mar 25.6	65.09	89.46	6625	227	267	0.003	279
D	Fragments	1968-23D,E											
D R	Cosmos 210	1968-24A	1968 Apr 3.46 7.84 days 1968 Apr 11.30	Sphere- cylinder 5530?	5 long? 2.4 dia	1968 Apr 3.8	81.39	90.27	6665	200	373	0.013	85
D	Cosmos 210 rocket	1968-24B	1968 Apr 3.46 8.75 days 1968 Apr 12.21	Cylinder 2500?	7.5 long 2.6 dia	1968 Apr 4.2	81.39	90.11	6658	200	360	0.012	82
D	Fragment	1968-24C											
D r	Apollo 6** Command module + Service module	1968-25A	1968 Apr 4.50 0.42 day 1968 Apr 4.92	Cone-cylinder 25000	10.36 long 3.91 dia	1968 Apr 4.5 1968 Apr 4.7	32.57 30.13	90.36 385.20	6676 17560	205 90?	392 22274	0.014 0.632	121 -
D	LEM Model 2 - Saturn IV B [Saturn 502]	1968-25B	1968 Apr 4.50 22.01 days 1968 Apr 26.51	Cylinder 24300?	24.2 long? 6.6 dia	1968 Apr 7.1 1968 Apr 14.1	32.59 32.54	90.17 89.69	6667 6644	200 189	378 342	0.013 0.011	149 231
D	Fragments	1968-25C-S											

*1968-23B and 23C attached to 1968-23A until orbit change.
**Before Apr 4.64, 1968-25A and B were joined.

	Name		Launch date, lifetime and descent date	Shape and weight (kg)	Size (m)	Date of orbital determination	Orbital inclination (deg)	Nodal period (min)	Semi major axis (km)	Perigee height (km)	Apogee height (km)	Orbital eccentricity	Argument of perigee (deg)
	OV1-13	1968-26A	1968 Apr 6.42 1000 years	Cylinder + hemisphere 107	1.40 long 0.69 dia	1968 Apr 6.6	100.05	199.72	11315	558	9316	0.387	184
	OV1-14	1968-26B	1968 Apr 6.42 1000 years	Cylinder + hemisphere 101	1.40 long 0.69 dia	1968 Apr 6.6	100.04	208.03	11620	571	9913	0.402	184
	OV1-14 rocket	1968-26C	1968 Apr 6.42 500 years	Cylinder 70?	2.05 long 0.72 dia	1968 Apr 6.6	100.04	208.4	11634	560	9957	0.404	-
	OV1-13 rocket	1968-26D	1968 Apr 6.42 500 years	Cylinder 70?	2.05 long 0.72 dia	1968 Apr 6.6	100.05	199.71	11315	553	9320	0.387	-
D	Luna 14 launcher	1968-27B	1968 Apr 7.42 2.39 days 1968 Apr 9.81	-	-	1968 Apr 7.5	51.78	88.78	6594	189	242	0.004	62
D	Luna 14 launcher rocket	1968-27C	1968 Apr 7.42 1.78 days 1968 Apr 9.20	Cylinder 2500?	7.5 long 2.6 dia	1968 Apr 7.6	51.79	88.67	6588	183	236	0.004	34
D	Cosmos 211	1968-28A	1968 Apr 9.48 214.74 days 1968 Nov 10.22	Ellipsoid 400?	1.8 long 1.2 dia	1968 Apr 9.5 1968 Jun 15.5 1968 Aug 31.2	81.90 81.8 81.8	102.27 99.93 96.38	7244 7133 6963	199 197 191	1532 1313 979	0.092 0.078 0.057	74 - -
D	Cosmos 211 rocket	1968-28B	1968 Apr 9.48 130.51 days 1968 Aug 17.99	Cylinder 1500?	8 long 1.65 dia	1968 Apr 9.5 1968 Jun 15.5	81.90 81.77	102.27 98.35	7244 7059	199 191	1532 1171	0.092 0.069	74 -

Space Vehicle: 1968-27A, Luna 14.

	Name		Launch date, lifetime and descent date	Shape and weight (kg)	Size (m)	Date of orbital determination	Orbital inclination (deg)	Nodal period (min)	Semi major axis (km)	Perigee height (km)	Apogee height (km)	Orbital eccentricity	Argument of perigee (deg)
D R	Cosmos 212*	1968-29A	1968 Apr 14.42 4.92 days 1968 Apr 19.34	Cylinder + 2 wings 6500?	7.5 long 2.2 dia	1968 Apr 14.5 1968 Apr 15.2 1968 Apr 18.6	51.84 51.66 51.60	88.55 88.25 88.72	6584 6569 6592	186 184 174	225 197 253	0.003 0.001 0.006	152 300 80
D	Cosmos 212 rocket	1968-29B	1968 Apr 14.42 2.17 days 1968 Apr 16.59	Cylinder 2500?	7.5 long 2.6 dia	1968 Apr 14.8	51.68	88.52	6583	203	207	0.0003	45
D R	Cosmos 213*	1968-30A	1968 Apr 15.40 5.01 days 1968 Apr 20.41	Cylinder + 2 wings 6500?	7.5 long 2.2 dia	1968 Apr 15.5 1968 Apr 16.8 1968 Apr 19.0	51.66 51.67 51.65	88.88 89.02 89.12	6599 6606 6611	188 195 193	254 261 272	0.005 0.005 0.006	116 90 90
D	Cosmos 213 rocket	1968-30B	1968 Apr 15.40 3.65 days 1968 Apr 19.05	Cylinder 2500?	7.5 long 2.6 dia	1968 Apr 16.7	51.65	88.86	6598	193	246	0.004	97
D	[Titan 3B Agena D]	1968-31A	1968 Apr 17.71 12 days 1968 Apr 29	Cylinder 3000?	8 long? 1.5 dia	1968 Apr 18.4 1968 Apr 22.0 1968 Apr 24.7	111.51 111.49 111.49	90.10 90.19 89.75	6659 6663 6641	134 132 130	427 438 396	0.022 0.023 0.020	125 119 115
D R	Cosmos 214	1968-32A	1968 Apr 18.44 7.96 days 1968 Apr 26.40	Sphere-cylinder 5530?	5 long? 2.4 dia	1968 Apr 19.1	81.40	90.25	6665	200	373	0.013	84
D	Cosmos 214 rocket	1968-32B	1968 Apr 18.44 11.47 days 1968 Apr 29.91	Cylinder 2500?	7.5 long 2.6 dia	1968 Apr 19.1	81.44	90.05	6657	186	372	0.014	81
D	Cosmos 215	1968-33A	1968 Apr 18.94 72.92 days 1968 Jun 30.86	Ellipsoid 400?	1.8 long 1.2 dia	1968 Apr 24.1 1968 May 23.5	48.41 48.41	91.03 90.46	6707 6679	255 238	403 363	0.011 0.009	134 -
D	Cosmos 215 rocket	1968-33B	1968 Apr 18.94 30.58 days 1968 May 19.52	Cylinder 1500?	8 long 1.65 dia	1968 Apr 21.9	48.41	90.94	6703	258	392	0.010	129

*Cosmos 212 and Cosmos 213 docked Apr 15.43 and separated Apr 15.59.

	Name		Launch date, lifetime and descent date	Shape and weight (kg)	Size (m)	Date of orbital determination	Orbital inclination (deg)	Nodal period (min)	Semi major axis (km)	Perigee height (km)	Apogee height (km)	Orbital eccentricity	Argument of perigee (deg)
D R	Cosmos 216	1968-34A	1968 Apr 20.44 7.98 days 1968 Apr 28.42	Sphere-cylinder 5530?	5 long? 2.4 dia	1968 Apr 20.7	51.84	89.12	6612	201	267	0.005	21
D	Cosmos 216 rocket	1968-34B	1968 Apr 20.44 2.95 days 1968 Apr 23.39	Cylinder 2500?	7.5 long 2.6 dia	1968 Apr 21.0	51.83	88.95	6603	185	264	0.006	19
D	Molniya 1H	1968-35A	1968 Apr 21.18 2109 days 1974 Jan 29	Windmill 1000?	3.4 long 1.6 dia	1968 Apr 29.2 1969 Feb 15.3 1971 Jan 16.5	64.85 65.2 65.3	713.12 717.73 717.69	26443 26555 26554	391 833 1531	39738 39521 38821	0.744 0.729 0.702	285 - -
D	Molniya 1H launcher	1968-35C	1968 Apr 21.18 38.45 days 1968 May 29.63	Irregular	-	1968 Apr 22.5	64.91	91.54	6730	244	460	0.016	66
D	Molniya 1H launcher rocket	1968-35B	1968 Apr 21.18 20.11 days 1968 May 11.29	Cylinder 2500?	7.5 long 2.6 dia	1968 Apr 23.7	64.97	90.69	6689	231	391	0.012	53
D	Molniya 1H rocket	1968-35D	1968 Apr 21.18 2270.77 days 1974 Jul 9.95	Cylinder 440	2.0 long 2.0 dia	1968 Nov 17.1 1969 Oct 1.0 1971 Feb 1.0	65.11 65.11 65.11	709.39 709.29 709.25	26349 26346 26345	680 1159 1616	39261 38777 38318	0.732 0.714 0.687	282 - -
D	Cosmos 217*	1968-36A	1968 Apr 24.67 2 days 1968 Apr 26	Cylinder?	10 long? 2.5 dia?	1968 Apr 24.8 1968 Apr 25.3	62.24 62.26	88.50 87.65	6581 6538	144 140	262 179	0.009 0.003	74 96
D R?	Cosmos 218	1968-37A	1968 Apr 25.03 0.07 day 1968 Apr 25.10	Cylinder	2 long? 1 dia?	1968 Apr 25.0	49.56	87.28	6521	123	162	0.003	23
D	Cosmos 218 rocket	1968-37B	1968 Apr 25.03 0.23 day 1968 Apr 25.26	Cylinder 1500?	8 long? 2.5 dia?	1968 Apr 25.1	49.59	87.42	6528	131	167	0.003	42
D	Cosmos 218 launch platform	1968-37C	1968 Apr 25.03 0.45 day 1968 Apr 25.48	Irregular	-	1968 Apr 25.3	49.60	87.48	6531	133	172	0.003	67

* Intended orbit announced by USSR: 62.2 deg, 93.4 min, 396-520 km height.

	Name		Launch date, lifetime and descent date	Shape and weight (kg)	Size (m)	Date of orbital determination	Orbital inclination (deg)	Nodal period (min)	Semi major axis (km)	Perigee height (km)	Apogee height (km)	Orbital eccentricity	Argument of perigee (deg)
D	Cosmos 219	1968-38A	1968 Apr 26.20 310.10 days 1969 Mar 2.30	Ellipsoid 400?	1.8 long 1.2 dia	1968 Apr 26.3 1968 Jul 31.2 1968 Dec 8.0	48.42 48.42 48.35	104.62 102.28 96.83	7358 7248 6988	215 213 203	1745 1526 1017	0.104 0.091 0.058	100 - -
D	Cosmos 219 rocket	1968-38B	1968 Apr 26.20 304.18 days 1969 Feb 24.38	Cylinder 1500?	8 long 1.65 dia	1968 Apr 26.3 1968 Jul 31.2 1968 Dec 8.0	48.42 48.42 48.42	104.38 102.04 96.82	7347 7237 6988	219 228 216	1718 1489 1003	0.102 0.087 0.056	100 - -
D	Fragment	1968-38C											
D r?	[Thorad Agena D]*	1968-39A	1968 May 1.90 14 days 1968 May 15	Cylinder 2000?	8 long? 1.5 dia	1968 May 2.8 1968 May 15.4	83.05 83.05	88.58 88.63	6582 6583	164 155	243 255	0.006 0.008	163 139
D	Agena D rocket*	1968-39B	1968 May 1.90 18.79 days 1968 May 20.69	Cylinder 700?	6 long? 1.5 dia	1968 May 17.0 1968 May 20.6	83.02 83.01	88.64 87.48	6584 6526	154 132	257 164	0.008 0.002	147 145
	Cosmos 220	1968-40A	1968 May 7.58 50 years	Cylinder + boom? 750?	1.4 long? 2.0 dia?	1968 May 7.7	74.10	99.15	7096	675	760	0.006	10
	Cosmos 220 rocket	1968-40B	1968 May 7.58 50 years	Cylinder 2200?	7.4 long 2.4 dia	1968 May 18.7	74.05	99.17	7097	678	759	0.006	352
	Fragment	1968-40C											
D	Iris** (ESRO 2)†	1968-41A	1968 May 17.09 1086.05 days 1971 May 8.14	12-sided cylinder 75	0.85 long 0.76 dia	1968 May 18.8 1969 Jun 16.0 1970 Aug 1.0	97.16 97.16 97.16	99.00 97.37 94.70	7088 7009 6881	334 325 308	1085 937 697	0.053 0.044 0.028	161 - -
D	Iris rocket	1968-41B	1968 May 17.09 682.83 days 1970 Mar 30.92	Cylinder 24	1.50 long 0.46 dia	1968 Jun 11.4 1969 Feb 15.3	97.21 97.17	98.85 97.29	7080 7005	331 323	1073 931	0.052 0.043	85 -
D	Fragment	1968-41C				1969 Oct 31.7	97.17	94.68	6880	311	692	0.028	-
	[Thor Burner 2]	1968-42A	1968 May 23.19 100 years	12-sided frustum 195	1.64 long 1.31 to 1.10 dia	1968 May 24.6	98.94	102.19	7239	817	904	0.006	59
	Burner 2 rocket	1968-42B	1968 May 23.19 80 years	Sphere-cone 66	1.32 long 0.94 dia	1968 Jun 14.7	98.88	102.21	7240	819	904	0.006	13

*Before May 15, 1968-39B was part of 1968-39A. **Infra-red interferometer spectrometer

†European Space Research Organisation.

	Name		Launch date, lifetime and descent date	Shape and weight (kg)	Size (m)	Date of orbital determination	Orbital inclination (deg)	Nodal period (min)	Semi major axis (km)	Perigee height (km)	Apogee height (km)	Orbital eccentricity	Argument of perigee (deg)
D	Cosmos 221	1968-43A	1968 May 24.30 463.84 days 1969 Aug 31.14	Ellipsoid 400?	1.8 long 1.2 dia	1968 May 25.0 1968 Oct 15.8 1969 Feb 15.3	48.41 48.41 48.37	108.30 105.13 101.15	7530 7382 7195	218 211 206	2086 1797 1428	0.124 0.107 0.085	103 - -
D	Cosmos 221 rocket	1968-43B	1968 May 24.30 324.29 days 1969 Apr 13.59	Cylinder 1500?	8 long 1.65 dia	1968 May 27.0 1968 Oct 31.2 1969 Feb 15.3	48.40 48.37 48.37	107.94 102.57 96.92	7513 7261 6993	218 213 205	2051 1553 1024	0.122 0.092 0.059	110 - -
D	Cosmos 222	1968-44A	1968 May 30.85 134 days 1968 Oct 11	Ellipsoid 400?	1.8 long 1.2 dia	1968 May 31.0 1968 Jul 31.2	70.91 70.91	92.28 91.44	6765 6725	285 255	488 438	0.015 0.014	95 -
D	Cosmos 222 rocket	1968-44B	1968 May 30.85 74.53 days 1968 Aug 13.38	Cylinder 1500?	8 long 1.65 dia	1968 May 31.0 1968 Jun 30.5	70.91 70.91	92.00 91.37	6752 6721	273 263	475 422	0.015 0.012	95 -
D R	Cosmos 223	1968-45A	1968 Jun 1.46 7.80 days 1968 Jun 9.26	Sphere-cylinder 5530?	5 long? 2.4 dia	1968 Jun 5.8	72.86	89.85	6645	200	333	0.010	45
D	Cosmos 223 rocket	1968-45B	1968 Jun 1.46 7.97 days 1968 Jun 9.43	Cylinder 2500?	7.5 long 2.6 dia	1968 Jun 2.8	72.86	89.72	6639	201	321	0.009	44
D R	Cosmos 224	1968-46A	1968 Jun 4.28 7.98 days 1968 Jun 12.26	Sphere-cylinder 5530?	5 long? 2.4 dia	1968 Jun 4.6	51.83	89.05	6608	203	256	0.004	31
D	Cosmos 224 rocket	1968-46B	1968 Jun 4.28 3.13 days 1968 Jun 7.41	Cylinder 2500?	7.5 long 2.6 dia	1968 Jun 4.6	51.82	88.85	6598	193	246	0.004	20
D	[Titan 3B Agena D]	1968-47A	1968 Jun 5.73 12.2 days 1968 Jun 17.9	Cylinder 3000?	8 long? 1.5 dia	1968 Jun 5.9 1968 Jun 17.1	110.52 110.52	90.31 89.51	6668 6629	123 122	456 381	0.025 0.020	118 110
D	Cosmos 225	1968-48A	1968 Jun 11.90 144 days 1968 Nov 2	Ellipsoid 400?	1.8 long 1.2 dia	1968 Jun 12.8 1968 Aug 31.0	48.40 48.36	92.15 91.31	6762 6721	255 246	512 439	0.019 0.014	118 -
D	Cosmos 225 rocket	1968-48B	1968 Jun 11.90 64.23 days 1968 Aug 15.13	Cylinder 1500?	8 long 1.65 dia	1968 Jun 12.7 1968 Jul 15.8	48.42 48.42	91.90 91.18	6750 6714	257 245	486 427	0.017 0.014	123 -
D	Fragments	1968-48C,D											

Name		Launch date, lifetime and descent date	Shape and weight (kg)	Size (m)	Date of orbital determination	Orbital inclination (deg)	Nodal period (min)	Semi major axis (km)	Perigee height (km)	Apogee height (km)	Orbital eccentricity	Argument of perigee (deg)
Cosmos 226	1968-49A	1968 Jun 12.55 16 years	Cylinder + 2 vanes 2000?	5 long? 1.5 dia?	1968 Jul 23.5	81.24	96.87	6987	579	639	0.004	-
Cosmos 226 rocket	1968-49B	1968 Jun 12.55 30 years	Cylinder 1440	3.8 long 2.6 dia	1968 Jul 23.5	81.25	97.10	6998	526	713	0.013	-
IDCSP 4-1	1968-50A	1968 Jun 13.59 > million yr	Polyhedron (26 faces) 45	0.8 long 0.9 dia	1968 Jun 15.4	0.19	1335.7	40178	33758	33841	0.001	153
IDCSP 4-2	1968-50B	1968 Jun 13.59 > million yr	Polyhedron (26 faces) 45	0.8 long 0.9 dia	1968 Jun 15.4	0.11	1335.5	40172	33725	33863	0.002	125
IDCSP 4-3	1968-50C	1968 Jun 13.59 > million yr	Polyhedron (26 faces) 45	0.8 long 0.9 dia	1968 Jun 15.5	0.10	1335.9	40181	33699	33907	0.003	123
IDCSP 4-4	1968-50D	1968 Jun 13.59 > million yr	Polyhedron (26 faces) 45	0.8 long 0.9 dia	1968 Jun 15.6	0.10	1338.0	40224	33737	33954	0.003	127
IDCSP 4-5	1968-50E	1968 Jun 13.59 > million yr	Polyhedron (26 faces) 45	0.8 long 0.9 dia	1968 Jun 15.4	0.19	1339.6	40256	33721	34035	0.004	121
IDCSP 4-6	1968-50F	1968 Jun 13.59 > million yr	Polyhedron (26 faces) 45	0.8 long 0.9 dia	1968 Jun 15.2	0.16	1342.0	40303	33724	34126	0.005	115
IDCSP 4-7	1968-50G	1968 Jun 13.59 > million yr	Polyhedron (26 faces) 45	0.8 long 0.9 dia	1968 Jun 15.6	0.17	1345.2	40367	33721	34256	0.007	123
IDCSP 4-8	1968-50H	1968 Jun 13.59 > million yr	Polyhedron (26 faces) 45	0.8 long 0.9 dia	1968 Jun 15.6	0.13	1350.6	40476	33752	34443	0.009	109
Transtage 12 [Titan 3C]	1968-50J	1968 Jun 13.59 > million yr	Cylinder 1500?	6 long? 3 dia	1968 Jun 15.5	0.14	1356.7	40598	33729	34710	0.012	125*

*Approximate orbit.

	Name		Launch date, lifetime and descent date	Shape and weight (kg)	Size (m)	Date of orbital determination	Orbital inclination (deg)	Nodal period (min)	Semi major axis (km)	Perigee height (km)	Apogee height (km)	Orbital eccentricity	Argument of perigee (deg)
D R	Cosmos 227	1968-51A	1968 Jun 18.26 7.99 days 1968 Jun 26.25	Sphere-cylinder 5530?	5 long? 2.4 dia	1968 Jun 21.6	51.81	89.06	6608	190	269	0.006	22
D	Cosmos 227 rocket	1968-51B	1968 Jun 18.26 3.04 days 1968 Jun 21.30	Cylinder 2500?	7.5 long 2.6 dia	1968 Jun 18.6	51.80	89.00	6604	179	272	0.007	1
D	[Thorad Agena D]	1968-52A	1968 Jun 20.91 25 days 1968 Jul 16	Cylinder 2000?	8 long? 1.5 dia?	1968 Jun 21.1	84.99	89.75	6638	193	326	0.010	163
D	Capsule*	1968-52B	1968 Jun 20.91 569.24 days 1970 Jan 11.15	Octagon? 60?	0.3 long? 0.9 dia?	1968 Jun 21.2 1969 Feb 15.3 1969 Aug 16.3	85.18 85.1 85.1	94.15 93.53 92.56	6856 6826 6777	437 417 378	519 478 419	0.006 0.004 0.003	350 - -
D R	Cosmos 228	1968-53A	1968 Jun 21.50 11.92 days 1968 Jul 3.42	Sphere-cylinder 5900?	5.9 long 2.4 dia	1968 Jun 22.5	51.62	89.00	6604	199	252	0.004	30
D	Cosmos 228 rocket	1968-53B	1968 Jun 21.50 3.49 days 1968 Jun 24.99	Cylinder 2500?	7.5 long 2.6 dia	1968 Jun 23.1	51.67	88.60	6585	194	220	0.002	15
D	Fragments	1968-53C-F											
D	Cosmic Ray Package A**	1968-53G	1968 Jun 21.50 15.87 days 1968 Jul 7.37	Ellipsoid 200?	0.9 long 1.9 dia	1968 Jul 7.0	51.61	87.50	6531	153	153	0	0
D R	Cosmos 229	1968-54A	1968 Jun 24.46 7.79 days 1968 Jul 4.25	Sphere-cylinder 5530?	5 long? 2.4 dia	1968 Jun 27.0	72.87	89.85	6645	207	327	0.009	56
D	Cosmos 229 rocket	1968-54B	1968 Jun 26.46 8.01 days 1968 Jul 4.47	Cylinder 2500?	7.5 long 2.6 dia	1968 Jun 26.8	72.84	89.75	6640	202	322	0.009	53

* 1968-52B ejected from 1968-52A on June 20.98.

** 1968-53G ejected from Cosmos 228 about July 1.3.

	Name		Launch date, lifetime and descent date	Shape and weight (kg)	Size (m)	Date of orbital determination	Orbital inclination (deg)	Nodal period (min)	Semi major axis (km)	Perigee height (km)	Apogee height (km)	Orbital eccentricity	Argument of perigee (deg)
	Explorer 38* (RAE A)**	1968-55A	1968 Jul 4.73 100000 years	Tubular cross 190	Arms of cross 229	1968 Jul 6.6 1968 Jul 7.4	120.64 120.64	156.71 224.41	9630 12234	642 5851	5862 5861	0.271 0.0004	162 148
	Explorer 38* rocket	1968-55B	1968 Jul 4.73 500 years	Cylinder 24	1.5 long 0.46 dia	1968 Jul 16.4	120.63	156.82	9635	636	5878	0.272	165
	Fragments	1968-55C,D											
D	Cosmos 230	1968-56A	1968 Jul 5.29 120.60 days 1968 Nov 2.89	Ellipsoid + 8 panels 400?	1.8 long 1.2 dia	1968 Jul 8.4 1968 Aug 31.2	48.40 48.37	92.75 92.12	6792 6760	285 277	543 487	0.019 0.016	131 -
D	Cosmos 230 rocket	1968-56B	1968 Jul 5.29 103.17 days 1968 Oct 16.46	Cylinder 1500?	8 long 1.65 dia	1968 Jul 5.4 1968 Aug 31.2	48.32 48.37	92.76 91.86	6793 6748	279 265	551 474	0.020 0.015	121 -
D	Molniya 1J	1968-57A	1968 Jul 5.64 1044 days 1971 May 15	Windmill 1000?	3.4 long 1.6 dia	1968 Jul 7.7 1969 Feb 15.3 1969 Nov 16.0	65.05 65.0 64.98	713.8 717.4 717.91	26480 26547 26560	401 573 435	39803 39764 39928	0.744 0.738 0.743	284 - -
D	Molniya 1J launcher rocket	1968-57B	1968 Jul 5.64 34.71 days 1968 Aug 9.35	Cylinder 2500?	7.5 long 2.6 dia	1968 Jul 6.0	65.03	91.35	6720	234	449	0.016	60
D	Molniya 1J launcher	1968-57C	1968 Jul 5.64 47 days 1968 Aug 22	Irregular	-	1968 Jul 6.7	65.01	91.65	6736	243	472	0.017	67
D	Fragments	1968-57D,E											
D	Molniya 1J rocket	1968-57F	1968 Jul 5.64 789 days 1970 Sep 2	Cylinder 440	2.0 long 2.0 dia	1968 Nov 18.5 1969 Oct 16.3 1970 Apr 1.0	64.95 64.97 64.97	711.0 710.9 710.28	26388 26387 26371	593 464 261	39427 39554 39724	0.736 0.741 0.748	282 - -
D R	Cosmos 231	1968-58A	1968 Jul 10.83 7.91 days 1968 Jul 18.74	Sphere- cylinder 5530?	5 long? 2.4 dia	1968 Jul 10.8	64.98	89.95	6650	199	345	0.011	45
D	Cosmos 231 rocket	1968-58B	1968 Jul 10.83 10.1 days 1968 Jul 20.9	Cylinder 2500?	7.5 long 2.6 dia	1968 Jul 12.8	64.96	89.52	6631	206	299	0.007	37

*1968-55A and 1968-55B remained attached until about Jul 7.25.

** Radio Astronomy Explorer.

	Name		Launch date, lifetime and descent date	Shape and weight (kg)	Size (m)	Date of orbital determination	Orbital inclination (deg)	Nodal period (min)	Semi major axis (km)	Perigee height (km)	Apogee height (km)	Orbital eccentricity	Argument of perigee (deg)
D	OV1-15 (Spades)*	1968-59A	1968 Jul 11.81 117.65 days 1968 Nov 6.46	Cylinder + hemispheres 215	1.47 long 0.66 dia	1968 Jul 12.8 1968 Aug 21.0 1968 Nov 1.0	89.88 89.86 89.78	104.82 101.10 91.88	7364 7188 6743	154 141 135	1818 1479 596	0.113 0.093 0.034	195 70 166
D	OV1-16 (Cannonball 1)	1968-59B	1968 Jul 11.81 38.67 days 1968 Aug 19.48	Sphere 273	0.58 dia	1968 Jul 11.9 1968 Aug 17.5	90.00 89.78	91.77 88.34	6737 6568	163 124	554 255	0.029 0.010	195 46
D	OV1-15 rocket	1968-59C	1968 Jul 11.81 35.62 days 1968 Aug 16.43	Cylinder 70?	2.05 long 0.72 dia	1968 Jul 14.1 1968 Aug 13.7	89.90 89.80	104.55 92.15	7350 6756	156 135	1788 621	0.111 0.036	194 85
D	OV1-16 rocket	1968-59D	1968 Jul 11.81 5 days 1968 Jul 16	Cylinder 70?	2.05 long 0.72 dia	1968 Jul 13.6	89.80	91.42	6721	141	544	0.030	194
D R	Cosmos 232	1968-60A	1968 Jul 16.55 7.73 days 1968 Jul 24.28	Sphere-cylinder 5530?	5 long? 2.4 dia	1968 Jul 16.6	65.32	89.85	6647	189	348	0.012	31
D	Cosmos 232 rocket	1968-60B	1968 Jul 16.55 10 days 1968 Jul 26	Cylinder 2500?	7.5 long 2.6 dia	1968 Jul 17.8	65.34	89.70	6639	201	321	0.009	24
D	Cosmos 233	1968-61A	1968 Jul 18.83 203.17 days 1969 Feb 7.00	Ellipsoid 400?	1.8 long 1.2 dia	1968 Jul 21.2 1968 Oct 31.2	81.94 81.94	102.05 98.22	7234 7051	198 191	1514 1154	0.091 0.068	68 -
D	Cosmos 233 rocket	1968-61B	1968 Jul 18.83 166.49 days 1969 Jan 1.32	Cylinder 1500?	8 long 1.65 dia	1968 Jul 24.8 1968 Oct 15.8	81.95 81.95	101.78 98.05	7222 7043	208 207	1479 1122	0.088 0.065	58 -

* Solar perturbation of atmospheric density experiments satellite

	Name		Launch date, lifetime and descent date	Shape and weight (kg)	Size (m)	Date of orbital determination	Orbital inclination (deg)	Nodal period (min)	Semi major axis (km)	Perigee height (km)	Apogee height (km)	Orbital eccentricity	Argument of perigee (deg)
D R	Cosmos 234	1968-62A	1968 Jul 30.29 6.04 days 1968 Aug 5.33	Sphere-cylinder 5530?	5 long? 2.4 dia	1968 Jul 30.4	51.83	89.42	6626	208	288	0.006	52
D	Cosmos 234 rocket	1968-62B	1968 Jul 30.29 8.48 days 1968 Aug 7.77	Cylinder 2500?	7.5 long 2.6 dia	1968 Jul 31.2	51.78	89.41	6625	207	287	0.006	36
	BMEWS 1* [Atlas Agena D]	1968-63A	1968 Aug 6.47 > million yr	Cylinder 700 full? 350 empty?	1.7 long? 1.4 dia?	1968 Aug 7	9.9	1436	42150	31680	39860	0.097	-
D	[Titan 3B Agena D]	1968-64A	1968 Aug 6.69 9 days 1968 Aug 16	Cylinder 3000?	8 long? 1.5 dia	1968 Aug 10.0	110.00	89.85	6647	142	395	0.019	107
D	[Thorad Agena D]	1968-65A	1968 Aug 7.90 19.45 days 1968 Aug 27.35	Cylinder 2000?	8 long? 1.5 dia	1968 Aug 9.7	82.11	88.60	6583	152	257	0.008	178
D	Explorer 39	1968-66A	1968 Aug 8.84 4701 days 1981 Jun 22	Inflated sphere 9.3	3.66 dia	1968 Aug 11.6 1977 Mar 1.0	80.66 80.66	118.25 114.40	7982 7807	670 684	2538 2174	0.117 0.095	168 -
	Explorer 40 (Injun 5)	1968-66B	1968 Aug 8.84 500 years	Hexagonal cylinder 69.6	0.74 long 0.76 dia	1968 Aug 11.6	80.67	118.33	7985	681	2533	0.116	168
	Explorer 39 rocket	1968-66C	1968 Aug 8.84 300 years	Cylinder 24	1.5 long 0.46 dia	1968 Aug 12.6	80.67	118.40	7987	682	2535	0.116	167
	Fragments	1968-66D-J											

*Ballistic Missile Early Warning System. Agena D rocket (1968-63B) is probably in an orbit similar to 1970-46B.

	Name		Launch date, lifetime and descent date	Shape and weight (kg)	Size (m)	Date of orbital determination	Orbital inclination (deg)	Nodal period (min)	Semi major axis (km)	Perigee height (km)	Apogee height (km)	Orbital eccentricity	Argument of perigee (deg)
D R	Cosmos 235	1968-67A	1968 Aug 9.29 7.95 days 1968 Aug 17.24	Sphere-cylinder 5530?	5 long? 2.4 dia	1968 Aug 15.9	51.81	89.27	6619	201	281	0.006	65
D	Cosmos 235 rocket	1968-67B	1968 Aug 9.29 5 days 1968 Aug 14	Cylinder 2500?	7.5 long 2.6 dia	1968 Aug 11.0	51.79	89.00	6606	188	267	0.006	13
D	ATS 4 [Atlas Centaur]	1968-68A	1968 Aug 10.94 67.72 days 1968 Oct 17.66	Cylinder 2600	10 long 3 dia	1968 Aug 22.8	29.04	93.92	6851	219	726	0.037	258
	Essa 7 (Tiros 17)	1968-69A	1968 Aug 16.48 10000 years	Cylinder 145	0.57 long 1.07 dia	1968 Aug 16.5	101.72	114.98	7832	1432	1476	0.003	142
	Essa 7 second stage	1968-69B	1968 Aug 16.48 5000 years	Cylinder 350?	4.9 long 1.43 dia	1968 Sep 16.2	101.72	114.89	7828	1426	1473	0.003	166
	Fragments	1968-69C-G											
	Cosmos 236	1968-70A	1968 Aug 27.48 20 years	Cylinder + vanes 850?	2 long? 1 dia?	1968 Sep 4.7	56.07	96.83	6987	588	630	0.003	94
D	Cosmos 236 rocket	1968-70B	1968 Aug 27.48 5155 days 1982 Oct 8	Cylinder 2200?	7.4 long 2.4 dia	1968 Aug 27.6	56.17	96.71	6981	575	631	0.004	92
D R	Cosmos 237	1968-71A	1968 Aug 27.52 7.79 days 1968 Sep 4.31	Sphere-cylinder 5530?	5 long? 2.4 dia	1968 Aug 31.4	65.42	89.70	6638	200	320	0.009	45
D	Cosmos 237 rocket	1968-71B	1968 Aug 27.52 7.95 days 1968 Sep 4.47	Cylinder 2500?	7.5 long 2.6 dia	1968 Aug 28.1	65.41	89.56	6632	194	313	0.009	40

	Name		Launch date, lifetime and descent date	Shape and weight (kg)	Size (m)	Date of orbital determination	Orbital inclination (deg)	Nodal period (min)	Semi major axis (km)	Perigee height (km)	Apogee height (km)	Orbital eccentricity	Argument of perigee (deg)
D R	Cosmos 238	1968-72A	1968 Aug 28.42 3.95 days 1968 Sep 1.37	Cylinder + 2 wings 6520?	7.5 long 2.2 dia	1968 Aug 29.0	51.68	88.43	6579	188	214	0.002	306
D	Cosmos 238 rocket	1968-72B	1968 Aug 28.42 1.55 days 1968 Aug 29.97	Cylinder 2500?	7.5 long 2.6 dia	1968 Aug 28.6	51.81	88.31	6573	162	228	0.005	320
D R	Cosmos 239	1968-73A	1968 Sep 5.29 7.99 days 1968 Sep 13.28	Sphere-cylinder 5530?	5 long? 2.4 dia	1968 Sep 7.0	51.80	89.17	6614	203	269	0.005	38
D	Cosmos 239 rocket	1968-73B	1968 Sep 5.29 3.97 days 1968 Sep 9.26	Cylinder 2500?	7.5 long 2.6 dia	1968 Sep 7.8	51.79	88.60	6586	175	241	0.005	43
D	[Titan 3B Agena D]	1968-74A	1968 Sep 10.77 15 days 1968 Sep 25	Cylinder 3000?	8 long? 1.5 dia?	1968 Sep 11.9	106.06	89.82	6643	125	404	0.021	126
D R	Cosmos 240	1968-75A	1968 Sep 14.28 7.01 days 1968 Sep 21.29	Sphere-cylinder 5530?	5 long? 2.4 dia	1968 Sep 14.6	51.83	89.29	6620	202	282	0.006	23
D	Cosmos 240 rocket	1968-75B	1968 Sep 14.28 3.39 days 1968 Sep 17.67	Cylinder 2500?	7.5 long 2.6 dia	1968 Sep 15.4	51.81	88.95	6603	179	271	0.007	16
D	Zond 5 launcher rocket	1968-76C	1968 Sep 14.90 3 days 1968 Sep 18	Cylinder 4000?	12 long? 4 dia	1968 Sep 16.3	51.52	88.52	6584	193	219	0.002	339
D	Zond 5 launcher	1968-76B	1968 Sep 14.90 1.43 days 1968 Sep 16.33	Irregular	-	1968 Sep 15.0	51.61	88.89	6600	129	314	0.014	348
D	Fragment	1968-76D											

Space Vehicle: Zond 5, 1968-76A, after passing near the Moon returned to Earth and was recovered on 1968 Sep 21.67.

	Name		Launch date, lifetime and descent date	Shape and weight (kg)	Size (m)	Date of orbital determination	Orbital inclination (deg)	Nodal period (min)	Semi major axis (km)	Perigee height (km)	Apogee height (km)	Orbital eccentricity	Argument of perigee (deg)
D R	Cosmos 241	1968-77A	1968 Sep 16.52 7.79 days 1968 Sep 24.31	Sphere-cylinder 5530?	5 long? 2.4 dia	1968 Sep 17.1	65.42	89.73	6640	202	322	0.009	52
D	Cosmos 241 rocket	1968-77B	1968 Sep 16.52 6.45 days 1968 Sep 22.97	Cylinder 2500?	7.5 long 2.6 dia	1968 Sep 19.4	65.41	89.20	6614	190	282	0.007	35
D	[Thor Agena D]	1968-78A	1968 Sep 18.90 19.25 days 1968 Oct 8.15	Cylinder 1500?	8 long? 1.5 dia	1968 Sep 19.0	83.02	90.12	6658	167	393	0.017	162
D	Capsule*	1968-78B	1968 Sep 18.90 374.68 days 1969 Sep 28.58	Octagon? 60?	0.3 long? 0.9 dia?	1968 Sep 20.3 1969 Feb 15.3 1969 Jul 16.3	83.22 83.22 83.13	94.75 94.12 92.54	6885 6855 6776	500 471 394	514 482 401	0.001 0.0007 0.0005	201 - -
D	Cosmos 242	1968-79A	1968 Sep 20.61 53.69 days 1968 Nov 13.30	Ellipsoid 400?	1.8 long 1.2 dia	1968 Sep 22.6	70.97	91.29	6717	272	406	0.010	59
D	Cosmos 242 rocket	1968-79B	1968 Sep 20.61 32.06 days 1968 Oct 22.67	Cylinder 1500?	8 long 1.65 dia	1968 Sep 24.1	70.97	91.10	6707	275	382	0.008	54
D R	Cosmos 243	1968-80A	1968 Sep 23.32 10.88 days 1968 Oct 4.20	Sphere-cylinder 5900?	5.9 long 2.4 dia	1968 Sep 23.9	71.29	89.54	6631	213	293	0.006	57
D	Cosmos 243 rocket	1968-80B	1968 Sep 23.32 6.0 days 1968 Sep 29.3	Cylinder 2500?	7.5 long 2.6 dia	1968 Sep 24.2	71.33	89.31	6619	201	281	0.006	35
D	Passive Microwave Package A**	1968-80C	1968 Sep 23.32 19 days 1968 Oct 12	Ellipsoid 200?	0.9 long 1.9 dia	1968 Oct 2.8 1968 Oct 12.0	71.33 71.29	89.38 87.86	6622 6546	203 154	284 182	0.006 0.002	31 346

*1968-78B ejected from 1968-78A on Sep 18.97.
**1968-80C ejected from 1968-80A about Oct 2.2.

	Name		Launch date, lifetime and descent date	Shape and weight (kg)	Size (m)	Date of orbital determination	Orbital inclination (deg)	Nodal period (min)	Semi major axis (km)	Perigee height (km)	Apogee height (km)	Orbital eccentricity	Argument of perigee (deg)
	OV2-5	1968-81A	1968 Sep 26.32 > million yr	Box + 4 paddles 204	0.59 long 0.59 wide 0.61 high	1968 Sep 27.0	2.9	1417.9	41844	35116	35816	0.008	-
D	ERS-28 (OV5-2)	1968-81B	1968 Sep 26.32 29 months? 1971 Feb?	Octahedron 10	0.28 side	1968 Sep 26.4 1969 Feb 22.5 1970 Apr 1.0	26.37 25.85 26.0	630.3 587.2 345.5	24364 23229 16311	184 162 158	35787 33539 19707	0.731 0.718 0.599	359 - -
	ERS-21 (OV5-4)	1968-81C	1968 Sep 26.32 > million yr	Octahedron 13	0.28 side	1968 Sep 30.5	3.0	1435.8	42159	35776	35785	0.0001	-
	LES 6	1968-81D	1968 Sep 26.32 > million yr	Cylinder 163	1.83 long 1.22 dia	1968 Sep 27.0	3.0	1431.2	42069	35597	35785	0.002	-
	Transtage 13 [Titan 3C]	1968-81E	1968 Sep 26.32 > million yr	Cylinder 1500?	6 long? 3 dia	1968 Sep 26.6	3.04	1425.3	41954	35408	35744	0.004	322
D	**Fragment**	**1968-81F**											
D R?	Cosmos 244	1968-82A	1968 Oct 2.57 0.06 day 1968 Oct 2.63	Cylinder?	2 long? 1 dia?	1968 Oct 2.6	49.57	87.33	6524	134	158	0.002	142
D	Cosmos 244 rocket	1968-82B	1968 Oct 2.57 0.3 day? 1968 Oct 2	Cylinder 1500?	8 long? 2.5 dia?	1968 Oct 2.7	49.57	87.34	6524	133	159	0.002	142
D	Cosmos 244 launch platform	1968-82C	1968 Oct 2.57 0.5 day? 1968 Oct 3	Irregular	-	1968 Oct 2.7	49.58	87.84	6549	149	193	0.003	70

	Name		Launch date, lifetime and descent date	Shape and weight (kg)	Size (m)	Date of orbital determination	Orbital inclination (deg)	Nodal period (min)	Semi major axis (km)	Perigee height (km)	Apogee height (km)	Orbital eccentricity	Argument of perigee (deg)
D	Cosmos 245	1968-83A	1968 Oct 3.54 104.45 days 1969 Jan 15.99	Ellipsoid 400?	1.8 long 1.2 dia	1968 Oct 3.7 1968 Nov 30.5	70.98 70.9	92.12 91.09	6757 6706	284 253	473 403	0.014 0.011	75 -
D	Cosmos 245 rocket	1968-83B	1968 Oct 3.54 43.71 days 1968 Nov 16.25	Cylinder 1500?	8 long 1.65 dia	1968 Oct 4.8	70.99	91.80	6742	283	445	0.012	78
D	Aurorae (ESRO 1)	1968-84A	1968 Oct 3.87 630.41 days 1970 Jun 26.28	Cylinder-cone 81	1.52 long 0.76 dia	1968 Oct 4.5 1969 Jun 16.0 1970 Feb 15.3	93.76 93.76 93.7	103.00 99.64 95.54	7276 7120 6923	258 252 243	1538 1231 846	0.088 0.069 0.044	164 - -
D	Aurorae rocket	1968-84B	1968 Oct 3.87 313.48 days 1969 Aug 13.35	Cylinder 24	1.50 long 0.46 dia	1968 Oct 5.1 1969 Feb 15.3 1969 Jun 16.0	93.74 93.74 93.6	103.04 99.46 94.60	7278 7111 6878	260 250 238	1540 1215 761	0.088 0.068 0.038	162 - -
D	Fragments	1968-84C,D											
D	Molniya 1K	1968-85A	1968 Oct 5.02 2841 days 1976 Jul 16	Windmill 1000?	3.4 long 1.6 dia	1968 Oct 7.1 1968 Nov 17.3 1972 Feb 1.0	64.87 65.03 65.25	712.0 718.2 728.0	26413 26566 26810	436 466 615	39633 39909 39248	0.742 0.742 0.702	285 284 -
D	Molniya 1K launcher rocket	1968-85B	1968 Oct 5.02 25.55 days 1968 Oct 30.57	Cylinder 2500?	7.5 long 2.6 dia	1968 Oct 9.4	64.96	91.07	6706	234	422	0.014	68
D	Molniya 1K launcher	1968-85C	1968 Oct 5.02 30.72 days 1968 Nov 4.74	Irregular	-	1968 Oct 11.8	64.96	91.43	6723	224	466	0.018	66
D	Molniya 1K rocket	1968-85D	1968 Oct 5.02 2405.92 days 1975 May 7.94	Cylinder 440	2.0 long 2.0 dia	1968 Nov 16.3 1969 Aug 16.3 1972 Jan 16.5	64.99 65.4 65.15	707.7 707.4 706.99	26306 26299 26289	466 366 1011	39390 39476 38811	0.740 0.743 0.719	284 - -

	Name		Launch date, lifetime and descent date	Shape and weight (kg)	Size (m)	Date of orbital determination	Orbital inclination (deg)	Nodal period (min)	Semi major axis (km)	Perigee height (km)	Apogee height (km)	Orbital eccentricity	Argument of perigee (deg)
D	[Thorad Agena D]	1968-86A	1968 Oct 5.47 902.33 days 1971 Mar 26.80	Cylinder 2000?	8 long? 1.5 dia	1968 Oct 8.6 1969 Oct 31.7 1970 Jul 1.3	74.97 74.97 74.97	94.55 93.99 93.14	6875 6848 6806	483 462 421	511 478 435	0.002 0.001 0.001	248 - -
D R	Cosmos 246	1968-87A	1968 Oct 7.51 4.76 days 1968 Oct 12.27	Sphere-cylinder 5530?	5 long? 2.4 dia	1968 Oct 9.1	65.37	89.18	6613	149	321	0.013	11
D	Cosmos 246 rocket	1968-87B	1968 Oct 7.51 1.14 days 1968 Oct 8.65	Cylinder 2500?	7.5 long 2.6 dia	1968 Oct 8.4	65.38	88.31	6570	133	251	0.009	12
D R	Cosmos 247	1968-88A	1968 Oct 11.50 7.74 days 1968 Oct 19.24	Sphere-cylinder 5530?	5 long? 2.4 dia	1968 Oct 12.4	65.39	89.94	6650	199	345	0.011	56
D	Cosmos 247 rocket	1968-88B	1968 Oct 11.50 7 days 1968 Oct 18	Cylinder 2500?	7.5 long 2.6 dia	1968 Oct 13.7	65.42	89.50	6629	191	311	0.009	36
D 3M R	Apollo 7*	1968-89A	1968 Oct 11.63 10.84 days 1968 Oct 22.47	Cone-cylinder 14690	10.36 long 3.91 dia	1968 Oct 12.9	31.63	89.78	6642	231	297	0.005	88
D	Saturn IV B* [Saturn 205]	1968-89B	1968 Oct 11.63 6.83 days 1968 Oct 18.46	Cylinder 13600	18.7 long 6.6 dia	1968 Oct 11.8	31.6	89.84	6645	227	307	0.006	-
D	Cosmos 248	1968-90A	1968 Oct 19.18 4147 days 1980 Feb 26	Cylinder? 1400?	4 long? 2 dia?	1968 Oct 20.1 1972 Feb 1.0	62.25 62.25	94.80 94.17	6887 6856	475 452	543 503	0.005 0.004	296 -
D	Fragments	1968-90B-E											

*1968-89A and B were joined together until Oct 11.75.

	Name		Launch date, lifetime and descent date	Shape and weight (kg)	Size (m)	Date of orbital determination	Orbital inclination (deg)	Nodal period (min)	Semi major axis (km)	Perigee height (km)	Apogee height (km)	Orbital eccentricity	Argument of perigee (deg)
	Cosmos 249*	1968-91A	1968 Oct 20.17 100 years	Cylinder?	4 long? 1.5 dia?	1968 Oct 21.9	62.35	112.13	7703	493	2157	0.108	76
D	Cosmos 249 rocket	1968-91B	1968 Oct 20.17 1 day 1968 Oct 21	Cylinder 1500?	8 long? 2.5 dia?	1968 Oct 20.6	62.27	88.38	6573	136	254	0.009	83
36d	Fragments	1968-91C-CK											
	[Thor Burner 2]	1968-92A	1968 Oct 23.19 100 years	12-sided Frustum 195	1.64 long 1.31 to 1.10 dia	1968 Oct 23.6	99.00	101.45	7204	797	855	0.004	34
	Burner 2 rocket	1968-92B	1968 Oct 23.19 80 years	Sphere-cone 66	1.32 long 0.94 dia	1968 Nov 12.3	98.99	101.45	7204	801	851	0.003	340
D R	Soyuz 2**	1968-93A	1968 Oct 25.38 2.95 days 1968 Oct 28.33	Sphere-cylinder + 2 wings 6520?	7.5 long 2.2 dia	1968 Oct 26.2	51.66	88.30	6568	170	210	0.003	310
D	Soyuz 2 rocket	1968-93B	1968 Oct 25.38 1.83 days 1968 Oct 27.21	Cylinder 2500?	7.5 long 2.6 dia	1968 Oct 25.8	51.72	88.48	6576	196	200	0.0003	164
D M R	Soyuz 3**	1968-94A	1968 Oct 26.36 3.94 days 1968 Oct 30.30	Sphere-cylinder + 2 wings 6575	7.5 long 2.2 dia	1968 Oct 27.3 1968 Oct 28.4 1968 Oct 29.4	51.66 51.66 51.66	88.30 88.70 88.87	6568 6588 6597	177 176 196	203 244 241	0.002 0.005 0.003	2 - -
D	Soyuz 3 rocket	1968-94B	1968 Oct 26.36 1.79 days 1968 Oct 28.15	Cylinder 2500?	7.5 long 2.6 dia	1968 Oct 27.6	51.66	87.99	6553	170	180	0.0007	312

* Cosmos 249 passed close to Cosmos 248 on 1968 Oct 20.32, then exploded.

** Soyuz 2 and 3 close approach occurred on 1968 Oct 26.4.

	Name		Launch date, lifetime and descent date	Shape and weight (kg)	Size (m)	Date of orbital determination	Orbital inclination (deg)	Nodal period (min)	Semi major axis (km)	Perigee height (km)	Apogee height (km)	Orbital eccentricity	Argument of perigee (deg)
D	Cosmos 250	1968-95A	1968 Oct 30.92 3395 days 1978 Feb 15	Cylinder + paddles? 900?	2 long? 1 dia?	1968 Nov 10.2 1970 Nov 25.0 1975 Feb 5.4	74.02 73.99 73.99	95.30 94.55 93.05	6910 6873 6802	522 486 419	542 504 428	0.001 0.001 0.001	325 - -
D	Cosmos 250 rocket	1968-95B	1968 Oct 30.92 3661 days 1978 Nov 8	Cylinder 2200?	7.4 long 2.4 dia	1968 Nov 9.4 1971 Apr 1.0 1977 Jul 1.0	74.01 74.00 74.00	95.20 94.47 93.00	6905 6869 6799	514 484 418	540 498 424	0.002 0.001 0	328 - -
D	Fragments	1968-95C-G											
D R	Cosmos 251	1968-96A	1968 Oct 31.38 11.86 days 1968 Nov 12.24	Sphere-cylinder 6300?	6.5 long? 2.4 dia	1968 Nov 3.1 1968 Nov 9.3 1968 Nov 9.7	64.87 64.95 64.92	88.99 89.62 89.68	6606 6637 6640	201 186 227	255 332 296	0.004 0.011 0.005	6 74 51
D	Cosmos 251 rocket	1968-96B	1968 Oct 31.38 2.56 days 1968 Nov 2.94	Cylinder 2500?	7.5 long 2.6 dia	1968 Nov 1.1	64.94	88.74	6593	197	233	0.003	5
D	Cosmos 251 engine*	1968-96E	1968 Oct 31.38 18.09 days 1968 Nov 18.47	Cone? 600? full	1.5 long? 2 dia?	Orbit similar to 1968-96A							
D	Fragments	1968-96C,D											
	Cosmos 252**	1968-97A	1968 Nov 1.02 200 years	Cylinder?	4 long? 2 dia?	1968 Nov 8.0	62.32	112.45	7718	531	2149	0.105	75
68d	Fragments	1968-97B-DX											
D	[Thor Agena D]	1968-98A	1968 Nov 3.90 19.99 days 1968 Nov 23.89	Cylinder 1500?	8 long? 1.5 dia	1968 Nov 7.7	82.15	88.90	6597	150	288	0.010	137
D	[Titan 3B Agena D]	1968-99A	1968 Nov 6.80 14 days 1968 Nov 20	Cylinder 3000?	8 long? 1.5 dia	1968 Nov 11.0 1968 Nov 19.8	106.0 105.97	89.73 89.03	6638 6604	130 134	390 318	0.020 0.014	127 107

* This engine, ejected from Cosmos 251 on 1968 Nov 12, retained the 1968-96A designation in the USA.

**Cosmos 252 passed close to Cosmos 248 on 1968 Nov 1.17, then exploded.

	Name		Launch date, lifetime and descent date	Shape and weight (kg)	Size (m)	Date of orbital determination	Orbital inclination (deg)	Nodal period (min)	Semi major axis (km)	Perigee height (km)	Apogee height (km)	Orbital eccentricity	Argument of perigee (deg)
D	TTS 2	1968-100B	1968 Nov 8.41 3967 days 1979 Sep 19	Octahedron 18	0.28 side	1968 Nov 14.3 1970 Nov 30.9	32.87 32.87	97.78 96.63	7037 6980	378 365	940 839	0.040 0.034	190 -
D	Pioneer 9 second stage	1968-100D	1968 Nov 8.41 3443 days 1978 Apr 13	Cylinder 400?	4.9 long 1.43 dia	1968 Nov 30.5 1970 Jul 1.0 1974 May 16.5	32.85 32.85 32.85	97.71 96.60 94.25	7034 6979 6864	374 365 345	937 836 626	0.040 0.034 0.020	- - -
D	Fragments	1968-100C,E											
D	Zond 6 launcher	1968-101B	1968 Nov 10.80 1.49 days 1968 Nov 12.29	-	-	1968 Nov 11.6	51.49	87.91	6553	175	175	0.000	137
D	Zond 6 launcher rocket	1968-101C	1968 Nov 10.80 3.16 days 1968 Nov 13.96	Cylinder 4000?	12 long? 4 dia	1968 Nov 11.3	51.49	88.63	6587	186	232	0.004	334
D	Fragments	1968-101D,E											
D R	Cosmos 253	1968-102A	1968 Nov 13.50 4.80 days 1968 Nov 18.30	Sphere-cylinder 5530?	5 long? 2.4 dia	1968 Nov 14.9	65.42	89.87	6645	200	333	0.010	53
D p	Cosmos 253 rocket*	1968-102B	1968 Nov 13.50 7.25 days 1968 Nov 20.75	Cylinder 2500?	7.5 long 2.6 dia	1968 Nov 15.7	65.40	89.54	6630	195	309	0.009	45
D	Proton 4	1968-103A	1968 Nov 16.49 249.96 days 1969 Jul 24.45	Cylinder 17000	3 long 4 dia	1968 Nov 16.8 1969 Feb 15.3	51.55 51.55	91.75 91.18	6741 6714	248 243	477 428	0.017 0.014	84 -
D	Proton 4 rocket	1968-103B	1968 Nov 16.49 69.73 days 1969 Jan 25.22	Cylinder 4000?	12 long? 4 dia	1968 Nov 18.2 1968 Dec 15.5	51.54 51.54	91.67 91.10	6738 6710	252 238	468 425	0.016 0.014	87 -

Space Vehicles: Pioneer 9 is 1968-100A; Zond 6 is 1968-101A; passed Moon, recovered on Earth 1968 Nov 17.59.

* Decay was observed over England and pieces were picked up at Southend.

	Name		Launch date, lifetime and descent date	Shape and weight (kg)	Size (m)	Date of orbital determination	Orbital inclination (deg)	Nodal period (min)	Semi major axis (km)	Perigee height (km)	Apogee height (km)	Orbital eccentricity	Argument of perigee (deg)
D R	Cosmos 254	1968-104A	1968 Nov 21.51 7.72 days 1968 Nov 29.23	Sphere-cylinder 5530?	5 long? 2.4 dia	1968 Nov 22.3	65.40	89.85	6644	197	335	0.010	50
D	Cosmos 254 rocket	1968-104B	1968 Nov 21.51 6.16 days 1968 Nov 27.67	Cylinder 2500?	7.5 long 2.6 dia	1968 Nov 23.1	65.40	89.55	6631	187	318	0.010	39
D R	Cosmos 255	1968-105A	1968 Nov 29.53 7.77 days 1968 Dec 7.30	Sphere-cylinder 5530?	5 long? 2.4 dia	1968 Nov 30.2	65.42	89.64	6635	197	317	0.009	44
D	Cosmos 255 rocket	1968-105B	1968 Nov 29.53 5.53 days 1968 Dec 5.06	Cylinder 2500?	7.5 long 2.6 dia	1968 Nov 30.7	65.40	89.42	6625	188	306	0.009	35
	Cosmos 256	1968-106A	1968 Nov 30.50 3000 years	Spheroid + 2 vanes? 650?	1.6 dia?	1968 Dec 1.2	74.05	109.45	7579	1175	1227	0.003	168
	Cosmos 256 rocket	1968-106B	1968 Nov 30.50 2000 years	Cylinder 2200?	7.4 long 2.4 dia	1968 Dec 2.8	74.04	109.33	7573	1168	1222	0.004	155
D	Cosmos 257	1968-107A	1968 Dec 3.62 91.54 days 1969 Mar 5.16	Ellipsoid 400?	1.8 long 1.2 dia	1968 Dec 4.6	70.94	91.97	6752	286	462	0.013	83
D	Cosmos 257 rocket	1968-107B	1968 Dec 3.62 43.44 days 1969 Jan 16.06	Cylinder 1500?	8 long 1.65 dia	1968 Dec 4.4	70.95	91.47	6727	275	423	0.011	84

	Name		Launch date, lifetime and descent date	Shape and weight (kg)	Size (m)	Date of orbital determination	Orbital inclination (deg)	Nodal period (min)	Semi major axis (km)	Perigee height (km)	Apogee height (km)	Orbital eccentricity	Argument of perigee (deg)
D	[Titan 3B Agena D]	1968-108A	1968 Dec 4.81 8 days 1968 Dec 12	Cylinder 3000?	8 long? 1.5 dia	1968 Dec 5.8 1968 Dec 11.1	106.24 106.17	93.30 91.94	6814 6747	136 134	736 604	0.044 0.035	56 127
D	Heos 1*	1968-109A	1968 Dec 5.79 2518 days 1975 Oct 28	16-sided cylinder 108	0.75 long 1.26 dia	1968 Dec 5.8 1970 Nov 7.6 1971 Dec 31.9	28.28 60.50 68.20	6750 6704.2 6704.1	118,300 117,778 117,782	418 20020 32430	223,440 202,780 190,380	0.943 0.776 0.671	268 - -
D	Heos 1 second stage	1968-109B	1968 Dec 5.79 21.84 days 1968 Dec 27.63	Cylinder 350?	4.9 long 1.43 dia	1968 Dec 8.8	28.3	93.66	6839	177	745	0.042	223
	OAO 2	1968-110A	1968 Dec 7.36 500 years	Octagonal cylinder 2012	3.05 long 2.15 dia	1968 Dec 10.4	35.00	100.16	7150	765	778	0.0009	315
	OAO 2 rocket	1968-110B	1968 Dec 7.36 200 years	Cylinder 1815	8.6 long 3.0 dia	1968 Dec 7.7	35.00	100.06	7145	717	817	0.007	239
D R	Cosmos 258	1968-111A	1968 Dec 10.35 7.90 days 1968 Dec 18.25	Sphere-cylinder 5530?	5 long? 2.4 dia	1968 Dec 11.0	64.98	89.59	6630	205	298	0.007	48
D	Cosmos 258 rocket	1968-111B	1968 Dec 10.35 6.82 days 1968 Dec 17.17	Cylinder 2500?	7.5 long 2.6 dia	1968 Dec 12.1	64.97	89.35	6620	202	282	0.006	32
D	[Thorad Agena D]	1968-112A	1968 Dec 12.93 15.65 days 1968 Dec 28.58	Cylinder 2000?	8 long? 1.5 dia	1968 Dec 13.6	81.02	88.67	6587	169	248	0.006	185
	Capsule	1968-112B	1968 Dec 12.93 10000 years	Octagon? 60?	0.3 long? 0.9 dia?	1968 Dec 14.7	80.33	114.45	7808	1391	1468	0.006	282
	Fragments	1968-112C-E											

* Highly eccentric orbit satellite (ESRO): Third stage Altair, 1968-109C (identical to 1968-55B) in orbit similar to 109A.

	Name		Launch date, lifetime and descent date	Shape and weight (kg)	Size (m)	Date of orbital determination	Orbital inclination (deg)	Nodal period (min)	Semi major axis (km)	Perigee height (km)	Apogee height (km)	Orbital eccentricity	Argument of perigee (deg)
D	Cosmos 259	1968-113A	1968 Dec 14.22 142.21 days 1969 May 5.43	Ellipsoid 400?	1.8 long 1.2 dia	1968 Dec 15.8 1969 Feb 22.5	48.40 48.38	100.22 97.45	7151 7018	215 208	1331 1072	0.078 0.062	107 -
D	Cosmos 259 rocket	1968-113B	1968 Dec 14.22 147.31 days 1969 May 10.53	Cylinder 1500?	8 long 1.65 dia	1968 Dec 15.1 1969 Feb 22.5	48.40 48.40	100.00 97.37	7140 7014	216 210	1308 1062	0.077 0.061	104 -
D	Fragment	1968-113C											
	Essa 8	1968-114A	1968 Dec 15.72 10000 years	Cylinder 132	0.56 long 1.07 dia	1968 Dec 16.4	101.90	114.70	7820	1410	1473	0.004	294
	Essa 8 second stage	1968-114B	1968 Dec 15.72 5000 years	Cylinder 350?	4.9 long 1.43 dia	1968 Dec 16.4	101.90	114.60	7815	1410	1460	0.003	283
	Fragments	1968-114C,D											
D	Cosmos 260*	1968-115A	1968 Dec 16.39 1666 days 1973 Jul 9	Windmill 1000?	3.4 long 1.6 dia	1968 Dec 16.9 1969 Nov 8.5 1971 Oct 16.5	64.93 65.0 65.0	712.36 712.26 712.20	26422 26420 26419	518 962 966	39570 39122 39115	0.739 0.722 0.722	285 - -
D	Cosmos 260 launcher	1968-115B	1968 Dec 16.39 52.21 days 1969 Feb 6.60	Irregular	-	1968 Dec 17.2	64.91	92.01	6753	240	510	0.020	71
D	Cosmos 260 launcher rocket	1968-115C	1968 Dec 16.39 32.71 days 1969 Jan 18.10	Cylinder 2500?	7.5 long 2.6 dia	1968 Dec 17.5	65.00	91.57	6730	230	473	0.018	69
D	Cosmos 260 rocket	1968-115D	1968 Dec 16.39 1740 days 1973 Sep 21	Cylinder 440	2.0 long 2.0 dia	1968 Dec 22.3 1969 Nov 8.5 1971 Oct 16.5	64.84 65.0 64.94	708.62 708.54 708.54	26330 26328 26328	494 966 963	39410 38934 38937	0.739 0.721 0.721	285 - -

*This is probably a Molniya satellite, replacing Molniya 1H (1968-35A).

Year of launch 1968, continued

	Name		Launch date, lifetime and descent date	Shape and weight (kg)	Size (m)	Date of orbital determination	Orbital inclination (deg)	Nodal period (min)	Semi major axis (km)	Perigee height (km)	Apogee height (km)	Orbital eccentricity	Argument of perigee (deg)
	Intelsat 3 F-2	1968-116A	1968 Dec 19.02 > million yr	Cylinder 293 full 137 empty	1.04 long 1.42 dia	1968 Dec 20.5 1977 May 13.0	0.7 6.47	1436 1632.52	42160 45927	35770 39365	35790 39732	0.0002 0.004	- 56†
D	Intelsat 3 F-2 third-stage	1968-116B	1968 Dec 19.02 1650 days 1973 Jun 26	Cylinder 24	1.5 long 0.46 dia	1972 May 1.0 1973 Jan 1.0	29.6 29.8	588.5 571.9	23263 22823	368 236	33401 32653	0.710 0.710	- -
D	Cosmos 261	1968-117A	1968 Dec 20.00 54.33 days 1969 Feb 12.33	Ellipsoid 400?	1.8 long 1.2 dia	1968 Dec 20.6 1969 Jan 12.7	71.03 71.03	93.08 92.01	6803 6751	207 195	642 551	0.032 0.026	85 41
D	Cosmos 261 rocket	1968-117B	1968 Dec 20.00 18.48 days 1969 Jan 7.48	Cylinder* 1500?	8 long* 1.65 dia	1968 Dec 22.8	71.00	92.85	6792	203	624	0.031	83
D	Fragments	1968-117C-Z											
D 3M R	Apollo 8	1968-118A	1968 Dec 21.54 6.12 days 1968 Dec 27.66	Cone-cylinder 28400 full 15640 empty	10.36 long 3.91 dia	1968 Dec 21.6 1968 Dec 21.66	32.60 30.71	88.15 24,400	6569 273,000	191 174	191 533,000	0 0.976	265 30
						In selenocentric orbit 1968 Dec 24.42 - Dec 25.25							
	Saturn IV B** [Saturn 503]	1968-118B	1968 Dec 21.54 Indefinite	Cylinder 11800	24.5 long 6.6 dia	1968 Dec 21.6	32.60	88.15	6569	191	191	0	265
						Entered heliocentric orbit 1968 Dec 21.67							
D	Cosmos 262	1968-119A	1968 Dec 26.41 204.20 days 1969 Jul 18.61	Ellipsoid 400?	1.8 long 1.2 dia	1968 Dec 27.78 1969 Feb 22.5	48.44 48.44	95.13 94.22	6907 6864	259 252	798 720	0.039 0.034	113 -
D	Cosmos 262 rocket	1968-119B	1968 Dec 26.41 124.87 days 1969 Apr 30.28	Cylinder 1500?	8 long 1.65 dia	1968 Dec 29.95 1969 Feb 22.5	48.45 48.45	94.90 93.60	6897 6834	257 252	781 660	0.038 0.030	124 -

* Before disintegration.

**Attached to 118A until Dec 21.67.

† Orbit raised when transmissions terminated.

	Name		Launch date, lifetime and descent date	Shape and weight (kg)	Size (m)	Date of orbital determination	Orbital inclination (deg)	Nodal period (min)	Semi major axis (km)	Perigee height (km)	Apogee height (km)	Orbital eccentricity	Argument of perigee (deg)
D	Venus 5 launcher rocket	1969-01B	1969 Jan 5.27 1.56 days 1969 Jan 6.83	Cylinder 2500?	7.5 long 2.6 dia	1969 Jan 5.3	51.82	88.65	6589	203	218	0.001	106
D	Venus 5 launcher	1969-01C	1969 Jan 5.27 1.96 days 1969 Jan 7.23	Irregular	-	1969 Jan 5.5	51.80	88.55	6584	186	225	0.003	133
D	Venus 6 launcher rocket	1969-02B	1969 Jan 10.24 0.9 day 1969 Jan 11.14	Cylinder 2500?	7.5 long 2.6 dia	1969 Jan 10.5	51.75	88.20	6567	184	193	0.001	297
D	Venus 6 launcher	1969-02C	1969 Jan 10.24 3.07 days 1969 Jan 13.31	Irregular	-	1969 Jan 11.3	51.66	88.51	6582	201	207	0.0005	118
D R	Cosmos 263	1969-03A	1969 Jan 12.51 7.72 days 1969 Jan 20.23	Sphere-cylinder 5530?	5 long? 2.4 dia	1969 Jan 13.8	65.43	89.74	6641	200	325	0.009	54
D	Cosmos 263 rocket	1969-03B	1969 Jan 12.51 5.49 days 1969 Jan 18.00	Cylinder 2500?	7.5 long 2.6 dia	1969 Jan 14.4	65.34	89.40	6624	174	317	0.011	43
D	Fragment	1969-03C											
D 1M R	Soyuz 4*	1969-04A	1969 Jan 14.32 2.95 days 1969 Jan 17.27	Sphere-cylinder + 2 wings 6625	7.5 long 2.2 dia	1969 Jan 14.5 1969 Jan 15.2	51.73 51.73	88.20 88.72	6566 6592	161 205	215 223	0.004 0.001	293 20
D	Soyuz 4 rocket	1969-04B	1969 Jan 14.32 0.78 day 1969 Jan 15.10	Cylinder 2500?	7.5 long 2.6 dia	1969 Jan 14.5	51.70	88.03	6558	147	212	0.005	276

Space Vehicles: Venus 5, 1969-01A; Venus 6, 1969-02A.

* Two crew members from Soyuz 5 transferred to Soyuz 4 during docking from Jan 16.35 to Jan 16.54 (See page 186.)

	Name		Launch date, lifetime and descent date	Shape and weight (kg)	Size (m)	Date of orbital determination	Orbital inclination (deg)	Nodal period (min)	Semi major axis (km)	Perigee height (km)	Apogee height (km)	Orbital eccentricity	Argument of perigee (deg)
D 3M R	Soyuz 5	1969-05A	1969 Jan 15.30 3.03 days 1969 Jan 18.33	Sphere-cylinder + 2 wings 6585	7.5 long 2.2 dia	1969 Jan 15.7	51.69	88.87	6600	210	233	0.002	66
D	Soyuz 5 rocket	1969-05B	1969 Jan 15.30 2.08 days 1969 Jan 17.38	Cylinder 2500?	7.5 long 2.6 dia	1969 Jan 15.5	51.69	88.50	6581	194	212	0.001	1
D	Fragment	1969-05C											
	OSO 5*	1969-06A	1969 Jan 22.70 16 years	Nonagonal box + vane 291	0.94 long 1.12 dia	1969 Jan 22.9	32.95	95.48	6927	536	561	0.002	241
D	OSO 5 rocket	1969-06B	1969 Jan 22.70 3972 days 1979 Dec 8	Cylinder 24	1.50 long 0.46 dia	1969 Feb 10.4 1972 Sep 1.0	32.99 32.99	95.51 94.72	6928 6889	538 502	562 520	0.002 0.001	65 -
D	[Titan 3B Agena D]	1969-07A	1969 Jan 22.80 12 days 1969 Feb 3	Cylinder 3000?	8 long? 1.5 dia	1969 Jan 23.2 1969 Jan 30.3	106.15 106.12	97.04 96.78	6994 6981	142 140	1090 1066	0.068 0.066	154 151
D	Fragment	1969-07B											
D R	Cosmos 264 **	1969-08A	1969 Jan 23.39 12.88 days 1969 Feb 5.27	Sphere-cylinder 6300?	6.5 long? 2.4 dia	1969 Jan 24.7 1969 Jan 29.7	69.94 69.92	89.57 89.73	6630 6639	209 213	295 308	0.006 0.007	47 53
D	Cosmos 264 rocket	1969-08B	1969 Jan 23.39 7.53 days 1969 Jan 30.92	Cylinder 2500?	7.5 long 2.6 dia	1969 Jan 25.2	69.94	89.41	6623	207	282	0.006	39
D	Cosmos 264 engine†	1969-08C	1969 Jan 23.39 21.48 days 1969 Feb 13.87	Cone 600? full	1.5 long? 2 dia?	1969 Feb 5.2	69.92	89.61	6633	207	303	0.007	33

* Orbiting Solar Observatory.

** Carried supplementary extragalactic gamma-ray experiment.

† 1969-08C ejected from 08A about Feb 4.4.

	Name		Launch date, lifetime and descent date	Shape and weight (kg)	Size (m)	Date of orbital determination	Orbital inclination (deg)	Nodal period (min)	Semi major axis (km)	Perigee height (km)	Apogee height (km)	Orbital eccentricity	Argument of perigee (deg)
	Isis 1*	1969-09A	1969 Jan 30.28 250 years	Polyhedron 241	1.07 long 1.27 dia	1969 Feb 4.2	88.42	128.42	8430	578	3526	0.175	161
	Isis 1 rocket	1969-09B	1969 Jan 30.28 150 years	Cylinder 24	1.50 long 0.46 dia	1969 Feb 1.7	88.42	128.30	8425	579	3515	0.174	165
D	[Thorad Agena D]**	1969-10A	1969 Feb 5.92 18.86 days 1969 Feb 24.78	Cylinder 2000?	8 long? 1.5 dia	1969 Feb 6.1	81.54	88.70	6587	178	239	0.005	161
	Capsule	1969-10B	1969 Feb 5.92 10000 years	Octagon? 60?	0.3 long? 0.9 dia?	1969 Feb 6.7	80.41	114.22	7796	1396	1441	0.003	39
	Fragment	1969-10C											
	Intelsat 3 F-3 †	1969-11A	1969 Feb 6.03 >million years	Cylinder 293 full 137 empty	1.04 long 1.42 dia	1969 Mar 17.0 1979 Dec 15	1.34 6.4	1436.4 1674	42173 46700	35782 39700	35808 40900	0.0003 0.013	74
	Intelsat 3 F-3 rocket	1969-11B	1969 Feb 6.03 20 years	Cylinder 24	1.50 long 0.46 dia	1972 May 1.0	29.5	613.9	23926	343	34753	0.719	-
D	Cosmos 265	1969-12A	1969 Feb 7.59 82.50 days 1969 May 1.09	Ellipsoid 400?	1.8 long 1.2 dia	1969 Feb 9.2	71.01	91.89	6745	275	458	0.014	71
D	Cosmos 265 rocket	1969-12B	1969 Feb 7.59 38.20 days 1969 Mar 17.79	Cylinder 1500?	8 long 1.65 dia	1969 Feb 9.2	71.01	91.67	6734	274	437	0.012	76

* International satellite for ionospheric studies (Canada).

** Thorad: long-tank thrust-augmented Thor.

† International telecommunications satellite. Intelsat 3 F-3 was raised about 4500 km in two manoeuvres on 13-14 December 1979

	Name		Launch date, lifetime and descent date	Shape and weight (kg)	Size (m)	Date of orbital determination	Orbital inclination (deg)	Nodal period (min)	Semi major axis (km)	Perigee height (km)	Apogee height (km)	Orbital eccentricity	Argument of perigee (deg)
	Tactical Comsat 1	1969-13A	1969 Feb 9.88 > million years	Cylinder +5 aerials 730	6.1 long 2.4 dia	1969 Apr 16.0	0.8	1436.0	42164	35768	35803	0.0004	-
	Tacsat 1 rocket [Titan 3C]	1969-13B	1969 Feb 9.88 > million years	Cylinder 1500?	6 long? 3.0 dia	1969 Feb 10	0.65	1446.5	42370	35940	36044	0.001	159
D R	Cosmos 266	1969-15A	1969 Feb 25.43 7.9 days 1969 Mar 5.3	Sphere-cylinder 5530?	5 long? 2.4 dia	1969 Feb 27.1	72.90	89.90	6647	202	336	0.010	48
D	Cosmos 266 rocket	1969-15B	1969 Feb 25.43 7.33 days 1969 Mar 4.76	Cylinder 2500?	7.5 long 2.6 dia	1969 Feb 27.2	72.89	89.62	6633	198	312	0.009	43
D	Fragment	1969-15C											
	Essa 9 *	1969-16A	1969 Feb 26.32 10000 years	Cylinder 145	0.56 long 1.07 dia	1969 Mar 3.4	101.79	115.28	7846	1427	1508	0.005	119
	Essa 9 rocket	1969-16B	1969 Feb 26.32 5000 years	Cylinder 24	1.50 long 0.46 dia	1969 Mar 1.3	101.79	115.21	7842	1423	1505	0.005	120
D R	Cosmos 267	1969-17A	1969 Feb 26.35 7.93 days 1969 Mar 6.28	Sphere-cylinder 5530?	5 long? 2.4 dia	1969 Feb 27.2	65.04	89.82	6645	205	329	0.009	50
D	Cosmos 267 rocket	1969-17B	1969 Feb 26.35 7.12 days 1969 Mar 5.47	Cylinder 2500?	7.5 long 2.6 dia	1969 Feb 27.2	64.99	89.63	6636	204	311	0.008	45

Space Vehicles: Mariner 6, 1969-14A; and Centaur rocket, 1969-14B.

* Environmental Science Services Administration.

	Name		Launch date, lifetime and descent date	Shape and weight (kg)	Size (m)	Date of orbital determination	Orbital inclination (deg)	Nodal period (min)	Semi major axis (km)	Perigee height (km)	Apogee height (km)	Orbital eccentricity	Argument of perigee (deg)
D 3M R	Apollo 9 (CM + SM)	1969-18A	1969 Mar 3.67 10.04 days 1969 Mar 13.71	Cone- cylinder 22030 full 11205 empty	10.36 long 3.91 dia	1969 Mar 4.5 1969 Mar 4.9 1969 Mar 9.5	32.57 33.84 33.63	88.64 91.46 88.49	6594 6733 6587	203 198 194	229 511 223	0.002 0.023 0.002	67 70 94
	Saturn IV B [Saturn 504]	1969-18B	1969 Mar 3.67 indefinite	Cylinder 13400	18.7 long 6.6 dia	1969 Mar 3.7 1969 Mar 3.9	32.57 32.57	88.15 118.6	6569 7999	191 191	191 3050	0 0.179	- -
							Entered heliocentric orbit 1969 Mar 3.92						
D	LEM 3* Ascent stage	1969-18C	1969 Mar 3.67 4617 days 1981 Oct 23	Box + 2 tanks 4450 full 2300 empty	2.52 high 3.76 wide 3.13 deep	1969 Mar 8.6 1971 Feb 1.0 1974 Mar 1.0	28.91 28.91 28.90	164.70 154.61 142.2	9959 9546 9024	227 234 225	6935 6102 5067	0.337 0.307 0.268	130 - -
D	LEM 3 Descent stage	1969-18D	1969 Mar 3.67 18.49 days 1969 Mar 22.16	Octagon + cone + legs 10075 full 1935 empty	1.57 high 3.13 wide	1969 Mar 9.4	33.63	89.25	6622	242	246	0.0004	256
D	[Titan 3B Agena D]	1969-19A	1969 Mar 4.81 14 days 1969 Mar 18	Cylinder 3000?	8 long? 1.5 dia	1969 Mar 5.2	92.00	90.50	6676	134	461	0.024	147
D	Cosmos 268	1969-20A	1969 Mar 5.55 429.78 days 1970 May 9.33	Ellipsoid 400?	1.8 long 1.2 dia	1969 Mar 5.9 1969 Jul 16.3 1969 Dec 16.3	48.40 48.37 48.3	109.14 105.08 100.29	7569 7380 7153	209 209 202	2173 1795 1348	0.130 0.107 0.080	102 - -
D	Cosmos 268 rocket	1969-20B	1969 Mar 5.55 342.52 days 1970 Feb 11.07	Cylinder 1500?	8 long 1.65 dia	1969 Mar 7.3 1969 Jun 16.0 1969 Oct 16.3	48.40 48.3 48.3	109.12 104.87 100.60	7568 7370 7169	221 219 212	2159 1764 1369	0.128 0.105 0.081	107 - -

* LEM attached to Apollo 9, separated from Saturn IV B on Mar 3.84. LEM is Lunar Excursion Module.
LEM with two crew members separated from Apollo 9 on Mar 7.53.
LEM ascent stage separated from descent stage on Mar 7.71; briefly re-docked with Apollo 9 on Mar 7.79.

	Name		Launch date, lifetime and descent date	Shape and weight (kg)	Size (m)	Date of orbital determination	Orbital inclination (deg)	Nodal period (min)	Semi major axis (km)	Perigee height (km)	Apogee height (km)	Orbital eccentricity	Argument of perigee (deg)
D	Cosmos 269	1969-21A	1969 Mar 5.73 3517 days 1978 Oct 21	Cylinder + paddles? 900?	2 long? 1 dia?	1969 Mar 9.5 1971 Dec 1.0	74.05 74.05	95.34 94.47	6912 6870	525 487	543 497	0.001 0.001	340 -
D	Cosmos 269 rocket	1969-21B	1969 Mar 5.73 3424 days 1978 Jul 20	Cylinder 2200?	7.4 long 2.4 dia	1969 Mar 9.5 1972 Jan 1.0	74.17 74.17	95.52 94.63	6922 6878	540 498	547 502	0.0005 0.0003	263 -
20d	Fragments	1969-21C-Y											
D R	Cosmos 270	1969-22A	1969 Mar 6.51 7.74 days 1969 Mar 14.25	Sphere- cylinder 5530?	5 long? 2.4 dia	1969 Mar 9.1	65.43	89.81	6644	200	331	0.010	53
D	Cosmos 270 rocket	1969-22B	1969 Mar 6.51 5.94 days 1969 Mar 12.45	Cylinder 2500?	7.5 long 2.6 dia	1969 Mar 9.4	65.41	89.22	6615	185	288	0.008	39
D R	Cosmos 271	1969-23A	1969 Mar 15.51 7.78 days 1969 Mar 23.29	Sphere- cylinder 5530?	5 long? 2.4 dia	1969 Mar 16.2	65.40	89.71	6638	196	324	0.010	47
D	Cosmos 271 rocket	1969-23B	1969 Mar 15.51 6.58 days 1969 Mar 22.09	Cylinder 2500?	7.5 long 2.6 dia	1969 Mar 19.2	65.40	88.71	6588	168	252	0.006	38
	Cosmos 272	1969-24A	1969 Mar 17.70 3000 years	Spheroid + paddles? 650?	1.6 dia?	1969 Mar 19.1	73.99	109.35	7574	1181	1211	0.002	260
	Cosmos 272 rocket	1969-24B	1969 Mar 17.70 2000 years	Cylinder 2200?	7.4 long 2.4 dia	1969 Mar 21.2	73.98	109.25	7569	1178	1203	0.002	231
	Fragment	1969-24C											

	Name		Launch date, lifetime and descent date	Shape and weight (kg)	Size (m)	Date of orbital determination	Orbital inclination (deg)	Nodal period (min)	Semi major axis (km)	Perigee height (km)	Apogee height (km)	Orbital eccentricity	Argument of perigee (deg)
D	OV1-17*	1969-25A	1969 Mar 18.32 352.23 days 1970 Mar 5.55	Cylinder + 2 hemi- spheres 142	1.40 long 0.69 dia	1969 Mar 28.7 1969 Jul 16.3 1969 Nov 16.0	99.18 99.1 99.1	93.21 92.59 91.83	6808 6778 6741	397 374 346	463 425 379	0.005 0.004 0.002	342 - -
D	OV1-18	1969-25B	1969 Mar 18.32 1258.68 days 1972 Aug 28.00	Cylinder + 2 hemi- spheres 125	1.40 long 0.69 dia	1969 Mar 20.9 1970 May 1.0 1971 Nov 1.0	98.86 98.86 98.86	95.15 94.35 92.92	6903 6864 6794	466 443 395	583 528 436	0.008 0.006 0.003	157 - -
	OV1-19	1969-25C	1969 Mar 18.32 300 years	Cylinder + 2 hemi- spheres 124	1.40 long 0.69 dia	1969 Mar 20.8	104.77	153.44	9493	466	5764	0.279	175
D	OV1-17A (Orbiscal 2)**	1969-25D	1969 Mar 18.32 6.23 days 1969 Mar 24.55	Cylinder 221	2.05 long 0.72 dia	1969 Mar 18.9	99.10	89.37	6620	175	309	0.010	152
	OV1-19 rocket	1969-25E	1969 Mar 18.32 150 years	Cylinder 70?	2.05 long 0.72 dia	1969 Mar 28.2	104.78	153.43	9493	463	5767	0.279	168
D	OV1-18 rocket	1969-25F	1969 Mar 18.32 3650 days 1979 Mar 16	Cylinder 70?	2.05 long 0.72 dia	1969 Mar 23.3 1972 Jan 16.5	98.85 98.85	95.15 94.35	6903 6864	465 441	584 530	0.009 0.006	149 -
D	Fragment	1969-25G											

* Orbiting Vehicle.

** Orbiting radio beacon ionospheric satellite - calibration.

	Name		Launch date, lifetime and descent date	Shape and weight (kg)	Size (m)	Date of orbital determination	Orbital inclination (deg)	Nodal period (min)	Semi major axis (km)	Perigee height (km)	Apogee height (km)	Orbital eccentricity	Argument of perigee (deg)
D	[Thorad Agena D]	1969-26A	1969 Mar 19.90 4.35 days 1969 Mar 24.25	Cylinder 2000?	8 long? 1.5 dia	1969 Mar 21.7	83.04	88.73	6588	179	241	0.005	137
D	Capsule*	1969-26B	1969 Mar 19.90 991.68 days 1971 Dec 6.58	Octagon? 60?	0.3 long? 0.9 dia?	1969 Mar 26.2 1970 Oct 31.7 1971 Jun 1.0	83.08 83.0 83.0	94.82 93.34 92.19	6886 6815 6758	504 436 378	513 438 382	0.0006 0.0001 0.0003	25 - -
D R	Cosmos 273	1969-27A	1969 Mar 22.51 7.73 days 1969 Mar 30.24	Sphere-cylinder 5530?	5 long? 2.4 dia	1969 Mar 26.4	65.43	89.78	6642	198	329	0.010	52
D	Cosmos 273 rocket	1969-27B	1969 Mar 22.51 6.40 days 1969 Mar 28.91	Cylinder 2500?	7.5 long 2.6 dia	1969 Mar 26.2	65.43	89.06	6606	188	267	0.006	68
D R	Cosmos 274**	1969-28A	1969 Mar 24.42 7.90 days 1969 Apr 1.32	Sphere-cylinder 5530?	5 long? 2.4 dia	1969 Mar 26.2	64.98	89.56	6631	206	300	0.007	49
D	Cosmos 274 rocket	1969-28B	1969 Mar 24.42 5.26 days 1969 Mar 29.68	Cylinder 2500?	7.5 long 2.6 dia	1969 Mar 26.0	65.00	89.26	6616	206	270	0.005	54
	Meteor 1	1969-29A	1969 Mar 26.52 60 years	Cylinder + 2 vanes 2200?	5 long? 1.5 dia?	1969 Mar 30.0	81.20	97.96	7038	633	687	0.004	80
	Meteor 1 rocket	1969-29B	1969 Mar 26.52 20 years	Cylinder 1440	3.8 long 2.6 dia	1969 Mar 31.1	81.21	98.14	7046	466	870	0.029	175
34d	Fragments	1969-29C-AP											

Space Vehicles: Mariner 7, 1969-30A; and Centaur rocket 1969-30B.

*1969-26B ejected from 1969-26A about Mar 19.97.

** Carried supplementary atmospheric-composition experiment.

	Name		Launch date, lifetime and descent date	Shape and weight (kg)	Size (m)	Date of orbital determination	Orbital inclination (deg)	Nodal period (min)	Semi major axis (km)	Perigee height (km)	Apogee height (km)	Orbital eccentricity	Argument of perigee (deg)
D	Cosmos 275	1969-31A	1969 Mar 28.67 315.61 days 1970 Feb 7.28	Ellipsoid 400?	1.8 long 1.2 dia	1969 Mar 30.0 1969 Sep 16.0	70.98 70.9	95.18 93.42	6905 6821	273 261	780 624	0.037 0.027	80 -
D	Cosmos 275 rocket	1969-31B	1969 Mar 28.67 171.76 days 1969 Sept 16.43	Cylinder 1500?	8 long 1.65 dia	1969 Mar 29.7 1969 Jun 16.0	70.96 70.9	95.07 93.30	6899 6815	274 260	768 613	0.036 0.026	82 -
D R	Cosmos 276	1969-32A	1969 Apr 4.43 6.90 days 1969 Apr 11.33	Sphere-cylinder 5530?	5 long? 2.4 dia	1969 Apr 6.2	81.36	90.25	6664	200	371	0.013	79
D	Cosmos 276 rocket	1969-32B	1969 Apr 4.43 10.07 days 1969 Apr 14.50	Cylinder 2500?	7.5 long 2.6 dia	1969 Apr 7.1	81.35	89.94	6648	203	337	0.010	73
D	Cosmos 277	1969-33A	1969 Apr 4.54 92.57 days 1969 Jul 6.11	Ellipsoid 400?	1.8 long 1.2 dia	1969 Apr 6.2	70.95	91.90	6745	268	466	0.015	87
D	Cosmos 277 rocket	1969-33B	1969 Apr 4.54 43.68 days 1969 May 18.22	Cylinder 1500?	8 long 1.65 dia	1969 Apr 6.8	70.94	91.77	6739	266	455	0.014	86
D R	Cosmos 278	1969-34A	1969 Apr 9.54 7.78 days 1969 Apr 17.32	Sphere-cylinder 5530?	5 long? 2.4 dia	1969 Apr 14.8	65.42	89.58	6632	198	310	0.008	44
D	Cosmos 278 rocket	1969-34B	1969 Apr 9.54 7 days 1969 Apr 16	Cylinder 2500?	7.5 long 2.6 dia	1969 Apr 10.2	65.40	89.49	6628	190	309	0.009	36

	Name		Launch date, lifetime and descent date	Shape and weight (kg)	Size (m)	Date of orbital determination	Orbital inclination (deg)	Nodal period (min)	Semi major axis (km)	Perigee height (km)	Apogee height (km)	Orbital eccentricity	Argument of perigee (deg)
D	Molniya IL	1969-35A	1969 Apr 11.11 1832 days 1974 Apr 17	Windmill + 6 vanes 1000?	3.4 long 1.6 dia	1969 Apr 22.1 1970 May 1.0 1971 May 16.5	64.94 65.07 65.15	713.50 717.67 717.44	26451 26554 26548	404 1060 1368	39741 39291 38972	0.744 0.720 0.708	285 - -
D	Molniya IL launcher rocket	1969-35B	1969 Apr 11.11 20.09 days 1969 May 1.20	Cylinder 2500?	7.5 long 2.6 dia	1969 Apr 12.8	65.02	90.94	6699	234	408	0.013	60
D	Molniya IL rocket	1969-35C	1969 Apr 11.11 1907.00 days 1974 Jul 1.11	Cylinder 440	2.0 long 2.0 dia	1969 Apr 24.5 1970 May 1.0 1971 Jun 1.0	64.98 65.07 65.1	710.28 710.28 710.24	26371 26371 26370	461 1037 1350	39524 39848 38633	0.741 0.719 0.707	285 - -
D	Molniya IL launcher	1969-35D	1969 Apr 11.11 24.03 days 1969 May 5.14	Irregular	-	1969 Apr 13.4	65.00	91.36	6720	222	461	0.018	65
	BMEWS 2† [Atlas Agena D]*	1969-36A	1969 Apr 13.10 > million years	Cylinder 700 full? 350 empty?	1.7 long? 1.4 dia?	1969 Apr 14	9.9?	1445	42350	32670	39270	0.078	-
	Nimbus 3	1969-37A	1969 Apr 14.33 800 years	Conical skeleton + 2 paddles 575	3.00 long 1.45 dia	1969 Apr 25.5	99.91	107.40	7483	1075	1135	0.004	215
	Secor 13** (EGRS 13)††	1969-37B	1969 Apr 14.33 2000 years	Rectangular box 20	0.33 x 0.28 x 0.23	1969 Apr 15.4	99.93	107.36	7481	1075	1130	0.004	230
	Nimbus 3 rocket	1969-37C	1969 Apr 14.33 1000 years	Cylinder 700?	6 long? 1.5 dia	1969 Apr 16.0	99.92	107.50	7488	1078	1141	0.004	-

* The Agena D rocket (1969-36B) is probably in an eccentric orbit like that of 1970-46B.

** Secor is Sequential collation of range.

† Ballistic Missile Early Warning Satellite.

†† Electronic Geodetic Ranging System.

	Name		Launch date, lifetime and descent date	Shape and weight (kg)	Size (m)	Date of orbital determination	Orbital inclination (deg)	Nodal period (min)	Semi major axis (km)	Perigee height (km)	Apogee height (km)	Orbital eccentricity	Argument of perigee (deg)
D R	Cosmos 279†	1969-38A	1969 Apr 15.35 7.98 days 1969 Apr 23.33	Sphere-cylinder 5530?	5 long? 2.4 dia	1969 Apr 18.4	51.74	89.04	6608	192	267	0.006	25
D	Cosmos 279 rocket	1969-38B	1969 Apr 15.35 2.76 days 1969 Apr 18.11	Cylinder 2500?	7.5 long 2.6 dia	1969 Apr 16.7	51.74	88.65	6589	169	253	0.006	9
D R	[Titan 3B Agena D]	1969-39A	1969 Apr 15.73 15 days 1969 Apr 30	Cylinder 3000?	8 long? 1.5 dia	1969 Apr 20.1 1969 Apr 30.5	108.76 108.76	89.96 89.03	6651 6605	135 130?	410 323?	0.021 0.015?	127 104
D R	Cosmos 280††	1969-40A	1969 Apr 23.42 12.86 days 1969 May 6.28	Sphere-cylinder 6300?	6.5 long? 2.4 dia	1969 Apr 25.1 1969 Apr 28.7	51.60 51.60	89.02 89.50	6607 6631	207 198	250 307	0.003 0.008	30 60
D	Cosmos 280 rocket	1969-40B	1969 Apr 23.42 3.38 days 1969 Apr 26.80	Cylinder 2500?	7.5 long 2.6 dia	1969 Apr 24.5	51.60	88.83	6598	196	243	0.003	38
D	Cosmos 280 engine*	1969-40C	1969 Apr 23.42 14.76 days 1969 May 8.18	Cone 600? full	1.5 long? 2 dia?	1969 May 5	51.58	89.06	6610	205	258	0.004	-
D	Fragment	1969-40D											
D	[Thorad Agena D]	1969-41A	1969 May 2.08 21.35 days 1969 May 23.43	Cylinder 2000?	8 long? 1.5 dia	1969 May 2.2	64.97	89.54	6631	179	326	0.011	98
D	Capsule**	1969-41B	1969 May 2.08 290.78 days 1970 Feb 16.86	Octagon? 60?	0.3 long? 0.9 dia?	1969 May 2.3 1969 Aug 16.3 1969 Nov 16.0	65.71 65.71 65.71	93.37 92.96 92.28	6815 6796 6763	401 393 361	473 442 408	0.005 0.004 0.003	39 - -
D	Fragment	1969-41C											

* 1969-40C was ejected from 1969-40A about May 5 (approximate orbit).
** 1969-41B was ejected from 1969-41A about May 2.15.

† Carried supplementary ion orientation system.
†† Carried supplementary charged-particles experiment.

	Name		Launch date, lifetime and descent date	Shape and weight (kg)	Size (m)	Date of orbital determination	Orbital inclination (deg)	Nodal period (min)	Semi major axis (km)	Perigee height (km)	Apogee height (km)	Orbital eccentricity	Argument of perigee (deg)
D R	Cosmos 281	1969-42A	1969 May 13.39 7.74 days 1969 May 21.13	Sphere-cylinder 5530?	5 long? 2.4 dia	1969 May 14.1	65.42	89.43	6625	191	303	0.008	33
D	Cosmos 281 rocket	1969-42B	1969 May 13.39 3.65 days 1969 May 17.04	Cylinder 2500?	7.5 long 2.6 dia	1969 May 14.1	65.40	89.28	6617	181	297	0.009	23
D R 3M	Apollo 10**	1969-43A	1969 May 18.70 8.00 days 1969 May 26.70	Cone-cylinder 28870 full 12460 empty	11.15 long 3.91 dia	1969 May 18.7 1969 May 18.9	32.56 33.2	88.03 2400	6562 273000	183 174	184 533000	0 0.976	107 30*
							In selenocentric orbit 1969 May 21.87 - May 24.42						
	Saturn IVB [Saturn 505]	1969-43B	1969 May 18.70 Indefinite	Cylinder 13600	18.7 long 6.6 dia	1969 May 18.7	32.56	88.03	6562	183	184	0	107
							Entered heliocentric orbit 1969 May 18.89						
	LEM 4 †	1969-43C	1969 May 18.70 Indefinite	Box + Octagon + legs 13993 full	4.1 high 3.76 wide 3.13 deep	1969 May 18.9	33.2	2400	273000	174	533000	0.976	30*
							Entered selenocentric orbit 1969 May 21.87						
D R	Cosmos 282	1969-44A	1969 May 20.36 7.7 days 1969 May 28.1	Sphere-cylinder 5530?	5 long? 2.4 dia	1969 May 21.8	65.40	89.73	6640	202	321	0.009	52
D	Cosmos 282 rocket	1969-44B	1969 May 20.36 8.32 days 1969 May 28.68	Cylinder 2500?	7.5 long 2.6 dia	1969 May 22.5	65.40	89.45	6626	206	290	0.006	41
	Intelsat 3 F-4	1969-45A	1969 May 22.08 > million years	Cylinder 293 full 137 empty	1.04 long 1.42 dia	1969 Jun 1.0 1969 Jul 1.0 1977 May 13.0	0.5 0.5 6.09	1418.9 1436.5 1636.33	41827 42172 45998	35226 35787 39487	35671 35801 39753	0.005 0.0002 0.003	- - 199
D	Intelsat 3 F-4 rocket	1969-45B	1969 May 22.08 2846 days 1977 Mar 7	Cylinder 24	1.50 long 0.46 dia	1972 May 1.0	28.5	640.9	24623	396	36093	0.725	-

* Approximate orbits. ** Apollo attached to LEM, separated from Saturn IVB on May 18.89.

† LEM with 2 crew members, separated from Apollo on May 22.80. LEM ascent stage jettisoned descent stage May 22.98; re-joined Apollo May 23.13; now in solar orbit.

	Name		Launch date, lifetime and descent date	Shape and weight (kg)	Size (m)	Date of orbital determination	Orbital inclination (deg)	Nodal period (min)	Semi major axis (km)	Perigee height (km)	Apogee height (km)	Orbital eccentricity	Argument of perigee (deg)
	OV5-5 (ERS 29)*	1969-46A	1969 May 23.33 million years	Octahedron 11	0.31 side	1969 May 24.3	33.03	3120.3	70736	17069	111647	0.668	184
	OV5-6 (ERS 26)	1969-46B	1969 May 23.33 million years	Tetrahedron 10	0.28 side	1969 May 24.3	32.86	3115.2	70658	16923	111636	0.670	184
	OV5-9	1969-46C	1969 May 23.33 million years	Modified Tetrahedron 13	0.28 side	1969 May 24.3	32.7	3115.4	70661	17046	111519	0.668	184
	Vela 9	1969-46D	1969 May 23.33 > million years	Icosahedron 259	1.27 dia	1969 May 24.5	32.8	6703	117933	110900	112210	0.006	-
	Vela 10	1969-46E	1969 May 23.33 > million years	Icosahedron 259	1.27 dia	1969 May 25.5	32.8	6709	117980	110920	112283	0.006	-
	Vela 9 rocket [Titan 3C]	1969-46F	1969 May 23.33 > million years	Cylinder 1500?	6 long? 3.0 dia		Orbit similar to 1969-46A						
D	Cosmos 283	1969-47A	1969 May 27.54 197.25 days 1969 Dec 10.79	Ellipsoid 400?	1.8 long 1.2 dia	1969 May 31.0 1969 Aug 31.7	81.94 81.9	101.95 98.48	7228 7065	199 195	1501 1178	0.090 0.070	64 -
D	Cosmos 283 rocket	1969-47B	1969 May 27.54 126.48 days 1969 Oct 1.02	Cylinder 1500?	8 long 1.65 dia	1969 May 30.5 1969 Aug 1.0	81.94 81.9	101.80 98.35	7221 7058	199 199	1486 1161	0.089 0.068	66 -

* Environment Research Satellite.

	Name		Launch date, lifetime and descent date	Shape and weight (kg)	Size (m)	Date of orbital determination	Orbital inclination (deg)	Nodal period (min)	Semi major axis (km)	Perigee height (km)	Apogee height (km)	Orbital eccentricity	Argument of perigee (deg)
D R	Cosmos 284	1969-48A	1969 May 29.29 7.95 days 1969 Jun 6.24	Sphere-cylinder 5530?	5 long? 2.4 dia	1969 May 31.5	51.76	89.45	6628	205	294	0.007	46
D	Cosmos 284 rocket	1969-48B	1969 May 29.29 6.02 days 1969 Jun 4.31	Cylinder 2500?	7.5 long 2.6 dia	1969 May 30.5	51.75	89.28	6620	197	286	0.007	30
D	Fragment	1969-48C											
D	Cosmos 285	1969-49A	1969 Jun 3.54 126.09 days 1969 Oct 7.63	Ellipsoid 400?	1.8 long 1.2 dia	1969 Jun 4.8 1969 Aug 16.3	71.03 70.98	92.16 91.11	6758 6707	267 249	493 409	0.017 0.012	85 -
D	Cosmos 285 rocket	1969-49B	1969 Jun 3.54 62.89 days 1969 Aug 5.43	Cylinder 1500?	8 long 1.65 dia	1969 Jun 3.9	71.03	92.06	6753	266	484	0.016	95
D R	[Titan 3B Agena D]	1969-50A	1969 Jun 3.70 11.2 days 1969 Jun 14.9	Cylinder 3000?	8 long? 1.5 dia	1969 Jun 4.1	110.00	90.04	6654	137	414	0.021	131
D	OGO 6*	1969-51A	1969 Jun 5.61 3781 days 1979 Oct 12	Box + booms 620	1.73 long 0.84 high 0.84 wide	1969 Jun 17.2 1971 Nov 1.0	82.00 82.00	99.71 98.46	7121 7062	397 391	1089 977	0.049 0.042	123 -
D	OGO 6 rocket	1969-51B	1969 Jun 5.61 4025 days 1980 Jun 12	Cylinder 700?	6.5 long 1.5 dia	1969 Jun 17.2 1972 Feb 1.0	82.00 82.00	99.66 98.43	7118 7061	396 402	1084 963	0.049 0.040	118 -
D R	Cosmos 286	1969-52A	1969 Jun 15.38 7.78 days 1969 Jun 23.16	Sphere-cylinder 5530?	5 long? 2.4 dia	1969 Jun 17.0	65.41	89.78	6642	200	327	0.010	51
D	Cosmos 286 rocket	1969-52B	1969 Jun 15.38 7.95 days 1969 Jun 23.33	Cylinder 2500?	7.5 long 2.6 dia	1969 Jun 17.2	65.40	89.43	6625	188	305	0.009	40

* Orbiting Geophysical Observatory.

	Name		Launch date, lifetime and descent date	Shape and weight (kg)	Size (m)	Date of orbital determination	Orbital inclination (deg)	Nodal period (min)	Semi major axis (km)	Perigee height (km)	Apogee height (km)	Orbital eccentricity	Argument of perigee (deg)
D	Explorer 41 (Imp 7)*	1969-53A	1969 Jun 21.37 1281 days 1972 Dec 23	Octagon + 4 vanes. 79	0.25 long 0.71 dia	1969 Jun 21.9 1970 Feb 15.3 1971 Jun 16.0	86.78 86.65 85.06	4843.5 4836.9 4840.9	94784 94743 94794	378 2083 3920	176,434 174,646 172,912	0.929 0.911 0.891	200 - -
D	Explorer 41 rocket	1969-53B	1969 Jun 21.37 42 months? 1972 Dec ?	Cylinder 24	1.57 long 0.50 dia	Orbit similar to 1969-53A							
D R	Cosmos 287	1969-54A	1969 Jun 24.29 7.96 days 1969 Jul 2.25	Sphere-cylinder 5530?	5 long? 2.4 dia	1969 Jun 25.1	51.77	88.95	6604	188	264	0.006	3
D	Cosmos 287 rocket	1969-54B	1969 Jun 24.29 2.83 days 1969 Jun 27.12	Cylinder 2500?	7.5 long 2.6 dia	1969 Jun 25.3	51.76	88.62	6588	171	249	0.006	351
D R	Cosmos 288	1969-55A	1969 Jun 27.30 7.98 days 1969 Jul 5.28	Sphere-cylinder 5530?	5 long? 2.4 dia	1969 Jun 28.0	51.76	89.17	6614	199	273	0.006	24
D	Cosmos 288 rocket	1969-55B	1969 Jun 27.30 5.29 days 1969 Jul 2.59	Cylinder 2500?	7.5 long 2.6 dia	1969 Jun 28.9	51.75	88.94	6604	193	258	0.005	29
D B R	Bios 3 capsule**	1969-56D	1969 Jun 29.14 8.85 days 1969 Jul 7.99	Blunt cone 259	1.2 long 1.02 dia	1969 Jul 1.1	33.56	91.86	6750	356	388	0.002	244
D	Bios 3 adapter	1969-56A	1969 Jun 29.14 205.43 days 1970 Jan 20.57	Cone-cylinder 440	1.8 long 1.45 dia	1969 Jul 1.1 1969 Oct 16.3	33.56 33.5	91.86 91.29	6750 6723	356 333	388 357	0.002 0.002	244 -
D	Bios 3 rocket	1969-56B	1969 Jun 29.14 155.82 days 1969 Dec 1.96	Cylinder 350?	4.9 long 1.43 dia	1969 Jul 1.8 1969 Sep 16.0	33.56 33.5	91.74 91.25	6745 6721	351 334	382 352	0.002 0.001	260 -
D	Fragment	1969-56C											

* Interplanetary monitoring platform.

** Before 1969 Jul 7.99, Bios capsule, which carried a monkey, was attached to Bios 3 adapter.

	Name		Launch date, lifetime and descent date	Shape and weight (kg)	Size (m)	Date of orbital determination	Orbital inclination (deg)	Nodal period (min)	Semi major axis (km)	Perigee height (km)	Apogee height (km)	Orbital eccentricity	Argument of perigee (deg)
D R	Cosmos 289	1969-57A	1969 Jul 10.38 4.80 days 1969 Jul 15.18	Sphere-cylinder 5530?	5 long? 2.4 dia	1969 Jul 14.1	65.40	89.69	6638	194	325	0.010	41
D	Cosmos 289 rocket	1969-57B	1969 Jul 10.38 5.69 days 1969 Jul 16.07	Cylinder 2500?	7.5 long 2.6 dia	1969 Jul 11.1	65.39	89.55	6631	191	314	0.009	36
D	Luna 15 launcher rocket	1969-58B	1969 Jul 13.13 3.70 days 1969 Jul 16.83	Cylinder 4000?	12 long? 4 dia	1969 Jul 13.4	51.55	88.72	6593	182	247	0.005	352
D	Luna 15 launcher	1969-58C	1969 Jul 13.13 3.60 days 1969 Jul 16.73	-	-	1969 Jul 13.3	51.50	88.71	6592	190	238	0.004	320
D 3M R	Apollo 11**	1969-59A	1969 Jul 16.56 8.14 days 1969 Jul 24.70	Cone-cylinder 28800 full 12050 empty	11.15 long 3.91 dia	1969 Jul 16.6 1969 Jul 16.8	32.51 33.3 In selenocentric orbit 1969 Jul 19.73 to Jul 22.21	88.03 24400	6562 273 000	183 174	184 533 000	0 0.976	183 30*
	Saturn IV B [Saturn 506]	1969-59B	1969 Jul 16.56 Indefinite	Cylinder 13300	18.7 long 6.6 dia	1969 Jul 16.6	32.51 Entered heliocentric orbit 1969 Jul 16.76	88.03	6562	183	184	0	183
D	LEM 5† Descent stage	1969-59D	1969 Jul 16.56 4.29 days 1969 Jul 20.85	Octagon + 4 legs 10243 full 2033 empty	1.57 high 3.13 wide	1969 Jul 16.8	33.3 Entered selenocentric orbit 1969 Jul 19.73 Landed on Moon 1969 July 20.85	24400	273 000	174	533 000	0.976	30*
	LEM 5 Ascent stage	1969-59C	1969 Jul 16.56 Indefinite	Box + 2 tanks 4818 full 2179 empty	2.52 high 3.76 wide 3.13 deep	1969 Jul 16.8	33.3 On Moon's surface 1969 Jul 20.85 to Jul 21.75 Now in selenocentric orbit	24400	273 000	174	533 000	0.976	30*

** Apollo attached to LEM, separated from Saturn IV B on Jul 16.76. First manned landing on Moon.
† LEM with 2 crew members, separated from Apollo on Jul 20.74
LEM ascent stage launched from Moon Jul 21.75; re-joined Apollo Jul 21.90.

* Approximate orbits.
Space Vehicle: Luna 15, 1969-58A

	Name		Launch date, lifetime and descent date	Shape and weight (kg)	Size (m)	Date of orbital determination	Orbital inclina-tion (deg)	Nodal period (min)	Semi major axis (km)	Perigee height (km)	Apogee height (km)	Orbital eccen-tricity	Argument of perigee (deg)
D R	Cosmos 290	1969-60A	1969 Jul 22.52 7.79 days 1969 Jul 30.31	Sphere-cylinder 5530?	5 long? 2.4 dia	1969 Jul 24.1	65.40	89.75	6641	194	332	0.010	44
D	Cosmos 290 rocket	1969-60B	1969 Jul 22.52 6.75 days 1969 Jul 29.27	Cylinder 2500?	7.5 long 2.6 dia	1969 Jul 24.1	65.41	89.52	6630	183	320	0.010	39
D	Molniya 1M	1969-61A	1969 Jul 22.54 696 days 1971 Jun 18	Windmill + 6 vanes 1000?	3.4 long 1.6 dia	1969 Jul 31.5 1969 Sep 16.0 1970 Oct 31.7	64.90 64.90 64.90	710.94 717.65 717.46	26387 26553 26549	499 525 316	39519 39825 40025	0.740 0.740 0.748	285 - -
D	Molniya 1M launcher rocket	1969-61B	1969 Jul 22.54 36.89 days 1969 Aug 28.43	Cylinder 2500?	7.5 long 2.6 dia	1969 Jul 23.9	64.93	91.73	6738	228	492	0.020	72
D	Molniya 1M launcher	1969-61C	1969 Jul 22.54 32.32 days 1969 Aug 23.86	Irregular	-	1969 Jul 23.9	64.93	91.90	6747	213	524	0.023	69
D	Molniya 1M rocket	1969-61D	1969 Jul 22.54 688 days 1971 Jun 10	Cylinder 440	2.0 long 2.0 dia	1969 Jul 29.5 1969 Dec 1.0 1970 Oct 31.7	64.89 64.88 64.88	707.80 707.84 707.58	26309 26311 26304	486 548 307	39376 39318 39545	0.739 0.737 0.746	285 - -
D	Fragments	1969-61E-G											
	[Thor Burner 2]	1969-62A	1969 Jul 23.19 80 years	12-faced Frustum 195	1.64 long 1.31 to 1.10 dia	1969 Jul 25.1	98.80	101.36	7200	788	856	0.005	2
	Burner 2 rocket	1969-62B	1969 Jul 23.19 80 years	Sphere-cone 66	1.32 long 0.94 dia	1969 Jul 23.6	98.78	101.36	7200	787	857	0.005	1

	Name		Launch date, lifetime and descent date	Shape and weight (kg)	Size (m)	Date of orbital determination	Orbital inclination (deg)	Nodal period (min)	Semi major axis (km)	Perigee height (km)	Apogee height (km)	Orbital eccentricity	Argument of perigee (deg)
D	[Thorad Agena D]	1969-63A	1969 Jul 24.06 30.44 days 1969 Aug 23.50	Cylinder 2000?	8 long? 1.5 dia	1969 Jul 27.9	74.98	88.49	6577	178	220	0.003	123
	Intelsat 3 F-5	1969-64A	1969 Jul 26.09 20 years	Cylinder 292 full	1.04 long 1.42 dia	1969 Jul 30.7 1973 Jan 1.0	30.33 30.33	146.42 138.4	9212 8863	271 268	5397 4701	0.278 0.250	204 -
D	Intelsat 3 F-5 rocket	1969-64B	1969 Jul 26.09 355.67 days 1970 Jul 16.76	Sphere-cone 66	1.32 long 0.94 dia	1969 Aug 4.9 1970 Apr 14.4	30.37 30.26	146.17 111.65	9201 7689	271 261	5375 2360	0.277 0.137	227 140
19d	Fragments	1969-64C-X											
D	[Thorad Agena D]	1969-65A	1969 Jul 31.44 1253.28 days 1973 Jan 4.72	Cylinder 2000?	8 long? 1.5 dia	1969 Aug 1.5 1970 Oct 1.3 1971 Nov 16.0	75.02 74.97 75.00	94.67 93.98 93.25	6880 6847 6812	462 439 410	541 498 457	0.006 0.004 0.003	238 - -
D	Cosmos 291	1969-66A	1969 Aug 6.24 33.70 days 1969 Sep 8.94	Cylinder?	4 long? 2 dia?	1969 Aug 7.2	62.24	91.46	6726	147	548	0.030	63
D	Cosmos 291 rocket	1969-66B	1969 Aug 6.24 5.11 days 1969 Aug 11.35	Cylinder 1500?	8 long? 2.5 dia?	1969 Aug 8.0	62.23	90.71	6690	143	481	0.025	63
D	Zond 7 launcher	1969-67B	1969 Aug 8.00 2.17 days 1969 Aug 10.17	-	-	1969 Aug 8.7	51.51	88.17	6565	183	191	0.001	139
D	Zond 7 launcher rocket	1969-67C	1969 Aug 8.00 4.37 days 1969 Aug 12.37	Cylinder 4000?	12 long? 4 dia	1969 Aug 10.1	51.52	88.46	6580	190	214	0.002	330
D	Fragment	1969-67D											

Space vehicle: Zond 7, 1969-67A, passed around the Moon Aug 11.18; was recovered on Earth at Aug 14.76

	Name		Launch date, lifetime and descent date	Shape and weight (kg)	Size (m)	Date of orbital determination	Orbital inclination (deg)	Nodal period (min)	Semi major axis (km)	Perigee height (km)	Apogee height (km)	Orbital eccentricity	Argument of perigee (deg)
D	OSO 6	1969-68A	1969 Aug 9.33 4228 days 1981 Mar 7	Nonagonal box + vane 290	0.94 long 1.12 dia	1969 Aug 10.8	32.96	94.95	6901	491	554	0.005	282
D	PAC 1* (OSO 6 rocket)	1969-68B	1969 Aug 9.33 2819 days 1977 Apr 28	Cylinder 470? payload 120	4.9 long 1.43 dia	1969 Aug 13.8 1971 Feb 1.0	32.95 32.95	94.89 94.23	6898 6865	485 450	554 514	0.005 0.004	311 -
T	ATS 5 †	1969-69A	1969 Aug 12.46 > million years	Cylinder 821 full 433 empty	1.83 long 1.52 dia	1969 Aug 23.2 1969 Nov 1.0	2.6 2.5	1463.8 1435.9	42705 42162	35760 35777	36894 35790	0.013 0.0002	291 -
	ATS 5 rocket	1969-69B	1969 Aug 12.46 200 000 years	Cylinder 1815	8.6 long 3.0 dia	1969 Sep 16.0	17.55	703.1	26192	2209	37419	0.672	185
	ATS 5** apogee motor	1969-69D	1969 Aug 12.46 > million years	Cone 388? full 39? empty	1.3 long? 0.5 dia?	1969 Sep 5	2.6	1463.8	42705	35760	36894	0.013	291
	Fragment	1969-69C											
	Cosmos 292	1969-70A	1969 Aug 13.92 100 years	Cylinder + paddles? 750?	2 long? 1 dia?	1969 Aug 28.8	74.06	99.96	7134	746	765	0.001	71
	Cosmos 292 rocket	1969-70B	1969 Aug 13.92 80 years	Cylinder 2200?	7.4 long 2.4 dia	1969 Aug 22.6	74.06	99.85	7129	738	763	0.002	58
	Fragment	1969-70C											

* Package Attitude Control.
** Separated from ATS 5 on 1969 Sep 5.

† Applications Technology Satellite.

	Name		Launch date, lifetime and descent date	Shape and weight (kg)	Size (m)	Date of orbital determination	Orbital inclination (deg)	Nodal period (min)	Semi major axis (km)	Perigee height (km)	Apogee height (km)	Orbital eccentricity	Argument of perigee (deg)
D R	Cosmos 293*	1969-71A	1969 Aug 16.50 11.93 days 1969 Aug 28.43	Sphere-cylinder 5900?	5.9 long? 2.4 dia	1969 Aug 16.8	51.77	89.08	6610	208	256	0.004	44
D	Cosmos 293 rocket	1969-71B	1969 Aug 16.50 3.93 days 1969 Aug 20.43	Cylinder 2500?	7.5 long 2.6 dia	1969 Aug 16.8	51.77	88.94	6603	202	248	0.003	31
D R	Cosmos 294	1969-72A	1969 Aug 19.54 7.79 days 1969 Aug 27.33	Sphere-cylinder 5530?	5 long? 2.4 dia	1969 Aug 20.3	65.40	89.79	6643	200	329	0.010	48
D	Cosmos 294 rocket	1969-72B	1969 Aug 19.54 7.62 days 1969 Aug 27.16	Cylinder 2500?	7.5 long 2.6 dia	1969 Aug 20.3	65.40	89.65	6636	195	320	0.009	43
D	Cosmos 295	1969-73A	1969 Aug 22.60 101.15 days 1969 Dec 1.75	Ellipsoid 400?	1.8 long 1.2 dia	1969 Aug 28.1	71.01	91.95	6748	270	469	0.015	77
D	Cosmos 295 rocket	1969-73B	1969 Aug 22.60 49.56 days 1969 Oct 11.16	Cylinder 1500?	8 long 1.65 dia	1969 Aug 27.4	71.01	91.80	6741	272	453	0.013	79
D R	[Titan 3B Agena D]	1969-74A	1969 Aug 22.67 16 days 1969 Sep 7	Cylinder 3000?	8 long? 1.5 dia	1969 Aug 24.7 1969 Aug 25.0 1969 Sep 1.1 1969 Sep 2.1	108.00 107.99 107.99 107.96	89.51 89.74 89.29 89.84	6628 6639 6618 6644	133 136 130 130	366 386 349 402	0.018 0.019 0.017 0.021	130 131 119 122

* Telemetry suggests Cosmos 293 carried a pickaback capsule, but none was apparently tracked or designated

	Name		Launch date, lifetime and descent date	Shape and weight (kg)	Size (m)	Date of orbital determination	Orbital inclination (deg)	Nodal period (min)	Semi major axis (km)	Perigee height (km)	Apogee height (km)	Orbital eccentricity	Argument of perigee (deg)
D R	Cosmos 296	1969-75A	1969 Aug 29.38 7.90 days 1969 Sep 6.28	Sphere-cylinder 5530?	5 long? 2.4 dia	1969 Aug 31.2	64.95	89.59	6633	207	302	0.007	50
D	Cosmos 296 rocket	1969-75B	1969 Aug 29.38 6.91 days 1969 Sep 5.29	Cylinder 2500?	7.5 long 2.6 dia	1969 Aug 30.4	64.95	89.42	6624	210	282	0.005	41
D R	Cosmos 297	1969-76A	1969 Sep 2.46 7.84 days 1969 Sep 10.30	Sphere-cylinder 5530?	5 long? 2.4 dia	1969 Sep 5.8	72.89	89.66	6635	205	309	0.008	47
D	Cosmos 297 rocket	1969-76B	1969 Sep 2.46 9 days 1969 Sep 11	Cylinder 2500?	7.5 long 2.6 dia	1969 Sep 3.1	72.89	89.53	6629	205	296	0.007	48
D R?	Cosmos 298	1969-77A	1969 Sep 15.67 0.06 day 1969 Sep 15.73	Cylinder	2 long? 1 dia?	1969 Sep 15.7	49.60	87.31	6523	127	162	0.003	-
D	Cosmos 298 rocket	1969-77B	1969 Sep 15.67 0.3 day? 1969 Sep 16	Cylinder 1500?	8 long? 2.5 dia?	1969 Sep 15.9	49.55	87.21	6518	123	156	0.003	24
D	Cosmos 298 launch platform	1969-77C	1969 Sep 15.67 0.5 day? 1969 Sep 16	Irregular	-	1969 Sep 16.0	49.55	87.45	6530	134	169	0.003	69
D	Fragments	1969-77D,E											

	Name		Launch date, lifetime and descent date	Shape and weight (kg)	Size (m)	Date of orbital determination	Orbital inclination (deg)	Nodal period (min)	Semi major axis (km)	Perigee height (km)	Apogee height (km)	Orbital eccentricity	Argument of perigee (deg)
D R	Cosmos 299	1969-78A	1969 Sep 18.36 3.98 days 1969 Sep 22.34	Sphere-cylinder 5530?	5 long? 2.4 dia	1969 Sep 19.4	64.97	89.41	6624	207	284	0.006	62
D	Cosmos 299 rocket	1969-78B	1969 Sep 18.36 5.69 days 1969 Sep 24.05	Cylinder 2500?	7.5 long 2.6 dia	1969 Sep 19.5	64.98	89.16	6612	206	261	0.004	54
D	[Thorad Agena D]	1969-79A	1969 Sep 22.88 19.74 days 1969 Oct 12.62	Cylinder 2000?	8 long? 1.5 dia	1969 Sep 24.3	85.03	88.83	6594	178	253	0.006	134
D	Capsule*	1969-79B	1969 Sep 22.88 600.84 days 1971 May 16.72	Octagon? 60?	0.3 long? 0.9 dia?	1969 Sep 29.7 1970 Oct 1.3 1971 Feb 1.0	85.16 85.16 85.16	94.51 93.23 92.22	6871 6808 6760	490 427 380	496 434 383	0.0005 0.0005 0.0002	305 - -
D	Cosmos 300**	1969-80A	1969 Sep 23.59 4 days 1969 Sep 27	Cylinder 19400? full	7 long 3.9 dia	1969 Sep 24.3 1969 Sep 27.4	51.52 51.52	88.16 87.54	6565 6533	184 148	189 162	0.0004 0.0011	193 317
D	Cosmos 300 rocket	1969-80B	1969 Sep 23.59 3.44 days 1969 Sep 27.03	Cylinder 4000?	12 long? 4 dia	1969 Sep 24.9	51.54	88.03	6558	179	181	0.0002	203
D R	Cosmos 301	1969-81A	1969 Sep 24.51 7.76 days 1969 Oct 2.27	Sphere-cylinder 5530?	5 long? 2.4 dia	1969 Sep 25.4	65.41	89.34	6620	195	289	0.007	34
D	Cosmos 301 rocket	1969-81B	1969 Sep 24.51 5.00 days 1969 Sep 29.51	Cylinder 2500?	7.5 long 2.6 dia	1969 Sep 25.6	65.40	89.13	6610	198	266	0.005	19

* 1969-79B ejected from 1969-79A on 1969 Sep 22.95.

**Probably an attempted lunar probe. (Payload 5600 kg?)

	Name		Launch date, lifetime and descent date	Shape and weight (kg)	Size (m)	Date of orbital determination	Orbital inclination (deg)	Nodal period (min)	Semi major axis (km)	Perigee height (km)	Apogee height (km)	Orbital eccentricity	Argument of perigee (deg)
D	Capsule	1969-82A	1969 Sep 30.57 394.65 days 1970 Oct 30.22	Octagon? 60?	0.3 long? 0.9 dia?	1969 Oct 7.1 1970 Mar 9.9 1970 Jul 1.3	69.64 69.64 69.64	93.91 93.28 92.38	6843 6813 6768	446 419 381	484 450 399	0.003 0.002 0.001	58 189 -
	Timation 2*	1969-82B	1969 Sep 30.57 900 years	Cylinder + boom? 57	1.1 long 0.4 dia	1970 Mar 3.1	70.02	103.48	7301	906	940	0.002	217
	[Thorad Agena D]	1969-82C	1969 Sep 30.57 600 years	-	-	1970 Feb 27.2	70.02	103.49	7302	906	941	0.002	222
	[Thorad Agena D]	1969-82D	1969 Sep 30.57 600 years	-	-	1970 Jan 3.2	70.02	103.49	7302	907	940	0.002	284
	[Thorad Agena D]	1969-82E	1969 Sep 30.57 600 years	-	-	1970 Feb 28.2	70.02	103.49	7302	906	941	0.002	221
	[Thorad Agena D]	1969-82F	1969 Sep 30.57 600 years	-	-	1970 Feb 3.4	70.03	103.49	7302	906	941	0.002	251
	[Thorad Agena D]	1969-82G	1969 Sep 30.57 600 years	-	-	1970 Feb 2.4	70.01	103.48	7301	906	940	0.002	254
	Tempsat 2	1969-82H	1969 Sep 30.57 750 years	Sphere (black) 14.5	0.41 dia	1970 Mar 2.2	70.02	103.48	7301	906	940	0.002	219

* Time/navigation satellite

1969-82 continued on page 208

	Name		Launch date, lifetime and descent date	Shape and weight (kg)	Size (m)	Date of orbital determination	Orbital inclination (deg)	Nodal period (min)	Semi major axis (km)	Perigee height (km)	Apogee height (km)	Orbital eccentricity	Argument of perigee (deg)
	SOICAL* cylinder	1969-82J	1969 Sep 30.57 20 years	Cylinder 2.4	1.02 long 0.25 dia	1970 Feb 23.8	70.02	103.46	7300	904	940	0.002	232
	SOICAL cone	1969-82K	1969 Sep 30.57 40 years	Cone 3.4	0.66 long 0.51 dia	1970 Feb 22.8	70.01	103.50	7302	903	945	0.003	228
	Agena D rocket	1969-82AB	1969 Sep 30.57 600 years	Cylinder 700?	6 long? 1.5 dia	1969 Oct 22.5	69.96	105.22	7383	918	1092	0.012	87
118d	Fragments	1969-82L-KL											
D	Boreas (ESRO 1B)**	1969-83A	1969 Oct 1.94 52.47 days 1969 Nov 23.41	Cone-cylinder 80	1.52 long 0.76 dia	1969 Oct 3.1	85.11	91.39	6718	291	389	0.007	313
D	Boreas rocket	1969-83B	1969 Oct 1.94 32.36 days 1969 Nov 3.30	Cylinder 24	1.50 long 0.46 dia	1969 Oct 5.9	85.13	91.36	6717	296	381	0.006	309
D	Fragments	1969-83C,D											
	Meteor 2	1969-84A	1969 Oct 6.07 40 years	Cylinder + 2 vanes 2200?	5 long? 1.5 dia?	1969 Oct 6.7	81.26	97.70	7025	613	681	0.005	131
	Meteor 2 rocket	1969-84B	1969 Oct 6.07 30 years	Cylinder 1440	3.8 long 2.6 dia	1969 Oct 8.4	81.24	97.86	7033	550	760	0.015	16[illegible]

* Space object identification calibration.

** European Space Research Organization.

	Name		Launch date, lifetime and descent date	Shape and weight (kg)	Size (m)	Date of orbital determination	Orbital inclination (deg)	Nodal period (min)	Semi major axis (km)	Perigee height (km)	Apogee height (km)	Orbital eccentricity	Argument of perigee (deg)
D 2M R	Soyuz 6*	1969-85A	1969 Oct 11.47 4.94 days 1969 Oct 16.41	Sphere-cylinder + 2 wings 6577	7.5 long 2.2 dia	1969 Oct 11.9	51.68	88.67	6590	192	231	0.003	3
D	Soyuz 6 rocket	1969-85B	1969 Oct 11.47 1.08 days 1969 Oct 12.55	Cylinder 2500?	7.5 long 2.6 dia	1969 Oct 11.8	51.67	88.15	6564	156	215	0.004	303
D 3M R	Soyuz 7	1969-86A	1969 Oct 12.45 4.94 days 1969 Oct 17.39	Sphere-cylinder + 2 wings 6570	7.5 long 2.2 dia	1969 Oct 14.5	51.65	88.77	6595	210	223	0.001	338
D	Soyuz 7 rocket	1969-86B	1969 Oct 12.45 2.09 days 1969 Oct 14.54	Cylinder 2500?	7.5 long 2.6 dia	1969 Oct 13.2	51.65	88.37	6575	191	202	0.001	348
D 2M R	Soyuz 8	1969-87A	1969 Oct 13.43 4.95 days 1969 Oct 18.38	Sphere-cylinder + 2 wings 6646	7.5 long 2.2 dia	1969 Oct 16.0	51.65	88.72	6592	201	227	0.002	40
D	Soyuz 8 rocket	1969-87B	1969 Oct 13.43 1.97 days 1969 Oct 15.40	Cylinder 2500?	7.5 long 2.6 dia	1969 Oct 14.6	51.66	88.13	6563	177	192	0.001	52
D	Intercosmos 1	1969-88A	1969 Oct 14.57 79.98 days 1970 Jan 2.55	Ellipsoid + 6 panels 400?	1.8 long 1.2 dia	1969 Oct 15.1	48.38	93.31	6818	254	626	0.027	112
D	Intercosmos 1 rocket	1969-88B	1969 Oct 14.57 63.06 days 1969 Dec 16.63	Cylinder 1500?	8 long 1.65 dia	1969 Oct 15.1	48.37	93.24	6815	249	624	0.027	111

* Soyuz 6, 7 and 8 closed to within a few hundred metres of each other in pairs at various times.

	Name		Launch date, lifetime and descent date	Shape and weight (kg)	Size (m)	Date of orbital determination	Orbital inclination (deg)	Nodal period (min)	Semi major axis (km)	Perigee height (km)	Apogee height (km)	Orbital eccentricity	Argument of perigee (deg)
D R	Cosmos 302	1969-89A	1969 Oct 17.49 7.78 days 1969 Oct 25.27	Sphere-cylinder 5530?	5 long? 2.4 dia	1969 Oct 18.8	65.41	89.69	6638	198	321	0.009	48
D	Cosmos 302 rocket	1969-89B	1969 Oct 17.49 7.11 days 1969 Oct 24.60	Cylinder 2500?	7.5 long 2.6 dia	1969 Oct 18.2	65.40	89.55	6631	196	310	0.009	44
D	Cosmos 303	1969-90A	1969 Oct 18.42 97.41 days 1970 Jan 23.83	Ellipsoid 400?	1.8 long 1.2 dia	1969 Oct 20.6	70.99	91.91	6746	270	466	0.015	80
D	Cosmos 303 rocket	1969-90B	1969 Oct 18.42 50.02 days 1969 Dec 7.44	Cylinder 1500?	8 long 1.65 dia	1969 Oct 20.1	71.00	91.80	6741	269	456	0.014	82
	Cosmos 304	1969-91A	1969 Oct 21.54 100 years	Cylinder + paddles? 750?	2 long? 1 dia?	1969 Oct 24.8	74.04	99.87	7130	742	761	0.001	320
	Cosmos 304 rocket	1969-91B	1969 Oct 21.54 80 years	Cylinder 2200?	7.4 long 2.4 dia	1969 Oct 24.5	74.05	99.74	7124	734	757	0.002	344
D	Cosmos 305*	1969-92A	1969 Oct 22.59 2.08 days 1969 Oct 24.67	Cylinder 19400? full	7 long 3.9 dia	1969 Oct 23.3	51.52	88.23	6569	175	206	0.002	293
D	Cosmos 305 rocket	1969-92B	1969 Oct 22.59 2.2 days 1969 Oct 24.8	Cylinder 4000?	12 long? 4 dia	1969 Oct 23.6	51.51	88.17	6565	171	203	0.002	285

* Possibly an attempted lunar probe. (Payload 5600 kg?)

	Name		Launch date, lifetime and descent date	Shape and weight (kg)	Size (m)	Date of orbital determination	Orbital inclination (deg)	Nodal period (min)	Semi major axis (km)	Perigee height (km)	Apogee height (km)	Orbital eccentricity	Argument of perigee (deg)
D R	Cosmos 306	1969-93A	1969 Oct 24.41 11.87 days 1969 Nov 5.28	Sphere-cylinder 5700?	5.0 long 2.4 dia	1969 Oct 25.1	64.97	89.64	6635	215	299	0.006	54
D	Cosmos 306 rocket	1969-93B	1969 Oct 24.41 5.49 days 1969 Oct 29.90	Cylinder 2500?	7.5 long 2.6 dia	1969 Oct 25.1	64.96	89.45	6626	203	293	0.007	45
D	Cosmos 307	1969-94A	1969 Oct 24.55 431.86 days 1970 Dec 30.41	Ellipsoid 400?	1.8 long 1.2 dia	1969 Oct 26.7 1970 Mar 31.9 1970 Aug 31.7	48.39 48.35 48.3	109.04 104.66 98.81	7564 7359 7083	214 210 201	2157 1752 1208	0.128 0.105 0.071	107 - -
D	Cosmos 307 rocket	1969-94B	1969 Oct 24.55 268.49 days 1970 Jul 20.04	Cylinder 1500?	8 long 1.65 dia	1969 Oct 26.6 1970 Jan 31.7 1970 May 1.0	48.41 48.3 48.3	108.78 104.18 98.82	7552 7337 7083	210 205 199	2138 1712 1211	0.128 0.103 0.071	107 - -
D	[Titan 3B Agena D]	1969-95A	1969 Oct 24.76 15 days 1969 Nov 8	Cylinder 3000?	8 long? 1.5 dia	1969 Oct 25.5	108.04	93.39	6816	136	740	0.044	144
D	Cosmos 308	1969-96A	1969 Nov 4.50 60.99 days 1970 Jan 4.49	Ellipsoid 400?	1.8 long 1.2 dia	1969 Nov 5.2	71.02	91.34	6718	271	408	0.010	68
D	Cosmos 308 rocket	1969-96B	1969 Nov 4.50 32.35 days 1969 Dec 6.85	Cylinder 1500?	8 long 1.65 dia	1969 Nov 5.5	71.02	91.20	6711	271	395	0.009	66

Year of launch 1969 continued

	Name		Launch date, lifetime and descent date	Shape and weight (kg)	Size (m)	Date of orbital determination	Orbital inclination (deg)	Nodal period (min)	Semi major axis (km)	Perigee height (km)	Apogee height (km)	Orbital eccentricity	Argument of perigee (deg)
	Azur (GRS A)†	1969-97A	1969 Nov 8.08 50 years	Cone-cylinder 71	1.13 long 0.76 dia	1969 Nov 10.5	102.96	122.00	8147	387	3150	0.170	158
	Azur rocket	1969-97B	1969 Nov 8.08 30 years	Cylinder 24	1.50 long 0.46 dia	1969 Nov 8.6	102.97	122.02	8148	390	3149	0.169	161
D	Fragments	1969-97C-F											
D R	Cosmos 309	1969-98A	1969 Nov 12.48 7.75 days 1969 Nov 20.23	Sphere-cylinder 5900?	5.9 long 2.4 dia	1969 Nov 14.2	65.40	89.99	6653	185	364	0.014	89
D	Cosmos 309 rocket	1969-98B	1969 Nov 12.48 10.31 days 1969 Nov 22.79	Cylinder 2500?	7.5 long 2.6 dia	1969 Nov 13.1	65.41	89.95	6651	192	353	0.012	97
D	Capsule	1969-98E	1969 Nov 12.48 18.09 days 1969 Nov 30.57	Ellipsoid 200?	0.9 long 1.9 dia	1969 Nov 19	65.4	89.85	6646	195	340	0.011	-
D	Fragments	1969-98C,D											
D 3M R	Apollo 12**	1969-99A	1969 Nov 14.68 10.19 days 1969 Nov 24.87	Cone-cylinder 28790, then 11250	11.15 long 3.91 dia	1969 Nov 14.7 1969 Nov 14.9	32.56 33.2	88.15 18150	6569 226087	183 207	199 439 210	0.001 0.971	-* 30*
							In selenocentric orbit 1969 Nov 18.16 to Nov 21.87						
	Saturn IV B [Saturn 507]	1969-99B	1969 Nov 14.68 Indefinite	Cylinder 13300	18.7 long 6.6 dia	1969 Nov 14.7 1969 Nov 14.9	32.56 31.6	88.15 60480	6569 518 800	183 163 100	199 861 800	0.001 0.673	-* 30*
D	LEM 6*** Ascent stage	1969-99C	1969 Nov 14.68 6.25 days 1969 Nov 20.93	Box + tanks 4774 full 2159 empty	2.52 high 3.76 wide 3.13 deep	1969 Nov 14.9	33.2	18150	226 087	207	439 210	0.971	30*
							On Moon's surface 1969 Nov 19.29 to Nov 20.60 Finally crashed on Moon 1969 Nov 20.93						
D	LEM 6 Descent stage	1969-99D	1969 Nov 14.68 4.61 days 1969 Nov 19.29	Octagon + cone + legs 10342 full 2211 empty	1.57 high 3.13 wide	1969 Nov 14.9	33.2	18150	226 087	207	439 210	0.971	30*
							Landed on Moon 1969 Nov 19.29						

** Apollo attached to LEM, separated from Saturn IV B on Nov 14.87.
*** LEM with 2 crew members, separated from Apollo on Nov 19.18.
Ascent stage relaunched from Moon Nov 20.60; briefly docked with Apollo Nov 20.75.

* Approximate orbits
† German Research Satellite

	Name		Launch date, lifetime and descent date	Shape and weight (kg)	Size (m)	Date of orbital determination	Orbital inclination (deg)	Nodal period (min)	Semi major axis (km)	Perigee height (km)	Apogee height (km)	Orbital eccentricity	Argument of perigee (deg)
D R	Cosmos 310	1969-100A	1969 Nov 15.36 7.92 days 1969 Nov 23.28	Sphere-cylinder 5530?	5 long? 2.4 dia	1969 Nov 16.2	65.00	89.91	6648	204	336	0.010	59
D	Cosmos 310 rocket	1969-100B	1969 Nov 15.36 7.91 days 1969 Nov 23.27	Cylinder 2500?	7.5 long 2.6 dia	1969 Nov 16.1	65.02	89.74	6640	204	319	0.009	52
	Skynet 1A	1969-101A	1969 Nov 22.03 >million years	Cylinder 243 full 126 empty	0.81 long 1.37 dia	1969 Dec 9.8 1970 Feb 16.0 1970 Jul 7.0 1970 Nov 25.6	2.40 2.2 1.88 1.56	1407.8 1436.0 1436.4 1436.4	41598 42162 42173 42172	34702 35779 35790 35794	35838 35788 35800 35794	0.012 0.0001 0.0001 0.0000	165 - 289 142
	Skynet 1A rocket	1969-101B	1969 Nov 22.03 20 years?	Sphere-cone 66	1.32 long 0.94 dia	1969 Nov 22.1 1970 Oct 1.0	27.73 27.19	656.9 632.6	25043 24411	245 215	37084 35850	0.735 0.730	181 -
D	Cosmos 311	1969-102A	1969 Nov 24.46 105.80 days 1970 Mar 10.26	Ellipsoid 400?	1.8 long 1.2 dia	1969 Nov 25.9	71.04	91.99	6749	273	469	0.014	79
D	Cosmos 311 rocket	1969-102B	1969 Nov 24.46 56.33 days 1970 Jan 19.79	Cylinder 1500?	8 long 1.65 dia	1969 Nov 25.2	71.03	91.88	6744	274	458	0.014	82
	Cosmos 312	1969-103A	1969 Nov 24.70 2500 years	Spheroid + paddles? 650?	1.6 dia?	1969 Dec 1.5	74.03	108.60	7540	1144	1179	0.002	339
	Cosmos 312 rocket	1969-103B	1969 Nov 24.70 1500 years	Cylinder 2200?	7.4 long 2.4 dia	1969 Nov 28.9	74.01	108.44	7532	1140	1168	0.002	6

	Name		Launch date, lifetime and descent date	Shape and weight (kg)	Size (m)	Date of orbital determination	Orbital inclination (deg)	Nodal period (min)	Semi major axis (km)	Perigee height (km)	Apogee height (km)	Orbital eccentricity	Argument of perigee (deg)
D R	Cosmos 313	1969-104A	1969 Dec 3.56 11.73 days 1969 Dec 15.29	Sphere-cylinder 5700?	5.0 long 2.4 dia	1969 Dec 4.1	65.40	89.07	6607	198	259	0.005	30
D	Cosmos 313 rocket	1969-104B	1969 Dec 3.56 3.26 days 1969 Dec 6.82	Cylinder 2500?	7.5 long 2.6 dia	1969 Dec 4.3	65.42	88.85	6596	190	245	0.004	10
D	[Thorad Agena D]	1969-105A	1969 Dec 4.90 36.26 days 1970 Jan 10.16	Cylinder 2000?	8 long? 1.5 dia	1969 Dec 6.4	81.48	88.61	6583	159	251	0.007	137
D	Cosmos 314	1969-106A	1969 Dec 11.54 100.63 days 1970 Mar 22.17	Ellipsoid 400?	1.8 long 1.2 dia	1969 Dec 13.3	71.01	91.93	6747	272	465	0.014	77
D	Cosmos 314 rocket	1969-106B	1969 Dec 11.54 46.93 days 1970 Jan 27.47	Cylinder 1500?	8 long 1.65 dia	1969 Dec 13.3	71.00	91.65	6733	273	437	0.012	79
D	Cosmos 315	1969-107A	1969 Dec 20.15 3382 days 1979 Mar 25	Cylinder + paddles? 900?	2 long? 1 dia?	1969 Dec 21.8 1972 May 1.0	74.04 74.04	95.26 94.58	6908 6876	518 489	542 506	0.002 0.001	344 -
D	Cosmos 315 rocket	1969-107B	1969 Dec 20.15 3390 days 1979 Apr 2	Cylinder 2200?	7.4 long 2.4 dia	1969 Dec 21.8 1972 May 16.5	74.04 74.04	95.12 94.48	6902 6871	511 481	537 505	0.002 0.002	8 -
D	Fragments	1969-107C-F											

	Name		Launch date, lifetime and descent date	Shape and weight (kg)	Size (m)	Date of orbital determination	Orbital inclination (deg)	Nodal period (min)	Semi major axis (km)	Perigee height (km)	Apogee height (km)	Orbital eccentricity	Argument of perigee (deg)
D P	Cosmos 316 †	1969-108A	1969 Dec 23.39 248.46 days 1970 Aug 28.85	Cylinder? 5000?	4 long? 1.5 dia	1969 Dec 24.5 1970 June 20.0 1970 Aug 28.0	49.50 49.48 49.45	102.82 95.20 87.90	7273 6910 6552	152 138 119	1638 926 226	0.102 0.057 0.008	83 43 354
D	Cosmos 316 rocket	1969-108C	1969 Dec 23.39 35.84 days 1970 Jan 28.23	Cylinder 1500?	8 long? 2.5 dia?	1969 Dec 26.0	49.49	102.15	7242	147	1581	0.099	88
D	Fragment	1969-108B											
D R	Cosmos 317	1969-109A	1969 Dec 23.58 12.72 days 1970 Jan 5.30	Sphere-cylinder 6300?	6.5 long? 2.4 dia	1969 Dec 24.4 1969 Dec 24.9	65.41 65.50	89.34 89.65	6621 6636	205 211	280 304	0.006 0.007	55 52
D	Cosmos 317 rocket	1969-109B	1969 Dec 23.58 5 days 1969 Dec 28	Cylinder 2500?	7.5 long 2.6 dia	1969 Dec 24.3	65.42	89.15	6611	201	265	0.005	45
D	Cosmos 317 engine*	1969-109C	1969 Dec 23.58 16 days 1970 Jan 8	Cone 600? full	1.5 long? 2 dia?	1970 Jan 3.7	65.40	89.08	6608	190	269	0.006	46
D	Fragment	1969-109D											
D	Intercosmos 2	1969-110A	1969 Dec 25.42 164.52 days 1970 Jun 7.94	Ellipsoid 400?	1.8 long 1.2 dia	1969 Dec 27.0 1970 Mar 31.9	48.40 48.38	98.48 95.08	7067 6904	200 193	1178 858	0.069 0.048	108 -
D	Intercosmos 2 rocket	1969-110B	1969 Dec 25.42 86.33 days 1970 Mar 21.75	Cylinder 1500?	8 long 1.65 dia	1969 Dec 27.9	48.40	98.14	7051	207	1139	0.066	112

† Pieces recovered in Oklahoma, Kansas and Texas.

* 1969-109C ejected from 1969-109A about 1970 Jan 3.4. (Cosmos 317 carried supplementary charged-particles experiment.)

	Name		Launch date, lifetime and descent date	Shape and weight (kg)	Size (m)	Date of orbital determination	Orbital inclination (deg)	Nodal period (min)	Semi major axis (km)	Perigee height (km)	Apogee height (km)	Orbital eccentricity	Argument of perigee (deg)
D R	Cosmos 318	1970-01A	1970 Jan 9.39 11.90 days 1970 Jan 21.29	Sphere-cylinder 5700?	5.0 long 2.4 dia	1970 Jan 10.8	64.97	89.29	6618	203	277	0.006	31
D	Cosmos 318 rocket	1970-01B	1970 Jan 9.39 6.21 days 1970 Jan 15.60	Cylinder 2500?	7.5 long 2.6 dia	1970 Jan 10.7	64.97	89.11	6610	200	263	0.005	29
D	[Titan 3B Agena D]	1970-02A	1970 Jan 14.78 18 days 1970 Feb 1	Cylinder 3000?	8 long? 1.5 dia	1970 Jan 15.3	109.96	89.69	6637	134	383	0.019	133
	Intelsat 3 F-6	1970-03A	1970 Jan 15.01 > million years	Cylinder 293 full 137 empty	1.04 long 1.42 dia	1970 Jan 15.7 1970 Feb 16.0 1977 May 13.0	28.04 0.9 5.16	629.71 1436.1 1465.69	24386 42165 42742	267 35773 36185	35748 35801 36542	0.727 0.0003 0.004	181 - 10*
	Intelsat 3 F-6 rocket	1970-03B	1970 Jan 15.01 80 years	Sphere-cone 66	1.32 long 0.94 dia	1970 Jan 15.7	28.04	629.70	24385	267	35747	0.727	181
D	Cosmos 319	1970-04A	1970 Jan 15.57 167.13 days 1970 Jul 1.70	Ellipsoid 400?	1.8 long 1.2 dia	1970 Jan 16.2 1970 Apr 1.0	81.96 81.96	102.03 98.97	7232 7088	200 195	1508 1224	0.090 0.073	74 -
D	Cosmos 319 rocket	1970-04B	1970 Jan 15.57 106.42 days 1970 May 1.99	Cylinder 1500?	8 long 1.65 dia	1970 Jan 16.2	81.96	101.85	7224	199	1493	0.090	74

* Transmissions terminated after orbit change

	Name		Launch date, lifetime and descent date	Shape and weight (kg)	Size (m)	Date of orbital determination	Orbital inclination (deg)	Nodal period (min)	Semi major axis (km)	Perigee height (km)	Apogee height (km)	Orbital eccentricity	Argument of perigee (deg)
D	Cosmos 320	1970-05A	1970 Jan 16.46 24.67 days 1970 Feb 10.13	Ellipsoid + annular tail? 400?	6.5 long? 1.2 dia?	1970 Jan 17.6	48.40	90.18	6665	247	326	0.006	139
D	Cosmos 320 rocket	1970-05B	1970 Jan 16.46 11.56 days 1970 Jan 28.02	Cylinder 1500?	8 long 1.65 dia	1970 Jan 18.5	48.41	90.19	6665	255	319	0.005	175
D	Fragments	1970-05C-G											
D	Cosmos 321*	1970-06A	1970 Jan 20.85 61.39 days 1970 Mar 23.24	Ellipsoid 400?	1.8 long 1.2 dia	1970 Jan 23.9	70.95	92.07	6754	272	479	0.015	73
D	Cosmos 321 rocket	1970-06B	1970 Jan 20.85 45.56 days 1970 Mar 7.41	Cylinder 1500?	8 long 1.65 dia	1970 Jan 24.1	70.95	91.84	6742	270	458	0.014	72
D	Fragments	1970-06C-D											
D R	Cosmos 322	1970-07A	1970 Jan 21.50 7.78 days 1970 Jan 29.28	Sphere-cylinder 5530?	5 long? 2.4 dia	1970 Jan 22.2	65.41	89.65	6635	195	319	0.009	41
D	Cosmos 322 rocket	1970-07B	1970 Jan 21.50 4.33 days 1970 Jan 25.83	Cylinder 2500?	7.5 long 2.6 dia	1970 Jan 22.2	65.40	89.46	6626	190	306	0.009	38

* Ionospheric and geomagnetic studies

	Name		Launch date, lifetime and descent date	Shape and weight (kg)	Size (m)	Date of orbital determination	Orbital inclination (deg)	Nodal period (min)	Semi major axis (km)	Perigee height (km)	Apogee height (km)	Orbital eccentricity	Argument of perigee (deg)
	ITOS 1*	1970-08A	1970 Jan 23.48 10000 years	Box + 3 panels 309	1.02 square 1.22 long	1970 Feb 14.3	102.00	115.10	7837	1436	1482	0.003	219
	Oscar 5†	1970-08B	1970 Jan 23.48 10000 years	Rectangular box 18	0.43 x 0.30 x 0.15	1970 Jan 25.4	101.96	115.08	7836	1435	1481	0.003	263
	ITOS 1 second stage	1970-08C	1970 Jan 23.48 5000 years	Cylinder 350?	4.9 long 1.43 dia	1970 Jan 25.4	101.95	115.11	7838	1441	1478	0.002	262
	SERT 2** (Agena D)	1970-09A	1970 Feb 4.12 800 years	Cylinder + 2 wings 1500	7.6 long 1.52 dia 12.2 span	1970 Feb 11.8 1970 Jul 23.0 1970 Nov 1.0	99.13 99.13 99.13	105.15 106.94 106.23	7378 7458 7427	997 1079 1045	1003 1081 1052	0.0004 0.0001 0.0005	328 - -
D R	Cosmos 323	1970-10A	1970 Feb 10.50 7.78 days 1970 Feb 18.28	Sphere- cylinder 5530?	5 long? 2.4 dia	1970 Feb 11.2	65.43	89.65	6636	201	314	0.009	49
D	Cosmos 323 rocket	1970-10B	1970 Feb 10.50 5.08 days 1970 Feb 15.58	Cylinder 2500?	7.5 long 2.6 dia	1970 Feb 12.2	65.41	89.27	6617	189	288	0.008	35
	Ohsumi †† [Lambda 4S]	1970-11A	1970 Feb 11.18 30 years	Cylinder 38 (payload 12)	1.00 long 0.48 dia	1970 Feb 14.7	31.07	144.20	9117	339	5138	0.263	131

* Improved Tiros Operational Satellite.
** Space Electric Rocket Test.
† Australian satellite
†† Japanese satellite

	Name		Launch date, lifetime and descent date	Shape and weight (kg)	Size (m)	Date of orbital determination	Orbital inclination (deg)	Nodal period (min)	Semi major axis (km)	Perigee height (km)	Apogee height (km)	Orbital eccentricity	Argument of perigee (deg)
	[Thor Burner 2]	1970-12A	1970 Feb 11.36 80 years	12-faced frustum 195	1.64 long 1.31 to 1.10 dia	1970 Feb 16.7	98.71	101.39	7201	773	874	0.007	12
	Burner 2 rocket	1970-12B	1970 Feb 11.36 60 years	Sphere-cone 66	1.32 long 0.94 dia	1970 Feb 18.4	98.74	101.38	7201	772	874	0.007	7
D	Molniya 1N	1970-13A	1970 Feb 19.79 2048 days 1975 Sep 29	Windmill + 6 vanes 1000?	3.4 long 1.6 dia	1970 Feb 26.2 1970 May 16.5 1973 Oct 16.5	65.44 65.57 65.1	703.13 717.73 717.95	26194 26556 26564	461 470 940	39170 39885 39426	0.739 0.742 0.724	285 - -
D	Molniya 1N launcher	1970-13B	1970 Feb 19.79 17.42 days 1970 Mar 9.21	Irregular	-	1970 Feb 20.6	65.42	91.43	6723	205	485	0.021	72
D	Molniya 1N launcher rocket	1970-13C	1970 Feb 19.79 20.07 days 1970 Mar 11.86	Cylinder 2500?	7.5 long 2.6 dia	1970 Feb 20.6	65.43	91.21	6713	220	449	0.017	70
D	Molniya 1N rocket	1970-13D	1970 Feb 19.79 2150 days 1976 Jan 9	Cylinder 440	2.0 long 2.0 dia	1970 Mar 2.0 1973 Oct 1.0	65.46 65.25	699.06 698.70	26093 26084	465 925	38964 38486	0.738 0.720	285 -
D	Cosmos 324	1970-14A	1970 Feb 27.73 84.53 days 1970 May 23.26	Ellipsoid 400?	1.8 long 1.2 dia	1970 Feb 28.4	71.03	91.97	6749	275	466	0.014	75
D	Cosmos 324 rocket	1970-14B	1970 Feb 27.73 42.75 days 1970 Apr 11.48	Cylinder 1500?	8 long 1.65 dia	1970 Mar 1.1	71.04	91.74	6737	277	441	0.012	73

	Name		Launch date, lifetime and descent date	Shape and weight (kg)	Size (m)	Date of orbital determination	Orbital inclination (deg)	Nodal period (min)	Semi major axis (km)	Perigee height (km)	Apogee height (km)	Orbital eccentricity	Argument of perigee (deg)
D R	Cosmos 325	1970-15A	1970 Mar 4.51 7.79 days 1970 Mar 12.30	Sphere-cylinder 5530?	5 long? 2.4 dia	1970 Mar 6.2	65.39	89.77	6642	200	327	0.010	49
D	Cosmos 325 rocket	1970-15B	1970 Mar 4.51 5.60 days 1970 Mar 10.11	Cylinder 2500?	7.5 long 2.6 dia	1970 Mar 6.2	65.40	89.50	6628	194	306	0.008	39
D	[Thorad Agena D]	1970-16A	1970 Mar 4.93 21.98 days 1970 Mar 26.91	Cylinder 2000?	8 long? 1.5 dia	1970 Mar 5.6	88.02	88.76	6590	167	257	0.007	121
D	Capsule*	1970-16B	1970 Mar 4.93 615.19 days 1971 Nov 10.12	Octagon? 60?	0.3 long? 0.9 dia?	1970 Mar 7.2 1970 Aug 31.7 1971 Apr 1.0	88.14 88.14 88.14	94.16 93.58 92.49	6856 6828 6773	442 420 376	514 479 414	0.005 0.004 0.003	10 - -
D	DIAL/WIKA**	1970-17A	1970 Mar 10.51 3131 days 1978 Oct 5	Octagonal door-knob 63	1.01 long 0.63 dia	1970 Mar 12.4 1971 Aug 1.0	5.44 5.44	104.20 102.14	7344 7247	301 304	1631 1434	0.091 0.078	112 -
D	DIAL rocket/ MIKA	1970-17B	1970 Mar 10.51 1644 days 1974 Sep 9	Cylinder + nozzle 120 (payload 52)	2.60 long 0.80 dia	1970 Mar 11.7 1971 Feb 1.0 1972 Oct 1.0	5.42 5.42 5.42	104.67 102.49 99.20	7367 7264 7108	313 303 293	1665 1468 1166	0.092 0.080 0.061	105 - -
D R	Cosmos 326	1970-18A	1970 Mar 13.34 7.88 days 1970 Mar 21.22	Sphere-cylinder 5530?	5 long? 2.4 dia	1970 Mar 13.8	81.35	90.20	6661	203	363	0.012	85
D	Cosmos 326 rocket	1970-18B	1970 Mar 13.34 9.98 days 1970 Mar 23.32	Cylinder 2500?	7.5 long 2.6 dia	1970 Mar 14.3	81.36	90.00	6651	202	346	0.011	83

* 1970-16B ejected from 1970-16A on 1970 Mar 5.00.
** DIamant B - ALLemand (French-German satellite).

	Name		Launch date, lifetime and descent date	Shape and weight (kg)	Size (m)	Date of orbital determination	Orbital inclination (deg)	Nodal period (min)	Semi major axis (km)	Perigee height (km)	Apogee height (km)	Orbital eccentricity	Argument of perigee (deg)
	Meteor 3	1970-19A	1970 Mar 17.47	Cylinder +	5 long?	1970 Mar 18.2	81.18	96.42	6964	537	635	0.007	236
			13½ years	2 vanes	1.5 dia?								
				2200?									
D	Meteor 3	1970-19B	1970 Mar 17.47	Cylinder	3.8 long	1970 Mar 18.8	81.17	96.55	6971	467	718	0.018	203
	rocket		4637 days	1440	2.6 dia								
			1982 Nov 26										
D	Fragment	1970-19C											
D	Cosmos 327	1970-20A	1970 Mar 18.61	Ellipsoid	1.8 long	1970 Mar 19.6	70.95	95.65	6928	280	819	0.039	87
			306.41 days	400?	1.2 dia	1970 Aug 31.7	70.95	93.64	6832	255	652	0.029	-
			1971 Jan 19.02										
D	Cosmos 327	1970-20B	1970 Mar 18.61	Cylinder	8 long	1970 Mar 19.6	70.94	95.57	6924	272	819	0.039	85
	rocket		192.44 days	1500?	1.65 dia	1970 Jul 1.3	70.94	93.39	6819	254	628	0.027	-
			1970 Sep 27.05										
	NATO 1 *	1970-21A	1970 Mar 20.99	Cylinder	0.81 long	1970 Mar 21.0	25.81	656.9	25043	281	37048	0.734	179
			> million	243 full	1.37 dia	1970 May 1.0	2.8	1403.4	41523	34429	35860	0.017	-
			years	117 empty		1975 May 1.0	1.7	1436.3	42168	35745	35834	0.001	-
	NATO 1	1970-21B	1970 Mar 20.99	Sphere-cone	1.32 long	1970 Mar 21.3	25.67	655.4	24993	295	36934	0.733	179
	rocket		100 years	66	0.94 dia								
	Fragment	1970-21C											
D	Cosmos 328**	1970-22A	1970 Mar 27.49	Sphere-	6.5 long?	1970 Mar 28.6	72.86	89.54	6629	203	299	0.007	56
R			12.77 days	cylinder	2.4 dia	1970 Mar 29.8	72.86	89.75	6639	206	316	0.008	59
			1970 Apr 9.26	6300?									
D	Cosmos 328	1970-22B	1970 Mar 27.49	Cylinder	7.5 long	1970 Mar 28.4	72.87	89.39	6622	204	284	0.006	53
	rocket		4.69 days	2500?	2.6 dia								
			1970 Apr 1.18										

* North Atlantic Treaty Organisation

** Cosmos 328 manoeuvred, but no jettisoned engine was tracked or designated

	Name		Launch date, lifetime and descent date	Shape and weight (kg)	Size (m)	Date of orbital determination	Orbital inclination (deg)	Nodal period (min)	Semi major axis (km)	Perigee height (km)	Apogee height (km)	Orbital eccentricity	Argument of perigee (deg)
D R	Cosmos 329	1970-23A	1970 Apr 3.36 11.87 days 1970 Apr 15.23	Sphere-cylinder 5700?	5.0 long 2.4 dia	1970 Apr 4.1	81.33	88.79	6591	198	228	0.002	319
D	Cosmos 329 rocket	1970-23B	1970 Apr 3.36 2.44 days 1970 Apr 5.80	Cylinder 2500?	7.5 long 2.6 dia	1970 Apr 4.3	81.32	88.52	6578	185	214	0.002	316
D	Cosmos 330	1970-24A	1970 Apr 7.47 3353 days 1979 Jun 12	Cylinder + paddles? 900?	2 long? 1 dia?	1970 Apr 8.6 1972 Sep 16.0	74.06 74.06	95.22 94.63	6907 6878	514 492	543 508	0.002 0.001	334 -
D	Cosmos 330 rocket	1970-24B	1970 Apr 7.47 3504 days 1979 Nov 10	Cylinder 2200?	7.4 long 2.4 dia	1970 Apr 10.2 1973 Feb 1.0	74.06 74.06	95.12 94.56	6902 6875	507 480	541 513	0.002 0.002	336 -
	Nimbus 4	1970-25A	1970 Apr 8.35 1200 years	Conical skeleton + 2 paddles 620	3.00 long 1.45 dia	1970 Apr 10.2	99.89	107.29	7476	1095	1100	0.0003	8
	TOPO 1	1970-25B	1970 Apr 8.35 2000 years	Rectangular box 18	0.36 x 0.30 x 0.23	1970 Apr 9.5	99.76	107.09	7466	1064	1111	0.003	193
	Nimbus 4 rocket	1970-25C	1970 Apr 8.35 1000 years	Cylinder* 700?	6 long? 1.5 dia	1970 Apr 10.2	99.89	106.86	7454	1066	1086	0.001	331
49d	Fragments	1970-25D-NR											
D R	Cosmos 331	1970-26A	1970 Apr 8.43 7.92 days 1970 Apr 16.35	Sphere-cylinder 5530?	5 long? 2.4 dia	1970 Apr 10.2	65.02	89.77	6641	206	320	0.009	45
D	Cosmos 331 rocket	1970-26B	1970 Apr 8.43 7.80 days 1970 Apr 16.23	Cylinder 2500?	7.5 long 2.6 dia	1970 Apr 10.1	65.01	89.60	6633	205	305	0.007	43

* Before explosion about 1970 Nov 4.

Year of launch 1970 continued

	Name		Launch date, lifetime and descent date	Shape and weight (kg)	Size (m)	Date of orbital determination	Orbital inclination (deg)	Nodal period (min)	Semi major axis (km)	Perigee height (km)	Apogee height (km)	Orbital eccentricity	Argument of perigee (deg)
	Vela 11	1970-27A	1970 Apr 8.46 > million years	Icosahedron 259	1.27 dia	1970 Apr 9.4	32.41	6729	118060	111210	112160	0.005	280
	Vela 12	1970-27B	1970 Apr 8.46 > million years	Icosahedron 259	1.27 dia	1970 Apr 11.6	32.52	6745	118230	111500	112210	0.003	47
	Vela 11 rocket [Titan 3C]	1970-27C	1970 Apr 8.46 > million years	Cylinder 1500?	6 long? 3.0 dia	1970 Apr 10.4	32.36	3005	68980	15040	110170	0.690	178*
	Cosmos 332	1970-28A	1970 Apr 11.71 100 years	Cylinder + paddles 750?	2 long? 1 dia?	1970 Apr 12.4	74.05	100.01	7136	755	761	0.0004	110
	Cosmos 332 rocket	1970-28B	1970 Apr 11.71 80 years	Cylinder 2200?	7.4 long 2.4 dia	1970 Apr 12.4	74.05	99.90	7131	744	762	0.001	34
D 3M R	Apollo 13**	1970-29A	1970 Apr 11.80 5.95 days 1970 Apr 17.75	Cone-cylinder 28890	11.15 long 3.91 dia	1970 Apr 11.8 1970 Apr 12.0	32.56 33.2	88.07 26320	6564 292520	186 200	186 572080	0 0.977	-* 30*
								Passed 252 km beyond Moon 1970 Apr 15.02					
D	Saturn IVB [Saturn 508]	1970-29B	1970 Apr 11.80 3.25 days 1970 Apr 15.05	Cylinder 13930	18.7 long 6.6 dia	1970 Apr 11.8 1970 Apr 12.0	32.56 33.2	88.07 26320	6564 292520	186 200	186 572080	0 0.977	-* 30*
								Crashed on Moon 1970 Apr 15.05					
D	LEM 7†	1970-29C	1970 Apr 11.80 5.95 days 1970 Apr 17.75	Box + octagon + 4 legs 15190, then 7890	4.1 high 3.76 wide 3.13 deep	1970 Apr 12.0	33.2	26320	292520	200	572080	0.977	30*
								Passed 252 km beyond Moon 1970 Apr 15.02. Decayed in Earth's atmosphere					

* Approximate orbits.

** Apollo attached to LEM, separated from Saturn IVB on Apr 11.98.

† Apollo command module jettisoned Service module on Apr 17.55; and jettisoned LEM on Apr 17.70.

	Name		Launch date, lifetime and descent date	Shape and weight (kg)	Size (m)	Date of orbital determination	Orbital inclination (deg)	Nodal period (min)	Semi major axis (km)	Perigee height (km)	Apogee height (km)	Orbital eccentricity	Argument of perigee (deg)
D R	Cosmos 333	1970-30A	1970 Apr 15.38 12.85 days 1970 Apr 28.23	Sphere-cylinder 6300?	6.5 long? 2.4 dia	1970 Apr 15.6	81.34	89.11	6607	219	239	0.001	80
D	Cosmos 333 rocket	1970-30B	1970 Apr 15.38 3.20 days 1970 Apr 18.58	Cylinder 2500?	7.5 long 2.6 dia	1970 Apr 15.7	81.36	88.99	6601	213	233	0.002	8
D	Cosmos 333 engine*	1970-30C	1970 Apr 15.38 16.64 days 1970 May 2.02	Cone 600? full	1.5 long? 2 dia?	1970 Apr 28.2	81.34	88.66	6585	195	218	0.002	331
D	Fragment	1970-30D											
D	[Titan 3B Agena D]	1970-31A	1970 Apr 15.66 21 days 1970 May 6	Cylinder 3000?	8 long? 1.5 dia	1970 Apr 16.6 1970 Apr 24.3	110.97 110.96	89.70 89.91	6637 6648	130 131	388 408	0.019 0.021	120 121
	Intelsat 3 F-7	1970-32A	1970 Apr 23.03 > million years	Cylinder 293 full 137 empty	1.04 long 1.42 dia	1970 May 17.9	0.21	1436.2	42167	35772	35805	0.0004	345
	Intelsat 3 F-7 rocket	1970-32B	1970 Apr 23.03 40 years	Sphere-cone 66	1.32 long 0.94 dia	1970 Apr 23.7 1971 Dec 1.0	28.04 28.1	643.2 629.5	24682 24329	272 193	36336 35708	0.731 0.730	181 -
D	Cosmos 334	1970-33A	1970 Apr 23.56 108.39 days 1970 Aug 9.95	Ellipsoid 400?	1.8 long 1.2 dia	1970 Apr 24.2	70.92	92.10	6755	272	482	0.016	81
D	Cosmos 334 rocket	1970-33B	1970 Apr 23.56 23.05 days 1970 May 16.61	Cylinder 1500?	8 long 1.65 dia	1970 Apr 24.3	70.92	91.95	6748	249	491	0.018	76
D	Fragments	1970-33C-E											

* 1970-30C ejected from 1970-30A about 1970 Apr 28.2.

	Name		Launch date, lifetime and descent date	Shape and weight (kg)	Size (m)	Date of orbital determination	Orbital inclination (deg)	Nodal period (min)	Semi major axis (km)	Perigee height (km)	Apogee height (km)	Orbital eccentricity	Argument of perigee (deg)
	China 1	1970-34A	1970 Apr 24.57 100 years	Spheroid? 173	1 dia?	1970 Apr 27.7	68.44	114.09	7792	441	2386	0.125	141
	China 1 rocket	1970-34B	1970 Apr 24.57 30 years	Cylinder	-	1970 Apr 30.0	68.45	114.09	7792	441	2386	0.125	139
	Fragment	1970-34C											
D	Cosmos 335*	1970-35A	1970 Apr 24.94 58.15 days 1970 Jun 22.09	Ellipsoid 400?	1.8 long 1.2 dia	1970 Apr 25.5	48.40	90.97	6704	250	401	0.011	127
D	Cosmos 335 rocket	1970-35B	1970 Apr 24.94 22.72 days 1970 May 17.66	Cylinder 1500?	8 long 1.65 dia	1970 Apr 25.5	48.44	90.88	6699	250	392	0.011	127
	Cosmos 336	1970-36A	1970 Apr 25.71 10000 years	Spheroid 40?	1.0 long? 0.8 dia?	1970 Apr 30.2	74.04	115.49	7855	1464	1490	0.002	199
	Cosmos 337	1970-36B	1970 Apr 25.71 10000 years	Spheroid 40?	1.0 long? 0.8 dia?	1970 Apr 27.1	74.05	116.27	7890	1470	1554	0.005	258
	Cosmos 338	1970-36C	1970 Apr 25.71 10000 years	Spheroid 40?	1.0 long? 0.8 dia?	1970 Apr 28.6	74.03	115.89	7873	1472	1518	0.003	250
	Cosmos 339	1970-36D	1970 Apr 25.71 9000 years	Spheroid 40?	1.0 long? 0.8 dia?	1970 Apr 30.2	74.04	115.10	7837	1446	1472	0.002	111
	Cosmos 340	1970-36E	1970 Apr 25.71 8000 years	Spheroid 40?	1.0 long? 0.8 dia?	1970 May 2.2	74.04	114.70	7819	1409	1473	0.004	100

* Ultra-violet radiation studies

1970-36 continued on page 226

	Name		Launch date, lifetime and descent date	Shape and weight (kg)	Size (m)	Date of orbital determination	Orbital inclination (deg)	Nodal period (min)	Semi major axis (km)	Perigee height (km)	Apogee height (km)	Orbital eccentricity	Argument of perigee (deg)
	Cosmos 341	1970-36F	1970 Apr 25.71 6000 years	Spheroid 40?	1.0 long? 0.8 dia?	1970 Apr 27.3	74.04	113.97	7786	1345	1471	0.008	98
	Cosmos 342	1970-36G	1970 Apr 25.71 5000 years	Spheroid 40?	1.0 long? 0.8 dia?	1970 Apr 27.3	74.04	113.62	7770	1313	1471	0.010	94
	Cosmos 343	1970-36H	1970 Apr 25.71 7000 years	Spheroid 40?	1.0 long? 0.8 dia?	1970 May 1.7	74.02	114.32	7802	1374	1474	0.006	97
	Cosmos 336 rocket	1970-36J	1970 Apr 25.71 20000 years	Cylinder 2200?	7.4 long 2.4 dia	1970 May 3.1	74.04	116.69	7910	1473	1590	0.007	257
	Meteor 4	1970-37A	1970 Apr 28.45 60 years	Cylinder + 2 vanes 2200?	5 long? 1.5 dia?	1970 Apr 29.5	81.23	98.12	7046	625	710	0.006	89
	Meteor 4 rocket	1970-37B	1970 Apr 28.45 60 years	Cylinder 1440	3.8 long 2.6 dia	1970 May 4.1	81.24	98.34	7056	571	785	0.015	136
D	Fragment	1970-37C											
D R	Cosmos 344	1970-38A	1970 May 12.43 7.85 days 1970 May 20.28	Sphere-cylinder 5530?	5 long? 2.4 dia	1970 May 13.2	72.90	89.83	6644	202	329	0.010	45
D	Cosmos 344 rocket	1970-38B	1970 May 12.43 7.61 days 1970 May 20.04	Cylinder 2500?	7.5 long 2.6 dia	1970 May 13.2	72.90	89.65	6635	204	309	0.008	34

	Name		Launch date, lifetime and descent date	Shape and weight (kg)	Size (m)	Date of orbital determination	Orbital inclination (deg)	Nodal period (min)	Semi major axis (km)	Perigee height (km)	Apogee height (km)	Orbital eccentricity	Argument of perigee (deg)
D R	Cosmos 345	1970-39A	1970 May 20.39 7.98 days 1970 May 28.37	Sphere-cylinder 5530?	5 long? 2.4 dia	1970 May 21.5	51.75	89.06	6609	192	270	0.006	13
D	Cosmos 345 rocket	1970-39B	1970 May 20.39 2.70 days 1970 May 23.09	Cylinder 2500?	7.5 long 2.6 dia	1970 May 20.8	51.76	88.97	6605	185	268	0.006	0
D	[Thorad Agena D]	1970-40A	1970 May 20.90 27.53 days 1970 Jun 17.43	Cylinder 2000?	8 long? 1.5 dia	1970 May 22.2	83.00	88.62	6583	162	247	0.006	141
D	Capsule* (Doppler Beacon 2)	1970-40B	1970 May 20.90 1387.24 days 1974 Mar 8.14	Octagon? 60?	0.3 long? 0.9 dia?	1970 May 22.4 1971 Jul 1.0 1973 Feb 1.0	83.12 83.12 83.12	94.59 93.92 92.61	6875 6843 6779	491 460 397	503 469 405	0.0009 0.0007 0.0006	61 - -
D 2M R	Soyuz 9	1970-41A	1970 Jun 1.79 17.71 days 1970 Jun 19.50	Sphere-cylinder + 2 wings 6500	7.5 long 2.2 dia	1970 Jun 2.0 1970 Jun 2.2 1970 Jun 3.5	51.64 51.66 51.67	88.47 89.06 89.48	6580 6609 6630	176 208 244	227 254 259	0.004 0.003 0.001	166 125 213
D	Soyuz 9 rocket	1970-41B	1970 Jun 1.79 1.83 days 1970 Jun 3.62	Cylinder 2500?	7.5 long 2.6 dia	1970 Jun 2.2	51.67	88.36	6574	194	198	0.0003	353
D R	Cosmos 346	1970-42A	1970 Jun 10.40 7.00 days 1970 Jun 17.40	Sphere-cylinder 5530?	5 long? 2.4 dia	1970 Jun 14.0	51.74	89.16	6614	197	274	0.006	36
D	Cosmos 346 rocket	1970-42B	1970 Jun 10.40 3.51 days 1970 Jun 13.91	Cylinder 2500?	7.5 long 2.6 dia	1970 Jun 12.6	51.74	88.46	6580	167	236	0.005	352

* 1970-40B ejected from 1970-40A about 1970 May 20.97.

	Name		Launch date, lifetime and descent date	Shape and weight (kg)	Size (m)	Date of orbital determination	Orbital inclination (deg)	Nodal period (min)	Semi major axis (km)	Perigee height (km)	Apogee height (km)	Orbital eccentricity	Argument of perigee (deg)
D	Cosmos 347	1970-43A	1970 Jun 12.40 513.18 days 1971 Nov 7.58	Ellipsoid 400?	1.8 long 1.2 dia	1970 Jun 15.5 1970 Dec 1.0 1971 May 16.5	48.41 48.3 48.3	107.90 103.60 98.20	7511 7310 7054	216 213 204	2050 1650 1148	0.122 0.098 0.067	110 - -
D	Cosmos 347 rocket	1970-43B	1970 Jun 12.40 247.51 days 1971 Feb 14.91	Cylinder 1500?	8 long 1.65 dia	1970 Jun 15.5 1970 Oct 1.0	48.41 48.37	107.77 103.56	7505 7308	215 209	2039 1650	0.121 0.099	110 -
D	Cosmos 348 *	1970-44A	1970 Jun 13.21 41.88 days 1970 Jul 25.09	Ellipsoid 400?	1.8 long 1.2 dia	1970 Jun 13.9	70.99	93.10	6804	201	651	0.033	90
D	Cosmos 348 rocket	1970-44B	1970 Jun 13.21 25.88 days 1970 Jul 9.09	Cylinder 1500?	8 long 1.65 dia	1970 Jun 15.5	70.98	92.78	6788	201	619	0.031	88
D	Fragment	1970-44C											
D R	Cosmos 349	1970-45A	1970 Jun 17.54 7.79 days 1970 Jun 25.33	Sphere-cylinder 5530?	5 long? 2.4 dia	1970 Jun 18.0	65.39	89.81	6644	199	332	0.010	46
D	Cosmos 349 rocket	1970-45B	1970 Jun 17.54 5.33 days 1970 Jun 22.87	Cylinder 2500?	7.5 long 2.6 dia	1970 Jun 17.8	65.40	89.67	6637	191	326	0.010	42
	BMEWS 3 [Atlas Agena D]	1970-46A	1970 Jun 19.48 20 years?	Cylinder 700 full?	1.7 long? 1.4 dia?	1970 Jul 15.2	28.21	588.85	23310	178	33685	0.719	197
	Agena D rocket	1970-46B	1970 Jun 19.48 20 years?	Cylinder 700?	6 long? 1.5 dia	1970 Sep 1.4 1972 Jul 1.0 1975 Mar 1.0	27.98 28.4 28.06	579.51 508.6 445.4	23041 21106 19317	171 194 224	33154 29262 25654	0.716 0.689 0.658	230 - -

* International atmospheric and auroral studies

	Name		Launch date, lifetime and descent date	Shape and weight (kg)	Size (m)	Date of orbital determination	Orbital inclination (deg)	Nodal period (min)	Semi major axis (km)	Perigee height (km)	Apogee height (km)	Orbital eccentricity	Argument of perigee (deg)
	Meteor 5	1970-47A	1970 Jun 23.60 400 years	Cylinder + 2 vanes 2200?	5 long? 1.5 dia?	1970 Jun 25.3	81.23	102.16	7238	831	888	0.004	248
	Meteor 5 rocket	1970-47B	1970 Jun 23.60 300 years	Cylinder 1440	3.8 long 2.6 dia	1970 Jun 27.1	81.23	102.34	7246	810	926	0.008	189
D	[Titan 3B Agena D]	1970-48A	1970 Jun 25.62 11 days 1970 Jul 6	Cylinder 3000?	8 long? 1.5 dia	1970 Jun 26.9 1970 Jun 28.7	108.87 108.87	89.70 89.74	6637 6639	129 128	389 394	0.020 0.020	118 119
D	Molniya 1P	1970-49A	1970 Jun 26.14 2061 days 1976 Feb 16	Windmill + 6 vanes 1000?	3.4 long 1.6 dia	1970 Jul 7.5 1970 Aug 1.0 1973 Feb 1.0	65.37 65.34 65.45	704.70 717.55 717.43	26232 26551 26548	448 473 1531	39260 39872 38808	0.740 0.742 0.702	285 - -
D	Molniya 1P rocket	1970-49D	1970 Jun 26.14 2076 days 1976 Mar 2	Cylinder 440	2.0 long 2.0 dia	1970 Jul 8.4 1973 Feb 1.0	65.39 65.43	700.30 700.30	26123 26123	469 1479	39021 38011	0.738 0.699	285 -
D	Molniya 1P launcher rocket	1970-49C	1970 Jun 26.14 21.70 days 1970 Jul 17.84	Cylinder 2500?	7.5 long 2.6 dia	1970 Jun 27.7	65.39	91.30	6717	213	464	0.019	63
D	Molniya 1P launcher	1970-49B	1970 Jun 26.14 29.54 days 1970 Jul 25.68	Irregular -	-	1970 Jun 27.7	65.40	91.40	6722	225	462	0.018	69
D R	Cosmos 350	1970-50A	1970 Jun 26.50 11.93 days 1970 Jul 8.43	Sphere-cylinder 5700?	5.0 long 2.4 dia	1970 Jun 27.5	51.73	89.04	6608	202	258	0.004	26
D	Cosmos 350 rocket	1970-50B	1970 Jun 26.50 2.61 days 1970 Jun 29.11	Cylinder 2500?	7.5 long 2.6 dia	1970 Jun 26.9	51.74	88.87	6600	190	253	0.005	9

	Name		Launch date, lifetime and descent date	Shape and weight (kg)	Size (m)	Date of orbital determination	Orbital inclination (deg)	Nodal period (min)	Semi major axis (km)	Perigee height (km)	Apogee height (km)	Orbital eccentricity	Argument of perigee (deg)
D	Cosmos 351	1970-51A	1970 Jun 27.32 108.15 days 1970 Oct 13.47	Ellipsoid 400?	1.8 long 1.2 dia	1970 Jun 29.4	70.99	91.93	6747	270	467	0.015	82
D	Cosmos 351 rocket	1970-51B	1970 Jun 27.32 52.2 days 1970 Aug 18.5	Cylinder 1500?	8 long 1.65 dia	1970 Jun 28.5	70.98	91.75	6738	270	450	0.013	81
D R	Cosmos 352	1970-52A	1970 Jul 7.44 7.95 days 1970 Jul 15.39	Sphere-cylinder 5530?	5 long? 2.4 dia	1970 Jul 8.2	51.78	89.46	6629	207	294	0.007	51
D	Cosmos 352 rocket	1970-52B	1970 Jul 7.44 4.59 days 1970 Jul 12.03	Cylinder 2500?	7.5 long 2.6 dia	1970 Jul 8.3	51.77	89.33	6622	187	301	0.009	32
D R	Cosmos 353	1970-53A	1970 Jul 9.57 11.72 days 1970 Jul 21.29	Sphere-cylinder 5700?	5.0 long 2.4 dia	1970 Jul 11.2	65.42	89.38	6622	204	284	0.006	58
D	Cosmos 353 rocket	1970-53B	1970 Jul 9.57 6.44 days 1970 Jul 16.01	Cylinder 2500?	7.5 long 2.6 dia	1970 Jul 10.3	65.44	89.23	6615	200	274	0.006	58?
D	[Thorad Agena D]	1970-54A	1970 Jul 23.06 26.99 days 1970 Aug 19.05	Cylinder 2000?	8 long? 1.5 dia	1970 Jul 25.2	60.00	90.04	6656	158	398	0.018	111
	Intelsat 3 F-8	1970-55A	1970 Jul 23.97 1 million years	Cylinder 137 empty	1.04 long 1.42 dia	1970 Jul 24.0 1970 Aug 1.0	27.98 13.3	642.7 1043	24666 34090	282 19400	36294 36030	0.730 0.244	189 -*
	Intelsat 3 F-8 rocket	1970-55B	1970 Jul 23.97 20 years?	Sphere-cone 66	1.32 long 0.94 dia	1970 Aug 6.6	27.98	642.7	24666	282	36294	0.730	189

* Approximate orbit (satellite lost).

	Name		Launch date, lifetime and descent date	Shape and weight (kg)	Size (m)	Date of orbital determination	Orbital inclination (deg)	Nodal period (min)	Semi major axis (km)	Perigee height (km)	Apogee height (km)	Orbital eccentricity	Argument of perigee (deg)
D R?	Cosmos 354	1970-56A	1970 Jul 28.92 0.06 day 1970 Jul 28.98	Cylinder	2 long? 1 dia?		Orbit similar to 1970-56C						
D	Cosmos 354 rocket	1970-56B	1970 Jul 28.92 0.36 day 1970 Jul 29.28	Cylinder 1500?	8 long? 2.5 dia?	1970 Jul 29.2	49.57	87.13	6514	114	157	0.003	158
D	Cosmos 354 launch platform	1970-56C	1970 Jul 28.92 0.49 day 1970 Jul 29.41	Irregular	-	1970 Jul 29.2	49.62	87.54	6534	134	178	0.003	58
D	Intercosmos 3	1970-57A	1970 Aug 7.13 121.31 days 1970 Dec 6.44	Ellipsoid 400?	1.8 long 1.2 dia	1970 Aug 9.3	48.41	99.70	7126	200	1295	0.077	110
D	Intercosmos 3 rocket	1970-57B	1970 Aug 7.13 102.57 days 1970 Nov 17.70	Cylinder 1500?	8 long 1.65 dia	1970 Aug 9.7	48.40	99.61	7122	201	1287	0.076	112
D R	Cosmos 355	1970-58A	1970 Aug 7.40 7.78 days 1970 Aug 15.18	Sphere-cylinder 5530?	5 long? 2.4 dia	1970 Aug 8.4	65.40	89.71	6639	199	322	0.009	45
D	Cosmos 355 rocket	1970-58B	1970 Aug 7.40 6.95 days 1970 Aug 14.35	Cylinder 2500?	7.5 long 2.6 dia	1970 Aug 9.1	65.39	89.45	6626	191	304	0.008	35
D	Cosmos 356 *	1970-59A	1970 Aug 10.84 53.11 days 1970 Oct 2.95	Ellipsoid 400?	1.8 long 1.2 dia	1970 Aug 11.3	81.96	92.62	6780	231	573	0.025	80
D	Cosmos 356 rocket	1970-59B	1970 Aug 10.84 52.00 days 1970 Oct 1.84	Cylinder 1500?	8 long 1.65 dia	1970 Aug 11.4	81.96	92.49	6773	236	554	0.023	79

* Atmospheric and auroral studies

	Name		Launch date, lifetime and descent date	Shape and weight (kg)	Size (m)	Date of orbital determination	Orbital inclination (deg)	Nodal period (min)	Semi major axis (km)	Perigee height (km)	Apogee height (km)	Orbital eccentricity	Argument of perigee (deg)
D	Venus 7 launcher rocket	1970-60B	1970 Aug 17.23 1.17 days 1970 Aug 18.40	Cylinder 2500?	7.5 long 2.6 dia	1970 Aug 17.5	51.75	88.28	6570	182	202	0.002	280
D	Venus 7 launcher	1970-60C	1970 Aug 17.23 1.68 days 1970 Aug 18.91	Irregular	-	1970 Aug 17.4	51.70	88.51	6582	174	233	0.004	349
D	[Titan 3B Agena D]	1970-61A	1970 Aug 18.62 16 days 1970 Sep 3	Cylinder 3000?	8 long? 1.5 dia	1970 Aug 21.2 1970 Aug 25.0	110.95 110.98	89.67 90.49	6636 6676	151 152	365 444	0.016 0.022	124 123
	Skynet 1B	1970-62A	1970 Aug 19.51 uncertain	Cylinder 243 full 129 empty	0.81 long 1.37 dia	1970 Aug 20.2	28.04	636.5	24534	270	36041	0.729	181*
	Skynet 1B rocket	1970-62B	1970 Aug 19.51 20 years?	Sphere-cone 66	1.32 long 0.94 dia	1970 Aug 20.2	28.04	636.5	24534	270	36041	0.729	181
D	Cosmos 357	1970-63A	1970 Aug 19.63 97.30 days 1970 Nov 24.93	Ellipsoid 400?	1.8 long 1.2 dia	1970 Aug 21.1	70.99	92.04	6752	272	476	0.015	73
D	Cosmos 357 rocket	1970-63B	1970 Aug 19.63 57 days 1970 Oct 15	Cylinder 1500?	8 long 1.65 dia	1970 Aug 22.0	70.98	91.81	6741	275	450	0.013	70
	Cosmos 358	1970-64A	1970 Aug 20.61 20 years	Cylinder + paddles? 900?	2 long? 1 dia?	1970 Aug 27.3	74.04	95.19	6905	515	539	0.002	316
D	Cosmos 358 rocket	1970-64B	1970 Aug 20.61 3243 days 1979 Jul 7	Cylinder 2200?	7.4 long 2.4 dia	1970 Aug 24.2 1973 Oct 1.0	74.03 74.03	95.08 94.40	6900 6867	505 471	539 507	0.003 0.003	325 -
D	Fragment	1970-64C											

Space Vehicle: Venus 7, 1970-60A. *Transfer orbit: present orbit unknown.

	Name		Launch date, lifetime and descent date	Shape and weight (kg)	Size (m)	Date of orbital determination	Orbital inclination (deg)	Nodal period (min)	Semi major axis (km)	Perigee height (km)	Apogee height (km)	Orbital eccentricity	Argument of perigee (deg)
D	Cosmos 359*	1970-65A	1970 Aug 22.22 76.04 days 1970 Nov 6.26	Sphere-cylinder 1180	3.5 long 1.2 dia	1970 Aug 24.3	51.13	95.57	6927	208	890	0.049	60
D	Cosmos 359 launcher rocket	1970-65B	1970 Aug 22.22 7.06 days 1970 Aug 29.28	Cylinder 2500?	7.5 long 2.6 dia	1970 Aug 23.5	51.80	89.49	6631	207	298	0.007	37
D	Cosmos 359 rocket	1970-65D	1970 Aug 22.22 410 days 1971 Oct 6	Cylinder 2400**	2.0 long 2.0 dia	1970 Sep 1.0 1971 Feb 1.0 1971 Jun 1.0	51.13 51.13 51.13	95.31 93.58 91.92	6914 6830 6749	205 202 195	867 701 546	0.048 0.037 0.026	79 288 -
D	Fragments	1970-65C, E											
D	[Thorad Agena D]	1970-66A	1970 Aug 26.12 1673 days 1975 Mar 26	Cylinder 2000?	8 long? 1.5 dia	1970 Aug 29.3 1972 Apr 1.0 1974 Feb 15.0	74.99 74.99 74.99	94.51 93.87 92.57	6872 6841 6778	484 457 398	504 469 402	0.001 0.001 0.0003	259 - -
	Navy Navigation Satellite 19	1970-67A	1970 Aug 27.56 1300 years	Octagon + 4 vanes + boom 58?	0.25 long 0.46 dia	1970 Aug 29.2	90.02	107.04	7466	955	1221	0.018	245
	Altair rocket	1970-67B	1970 Aug 27.56 700 years	Cylinder 24	1.5 long 0.46 dia	1970 Sep 6.9	90.04	107.05	7467	952	1225	0.018	218
1d	Fragments	1970-67C-E											
D R	Cosmos 360	1970-68A	1970 Aug 29.36 9.93 days 1970 Sep 8.29	Sphere-cylinder 6300?	6.5 long? 2.4 dia	1970 Sep 2.8	64.99	89.64	6635	209	305	0.007	37
D	Cosmos 360 rocket	1970-68B	1970 Aug 29.36 4.34 days 1970 Sep 2.70	Cylinder 2500?	7.5 long 2.6 dia	1970 Aug 31.1	65.00	88.91	6600	187	256	0.005	26
D	Cosmos 360 engine ***	1970-68C	1970 Aug 29.36 11 days 1970 Sep 9	Cone 600? full	1.5 long? 2 dia?	1970 Sep 7.7	64.93	88.61	6585	193	220	0.002	85
D	Fragments	1970-68D, E											

* Cosmos 359 was probably an attempted Venus probe.

**Mass before incomplete burn was approximately 5400kg

*** 1970-68C ejected from 1970-68A about 1970 Sep 7.4.

Year of launch 1970 continued

	Name		Launch date, lifetime and descent date	Shape and weight (kg)	Size (m)	Date of orbital determination	Orbital inclination (deg)	Nodal period (min)	Semi major axis (km)	Perigee height (km)	Apogee height (km)	Orbital eccentricity	Argument of perigee (deg)
	BMEWS 4 [Atlas Agena D]	1970-69A	1970 Sep 1.04 > million years	Cylinder 700 full? 350 empty?	1.7 long? 1.4 dia?	1970 Sep 1.0 1970 Oct 1.0	28.50 10.3	88.03 1441.9	6563 42279	179 31947	190 39855	0.001 0.094	242 -
	Agena D rocket	1970-69B	1970 Sep 1.04 20 years?	Cylinder 700?	6 long? 1.5 dia	Orbit similar to 1970-46B							
	[Thor Burner 2]	1970-70A	1970 Sep 3.36 80 years	12-faced frustum 195	1.64 long 1.31 to 1.10 dia	1970 Sep 4.9	98.73	101.30	7197	764	874	0.008	21
	Burner 2 rocket	1970-70B	1970 Sep 3.36 60 years	Sphere-cone 66	1.32 long 0.94 dia	1970 Sep 9.2	98.75	101.29	7197	765	872	0.007	7
D R	Cosmos 361	1970-71A	1970 Sep 8.44 12.8 days 1970 Sep 21.2	Sphere-cylinder 6300?	6.5 long? 2.4 dia	1970 Sep 9.1 1970 Sep 13.2	72.87 72.87	89.59 90.09	6632 6656	209 203	298 353	0.007 0.011	56 37
D	Cosmos 361 rocket	1970-71B	1970 Sep 8.44 5.57 days 1970 Sep 14.01	Cylinder 2500?	7.5 long 2.6 dia	1970 Sep 8.7	72.85	89.47	6626	204	291	0.007	54
D	Cosmos 361 engine*	1970-71C	1970 Sep 8.44 27 days 1970 Oct 5	Cone 600? full	1.5 long? 2 dia?	1970 Sep 20.2	72.87	89.95	6650	209	334	0.009	25
D	Fragment	1970-71D											
D	Luna 16 launcher rocket	1970-72B	1970 Sep 12.56 3.19 days 1970 Sep 15.75	Cylinder 4000?	12 long? 4 dia	1970 Sep 12.9	51.53	88.70	6591	185	241	0.004	344
D	Luna 16 launcher	1970-72C	1970 Sep 12.56 3.29 days 1970 Sep 15.85	-	-	1970 Sep 12.9	51.50	88.71	6592	186	241	0.004	334

Space Vehicle: Luna 16, 1970-72A.

* 1970-71C ejected from 1970-71A about 1970 Sep 20.2.

	Name		Launch date, lifetime and descent date	Shape and weight (kg)	Size (m)	Date of orbital determination	Orbital inclination (deg)	Nodal period (min)	Semi major axis (km)	Perigee height (km)	Apogee height (km)	Orbital eccentricity	Argument of perigee (deg)
D	Cosmos 362	1970-73A	1970 Sep 16.50 392.40 days 1971 Oct 13.90	Ellipsoid 400?	1.8 long 1.2 dia	1970 Sep 18.0 1971 Apr 1.0	70.96 70.96	95.65 93.26	6928 6812	270 255	829 613	0.040 0.026	82 -
D	Cosmos 362 rocket	1970-73B	1970 Sep 16.50 193.33 days 1971 Mar 28.83	Cylinder 1500?	8 long 1.65 dia	1970 Sep 17.0 1971 Jan 1.0	70.96 70.96	95.51 93.27	6921 6813	271 252	815 617	0.039 0.027	83 252
D R	Cosmos 363	1970-74A	1970 Sep 17.35 11.86 days 1970 Sep 29.21	Sphere-cylinder 5700?	5.0 long 2.4 dia	1970 Sep 18.1	65.01	89.53	6629	208	294	0.007	62
D	Cosmos 363 rocket	1970-74B	1970 Sep 17.35 5.44 days 1970 Sep 22.79	Cylinder 2500?	7.5 long 2.6 dia	1970 Sep 18.1	65.00	89.33	6619	205	277	0.006	54
D R	Cosmos 364	1970-75A	1970 Sep 22.54 9.76 days 1970 Oct 2.30	Sphere-cylinder 6300?	6.5 long? 2.4 dia	1970 Sep 25.7 1970 Oct 1.5	65.41 65.43	89.49 89.50	6628 6628	202 196	297 304	0.007 0.008	75 52
D	Cosmos 364 rocket	1970-75B	1970 Sep 22.54 5.96 days 1970 Sep 28.50	Cylinder 2500?	7.5 long 2.6 dia	1970 Sep 23.2	65.40	89.50	6628	204	296	0.007	59
D	Cosmos 364 engine*	1970-75C	1970 Sep 22.54 17 days 1970 Oct 9	Cone 600? full	1.5 long? 2 dia?	1970 Oct 5.2	65.39	91.16	6710	222	442	0.016	2

* 1970-75C ejected from 1970-75A about 1970 Oct 1.5.

	Name		Launch date, lifetime and descent date	Shape and weight (kg)	Size (m)	Date of orbital determination	Orbital inclina-tion (deg)	Nodal period (min)	Semi major axis (km)	Perigee height (km)	Apogee height (km)	Orbital eccen-tricity	Argument of perigee (deg)
D R?	Cosmos 365	1970-76A	1970 Sep 25.59 0.06 day 1970 Sep 25.65	Cylinder	2 long? 1 dia?				Orbit similar to 1970-76B				
D	Cosmos 365 launch platform	1970-76B	1970 Sep 25.59 0.43 day 1970 Sep 26.02	Irregular	-	1970 Sep 25.9	49.66	87.49	6532	133	174	0.003	174
D	Cosmos 365 rocket	1970-76C	1970 Sep 25.59 0.3 day? 1970 Sep 25	Cylinder 1500?	8 long? 2.5 dia?	1970 Sep 25.7	49.66	87.20	6517	117	161	0.003	-
D	Molniya 1Q	1970-77A	1970 Sep 29.35 1999 days 1976 Mar 20	Windmill + 6 vanes 1000?	3.4 long 1.6 dia	1970 Oct 1.0 1970 Nov 1.0 1971 Sep 16.0	65.5 65.3 65.75	706.18 717.57 717.63	26268 26551 26553	480 723 400	39300 39623 39949	0.739 0.733 0.745	285 284 -
D	Molniya 1Q launcher	1970-77C	1970 Sep 29.35 24.08 days 1970 Oct 23.43	Irregular	-	1970 Sep 30.9	65.43	91.50	6727	228	469	0.018	74
D	Molniya 1Q launcher rocket	1970-77B	1970 Sep 29.35 16.77 days 1970 Oct 16.12	Cylinder 2500?	7.5 long 2.6 dia	1970 Oct 1.6	65.39	90.99	6702	205	443	0.018	57
D	Molniya 1Q rocket	1970-77D	1970 Sep 29.35 610 days 1972 May 31	Cylinder 440	2.0 long 2.0 dia	1970 Nov 1.0 1971 Apr 1.0 1971 Sep 1.0	65.41 65.6 65.8	704.67 704.43 704.07	26231 26225 26216	490 358 236	39215 39335 39439	0.738 0.743 0.748	- - -
D	Fragment	1970-77E											
D R	Cosmos 366	1970-78A	1970 Oct 1.35 11.9 days 1970 Oct 13.3	Sphere- cylinder 5700?	5.0 long 2.4 dia	1970 Oct 2.5	64.96	89.48	6628	204	295	0.007	39
D	Cosmos 366 rocket	1970-78B	1970 Oct 1.35 5 days 1970 Oct 6	Cylinder 2500?	7.5 long 2.6 dia	1970 Oct 3.1	64.96	89.13	6610	177	286	0.008	42

	Name		Launch date, lifetime and descent date	Shape and weight (kg)	Size (m)	Date of orbital determination	Orbital inclination (deg)	Nodal period (min)	Semi major axis (km)	Perigee height (km)	Apogee height (km)	Orbital eccentricity	Argument of perigee (deg)
	Cosmos 367 †	1970-79A	1970 Oct 3.44 600 years	Cone-cylinder	6 long? 2 dia?	1970 Oct 6.8	65.28	104.53	7351	922	1024	0.007	312
D	Cosmos 367 rocket	1970-79B	1970 Oct 3.44 2.93 days 1970 Oct 6.37	Cylinder 1500?	8 long? 2.5 dia?	1970 Oct 4.9	65.21	89.21	6614	226	246	0.002	227
D	Cosmos 367 platform	1970-79C	1970 Oct 3.44 27.69 days 1970 Oct 31.13	Irregular	-	1970 Oct 4.8	65.09	89.60	6633	246	264	0.001	258
D	Fragments	1970-79D,E											
D B R	Cosmos 368	1970-80A	1970 Oct 8.53 5.98 days 1970 Oct 14.51	Sphere-cylinder 5900?	5.9 long 2.4 dia	1970 Oct 9.7	64.99	90.56	6680	204	400	0.015	56
D	Cosmos 368 rocket	1970-80B	1970 Oct 8.53 11.60 days 1970 Oct 20.13	Cylinder 2500?	7.5 long 2.6 dia	1970 Oct 9.7	65.00	90.43	6674	203	389	0.014	52
D	Capsule*	1970-80E	1970 Oct 8.53 27 days 1970 Nov 4	Ellipsoid 200?	0.9 long 1.9 dia	1970 Nov 1.0	64.9	88.70	6589	178	243	0.005	-
D	Fragments	1970-80C,D											
D	Cosmos 369	1970-81A	1970 Oct 8.63 106.20 days 1971 Jan 22.83	Ellipsoid 400?	1.8 long 1.2 dia	1970 Oct 9.6	70.93	92.32	6766	269	506	0.018	86
D	Cosmos 369 rocket	1970-81B	1970 Oct 8.63 52.76 days 1970 Nov 30.39	Cylinder 1500?	8 long 1.65 dia	1970 Oct 10.9	70.93	92.15	6757	271	486	0.016	84

*1970-80E ejected from 1970-80A on 1970 Oct 13

†1970-79B and 79C attached to 1970-79A until orbit change

	Name		Launch date, lifetime and descent date	Shape and weight (kg)	Size (m)	Date of orbital determination	Orbital inclination (deg)	Nodal period (min)	Semi major axis (km)	Perigee height (km)	Apogee height (km)	Orbital eccentricity	Argument of perigee (deg)
D R	Cosmos 370	1970-82A	1970 Oct 9.46 12.84 days 1970 Oct 22.30	Sphere-cylinder 6300?	6.5 long? 2.4 dia	1970 Oct 10.9	64.92	89.40	6623	202	288	0.006	33
D	Cosmos 370 rocket	1970-82B	1970 Oct 9.46 4.13 days 1970 Oct 13.59	Cylinder 2500?	7.5 long 2.6 dia	1970 Oct 11.7	64.93	88.77	6592	187	241	0.004	9
D	Cosmos 370 engine*	1970-82C	1970 Oct 9.46 18 days 1970 Oct 27	Cone 600? full	1.5 long? 2 dia?	1970 Oct 22.3	64.92	88.74	6591	182	243	0.005	-
	Cosmos 371	1970-83A	1970 Oct 12.58 100 years	Cylinder + paddles? 750?	2 long? 1 dia?	1970 Oct 31.0	74.00	99.92	7132	750	758	0.0005	16
	Cosmos 371 rocket	1970-83B	1970 Oct 12.58 80 years	Cylinder 2200?	7.4 long 2.4 dia	1970 Oct 12.9	74.00	99.81	7127	743	755	0.0009	107
D	Intercosmos 4	1970-84A	1970 Oct 14.48 95.20 days 1971 Jan 17.68	Ellipsoid + 6 panels 400?	1.8 long 1.2 dia	1970 Oct 16.9	48.41	93.56	6830	255	649	0.029	118
D	Intercosmos 4 rocket	1970-84B	1970 Oct 14.48 63.63 days 1970 Dec 17.11	Cylinder 1500?	8 long 1.65 dia	1970 Oct 16.7	48.41	93.29	6817	259	619	0.026	117
	Meteor 6	1970-85A	1970 Oct 15.48 60 years	Cylinder + 2 vanes 2200?	5 long? 1.5 dia?	1970 Oct 25.1	81.21	97.49	7015	626	648	0.002	56
	Meteor 6 rocket	1970-85B	1970 Oct 15.48 60 years	Cylinder 1440	3.8 long 2.6 dia	1970 Oct 22.3	81.23	97.62	7022	551	736	0.013	155
	Fragment	1970-85C											

*1970-82C ejected from 1970-82A on 1970 Oct 21 or 22.

	Name		Launch date, lifetime and descent date	Shape and weight (kg)	Size (m)	Date of orbital determination	Orbital inclination (deg)	Nodal period (min)	Semi major axis (km)	Perigee height (km)	Apogee height (km)	Orbital eccentricity	Argument of perigee (deg)
	Cosmos 372	1970-86A	1970 Oct 16.63 100 years	Cylinder + paddles? 750?	2 long? 1 dia?	1970 Oct 29.2	74.06	100.80	7174	785	806	0.001	9
	Cosmos 372 rocket	1970-86B	1970 Oct 16.63 80 years	Cylinder 2200?	7.4 long 2.4 dia	1970 Oct 29.2	74.06	100.70	7169	776	806	0.002	7
	Fragments	1970-86C,D											
D	Cosmos 373	1970-87A	1970 Oct 20.24 3427 days 1980 Mar 8	Cylinder?	4 long? 2 dia?	1970 Oct 20.5 1970 Oct 31.1	62.93 62.92	94.77 94.83	6886 6889	472 466	544 556	0.005 0.006	290 311
D	Cosmos 373 rocket	1970-87B	1970 Oct 20.24 90 days 1971 Jan 18	Cylinder 1500?	8 long? 2.5 dia?	1970 Oct 23.2	62.26	95.45	6919	145	937	0.057	54
D,	Zond 8 launcher rocket	1970-88B	1970 Oct 20.83 6 days 1970 Oct 26	Cylinder 4000?	12 long? 4 dia	1970 Oct 21.6	51.51	88.68	6591	202	223	0.002	338
	Cosmos 374*	1970-89A	1970 Oct 23.18 150 years	Cylinder?	4 long? 2 dia?	1970 Oct 30.2	62.95	112.26	7709	521	2141	0.105	61
	Cosmos 374 rocket	1970-89B	1970 Oct 23.18 100 years	Cylinder 1500?	8 long? 2.5 dia?	1970 Oct 28.5	62.93	111.83	7690	517	2106	0.103	62
48d	Fragments	1970-89C-CU											
D	[Titan 3B Agena D]	1970-90A	1970 Oct 23.74 19 days 1970 Nov 11	Cylinder 3000?	8 long? 1.5 dia	1970 Oct 24.3	111.06	89.83	6644	135	396	0.020	137

Space Vehicle: Zond 8, 1970-88A, passed 1120 km beyond Moon on Oct 24.1; recovered on Earth Oct 27.58

*1970-89A passed close to 1970-87A on Oct 23.34, then exploded.

	Name		Launch date, lifetime and descent date	Shape and weight (kg)	Size (m)	Date of orbital determination	Orbital inclination (deg)	Nodal period (min)	Semi major axis (km)	Perigee height (km)	Apogee height (km)	Orbital eccentricity	Argument of perigee (deg)
	Cosmos 375*	1970-91A	1970 Oct 30.09 150 years	Cylinder?	4 long? 2 dia?	1970 Nov 3.0	62.82	111.82	7689	524	2098	0.102	56
	Cosmos 375 rocket	1970-91B	1970 Oct 30.09 100 years	Cylinder 1500?	8 long? 2.5 dia?	1970 Nov 3.9	62.78	111.52	7674	526	2066	0.100	56
10d	Fragments	1970-91C-AQ											
D R	Cosmos 376	1970-92A	1970 Oct 30.56 12.71 days 1970 Nov 12.27	Sphere-cylinder 6300?	6.5 long? 2.4 dia	1970 Oct 31.9 1970 Nov 10.3	65.38 65.42	89.43 89.23	6625 6616	207 203	286 272	0.006 0.005	73 65
D	Cosmos 376 rocket	1970-92B	1970 Oct 30.56 5.61 days 1970 Nov 5.17	Cylinder 2500?	7.5 long 2.6 dia	1970 Oct 31.8	65.40	89.22	6614	209	263	0.004	62
D	Cosmos 376 engine**	1970-92C	1970 Oct 30.56 21 days 1970 Nov 20	Cone 600? full	1.5 long? 2 dia?	1970 Nov 15.6	65.36	88.67	6588	192	228	0.003	70
	IMEWS 1† [Titan 3C]	1970-93A	1970 Nov 6.44 >million years	Cylinder + 4 panels 820?	6 long? 2.5 dia?	1970 Nov 7.1 1970 Dec 1.0	26.29 7.8	635.1 1197.1	24473 37346	300 26050	35890 35886	0.727 0.132	180 -
	Transtage	1970-93B	1970 Nov 6.44 > million years	Cylinder 1500?	6 long? 3 dia	Orbit similar to 1970-93A							
D	Titan 3C second stage	1970-93C	1970 Nov 6.44 1.51 days 1970 Nov 7.95	Cylinder 1900	6 long 3.0 dia	1970 Nov 7.1	28.6	89.15	6617	147	331	0.014	120

* 1970-91A passed close to 1970-87A on Oct 30.25.
** Engine probably ejected about Nov 10.

† Integrated Missile Early Warning Satellite.

	Name		Launch date, lifetime and descent date	Shape and weight (kg)	Size (m)	Date of orbital determination	Orbital inclination (deg)	Nodal period (min)	Semi major axis (km)	Perigee height (km)	Apogee height (km)	Orbital eccentricity	Argument of perigee (deg)
D B	Orbiting Frog Otolith 1 [Scout]	1970-94A	1970 Nov 9.25 180.96 days 1971 May 9.21	Octagonal cylinder 133	1.19 long 0.76 dia	1970 Nov 11.1 1971 Feb 1.0	37.41 37.41	92.64 91.86	6789 6750	304 278	518 466	0.016 0.014	134 -
D	Radiation Meteoroid - OFO 1 rocket	1970-94B	1970 Nov 9.25 90.15 days 1971 Feb 7.40	Cylinder 45 (payload 21)	1.68 long 0.76 dia	1970 Nov 11.1	37.41	92.71	6793	303	526	0.016	133
D	Fragments	1970-94C-E											
D	Luna 17 launcher rocket	1970-95B	1970 Nov 10.61 2.96 days 1970 Nov 13.57	Cylinder 4000?	12 long? 4 dia	1970 Nov 11.5	51.53	88.57	6584	184	228	0.003	299
D	Luna 17 launcher	1970-95C	1970 Nov 10.61 3.00 days 1970 Nov 13.61	-	-	1970 Nov 10.8	51.55	88.73	6593	192	237	0.003	296
D R	Cosmos 377	1970-96A	1970 Nov 11.39 11.9 days 1970 Nov 23.3	Sphere-cylinder 5700?	5.0 long 2.4 dia	1970 Nov 11.8	64.99	89.40	6623	204	286	0.006	44
D	Cosmos 377 rocket	1970-96B	1970 Nov 11.39 5.49 days 1970 Nov 16.88	Cylinder 2500?	7.5 long 2.6 dia	1970 Nov 11.4	64.98	89.36	6621	201	285	0.006	39
D	Cosmos 378*	1970-97A	1970 Nov 17.77 638.86 days 1972 Aug 17.63	Octagonal ellipsoid? 400?	1.8 long? 1.5 dia?	1970 Nov 20.0 1971 Oct 16.5	74.00 74.00	104.88 99.95	7366 7133	234 227	1742 1283	0.102 0.074	122 -
D	Cosmos 378 rocket	1970-97B	1970 Nov 17.77 682.98 days 1972 Sep 30.75	Cylinder 2200?	7.4 long 2.4 dia	1970 Nov 20.0 1971 Nov 16.0	74.00 74.00	104.75 99.99	7360 7135	233 222	1730 1292	0.102 0.075	121 -

Space Vehicle: Luna 17, 1970-95A

* Ionospheric studies

	Name		Launch date, lifetime and descent date	Shape and weight (kg)	Size (m)	Date of orbital determination	Orbital inclination (deg)	Nodal period (min)	Semi major axis (km)	Perigee height (km)	Apogee height (km)	Orbital eccentricity	Argument of perigee (deg)
D	[Thorad Agena D]	1970-98A	1970 Nov 18.89 22.78 days 1970 Dec 11.67	Cylinder 2000?	8 long? 1.5 dia	1970 Nov 21.0	82.99	88.70	6587	185	232	0.004	164
D	Capsule*	1970-98B	1970 Nov 18.89 2492 days 1977 Sep 14	Octagon? 60?	0.3 long? 0.9 dia?	1970 Nov 20.6 1972 Sep 1.0	83.18 83.18	94.63 93.95	6877 6844	487 455	511 477	0.002 0.002	258 -
	Cosmos 379	1970-99A	1970 Nov 24.22 14 years	-	-	1970 Nov 24.9 1970 Nov 25.9 1970 Nov 30.2 1973 Jun 1.0	51.62 51.62 51.69 51.70	88.67 98.73 259.64 225.2	6590 7078 13483 12262	192 190 175 174	232 1210 14035 11593	0.003 0.072 0.514 0.466	80 62 72 -
D	Cosmos 379 rocket	1970-99B	1970 Nov 24.22 2.34 days 1970 Nov 26.56	Cylinder 2500?	7.5 long 2.6 dia	1970 Nov 24.9	51.61	88.46	6580	189	214	0.002	59
D	Cosmos 379 platform	1970-99C	1970 Nov 24.22 81 days 1971 Feb 13	-	-	1970 Nov 27.4	51.52	98.58	7071	187	1198	0.071	67
D	Fragment	1970-99D											
D	Cosmos 380	1970-100A	1970 Nov 24.46 205.43 days 1971 Jun 17.89	Ellipsoid 400?	1.8 long 1.2 dia	1970 Nov 25.2 1971 Mar 1.0	81.95 81.95	102.15 98.39	7238 7059	199 195	1520 1167	0.091 0.069	73 -
D	Cosmos 380 rocket	1970-100B	1970 Nov 24.46 135.76 days 1971 Apr 9.22	Cylinder 1500?	8 long 1.65 dia	1970 Nov 25.8 1971 Feb 1.0	81.96 81.96	101.93 98.40	7227 7060	197 193	1501 1170	0.090 0.069	71 -

* 1970-98B ejected from 1970-98A on 1970 Nov 18.96.

	Name		Launch date, lifetime and descent date	Shape and weight (kg)	Size (m)	Date of orbital determination	Orbital inclination (deg)	Nodal period (min)	Semi major axis (km)	Perigee height (km)	Apogee height (km)	Orbital eccentricity	Argument of perigee (deg)
D	Molniya IR	1970-101A	1970 Nov 27.66 1824 days 1975 Nov 25	Windmill + 6 vanes 1000?	3.4 long 1.6 dia	1970 Nov 29.2 1971 Jan 1.0 1973 Feb 1.0	65.48 65.49 65.50	707.09 717.81 717.73	26292 26557 26555	471 463 1379	39356 39895 38975	0.740 0.742 0.708	285 - -
D	Molniya IR launcher rocket	1970-101B	1970 Nov 27.66 13.69 days 1970 Dec 11.35	Cylinder 2500?	7.5 long 2.6 dia	1970 Nov 28.8	65.42	90.61	6683	213	397	0.014	51
D	Molniya IR launcher	1970-101C	1970 Nov 27.66 20.08 days 1970 Dec 17.74	Irregular	-	1970 Nov 29.5	65.39	91.02	6703	216	434	0.016	60
D	Molniya IR rocket	1970-101D	1970 Nov 27.66 1862 days 1976 Jan 2	Cylinder 440	2.0 long 2.0 dia	1970 Nov 30.6 1973 Feb 1.0	65.41 65.46	702.99 702.97	26190 26190	412 1350	39212 38273	0.741 0.705	285 -
	Cosmos 381*	1970-102A	1970 Dec 2.17 1200 years	Cylinder + boom 700?	1.4 long 2.0 dia	1970 Dec 7.6	74.04	104.93	7369	968	1013	0.003	270
	Cosmos 381 rocket	1970-102B	1970 Dec 2.17 600 years	Cylinder 2200?	7.4 long 2.4 dia	1970 Dec 4.6	74.03	104.82	7364	967	1004	0.002	263
	Fragments	1970-102C-F											
	Cosmos 382	1970-103A	1970 Dec 2.69 1000 years	Sphere?	5 dia?	1970 Dec 4.5 1970 Dec 7.4 1970 Dec 8.0	51.54 51.55 55.87	142.82 158.93 171.06	9053 9722 10,208	305 1615 2577	5045 5072 5082	0.262 0.178 0.123	261 258 248
	Cosmos 382 rocket	1970-103B	1970 Dec 2.69 30000 years	Cylinder 4000?	12 long? 4 dia	1970 Dec 5.1 1971 Jan 1.0	51.53 51.54	144.07 158.74	9105 9714	409 1590	5045 5081	0.255 0.180	261 285
	Cosmos 382 platform	1970-103C	1970 Dec 2.69 30000 years	-	-	1970 Dec 14.4	51.59	159.07	9727	1614	5084	0.178	266
2d	Fragments	1970-103D-G											

* Topside ionospheric sounder

	Name		Launch date, lifetime and descent date	Shape and weight (kg)	Size (m)	Date of orbital determination	Orbital inclination (deg)	Nodal period (min)	Semi major axis (km)	Perigee height (km)	Apogee height (km)	Orbital eccentricity	Argument of perigee (deg)
D R	Cosmos 383*	1970-104A	1970 Dec 3.58 12.69 days 1970 Dec 16.27	Sphere-cylinder 6300?	6.5 long? 2.4 dia	1970 Dec 4.7 1970 Dec 14.1	65.41 65.42	89.33 89.53	6620 6630	204 181	279 323	0.006 0.011	57 53
D	Cosmos 383 rocket	1970-104B	1970 Dec 3.58 6.12 days 1970 Dec 9.70	Cylinder 2500?	7.5 long 2.6 dia	1970 Dec 4.4	65.40	89.15	6611	200	266	0.005	55
D R	Cosmos 384	1970-105A	1970 Dec 10.47 11.80 days 1970 Dec 22.27	Sphere-cylinder 5900?	5.9 long 2.4 dia	1970 Dec 11.5	72.88	89.46	6626	203	292	0.007	57
D	Cosmos 384 rocket	1970-105B	1970 Dec 10.47 5.31 days 1970 Dec 15.78	Cylinder 2500?	7.5 long 2.6 dia	1970 Dec 12.4	72.89	89.16	6610	195	268	0.005	41
D	Capsule**	1970-105E	1970 Dec 10.47 17 days 1970 Dec 27	Ellipsoid 200?	0.9 long 1.9 dia	1970 Dec 18.3	72.88	89.27	6616	204	271	0.005	44
D	Fragments	1970-105C,D											
	NOAA 1 † (ITOS)	1970-106A	1970 Dec 11.48 10000 years	Box 306	1.25 long 1.02 square	1970 Dec 20.9	101.94	114.93	7829	1429	1473	0.003	237
	NOAA 1 second stage (CEP 1)††	1970-106B	1970 Dec 11.48 5000 years	Cylinder 350?	4.9 long 1.43 dia	1970 Dec 24.9	101.92	114.91	7828	1425	1475	0.003	226
	Fragment	1970-106C											

* Cosmos 383 manoeuvred, but no jettisoned engine was tracked or designated.

** 1970-105E ejected from 1970-105A on 1970 Dec 17.6

† National Oceanic and Atmospheric Administration

†† Cylindrical Electrostatic Probe

	Name		Launch date, lifetime and descent date	Shape and weight (kg)	Size (m)	Date of orbital determination	Orbital inclination (deg)	Nodal period (min)	Semi major axis (km)	Perigee height (km)	Apogee height (km)	Orbital eccentricity	Argument of perigee (deg)
D	Explorer 42 (SAS 1) *	1970-107A	1970 Dec 12.45 3036 days 1979 Apr 5	Cylinder + 4 paddles 143	1.16 long 0.56 dia	1970 Dec 12.7	3.04	95.30	6921	522	563	0.003	0
D	Explorer 42 rocket	1970-107B	1970 Dec 12.45 9 years ?	Cylinder 24	1.50 long 0.46 dia	1970 Dec 12.4	2.91	95.22	6917	529	549	0.002	352
	Cosmos 385	1970-108A	1970 Dec 12.54 1200 years	Cylinder + boom? 700?	1.4 long 2.0 dia	1970 Dec 15.9	74.02	104.75	7360	978	986	0.0005	312
	Cosmos 385 rocket	1970-108B	1970 Dec 12.54 600 years	Cylinder 2200?	7.4 long 2.4 dia	1970 Dec 15.3	74.02	104.64	7355	974	979	0.0003	129
D L	Peole 1**	1970-109A	1970 Dec 12.54 3474 days 1980 Jun 16	Octahedron 70	0.55 long 0.70 dia	1971 Jan 12.8	15.00	97.17	7010	517	747	0.016	248
	Peole 1 rocket	1970-109B	1970 Dec 12.54 50 years	Cylinder 68	1.60 long? 0.65 dia	1970 Dec 24.7	15.00	98.43	7070	635	749	0.008	319
D	Fragments	1970-109C-F											

* Small Astronomical Satellite. ** Préliminaire à Eole

	Name		Launch date, lifetime and descent date	Shape and weight (kg)	Size (m)	Date of orbital determination	Orbital inclination (deg)	Nodal period (min)	Semi major axis (km)	Perigee height (km)	Apogee height (km)	Orbital eccentricity	Argument of perigee (deg)
D R	Cosmos 386	1970-110A	1970 Dec 15.42 12.9 days 1970 Dec 28.3	Sphere-cylinder 6300?	6.5 long? 2.4 dia	1970 Dec 16.5	64.99	89.40	6624	215	276	0.005	52
D	Cosmos 386 rocket	1970-110B	1970 Dec 15.42 3.88 days 1970 Dec 19.30	Cylinder 2500?	7.5 long 2.6 dia	1970 Dec 16.2	65.01	88.91	6599	196	245	0.004	8
D	Cosmos 386 engine*	1970-110E	1970 Dec 15.42 19 days 1971 Jan 3	Cone 600? full	1.5 long? 2 dia?	1970 Dec 29.6	64.99	89.77	6642	213	315	0.008	40
D	Fragments	1970-110C,D,F											
D	Cosmos 387	1970-111A	1970 Dec 16.19 3321 days 1980 Jan 19	Cylinder + paddles? 900?	2 long? 1 dia?	1970 Dec 24.0	74.01	95.31	6911	528	538	0.0008	6
D	Cosmos 387 rocket	1970-111B	1970 Dec 16.19 3387 days 1980 Mar 25	Cylinder 2200?	7.4 long 2.4 dia	1970 Dec 24.0	74.01	95.13	6902	513	535	0.002	31
D	Fragments	1970-111C-E											
D	Cosmos 388	1970-112A	1970 Dec 18.40 143.40 days 1971 May 10.80	Ellipsoid 400?	1.8 long 1.2 dia	1970 Dec 20.3 1971 Mar 1.0	70.95 70.95	92.32 91.51	6766 6726	271 257	505 439	0.017 0.014	78 -
D	Cosmos 388 rocket	1970-112B	1970 Dec 18.40 61.35 days 1971 Feb 17.75	Cylinder 1500?	8 long 1.65 dia	1970 Dec 20.4	70.96	92.18	6759	268	494	0.017	79

*1970-110E ejected from 1970-110A about 1970 Dec 28.

	Name		Launch date, lifetime and descent date	Shape and weight (kg)	Size (m)	Date of orbital determination	Orbital inclination (deg)	Nodal period (min)	Semi major axis (km)	Perigee height (km)	Apogee height (km)	Orbital eccentricity	Argument of perigee (deg)
	Cosmos 389	1970-113A	1970 Dec 18.68 60 years	Cylinder + 2 vanes 2500?	5 long? 1.5 dia?	1970 Dec 25.3	81.19	98.06	7043	642	687	0.003	270
	Cosmos 389 rocket	1970-113B	1970 Dec 18.68 60 years	Cylinder 1440	3.8 long 2.6 dia	1970 Dec 29.7	81.20	98.08	7044	602	729	0.009	172
D	Fragment	1970-113C											
D	Molniya 1S	1970-114A	1970 Dec 25.16 31 months 1973 Jul	Windmill + 6 vanes 1000?	3.4 long 1.6 dia	1970 Dec 28.7 1971 Feb 1.0 1972 Jul 1.0	64.99 64.99 65.00	711.80 717.75 720.96	26408 26556 26635	495 474 429	39565 39881 40085	0.740 0.742 0.744	285 - -
D	Molniya 1S launcher rocket	1970-114B	1970 Dec 25.16 29 days 1971 Jan 23	Cylinder 2500?	7.5 long 2.6 dia	1970 Dec 26.7	65.01	91.10	6707	230	428	0.015	63
D	Molniya 1S launcher	1970-114C	1970 Dec 25.16 47.12 days 1971 Feb 10.28	Irregular	-	1970 Dec 28.5	64.98	91.65	6734	240	472	0.017	70
D	Molniya 1S rocket	1970-114F	1970 Dec 25.16 799 days 1973 Mar 3	Cylinder 440	2.0 long 2.0 dia	1971 Jan 3.6 1971 Sep 1.0 1972 Jun 1.0	65.02 64.99 65.02	709.66 709.66 709.48	26356 26356 26351	366 500 327	39589 39455 39619	0.744 0.739 0.745	285 - -
D	Fragments	1970-114D,E,G											

	Name		Launch date, lifetime and descent date	Shape and weight (kg)	Size (m)	Date of orbital determination	Orbital inclination (deg)	Nodal period (min)	Semi major axis (km)	Perigee height (km)	Apogee height (km)	Orbital eccentricity	Argument of perigee (deg)
D R	Cosmos 390	1971-01A	1971 Jan 12.40 12.83 days 1971 Jan 25.23	Sphere-cylinder 6300?	6.5 long? 2.4 dia	1971 Jan 14.1	65.01	89.28	6618	204	275	0.005	49
D	Cosmos 390 rocket	1971-01B	1971 Jan 12.40 5.05 days 1971 Jan 17.45	Cylinder 2500?	7.5 long 2.6 dia	1971 Jan 13.2	65.02	89.13	6610	202	262	0.004	44
D	Cosmos 390 engine*	1971-01C	1971 Jan 12.40 20 days 1971 Feb 1	Cone 600? full	1.5 long? 2 dia?	1971 Jan 23.8	65.02	89.06	6607	204	253	0.004	61
D	Fragments	1971-01D-F											
D	Cosmos 391	1971-02A	1971 Jan 14.50 402.57 days 1972 Feb 21.07	Ellipsoid 400?	1.8 long 1.2 dia	1971 Jan 15.5 1971 Aug 1.0	70.91 70.91	95.31 93.54	6913 6827	267 256	803 641	0.039 0.028	86 -
D	Cosmos 391 rocket	1971-02B	1971 Jan 14.50 218.62 days 1971 Aug 21.12	Cylinder 1500?	8 long 1.65 dia	1971 Jan 15.8 1971 May 1.0	70.92 70.92	95.20 93.33	6908 6816	267 253	792 623	0.038 0.027	86 -
	Meteor 7	1971-03A	1971 Jan 20.48 60 years	Cylinder + 2 vanes 2200?	5 long? 1.5 dia?	1971 Feb 1.0	81.21	97.60	7021	629	656	0.002	5
	Meteor 7 rocket	1971-03B	1971 Jan 20.48 60 years	Cylinder 1440	3.8 long 2.6 dia	1971 Jan 28.6	81.22	97.76	7029	564	737	0.012	139
2d	Fragments**	1971-00A-E											

*1971-01C ejected from 1971-01A about 1971 Jan 23.7.

**These unidentified fragments were discovered in orbit, and catalogued on 1971 Feb 9 (A,B,C) and 1971 Jun 30 (D,E). 1971-00D decayed 1973 Jun 26 and 1971-00C decayed on 1982 May 7.

	Name		Launch date, lifetime and descent date	Shape and weight (kg)	Size (m)	Date of orbital determination	Orbital inclina-tion (deg)	Nodal period (min)	Semi major axis (km)	Perigee height (km)	Apogee height (km)	Orbital eccen-tricity	Argument of perigee (deg)
D R	Cosmos 392	1971-04A	1971 Jan 21.36 11.83 days 1971 Feb 2.19	Sphere-cylinder 5700?	5.0 long 2.4 dia	1971 Jan 23.2	64.99	89.32	6619	204	278	0.006	45
D	Cosmos 392 rocket	1971-04B	1971 Jan 21.36 5 days 1971 Jan 26	Cylinder 2500?	7.5 long 2.6 dia	1971 Jan 22.4	65.00	89.17	6612	198	270	0.005	20
D	[Titan 3B Agena D]	1971-05A	1971 Jan 21.77 19 days 1971 Feb 9	Cylinder 3000?	8 long? 1.5 dia	1971 Jan 21.9	110.86	90.09	6657	139	418	0.021	144
D	Fragments	1971-05B,C											
	Intelsat 4 F-2	1971-06A	1971 Jan 26.03 > million years	Cylinder 1410 full 707 empty	2.82 long 2.39 dia	1971 Jan 26.1 1971 Feb 17.2 1971 Apr 1.0	28.25 0.55 0.55	638.7 1450.8 1436.1	24553 42453 42165	548 35801 35779	35801 36349 35794	0.718 0.006 0.0002	179 59 -
	Intelsat 4 F-2 rocket	1971-06B	1971 Jan 26.03 7000 years	Cylinder 1815	8.6 long 3.0 dia	1971 Feb 4.3	28.20	654.5	24966	597	36578	0.721	184
D	Cosmos 393	1971-07A	1971 Jan 26.53 140.86 days 1971 Jun 16.39	Ellipsoid 400?	1.8 long 1.2 dia	1971 Jan 28.4 1971 Apr 1.0	71.03 71.03	92.13 91.29	6757 6716	272 257	485 418	0.016 0.012	77 -
D	Cosmos 393 rocket	1971-07B	1971 Jan 26.53 63.81 days 1971 Mar 31.34	Cylinder 1500?	8 long 1.65 dia	1971 Jan 27.9	71.03	92.03	6752	271	476	0.015	82

Year of launch 1971, continued

	Name		Launch date, lifetime and descent date	Shape and weight (kg)	Size (m)	Date of orbital determination	Orbital inclination (deg)	Nodal period (min)	Semi major axis (km)	Perigee height (km)	Apogee height (km)	Orbital eccentricity	Argument of perigee (deg)
D 3M R	Apollo 14**	1971-08A	1971 Jan 31.88 9.00 days 1971 Feb 9.88	Cone-cylinder 29229 initially	11.15 long 3.91 dia	1971 Jan 31.9 1971 Feb 1.1	32.56 33.2	88.07 26320	6564 292520	186 200	186 572080	0 0.977	- * 30 *
							In selenocentric orbit 1971 Feb 4.29 to Feb 7.07						
D	Saturn IV B [Saturn 509]	1971-08B	1971 Jan 31.88 3.44 days 1971 Feb 4.32	Cylinder 13990	18.7 long 6.6 dia	1971 Jan 31.9 1971 Feb 1.1	32.56 33.2	88.07 26320	6564 292520	186 200	186 572080	0 0.977	- * 30 *
							Crashed on Moon 1971 Feb 4.32						
D	LEM 8 descent stage	1971-08D	1971 Jan 31.88 4.51 days 1971 Feb 5.39	Octagon + 4 legs 10420 full 2139 empty	1.57 high 3.13 wide	1971 Feb 1.1	33.2	26320	292520	200	572080	0.977	30 *
							Entered selenocentric orbit 1971 Feb 4.29 Landed on Moon 1971 Feb 5.39						
D	LEM 8† ascent stage	1971-08C	1971 Jan 31.88 6.15 days 1971 Feb 7.03	Box + tanks 4857 full 2128 empty	2.52 high 3.76 wide 3.13 deep	1971 Feb 1.1	33.2	26320	292520	200	572080	0.977	30 *
							On Moon's surface 1971 Feb 5.39 to Feb 6.78 Finally crashed on Moon 1971 Feb 7.03						
	NATO 2	1971-09A	1971 Feb 3.07 > million years	Cylinder 243 full 129 empty	0.81 long 1.37 dia	1971 Feb 15.0 1971 Mar 15.0 1975 May 1.0	27.83 2.8 0.8	587.5 1403.4 1436.3	23238 41523 42168	299 34429 35778	33420 35860 35802	0.713 0.017 0.0003	249 * - * -
	NATO 2 rocket	1971-09B	1971 Feb 3.07 20 years ?	Sphere-cone 66	1.32 long 0.94 dia	1971 Mar 1.0	25.9	665.0	25236	283	37433	0.736	-
1d	Fragments	1971-09C,D											
	Cosmos 394	1971-10A	1971 Feb 9.79 40 years	Cylinder?	4 long? 2 dia?	1971 Feb 12.2	65.84	96.54	6971	572	614	0.003	352
	Cosmos 394 rocket	1971-10B	1971 Feb 9.79 13 years	Cylinder 2200?	7.4 long 2.4 dia	1971 Feb 16.3	65.84	96.43	6966	564	612	0.003	357
D	Fragment	1971-10C											

*Approximate orbits.
**Apollo attached to LEM, separated from Saturn IVB on Feb 1.13.
† LEM with two crew members, separated from Apollo on Feb 5.20.
Ascent stage relaunched from Moon Feb 6.78; briefly docked with Apollo Feb 6.86.

	Name		Launch date, lifetime and descent date	Shape and weight (kg)	Size (m)	Date of orbital determination	Orbital inclination (deg)	Nodal period (min)	Semi major axis (km)	Perigee height (km)	Apogee height (km)	Orbital eccentricity	Argument of perigee (deg)
	Tansei 1* [Mu 4S]	1971-11A	1971 Feb 16.17 1000 years	26-sided cylinder 62	0.83 long 0.71 dia	1971 Mar 1.7	29.66	105.95	7422	984	1103	0.008	219
	Tansei 1* rocket	1971-11B	1971 Feb 16.17 500 years	Sphere-cone 90?	1.86 long 0.79 dia	1971 May 1.0	29.66	104.74	7361	973	993	0.001	-
D	Fragment	1971-11C											
	[Thor Burner 2]	1971-12A	1971 Feb 17.16 80 years	12-faced frustum 195	1.64 long 1.31 to 1.10 dia	1971 Feb 28.0	98.83	100.86	7176	763	833	0.005	327
	Burner 2 rocket	1971-12B	1971 Feb 17.16 60 years	Sphere-cone 66	1.32 long 0.94 dia	1971 Feb 19.0	98.78	100.96	7181	779	827	0.003	0
	Calsphere 3**	1971-12C	1971 Feb 17.16 25 years	Sphere (Aluminium) 0.73	0.26 dia	1971 Feb 28.0	98.84	100.89	7178	765	834	0.005	331
	Calsphere 4	1971-12D	1971 Feb 17.16 25 years	Sphere (Aluminium) 0.73	0.26 dia	1971 Feb 19.0	98.84	100.86	7176	763	833	0.005	353
	Calsphere 5	1971-12E	1971 Feb 17.16 25 years	Sphere (Gold) 0.73	0.26 dia	1971 Feb 27.9	98.82	100.95	7181	773	832	0.004	328
D	Cosmos 395	1971-13A	1971 Feb 17.88 3336 days 1980 Apr 6	Cylinder + paddles? 900?	2 long? 1 dia?	1971 Mar 2.3	74.04	95.41	6916	529	546	0.001	48
D	Cosmos 395 rocket	1971-13B	1971 Feb 17.88 3440 days 1980 Jul 19	Cylinder 2200?	7.4 long 2.4 dia	1971 Feb 25.4	74.03	95.30	6910	519	545	0.002	48
D	Fragments	1971-13C-E											

*Japanese satellite.

** Calibration sphere.

	Name		Launch date, lifetime and descent date	Shape and weight (kg)	Size (m)	Date of orbital determination	Orbital inclination (deg)	Nodal period (min)	Semi major axis (km)	Perigee height (km)	Apogee height (km)	Orbital eccentricity	Argument of perigee (deg)
D R	Cosmos 396	1971-14A	1971 Feb 18.59 12.70 days 1971 Mar 3.29	Sphere-cylinder 6300?	6.5 long? 2.4 dia	1971 Feb 20.4 1971 Feb 25.8	65.42 65.42	89.40 89.30	6624 6619	205 189	286 292	0.006 0.008	63 37
D	Cosmos 396 rocket	1971-14B	1971 Feb 18.59 6.98 days 1971 Feb 25.57	Cylinder 2500?	7.5 long 2.6 dia	1971 Feb 20.4	65.40	89.18	6613	206	263	0.004	61
D	Cosmos 396 engine*	1971-14E	1971 Feb 18.59 18 days 1971 Mar 8	Cone 600? full	1.5 long? 2 dia?	1971 Mar 3.4	65.42	89.00	6604	179	272	0.007	33
D	Fragments	1971-14C,D,F											
	Cosmos 397**	1971-15A	1971 Feb 25.47 150 years	Cylinder?	4 long? 2 dia?	1971 Mar 6.6	65.73	113.51	7766	574	2202	0.105	47
D	Cosmos 397 rocket	1971-15B	1971 Feb 25.47 6.78 days 1971 Mar 4.25	Cylinder 1500?	8 long? 2.5 dia?	1971 Feb 26.9	65.10	92.11	6757	144	613	0.035	53
15d	Fragments	1971-15C-CM											
	Cosmos 398	1971-16A	1971 Feb 26.22 20 years	-	-	1971 Feb 28.1 1971 Feb 28.4	51.61 51.59	88.86 216.13	6599 11931	189 203	252 10903	0.005 0.449	75 81
D	Cosmos 398 rocket	1971-16B	1971 Feb 26.22 3.12 days 1971 Mar 1.34	Cylinder 2500?	7.5 long 2.6 dia	1971 Feb 26.9	51.61	88.69	6591	186	239	0.004	58
D	Cosmos 398 platform	1971-16C	1971 Feb 26.22 75 days 1971 May 12	-	-	1971 Mar 6.8	51.60	98.47	7065	186	1188	0.071	89
D	Fragments	1971-16D,E											

*1971-14E ejected from 1971-14A about 1971 Mar 3.3.

**1971-15A passed close to 1971-10A about 1971 Feb 25.60, then exploded.

	Name		Launch date, lifetime and descent date	Shape and weight (kg)	Size (m)	Date of orbital determination	Orbital inclination (deg)	Nodal period (min)	Semi major axis (km)	Perigee height (km)	Apogee height (km)	Orbital eccentricity	Argument of perigee (deg)
D	Cosmos 399	1971-17A	1971 Mar 3.40	Sphere-	6.5 long?	1971 Mar 4.6	65.00	89.34	6620	201	283	0.006	36
R			13.84 days	cylinder	2.4 dia	1971 Mar 10.9	64.99	90.86	6695	196	438	0.018	-
			1971 Mar 17.24	6300?									
D	Cosmos 399 rocket	1971-17B	1971 Mar 3.40 5.09 days 1971 Mar 8.49	Cylinder 2500?	7.5 long 2.6 dia	1971 Mar 4.4	65.01	89.14	6611	199	266	0.005	30
D	Cosmos 399 engine*	1971-17C	1971 Mar 3.40 22 days 1971 Mar 25	Cone 600? full	1.5 long? 2 dia?	1971 Mar 15.6	64.99	90.70	6687	195	423	0.017	34
D	Fragments	1971-17D,E											
D	China 2**	1971-18A	1971 Mar 3.51	Spheroid?	1 dia?	1971 Mar 5.0	69.90	106.18	7427	268	1830	0.105	191
			3028 days 1979 Jun 17	221		1972 Oct 16.5	69.89	103.88	7320	262	1622	0.093	-
D	China 2 rocket	1971-18B	1971 Mar 3.51	Cylinder	-	1971 Mar 13.5	69.91	106.10	7423	265	1825	0.105	180
			1811 days			1972 Mar 1.0	69.89	103.82	7317	267	1611	0.092	-
			1976 Feb 16			1973 Nov 1.0	69.89	99.26	7099	256	1186	0.065	-
D	Explorer 43 (Imp 8)	1971-19A	1971 Mar 13.68	16-sided	1.82 long	1971 Mar 17.8	28.80	5956.1	108843	353	204577	0.938	303
			1299 days 1974 Oct 2	cylinder 288	1.36 dia	1973 Jan 1.0	39.90	5979.5	109128	13755	191745	0.815	-
D	Explorer 43 second stage	1971-19B	1971 Mar 13.68 70.25 days 1971 May 22.93	Cylinder 350	4.9 long 1.43 dia	1971 Mar 13.8	28.74	92.22	6768	237	543	0.023	297
D	Explorer 43 third stage	1971-19C	1971 Mar 13.68 43 months? 1974 Oct?	Sphere-cone 66	1.32 long 0.94 dia	1971 Mar 13.7	28.75	5628	104783	235	196575	0.937	303
D	Fragments	1971-19D,E											

* 1971-17C ejected from 1971-17A about 1971 Mar 15.2.

**1971-18A and 1971-18B were probably joined until 1971 Mar 11.

	Name		Launch date, lifetime and descent date	Shape and weight (kg)	Size (m)	Date of orbital determination	Orbital inclination (deg)	Nodal period (min)	Semi major axis (km)	Perigee height (km)	Apogee height (km)	Orbital eccentricity	Argument of perigee (deg)
	Cosmos 400	1971-20A	1971 Mar 18.91 1200 years	Cylinder?	4 long? 2 dia?	1971 Mar 20.3	65.83	104.99	7373	983	1006	0.002	267
	Cosmos 400 rocket	1971-20B	1971 Mar 18.91 600 years	Cylinder 2200?	7.4 long 2.4 dia	1971 Mar 27.9	65.82	104.88	7367	983	995	0.001	250
	Fragment	1971-20C											
	SDS-A † [Titan 3B Agena D]	1971-21A	1971 Mar 21.16 20 years	Cylinder?	-	1971 Mar 21.4	63.19	596.7	23473	390	33800	0.711	-
	SDS-A rocket	1971-21B	1971 Mar 21.16 20 years	Cylinder 700?	6 long? 1.5 dia	1971 May 19.6 1973 Dec 1.0 1975 Jan 1.0	63.19 63.0 63.02	700.5 700.1 700.1	26128 26118 26120	310 1148 931	39190 38332 38552	0.744 0.712 0.720	270 - -
D	[Thorad Agena D]	1971-22A	1971 Mar 24.88 18.81 days 1971 Apr 12.69	Cylinder 2000?	8 long? 1.5 dia	1971 Mar 25.5	81.52	88.56	6580	157	246	0.007	136
D R	Cosmos 401	1971-23A	1971 Mar 27.46 12.81 days 1971 Apr 9.27	Sphere-cylinder 6300?	6.5 long? 2.4 dia	1971 Mar 31.1 1971 Apr 2.4 1971 Apr 3.7	72.83 72.84 72.83	89.26 89.85 90.40	6616 6645 6672	185 185 186	290 348 401	0.008 0.012 0.016	45 49 50
D	Cosmos 401 rocket	1971-23B	1971 Mar 27.46 5.94 days 1971 Apr 2.40	Cylinder 2500?	7.5 long 2.6 dia	1971 Mar 28.7	72.84	89.31	6618	197	283	0.006	47
D	Cosmos 401 engine*	1971-23C	1971 Mar 27.46 20 days 1971 Apr 16	Cone 600? full	1.5 long? 2 dia?	1971 Apr 8.7	72.83	90.26	6665	183	390	0.015	39
D	Fragments	1971-23D,E											

*1971-23C ejected from 1971-23A about 1971 Apr 8.4.

† Satellite Data System.

	Name		Launch date, lifetime and descent date	Shape and weight (kg)	Size (m)	Date of orbital determination	Orbital inclination (deg)	Nodal period (min)	Semi major axis (km)	Perigee height (km)	Apogee height (km)	Orbital eccentricity	Argument of perigee (deg)
	Isis 2	1971-24A	1971 Apr 1.12 8000 years	Polyhedron 264	1.22 long 1.27 dia	1971 Apr 9.9	88.15	113.67	7772	1358	1429	0.005	120
	Isis 2 rocket	1971-24B	1971 Apr 1.12 5000 years	Cylinder 24	1.50 long 0.46 dia	1971 Apr 4.1	88.16	113.63	7770	1355	1428	0.005	129
	Fragment	1971-24C											
	Cosmos 402 †	1971-25A	1971 Apr 1.48 600 years	Cone-cylinder	6 long? 2 dia?	1971 Apr 1.5 1971 Apr 9.6	64.97 64.98	89.71 104.94	6639 7370	247 948	274 1036	0.002 0.006	231 264
D	Cosmos 402 platform	1971-25B	1971 Apr 1.48 35.10 days 1971 May 6.58	Irregular	-	1971 Apr 3.6	64.97	89.59	6633	247	263	0.001	229
D	Cosmos 402 rocket	1971-25C	1971 Apr 1.48 5 days 1971 Apr 6	Cylinder 1500?	8 long? 2.5 dia?	1971 Apr 2.3	64.96	89.46	6627	239	258	0.001	271
D	Fragment	1971-25D											
D R	Cosmos 403	1971-26A	1971 Apr 2.35 11.8 days 1971 Apr 14.2	Sphere-cylinder 5700?	5.0 long 2.4 dia	1971 Apr 4.1	81.34	88.96	6600	214	230	0.001	1
D	Cosmos 403 rocket	1971-26B	1971 Apr 2.35 3.27 days 1971 Apr 5.62	Cylinder 2500?	7.5 long 2.6 dia	1971 Apr 3.1	81.33	88.81	6593	201	228	0.002	346
D	Cosmos 404*	1971-27A	1971 Apr 4.60 <0.4 day ? 1971 Apr 4?	Cylinder?	4 long? 2 dia?	1971 Apr 4.7 1971 Apr 5.4	65.74 65.15	103.12 94.22	7284 6862	802 169	1010 799	0.014 0.046	245 50
D	Cosmos 404 rocket	1971-27B	1971 Apr 4.60 7.44 days 1971 Apr 12.04	Cylinder 1500?	8 long? 2.5 dia?	1971 Apr 5.6	65.08	92.34	6768	148	632	0.036	53
D	Fragments	1971-27C,D											

*Cosmos 404 passed close to Cosmos 400 about 1971 Apr 4.63, then de-orbited over Pacific Ocean?

† 1971-25B and 25C attached to 1971-25A until orbit change.

	Name		Launch date, lifetime and descent date	Shape and weight (kg)	Size (m)	Date of orbital determination	Orbital inclination (deg)	Nodal period (min)	Semi major axis (km)	Perigee height (km)	Apogee height (km)	Orbital eccentricity	Argument of perigee (deg)
	Cosmos 405	1971-28A	1971 Apr 7.30 60 years	Cylinder + 2 vanes 2500?	5 long? 1.5 dia?	1971 Apr 8.0	81.24	98.33	7056	673	683	0.001	139
	Cosmos 405 rocket	1971-28B	1971 Apr 7.30 60 years	Cylinder 1440	3.8 long 2.6 dia	1971 Apr 10.3	81.25	98.43	7061	616	749	0.009	180
1d	Fragments	1971-28C,D											
D R	Cosmos 406	1971-29A	1971 Apr 14.34 9.9 days 1971 Apr 24.2	Sphere- cylinder 6300?	6.5 long? 2.4 dia	1971 Apr 14.4	81.31	89.16	6610	217	246	0.002	33
D	Cosmos 406 rocket	1971-29B	1971 Apr 14.34 3.62 days 1971 Apr 17.96	Cylinder 2500?	7.5 long 2.6 dia	1971 Apr 14.7	81.32	89.02	6603	211	238	0.002	20
D	Cosmos 406 engine*	1971-29E	1971 Apr 14.34 14 days 1971 Apr 28	Cone 600? full	1.5 long? 2 dia?	1971 Apr 25.8	81.31	88.31	6568	184	195	0.001	-
D	Fragments	1971-29C,D,F											
D	Tournesol 1 [Diamant B]	1971-30A	1971 Apr 15.40 3210 days 1980 Jan 28	Cylinder + 4 vanes 96	0.80 long 0.70 dia	1971 Apr 19.5	46.37	96.16	6955	457	697	0.017	63
D	Tournesol 1 rocket	1971-30B	1971 Apr 15.40 3007 days 1979 Jul 9	Cylinder 68	1.60 long? 0.65 dia	1971 Apr 20.9	46.38	96.09	6952	448	699	0.018	67
D	Fragments	1971-30C-H											

* 1971-29E ejected from 1971-29A on 1971 Apr 22.

	Name		Launch date, lifetime and descent date	Shape and weight (kg)	Size (m)	Date of orbital determination	Orbital inclination (deg)	Nodal period (min)	Semi major axis (km)	Perigee height (km)	Apogee height (km)	Orbital eccentricity	Argument of perigee (deg)
	Meteor 8	1971-31A	1971 Apr 17.49 30 years	Cylinder + 2 vanes 2200?	5 long? 1.5 dia?	1971 Apr 18.9	81.24	97.17	7000	610	633	0.002	280
	Meteor 8 rocket	1971-31B	1971 Apr 17.49 30 years	Cylinder 1440	3.8 long 2.6 dia	1971 Apr 19.6	81.24	97.36	7009	554	708	0.011	177
D B	Salyut 1*	1971-32A	1971 Apr 19.07 175 days 1971 Oct 11	Cylinder + 4 wings 18500	14 long 4.15 max dia 2.0 min dia	1971 Apr 20.1 1971 Apr 28.8	51.56 51.57	88.53 89.67	6583 6639	200 251	210 271	0.0008 0.0014	341 188
D	Salyut 1 rocket	1971-32B	1971 Apr 19.07 1 day 1971 Apr 20	Cylinder 4000?	12 long? 4 dia	1971 Apr 19.4	51.55	88.30	6572	176	211	0.003	269
D	Fragments	1971-32C-G											
D	[Titan 3B Agena D]	1971-33A	1971 Apr 22.65 21 days 1971 May 13	Cylinder 3000?	8 long? 1.5 dia	1971 Apr 23.2	110.93	89.85	6645	132	401	0.020	127
D 3M R	Soyuz 10**	1971-34A	1971 Apr 23.00 1.99 days 1971 Apr 24.99	Sphere-cylinder + 2 wings. 6575?	7.5 long 2.2 dia	1971 Apr 23.7 1971 Apr 24.4	51.60 51.56	89.11 88.65	6612 6589	209 190	258 231	0.004 0.003	102 119
D	Soyuz 10 rocket	1971-34B	1971 Apr 23.00 2.74 days 1971 Apr 25.74	Cylinder 2500?	7.5 long 2.6 dia	1971 Apr 24.2	51.59	88.42	6578	194	205	0.0008	37
	Cosmos 407	1971-35A	1971 Apr 23.48 120 years	Cylinder + paddles 750?	2 long? 1 dia?	1971 Apr 28.6	74.06	100.99	7183	791	819	0.002	83
	Cosmos 407 rocket	1971-35B	1971 Apr 23.48 100 years	Cylinder 2200?	7.4 long 2.4 dia	1971 Apr 27.5	74.06	100.90	7179	783	818	0.002	74
	Fragments	1971-35C-L											

* De-orbited over Pacific Ocean.

**Soyuz 10 docked with Salyut 1 from Apr 24.07 to Apr 24.30.

	Name		Launch date, lifetime and descent date	Shape and weight (kg)	Size (m)	Date of orbital determination	Orbital inclination (deg)	Nodal period (min)	Semi major axis (km)	Perigee height (km)	Apogee height (km)	Orbital eccentricity	Argument of perigee (deg)
D	San Marco 3	1971-36A	1971 Apr 24.31 218.88 days 1971 Nov 29.19	Sphere 164	0.76 dia	1971 Apr 28.0 1971 Sep 1.0	3.23 3.23	93.82 92.17	6848 6766	222 214	718 562	0.036 0.026	4 -
D	San Marco 3 rocket	1971-36B	1971 Apr 24.31 42 days 1971 Jun 5	Cylinder 24	1.50 long 0.46 dia	1971 May 3.0	3.25	93.42	6828	209	691	0.035	-*
D	Ccsmos 408	1971-37A	1971 Apr 24.47 248.80 days 1971 Dec 29.27	Ellipsoid 400?	1.8 long 1.2 dia	1971 Apr 25.2 1971 Aug 31.7	81.83 81.83	102.10 97.68	7235 7024	200 194	1514 1098	0.091 0.064	72 -
D	Cosmos 408 rocket	1971-37B	1971 Apr 24.47 158.23 days 1971 Sep 29.70	Cylinder 1500?	8 long 1.65 dia	1971 Apr 25.9 1971 Jul 16.5	81.83 81.83	101.89 97.95	7225 7038	201 194	1493 1125	0.089 0.066	70 -
	Cosmos 409	1971-38A	1971 Apr 28.61 3000 years	Spheroid + paddles? 650?	1.6 dia?	1971 May 2.3	74.01	109.36	7575	1177	1216	0.003	218
	Cosmos 409 rocket	1971-38B	1971 Apr 28.61 2000 years	Cylinder 2200?	7.4 long 2.4 dia	1971 Apr 30.9	74.00	109.23	7569	1173	1208	0.002	204
	IMEWS 2 [Titan 3C]	1971-39A	1971 May 5.33 >million years	Cylinder + 4 panels 820?	6 long? 2.5 dia?	1971 May 5.3 1971 Jun 1.0	26.36 0.87	630.95 1434.0	24419 42124	295 35651	35787 35840	0.727 0.002	180 -
	Transtage	1971-39B	1971 May 5.33 >million years	Cylinder 1500?	6 long? 3 dia	1971 May 5.3 1971 Jun 1.0	26.36 0.87	630.95 1434.0	24419 42124	295 35651	35787 35840	0.727 0.002	180 -
D	Titan 3C second stage	1971-39C	1971 May 5.33 1 day 1971 May 6	Cylinder 1900	6 long 3.0 dia	1971 May 5.3	28.5	89.91	6655	148	406	0.019	114

*Approximate orbit.

	Name		Launch date, lifetime and descent date	Shape and weight (kg)	Size (m)	Date of orbital determination	Orbital inclination (deg)	Nodal period (min)	Semi major axis (km)	Perigee height (km)	Apogee height (km)	Orbital eccentricity	Argument of perigee (deg)
D R	Cosmos 410	1971-40A	1971 May 6.27 11.90 days 1971 May 18.17	Sphere-cylinder 5900?	5.9 long 2.4 dia	1971 May 7.4	64.96	89.35	6621	205	280	0.006	43
D	Cosmos 410 rocket	1971-40B	1971 May 6.27 5.19 days 1971 May 11.46	Cylinder 2500?	7.5 long 2.6 dia	1971 May 6.8	64.96	89.21	6614	202	270	0.005	37
D	Excess Radiation Package A*	1971-40C	1971 May 6.27 19 days 1971 May 25	Ellipsoid 200?	0.9 long 1.9 dia	1971 May 17.7	64.95	89.11	6609	197	264	0.005	36
D	Fragment	1971-40D											
	Cosmos 411	1971-41A	1971 May 7.60 5000 years	Spheroid 40?	1.0 long? 0.8 dia?	1971 May 10.0	74.03	113.91	7783	1318	1492	0.011	100
	Cosmos 412	1971-41B	1971 May 7.60 10000 years	Spheroid 40?	1.0 long? 0.8 dia?	1971 May 11.3	74.04	116.20	7888	1482	1537	0.003	219
	Cosmos 413	1971-41C	1971 May 7.60 10000 years	Spheroid 40?	1.0 long? 0.8 dia?	1971 May 8.6	74.04	115.84	7871	1476	1509	0.002	205
	Cosmos 414	1971-41D	1971 May 7.60 9000 years	Spheroid 40?	1.0 long? 0.8 dia?	1971 May 9.9	74.02	115.16	7840	1428	1496	0.004	123
	Cosmos 415	1971-41E	1971 May 7.60 9000 years	Spheroid 40?	1.0 long? 0.8 dia?	1971 May 10.9	74.01	115.50	7856	1452	1503	0.003	145
	Cosmos 416	1971-41F	1971 May 7.60 7000 years	Spheroid 40?	1.0 long? 0.8 dia?	1971 May 11.3	74.02	114.54	7812	1373	1494	0.008	105

*1971-40C ejected from 1971-40A on 1971 May 17.

1971-41 continued on page 260

	Name		Launch date, lifetime and descent date	Shape and weight (kg)	Size (m)	Date of orbital determination	Orbital inclination (deg)	Nodal period (min)	Semi major axis (km)	Perigee height (km)	Apogee height (km)	Orbital eccentricity	Argument of perigee (deg)
	Cosmos 417	1971-41G	1971 May 7.60 6000 years	Spheroid 40?	1.0 long? 0.8 dia?	1971 May 10.0	74.01	114.23	7798	1344	1495	0.010	106
	Cosmos 418	1971-41H	1971 May 7.60 8000 years	Spheroid 40?	1.0 long? 0.8 dia?	1971 May 11.2	74.01	114.85	7826	1401	1495	0.006	111
	Cosmos 411 rocket	1971-41J	1971 May 7.60 20000 years	Cylinder 2200?	7.4 long 2.4 dia	1971 May 9.9	74.04	116.87	7918	1487	1592	0.007	237
D	Cosmos 419*	1971-42A	1971 May 10.71 2.13 days 1971 May 12.84	Cone-cylinder? 23400? full	16 long? 4 dia	1971 May 11.6	51.53	87.47	6530	145	159	0.001	297
D R	Cosmos 420	1971-43A	1971 May 18.34 10.93 days 1971 May 29.27	Sphere-cylinder 6300?	6.5 long? 2.4 dia	1971 May 19.5	51.75	89.00	6606	199	257	0.004	21
D	Cosmos 420 rocket	1971-43B	1971 May 18.34 2.72 days 1971 May 21.06	Cylinder 2500?	7.5 long 2.6 dia	1971 May 19.0	51.79	88.60	6586	186	230	0.003	353
D	Cosmos 420 engine **	1971-43C	1971 May 18.34 16 days 1971 Jun 3	Cone 600? full	1.5 long? 2 dia?	1971 May 28.5	51.77	88.81	6597	197	240	0.003	62
D	Fragments	1971-43D,E											
D	Cosmos 421	1971-44A	1971 May 19.43 172.75 days 1971 Nov 8.18	Ellipsoid 400?	1.8 long 1.2 dia	1971 May 19.8 1971 Aug 16.5	70.96 70.96	91.99 91.23	6749 6713	273 259	469 410	0.014 0.011	75 -
D	Cosmos 421 rocket	1971-44B	1971 May 19.43 96.33 days 1971 Aug 23.76	Cylinder 1500?	8 long 1.65 dia	1971 May 19.7	70.98	91.85	6743	274	455	0.013	73

* Cosmos 419 was probably an attempted Mars probe.

**1971-43C ejected from 1971-43A about 1971 May 28.4

	Name		Launch date, lifetime and descent date	Shape and weight (kg)	Size (m)	Date of orbital determination	Orbital inclination (deg)	Nodal period (min)	Semi major axis (km)	Perigee height (km)	Apogee height (km)	Orbital eccentricity	Argument of perigee (deg)
D	Mars 2 launcher rocket	1971-45B	1971 May 19.68 2 days 1971 May 21	Cylinder 4000?	12 long? 4 dia	1971 May 20.7	51.52	87.52	6533	137	172	0.003	24
D	Mars 2 launcher	1971-45C	1971 May 19.68 2 days 1971 May 21	-	-	1971 May 20.7	51.52	87.53	6533	137	173	0.003	24
	Cosmos 422	1971-46A	1971 May 22.03 1200 years	Cylinder + boom? 700?	1.4 long 2.0 dia	1971 May 23.6	74.03	105.10	7377	988	1010	0.001	242
	Cosmos 422 rocket	1971-46B	1971 May 22.03 600 years	Cylinder 2200?	7.4 long 2.4 dia	1971 May 22.4	74.03	104.99	7372	983	1004	0.001	210
D	Cosmos 423	1971-47A	1971 May 27.50 183.16 days 1971 Nov 26.66	Ellipsoid 400?	1.8 long 1.2 dia	1971 May 28.8 1971 Aug 31.7	71.03 71.03	92.15 91.34	6758 6718	272 257	487 423	0.016 0.012	74 -
D	Cosmos 423 rocket	1971-47B	1971 May 27.50 93.96 days 1971 Aug 29.46	Cylinder 1500?	8 long 1.65 dia	1971 May 28.8	71.03	91.91	6746	274	461	0.014	72
D R	Cosmos 424	1971-48A	1971 May 28.44 12.71 days 1971 Jun 10.15	Sphere-cylinder 6300?	6.5 long? 2.4 dia	1971 May 30.6 1971 Jun 1.6	65.40 65.41	89.36 89.40	6621 6623	204 177	282 313	0.006 0.010	65 59
D	Cosmos 424 rocket	1971-48B	1971 May 28.44 6.15 days 1971 Jun 3.59	Cylinder 2500?	7.5 long 2.6 dia	1971 May 30.3	65.40	89.15	6611	196	270	0.006	46

Space vehicle: Mars 2, 1971-45A

1971-48 continued on page 262

	Name		Launch date, lifetime and descent date	Shape and weight (kg)	Size (m)	Date of orbital determination	Orbital inclination (deg)	Nodal period (min)	Semi major axis (km)	Perigee height (km)	Apogee height (km)	Orbital eccentricity	Argument of perigee (deg)
D	Cosmos 424 engine*	1971-48C	1971 May 28.44 18 days 1971 Jun 15	Cone 600? full	1.5 long? 2 dia?	1971 Jun 9.6	65.41	89.15	6611	175	291	0.009	59
D	Fragments	1971-48D,E											
D	Mars 3 launcher rocket	1971-49B	1971 May 28.64 3 days 1971 May 31	Cylinder 4000?	12 long? 4 dia	1971 May 29.9	51.57	88.21	6567	139	239	0.008	100
D	Mars 3 launcher	1971-49C	1971 May 28.64 3 days 1971 May 31	-	-	1971 May 29.9	51.57	88.17	6565	140	234	0.007	66
D	Fragment	1971-49D											
D	Cosmos 425	1971-50A	1971 May 29.16 3153 days 1980 Jan 15	Cylinder + paddles? 900?	2 long? 1 dia?	1971 Jun 6.9	74.03	95.24	6908	506	553	0.003	316
D	Cosmos 425 rocket	1971-50B	1971 May 29.16 3105 days 1979 Nov 28	Cylinder 2200?	7.4 long 2.4 dia	1971 Jun 6.9	74.04	95.20	6906	499	556	0.004	310
D	Fragments	1971-50C-F											
	Cosmos 426	1971-52A	1971 Jun 4.76 30 years	Octagonal ellipsoid? 400?	1.8 long? 1.5 dia?	1971 Jun 6.6	74.03	109.29	7571	389	1997	0.106	132
	Cosmos 426 rocket	1971-52B	1971 Jun 4.76 30 years	Cylinder 2200?	7.4 long 2.4 dia	1971 Jun 6.3	74.03	109.17	7565	389	1985	0.106	132
D	Fragments	1971-52C-F											

*1971-48C ejected from 1971-48A about 1971 Jun 9.5.

Space Vehicles: Mars 3, 1971-49A
Mariner 9, 1971-51A; and Centaur rocket, 1971-51B

	Name		Launch date, lifetime and descent date	Shape and weight (kg)	Size (m)	Date of orbital determination	Orbital inclination (deg)	Nodal period (min)	Semi major axis (km)	Perigee height (km)	Apogee height (km)	Orbital eccentricity	Argument of perigee (deg)
D 3M R	Soyuz 11*	1971-53A	1971 Jun 6.21 23.76 days 1971 Jun 29.97	Sphere-cylinder + 2 wings 6790?	7.5 long 2.2 dia	1971 Jun 6.4 1971 Jun 6.6 1971 Jun 9.9	51.57 51.60 51.56	88.41 88.30 89.66	6577 6571 6638	189 177 256	209 209 264	0.002 0.002 0.001	285 291 74
D	Soyuz 11 rocket	1971-53B	1971 Jun 6.21 1.64 days 1971 Jun 7.85	Cylinder 2500?	7.5 long 2.6 dia	1971 Jun 6.5	51.63	88.31	6572	176	211	0.003	286
D	SESP-1† [Thor Burner 2]	1971-54A	1971 Jun 8.59 3890 days 1982 Jan 31	Sphere-cone? 260?	3.0 long? 1.31 dia?	1971 Jun 9.0	90.22	95.95	6941	545	581	0.003	210
D R	Cosmos 427	1971-55A	1971 Jun 11.42 11.8 days 1971 Jun 23.2	Sphere-cylinder 6300?	6.5 long? 2.4 dia	1971 Jun 12.3 1971 Jun 22.9	72.84 72.85	89.70 89.48	6637 6626	204 177	314 319	0.008 0.011	58 46
D	Cosmos 427 rocket	1971-55B	1971 Jun 11.42 8.88 days 1971 Jun 20.30	Cylinder 2500?	7.5 long 2.6 dia	1971 Jun 11.5	72.80	89.66	6635	205	309	0.008	59
D	Cosmos 427 engine**	1971-55E	1971 Jun 11.42 17 days 1971 Jun 28	Cone 600? full	1.5 long? 2 dia?	1971 Jun 23.6	72.85	89.36	6620	176	308	0.010	47
D	Fragments	1971-55C, D											
D	[Titan 3D]	1971-56A	1971 Jun 15.78 52 days 1971 Aug 6	Cylinder 13300? full	15 long 3.0 dia	1971 Jun 16.1 1971 Jul 3.2	96.41 96.39	89.38 89.31	6620 6617	184 183	300 294	0.009 0.008	170 166
D	Titan 3D rocket	1971-56B	1971 Jun 15.78 4.65 days 1971 Jun 20.43	Cylinder 1900	6 long 3.0 dia	1971 Jun 17.4	96.39	88.93	6599	179	262	0.006	170

*Soyuz 11 docked with Salyut 1 from 1971 Jun 7.32 to Jun 29.77. Crew died from depressurisation after jettisoning orbital module Jun 29.95.

**1971-55E ejected from 1971-55A on 1971 Jun 23.

† Space Experiments Support Program.

	Name		Launch date, lifetime and descent date	Shape and weight (kg)	Size (m)	Date of orbital determination	Orbital inclination (deg)	Nodal period (min)	Semi major axis (km)	Perigee height (km)	Apogee height (km)	Orbital eccentricity	Argument of perigee (deg)
D R	Cosmos 428	1971-57A	1971 Jun 24.34 11.93 days 1971 Jul 6.27	Sphere-cylinder 5900?	5.9 long 2.4 dia	1971 Jun 25.0	51.76	89.07	6610	206	257	0.004	39
D	Cosmos 428 rocket	1971-57B	1971 Jun 24.34 4.33 days 1971 Jun 28.67	Cylinder 2500?	7.5 long 2.6 dia	1971 Jun 25.7	51.74	88.79	6596	194	241	0.004	21
D	Cosmic Ray Package B*	1971-57G	1971 Jun 24.34 19 days 1971 Jul 13	Ellipsoid 200?	0.9 long 1.9 dia	1971 Jul 5.9	51.76	88.82	6597	199	239	0.003	80
D	Fragments	1971-57C-F											
D	Explorer 44 (SR 10)	1971-58A	1971 Jul 8.96 3082 days 1979 Dec 15	12-sided cylinder + 4 vanes 118	0.58 long 0.76 dia	1971 Jul 9.1	51.06	95.23	6911	433	632	0.014	278
D	Explorer 44 rocket	1971-58B	1971 Jul 8.96 1665 days 1976 Jan 28	Cylinder 24	1.50 long 0.46 dia	1971 Jul 11.1 1972 Sep 16.0	51.06 51.06	95.23 94.48	6911 6875	435 418	630 575	0.014 0.011	284 -
D	Fragments	1971-58C, D											
	Meteor 9	1971-59A	1971 Jul 16.07 60 years	Cylinder + 2 vanes 2200?	5 long? 1.5 dia?	1971 Jul 16.2	81.19	97.29	7006	614	642	0.002	329
	Meteor 9 rocket	1971-59B	1971 Jul 16.07 60 years	Cylinder 1440	3.8 long 2.6 dia	1971 Jul 20.3	81.21	97.53	7017	559	719	0.011	174
D	Fragment	1971-59C											

*1971-57G ejected from 1971-57A on 1971 Jul 5.

	Name		Launch date, lifetime and descent date	Shape and weight (kg)	Size (m)	Date of orbital determination	Orbital inclination (deg)	Nodal period (min)	Semi major axis (km)	Perigee height (km)	Apogee height (km)	Orbital eccentricity	Argument of perigee (deg)
D	[Thorad Agena D]	1971-60A	1971 Jul 16.45 **2603 days** **1978 Aug 31**	Cylinder 2000?	8 long? 1.5 dia	1971 Jul 18.9 1973 Nov 1.0	75.00 74.99	94.59 94.00	6876 6848	488 462	508 477	0.001 0.001	243 -
D R	Cosmos 429	1971-61A	1971 Jul 20.42 12.9 days 1971 Aug 2.3	Sphere- cylinder **6300?**	**6.5 long?** **2.4 dia**	1971 Jul 22.6 1971 Jul 24.2	51.76 51.76	88.98 88.81	6605 6597	202 179	252 258	0.004 0.006	34 66
D	Cosmos 429 rocket	1971-61B	1971 Jul 20.42 2.94 days 1971 Jul 23.36	Cylinder 2500?	7.5 long 2.6 dia	1971 Jul 21.2	51.76	88.76	6594	185	247	0.005	19
D	**Cosmos 429 engine***	1971-61E	1971 Jul 20.42 15 days 1971 Aug 4	**Cone** **600? full**	**1.5 long?** 2 dia?	1971 Aug 3.2	51.81	89.89	6651	185	360	0.013	37
D	Fragments	1971-61C,D											
D R	Cosmos 430	1971-62A	1971 Jul 23.46 12.7 days 1971 Aug 5.2	Sphere- cylinder **6300?**	**6.5 long?** **2.4 dia**	1971 Jul 24.4 1971 Jul 26.5	65.41 65.40	89.54 89.05	6630 6606	199 188	305 267	0.008 0.006	50 76
D	Cosmos 430 rocket	1971-62B	1971 Jul 23.46 5.70 days 1971 Jul 29.16	Cylinder 2500?	7.5 long 2.6 dia	1971 Jul 24.8	65.41	89.17	6612	193	274	0.006	29
D	**Cosmos 430 engine****	1971-62F	1971 Jul 23.46 16.74 days 1971 Aug 9.20	**Cone** **600? full**	**1.5 long?** 2 dia?	1971 Aug 5.3	65.41	88.68	6587	181	237	0.004	66
D	Fragments	1971-62C-E,G											

* 1971-61E ejected from 1971-61A about 1971 Aug 1.
** 1971-62F ejected from 1971-62A about 1971 Aug 4.

	Name		Launch date, lifetime and descent date	Shape and weight (kg)	Size (m)	Date of orbital determination	Orbital inclination (deg)	Nodal period (min)	Semi major axis (km)	Perigee height (km)	Apogee height (km)	Orbital eccentricity	Argument of perigee (deg)
D 3M R	Apollo 15**	1971-63A	1971 Jul 26.57 12.30 days 1971 Aug 7.87	Cone-cylinder 30340 initially	11.15 long 3.91 dia	1971 Jul 26.6 1971 Jul 26.8	32.56 33.2	87.77 26320	6549 292520	169 200	173 572080	0.0003 0.977	-* 30*
							In selenocentric orbit 1971 Jul 29.84 to Aug 4.89						
D	Saturn IVB [Saturn 510]	1971-63B	1971 Jul 26.57 3.30 days 1971 Jul 29.87	Cylinder 13990	18.7 long 6.6 dia	1971 Jul 26.6 1971 Jul 26.8	32.56 33.2	87.77 26320	6549 292520	169 200	173 572080	0.0003 0.977	-* 30*
							Crashed on Moon 1971 Jul 29.87						
D	LEM 10 descent stage	1971-63E	1971 Jul 26.57 4.36 days 1971 Jul 30.93	Octagon + 4 legs 11404 full? 2803 empty	1.57 high 3.13 wide	1971 Jul 26.8	33.2	26320	292520	200	572080	0.977	30*
							Entered selenocentric orbit 1971 Jul 29.84 Landed on Moon 1971 Jul 30.93						
D	LEM 10 † ascent stage	1971-63C	1971 Jul 26.57 7.56 days 1971 Aug 3.13	Box + tanks 5030 full? 2127 empty	2.52 high 3.76 wide 3.13 deep	1971 Jul 26.8	33.2	26320	292520	200	572080	0.977	30*
							On Moon's surface 1971 Jul 30.93 to Aug 2.72 Finally crashed on Moon 1971 Aug 3.13						
D	Molniya 1T	1971-64A	1971 Jul 28.15 2183 days 1977 Jul 19	Windmill + 6 vanes 1000?	3.4 long 1.6 dia	1971 Aug 1.6 1971 Sep 1.0	65.37 65.43	704.99 717.75	26239 26556	468 478	39254 39877	0.739 0.742	285 -
D	Molniya 1T launcher	1971-64B	1971 Jul 28.15 32.20 days 1971 Aug 29.35	Irregular	-	1971 Jul 28.6	65.42	91.29	6716	217	459	0.018	63
D	Molniya 1T launcher rocket	1971-64C	1971 Jul 28.15 26.96 days 1971 Aug 24.11	Cylinder 2500?	7.5 long 2.6 dia	1971 Jul 30.0	65.40	91.28	6716	202	473	0.020	63
D	Molniya 1T rocket	1971-64D	1971 Jul 28.15 2209 days 1977 Aug 14	Cylinder 440	2.0 long 2.0 dia	1971 Aug 1.6	65.37	700.62	26131	442	39064	0.739	285
D	Fragments	1971-64E,F											

** Apollo attached to LEM separated from Saturn IVB on Jul 26.74.

† LEM with two crew members separated from Apollo on Jul 30.74
Ascent stage relaunched from Moon Aug 2.72; briefly docked with Apollo Aug 2.80. **(Apollo 15 subsatellite, 1971-63D, in selenocentric orbit)**

*Approximate orbits.

	Name		Launch date, lifetime and descent date	Shape and weight (kg)	Size (m)	Date of orbital determination	Orbital inclination (deg)	Nodal period (min)	Semi major axis (km)	Perigee height (km)	Apogee height (km)	Orbital eccentricity	Argument of perigee (deg)
D R	Cosmos 431	1971-65A	1971 Jul 30.36 11.91 days 1971 Aug 11.27	Sphere-cylinder 5700?	5.0 long 2.4 dia	1971 Jul 31.5	51.77	88.95	6604	194	257	0.005	21
D	Cosmos 431 rocket	1971-65B	1971 Jul 30.36 5.06 days 1971 Aug 4.42	Cylinder 2500?	7.5 long 2.6 dia	1971 Jul 31.3	51.78	88.84	6598	202	237	0.003	0
D	Fragment	1971-65C											
D R	Cosmos 432	1971-66A	1971 Aug 5.42 12.91 days 1971 Aug 18.33	Sphere-cylinder 6300?	6.5 long? 2.4 dia	1971 Aug 6.7	51.74	88.97	6605	194	259	0.005	73
D	Cosmos 432 rocket	1971-66B	1971 Aug 5.42 3.91 days 1971 Aug 9.33	Cylinder 2500?	7.5 long 2.6 dia	1971 Aug 5.7	51.77	88.88	6600	191	253	0.005	15
D	Cosmos 432 engine*	1971-66D	1971 Aug 5.42 16 days 1971 Aug 21	Cone 600? full	1.5 long? 2 dia ?	1971 Aug 18.4	51.74	88.62	6587	187	231	0.003	-
D	Fragments	1971-66C,E											
D	OV1-20	1971-67A	1971 Aug 7.00 22.00 days 1971 Aug 29.00	Cylinder 70?	2.05 long? 0.72 dia?	1971 Aug 7.0	92.00	106.16	7423	133	1957	0.123	173
	OV1-21 rocket	1971-67B	1971 Aug 7.00 80 years	Cone-cylinder 70?	2.05 long 0.72 dia	1971 Aug 9.5	87.64	102.01	7231	792	914	0.008	216
D	Cannonball 2	1971-67C	1971 Aug 7.00 177.72 days 1972 Jan 31.72	Sphere 364	0.66 dia	1971 Aug 7.4	92.01	106.29	7430	133	1970	0.124	172
D	Musketball	1971-67D	1971 Aug 7.00 43.60 days 1971 Sep 19.60	Sphere 61	0.30 dia	1971 Aug 7.2	87.61	94.87	6889	137	884	0.054	167

*1971-66D ejected from 1971-66A about 1971 Aug 17.

1971-67 launch continued on page 268.

	Name		Launch date, lifetime and descent date	Shape and weight (kg)	Size (m)	Date of orbital determination	Orbital inclina-tior (deg)	Nodal period (min)	Semi major axis (km)	Perigee height (km)	Apogee height (km)	Orbital eccen-tricity	Argument of perigee (deg)
	Rigid Sphere 2 (LCS 4) *	1971-67E	1971 Aug 7.00 75 years	Sphere (magnesium) 37	1.12 dia	1971 Aug 9.2	87.62	102.03	7232	795	913	0.008	219
D	Balloon (Mylar)	1971-67F	1971 Aug 7.00 309 days 1972 Jun 11	Inflated sphere 0.8	2.13 dia	1971 Aug 8.3 1972 Apr 10.6	87.61 87.62	101.86 98.77	7224 7082	777 684	914 724	0.009 0.003	214 119
D	Grid sphere 2	1971-67G	1971 Aug 7.00 2780 days 1979 Mar 18	Spherical skeleton 4.0	2.13 dia	1971 Aug 8.7	87.64	101.93	7227	783	915	0.009	216
D	Grid sphere 1	1971-67H	1971 Aug 7.00 3009 days 1979 Nov 2	Spherical skeleton 6.2	2.13 dia	1971 Sep 1.0	87.63	101.92	7227	777	920	0.009	-
	Apogee rocket	1971-67K	1971 Aug 7.00 75 years	Cone + nozzle 30?	1.1 long 1.2 dia	1971 Sep 1.0	87.63	102.05	7233	792	918	0.009	-
D	Rigid Sphere 1	1971-67P	1971 Aug 7.00 3678 days 1981 Sep 1	Sphere (aluminium) 1.6	0.61 dia	1971 Sep 1.0	87.63	101.98	7230	786	917	0.009	-
	Fragments	1971-67J,L-N											
D R?	Cosmos 433	1971-68A	1971 Aug 8.99 0.06 days 1971 Aug 9.05	Cylinder	2 long? 1 dia?	1971 Aug 9.0	49.41	88.54	6584	112	299	0.014	-
D	Cosmos 433 launch platform	1971-68B	1971 Aug 8.99 1.79 days 1971 Aug 10.78	Irregular	-	1971 Aug 9.2	49.41	88.55	6584	112	300	0.014	169
D	Cosmos 433 rocket	1971-68C	1971 Aug 8.99 1.04 days 1971 Aug 10.03	Cylinder 1500?	8 long? 2.5 dia?	1971 Aug 9.8	49.50	87.58	6536	142	174	0.003	113

* Lincoln Calibration Satellite

	Name		Launch date, lifetime and descent date	Shape and weight (kg)	Size (m)	Date of orbital determination	Orbital inclination (deg)	Nodal period (min)	Semi major axis (km)	Perigee height (km)	Apogee height (km)	Orbital eccentricity	Argument of perigee (deg)
D	Cosmos 434	1971-69A	1971 Aug 12.23 3664 days 1981 Aug 23	-	-	1971 Aug 12.9	51.60	88.98	6606	188	267	0.006	85
						1971 Aug 16.4			Orbit similar to 1971-69 D				
						1971 Aug 27.3	51.60	228.24	12373	186	11804	0.469	93
D	Cosmos 434 rocket	1971-69B	1971 Aug 12.23 6 days 1971 Aug 18	Cylinder 2500?	7.5 long 2.6 dia	1971 Aug 12.9	51.59	88.98	6606	194	261	0.005	86
D	Cosmos 434 platform	1971-69D	1971 Aug 12.23 77 days 1971 Oct 28	-	-	1971 Aug 17.5	51.60	99.90	7137	189	1328	0.080	92
D	Fragment	1971-69C											
D	[Titan 3B Agena D]	1971-70A	1971 Aug 12.59 22 days 1971 Sep 3	Cylinder 3000?	8 long? 1.5 dia	1971 Aug 13.2 1971 Aug 30.2	111.00 110.96	90.13 89.98	6659 6651	137 136	424 410	0.022 0.021	134 128
D	Fragment	1971-70B											
	Eole 1 *	1971-71A	1971 Aug 16.78 80 years	Cone-octagon 84	0.58 long 0.71 dia	1971 Aug 28.1	50.16	100.62	7169	677	904	0.016	353
	Eole 1 rocket	1971-71B	1971 Aug 16.78 60 years	Cylinder 24	1.50 long 0.46 dia	1971 Aug 18.3	50.18	100.55	7166	667	908	0.017	322
	Fragments	1971-71C,D											
D	Cosmos 435	1971-72A	1971 Aug 27.46 153.60 days 1972 Jan 28.06	Ellipsoid 400?	1.8 long 1.2 dia	1971 Aug 28.0 1971 Nov 16.0	70.96 70.96	92.09 91.28	6755 6715	271 257	482 417	0.016 0.012	79 -
D	Cosmos 435 rocket	1971-72B	1971 Aug 27.46 84.67 days 1971 Nov 20.13	Cylinder 1500?	8 long 1.65 dia	1971 Aug 28.0	70.98	91.96	6748	272	468	0.015	80

*French Cooperative Applications Satellite 1, launched by NASA. To study southern-hemisphere winds: 'Eole' is god of the winds.

	Name		Launch date, lifetime and descent date	Shape and weight (kg)	Size (m)	Date of orbital determination	Orbital inclination (deg)	Nodal period (min)	Semi major axis (km)	Perigee height (km)	Apogee height (km)	Orbital eccentricity	Argument of perigee (deg)
D	Luna 18 launcher rocket	1971-73D	1971 Sep 2.57 4.56 days 1971 Sep 7.13	Cylinder 4000?	12 long? 4 dia	1971 Sep 3.7	51.56	88.64	6588	193	227	0.003	334
D	Luna 18 launcher	1971-73C	1971 Sep 2.57 4.64 days 1971 Sep 7.21	-	-	1971 Sep 3.5	51.57	88.72	6592	186	242	0.004	327
D	Fragment	1971-73E											
D	Cosmos 436	1971-74A	1971 Sep 7.06 3041 days 1980 Jan 4	Cylinder + paddles? 900?	2 long ? 1 dia?	1971 Sep 19.9	74.04	95.18	6905	509	545	0.003	309
D	Cosmos 436 rocket	1971-74B	1971 Sep 7.06 3110 days 1980 Mar 13	Cylinder 2200?	7.4 long 2.4 dia	1971 Sep 19.8	74.04	95.03	6898	502	537	0.002	329
D	Fragments*	1971-74C-Q											
D	Cosmos 437	1971-75A	1971 Sep 10.15 3123 days 1980 Mar 29	Cylinder + paddles? 900?	2 long? 1 dia?	1971 Sep 10.4	74.05	95.31	6911	519	548	0.002	345
D	Cosmos 437 rocket	1971-75B	1971 Sep 10.15 3040 days 1980 Jan 6	Cylinder 2200?	7.4 long 2.4 dia	1971 Sep 10.3	74.05	95.18	6905	508	545	0.003	170
D	Fragment	1971-75C											
D	[Thorad Agena D]	1971-76A	1971 Sep 10.90 25.02 days 1971 Oct 5.92	Cylinder 2000?	8 long? 1.5 dia	1971 Sep 11.6	74.95	88.48	6578	156	244	0.007	156
D	Capsule	1971-76B	1971 Sep 10.90 1607 days 1976 Feb 3	Octagon? 60?	0.3 long? 0.9 dia?	1971 Sep 11.5 1972 Nov 16.0	75.07 75.07	94.60 93.94	6877 6845	492 456	507 477	0.001 0.002	236 -
D	Fragment	1971-76C											

Space Vehicle: Luna 18, 1971-73A; and fragment, 1971-73B.
*Fragments designated 1971-74H to Q probably belong to the 1971-75, -103, and -114 launches.

	Name		Launch date, lifetime and descent date	Shape and weight (kg)	Size (m)	Date of orbital determination	Orbital inclination (deg)	Nodal period (min)	Semi major axis (km)	Perigee height (km)	Apogee height (km)	Orbital eccentricity	Argument of perigee (deg)
D R	Cosmos 438	1971-77A	1971 Sep 14.54 12.72 days 1971 Sep 27.26	Sphere-cylinder 6300?	6.5 long? 2.4 dia	1971 Sep 15.2 1971 Sep 21.7	65.40 65.40	89.54 89.44	6630 6625	208 175	296 319	0.007 0.011	72 62
D	Cosmos 438 rocket	1971-77B	1971 Sep 14.54 8 days 1971 Sep 22	Cylinder 2500?	7.5 long 2.6 dia	1971 Sep 15.5	65.39	89.34	6620	196	288	0.007	63
D	Cosmos 438 engine*	1971-77F	1971 Sep 14.54 15.55 days 1971 Sep 30.09	Cone 600? full	1.5 long? 2 dia?	1971 Sep 27.7	65.40	88.80	6593	170	260	0.007	62
D	Fragments	1971-77C-E											
D R	Cosmos 439	1971-78A	1971 Sep 21.50 10.74 days 1971 Oct 2.24	Sphere-cylinder 5700?	5.0 long 2.4 dia	1971 Sep 23.4	65.41	89.41	6624	207	284	0.006	89
D	Cosmos 439 rocket	1971-78B	1971 Sep 21.50 6.46 days 1971 Sep 27.96	Cylinder 2500?	7.5 long 2.6 dia	1971 Sep 22.4	65.41	89.28	6617	203	275	0.005	77
D	Cosmos 440	1971-79A	1971 Sep 24.44 401.18 days 1972 Oct 29.62	Ellipsoid 400?	1.8 long 1.2 dia	1971 Sep 30.9 1972 Feb 1.0 1972 Jun 16.0	71.00 70.99 70.99	95.21 94.38 92.66	6907 6866 6783	272 267 254	785 709 555	0.037 0.032 0.022	71 - -
D	Cosmos 440 rocket	1971-79B	1971 Sep 24.44 229.38 days 1972 May 10.82	Cylinder 1500?	8 long 1.65 dia	1971 Sep 28.5 1972 Jan 1.0	71.00 71.00	95.19 93.98	6906 6847	270 265	785 672	0.037 0.030	74 -

*1971-77F ejected from 1971-77A about 1971 Sep 27.2.

	Name		Launch date, lifetime and descent date	Shape and weight (kg)	Size (m)	Date of orbital determination	Orbital inclination (deg)	Nodal period (min)	Semi major axis (km)	Perigee height (km)	Apogee height (km)	Orbital eccentricity	Argument of perigee (deg)
	Shinsei** [Mu 4S]	1971-80A	1971 Sep 28.17 5000 years	26-faced polyhedron 65	0.75 long 0.71 dia	1971 Oct 1.1	32.06	112.92	7745	869	1865	0.064	136
	Shinsei rocket	1971-80B	1971 Sep 28.17 4000 years	Sphere-cone 90?	1.86 long 0.79 dia	1971 Sep 30.9	32.05	111.75	7689	867	1755	0.058	137
D R	Cosmos 441	1971-81A	1971 Sep 28.32 11.91 days 1971 Oct 10.23	Sphere-cylinder 6300?	6.5 long? 2.4 dia	1971 Sep 28.8 1971 Oct 1.7	65.02 65.04	89.21 89.00	6614 6604	204 173	268 278	0.005 0.008	33 57
D	Cosmos 441 rocket	1971-81B	1971 Sep 28.32 5.09 days 1971 Oct 3.41	Cylinder 2500?	7.5 long 2.6 dia	1971 Sep 28.7	65.03	89.02	6605	206	247	0.003	35
D	Cosmos 441 engine*	1971-81E	1971 Sep 28.32 15.25 days 1971 Oct 13.57	Cone 600? full	1.5 long? 2 dia?	1971 Oct 11.6	65.04	88.66	6587	174	244	0.005	53
D	Fragments	1971-81C,D,F,G											
D	Luna 19 launcher	1971-82B	1971 Sep 28.42 3.45 days 1971 Oct 1.87	-	-	1971 Sep 28.5	51.58	88.76	6594	172	260	0.007	348
D	Luna 19 launcher rocket	1971-82D	1971 Sep 28.42 3.39 days 1971 Oct 1.81	Cylinder 4000?	12 long? 4 dia	1971 Sep 29.1	51.59	88.44	6578	198	202	0.0003	357

*1971-81E ejected from 1971-81A about 1971 Oct 10.

** Japanese satellite

Space Vehicle: Luna 19, 1971-82A; and fragment, 1971-82C.

	Name		Launch date, lifetime and descent date	Shape and weight (kg)	Size (m)	Date of orbital determination	Orbital inclination (deg)	Nodal period (min)	Semi major axis (km)	Perigee height (km)	Apogee height (km)	Orbital eccentricity	Argument of perigee (deg)
D	OSO 7	1971-83A	1971 Sep 29.41 1013.90 days 1974 Jul 9.31	Nonagonal box + vane 635	1.71 long? 1.42 dia?	1971 Oct 2.9 1972 Jun 1.0 1973 May 16.5	33.13 33.13 33.13	93.40 92.84 91.86	6825 6798 6749	323 315 298	571 525 443	0.018 0.015 0.011	91 - -
D	TTS 3 †	1971-83B	1971 Sep 29.41 2547 days 1978 Sep 19	Octahedron 20	0.30 side	1971 Oct 3.2 1973 May 1.0	33.09 33.09	94.17 93.58	6863 6834	398 385	572 527	0.013 0.010	85 -
D	OSO 7 second stage	1971-83C	1971 Sep 29.41 847.61 days 1974 Jan 24.02	Cylinder 350?	4.9 long 1.43 dia	1971 Oct 1.1 1972 Jul 1.0 1973 Apr 16.0	33.08 33.08 33.08	93.88 93.17 92.13	6849 6814 6762	371 358 333	570 514 435	0.015 0.011 0.007	70 - -
D	Fragment	1971-83D											
D R	Cosmos 442	1971-84A	1971 Sep 29.48 12.75 days 1971 Oct 12.23	Sphere- cylinder 6300?	6.5 long? 2.4 dia	1971 Oct 2.9 1971 Oct 6.3	72.86 72.81	89.47 89.62	6626 6634	182 179	313 333	0.010 0.012	58 56
D	Cosmos 442 rocket	1971-84B	1971 Sep 29.48 6.51 days 1971 Oct 5.99	Cylinder 2500?	7.5 long 2.6 dia	1971 Oct 2.8	72.85	89.01	6603	195	254	0.004	74
D	Cosmos 442 engine*	1971-84D	1971 Sep 29.48 18.38 days 1971 Oct 17.86	Cone 600? full	1.5 long? 2 dia?	1971 Oct 12.2	72.85	89.37	6621	177	309	0.010	41
D	Fragments	1971-84C,E											
D R	Cosmos 443	1971-85A	1971 Oct 7.52 11.67 days 1971 Oct 19.19	Sphere- cylinder 5900?	5.9 long 2.4 dia	1971 Oct 8.9	65.40	89.55	6631	204	301	0.007	61
D	Cosmos 443 rocket	1971-85B	1971 Oct 7.52 6.47 days 1971 Oct 13.99	Cylinder 2500?	7.5 long 2.6 dia	1971 Oct 8.4	65.39	89.42	6624	199	293	0.007	52

*1971-84D ejected from 1971-84A about 1971 Oct 12.

† Test and Training Satellite.

1971-85 continued on page 274

	Name		Launch date, lifetime and descent date	Shape and weight (kg)	Size (m)	Date of orbital determination	Orbital inclination (deg)	Nodal period (min)	Semi major axis (km)	Perigee height (km)	Apogee height (km)	Orbital eccentricity	Argument of perigee (deg)
D	Excess Radiation Package B*	1971-85F	1971 Oct 7.52 23 days 1971 Oct 30	Ellipsoid 200?	0.9 long 1.9 dia	1971 Oct 26.2	65.40	88.61	6584	186	225	0.003	-
D	Fragments	1971-85C-E											
	Cosmos 444	1971-86A	1971 Oct 13.57 6000 years	Spheroid 40?	1.0 long? 0.8 dia?	1971 Oct 13.9	74.03	114.16	7795	1324	1509	0.012	108
	Cosmos 445	1971-86B	1971 Oct 13.57 7000 years	Spheroid 40?	1.0 long? 0.8 dia?	1971 Oct 19.9	74.03	114.53	7811	1353	1513	0.010	104
	Cosmos 446	1971-86C	1971 Oct 13.57 8000 years	Spheroid 40?	1.0 long? 0.8 dia?	1971 Oct 15.0	74.03	114.88	7827	1384	1513	0.008	117
	Cosmos 447	1971-86D	1971 Oct 13.57 9000 years	Spheroid 40?	1.0 long? 0.8 dia?	1971 Oct 16.9	74.03	115.21	7843	1414	1515	0.006	122
	Cosmos 448	1971-86E	1971 Oct 13.57 9000 years	Spheroid 40?	1.0 long? 0.8 dia?	1971 Oct 14.0	74.03	115.58	7860	1441	1522	0.005	130
	Cosmos 449	1971-86F	1971 Oct 13.57 10000 years	Spheroid 40?	1.0 long? 0.8 dia?	1971 Oct 16.2	74.04	116.33	7892	1484	1544	0.004	194
	Cosmos 450	1971-86G	1971 Oct 13.57 10000 years	Spheroid 40?	1.0 long? 0.8 dia?	1971 Oct 14.8	74.03	115.94	7876	1465	1530	0.004	168
	Cosmos 451	1971-86H	1971 Oct 13.57 10000 years	Spheroid 40?	1.0 long? 0.8 dia?	1971 Oct 21.2	74.03	116.73	7911	1492	1574	0.005	212
	Cosmos 444 rocket	1971-86J	1971 Oct 13.57 20000 years	Cylinder 2200?	7.4 long 2.4 dia	1971 Oct 14.8	74.03	117.43	7943	1501	1628	0.008	247

*1971-85F ejected from 1971-85A about 1971 Oct 19

	Name		Launch date, lifetime and descent date	Shape and weight (kg)	Size (m)	Date of orbital determination	Orbital inclination (deg)	Nodal period (min)	Semi major axis (km)	Perigee height (km)	Apogee height (km)	Orbital eccentricity	Argument of perigee (deg)
	DMSP 1 † [Thor Burner 2]	1971-87A	1971 Oct 14.40 80 years	12-faced frustum 195	1.64 long 1.31 to 1.10 dia	1971 Oct 14.9	98.96	101.68	7215	796	877	0.006	231
	Burner 2 rocket	1971-87B	1971 Oct 14.40 60 years	Sphere-cone 66	1.32 long 0.94 dia	1971 Oct 14.4	98.94	101.78	7220	796	888	0.006	230
D R	Cosmos 452	1971-88A	1971 Oct 14.38 12.83 days 1971 Oct 27.21	Sphere-cylinder 6300?	6.5 long? 2.4 dia	1971 Oct 15.4 1971 Oct 19.5	64.97 64.98	89.07 89.41	6607 6624	198 176	260 316	0.005 0.011	17 50
D	Cosmos 452 rocket	1971-88B	1971 Oct 14.38 4.48 days 1971 Oct 18.86	Cylinder 2500?	7.5 long 2.6 dia	1971 Oct 15.8	64.98	88.70	6589	189	233	0.003	359
D	Cosmos 452 engine*	1971-88E	1971 Oct 14.38 16 days 1971 Oct 30	Cone 600? full	1.5 long? 2 dia?	1971 Oct 28.1	64.97	90.07	6657	196	362	0.012	18
D	Fragments	1971-88C,D,F											
	ASTEX** [Thorad Agena D]	1971-89A	1971 Oct 17.57 200 years	Cylinder + 2 wings 1500?	9.6 long? 1.5 dia 9.8 span	1971 Oct 27.0	92.72	100.65	7166	773	803	0.002	348
D	Cosmos 453	1971-90A	1971 Oct 19.53 151.55 days 1972 Mar 19.08	Ellipsoid 400?	1.8 long 1.2 dia	1971 Oct 26.2 1972 Jan 1.0	71.00 71.00	92.19 91.47	6760 6725	271 259	493 434	0.016 0.013	67 -
D	Cosmos 453 rocket	1971-90B	1971 Oct 19.53 73.61 days 1972 Jan 1.14	Cylinder 1500?	8 long 1.65 dia	1971 Oct 24.9	71.00	92.08	6754	270	482	0.016	73
D	Fragment	1971-90C											

* 1971-88E ejected from 1971-88A on 1971 Oct 26.

** USAF Advanced Space Technology Experiments.

† USAF Defense Meteorological Satellite Program

	Name		Launch date, lifetime and descent date	Shape and weight (kg)	Size (m)	Date of orbital determination	Orbital inclination (deg)	Nodal period (min)	Semi major axis (km)	Perigee height (km)	Apogee height (km)	Orbital eccentricity	Argument of perigee (deg)
D	ITOS B* second stage	1971-91A	1971 Oct 21.48 274 days 1972 Jul 21	Cylinder 350?	4.9 long 1.43 dia	1971 Oct 23.9 1972 Mar 16.5	102.62 102.59	102.53 99.63	7255 7117	279 292	1474 1186	0.082 0.063	149 -
3d	Fragments	1971-91B-E											
D	[Titan 3B Agena D]	1971-92A	1971 Oct 23.72 25 days 1971 Nov 17	Cylinder 3000?	8 long ? 1.5 dia	1971 Oct 24.3 1971 Oct 31.4	110.94 110.96	90.02 89.95	6653 6650	134 133	416 410	0.021 0.021	135 133
D	Fragment	1971-92B											
	Prospero** [Black Arrow]	1971-93A	1971 Oct 28.17 150 years	26-faced polyhedron 66	0.70 long 1.12 dia	1971 Nov 1.0	82.06	106.53	7443	547	1582	0.069	329
	Waxwing rocket	1971-93B	1971 Oct 28.17 150 years	Cylinder + nozzle 47	1.36 long 0.71 dia	1971 Oct 30.5	82.05	106.62	7447	546	1593	0.070	333
D	Fragment***	1971-93C											
D R	Cosmos 454	1971-94A	1971 Nov 2.60 13.7 days 1971 Nov 16.3	Sphere- cylinder 6300?	6.5 long? 2.4 dia	1971 Nov 2.9 1971 Nov 4.9	65.42 65.45	89.14 90.03	6611 6655	203 203	262 350	0.004 0.011	53 44
D	Cosmos 454 rocket	1971-94B	1971 Nov 2.60 4.12 days 1971 Nov 6.72	Cylinder 2500?	7.5 long 2.6 dia	1971 Nov 2.9	65.44	89.05	6606	199	257	0.004	39
D	Cosmos 454 engine†	1971-94D	1971 Nov 2.60 22.45 days 1971 Nov 25.05	Cone 600? full	1.5 long? 2 dia?	1971 Nov 16.1	65.45	89.88	6647	202	336	0.010	35
D	Fragments	1971-94C,E,F											

*The satellite ITOS B failed to achieve orbit

†1971-94D ejected from 94A on 1971 Nov 14.

**UK technological satellite, known as X3 before launch.

***This object is probably an aerial about 1 m long.

	Name		Launch date, lifetime and descent date	Shape and weight (kg)	Size (m)	Date of orbital determination	Orbital inclination (deg)	Nodal period (min)	Semi major axis (km)	Perigee height (km)	Apogee height (km)	Orbital eccentricity	Argument of perigee (deg)
	DSCS 1 † [Titan 3C]	1971-95A	1971 Nov 3.13 > million years	Cylinder + 2 dishes 522	1.83 long 2.74 dia	1971 Nov 25.9	2.70	1435.2	42148	35065	36475	0.017	201
	DSCS 2	1971-95B	1971 Nov 3.13 > million years	Cylinder + 2 dishes 522	1.83 long 2.74 dia	1971 Nov 27.9	2.28	1438.0	42202	35349	36299	0.011	227
	DSCS 1 rocket	1971-95C	1971 Nov 3.13 > million years	Cylinder 1500?	6 long? 3.0 dia	1971 Nov 3.4	2.63	1481.3	43043	36089	37240	0.013	199
	Explorer 45 (SSS-1)*	1971-96A	1971 Nov 15.24 **30** years	Octagon 52	0.76 long 0.68 dia	1971 Nov 18.0	3.58	466.85	19942	233	26895	0.669	261
	Explorer 45 rocket	1971-96B	1971 Nov 15.24 20 years	Cylinder 24	1.50 long 0.46 dia	1972 Apr 15.1	3.27	457.7	19677	275	26323	0.662	81
D	Cosmos 455	1971-97A	1971 Nov 17.47 143.55 days 1972 Apr 9.02	Ellipsoid 400?	1.8 long 1.2 dia	1971 Nov 19.9 1972 Feb 1.0	71.00 71.00	92.19 91.46	6760 6724	272 259	491 433	0.016 0.013	73 -
D	Cosmos 455 rocket	1971-97B	1971 Nov 17.47 82.60 days 1972 Feb 8.07	Cylinder 1500?	8 long 1.65 dia	1971 Nov 18.0	70.98	92.03	6752	274	473	0.015	77
D R	Cosmos 456	1971-98A	1971 Nov 19.50 12.8 days 1971 Dec 2.3	Sphere-cylinder 6300?	6.5 long? 2.4 dia	1971 Nov 20.9 1971 Nov 26.4	72.86 72.87	89.34 89.98	6619 6651	178 186	304 360	0.010 0.013	55 64
D	Cosmos 456 rocket	1971-98B	1971 Nov 19.50 6.25 days 1971 Nov 25.75	Cylinder 2500?	7.5 long 2.6 dia	1971 Nov 19.8	72.87	89.50	6627	201	297	0.007	57

*Small scientific satellite

†Defence Satellite Communication System

1971-98 continued on page 278

	Name		Launch date, lifetime and descent date	Shape and weight (kg)	Size (m)	Date of orbital determination	Orbital inclination (deg)	Nodal period (min)	Semi major axis (km)	Perigee height (km)	Apogee height (km)	Orbital eccentricity	Argument of perigee (deg)
D	Cosmos 456 engine*	1971-98E	1971 Nov 19.50 21.27 days 1971 Dec 10.77	Cone 600? full	1.5 long? 2 dia?	1971 Dec 2.4	72.84	89.75	6640	187	336	0.011	52
D	Fragments	1971-98C,D,F-H											
	Cosmos 457	1971-99A	1971 Nov 20.75 3000 years	Spheroid + paddles? 650?	1.6 dia?	1971 Nov 23.5	74.04	109.50	7581	1185	1221	0.002	224
	Cosmos 457 rocket	1971-99B	1971 Nov 20.75 2000 years	Cylinder 2200?	7.4 long 2.4 dia	1971 Nov 27.6	74.03	109.37	7575	1182	1212	0.002	200
D	Molniya 2A	1971-100A	1971 Nov 24.40 28 months? 1974 Mar?	Windmill + 6 vanes? 1250?	4.2 long? 1.6 dia?	1971 Nov 24.4 1972 Feb 1.0 1972 Aug.1.0	65.47 65.47 65.47	712.03 717.71 717.73	26414 26555 26555	517 466 567	39554 39887 39787	0.739 0.742 0.739	285 - -
D	Molniya 2A launcher rocket	1971-100B	1971 Nov 24.40 25.13 days 1971 Dec 19.53	Cylinder 2500?	7.5 long 2.6 dia	1971 Nov 24.7	65.42	91.06	6705	218	436	0.016	60
D	Molniya 2A launcher	1971-100C	1971 Nov 24.40 36.33 days 1971 Dec 30.73	Irregular	-	1971 Nov 24.7	65.43	91.42	6723	225	465	0.018	68
D	Molniya 2A rocket	1971-100E	1971 Nov 24.40 773 days 1974 Jan 5	Cylinder 440	2.0 long 2.0 dia	1972 Jan 1.0 1972 Aug 1.0 1973 Jun 1.0	65.47 65.47 65.47	703.63 703.53 703.41	26206 26203 26200	450 551 348	39205 39099 39296	0.739 0.736 0.743	- - -
D	Fragment	1971-100D											
D	Cosmos 458	1971-101A	1971 Nov 29.43 142.90 days 1972 Apr 20.33	Ellipsoid 400?	1.8 long 1.2 dia	1971 Nov 29.7 1972 Feb 15.5	70.96 70.96	92.25 91.52	6763 6727	272 258	497 440	0.017 0.014	80 -
D	Cosmos 458 rocket	1971-101B	1971 Nov 29.43 77.49 days 1972 Feb 14.92	Cylinder 1500?	8 long 1.65 dia	1971 Nov 30.1	70.96	92.07	6754	270	481	0.016	84

*1971-98E ejected from 1971-98A on 1971 Dec 1.

	Name		Launch date, lifetime and descent date	Shape and weight (kg)	Size (m)	Date of orbital determination	Orbital inclination (deg)	Nodal period (min)	Semi major axis (km)	Perigee height (km)	Apogee height (km)	Orbital eccentricity	Argument of perigee (deg)
D	Cosmos 459	1971-102A	1971 Nov 29.73 27.92 days 1971 Dec 27.65	Cylinder?	4 long? 2 dia?	1971 Dec 3.2	65.81	89.34	6620	224	260	0.003	3
D	Cosmos 459 rocket	1971-102B	1971 Nov 29.73 11.29 days 1971 Dec 11.02	Cylinder 2200?	7.4 long 2.4 dia	1971 Nov 30.2	65.87	89.34	6621	223	262	0.003	4
D	Fragment	1971-102C											
D	Cosmos 460	1971-103A	1971 Nov 30.70 3018 days 1980 Mar 5	Cylinder + paddles? 900?	2 long? 1 dia?	1971 Nov 30.9	74.01	95.25	6908	528	532	0.0003	328
D	Cosmos 460 rocket	1971-103B	1971 Nov 30.70 3078 days 1980 May 4	Cylinder 2200?	7.4 long 2.4 dia	1971 Dec 1.9	74.04	95.14	6903	508	541	0.002	349
D	Fragment	1971-103C											
D	Intercosmos 5	1971-104A	1971 Dec 2.35 126.84 days 1972 Apr 7.19	Ellipsoid 400?	1.8 long 1.2 dia	1971 Dec 4.0	48.42	98.49	7068	198	1181	0.070	110
D	Intercosmos 5 rocket	1971-104B	1971 Dec 2.35 90.67 days 1972 Mar 2.02	Cylinder 1500?	8 long 1.65 dia	1971 Dec 5.4	48.43	98.21	7054	199	1152	0.068	116
D	Fragment	1971-104C											
D	Cosmos 461*	1971-105A	1971 Dec 2.73 2638 days 1979 Feb 21	Cylinder? 950?	-	1971 Dec 5.5	69.23	94.61	6878	488	511	0.002	352
D	Cosmos 461 rocket	1971-105B	1971 Dec 2.73 2705 days 1979 Apr 29	Cylinder 2200?	7.4 long 2.4 dia	1971 Dec 6.6	69.23	94.45	6870	476	507	0.002	11
D	Fragments	1971-105C,D											

* Gamma radiation studies

	Name		Launch date, lifetime and descent date	Shape and weight (kg)	Size (m)	Date of orbital determination	Orbital inclination (deg)	Nodal period (min)	Semi major axis (km)	Perigee height (km)	Apogee height (km)	Orbital eccentricity	Argument of perigee (deg)
D	Cosmos 462*	1971-106A	1971 Dec 3.55 1218.35 days 1975 Apr 4.90	Cylinder?	4 long? 2 dia?	1971 Dec 5.9 1972 Aug 1.0 1973 Oct 1.0	65.75 65.75 65.75	105.43 103.06 98.80	7393 7282 7081	230 212 220	1800 1595 1185	0.106 0.095 0.068	53 - -
D	Cosmos 462 rocket	1971-106C	1971 Dec 3.55 33.23 days 1972 Jan 5.78	Cylinder 1500?	8 long? 2.5 dia?	1971 Dec 5.0	62.31	101.96	7230	143	1561	0.098	57
D	Fragments	1971-106B, D-AE											
D R	Cosmos 463	1971-107A	1971 Dec 6.41 4.96 days 1971 Dec 11.37	Sphere-cylinder 6300?	6.5 long? 2.4 dia	1971 Dec 6.6 1971 Dec 9.6	64.97 64.97	89.24 89.27	6616 6617	202 182	273 296	0.005 0.009	37 45
D	Cosmos 463 rocket	1971-107B	1971 Dec 6.41 5.39 days 1971 Dec 11.80	Cylinder 2500?	7.5 long 2.6 dia	1971 Dec 6.8	64.99	89.18	6613	202	267	0.005	38
D	Cosmos 463 engine**	1971-107D	1971 Dec 6.41 9.57 days 1971 Dec 15.98	Cone 600? full	1.5 long? 2 dia?	1971 Dec 10.3	64.97	89.17	6612	180	288	0.008	48
D	Fragment	1971-107C											
D R	Cosmos 464	1971-108A	1971 Dec 10.46 5.8 days 1971 Dec 16.3	Sphere-cylinder 6300?	6.5 long? 2.4 dia	1971 Dec 11.0 1971 Dec 14.9	72.84 72.86	90.34 89.23	6669 6614	206 178	375 293	0.013 0.009	72 30
D	Cosmos 464 rocket	1971-108B	1971 Dec 10.46 14.25 days 1971 Dec 24.71	Cylinder 2500?	7.5 long 2.6 dia	1971 Dec 11.0	72.85	90.26	6665	205	369	0.012	69
D	Cosmos 464 engine †	1971-108E	1971 Dec 10.46 9.23 days 1971 Dec 19.69	Cone 600? full	1.5 long? 2 dia?	1971 Dec 15.4	72.86	89.09	6607	166	291	0.009	23
D	Fragments	1971-108C, D											

*Cosmos 462 passed close to Cosmos 459 on 1971 Dec 3.70, then exploded.

**1971-107D ejected from 1971-107A on 1971 Dec 10.

† 1971-108E ejected from 1971-108A on 1971 Dec 15.

	Name		Launch date, lifetime and descent date	Shape and weight (kg)	Size (m)	Date of orbital determination	Orbital inclination (deg)	Nodal period (min)	Semi major axis (km)	Perigee height (km)	Apogee height (km)	Orbital eccentricity	Argument of perigee (deg)
D	Ariel 4*	1971-109A	1971 Dec 11.86 2557.8 days 1978 Dec 12.7	Cylinder + 4 paddles 99.5	0.91 long 0.76 dia	1971 Dec 12.6 1973 Nov 1.0	82.99 82.99	95.35 94.74	6913 6883	477 457	593 552	0.008 0.007	242 -
D	Ariel 4 rocket	1971-109B	1971 Dec 11.86 2433 days 1978 Aug 9	Cylinder 24	1.5 long 0.46 dia	1971 Dec 12.6 1973 Aug 1.0	83.00 82.99	95.33 94.72	6912 6882	477 457	591 550	0.008 0.007	244 -
D	Fragments	1971-109C,D											
	[Thorad Agena D]	1971-110A	1971 Dec 14.51 700 years	-	-	1971 Dec 26.5	70.00	104.93	7369	983	999	0.001	235
	Agena D rocket	1971-110B	1971 Dec 14.51 600 years	Cylinder 700?	6 long? 1.5 dia	1971 Dec 29.2	70.02	104.17	7333	943	967	0.002	297
	[Thorad Agena D] **	1971-110C	1971 Dec 14.51 700 years	Box + aerials?	0.3 x 0.9 x 2.4?	1971 Dec 26.4	70.01	104.93	7369	983	999	0.001	242
	[Thorad Agena D] **	1971-110D	1971 Dec 14.51 700 years	Box + aerials?	0.3 x 0.9 x 2.4?	1972 Jan 1.5	70.01	104.90	7368	982	997	0.001	221
	[Thorad Agena D] **	1971-110E	1971 Dec 14.51 700 years	Box + aerials?	0.3 x 0.9 x 2.4?	1972 Jan 1.0	70.01	104.89	7367	981	997	0.001	-
	Cosmos 465	1971-111A	1971 Dec 15.19 1200 years	Cylinder + boom? 700?	1.4 long 2.0 dia	1971 Dec 29.2	74.03	104.94	7369	970	1012	0.003	260
	Cosmos 465 rocket	1971-111B	1971 Dec 15.19 600 years	Cylinder 2200?	7.4 long 2.4 dia	1971 Dec 26.9	74.03	104.83	7364	970	1002	0.002	263

*British satellite, known as UK4 before launch.

** SSU precursors

	Name		Launch date, lifetime and descent date	Shape and weight (kg)	Size (m)	Date of orbital determination	Orbital inclination (deg)	Nodal period (min)	Semi major axis (km)	Perigee height (km)	Apogee height (km)	Orbital eccentricity	Argument of perigee (deg)
D R	Cosmos 466	1971-112A	1971 Dec 16.41 10.9 days 1971 Dec 27.3	Sphere-cylinder 6300?	6.5 long? 2.4 dia	1971 Dec 16.8 1971 Dec 23.6	65.01 64.99	89.39 89.94	6623 6650	209 179	280 365	0.005 0.014	43 51
D	Cosmos 466 rocket	1971-112B	1971 Dec 16.41 4.63 days 1971 Dec 21.04	Cylinder 2500?	7.5 long 2.6 dia	1971 Dec 17.3	65.01	89.20	6614	193	278	0.006	25
D	Cosmos 466 engine*	1971-112C	1971 Dec 16.41 17.66 days 1972 Jan 3.07	Cone 600? full	1.5 long? 2 dia?	1971 Dec 27.7	65.01	89.78	6642	173	355	0.014	48
D	Fragment	1971-112D											
D	Cosmos 467	1971-113A	1971 Dec 17.45 122.80 days 1972 Apr 18.25	Ellipsoid 400?	1.8 long 1.2 dia	1971 Dec 23.6 1972 Feb 15.5	71.00 71.00	91.95 91.44	6748 6723	267 260	472 430	0.015 0.013	78 -
D	Cosmos 467 rocket	1971-113B	1971 Dec 17.45 62.38 days 1972 Feb 17.83	Cylinder 1500?	8 long 1.65 dia	1971 Dec 20.4	71.01	91.80	6740	270	454	0.014	80
	Cosmos 468	1971-114A	1971 Dec 17.54 120 years	Cylinder + paddles? 750?	2 long? 1 dia?	1971 Dec 27.2	74.03	100.83	7176	786	809	0.002	22
	Cosmos 468 rocket	1971-114B	1971 Dec 17.54 100 years	Cylinder 2200?	7.4 long 2.4 dia	1971 Dec 26.1	74.03	100.76	7172	775	813	0.003	16
	Fragments	1971-114C-E											

* 1971-112C ejected from 1971-112A about 1971 Dec 26.

	Name		Launch date, lifetime and descent date	Shape and weight (kg)	Size (m)	Date of orbital determination	Orbital inclination (deg)	Nodal period (min)	Semi major axis (km)	Perigee height (km)	Apogee height (km)	Orbital eccentricity	Argument of perigee (deg)
D	Molniya 1U	1971-115A	1971 Dec 19.96 1942 days 1977 Apr 13	Windmill + 6 vanes 1000?	3.4 long 1.6 dia	1971 Dec 29.3 1972 Feb 1.0	65.42 65.42	703.28 717.79	26197 26557	499 417	39139 39940	0.738 0.744	285 -
D	Molniya 1U rocket	1971-115B	1971 Dec 19.96 2104 days 1977 Sep 22	Cylinder 440	2.0 long 2.0 dia	1971 Dec 31.2	65.45	699.22	26097	451	38986	0.738	285
D	Molniya 1U launcher rocket	1971-115C	1971 Dec 19.96 38 days 1972 Jan 26	Cylinder 2500?	7.5 long 2.6 dia	1971 Dec 21.7	65.41	91.56	6730	222	481	0.019	72
D	Molniya 1U launcher	1971-115D	1971 Dec 19.96 39 days 1972 Jan 27	Irregular	-	1971 Dec 21.7	65.37	91.64	6734	222	489	0.020	71
	Intelsat 4 F-3	1971-116A	1971 Dec 20.05 >million years	Cylinder 1410 full 707 empty	2.82 long 2.39 dia	1971 Dec 20.1 1972 Jan 1.0	28.23 0.4	640.3 1436.2	24616 42167	550 35749	35926 35828	0.719 0.001	179 -
	Intelsat 4 F-3 rocket	1971-116B	1971 Dec 20.05 6000 years	Cylinder 1815	8.6 long 3.0 dia	1971 Dec 20.1	28.23	640.3	24616	550	35926	0.719	179**
	Cosmos 469	1971-117A	1971 Dec 25.48 600 years	Cone-cylinder	6 long? 2 dia?	1971 Dec 26.0 1972 Jan 15.6	64.94 64.50	89.60 104.74	6634 7360	249 941	262 1023	0.001 0.006	286 346
D	Cosmos 469* platform	1971-117B	1971 Dec 25.48 45.60 days 1972 Feb 9.08	Irregular	-	1972 Jan 4.4	64.96	89.57	6632	247	261	0.001	301
D	Cosmos 469* rocket	1971-117C	1971 Dec 25.48 13.16 days 1972 Jan 7.64	Cylinder 1500?	8 long? 2.5 dia?	1972 Jan 4.1	64.96	89.48	6628	244	255	0.001	305

*1971-117C, and probably 1971-117B, attached to 1971-117A until orbit change on 1972 Jan 4.29.

**Approximate orbit.

	Name		Launch date, lifetime and descent date	Shape and weight (kg)	Size (m)	Date of orbital determination	Orbital inclination (deg)	Nodal period (min)	Semi major axis (km)	Perigee height (km)	Apogee height (km)	Orbital eccentricity	Argument of perigee (deg)
D R	Cosmos 470	1971-118A	1971 Dec 27.59 9.76 days 1972 Jan 6.35	Sphere-cylinder 5900?	5.9 long 2.4 dia	1971 Dec 28.6	65.42	89.03	6605	194	260	0.005	18
D	Cosmos 470 rocket	1971-118B	1971 Dec 27.59 2.95 days 1971 Dec 30.54	Cylinder 2500?	7.5 long 2.6 dia	1971 Dec 28.6	65.43	88.75	6591	177	249	0.005	17
D	Capsule*	1971-118C	1971 Dec 27.59 10.02 days 1972 Jan 6.61	Ellipsoid 200?	0.9 long 1.9 dia	1972 Jan 6.3	65.43	88.43	6575	183	211	0.002	5
D	Fragments	1971-118D, E											
	Aureole 1**	1971-119A	1971 Dec 27.79 70 years	Octagonal ellipsoid 400?	1.8 long? 1.5 dia?	1972 Jan 1.9	73.98	114.65	7817	400	2477	0.133	98
	Aureole 1 rocket	1971-119B	1971 Dec 27.79 50 years	Cylinder 2200?	7.4 long 2.4 dia	1972 Jan 3.0	74.00	114.51	7810	394	2469	0.133	96
D	Fragments	1971-119C-E											
	Meteor 10	1971-120A	1971 Dec 29.45 500 years	Cylinder + 2 vanes 2200?	5 long? 1.5 dia?	1972 Jan 2.2 1972 Feb 17.5	81.25 81.25	102.66 102.32	7262 7245	878 859	889 874	0.001 0.001	221 - †
	Meteor 10 rocket	1971-120B	1971 Dec 29.45 400 years	Cylinder 1440	3.8 long 2.6 dia	1972 Jan 2.6	81.26	102.72	7264	845	927	0.006	168
	Fragments	1971-120C-E											

*1971-118C ejected from 1971-118A about 1972 Jan 6. **French satellite, launched by USSR. † Meteor 10 carried orbital adjustment motor.

	Name		Launch date, lifetime and descent date	Shape and weight (kg)	Size (m)	Date of orbital determination	Orbital inclination (deg)	Nodal period (min)	Semi major axis (km)	Perigee height (km)	Apogee height (km)	Orbital eccentricity	Argument of perigee (deg)
D R	Cosmos 471	1972-01A	1972 Jan 12.42 12.9 days 1972 Jan 25.3	Sphere-cylinder 6300?	6.5 long? 2.4 dia	1972 Jan 12.8 1972 Jan 23.2	64.99 64.99	89.66 88.96	6637 6602	201 178	316 270	0.009 0.007	57 49
D	Cosmos 471 rocket	1972-01B	1972 Jan 12.42 7.84 days 1972 Jan 20.26	Cylinder 2500?	7.5 long 2.6 dia	1972 Jan 12.8	65.01	89.58	6633	198	311	0.008	54
D	Cosmos 471 engine*	1972-01G	1972 Jan 12.42 14 days 1972 Jan 26	Cone 600? full	1.5 long? 2 dia?	1972 Jan 26.2	64.98	88.50	6579	181	220	0.003	15
D	Fragments	1972-01C-F											
D	[Titan 3D]	1972-02A	1972 Jan 20.77 40 days 1972 Feb 29	Cylinder 13300? full	15 long 3.0 dia	1972 Jan 21.3 1972 Jan 31.8	97.00 97.00	89.41 89.47	6622 6625	157 149	331 344	0.013 0.015	140 126
D	Titan 3D rocket	1972-02B	1972 Jan 20.77 2.70 days 1972 Jan 23.47	Cylinder 1900	6 long 3.0 dia	1972 Jan 20.9	96.99	89.23	6613	163	306	0.011	139
D	Capsule	1972-02D	1972 Jan 20.77 2644 days 1979 Apr 17	Octagon 60?	0.3 long? 0.9 dia?	1972 Jan 31.4	96.59	94.86	6889	472	549	0.006	129
D	Fragment	1972-02C											
	Intelsat 4 F-4	1972-03A	1972 Jan 23.01 > million years	Cylinder + 2 aerials 1410 full	2.82 long 2.39 dia	1972 Jan 25.9 1972 Apr 16.0	28.22 0.4	654.5 1436.1	24967 42166	562 35781	36615 35794	0.722 0.0002	181** -
	Intelsat 4 F-4 rocket	1972-03B	1972 Jan 23.01 6000 years	Cylinder 1815	8.6 long 3.0 dia	1972 Jan 26.9	28.22	654.5	24967	562	36615	0.722	181

* 1972-01G ejected from 1972-01A about 1972 Jan 23

** Approximate orbit

	Name		Launch date, lifetime and descent date	Shape and weight (kg)	Size (m)	Date of orbital determination	Orbital inclination (deg)	Nodal period (min)	Semi major axis (km)	Perigee height (km)	Apogee height (km)	Orbital eccentricity	Argument of perigee (deg)
D	Cosmos 472	1972-04A	1972 Jan 25.47 206.48 days 1972 Aug 18.95	Ellipsoid 400?	1.8 long 1.2 dia	1972 Jan 28.2 1972 May 1.0	82.01 81.99	102.26 98.37	7243 7058	194 186	1536 1173	0.093 0.070	67 -
D	Cosmos 472 rocket	1972-04B	1972 Jan 25.47 101.58 days 1972 May 6.05	Cylinder 1500?	8 long 1.65 dia	1972 Jan 27.9	82.00	102.15	7238	193	1526	0.092	69
D	Heos 2*	1972-05A	1972 Jan 31.72 913.73 days 1974 Aug 2.45	16-faced cylinder 117	0.75 long 1.33 dia	1972 Feb 3.5 1973 Jul 1.0	89.91 88.0	7477.1 7496.9	126663 126894	405 5442	240164 235589	0.946 0.907	310 -
D	Heos 2 second stage	1972-05B	1972 Jan 31.72 2431 days 1978 Sep 27	Cylinder 350?	4.9 long 1.43 dia	1972 Jan 31.8 1973 May 16.5	89.81 89.79	99.25 97.91	7100 7036	320 310	1123 1005	0.057 0.049	279 -
D	Heos 2 rocket	1972-05C	1972 Jan 31.72 31 months? 1974 Aug?	Cylinder 24	1.50 long 0.46 dia	Orbit similar to 1972-05A							
D	Fragments	1972-05D,E											
D R	Cosmos 473	1972-06A	1972 Feb 3.36 11.90 days 1972 Feb 15.26	Sphere-cylinder 5700?	5.0 long 2.4 dia	1972 Feb 4.0	65.01	89.68	6638	205	314	0.008	51
D	Cosmos 473 rocket	1972-06B	1972 Feb 3.36 8.24 days 1972 Feb 11.60	Cylinder 2500?	7.5 long 2.6 dia	1972 Feb 3.7	65.04	89.59	6633	203	307	0.008	46
D	Luna 20 launcher	1972-07C	1972 Feb 14.14 3.57 days 1972 Feb 17.71	-	-	1972 Feb 14.5	51.51	88.73	6593	191	238	0.004	317
D	Luna 20 launcher rocket	1972-07D	1972 Feb 14.14 3.78 days 1972 Feb 17.92	Cylinder 4000?	12 long? 4 dia	1972 Feb 15.0	51.48	88.64	6588	188	232	0.003	322

Space Vehicle: Luna 20, 1972-07A; Luna 20 rocket, 1972-07B

* Highly eccentric orbit satellite, launched for ESRO by NASA

	Name		Launch date, lifetime and descent date	Shape and weight (kg)	Size (m)	Date of orbital determination	Orbital inclination (deg)	Nodal period (min)	Semi major axis (km)	Perigee height (km)	Apogee height (km)	Orbital eccentricity	Argument of perigee (deg)
D R	Cosmos 474	1972-08A	1972 Feb 16.40 12.9 days 1972 Feb 29.3	Sphere-cylinder 6300?	6.5 long 2.4 dia	1972 Feb 17.4	64.97	89.79	6643	213	317	0.008	72
D	Cosmos 474 rocket	1972-08B	1972 Feb 16.40 7.50 days 1972 Feb 23.90	Cylinder 2500?	7.5 long 2.6 dia	1972 Feb 17.9	64.98	89.63	6635	202	312	0.008	48
D	Cosmos 474 engine*	1972-08E	1972 Feb 16.40 14 days 1972 Mar 1	Cone 600? full	1.5 long? 2 dia?	1972 Feb 29.3	64.95	90.45	6676	195	401	0.015	9
D	Fragments	1972-08C, D											
	Cosmos 475	1972-09A	1972 Feb 25.40 1200 years	Cylinder + boom? 700?	1.4 long 2.0 dia	1972 Feb 25.8	74.08	104.81	7363	970	1000	0.002	287
	Cosmos 475 rocket	1972-09B	1972 Feb 25.40 600 years	Cylinder 2200?	7.4 long 2.4 dia	1972 Feb 25.5	74.09	104.69	7357	964	994	0.002	291
	IMEWS 3 [Titan 3C]	1972-10A	1972 Mar 1.40 >million years	Cylinder + 4 panels 820?	6 long? 2.5 dia?	1972 Mar 1.5 1972 Apr 1.0	28.58 0.2	89.47 1429.9	6633 42067	153 35416	357 35962	0.015 0.006	114 -
	Transtage	1972-10B	1972 Mar 1.40 >million years	Cylinder 1500?	6 long? 3 dia	1972 Apr 1.0	0.2	1429.9	42067	35416	35962	0.006	-
D	Titan 3C second stage	1972-10C	1972 Mar 1.40 2.00 days 1972 Mar 3.40	Cylinder 1900	6 long 3.0 dia	1972 Mar 2.4	28.63	88.88	6604	151	300	0.011	123

* 1972-08E ejected from 1972-08A about 1972 Feb 28.

	Name		Launch date, lifetime and descent date	Shape and weight (kg)	Size (m)	Date of orbital determination	Orbital inclination (deg)	Nodal period (min)	Semi major axis (km)	Perigee height (km)	Apogee height (km)	Orbital eccentricity	Argument of perigee (deg)
	Cosmos 476	1972-11A	1972 Mar 1.47 60 years	Cylinder + 2 vanes? 2500?	5 long? 1.5 dia?	1972 Mar 5.9	81.23	97.24	7003	617	633	0.001	180
	Cosmos 476 rocket	1972-11B	1972 Mar 1.47 60 years	Cylinder 1440	3.8 long 2.6 dia	1972 Mar 5.9	81.23	97.35	7009	569	692	0.009	160
D	Pioneer 10 second stage	1972-12C	1972 Mar 3.08 168 days 1972 Aug 18	Cylinder 1815	8.6 long 3.0 dia	1972 Apr 1.0	31.6	1523.2	43852	158	74789	0.851	-
D R	Cosmos 477	1972-13A	1972 Mar 4.42 11.8 days 1972 Mar 16.2	Sphere-cylinder 5900?	5.9 long 2.4 dia	1972 Mar 5.0	72.85	89.60	6632	202	306	0.008	65
D	Cosmos 477 rocket	1972-13B	1972 Mar 4.42 5.99 days 1972 Mar 10.41	Cylinder 2500?	7.5 long 2.6 dia	1972 Mar 4.8	72.84	89.48	6626	202	294	0.007	62
D	Excess Radiation Package C	1972-13E	1972 Mar 4.42 19.07 days 1972 Mar 23.49	Ellipsoid 200?	0.9 long 1.9 dia	1972 Mar 14.7	72.85	89.43	6624	199	292	0.007	49
D	Fragments	1972-13C,D,F											

Space vehicle: Pioneer 10, 1972-12A; Burner 2 rocket, 1972-12B

	Name		Launch date, lifetime and descent date	Shape and weight (kg)	Size (m)	Date of orbital determination	Orbital inclination (deg)	Nodal period (min)	Semi major axis (km)	Perigee height (km)	Apogee height (km)	Orbital eccentricity	Argument of perigee (deg)
D	TD 1A * (ESRO)	1972-14A	1972 Mar 12.08 2859 days 1980 Jan 9	Box + paddles 472	2.12 x 0.99 x 0.89	1972 Mar 12.9	97.55	95.41	6916	524	551	0.002	323
D	TD 1A second stage	1972-14B	1972 Mar 12.08 3133 days 1980 Oct 9	Cylinder 350?	4.9 long 1.43 dia	1972 Mar 12.9	97.54	95.34	6912	525	543	0.001	323
D	Fragments	1972-14C-E											
D R	Cosmos 478	1972-15A	1972 Mar 15.54 12.68 days 1972 Mar 28.22	Sphere-cylinder 6300?	6.5 long? 2.4 dia	1972 Mar 16.2 1972 Mar 23.8	65.39 65.40	89.48 89.65	6628 6636	204 177	295 339	0.007 0.012	75 58
D	Cosmos 478 rocket	1972-15B	1972 Mar 15.54 6.33 days 1972 Mar 21.87	Cylinder 2500?	7.5 long 2.6 dia	1972 Mar 15.8	65.41	89.46	6627	177	320	0.011	54
D	Cosmos 478 engine**	1972-15F	1972 Mar 15.54 15 days 1972 Mar 30	Cone 600? full	1.5 long? 2 dia?	1972 Mar 29.7	65.42	89.83	6645	189	345	0.012	19
D	Fragments	1972-15C-E											
D	[Titan 3B Agena D]	1972-16A	1972 Mar 17.71 25 days 1972 Apr 11	Cylinder 3000?	8 long? 1.5 dia	1972 Mar 18.2	110.98	89.91	6648	131	409	0.021	149

* Thor Delta

** 1972-15F ejected from 1972-15A about 1972 Mar 25

	Name		Launch date, lifetime and descent date	Shape and weight (kg)	Size (m)	Date of orbital determination	Orbital inclination (deg)	Nodal period (min)	Semi major axis (km)	Perigee height (km)	Apogee height (km)	Orbital eccentricity	Argument of perigee (deg)
D	Cosmos 479	1972-17A	1972 Mar 22.86 2944 days 1980 Apr 13	Cylinder + paddles? 900?	2 long? 1 dia?	1972 Apr 2.5	74.06	95.20	6906	514	542	0.002	306
D	Cosmos 479 rocket	1972-17B	1972 Mar 22.86 2973 days 1980 May 12	Cylinder 2200?	7.4 long 2.4 dia	1972 Apr 1.5	74.07	95.10	6901	506	540	0.003	324
D	Fragments	1972-17C-E											
	DMSP 2 [Thor Burner 2]	1972-18A	1972 Mar 24.37 100 years	12-sided frustum 195	1.64 long 1.31 to 1.10 dia	1972 Mar 28.1	98.80	101.83	7222	803	885	0.006	217
	Burner 2 rocket	1972-18B	1972 Mar 24.37 80 years	Sphere-cone 66	1.32 long 0.94 dia	1972 Mar 27.5	98.80	101.72	7217	801	876	0.005	221
	Cosmos 480	1972-19A	1972 Mar 25.10 3000 years	Spheroid + paddles? 650?	1.6 dia?	1972 Mar 26.3	82.97	109.21	7567	1175	1203	0.002	229
	Cosmos 480 rocket	1972-19B	1972 Mar 25.10 2000 years	Cylinder 2200?	7.4 long 2.4 dia	1972 Mar 27.7	82.97	109.07	7561	1171	1194	0.001	207
D	Cosmos 481 *	1972-20A	1972 Mar 25.45 161.39 days 1972 Sep 2.84	Ellipsoid 400?	1.8 long 1.2 dia	1972 Mar 26.2 1972 Jun 1.0	71.03 71.03	92.40 91.52	6770 6727	269 255	514 443	0.018 0.014	82 -
D	Cosmos 481 rocket	1972-20B	1972 Mar 25.45 77.70 days 1972 Jun 11.15	Cylinder 1500?	8 long 1.65 dia	1972 Mar 26.3	71.03	92.26	6763	270	499	0.017	83

* Geomagnetic studies

	Name		Launch date, lifetime and descent date	Shape and weight (kg)	Size (m)	Date of orbital determination	Orbital inclination (deg)	Nodal period (min)	Semi major axis (km)	Perigee height (km)	Apogee height (km)	Orbital eccentricity	Argument of perigee (deg)
D	Venus 8 launcher rocket	1972-21B	1972 Mar 27.18 2.26 days 1972 Mar 29.44	Cylinder 2500?	7.5 long 2.6 dia	1972 Mar 27.9	51.78	88.57	6585	191	222	0.002	84
D	Venus 8 launcher	1972-21C	1972 Mar 27.18 2.73 days 1972 Mar 29.91	Irregular	-	1972 Mar 28.0	51.77	88.79	6596	194	241	0.004	110
	Meteor 11	1972-22A	1972 Mar 30.59 500 years	Cylinder + 2 vanes 2200?	5 long? 1.5 dia?	1972 Apr 1.6	81.23	102.59	7258	868	891	0.002	257
	Meteor 11 rocket	1972-22B	1972 Mar 30.59 400 years	Cylinder 1440	3.8 long 2.6 dia	1972 Apr 2.5	81.24	102.72	7264	840	932	0.006	172
D	Cosmos 482*	1972-23A	1972 Mar 31.17 3322 days 1981 May 5	Sphere-cylinder 1180	3.5 long 1.2 dia	1972 Mar 31.8 1973 Dec 16.5	52.22 52.19	201.44 185.50	11383 10777	205 209	9805 8589	0.422 0.389	42 -
D	Cosmos 482 launcher rocket	1972-23B	1972 Mar 31.17 1.74 days 1972 Apr 1.91	Cylinder 2500?	7.5 long 2.6 dia	1972 Mar 31.5	51.78	88.54	6584	196	215	0.001	116
D p	Cosmos 482 launcher	1972-23C	1972 Mar 31.17 2.38 days 1972 Apr 2.55	Irregular	-	1972 Mar 31.9	51.74	88.59	6586	179	237	0.004	125
	Cosmos 482 rocket	1972-23D	1972 Mar 31.17 10.9 years	Cylinder 5410? full	2.0 long 2.0 dia	1972 Apr 3.2	52.16	200.93	11365	207	9767	0.421	44
	Fragment	1972-23E											

Space vehicle: Venus 8, 1972-21A

*Probably an attempted Venus probe.

	Name		Launch date, lifetime and descent date	Shape and weight (kg)	Size (m)	Date of orbital determination	Orbital inclination (deg)	Nodal period (min)	Semi major axis (km)	Perigee height (km)	Apogee height (km)	Orbital eccentricity	Argument of perigee (deg)
D R	Cosmos 483	1972-24A	1972 Apr 3.43 11.82 days 1972 Apr 15.25	Sphere-cylinder 6300?	6.5 long? 2.4 dia	1972 Apr 3.8	72.81	89.74	6639	209	313	0.008	73
D	Cosmos 483 rocket	1972-24B	1972 Apr 3.43 8.47 days 1972 Apr 11.90	Cylinder 2500?	7.5 long 2.6 dia	1972 Apr 4.2	72.80	89.64	6634	202	310	0.008	63
D	Cosmos 483 engine*	1972-24D	1972 Apr 3.43 15.95 days 1972 Apr 19.38	Cone 600? full	1.5 long? 2 dia?	1972 Apr 13.9	72.82	89.47	6626	180	315	0.010	33
D	Fragments	1972-24C,E											
D	Molniya 1v	1972-25A	1972 Apr 4.86 666 days 1974 Jan 30	Windmill + 6 vanes 1000?	3.4 long 1.6 dia	1972 Apr 5.0 1972 May 1.0 1973 Mar 16.5	65.6 65.53 65.7	705.35 717.69 717.65	26248 26554 26553	480 442 360	39260 39910 39990	0.739 0.743 0.746	285 285 -
D	Molniya 1v rocket	1972-25G	1972 Apr 4.86 703 days 1974 Mar 8	Cylinder 440	2.0 long 2.0 dia	1972 May 1.0 1972 Nov 1.0 1973 Mar 16.5	65.5 65.6 65.7	700.02 700.00 698.5	26116 26116 26078	454 338 214	39022 39137 39186	0.738 0.743 0.747	- - -
D	Molniya 1v launcher rocket	1972-25C	1972 Apr 4.86 23.70 days 1972 Apr 28.56	Cylinder 2500?	7.5 long 2.6 dia	1972 Apr 6.1	65.54	91.28	6716	222	454	0.017	66
D	Molniya 1v launcher	1972-25D	1972 Apr 4.86 33.29 days 1972 May 8.15	Irregular	-	1972 Apr 6.2	65.52	91.57	6730	231	473	0.018	70
D	SRET 1†	1972-25B	1972 Apr 4.86 693 days? 1974 Feb 26?	Octahedron 15.4	0.56 dia	1972 May 1.0	65.6	704.70	26232	458	39250	0.739	-
D	Fragments	1972-25E,F											

* 1972-24D ejected from 1972-24A on 1972 Apr 13

† French "Satellite for Research on Environment and Technology".

	Name		Launch date, lifetime and descent date	Shape and weight (kg)	Size (m)	Date of orbital determination	Orbital inclination (deg)	Nodal period (min)	Semi-major axis (km)	Perigee height (km)	Apogee height (km)	Orbital eccentricity	Argument of perigee (deg)
D R	Cosmos 484	1972-26A	1972 Apr 6.34 11.87 days 1972 Apr 18.21	Sphere-cylinder 5900?	5.9 long 2.4 dia	1972 Apr 7.3	81.30	88.73	6588	196	224	0.002	302
D	Cosmos 484 rocket	1972-26B	1972 Apr 6.34 2.08 days 1972 Apr 8.42	Cylinder 2500?	7.5 long 2.6 dia	1972 Apr 6.9	81.31	88.58	6581	184	221	0.003	333
D	Solar Radiation Package A	1972-26C	1972 Apr 6.34 12.52 days 1972 Apr 18.86	Ellipsoid 200?	0.9 long 1.9 dia	1972 Apr 18.6	81.30	88.01	6553	170	179	0.001	-
D R	Intercosmos 6	1972-27A	1972 Apr 7.42 4.0 days 1972 Apr 11.4	Cylinder? 5700? (payload 1070)	5.0 long 2.4 dia	1972 Apr 7.7	51.78	88.94	6604	203	248	0.003	24
D	Intercosmos 6 rocket	1972-27B	1972 Apr 7.42 2.85 days 1972 Apr 10.27	Cylinder 2500?	7.5 long 2.6 dia	1972 Apr 7.8	51.73	88.79	6596	192	244	0.004	4
D	Cosmos 485	1972-28A	1972 Apr 11.46 140.77 days 1972 Aug 30.23	Ellipsoid 400?	1.8 long 1.2 dia	1972 Apr 11.6 1972 Jun.16.0	70.99 70.99	92.05 91.14	6753 6709	271 254	479 408	0.015 0.012	87 -
D	Cosmos 485 rocket	1972-28B	1972 Apr 11.46 65.90 days 1972 Jun 16.36	Cylinder 1500?	8 long 1.65 dia	1972 Apr 11.5	70.98	91.85	6743	276	453	0.013	87
D	Prognoz 1	1972-29A	1972 Apr 14.04 9 years 1981 Apr?	Spheroid + 4 vanes 845	1.8 dia	1972 Apr 14.1 1972 May 1.0	64.92 65.0	91.22 5782.1	6713 106714	224 1005	446 199 667	0.016 0.931	60 -
D	Prognoz 1 launcher rocket	1972-29B	1972 Apr 14.04 27.56 days 1972 May 11.60	Cylinder 2500?	7.5 long 2.6 dia	1972 Apr 14.8	64.97	91.20	6712	227	440	0.016	61

1972-29 continued on page 294

	Name		Launch date, lifetime and descent date	Shape and weight (kg)	Size (m)	Date of orbital determination	Orbital inclination (deg)	Nodal period (min)	Semi major axis (km)	Perigee height (km)	Apogee height (km)	Orbital eccentricity	Argument of perigee (deg)
D	Prognoz 1 launcher	1972-29C	1972 Apr 14.04 33.29 days 1972 May 17.33	Irregular	-	1972 Apr 15.9	64.96	91.40	6 722	228	459	0.017	65
D	Prognoz 1 rocket	1972-29F	1972 Apr 14.04 9 years 1981 Apr?	Cylinder 440	2.0 long 2.0 dia		Orbit similar to 1972-29A						
D	Fragments	1972-29D,E											
D R	Cosmos 486	1972-30A	1972 Apr 14.34 12.9 days 1972 Apr 27.2	Sphere-cylinder 6300?	6.5 long? 2.4 dia	1972 Apr 15.9 1972 Apr 20.3	81.33 81.33	88.64 89.09	6 584 6 606	178 186	234 270	0.004 0.006	67 65
D	Cosmos 486 rocket	1972-30B	1972 Apr 14.34 3.59 days 1972 Apr 17.93	Cylinder 2500?	7.5 long 2.6 dia	1972 Apr 14.4	81.32	89.04	6 604	209	243	0.003	21
D	Cosmos 486 engine	1972-30F	1972 Apr 14.34 14 days 1972 Apr 28	Cone 600? full	1.5 long? 2 dia?	1972 Apr 28.0	81.25	88.83	6 594	188	243	0.004	1
D	Fragments	1972-30C-E											
D R	Apollo 16**	1972-31A	1972 Apr 16.75 11.07 days	Cone-cylinder	11.15 long 3.91 dia	1972 Apr 16.8 1972 Apr 17.0	32.56 33.2	87.82 26320	6 552 292 520	169 200	178 572 080	0.0007 0.977	-* 30*
3M			1972 Apr 27.82	30358 full			In selenocentric orbit 1972 Apr 19.85 to Apr 25.09						
D	Saturn IVB [Saturn 511]	1972-31B	1972 Apr 16.75 3.13 days	Cylinder 13970	18.7 long 6.6 dia	1972 Apr 16.8 1972 Apr 17.0	32.56 33.2	87.82 26320	6 552 292 520	169 200	178 572 080	0.0007 0.977	-* 30*
			1972 Apr 19.88				Crashed on Moon 1972 Apr 19.88						
D	LEM 11 descent stage	1972-31E	1972 Apr 16.75 4.35 days 1972 Apr 21.10	Octagon + legs. 11398 2759 empty	1.57 high 3.13 wide	1972 Apr 17.0	33.2	26320	292 520	200	572 080	0.977	30*
							Entered selenocentric orbit 1972 Apr 19.85 Landed on Moon 1972 Apr 21.10						
	LEM 11*** ascent stage	1972-31C	1972 Apr 16.75 Indefinite	Box + tanks 5040 full 2134 empty	2.52 high 3.76 wide 3.13 deep	1972 Apr 17.0	33.2	26320	292 520	200	572 080	0.977	30*
							On Moon's surface 1972 Apr 21.10 to Apr 24.06 Now in selenocentric orbit.						

** Apollo attached to LEM, separated from Saturn IVB on Apr 16.93.
*** LEM with two crew members separated from Apollo on Apr 20.76.
Ascent stage relaunched from Moon Apr 24.06; briefly docked with Apollo Apr 24.14

* Approximate orbits.

	Name		Launch date, lifetime and descent date	Shape and weight (kg)	Size (m)	Date of orbital determination	Orbital inclination (deg)	Nodal period (min)	Semi major axis (km)	Perigee height (km)	Apogee height (km)	Orbital eccentricity	Argument of perigee (deg)
D	[Thorad Agena D]	1972-32A	1972 Apr 19.91 22.77 days 1972 May 12.68	Cylinder 2000?	8 long? 1.5 dia	1972 Apr 20.5	81.48	88.85	6594	155	277	0.009	171
D	Cosmos 487	1972-33A	1972 Apr 21.50 155.59 days 1972 Sep 24.09	Ellipsoid 400?	1.8 long 1.2 dia	1972 Apr 23.1	70.97	92.29	6765	268	505	0.018	86
						1972 Jul 1.0	70.97	91.43	6723	255	435	0.013	-
D	Cosmos 487 rocket	1972-33B	1972 Apr 21.50 73.04 days 1972 Jul 3.54	Cylinder 1500?	8 long 1.65 dia	1972 Apr 23.9	70.96	92.14	6757	267	491	0.017	87
D	Fragment	1972-33C											
D	Cosmos 488	1972-34A	1972 May 5.48 12.7 days 1972 May 18.2	Sphere-cylinder 6300?	6.5 long? 2.4 dia	1972 May 6.0	65.41	89.50	6629	207	294	0.007	65
R						1972 May 14.6	65.41	89.79	6643	176	353	0.013	44
D	Cosmos 488 rocket	1972-34B	1972 May 5.48 6.12 days 1972 May 11.60	Cylinder 2500?	7.5 long 2.6 dia	1972 May 6.0	65.40	89.40	6624	201	290	0.007	54
D	Cosmos 488 engine*	1972-34E	1972 May 5.48 18 days 1972 May 23	Cone 600? full	1.5 long? 2 dia?	1972 May 17.2	65.41	89.65	6636	178	338	0.012	-
D	Fragments	1972-34C,D,F-H											
	Cosmos 489	1972-35A	1972 May 6.47 1200 years	Cylinder + boom? 700?	1.4 long 2.0 dia	1972 May 6.9	74.02	104.82	7364	969	1002	0.002	282
	Cosmos 489 rocket	1972-35B	1972 May 6.47 600 years	Cylinder 2200?	7.4 long 2.4 dia	1972 May 8.1	74.02	104.57	7356	965	991	0.002	286

* 1972-34E ejected from 1972-34A about 1972 May 17.

	Name		Launch date, lifetime and descent date	Shape and weight (kg)	Size (m)	Date of orbital determination	Orbital inclination (deg)	Nodal period (min)	Semi major axis (km)	Perigee height (km)	Apogee height (km)	Orbital eccentricity	Argument of perigee (deg)
D R	Cosmos 490	1972-36A	1972 May 17.43 11.73 days 1972 May 29.16	Sphere-cylinder 5900?	5.9 long 2.4 dia	1972 May 17.8	65.42	89.39	6623	205	285	0.006	63
D	Cosmos 490 rocket	1972-36B	1972 May 17.43 5.77 days 1972 May 23.20	Cylinder 2500?	7.5 long 2.6 dia	1972 May 17.8	65.40	89.30	6619	204	277	0.005	58
D	Cosmic Ray Package C*	1972-36D	1972 May 17.43 20.35 days 1972 Jun 6.78	Ellipsoid 200?	0.9 long 1.9 dia	1972 May 27.8	65.39	89.19	6613	202	268	0.005	59
D	Fragment	1972-36C											
D	Molniya 2B	1972-37A	1972 May 19.61 1768 days 1977 Mar 22	Windmill + 6 vanes 1250?	4.2 long? 1.6 dia?	1972 May 24.6 1972 Jul 1.0 1973 Jul 1.0	65.42 65.42 65.8	705.11 720.11 716.19	26243 26614 26517	440 408 251	39290 40064 40027	0.740 0.745 0.750	285 - -
D	Molniya 2B launcher rocket	1972-37B	1972 May 19.61 20.03 days 1972 Jun 8.64	Cylinder 2500?	7.5 long 2.6 dia	1972 May 19.8	65.41	90.85	6695	217	417	0.015	62
D	Molniya 2B launcher	1972-37C	1972 May 19.61 20.77 days 1972 Jun 9.38	Irregular -	-	1972 May 19.8	65.40	91.21	6713	207	463	0.019	64
D	Molniya 2B rocket	1972-37G	1972 May 19.61 1839 days 1977 Jun 1	Cylinder 440	2.0 long 2.0 dia	1972 Jul 1.0 1973 Jun 1.0	65.3 65.7	704.07 703.38	26216 26198	453 323	39223 39317	0.739 0.744	- -
D	Fragments	1972-37D-F											

* 1972-36D ejected from 1972-36A about 1972 May 27.

	Name		Launch date, lifetime and descent date	Shape and weight (kg)	Size (m)	Date of orbital determination	Orbital inclination (deg)	Nodal period (min)	Semi major axis (km)	Perigee height (km)	Apogee height (km)	Orbital eccentricity	Argument of perigee (deg)
D R	Cosmos 491	1972-38A	1972 May 25.28 13.9 days 1972 Jun 8.1	Sphere-cylinder 6300?	6.5 long? 2.4 dia	1972 May 26.6 1972 May 30.4	64.98 64.98	88.95 89.92	6601 6649	177 173	269 369	0.007 0.015	57 54
D	Cosmos 491 rocket	1972-38B	1972 May 25.28 5.47 days 1972 May 30.75	Cylinder 2500?	7.5 long 2.6 dia	1972 May 26.3	64.96	89.14	6611	194	271	0.006	24
D	Cosmos 491 engine*	1972-38E	1972 May 25.28 17.25 days 1972 Jun 11.53	Cone 600? full	1.5 long? 2 dia?	1972 Jun 7.4	64.97	89.49	6628	173	327	0.012	51
D	Fragments	1972-38C,D,F											
D	[Thorad Agena D]	1972-39A	1972 May 25.78 10.20 days 1972 Jun 4.98	Cylinder 2000?	8 long? 1.5 dia	1972 May 26.0	96.34	89.17	6610	158	305	0.011	143
D R	Cosmos 492	1972-40A	1972 Jun 9.30 12.86 days 1972 Jun 22.16	Sphere-cylinder 6300?	6.5 long? 2.4 dia	1972 Jun 9.7 1972 Jun 15.1	64.97 64.97	89.77 89.34	6642 6621	205 173	323 312	0.009 0.010	54 55
D	Cosmos 492 rocket	1972-40B	1972 Jun 9.30 9.10 days 1972 Jun 18.40	Cylinder 2500?	7.5 long 2.6 dia	1972 Jun 10.4	64.96	89.56	6632	201	306	0.008	39
D	Cosmos 492 engine**	1972-40D	1972 Jun 9.30 18.50 days 1972 Jun 27.80	Cone 600? full	1.5 long? 2 dia?	1972 Jun 21.6	64.96	89.35	6621	180	306	0.010	55
D	Fragments	1972-40C,E											

* 1972-38E ejected from 1972-38A on 1972 Jun 7.

** 1972-40D ejected from 1972-40A on 1972 Jun 21.

	Name		Launch date, lifetime and descent date	Shape and weight (kg)	Size (m)	Date of orbital determination	Orbital inclination (deg)	Nodal period (min)	Semi major axis (km)	Perigee height (km)	Apogee height (km)	Orbital eccentricity	Argument of perigee (deg)
	Intelsat 4 F-5	1972-41A	1972 Jun 13.91 >million years	Cylinder 1410 full 720 empty	2.82 long 2.39 dia	1972 Jun 14.0 1972 Jul 23.0	27.00 0.15	651.4 1436.2	24937 42166	548 35782	36570 35794	0.722 0.0001	189* -
	Intelsat 4 F-5 rocket	1972-41B	1972 Jun 13.91 6000 years	Cylinder 1815	8.6 long 3.0 dia	1972 Jun 30.9	27.00	651.4	24937	548	36570	0.722	189
D R	Cosmos 493	1972-42A	1972 Jun 21.27 11.89 days 1972 Jul 3.16	Sphere-cylinder 5700?	5.0 long 2.4 dia	1972 Jun 21.5	64.98	89.25	6617	203	274	0.005	33
D	Cosmos 493 rocket	1972-42B	1972 Jun 21.27 4.44 days 1972 Jun 25.71	Cylinder 2500?	7.5 long 2.6 dia	1972 Jun 22.0	64.98	89.06	6607	195	263	0.005	26
D	Fragment	1972-42C											
	Cosmos 494	1972-43A	1972 Jun 23.39 120 years	Cylinder + paddles? 750?	2 long? 1 dia?	1972 Jun 25.8	74.06	100.83	7175	790	804	0.001	51
	Cosmos 494 rocket	1972-43B	1972 Jun 23.39 100 years	Cylinder 2200?	7.4 long 2.4 dia	1972 Jun 29.5	74.06	100.70	7169	779	803	0.002	25
1d	Fragments	1972-43C-E											
D R	Cosmos 495	1972-44A	1972 Jun 23.48 12.7 days 1972 Jul 6.2	Sphere-cylinder 6300?	6.5 long? 2.4 dia	1972 Jun 24.5 1972 Jun 27.7	65.41 65.41	89.30 88.90	6618 6598	202 150	278 290	0.006 0.011	48 48
D	Cosmos 495 rocket	1972-44B	1972 Jun 23.48 5.61 days 1972 Jun 29.09	Cylinder 2500?	7.5 long 2.6 dia	1972 Jun 24.5	65.40	89.13	6610	195	268	0.006	40
D	Cosmos 495 engine**	1972-44D	1972 Jun 23.48 14.98 days 1972 Jul 8.46	Cone 600? full	1.5 long? 2 dia?	1972 Jul 7.0	65.41	88.98	6602	165	283	0.009	-
D	Fragments	1972-44C,E											

** 1972-44D ejected from 1972-44A about 1972 Jul 5.

* Approximate orbit.

	Name		Launch date, lifetime and descent date	Shape and weight (kg)	Size (m)	Date of orbital determination	Orbital inclination (deg)	Nodal period (min)	Semi major axis (km)	Perigee height (km)	Apogee height (km)	Orbital eccentricity	Argument of perigee (deg)
D R	Cosmos 496	1972-45A	1972 Jun 26.62 6.0 days 1972 Jul 2.6	Sphere-cylinder + 2 wings 6570?	7.5 long 2.2 dia	1972 Jun 27.6	51.61	89.53	6632	187	321	0.010	85
D	Cosmos 496 rocket	1972-45B	1972 Jun 26.62 5.96 days 1972 Jul 2.58	Cylinder 2500?	7.5 long 2.6 dia	1972 Jun 26.9	51.59	89.49	6630	185	319	0.010	77
D	Fragment	1972-45C											
D	Prognoz 2	1972-46A	1972 Jun 29.16 10.3 years 1982 Oct?	Spheroid + 4 vanes 845	1.8 dia?		Initial orbit similar to 1972-46C						
						1972 Aug 1.0	65.3	5849.2	107 539	517	201 804	0.936	-
D	Prognoz 2 launcher rocket	1972-46B	1972 Jun 29.16 33.51 days 1972 Aug 1.67	Cylinder 2500?	7.5 long 2.6 dia	1972 Jul 1.1	64.96	91.08	6707	230	427	0.015	64
D P	Prognoz 2* launcher	1972-46C	1972 Jun 29.16 45.08 days 1972 Aug 13.24	Irregular	-	1972 Jul 1.6	64.94	91.57	6730	235	469	0.017	66
D	Prognoz 2 rocket	1972-46F	1972 Jun 29.16 10.3 years 1982 Oct?	Cylinder 440	2.0 long 2.0 dia		Orbit similar to 1972-46A second orbit						
D	Fragments	1972-46D,E											
D	Intercosmos 7	1972-47A	1972 Jun 30.25 97.40 days 1972 Oct 5.65	Ellipsoid + 6 panels? 400?	1.8 long 1.2 dia	1972 Jul 1.8	48.41	92.60	6784	260	551	0.021	113
D	Intercosmos 7 rocket	1972-47B	1972 Jun 30.25 90.80 days 1972 Sep 29.05	Cylinder 1500?	8 long 1.65 dia	1972 Jul 1.8	48.42	92.35	6771	265	521	0.019	114
D	Fragment	1972-47C											

* Pieces (spheres) picked up in Australia.

	Name		Launch date, lifetime and descent date	Shape and weight (kg)	Size (m)	Date of orbital determination	Orbital inclination (deg)	Nodal period (min)	Semi major axis (km)	Perigee height (km)	Apogee height (km)	Orbital eccentricity	Argument of perigee (deg)
D	Cosmos 497	1972-48A	1972 Jun 30.39 494.80 days 1973 Nov 7.19	Ellipsoid 400?	1.8 long 1.2 dia	1972 Jul 2.2 1972 Dec 16.5 1973 Jun 1.0	70.98 70.98 70.98	95.23 93.86 92.37	6908 6841 6769	271 265 252	788 661 529	0.037 0.029 0.020	80 - -
D	Cosmos 497 rocket	1972-48B	1972 Jun 30.39 259.88 days 1973 Mar 17.27	Cylinder 1500?	8 long 1.65 dia	1972 Jul 2.2 1972 Oct 1.0 1973 Jan 1.0	71.00 70.99 70.98	95.05 93.88 92.27	6899 6842 6764	273 263 251	768 664 520	0.036 0.029 0.020	80 - -
D	Fragment	1972-48C											
	Meteor 12	1972-49A	1972 Jun 30.79 500 years	Cylinder + 2 vanes 2200?	5 long? 1.5 dia?	1972 Jul 2.5	81.22	102.95	7275	889	905	0.001	30
	Meteor 12 rocket	1972-49B	1972 Jun 30.79 400 years	Cylinder 1440	3.8 long 2.6 dia	1972 Jul 2.9	81.22	103.05	7280	865	939	0.004	177
D	Cosmos 498	1972-50A	1972 Jul 5.40 143.26 days 1972 Nov 25.66	Ellipsoid 400?	1.8 long 1.2 dia	1972 Jul 6.7 1972 Sep 16.0	70.95 70.95	92.12 91.26	6757 6715	267 256	490 418	0.017 0.012	70 -
D	Cosmos 498 rocket	1972-50B	1972 Jul 5.40 83.80 days 1972 Sep 27.20	Cylinder 1500?	8 long 1.65 dia	1972 Jul 5.8	70.95	91.95	6748	273	467	0.014	81
D R	Cosmos 499	1972-51A	1972 Jul 6.45 10.88 days 1972 Jul 17.33	Sphere-cylinder 6300?	6.5 long? 2.4 dia	1972 Jul 6.7	51.77	89.31	6622	204	283	0.006	39
D	Cosmos 499 rocket	1972-51B	1972 Jul 6.45 3.77 days 1972 Jul 10.22	Cylinder 2500?	7.5 long 2.6 dia	1972 Jul 7.1	51.76	89.18	6615	205	269	0.005	44
D	Cosmos 499 engine†	1972-51E	1972 Jul 6.45 14.13 days 1972 Jul 20.58	Cone 600? full	1.5 long? 2 dia?	1972 Jul 17.0	51.76	88.84	6598	171	269	0.007	76
D	Fragments	1972-51C,D											

† 1972-51E ejected from 1972-51A on 1972 Jul 16.

	Name		Launch date, lifetime and descent date	Shape and weight (kg)	Size (m)	Date of orbital determination	Orbital inclination (deg)	Nodal period (min)	Semi major axis (km)	Perigee height (km)	Apogee height (km)	Orbital eccentricity	Argument of perigee (deg)
D	[Titan 3D]	1972-52A	1972 Jul 7.74 68 days 1972 Sep 13	Cylinder 13300? full	15 long 3.0 dia	1972 Jul 8.0	96.88	88.77	6591	174	251	0.006	162
D	Titan 3D rocket	1972-52B	1972 Jul 7.74 1.92 days 1972 Jul 9.66	Cylinder 1900	6 long 3.0 dia	1972 Jul 8.9	96.88	88.07	6556	142	213	0.005	158
D	Capsule	1972-52C	1972 Jul 7.74 2129 days 1978 May 6	Octagon 60?	0.3 long? 0.9 dia?	1972 Jul 9.6	96.15	94.66	6879	497	504	0.0005	238
D	Cosmos 500	1972-53A	1972 Jul 10.68 2819 days 1980 Mar 29	Cylinder + paddles? 900?	2 long? 1 dia?	1972 Jul 11.5	74.07	95.18	6905	505	549	0.003	346
D	Cosmos 500 rocket	1972-53B	1972 Jul 10.68 2848 days 1980 Apr 27	Cylinder 2200?	7.4 long 2.4 dia	1972 Jul 12.2	74.06	95.08	6900	500	544	0.003	345
D	Fragments	1972-53C-F											
D	Cosmos 501	1972-54A	1972 Jul 12.25 666.63 days 1974 May 9.88	Ellipsoid 400?	1.8 long 1.2 dia	1972 Jul 12.5 1972 Dec 16.5 1973 Sep 1.0	48.52 48.45 48.38	109.21 105.60 100.01	7573 7406 7140	221 213 206	2168 1843 1318	0.129 0.110 0.078	105 - -
D	Cosmos 501 rocket	1972-54B	1372 Jul 12.25 333 days 1973 Jun 10	Cylinder 1500?	8 long 1.65 dia	1972 Jul 13.4 1972 Oct 16.5 1973 Feb 15.0	48.45 48.45 48.35	108.75 105.28 99.68	7550 7390 7124	215 212 207	2129 1812 1284	0.127 0.108 0.076	103 - -
D	Fragments	1972-54C-J											

	Name		Launch date, lifetime and descent date	Shape and weight (kg)	Size (m)	Date of orbital determination	Orbital inclination (deg)	Nodal period (min)	Semi major axis (km)	Perigee height (km)	Apogee height (km)	Orbital eccentricity	Argument of perigee (deg)
D R	Cosmos 502	1972-55A	1972 Jul 13.61 11.7 days 1972 Jul 25.3	Sphere-cylinder 5900?	5.9 long 2.4 dia	1972 Jul 14.3	65.40	89.16	6611	204	262	0.004	47
D	Cosmos 502 rocket	1972-55B	1972 Jul 13.61 5 days 1972 Jul 18	Cylinder 2500?	7.5 long 2.6 dia	1972 Jul 14.3	65.40	89.00	6603	202	248	0.004	24
D	Capsule*	1972-55E	1972 Jul 13.61 13 days 1972 Jul 26	Ellipsoid 200?	0.9 long 1.9 dia	1972 Jul 26.7	65.39	88.54	6580	171	233	0.005	2
D	Fragments	1972-55C,D											
D R	Cosmos 503	1972-56A	1972 Jul 19.58 12.70 days 1972 Aug 1.28	Sphere-cylinder 6300?	6.5 long? 2.4 dia	1972 Jul 19.9 1972 Jul 25.5	65.43 65.41	89.40 89.33	6623 6620	202 169	288 314	0.006 0.011	59 51
D	Cosmos 503 rocket	1972-56B	1972 Jul 19.58 7.49 days 1972 Jul 27.07	Cylinder 2500?	7.5 long 2.6 dia	1972 Jul 20.1	65.40	89.32	6619	207	275	0.005	53
D	Cosmos 503 engine**	1972-56E	1972 Jul 19.58 14.70 days 1972 Aug 3.28	Cone 600? full	1.5 long? 2 dia?	1972 Jul 31.4	65.46	89.08	6607	163	295	0.010	53
D	Fragments	1972-56C,D											

* 1972-55E ejected from 1972-55A on 1972 Jul 25.
** 1972-56E ejected from 1972-56A on 1972 Jul 31.

Name		Launch date, lifetime and descent date	Shape and weight (kg)	Size (m)	Date of orbital determination	Orbital inclination (deg)	Nodal period (min)	Semi major axis (km)	Perigee height (km)	Apogee height (km)	Orbital eccentricity	Argument of perigee (deg)
Cosmos 504	1972-57A	1972 Jul 20.75 5000 years	Spheroid 40?	1.0 long? 0.8 dia?	1972 Jul 24.1	74.02	114.03	7789	1324	1498	0.011	89
Cosmos 505	1972-57B	1972 Jul 20.75 6000 years	Spheroid 40?	1.0 long? 0.8 dia?	1972 Jul 29.9	74.03	114.37	7804	1354	1498	0.009	84
Cosmos 506	1972-57C	1972 Jul 20.75 7000 years	Spheroid 40?	1.0 long? 0.8 dia?	1972 Jul 25.8	74.02	114.70	7819	1384	1498	0.007	89
Cosmos 507	1972-57D	1972 Jul 20.75 8000 years	Spheroid 40?	1.0 long? 0.8 dia?	1972 Jul 24.2	74.02	115.03	7834	1414	1498	0.005	94
Cosmos 508	1972-57E	1972 Jul 20.75 9000 years	Spheroid 40?	1.0 long? 0.8 dia?	1972 Jul 24.2	74.02	115.37	7850	1446	1497	0.003	95
Cosmos 509	1972-57F	1972 Jul 20.75 10000 years	Spheroid 40?	1.0 long? 0.8 dia?	1972 Jul 26.2	74.02	115.73	7866	1475	1501	0.002	124
Cosmos 510	1972-57G	1972 Jul 20.75 10000 years	Spheroid 40?	1.0 long? 0.8 dia?	1972 Jul 29.7	74.02	116.10	7883	1497	1512	0.001	235
Cosmos 511	1972-57H	1972 Jul 20.75 10000 years	Spheroid 40?	1.0 long? 0.8 dia?	1972 Jul 26.2	74.03	116.48	7900	1496	1548	0.003	246
Cosmos 504 rocket	1972-57J	1972 Jul 20.75 20000 years	Cylinder 2200?	7.4 long 2.4 dia	1972 Jul 22.9	74.03	117.08	7927	1500	1598	0.006	265

	Name		Launch date, lifetime and descent date	Shape and weight (kg)	Size (m)	Date of orbital determination	Orbital inclination (deg)	Nodal period (min)	Semi major axis (km)	Perigee height (km)	Apogee height (km)	Orbital eccentricity	Argument of perigee (deg)
	Landsat 1 (ERTS*)	1972-58A	1972 Jul 23.75 100 years	Conical skeleton + 2 paddles 816	3.0 long 1.45 dia	1972 Jul 30.9	99.12	103.27	7290	903	921	0.001	284
	Landsat 1 second stage **	1972-58B	1972 Jul 23.75 disintegrated	Cylinder + annulus 350?	6.4 long 1.52 and 2.44 dia	1972 Jul 30.2	98.51	100.34	7152	637	910	0.019	221
130d	Fragments	1972-58C-HZ											
D R	Cosmos 512	1972-59A	1972 Jul 28.43 11.7 days 1972 Aug 9.1	Sphere-cylinder 5700?	5.0 long 2.4 dia	1972 Jul 30.2	65.39	89.25	6616	203	273	0.005	47
D	Cosmos 512 rocket	1972-59B	1972 Jul 28.43 4.38 days 1972 Aug 1.81	Cylinder 2500?	7.5 long 2.6 dia	1972 Jul 29.3	65.38	89.06	6607	190	267	0.006	35
D R	Cosmos 513	1972-60A	1972 Aug 2.35 12.86 days 1972 Aug 15.21	Sphere-cylinder 6300?	6.5 long? 2.4 dia	1972 Aug 3.5 1972 Aug 4.1	64.97 64.97	89.73 89.36	6640 6621	203 174	320 312	0.009 0.010	44 56
D	Cosmos 513 rocket	1972-60B	1972 Aug 2.35 6.92 days 1972 Aug 9.27	Cylinder 2500?	7.5 long 2.6 dia	1972 Aug 3.2	64.98	89.56	6631	203	303	0.008	39
D	Cosmos 513 engine †	1972-60E	1972 Aug 2.35 22 days 1972 Aug 24	Cone 600? full	1.5 long? 2 dia?	1972 Aug 15.1	64.95	89.33	6620	161	322	0.012	55
D	Fragments	1972-60C,D,F											

* Earth Resources Technology Satellite.

† 1972-60E ejected from 1972-60A on 1972 Aug 14.

** Landsat 1 second stage disintegrated on 1975 May 22.77 near 33.3 deg South, 45.1 deg East. 1972-58B is now a fragment.

	Name		Launch date, lifetime and descent date	Shape and weight (kg)	Size (m)	Date of orbital determination	Orbital inclination (deg)	Nodal period (min)	Semi major axis (km)	Perigee height (km)	Apogee height (km)	Orbital eccentricity	Argument of perigee (deg)
D	Explorer 46 (MTS*)	1972-61A	1972 Aug 13.63 2637 days 1979 Nov 2	Cylinder + 4 panels 175	3.2 long 0.5 dia?	1972 Sep 4.5	37.70	97.65	7030	492	811	0.023	267
D	Fragments	1972-61B,C											
	Cosmos 514	1972-62A	1972 Aug 16.64 1200 years	Cylinder? 700?	1.3 long? 1.9 dia?	1972 Sep 3.1	82.97	104.43	7345	958	975	0.001	100
	Cosmos 514 rocket	1972-62B	1972 Aug 16.64 600 years	Cylinder 2200?	7.4 long 2.4 dia	1972 Sep 4.8	82.97	104.37	7342	952	975	0.002	94
	Fragments	1972-62C,D											
D R	Cosmos 515	1972-63A	1972 Aug 18.42 12.76 days 1972 Aug 31.18	Sphere-cylinder 6300?	6.5 long? 2.4 dia	1972 Aug 19.4	72.86	89.66	6635	189	325	0.010	50
D	Cosmos 515 rocket	1972-63B	1972 Aug 18.42 4.81 days 1972 Aug 23.23	Cylinder 2500?	7.5 long 2.6 dia	1972 Aug 19.5	72.88	89.13	6609	207	254	0.004	31
D	Cosmos 515 engine**	1972-63D	1972 Aug 18.42 17 days 1972 Sep 4	Cone 600? full	1.5 long? 2 dia?	1972 Sep 1.0	72.87	88.82	6593	174	256	0.006	-
D	Fragments	1972-63C,E											
D	Denpa† [Mu 4S]	1972-64A	1972 Aug 19.11 2830 days 1980 May 19	Octagonal cylinder 75	0.68 long 0.71 dia	1972 Sep 2.1	31.03	156.85	9646	245	6291	0.313	175
D	Denpa rocket	1972-64B	1972 Aug 19.11 3174 days 1981 Apr 28	Sphere-cone 90	1.86 long 0.79 dia	1973 Jan 2.3	31.02	157.55	9675	235	6358	0.316	275

* Meteoroid Technology Satellite.
** 1972-63D ejected from 1972-63A on 1972 Aug 29.

† Japanese satellite.

	Name		Launch date, lifetime and descent date	Shape and weight (kg)	Size (m)	Date of orbital determination	Orbital inclination (deg)	Nodal period (min)	Semi major axis (km)	Perigee height (km)	Apogee height (km)	Orbital eccentricity	Argument of perigee (deg)
	Copernicus (OAO 3)*	1972-65A	1972 Aug 21.44 500 years	Octagonal cylinder 2220	3.05 long 2.15 dia	1972 Aug 22.1	35.01	99.49	7118	736	744	0.0006	293
	Copernicus rocket	1972-65B	1972 Aug 21.44 200 years	Cylinder 1815	8.6 long 3.0 dia	1972 Sep 4.4	35.02	99.44	7116	695	780	0.006	350
D	Fragment	1972-65C											
	Cosmos 516**	1972-66A	1972 Aug 21.44 600 years	Cone-cylinder	6 long? 2 dia?	1972 Sep 1.5 1972 Oct 1.7	64.98 64.82	89.64 104.57	6635 7353	251 920	263 1030	0.0009 0.007	260 343
D	Cosmos 516 rocket	1972-66B	1972 Aug 21.44 35.26 days 1972 Sep 25.70	Cylinder 1500?	8 long? 2.5 dia?	1972 Sep 23.1	64.98	89.44	6625	239	255	0.001	312
D	Cosmos 516 platform	1972-66C	1972 Aug 21.44 60 days 1972 Oct 20	Irregular	-	1972 Sep 24.9	64.98	89.52	6629	243	259	0.001	270
D R	Cosmos 517	1972-67A	1972 Aug 30.35 11.86 days 1972 Sep 11.21	Sphere-cylinder 5700?	5.0 long 2.4 dia	1972 Aug 31.1	64.98	89.42	6624	204	288	0.006	39
D	Cosmos 517 rocket	1972-67B	1972 Aug 30.35 4.51 days 1972 Sep 3.86	Cylinder 2500?	7.5 long 2.6 dia	1972 Aug 31.2	65.00	89.15	6611	197	268	0.005	25
D	Fragment	1972-67C											
D	[Titan 3B Agena D]	1972-68A	1972 Sep 1.74 29 days 1972 Sep 30	Cylinder 3000?	8 long? 1.5 dia	1972 Sep 2.0	110.50	89.71	6638	140	380	0.018	147

* Orbiting Astronomical Observatory

** 1972-66B and 66C attached to 1972-66A until orbit change about 1972 Sep 21.93

	Name		Launch date, lifetime and descent date	Shape and weight (kg)	Size (m)	Date of orbital determination	Orbital inclination (deg)	Nodal period (min)	Semi major axis (km)	Perigee height (km)	Apogee height (km)	Orbital eccentricity	Argument of perigee (deg)
	Triad 1 [Scout]	1972-69A	1972 Sep 2.74 90 years	Dumb-bell 94	7.3 long 0.59 dia	1972 Sep 4.5	90.14	100.68	7168	716	863	0.010	352
	Altair rocket	1972-69B	1972 Sep 2.74 70 years	Cylinder 24	1.50 long 0.46 dia	1972 Sep 4.2	90.13	100.70	7169	738	843	0.007	338
	Fragment	1972-69C											
D R	Cosmos 518	1972-70A	1972 Sep 15.40 8.85 days 1972 Sep 24.25	Sphere-cylinder 5900?	5.9 long 2.4 dia	1972 Sep 16.2	72.84	89.64	6634	204	307	0.008	59
D	Cosmos 518 rocket	1972-70B	1972 Sep 15.40 7.10 days 1972 Sep 22.50	Cylinder 2500?	7.5 long 2.6 dia	1972 Sep 16.2	72.83	89.46	6625	199	295	0.007	47
D	Capsule*	1972-70C	1972 Sep 15.40 12 days 1972 Sep 27	Ellipsoid 200?	0.9 long 1.9 dia	1972 Sep 23.7	72.85	89.30	6617	200	278	0.006	37
D R	Cosmos 519	1972-71A	1972 Sep 16.35 9.90 days 1972 Sep 26.25	Sphere-cylinder 6300?	6.5 long? 2.4 dia	1972 Sep 18.2	71.33	90.19	6662	207	360	0.012	47
D	Cosmos 519 rocket	1972-71B	1972 Sep 16.35 8.35 days 1972 Sep 24.70	Cylinder 2500?	7.5 long 2.6 dia	1972 Sep 18.0	71.36	89.56	6630	201	302	0.008	41
D	Cosmos 519 engine**	1972-71D	1972 Sep 16.35 28 days 1972 Oct 14	Cone 600? full	1.5 long? 2 dia?	1972 Oct 1.0	71.2	89.76	6640	200	323	0.009	-
D	Fragments	1972-71C,E											

*1972-70C ejected from 1972-70A on 1972 Sep 23.
**1972-71D ejected from 1972-71A on 1972 Sep 25.

	Name		Launch date, lifetime and descent date	Shape and weight (kg)	Size (m)	Date of orbital determination	Orbital inclination (deg)	Nodal period (min)	Semi major axis (km)	Perigee height (km)	Apogee height (km)	Orbital eccentricity	Argument of perigee (deg)
	Cosmos 520	1972-72A	1972 Sep 19.81 30 years	Windmill + 6 vanes 1250?	4.2 long? 1.6 dia?	1972 Sep 20.0 1972 Dec 30.7	62.88 62.89	93.52 715.02	6826 26488	227 750	669 39470	0.032 0.731	113 319
D	Cosmos 520 launcher rocket	1972-72B	1972 Sep 19.81 30.64 days 1972 Oct 20.45	Cylinder 2500?	7.5 long 2.6 dia	1972 Sep 21.3	62.89	92.92	6796	187	649	0.034	118
D	Cosmos 520 launcher	1972-72C	1972 Sep 19.81 47 days 1972 Nov 5	Irregular	-	1972 Sep 22.4	62.82	92.78	6789	214	607	0.029	119
	Cosmos 520 rocket	1972-72E	1972 Sep 19.81 30 years	Cylinder 440	2.0 long 2.0 dia	1973 Jan 1.0	62.89	706.77	26283	750	39059	0.729	-
D	Fragment	1972-72D											
	Explorer 47 (Imp 9)	1972-73A	1972 Sep 23.06 > million years	16-sided cylinder 390 full	1.58 long 1.35 dia	1972 Sep 23.1 1972 Sep 25.6	28.63 17.23	7349.9 17670	125 223 224 740	260 201 100	237 429 235 600	0.947 0.077	141 76
D	Explorer 47 second stage	1972-73B	1972 Sep 23.06 3 days 1972 Sep 26	Cylinder 350?	4.9 long 1.43 dia	1972 Sep 23.1	28.8	90.02	6660	153	411	0.019	-
D	Explorer 47 third stage	1972-73C	1972 Sep 23.06 1245 days? 1976 Feb 20?	Sphere-cone 66	1.32 long 0.94 dia	1972 Oct 1.0	28.8	7780.3	130 064	246	247 125	0.949	142
D	Fragments	1972-73D,E											
	Cosmos 521	1972-74A	1972 Sep 29.84 1200 years	Cylinder?	4 long? 2 dia?	1972 Sep 30.2	65.89	104.97	7372	965	1022	0.004	194
	Cosmos 521 rocket	1972-74B	1972 Sep 29.84 600 years	Cylinder 2200?	7.4 long 2.4 dia	1972 Sep 30.3	65.83	104.87	7367	975	1002	0.002	281
	Fragment	1972-74C											

	Name		Launch date, lifetime and descent date	Shape and weight (kg)	Size (m)	Date of orbital determination	Orbital inclination (deg)	Nodal period (min)	Semi major axis (km)	Perigee height (km)	Apogee height (km)	Orbital eccentricity	Argument of perigee (deg)
D	Molniya 2C	1972-75A	1972 Sep 30.85 1930 days 1978 Jan 12	Windmill + 6 vanes? 1250?	4.2 long? 1.6 dia?	1972 Oct 9.2 1972 Nov 1.0	65.63 65.3	703.2 717.50	26192 26550	392 551	39240 39792	0.742 0.739	284 -
D	Molniya 2C launcher rocket	1972-75B	1972 Sep 30.85 26 days 1972 Oct 26	Cylinder 2500?	7.5 long 2.6 dia	1972 Oct 1.3	65.42	91.34	6719	216	466	0.019	68
D	Molniya 2C launcher	1972-75C	1972 Sep 30.85 32.14 days 1972 Nov 1.99	Irregular	-	1972 Oct 3.1	65.39	91.48	6726	230	466	0.017	71
D	Molniya 2C rocket	1972-75D	1972 Sep 30.85 1970 days 1978 Feb 21	Cylinder 440	2.0 long 2.0 dia	1973 Jan 1.0	65.4	700.58	26130	517	38987	0.736	-
	Radcat* [Atlas Burner 2]	1972-76A	1972 Oct 2.84 40 years	Cylinder 208	12.2 long 3.05 dia	1972 Oct 5.7	98.44	99.64	7118	731	749	0.001	265
	Radsat**	1972-76B	1972 Oct 2.84 40 years	726	- 1.7 dia?	1972 Oct 5.7	98.45	99.69	7121	732	753	0.001	245
	Burner 2 rocket	1972-76C	1972 Oct 2.84 50 years	Sphere-cone 66	1.32 long 0.94 dia	1972 Oct 20.9	98.44	99.66	7119	731	751	0.001	202
1d	Fragments	1972-76D, E											
D R	Cosmos 522	1972-77A	1972 Oct 4.50 12.78 days 1972 Oct 17.28	Sphere-cylinder 6300?	6.5 long? 2.4 dia	1972 Oct 6.4 1972 Oct 15.3	72.83 72.84	89.74 89.99	6639 6651	206 178	316 368	0.008 0.014	62 37
D	Cosmos 522 rocket	1972-77B	1972 Oct 4.50 9.66 days 1972 Oct 14.16	Cylinder 2500?	7.5 long 2.6 dia	1972 Oct 6.4	72.84	89.43	6624	198	294	0.007	56

*Radar calibration target.
**Contains experiments on ultraviolet and gamma radiation.

1972-77 continued on page 310

	Name		Launch date, lifetime and descent date	Shape and weight (kg)	Size (m)	Date of orbital determination	Orbital inclination (deg)	Nodal period (min)	Semi major axis (km)	Perigee height (km)	Apogee height (km)	Orbital eccentricity	Argument of perigee (deg)
D	Cosmos 522 engine*	1972-77C	1972 Oct 4.50 19.51 days 1972 Oct 24.01	Cone 600? full	1.5 long? 2 dia?	1972 Oct 16.3	72.82	90.03	6653	180	370	0.014	35
D	Fragment	1972-77D											
D	Cosmos 523	1972-78A	1972 Oct 5.48 153.49 days 1973 Mar 7.97	Ellipsoid 400?	1.8 long 1.2 dia	1972 Oct 8.2 1972 Dec 16.5	71.03 71.03	92.09 91.22	6755 6713	272 257	481 413	0.015 0.012	73 -
D	Cosmos 523 rocket	1972-78B	1972 Oct 5.48 62.92 days 1972 Dec 7.40	Cylinder 1500?	8 long 1.65 dia	1972 Oct 6.0	71.03	91.88	6744	272	460	0.014	81
D	Fragments	1972-78C-M											
D	[Titan 3D]	1972-79A	1972 Oct 10.75 90 days 1973 Jan 8	Cylinder 13300? full	15 long 3.0 dia	1972 Oct 11.5 1972 Oct 21.5	96.47 96.46	88.93 88.84	6599 6594	160 164	281 267	0.009 0.008	186 157
D	Titan 3D rocket	1972-79B	1972 Oct 10.75 2.09 days 1972 Oct 12.84	Cylinder 1900	6 long 3.0 dia	1972 Oct 11.4	96.47	88.59	6582	150	257	0.008	185
	Capsule	1972-79C	1972 Oct 10.75 10000 years	Octagon 60?	0.3 long? 0.9 dia?	1973 Sep 8.9	95.62	114.79	7824	1423	1469	0.003	204
	Fragments	1972-79D,E											
D	Cosmos 524	1972-80A	1972 Oct 11.56 164.64 days 1973 Mar 25.20	Ellipsoid 400?	1.8 long 1.2 dia	1972 Oct 12.8 1973 Jan 1.0	70.99 70.99	92.33 91.37	6767 6721	267 251	512 434	0.018 0.014	86 -
D	Cosmos 524 rocket	1972-80B	1972 Oct 11.56 69.92 days 1972 Dec 20.48	Cylinder 1500?	8 long 1.65 dia	1972 Oct 12.8	70.99	92.08	6754	265	488	0.017	89

*1972-77C ejected from 1972-77A on 1972 Oct 16.

	Name		Launch date, lifetime and descent date	Shape and weight (kg)	Size (m)	Date of orbital determination	Orbital inclination (deg)	Nodal period (min)	Semi major axis (km)	Perigee height (km)	Apogee height (km)	Orbital eccentricity	Argument of perigee (deg)
D	Molniya 1W	1972-81A	1972 Oct 14.26 1844 days 1977 Nov 1	Windmill + 6 vanes 1000?	3.4 long 1.6 dia	1972 Oct 14.4 1972 Oct 16.0 1972 Nov 2.0	65.41 65.3 65.30	91.61 706.18 717.81	6732 26268 26557	226 480 636	481 39300 39722	0.019 0.739 0.736	65 285* 284
D	Molniya 1W launcher rocket	1972-81B	1972 Oct 14.26 21.65 days 1972 Nov 4.91	Cylinder 2500?	7.5 long 2.6 dia	1972 Oct 15.3	65.42	91.30	6717	211	467	0.019	60
D	Molniya 1W launcher	1972-81C	1972 Oct 14.26 33.23 days 1972 Nov 16.49	Irregular	-	1972 Oct 15.2	65.39	91.65	6734	223	488	0.020	72
D	Molniya 1W rocket	1972-81E	1972 Oct 14.26 883.47 days 1975 Mar 16.73	Cylinder 440	2.0 long 2.0 dia	1972 Dec 12.3	65.53	701.40	26152	474	39074	0.738	284
D	Fragment	1972-81D											
	NOAA 2 (ITOS)	1972-82A	1972 Oct 15.72 10000 years	Box 334	1.25 long 1.02 square	1972 Oct 16.1	101.77	115.01	7833	1451	1458	0.0004	223
	Oscar 6	1972-82B	1972 Oct 15.72 10000 years	Rectangular box 16	0.43 x 0.30 x 0.15	1972 Oct 16.4	101.76	115.01	7833	1450	1459	0.0006	239
	NOAA 2 second stage	1972-82C	1972 Oct 15.72 1000 years	Cylinder 350?	4.9 long 1.43 dia	1972 Oct 23.9	102.80	109.36	7575	918	1475	0.037	306
D R	Cosmos 525	1972-83A	1972 Oct 18.50 10.71 days 1972 Oct 29.21	Sphere-cylinder 5900?	5.9 long 2.4 dia	1972 Oct 19.0	65.39	89.25	6616	207	269	0.005	66
D	Cosmos 525 rocket	1972-83B	1972 Oct 18.50 4.25 days 1972 Oct 22.75	Cylinder 2500?	7.5 long 2.6 dia	1972 Oct 19.2	65.39	89.14	6611	198	267	0.005	43

*Approximate orbit.

1972-83 continued on page 312

	Name		Launch date, lifetime and descent date	Shape and weight (kg)	Size (m)	Date of orbital determination	Orbital inclination (deg)	Nodal period (min)	Semi major axis (km)	Perigee height (km)	Apogee height (km)	Orbital eccentricity	Argument of perigee (deg)
D	Capsule*	1972-83C	1972 Oct 18.50 14.47 days 1972 Nov 1.97	Ellipsoid 200?	0.9 long 1.9 dia	1972 Oct 27.3	65.38	89.01	6604	192	260	0.005	38
D	Fragment	1972-83D											
D	Cosmos 526	1972-84A	1972 Oct 25.45 165 days 1973 Apr 8	Ellipsoid 400?	1.8 long 1.2 dia	1972 Oct 26.1 1973 Jan 16.5	70.96 70.96	92.15 91.26	6758 6715	273 256	486 418	0.016 0.012	73 -
D	Cosmos 526 rocket	1972-84B	1972 Oct 25.45 75.16 days 1973 Jan 8.61	Cylinder 1500?	8 long 1.65 dia	1972 Oct 26.1	70.96	91.93	6747	273	465	0.014	71
	Meteor 13	1972-85A	1972 Oct 26.92 500 years	Cylinder + 2 vanes 2200?	5 long? 1.5 dia?	1972 Oct 28.5	81.27	102.57	7257	867	891	0.002	264
	Meteor 13 rocket	1972-85B	1972 Oct 26.92 400 years	Cylinder 1440	3.8 long 2.6 dia	1972 Oct 28.8	81.27	102.67	7262	841	927	0.006	185
D R	Cosmos 527	1972-86A	1972 Oct 31.57 12.7 days 1972 Nov 13.3	Sphere-cylinder 6300?	6.5 long? 2.4 dia	1972 Oct 31.8 1972 Nov 2.3	65.37 65.40	89.62 89.11	6635 6609	207 178	306 284	0.007 0.008	80 47
D	Cosmos 527 rocket	1972-86B	1972 Oct 31.57 7.95 days 1972 Nov 8.52	Cylinder 2500?	7.5 long 2.6 dia	1972 Oct 31.8	65.39	89.51	6629	207	295	0.007	75
D	Cosmos 527 engine**	1972-86E	1972 Oct 31.57 18 days 1972 Nov 18	Cone 600? full	1.5 long? 2 dia?	1972 Nov 12.7	65.43	89.53	6630	181	323	0.011	46
D	Fragments	1972-86C,D,F											

* 1972-83C ejected from 1972-83A on 1972 Oct 26.43
** 1972-86E ejected from 1972-86A on 1972 Nov 12.

Name		Launch date, lifetime and descent date	Shape and weight (kg)	Size (m)	Date of orbital determination	Orbital inclination (deg)	Nodal period (min)	Semi major axis (km)	Perigee height (km)	Apogee height (km)	Orbital eccentricity	Argument of perigee (deg)
Cosmos 528	1972-87A	1972 Nov 1.08 7000 years	Spheroid 40?	1.0 long? 0.8 dia?	1972 Nov 4.6	74.03	114.21	7797	1368	1470	0.007	93
Cosmos 529	1972-87B	1972 Nov 1.08 8000 years	Spheroid 40?	1.0 long? 0.8 dia?	1972 Nov 4.6	74.03	114.61	7815	1404	1470	0.004	96
Cosmos 530	1972-87C	1972 Nov 1.08 6000 years	Spheroid 40?	1.0 long? 0.8 dia?	1972 Nov 3.3	74.03	113.85	7781	1336	1469	0.009	92
Cosmos 531	1972-87D	1972 Nov 1.08 9000 years	Spheroid 40?	1.0 long? 0.8 dia?	1972 Nov 3.9	74.03	114.83	7825	1423	1471	0.003	110
Cosmos 532	1972-87E	1972 Nov 1.08 4000 years	Spheroid 40?	1.0 long? 0.8 dia?	1972 Nov 3.3	74.04	113.49	7764	1302	1470	0.011	90
Cosmos 533	1972-87F	1972 Nov 1.08 5000 years	Spheroid 40?	1.0 long? 0.8 dia?	1972 Nov 4.3	74.03	113.68	7773	1319	1470	0.010	95
Cosmos 534	1972-87G	1972 Nov 1.08 6000 years	Spheroid 40?	1.0 long? 0.8 dia?	1972 Nov 5.5	74.04	114.03	7789	1351	1470	0.008	91
Cosmos 535	1972-87H	1972 Nov 1.08 8000 years	Spheroid 40?	1.0 long? 0.8 dia?	1972 Nov 2.2	74.04	114.42	7806	1385	1471	0.006	102
Cosmos 528 rocket	1972-87J	1972 Nov 1.08 20000 years	Cylinder 2200?	7.4 long 2.4 dia	1972 Nov 3.6	74.03	116.70	7910	1470	1594	0.008	267

	Name		Launch date, lifetime and descent date	Shape and weight (kg)	Size (m)	Date of orbital determination	Orbital inclination (deg)	Nodal period (min)	Semi major axis (km)	Perigee height (km)	Apogee height (km)	Orbital eccentricity	Argument of perigee (deg)
D	Cosmos 536	1972-88A	1972 Nov 3.07 2816 days 1980 Jul 20	Cylinder + paddles? 900?	2 long? 1 dia?	1972 Nov 4.2	74.02	95.27	6909	518	544	0.002	15
D	Cosmos 536 rocket	1972-88B	1972 Nov 3.07 2815 days 1980 Jul 19	Cylinder 2200?	7.4 long 2.4 dia	1972 Nov 4.3	74.02	95.16	6904	508	543	0.002	11
1d	Fragments	1972-88C,D											
	DMSP 3 [Thor Burner 2]	1972-89A	1972 Nov 9.21 80 years	12-faced frustum 195	1.64 long 1.31 to 1.10 dia	1972 Nov 20.6	98.65	101.80	7221	813	872	0.004	200
	Burner 2 rocket	1972-89B	1972 Nov 9.21 60 years	Sphere-cone 66	1.32 long 0.94 dia	1972 Nov 10.7	98.64	101.82	7222	816	871	0.004	221
	Telesat 1 (Anik 1)	1972-90A	1972 Nov 10.05 >million years	Cylinder 562 full 295 empty	1.52 long 1.83 dia	1972 Dec 1.0 1973 Jan 1.0	0.4 0.1	1455.5 1436.0	42543 42164	35822 35780	36508 35791	0.008 0.0001	- -
D	Telesat 1 second stage	1972-90B	1972 Nov 10.05 178.31 days 1973 May 7.36	Cylinder + annulus 350?	6.4 long 1.52 and 2.44 dia	1972 Nov 10.1 1973 Feb 15.0	28.57 28.57	103.78 98.45	7319 7066	203 181	1678 1194	0.101 0.072	183 -
	Telesat 1 third stage	1972-90C	1972 Nov 10.05 20 years?	Sphere-cone 66	1.32 long 0.94 dia	1972 Nov 10.1	28.57	645.3	24734	203	36508	0.734	181*
D	Explorer 48 (SAS 2)	1972-91A	1972 Nov 15.93 1374? days 1976 Aug 20?	Cylinder + 4 paddles 186	1.29 long 0.55 dia	1972 Nov 17.0	1.90	95.20	6916	444	632	0.014	79
D	Explorer 48 rocket	1972-91B	1972 Nov 15.93 2358 days 1979 May 1	Cylinder 24	1.50 long 0.46 dia	1972 Nov 16.7 1973 Aug 27.8	1.84 1.90	95.12 94.68	6912 6890	450 430	618 594	0.013 0.012	71* 76

* Approximate orbits. Telesat 1 is Canadian.

	Name		Launch date, lifetime and descent date	Shape and weight (kg)	Size (m)	Date of orbital determination	Orbital inclination (deg)	Nodal period (min)	Semi major axis (km)	Perigee height (km)	Apogee height (km)	Orbital eccentricity	Argument of perigee (deg)
D	ESRO 4	1972-92A	1972 Nov 22.01 509.20 days 1974 Apr 15.21	Cylinder 114	0.90 long 0.76 dia	1972 Nov 22.9 1973 May 1.0 1973 Oct 1.0	91.11 91.09 91.09	99.00 96.99 95.15	7087 6989 6902	245 240 234	1173 982 814	0.065 0.053 0.042	174 - -
D	ESRO 4 rocket	1972-92B	1972 Nov 22.01 229.77 days 1973 Jul 9.78	Cylinder 24	1.50 long 0.46 dia	1972 Nov 24.1 1973 Mar 16.5	91.11 91.09	98.93 95.92	7084 6939	244 232	1168 889	0.065 0.047	169 -
D R	Cosmos 537	1972-93A	1972 Nov 25.38 11.8 days 1972 Dec 7.2	Sphere-cylinder 5700?	5.0 long 2.4 dia	1972 Nov 26.2	64.95	89.59	6633	204	305	0.008	41
D	Cosmos 537 rocket	1972-93B	1972 Nov 25.38 6.93 days 1972 Dec 2.31	Cylinder 2500?	7.5 long 2.6 dia	1972 Nov 25.9	64.97	89.49	6628	201	298	0.007	35
D	Intercosmos 8	1972-94A	1972 Nov 30.91 91.76 days 1973 Mar 2.67	Ellipsoid + 6 panels? 400?	1.8 long 1.2 dia	1972 Dec 2.3	71.01	93.11	6805	204	649	0.033	85
D	Intercosmos 8 rocket	1972-94B	1972 Nov 30.91 42.70 days 1973 Jan 12.61	Cylinder 1500?	8 long 1.65 dia	1972 Dec 2.3	71.01	92.95	6797	206	631	0.031	87
D	Fragment	1972-94C											

	Name		Launch date, lifetime and descent date	Shape and weight (kg)	Size (m)	Date of orbital determination	Orbital inclination (deg)	Nodal period (min)	Semi major axis (km)	Perigee height (km)	Apogee height (km)	Orbital eccentricity	Argument of perigee (deg)
D	Molniya 1X	1972-95A	1972 Dec 2.20 1166 days 1976 Feb 11	Windmill + 6 vanes 1000?	3.4 long 1.6 dia	1972 Dec 31.9 1973 Sep 1.0	65.01 65.01	717.70 717.74	26554 26555	555 582	39797 39772	0.739 0.738	285 -
D	Molniya 1X launcher rocket	1972-95B	1972 Dec 2.20 39.72 days 1973 Jan 10.92	Cylinder 2500?	7.5 long 2.6 dia	1972 Dec 4.2	64.98	91.23	6714	230	441	0.016	64
D	Molniya 1X launcher	1972-95C	1972 Dec 2.20 62.08 days 1973 Feb 2.28	Irregular	-	1972 Dec 3.1	64.98	91.93	6749	210	531	0.024	60
D	Molniya 1X rocket	1972-95F	1972 Dec 2.20 815 days 1975 Feb 25	Cylinder 440	2.0 long 2.0 dia	1973 Mar 1.0 1973 Sep 1.0	65.00 64.99	699.14 699.18	26094 26095	545 571	38887 38863	0.735 0.734	- -
D	Fragments	1972-95D, E											
D R 3M	Apollo 17**	1972-96A	1972 Dec 7.23 12.58 days 1972 Dec 19.81	Cone-cylinder 30340 full	11.15 long 3.91 dia	1972 Dec 7.3 1972 Dec 7.5	32.56 33.2 In selenocentric orbit 1972 Dec 10.83 to Dec 16.98	87.82 26320	6552 292 520	169 200	178 572 080	0.0007 0.977	- * 30*
D	Saturn IVB [Saturn 512]	1972-96B	1972 Dec 7.23 3.63 days 1972 Dec 10.86	Cylinder 13930	18.7 long 6.6 dia	1972 Dec 7.3 1972 Dec 7.5	32.56 33.2 Crashed on Moon 1972 Dec 10.86	87.82 26320	6552 292 520	169 200	178 572 080	0.0007 0.977	- * 30*
D	LEM 12 descent stage	1972-96D	1972 Dec 7.23 4.60 days 1972 Dec 11.83	Octagon + legs 11390 full 2792 empty	1.57 high 3.13 wide	1972 Dec 7.5	33.2 Entered selenocentric orbit 1972 Dec 10.83 Landed on Moon 1972 Dec 11.83	26320	292 520	200	572 080	0.977	30*
D	LEM 12 † ascent stage	1972-96C	1972 Dec 7.23 8.06 days 1972 Dec 15.29	Box + tanks 5050 full 2145 empty	2.52 high 3.76 wide 3.13 deep	1972 Dec 7.5	33.2 On Moon's surface 1972 Dec 11.83 to Dec 14.96 Finally crashed on Moon 1972 Dec 15.29	26320	292 520	200	572 080	0.977	30*

** Apollo attached to LEM separated from Saturn IVB on Dec 7.41.

† LEM with two crew members separated from Apollo on Dec 11.72.

Ascent stage relaunched from Moon Dec 14.96; briefly docked with Apollo Dec 15.04.

* Approximate orbits.

	Name		Launch date, lifetime and descent date	Shape and weight (kg)	Size (m)	Date of orbital determination	Orbital inclination (deg)	Nodal period (min)	Semi major axis (km)	Perigee height (km)	Apogee height (km)	Orbital eccentricity	Argument of perigee (deg)
T	Nimbus 5	1972-97A	1972 Dec 11.33 1600 years	Conical skeleton 770	3.00 long 1.45 dia	1972 Dec 14.0	99.95	107.25	7474	1089	1102	0.0008	230
	Nimbus 5 second stage	1972-97B	1972 Dec 11.33 1600 years	Cylinder + annulus 350?	6.4 long 1.52 and 2.44 dia	1972 Dec 11.7 1973 Jan 1.0	99.93 99.8	107.34 111.8	7478 7690	1088 1105	1112 1518	0.002 0.027	214 -
D	Molniya 2D	1972-98A	1972 Dec 12.29 771 days 1975 Jan 22	Windmill + 6 vanes? 1250?	4.2 long? 1.6 dia?	1972 Dec 12.3 1973 Jan 2.5 1974 Feb 1.0	65.26 65.31 65.42	706.48 717.67 717.53	26276 26554 26550	495 465 358	39300 39886 39986	0.738 0.742 0.746	285 285 -
D	Molniya 2D launcher rocket	1972-98B	1972 Dec 12.29 31.71 days 1973 Jan 13.00	Cylinder 2500?	7.5 long 2.6 dia	1972 Dec 16.0	65.41	91.14	6709	218	444	0.017	63
D	Molniya 2D launcher	1972-98C	1972 Dec 12.29 38.24 days 1973 Jan 19.53	Irregular	-	1972 Dec 16.0	65.39	91.39	6721	220	467	0.018	64
D	Molniya 2D rocket	1972-98D	1972 Dec 12.29 664 days 1974 Oct 7	Cylinder 440	2.0 long 2.0 dia	1973 Jan 31.5 1973 Jul 1.0	65.38 65.40	702.02 701.90	26166 26163	441 484	39135 39086	0.739 0.738	284 -
D R	Cosmos 538	1972-99A	1972 Dec 14.58 12.7 days 1972 Dec 27.3	Sphere-cylinder 6300?	6.5 long? 2.4 dia	1972 Dec 15.6 1972 Dec 16.5	65.40 65.43	89.38 89.16	6622 6611	205 184	283 282	0.006 0.007	60 58
D	Cosmos 538 rocket	1972-99B	1972 Dec 14.58 8.24 days 1972 Dec 22.82	Cylinder 2500?	7.5 long 2.6 dia	1972 Dec 15.1	65.40	89.24	6615	199	275	0.006	48

1972-99 continued on page 318

	Name		Launch date, lifetime and descent date	Shape and weight (kg)	Size (m)	Date of orbital determination	Orbital inclination (deg)	Nodal period (min)	Semi major axis (km)	Perigee height (km)	Apogee height (km)	Orbital eccentricity	Argument of perigee (deg)
D	Cosmos 538 engine*	1972-99E	1972 Dec 14.58 21 days 1973 Jan 4	Cone 600? full	1.5 long? 2 dia?	1972 Dec 30.2	65.40	89.12	6609	177	285	0.008	41
D	Fragments	1972-99C,D											
D	Aeros 1 (GRS B)†	1972-100A	1972 Dec 16.48 248.91 days 1973 Aug 22.39	Cylinder 127	0.74 long 0.91 dia	1972 Dec 18.1 1973 Apr 1.0	96.94 96.94	95.57 93.65	6923 6830	223 217	867 687	0.047 0.034	162 -
D	Aeros 1 rocket	1972-100B	1972 Dec 16.48 71.45 days 1973 Feb 25.93	Cylinder 24	1.50 long 0.46 dia	1972 Dec 18.4	96.94	95.50	6920	224	859	0.046	161
	BMEWS 5 [Atlas Agena D]	1972-101A	1972 Dec 20.07? > million years	Cylinder? 700 full? 350 empty?	1.7 long? 1.4 dia?	1973 Jan 1.0	9.7	1440.4	42248	31012	40728	0.115	-
	Agena D rocket	1972-101B	1972 Dec 20.07? 20 years?	Cylinder 700?	6 long? 1.5 dia?	1973 Feb 1.0	28.12	583.76	23163	200	33370	0.716	210**
	Cosmos 539	1972-102A	1972 Dec 21.08 5000 years	- 500?	-	1972 Dec 28.1	74.02	112.98	7741	1343	1383	0.003	225
	Cosmos 539 rocket	1972-102B	1972 Dec 21.08 4000 years	Cylinder 2200?	7.4 long 2.4 dia	1972 Dec 28.2	74.02	112.85	7735	1339	1374	0.002	205

*1972-99E ejected from 1972-99A about 1972 Dec 26.

**Aproximate orbit.

†German Research Satellite.

	Name		Launch date, lifetime and descent date	Shape and weight (kg)	Size (m)	Date of orbital determination	Orbital inclination (deg)	Nodal period (min)	Semi major axis (km)	Perigee height (km)	Apogee height (km)	Orbital eccentricity	Argument of perigee (deg)
D	[Titan 3B Agena D]	1972-103A	1972 Dec 21.74 33 days 1973 Jan 23	Cylinder 3000?	8 long? 1.5 dia	1972 Dec 25.7 1972 Dec 25.9	110.45 110.44	89.68 89.81	6637 6643	139 132	378 398	0.018 0.020	151 153
	Cosmos 540	1972-104A	1972 Dec 25.96 120 years	Cylinder + paddles 750?	2 long? 1 dia?	1972 Dec 29.2	74.08	100.79	7174	781	810	0.002	348
	Cosmos 540 rocket	1972-104B	1972 Dec 25.96 100 years	Cylinder 2200?	7.4 long 2.4 dia	1972 Dec 27.1	74.07	100.60	7164	769	803	0.002	19
	Fragments	1972-104C-E											
D R	Cosmos 541	1972-105A	1972 Dec 27.44 11.9 days 1973 Jan 8.3	Sphere-cylinder 5900?	5.9 long 2.4 dia	1972 Dec 28.7	81.31	90.21	6662	221	346	0.009	72
D	Cosmos 541 rocket	1972-105B	1972 Dec 27.44 16.33 days 1973 Jan 12.77	Cylinder 2500?	7.5 long 2.6 dia	1972 Dec 28.3	81.33	90.10	6656	218	338	0.009	63
D	Capsule*	1972-105F	1972 Dec 27.44 19 days 1973 Jan 15	Ellipsoid 200?	0.9 long 1.9 dia	1973 Jan 9.2	81.32	89.97	6650	220	323	0.008	25
D	Fragments	1972-105C-E											
	Cosmos 542	1972-106A	1972 Dec 28.46 11 years	Cylinder + 2 vanes? 2500?	5 long? 1.5 dia?	1972 Dec 31.0	81.22	96.38	6962	527	641	0.008	265
	Cosmos 542 rocket	1972-106B	1972 Dec 28.46 15 years	Cylinder 1440	3.8 long 2.6 dia	1972 Dec 31.2	81.21	96.40	6963	509	661	0.011	224

*1972-105F ejected from 1972-105A about 1973 Jan 8.

	Name		Launch date, lifetime and descent date	Shape and weight (kg)	Size (m)	Date of orbital determination	Orbital inclination (deg)	Nodal period (min)	Semi major axis (km)	Perigee height (km)	Apogee height (km)	Orbital eccentricity	Argument of perigee (deg)
D	Luna 21 launcher rocket	1973-01B	1973 Jan 8.29 4 days 1973 Jan 12	Cylinder 4000?	12 long? 4 dia	1973 Jan 9.0	51.55	88.68	6591	190	235	0.003	318
D	Luna 21 launcher	1973-01C	1973 Jan 8.29 5 days 1973 Jan 13	-	-	1973 Jan 9.3	51.55	88.62	6588	183	236	0.004	324
D R	Cosmos 543	1973-02A	1973 Jan 11.42 12.9 days 1973 Jan 24.3	Sphere-cylinder 6300?	6.5 long? 2.4 dia	1973 Jan 12.0 1973 Jan 13.6	64.98 64.97	89.62 89.28	6634 6618	203 175	309 304	0.008 0.010	54 47
D	Cosmos 543 rocket	1973-02B	1973 Jan 11.42 11 days 1973 Jan 22	Cylinder 2500?	7.5 long 2.6 dia	1973 Jan 12.1	64.98	89.51	6629	203	299	0.007	53
D	Cosmos 543 engine*	1973-02C	1973 Jan 11.42 16.76 days 1973 Jan 28.18	Cone 600? full	1.5 long? 2 dia?	1973 Jan 23.7	64.98	89.06	6607	167	290	0.009	40
D	Fragment	1973-02D											
D	Cosmos 544	1973-03A	1973 Jan 20.15 2703 days 1980 Jun 15	Cylinder + paddles? 900?	2 long? 1 dia?	1973 Jan 26.2	74.03	95.23	6907	510	548	0.003	345
D	Cosmos 544 rocket	1973-03B	1973 Jan 20.15 2675 days 1980 May 18	Cylinder 2200?	7.4 long 2.4 dia	1973 Jan 20.8	74.03	95.11	6901	501	545	0.003	0
D	Fragments	1973-03C,D											

Space Vehicle: Luna 21, 1973-01A

*1973-02C ejected from 1973-02A on 1973 Jan 23

	Name		Launch date, lifetime and descent date	Shape and weight (kg)	Size (m)	Date of orbital determination	Orbital inclination (deg)	Nodal period (min)	Semi major axis (km)	Perigee height (km)	Apogee height (km)	Orbital eccentricity	Argument of perigee (deg)
D	Cosmos 545	1973-04A	1973 Jan 24.49 188.41 days 1973 Jul 31.90	Ellipsoid 400?	1.8 long 1.2 dia	1973 Jan 25.2 1973 May 1.0	71.00 71.00	92.20 91.24	6760 6714	269 252	495 420	0.017 0.013	84 -
D	Cosmos 545 rocket	1973-04B	1973 Jan 24.49 86.75 days 1973 Apr 21.24	Cylinder 1500?	8 long 1.65 dia	1973 Jan 25.2	71.01	92.03	6752	272	475	0.015	85
	Cosmos 546	1973-05A	1973 Jan 26.49 50 years	-	-	1973 Jan 29.6	50.66	96.51	6973	575	614	0.003	46
	Cosmos 546 rocket	1973-05B	1973 Jan 26.49 13 years	Cylinder 2200?	7.4 long 2.4 dia	1973 Jan 29.0	50.66	96.42	6968	566	614	0.003	41
D R	Cosmos 547	1973-06A	1973 Feb 1.36 11.8 days 1973 Feb 13.2	Sphere-cylinder 5700?	5.0 long 2.4 dia	1973 Feb 2.2	64.97	89.63	6635	203	310	0.008	51
D	Cosmos 547 rocket	1973-06B	1973 Feb 1.36 8.5 days 1973 Feb 9.9	Cylinder 2500?	7.5 long 2.6 dia	1973 Feb 2.2	64.97	89.43	6625	204	289	0.006	41
D	Fragment	1973-06C											

	Name		Launch date, lifetime and descent date	Shape and weight (kg)	Size (m)	Date of orbital determination	Orbital inclination (deg)	Nodal period (min)	Semi-major axis (km)	Perigee height (km)	Apogee height (km)	Orbital eccentricity	Argument of perigee (deg)
D	Molniya 1Y	1973-07A	1973 Feb 3.25 1723 days 1977 Oct 23	Windmill + 6 vanes 1000?	3.4 long 1.6 dia	1973 Feb 8.7 1973 Mar 1.0	65.00 65.00	703.15 717.65	26194 26553	470 578	39164 39772	0.739 0.738	285 -
D	Molniya 1Y launcher rocket	1973-07B	1973 Feb 3.25 37.83 days 1973 Mar 13.08	Cylinder 2500?	7.5 long 2.6 dia	1973 Feb 6.0	64.97	91.23	6714	229	442	0.016	64
D	Molniya 1Y launcher	1973-07C	1973 Feb 3.25 42.79 days 1973 Mar 18.04	Irregular	-	1973 Feb 5.1	64.96	91.53	6728	228	472	0.018	67
D	Molniya 1Y rocket	1973-07E	1973 Feb 3.25 2008 days 1978 Aug 4	Cylinder 440	2.0 long 2.0 dia	1973 Mar 1.0	65.10	702.53	26179	508	39093	0.737	-
D	Fragment	1973-07D											
D R	Cosmos 548	1973-08A	1973 Feb 8.56 12.7 days 1973 Feb 21.3	Sphere-cylinder 6300?	6.5 long? 2.4 dia	1973 Feb 9.2 1973 Feb 14.1	65.38 65.39	89.55 89.37	6631 6622	205 171	300 317	0.007 0.011	72 48
D	Cosmos 548 rocket	1973-08B	1973 Feb 8.56 8.12 days 1973 Feb 16.68	Cylinder 2500?	7.5 long 2.6 dia	1973 Feb 9.0	65.38	89.40	6624	202	289	0.007	60
D	Cosmos 548 engine *	1973-08F	1973 Feb 8.56 18 days 1973 Feb 26	Cone 600? full	1.5 long? 2 dia?	1973 Feb 21.6	65.42	90.61	6684	159	453	0.022	7
D	Fragments	1973-08C-E,G											

* 1973-08F ejected from 1973-08A about 1973 Feb 20.

Year of launch 1973 continued

	Name		Launch date, lifetime and descent date	Shape and weight (kg)	Size (m)	Date of orbital determination	Orbital inclination (deg)	Nodal period (min)	Semi major axis (km)	Perigee height (km)	Apogee height (km)	Orbital eccentricity	Argument of perigee (deg)
D	Prognoz 3	1973-09A	1973 Feb 15.05 4 years?	Spheroid + 4 vanes 845	1.8 dia?	1973 Feb 15.0	65.0	5783	106670	590	200000	0.935	- *
D	Prognoz 3 launcher rocket	1973-09B	1973 Feb 15.05 35.69 days 1973 Mar 22.74	Cylinder 2500?	7.5 long 2.6 dia	1973 Feb 15.9	65.00	91.24	6714	229	442	0.016	64
D	Prognoz 3 launcher	1973-09C	1973 Feb 15.05 36.25 days 1973 Mar 23.30	Irregular	-	1973 Feb 15.9	65.03	91.53	6728	216	484	0.020	67
D	Prognoz 3 rocket	1973-09D	1973 Feb 15.05 4 years?	Cylinder 440	2.0 long 2.0 dia		Orbit similar to 1973-09A						
D	Cosmos 549	1973-10A	1973 Feb 28.19 2678 days 1980 Jun 29	Cylinder + paddles? 900?	2 long? 1 dia?	1973 Mar 1.1	74.02	95.23	6907	513	545	0.002	357
D	Cosmos 549 rocket	1973-10B	1973 Feb 28.19 2758 days 1980 Sep 17	Cylinder 2200?	7.4 long 2.4 dia	1973 Mar 1.6	74.02	95.11	6901	504	542	0.003	4
D	Fragment	1973-10C											
D R	Cosmos 550	1973-11A	1973 Mar 1.53 9.8 days 1973 Mar 11.3	Sphere- cylinder 6300?	6.5 long? 2.4 dia	1973 Mar 2.2 1973 Mar 10.8	65.42 65.42	89.73 89.29	6640 6618	206 183	317 296	0.008 0.008	78 55
D	Cosmos 550 rocket	1973-11B	1973 Mar 1.53 8.27 days 1973 Mar 9.80	Cylinder 2500?	7.5 long 2.6 dia	1973 Mar 2.2	65.42	89.63	6635	204	309	0.008	70
D	Cosmos 550 engine	1973-11C	1973 Mar 1.53 17 days 1973 Mar 18	Cone 600? full	1.5 long? 2 dia?	1973 Mar 13.7	65.42	89.24	6615	184	290	0.008	-
D	Fragments	1973-11D-F											

* Approximate orbit.

	Name		Launch date, lifetime and descent date	Shape and weight (kg)	Size (m)	Date of orbital determination	Orbital inclination (deg)	Nodal period (min)	Semi major axis (km)	Perigee height (km)	Apogee height (km)	Orbital eccentricity	Argument of perigee (deg)
D R	Cosmos 551	1973-12A	1973 Mar 6.39 13.8 days 1973 Mar 20.2	Sphere-cylinder 6300?	6.5 long? 2.4 dia	1973 Mar 6.8 1973 Mar 9.5 1973 Mar 15.3	65.00 65.00 65.01	89.52 89.40 90.18	6629 6623 6662	206 173 170	296 317 398	0.007 0.011 0.017	49 58 55
D	Cosmos 551 rocket	1973-12B	1973 Mar 6.39 6.74 days 1973 Mar 13.13	Cylinder 2500?	7.5 long 2.6 dia	1973 Mar 6.9	65.00	89.39	6623	203	286	0.006	50
D	Cosmos 551 engine*	1973-12D	1973 Mar 6.39 15.86 days 1973 Mar 22.25	Cone 600? full	1.5 long? 2 dia?	1973 Mar 16.5	65.02	89.96	6651	170	376	0.015	53
D	Fragments	1973-12C,E											
	BMEWS 6 [Atlas Agena D]	1973-13A	1973 Mar 6.5 > million years	Cylinder? 700 full? 350 empty?	1.7 long? 1.4 dia?	1973 Apr 1.0	0.2	1435.1	42145	35679	35855	0.002	-
D	Agena D rocket	1973-13B	1973 Mar 6.5 455 days 1974 Jun 4	Cylinder 700?	6 long? 1.5 dia	1973 Mar 19.7	28.27	587.96	23284	228	33584	0.717	190
D	[Titan 3D]	1973-14A	1973 Mar 9.88 71 days 1973 May 19	Cylinder 13300? full	15 long 3.0 dia	1973 Mar 10.7	95.70	88.76	6589	152	270	0.009	140
D	Titan 3D rocket	1973-14B	1973 Mar 9.88 1.65 days 1973 Mar 11.53	Cylinder 1900	6 long 3.0 dia	1973 Mar 10.7	95.68	88.27	6565	151	222	0.005	133
	Meteor 14	1973-15A	1973 Mar 20.47 500 years	Cylinder + 2 vanes 2200?	5 long? 1.5 dia?	1973 Mar 20.6	81.27	102.64	7261	873	892	0.001	269
	Meteor 14 rocket	1973-15B	1973 Mar 20.47 400 years	Cylinder 1440	3.8 long 2.6 dia	1973 Mar 21.6	81.27	102.77	7267	844	933	0.006	175

* 1973-12D ejected from 1973-12A on 1973 Mar 16.

	Name		Launch date, lifetime and descent date	Shape and weight (kg)	Size (m)	Date of orbital determination	Orbital inclination (deg)	Nodal period (min)	Semi major axis (km)	Perigee height (km)	Apogee height (km)	Orbital eccentricity	Argument of perigee (deg)
D R	Cosmos 552	1973-16A	1973 Mar 22.42 11.76 days 1973 Apr 3.18	Sphere-cylinder 5900?	5.9 long 2.4 dia	1973 Mar 22.8	72.84	89.68	6636	204	312	0.008	61
D	Cosmos 552 rocket	1973-16B	1973 Mar 22.42 8.54 days 1973 Mar 30.96	Cylinder 2500?	7.5 long 2.6 dia	1973 Mar 23.0	72.84	89.52	6628	206	294	0.007	58
D	Capsule*	1973-16C	1973 Mar 22.42 17.59 days 1973 Apr 9.01	Ellipsoid 200?	0.9 long 1.9 dia	1973 Mar 26.2	72.85	89.61	6633	199	310	0.008	51
D	Salyut 2	1973-17A	1973 Apr 3.38 55.11 days 1973 May 28.49	Cylinder + 4 wings 18500	14 long 4.15 max dia 2.0 min dia	1973 Apr 4.0 1973 Apr 5.7 1973 Apr 8.6	51.56 51.57 51.57	88.99 89.42 89.81	6606 6626 6646	207 237 257	248 259 278	0.003 0.002 0.002	59 226 81
D	Salyut 2 rocket	1973-17B	1973 Apr 3.38 3 days 1973 Apr 6	Cylinder 4000?	12 long? 4 dia	1973 Apr 3.6	51.48	88.82	6597	194	244	0.004	2
D	Fragments**	1973-17C-AB											
D	Molniya 2E	1973-18A	1973 Apr 5.47 2102 days 1979 Jan 6	Windmill + 6 vanes 1250?	4.2 long? 1.6 dia?	1973 Apr 18.2 1973 Apr 18.7	65.49 65.24	702.19 717.73	26170 26555	477 532	39107 39822	0.738 0.740	285 285
D	Molniya 2E launcher rocket	1973-18B	1973 Apr 5.47 19.05 days 1973 Apr 24.52	Cylinder 2500?	7.5 long 2.6 dia	1973 Apr 6.2	65.39	91.53	6728	188	511	0.024	70
D	Molniya 2E launcher	1973-18C	1973 Apr 5.47 25.40 days 1973 Apr 30.87	Irregular	-	1973 Apr 6.0	65.40	91.25	6715	217	456	0.018	72

* 1973-16C ejected from 1973-16A about 1973 Mar 26. ** Fragments designated on 1973 Apr 4.45

1973-18 continued on page 326

	Name		Launch date, lifetime and descent date	Shape and weight (kg)	Size (m)	Date of orbital determination	Orbital inclination (deg)	Nodal period (min)	Semi major axis (km)	Perigee height (km)	Apogee height (km)	Orbital eccentricity	Argument of perigee (deg)
D	Molniya 2E rocket	1973-18D	1973 Apr 5.47 68 months? 1978 Dec?	Cylinder 440	2.0 long 2.0 dia	1973 May 2.2	65.44	699.04	26092	467	38961	0.738	284
D	Pioneer 11 second stage	1973-19C	1973 Apr 6.09 577 days 1974 Nov 4	Cylinder 1815	8.6 long 3.0 dia	1973 May 1.0	34.9	2342.0	58420	161	103922	0.888	-
D	Cosmos 553	1973-20A	1973 Apr 12.50 213.45 days 1973 Nov 11.95	Ellipsoid 400?	1.8 long 1.2 dia	1973 Apr 13.0 1973 Jul 31.9	70.96 70.96	92.22 91.38	6761 6721	272 256	494 429	0.016 0.013	80 -
D	Cosmos 553 rocket	1973-20B	1973 Apr 12.50 116.96 days 1973 Aug 7.46	Cylinder 1500?	8 long 1.65 dia	1973 Apr 13.0	70.96	92.10	6755	272	482	0.016	80
D	Cosmos 554*	1973-21A	1973 Apr 19.38 38 days 1973 May 27	Sphere-cylinder 6300?	6.5 long? 2.4 dia	1973 Apr 19.7 1973 Apr 21.7 1973 Apr 25.8	72.85 72.85 72.85	89.50 89.58 90.04	6627 6631 6654	194 171 171	304 335 380	0.008 0.012 0.016	45 57 48
D	Cosmos 554 rocket	1973-21B	1973 Apr 19.38 7.17 days 1973 Apr 26.55	Cylinder 2500?	7.5 long 2.6 dia	1973 Apr 19.6	72.85	89.47	6626	202	293	0.007	58
D	Cosmos 554 engine**	1973-21E	1973 Apr 19.38 12 days 1973 May 1	Cone 600? full	1.5 long 2 dia?	1973 Apr 27.0	72.85	89.80	6642	169	359	0.014	43
D	Fragments	1973-21C,D,F-HE											
D	Intercosmos 9 (Copernicus 500)	1973-22A	1973 Apr 19.43 179.46 days 1973 Oct 15.89	Ellipsoid 400?	1.8 long 1.2 dia	1973 Apr 21.0 1973 Jul 16.5	48.42 48.39	102.12 97.73	7241 7030	199 192	1526 1111	0.092 0.065	108 -
D	Intercosmos 9 rocket	1973-22B	1973 Apr 19.43 181.37 days 1973 Oct 17.80	Cylinder 1500?	8 long 1.65 dia	1973 Apr 21.0 1973 Jul 16.5	48.41 48.39	102.12 97.71	7241 7029	199 194	1526 1107	0.092 0.065	108 -

Space Vehicle: Pioneer 11, 1973-19A
Burner 2 rocket, 1973-19B

* 1973-21A disintegrated about 1973 May 7.1

** 1973-21E ejected from 1973-21A on 1973 Apr 26

	Name		Launch date, lifetime and descent date	Shape and weight (kg)	Size (m)	Date of orbital determination	Orbital inclination (deg)	Nodal period (min)	Semi major axis (km)	Perigee height (km)	Apogee height (km)	Orbital eccentricity	Argument of perigee (deg)
	Telesat 2 (Anik 2)	1973-23A	1973 Apr 20.98 > million years	Cylinder 565 full 295 empty	1.52 long 1.83 dia	1973 May 1.0 1973 Jun 1.0	0.1 0.1	1430.7 1436.2	42035 42167	35604 35786	35709 35792	0.001 0.0001	- -
D	Telesat 2 second stage	1973-23B	1973 Apr 20.98 911 days 1975 Oct 18	Cylinder + annulus 350?	6.4 long 1.52 and 2.44 dia	1973 Apr 21.9 1973 Dec 1.0	29.48 29.48	115.80 111.30	7877 7673	215 212	2783 2377	0.163 0.141	188 -
	Telesat 2 third stage	1973-23C	1973 Apr 20.98 20 years?	Sphere-cone 66	1.32 long 0.94 dia	1973 Apr 21.9	29.48	645.5	24740	215	36508	0.734	188*
D R	Cosmos 555	1973-24A	1973 Apr 25.45 11.9 days 1973 May 7.3	Sphere-cylinder 5900?	5.9 long 2.4 dia	1973 Apr 25.6	81.33	89.02	6603	216	233	0.001	3
D	Cosmos 555 rocket	1973-24B	1973 Apr 25.45 3.22 days 1973 Apr 28.67	Cylinder 2500?	7.5 long 2.6 dia	1973 Apr 25.5	81.32	88.93	6598	209	231	0.002	334
D	Capsule**	1973-24D	1973 Apr 25.45 14.39 days 1973 May 9.84	Ellipsoid 200?	0.9 long 1.9 dia	1973 May 4.7	81.33	88.84	6594	209	222	0.001	327
D	Fragment	1973-24C											
D R	Cosmos 556	1973-25A	1973 May 5.29 8.9 days 1973 May 14.2	Sphere-cylinder 6300?	6.5 long? 2.4 dia	1973 May 5.6 1973 May 6.4	81.33 81.32	88.97 89.31	6600 6617	218 210	225 267	0.0005 0.004	18 5
D	Cosmos 556 rocket	1973-25B	1973 May 5.29 3.69 days 1973 May 8.98	Cylinder 2500?	7.5 long 2.6 dia	1973 May 6.3	81.32	88.77	6590	207	216	0.0006	353

*Approximate orbit; Telesat 2 is Canadian

**1973-24D ejected from 1973-24A on 1973 May 4

1973-25 continued on page 328

	Name		Launch date, lifetime and descent date	Shape and weight (kg)	Size (m)	Date of orbital determination	Orbital inclination (deg)	Nodal period (min)	Semi major axis (km)	Perigee height (km)	Apogee height (km)	Orbital eccentricity	Argument of perigee (deg)
D	Cosmos 556 engine*	1973-25C	1973 May 5.29 18.90 days 1973 May 24.19	Cone 600? full	1.5 long? 2 dia?	1973 May 13.8	81.32	89.19	6611	210	255	0.003	326
D	Fragments	1973-25D,E											
D	Cosmos 557**	1973-26A	1973 May 11.02 11.11 days 1973 May 22.13	Cylinder + 4 wings 18500	14 long 4.15 max dia 2.0 min dia	1973 May 12.2	51.59	89.01	6607	214	243	0.002	83
D	Cosmos 557 rocket	1973-26B	1973 May 11.02 6.01 days 1973 May 17.03	Cylinder 4000?	12 long? 4 dia	1973 May 12.2	51.60	88.79	6596	209	226	0.001	57
D P	Skylab 1† [Saturn 513]	1973-27A	1973 May 14.73 2248.96 days 1979 Jul 11.69	Cylinders + 4 vanes 74783	25.64 long 6.6 dia 27.4 span	1973 May 16.5	50.04	93.18	6811	427	439	0.0009	285
D	Skylab 1 rocket	1973-27B	1973 May 14.73 606.60 days 1975 Jan 11.33	Cylinder 35400?	24.8 long 10.0 dia	1973 May 15.0 1974 Feb 1.0	50.05 50.05	92.51 91.99	6779 6755	363 347	439 407	0.006 0.004	249 -
D	Fragments	1973-27C-Z											
D	[Titan 3B Agena D]	1973-28A	1973 May 16.69 28 days 1973 Jun 13	Cylinder 3000?	8 long? 1.5 dia	1973 May 16.8 1973 May 17.7	110.49 110.51	89.39 89.89	6622 6647	136 139	352 399	0.016 0.020	143 141
D	Cosmos 558	1973-29A	1973 May 17.56 219.33 days 1973 Dec 22.89	Ellipsoid 400?	1.8 long 1.2 dia	1973 May 17.8 1973 Sep 1.0	70.98 70.98	92.26 91.54	6763 6729	269 256	501 445	0.017 0.014	86 -
D	Cosmos 558 rocket	1973-29B	1973 May 17.56 116.33 days 1973 Sep 10.89	Cylinder 1500?	8 long 1.65 dia	1973 May 17.8	70.98	92.15	6758	268	491	0.016	89

* 1973-25C ejected from 1973-25A on 1973 May 13. ** Cosmos 557 was probably an intended Salyut.

† Decay was observed and pieces were picked up in SW Australia.

	Name		Launch date, lifetime and descent date	Shape and weight (kg)	Size (m)	Date of orbital determination	Orbital inclination (deg)	Nodal period (min)	Semi major axis (km)	Perigee height (km)	Apogee height (km)	Orbital eccentricity	Argument of perigee (deg)
D	Cosmos 559	1973-30A	1973 May 18.46 4.8 days 1973 May 23.3	Sphere-cylinder 6300?	6.5 long? 2.4 dia	1973 May 18.6	65.41	89.79	6643	204	325	0.009	68
R													
D	Cosmos 559 rocket	1973-30B	1973 May 18.46 10 days 1973 May 28	Cylinder 2500?	7.5 long 2.6 dia	1973 May 18.9	65.41	89.74	6640	205	319	0.009	75
D	Cosmos 559 engine**	1973-30D	1973 May 18.46 24.87 days 1973 Jun 12.33	Cone 600? full	1.5 long? 2 dia?	1973 May 22.8	65.39	89.78	6642	211	317	0.008	84
D	Fragment	1973-30C											
D	Cosmos 560	1973-31A	1973 May 23.44 12.8 days 1973 Jun 5.2	Sphere-cylinder 6300?	6.5 long? 2.4 dia	1973 May 24.1	72.85	89.68	6637	203	314	0.008	68
R						1973 May 25.3	72.84	89.41	6623	181	309	0.010	60
D	Cosmos 560 rocket	1973-31B	1973 May 23.44 8.85 days 1973 Jun 1.29	Cylinder 2500?	7.5 long 2.6 dia	1973 May 24.1	72.84	89.54	6630	201	302	0.008	52
D	Cosmos 560 engine	1973-31D	1973 May 23.44 20.17 days 1973 Jun 12.61	Cone 600? full	1.5 long? 2 dia?	1973 Jun 5.7	72.85	89.23	6614	175	297	0.009	29
D	Fragments	1973-31C,E											
D	Skylab 2† [Saturn 206]	1973-32A	1973 May 25.54 28.04 days 1973 Jun 22.58	Cone-cylinder 13780	10.36 long 3.91 dia	1973 May 25.5	50.04	89.59	6636	156	359	0.015	-
3M						1973 May 25.7	50.04	91.81	6744	359	373	0.001	-
R						1973 May 31.9	50.03	93.17	6811	425	440	0.001	325
D	Skylab 2 rocket	1973-32B	1973 May 25.54 < ½ day 1973 May 25	Cylinder 13600?	18.7 long 6.6 dia	1973 May 25.5	50.04	89.52	6632	156	352	0.015	-

† Skylab 2 rendezvous with Skylab 1 on 1973 May 25.86; docked 1973 May 26.16; undocked 1973 Jun 22.37

** 1973-30D ejected from 1973-30A on 1973 May 22.

	Name		Launch date, lifetime and descent date	Shape and weight (kg)	Size (m)	Date of orbital determination	Orbital inclination (deg)	Nodal period (min)	Semi major axis (km)	Perigee height (km)	Apogee height (km)	Orbital eccentricity	Argument of perigee (deg)
D R	Cosmos 561	1973-33A	1973 May 25.57 11.67 days 1973 Jun 6. 24	Sphere- cylinder 5900?	5.9 long 2.4 dia	1973 May 26.6	65.41	89.51	6629	206	295	0.007	68
D	Cosmos 561 rocket	1973-33B	1973 May 25.57 7.72 days 1973 Jun 2.29	Cylinder 2500?	7.5 long 2.6 dia	1973 May 26.7	65.40	89.34	6620	200	284	0.006	59
D	Capsule*	1973-33D	1973 May 25.57 25.61 days 1973 Jun 20.18	Ellipsoid 200?	0.9 long 1.9 dia	1973 Jun 3.9	65.39	89.39	6623	206	283	0.006	69
D	Fragment	1973-33C											
	Meteor 15	1973-34A	1973 May 29.43 500 years	Cylinder + 2 vanes, 2200?	5 long? 1.5 dia?	1973 May 30.8	81.22	102.48	7253	853	896	0.003	284
	Meteor 15 rocket	1973-34B	1973 May 29.43 400 years	Cylinder 1440	3.8 long 2.6 dia	1973 May 31.5	81.23	102.72	7264	852	920	0.005	178
D	Cosmos 562	1973-35A	1973 Jun 5.48 215.91 days 1974 Jan 7.39	Ellipsoid 400?	1.8 long 1.2 dia	1973 Jun 5.7 1973 Sep 16.0	70.98 70.98	92.13 91.46	6757 6725	270 259	487 434	0.016 0.013	85 -
D	Cosmos 562 rocket	1973-35B	1973 Jun 5.48 116.23 days 1973 Sep 29.71	Cylinder 1500?	8 long 1.65 dia	1973 Jun 5.6	70.99	92.04	6752	264	484	0.016	80
D R	Cosmos 563	1973-36A	1973 Jun 6.48 11.72 days 1973 Jun 18.20	Sphere- cylinder 6300?	6.5 long? 2.4 dia	1973 Jun 7.0 1973 Jun 8.6	65.40 65.40	89.53 89.17	6630 6612	206 177	298 291	0.007 0.009	64 64
D	Cosmos 563 rocket	1973-36B	1973 Jun 6.48 7 days 1973 Jun 13	Cylinder 2500?	7.5 long 2.6 dia	1973 Jun 7.0	65.40	89.40	6624	200	292	0.007	57

*1973-33D ejected from 1973-33A on 1973 Jun 3

1973-36 continued on page 331

	Name		Launch date, lifetime and descent date	Shape and weight (kg)	Size (m)	Date of orbital determination	Orbital inclination (deg)	Nodal period (min)	Semi major axis (km)	Perigee height (km)	Apogee height (km)	Orbital eccentricity	Argument of perigee (deg)
D	Cosmos 563 engine*	1973-36D	1973 Jun 6.48 19 days 1973 Jun 25	Cone 600? full	1.5 long? 2 dia?	1973 Jun 20.8	65.40	89.00	6604	173	278	0.008	-
D	Fragments	1973-36C, E											
	Cosmos 564	1973-37A	1973 Jun 8.65 8000 years	Spheroid 40?	1.0 long? 0.8 dia?	1973 Jun 9.9	74.03	114.68	7818	1395	1484	0.006	112
	Cosmos 565	1973-37B	1973 Jun 8.65 9000 years	Spheroid 40?	1.0 long? 0.8 dia?	1973 Jun 10.2	74.01	115.36	7849	1450	1492	0.003	155
	Cosmos 566	1973-37C	1973 Jun 8.65 9000 years	Spheroid 40?	1.0 long? 0.8 dia?	1973 Jun 11.0	74.01	115.12	7838	1435	1485	0.003	126
	Cosmos 567	1973-37D	1973 Jun 8.65 9000 years	Spheroid 40?	1.0 long? 0.8 dia?	1973 Jun 11.0	74.01	114.88	7828	1414	1486	0.005	122
	Cosmos 568	1973-37E	1973 Jun 8.65 8000 years	Spheroid 40?	1.0 long? 0.8 dia?	1973 Jun 10.6	74.02	114.43	7808	1378	1482	0.007	107
	Cosmos 569	1973-37F	1973 Jun 8.65 7000 years	Spheroid 40?	1.0 long? 0.8 dia?	1973 Jun 10.2	74.02	114.23	7799	1359	1482	0.008	106
	Cosmos 570	1973-37G	1973 Jun 8.65 6000 years	Spheroid 40?	1.0 long? 0.8 dia?	1973 Jun 10.2	74.02	114.03	7789	1341	1481	0.009	105
	Cosmos 571	1973-37H	1973 Jun 8.65 6000 years	Spheroid 40?	1.0 long? 0.8 dia?	1973 Jun 10.9	74.03	113.81	7779	1321	1481	0.010	98
	Cosmos 564 rocket	1973-37J	1973 Jun 8.65 20000 years	Cylinder 2200?	7.4 long 2.4 dia	1973 Jun 10.9	74.02	116.95	7921	1479	1606	0.008	254

* 1973-36D ejected from 1973-36A on 1973 Jun 17

	Name		Launch date, lifetime and descent date	Shape and weight (kg)	Size (m)	Date of orbital determination	Orbital inclination (deg)	Nodal period (min)	Semi major axis (km)	Perigee height (km)	Apogee height (km)	Orbital eccentricity	Argument of perigee (deg)
D R	Cosmos 572	1973-38A	1973 Jun 10.43 12.86 days 1973 Jun 23.29	Sphere-cylinder 6300?	6.5 long? 2.4 dia	1973 Jun 10.6 1973 Jun 15.0	51.66 51.64	89.32 88.89	6622 6601	206 174	281 271	0.006 0.007	57 68
D	Cosmos 572 rocket	1973-38B	1973 Jun 10.43 5.90 days 1973 Jun 16.33	Cylinder 2500?	7.5 long 2.6 dia	1973 Jun 10.6	51.64	89.22	6617	199	280	0.006	39
D	Cosmos 572 engine*	1973-38C	1973 Jun 10.43 16 days 1973 Jun 26	Cone 600? full	1.5 long? 2 dia?	1973 Jun 23.0	51.65	88.95	6604	171	280	0.008	91
D	Fragment	1973-38D											
D	Explorer 49 third stage	1973-39B	1973 Jun 10.59 5 years? 1978?	Sphere-cone 66	1.32 long 0.94 dia	1973 Jun 10.6	29.11	15013	201591	182	390244	0.967	118**
D	Explorer 49 second stage	1973-39C	1973 Jun 10.59 233.10 days 1974 Jan 29.69	Cylinder 600?	5.2 long 2.44 dia	1973 Jun 11.7 1973 Oct 1.0	29.34 29.34	107.80 102.30	7506 7249	182 177	2074 1565	0.126 0.096	125 -
2d	Fragments	1973-39D,E,H											
T	IMEWS 4 [Titan 3C]	1973-40A	1973 June 12.4? > million years	Cylinder + 4 panels 820?	6 long? 2.5 dia?	1973 Jun 12.7 1973 Jul 1.0	26.33 0.3	633.0 1435.9	24446 42160	297 35777	35839 35786	0.727 0.0001	180** -
	Transtage	1973-40B	1973 Jun 12.4? > million years	Cylinder 1500?	6 long? 3 dia	1973 Jul 1.0 1980 Aug 5.5	0.3 5.47	1435.9 1445.5	42160 42350	35777 35786	35786 36157	0.0001 0.004	- 1

* 1973-38C ejected from 1973-38A on 1973 Jun 22

Space Vehicle: Explorer 49 (RAE 2), 1973-39A;
Explorer 49 retrorocket, 1973-39F;
Fragment, 1973-39G.

** Approximate orbits

	Name		Launch date, lifetime and descent date	Shape and weight (kg)	Size (m)	Date of orbital determination	Orbital inclination (deg)	Nodal period (min)	Semi major axis (km)	Perigee height (km)	Apogee height (km)	Orbital eccentricity	Argument of perigee (deg)
D R	Cosmos 573	1973-41A	1973 Jun 15.25 2.0 days 1973 Jun 17.3	Sphere-cylinder 6570?	7.5 long 2.2 dia	1973 Jun 16.3	51.58	89.46	6629	192	309	0.009	77
D	Cosmos 573 rocket	1973-41B	1973 Jun 15.25 5.95 days 1973 Jun 21.20	Cylinder 2500?	7.5 long 2.6 dia	1973 Jun 16.3	51.60	89.26	6619	189	292	0.008	77
D	Fragment	1973-41C											
	Cosmos 574	1973-42A	1973 Jun 20.26 1400 years	Cylinder? 700?	1.3 long? 1.9 dia?	1973 Jun 23.2	82.94	105.14	7378	985	1014	0.002	281
	Cosmos 574 rocket	1973-42B	1973 Jun 20.26 750 years	Cylinder 2200?	7.4 long 2.4 dia	1973 Jun 22.2	82.95	105.03	7373	984	1005	0.001	261
D R	Cosmos 575	1973-43A	1973 Jun 21.56 11.71 days 1973 Jul 3.27	Sphere-cylinder 5700?	5.0 long 2.4 dia	1973 Jun 23.0	65.41	89.25	6616	204	271	0.005	50
D	Cosmos 575 rocket	1973-43B	1973 Jun 21.56 5.68 days 1973 Jun 27.24	Cylinder 2500?	7.5 long 2.6 dia	1973 Jun 22.5	65.39	89.10	6609	196	265	0.005	34
D R	Cosmos 576	1973-44A	1973 Jun 27.50 11.79 days 1973 Jul 9.29	Sphere-cylinder 5900?	5.9 long 2.4 dia	1973 Jun 28.6	72.86	89.88	6646	204	332	0.010	63
D	Cosmos 576 rocket	1973-44B	1973 Jun 27.50 9.81 days 1973 Jul 7.31	Cylinder 2500?	7.5 long 2.6 dia	1973 Jun 28.7	72.85	89.66	6635	199	315	0.009	52

1973-44 continued on page 334

	Name		Launch date, lifetime and descent date	Shape and weight (kg)	Size (m)	Date of orbital determination	Orbital inclination (deg)	Nodal period (min)	Semi major axis (km)	Perigee height (km)	Apogee height (km)	Orbital eccentricity	Argument of perigee (deg)
D	Capsule*	1973-44H	1973 Jun 27.50 17 days 1973 Jul 14	Ellipsoid 200?	0.9 long 1.9 dia	1973 Jul 10.1	72.90	91.42	6723	212	478	0.020	359
D	Fragments	1973-44C-G											
D	Molniya 2F	1973-45A	1973 Jul 11.42 1822 days? 1978 Jul 7?	Windmill + 6 vanes 1250?	4.2 long? 1.6 dia?	1973 Jul 27.1 1973 Sep 1.0	65.41 65.58	705.06 717.81	26241 26557	441 422	39285 39936	0.740 0.744	285 -
D	Molniya 2F launcher	1973-45B	1973 Jul 11.42 41.74 days 1973 Aug 22.16	Irregular	-	1973 Jul 11.7	65.41	91.36	6720	221	463	0.018	67
D	Molniya 2F launcher rocket	1973-45C	1973 Jul 11.42 23.00 days 1973 Aug 3.42	Cylinder 2500?	7.5 long 2.6 dia	1973 Jul 12.4	65.42	91.34	6719	197	484	0.021	67
D	Molniya 2F rocket	1973-45D	1973 Jul 11.42 1851 days 1978 Aug 5	Cylinder 440	2.0 long 2.0 dia	1973 Jul 22.7	65.40	702.58	26179	439	39163	0.740	285
D	[Titan 3D]	1973-46A	1973 Jul 13.85 91 days 1973 Oct 12	Cylinder 13300? full	15 long 3.0 dia	1973 Jul 15.7	96.21	88.77	6591	156	269	0.009	145
D	Titan 3D rocket	1973-46B	1973 Jul 13.85 1.97 days 1973 Jul 15.82	Cylinder 1900	6 long 3.0 dia	1973 Jul 15.1	96.21	88.11	6558	145	215	0.005	143

* 1973-44H ejected from 1973-44A on 1973 Jul 9

	Name		Launch date, lifetime and descent date	Shape and weight (kg)	Size (m)	Date of orbital determination	Orbital inclination (deg)	Nodal period (min)	Semi major axis (km)	Perigee height (km)	Apogee height (km)	Orbital eccentricity	Argument of perigee (deg)
D	Mars 4 launcher rocket	1973-47B	1973 Jul 21.81 1.45 days 1973 Jul 23.26	Cylinder 4000?	12 long? 4 dia	1973 Jul 22.5	51.52	87.94	6553	156	194	0.003	315
D	Mars 4 launcher	1973-47C	1973 Jul 21.81 5 days 1973 Jul 26	Irregular	-	1973 Jul 22.8	51.52	87.70	6541	147	179	0.002	321
D R	Cosmos 577	1973-48A	1973 Jul 25.48 12.71 days 1973 Aug 7.19	Sphere-cylinder 6300?	6.5 long? 2.4 dia	1973 Jul 25.7 1973 Jul 28.6	65.39 65.40	89.45 89.33	6626 6620	207 171	289 313	0.006 0.011	56 54
D	Cosmos 577 rocket	1973-48B	1973 Jul 25.48 6.59 days 1973 Aug 1.07	Cylinder 2500?	7.5 long 2.6 dia	1973 Jul 27.3	65.39	89.17	6612	192	276	0.006	40
D	Cosmos 577 engine*	1973-48 D	1973 Jul 25.48 18 days 1973 Aug 12	Cone 600? full	1.5 long? 2 dia?	1973 Aug 6.8	65.41	89.03	6605	172	282	0.008	50
D	Fragments	1973-48C, E											
D	Mars 5 launcher rocket	1973-49B	1973 Jul 25.79 1.47 days 1973 Jul 27.26	Cylinder 4000?	12 long? 4 dia	1973 Jul 26.6	51.55	87.86	6549	153	189	0.003	326
D	Mars 5 launcher	1973-49C	1973 Jul 25.79 2 days 1973 Jul 27	Irregular -	-	1973 Jul 26.5	51.55	87.77	6545	159	174	0.001	-
D 3M R	Skylab 3** [Saturn 207]	1973-50A	1973 Jul 28.47 59.46 days 1973 Sep 25.93	Cone-cylinder 13860	10.36 long 3.91 dia	1973 Jul 28.5 1973 Jul 28.9	50.03 50.03	88.33 93.15	6573 6810	159 425	230 438	0.005 0.001	- -
D	Skylab 3 rocket	1973-50B	1973 Jul 28.47 0.24 day 1973 Jul 28.71	Cylinder 13600?	18.7 long 6.6 dia	1973 Jul 28.5	50.03	88.26	6569	159	223	0.005	56

* 1973-48 D was ejected from 1973-48A on 1973 Aug 6

Space Vehicles: Mars 4, 1973-47A; and Mars 5, 1973-49A

** Skylab 3 docked with Skylab 1 on 1973 Jul 28.82

	Name		Launch date, lifetime and descent date	Shape and weight (kg)	Size (m)	Date of orbital determination	Orbital inclination (deg)	Nodal period (min)	Semi major axis (km)	Perigee height (km)	Apogee height (km)	Orbital eccentricity	Argument of perigee (deg)
D R	Cosmos 578	1973-51A	1973 Aug 1.59 11.7 days 1973 Aug 13.3	Sphere-cylinder 5700?	5.0 long 2.4 dia	1973 Aug 2.2	65.38	89.41	6624	200	292	0.007	47
D	Cosmos 578 rocket	1973-51B	1973 Aug 1.59 7.72 days 1973 Aug 9.31	Cylinder 2500?	7.5 long 2.6 dia	1973 Aug 2.4	65.38	89.27	6617	200	278	0.006	48
D	Fragments	1973-51C-E											
D	Mars 6 launcher rocket	1973-52B	1973 Aug 5.74 1.52 days 1973 Aug 7.26	Cylinder 4000?	12 long? 4 dia	1973 Aug 6.5	51.54	87.92	6552	155	193	0.003	320
D	Mars 6 launcher	1973-52C	1973 Aug 5.74 2 days 1973 Aug 7	Irregular -	-	1973 Aug 6.5	51.5	87.91	6552	154	193	0.003	-
D	Mars 7 launcher rocket	1973-53B	1973 Aug 9.71 1.46 days 1973 Aug 11.17	Cylinder 4000?	12 long? 4 dia	1973 Aug 10.4	51.51	87.91	6552	154	193	0.003	328
D	Mars 7 launcher	1973-53C	1973 Aug 9.71 2 days 1973 Aug 11	Irregular -	-	Orbit similar to 1973-53B							
T	DMSP 4 [Thor Burner 2]	1973-54A	1973 Aug 17.20 80 years	12-faced frustum 195	1.64 long 1.31 to 1.10 dia	1973 Aug 17.5	98.86	101.58	7210	811	852	0.003	242
	Burner 2 rocket	1973-54B	1973 Aug 17.20 60 years	Sphere-cone 66	1.32 long 0.94 dia	1973 Aug 18.4	98.84	101.54	7208	808	851	0.003	245

Space Vehicles: Mars 6, 1973-52A; and Mars 7, 1973-53A.

	Name		Launch date, lifetime and descent date	Shape and weight (kg)	Size (m)	Date of orbital determination	Orbital inclination (deg)	Nodal period (min)	Semi major axis (km)	Perigee height (km)	Apogee height (km)	Orbital eccentricity	Argument of perigee (deg)
D R	Cosmos 579	1973-55A	1973 Aug 21.52 12.7 days 1973 Sep 3.2	Sphere-cylinder 6300?	6.5 long? 2.4 dia	1973 Aug 22.6 1973 Aug 24.7	65.41 65.42	89.27 89.38	6617 6623	196 175	282 314	0.007 0.010	58 53
D	Cosmos 579 rocket	1973-55B	1973 Aug 21.52 6.64 days 1973 Aug 28.16	Cylinder 2500?	7.5 long 2.6 dia	1973 Aug 23.2	65.41	89.22	6615	194	279	0.006	45
D	Cosmos 579 engine*	1973-55D	1973 Aug 21.52 14.86 days 1973 Sep 5.38	Cone 600? full	1.5 long? 2 dia?	1973 Sep 3.0	65.43	88.65	6586	169	247	0.006	49
D	Fragments	1973-55C,E											
	SDS-B [Titan 3B Agena D]	1973-56A	1973 Aug 21.67 100 years?	Cylinder?	-	1973 Sep 1.0	63.29	705.68	26256	460	39296	0.740	269
	SDS-B rocket	1973-56B	1973 Aug 21.67 100 years?	Cylinder 700?	6 long? 1.5 dia	1973 Sep 1.4 1975 Jan 1.0	63.27 63.21	699.80 697.4	26111 26051	360 808	39105 38538	0.742 0.724	269 -
B	Cosmos 580	1973-57A	1973 Aug 22.48 221.70 days 1974 Apr 1.18	Ellipsoid 400?	1.8 long 1.2 dia	1973 Aug 23.9 1973 Dec 16.5	71.00 70.99	92.22 91.23	6761 6714	273 256	493 415	0.016 0.012	76 -
D	Cosmos 580 rocket	1973-57B	1973 Aug 22.48 100.48 days 1973 Nov 30.96	Cylinder 1500?	8 long 1.65 dia	1973 Aug 23.6	71.00	92.08	6754	273	479	0.015	75
T	Intelsat 4 F-7	1973-58A	1973 Aug 23.96 > million years	Cylinder 1410 full 720 empty	2.82 long 2.39 dia	1973 Aug 24.0 1973 Aug 25.5 1973 Oct 1.0	27.38 0.4 0.3	657.0 1432.7 1436.3	25026 42111 42169	570 35539 35784	36726 35927 35797	0.722 0.005 0.0002	180 - -
	Intelsat 4 F-7 rocket	1973-58B	1973 Aug 23.96 6000 years	Cylinder 1815	8.6 long 3.0 dia	1973 Aug 28.7	27.50	655.2	24983	597	36612	0.721	182

* 1973-55D was ejected from 1973-55A on 1973 Sep 2.

	Name		Launch date, lifetime and descent date	Shape and weight (kg)	Size (m)	Date of orbital determination	Orbital inclination (deg)	Nodal period (min)	Semi major axis (km)	Perigee height (km)	Apogee height (km)	Orbital eccentricity	Argument of perigee (deg)
D R	Cosmos 581	1973-59A	1973 Aug 24.47 12.8 days 1973 Sep 6.3	Sphere-cylinder 6300?	6.5 long? 2.4 dia	1973 Aug 25.3 1973 Aug 26.6	51.62 51.61	89.40 89.00	6626 6606	208 172	288 284	0.006 0.008	44 62
D	Cosmos 581 rocket	1973-59B	1973 Aug 24.47 6.66 days 1973 Aug 31.13	Cylinder 2500?	7.5 long 2.6 dia	1973 Aug 25.6	51.62	89.15	6614	202	269	0.005	34
D	Cosmos 581 engine*	1973-59E	1973 Aug 24.47 15.78 days 1973 Sep 9.25	Cone 600? full	1.5 long? 2 dia?	1973 Sep 5.6	51.62	88.86	6599	176	266	0.007	103
D	Fragments	1973-59C,D											
D	Cosmos 582	1973-60A	1973 Aug 28.42 2565 days 1980 Sep 5	Cylinder + paddles? 900?	2 long? 1 dia?	1973 Aug 31.3	74.04	95.27	6909	519	543	0.002	359
D	Cosmos 582 rocket	1973-60B	1973 Aug 28.42 2683 days 1981 Jan 1	Cylinder 2200?	7.4 long 2.4 dia	1973 Sep 1.0	74.04	95.19	6905	510	544	0.003	14
D	Fragments	1973-60C-E											
D	Molniya 1Z	1973-61A	1973 Aug 30.01 2049 days? 1979 Apr 10?	Windmill + 6 vanes 1000?	3.4 long 1.6 dia	1973 Sep 6.2	65.47	717.77	26556	463	39893	0.742	284
D	Molniya 1Z launcher rocket	1973-61B	1973 Aug 30.01 32.77 days 1973 Oct 1.78	Cylinder 2500?	7.5 long 2.6 dia	1973 Aug 31.0	65.43	91.30	6717	219	458	0.018	65
D	Molniya 1Z launcher	1973-61C	1973 Aug 30.01 39.48 days 1973 Oct 8.49	Irregular	-	1973 Aug 30.9	65.40	91.57	6730	223	481	0.019	69

* 1973-59E ejected from 1973-59A on 1973 Sep 5.

1973-61 continued on page 339

	Name		Launch date, lifetime and descent date	Shape and weight (kg)	Size (m)	Date of orbital determination	Orbital inclination (deg)	Nodal period (min)	Semi major axis (km)	Perigee height (km)	Apogee height (km)	Orbital eccentricity	Argument of perigee (deg)
D	Molniya 1Z rocket	1973-61F	1973 Aug 30.01 2038 days? 1979 Mar 30?	Cylinder 440	2.0 long 2.0 dia	1973 Oct 1.0	65.3	678.2	25572	435	37953	0.733	-
D	Fragments	1973-61D,E											
D R	Cosmos 583	1973-62A	1973 Aug 30.44 12.9 days 1973 Sep 12.3	Sphere-cylinder 5700?	5.0 long 2.4 dia	1973 Aug 31.6	64.92	89.52	6629	204	298	0.007	46
D	Cosmos 583 rocket	1973-62B	1973 Aug 30.44 6.74 days 1973 Sep 6.18	Cylinder 2500?	7.5 long 2.6 dia	1973 Aug 31.2	64.92	89.35	6621	200	285	0.006	37
D	Fragment	1973-62C											
D R	Cosmos 584	1973-63A	1973 Sep 6.45 13.74 days 1973 Sep 20.19	Sphere-cylinder 6300?	6.5 long? 2.4 dia	1973 Sep 7.5 1973 Sep 11.0 1973 Sep 15.4	72.85 72.84 72.85	89.95 89.21 89.70	6649 6613 6637	205 204 204	336 265 314	0.010 0.005 0.008	63 65 60
D	Cosmos 584 rocket	1973-63B	1973 Sep 6.45 10.70 days 1973 Sep 17.15	Cylinder 2500?	7.5 long 2.6 dia	1973 Sep 7.2	72.85	89.83	6643	202	327	0.009	60
D	Cosmos 584 engine*	1973-63E	1973 Sep 6.45 22.44 days 1973 Sep 28.89	Cone 600? full	1.5 long? 2 dia?	1973 Sep 19.5	72.85	89.60	6632	203	305	0.008	47
D	Fragments	1973-63C,D,F,G											
	Cosmos 585	1973-64A	1973 Sep 8.07 6000 years	- 500?	-	1973 Sep 8.4	73.99	113.63	7770	1368	1416	0.003	237
	Cosmos 585 rocket	1973-64B	1973 Sep 8.07 4000 years	Cylinder 2200?	7.4 long 2.4 dia	1973 Sep 8.7	73.97	113.49	7763	1375	1395	0.001	218
	Fragment	1973-64C											

*1973-63E ejected from 1973-63A on 1973 Sep 19

	Name		Launch date, lifetime and descent date	Shape and weight (kg)	Size (m)	Date of orbital determination	Orbital inclination (deg)	Nodal period (min)	Semi major axis (km)	Perigee height (km)	Apogee height (km)	Orbital eccentricity	Argument of perigee (deg)
	Cosmos 586	1973-65A	1973 Sep 14.02 1200 years	Cylinder? 700?	1.3 long? 1.9 dia?	1973 Sep 16.9	82.94	104.89	7368	971	1009	0.003	267
	Cosmos 586 rocket	1973-65B	1973 Sep 14.02 600 years	Cylinder 2200?	7.4 long 2.4 dia	1973 Sep 16.1	82.95	104.75	7361	969	997	0.002	260
D R	Cosmos 587	1973-66A	1973 Sep 21.55 12.8 days 1973 Oct 4.3	Sphere-cylinder 6300?	6.5 long? 2.4 dia	1973 Sep 22.5 1973 Sep 26.6	65.42 65.44	89.55 89.18	6631 6612	205 174	300 294	0.007 0.009	66 57
D	Cosmos 587 rocket	1973-66B	1973 Sep 21.55 6.49 days 1973 Sep 28.04	Cylinder 2500?	7.5 long 2.6 dia	1973 Sep 22.2	65.41	89.41	6624	201	290	0.007	60
D	Cosmos 587 engine*	1973-66D	1973 Sep 21.55 17 days 1973 Oct 8	Cone 600? full	1.5 long? 2 dia?	1973 Oct 3.3	65.42	89.40	6623	170	320	0.011	50
D	Fragments	1973-66C,E											
D 2M R	Soyuz 12	1973-67A	1973 Sep 27.51 1.97 days 1973 Sep 29.48	Sphere-cylinder 6570?	7.5 long 2.2 dia	1973 Sep 27.6 1973 Sep 27.8	51.58 51.58	88.54 91.20	6583 6713	181 326	229 344	0.004 0.001	10 301
D	Soyuz 12 rocket	1973-67B	1973 Sep 27.51 2.02 days 1973 Sep 29.53	Cylinder 2500?	7.5 long 2.6 dia	1973 Sep 27.8	51.58	88.49	6581	186	219	0.003	70
D	Soyuz 12 orbital module	1973-67C	1973 Sep 27.51 116 days 1974 Jan 21	Spheroid 1200?	2.5 long 2.2 dia	1973 Oct 1.0	51.58	91.07	6707	311	346	0.003	-
D	Fragment	1973-67D											

*1973-66D ejected from 1973-66A on 1973 Oct 3

	Name		Launch date, lifetime and descent date	Shape and weight (kg)	Size (m)	Date of orbital determination	Orbital inclination (deg)	Nodal period (min)	Semi major axis (km)	Perigee height (km)	Apogee height (km)	Orbital eccentricity	Argument of perigee (deg)
D	[Titan 3B Agena D]	1973-68A	1973 Sep 27.72 32 days 1973 Oct 29	Cylinder 3000?	8 long? 1.5 dia	1973 Sep 28.3	110.48	89.67	6636	131	385	0.019	146
	Cosmos 588	1973-69A	1973 Oct 2.91 10000 years	Spheroid 40?	1.0 long? 0.8 dia?	1973 Oct 3.0	74.00	115.37	7851	1451	1494	0.003	150
	Cosmos 589	1973-69B	1973 Oct 2.91 9000 years	Spheroid 40?	1.0 long? 0.8 dia?	1973 Oct 3.0	74.01	114.95	7831	1419	1487	0.004	123
	Cosmos 590	1973-69C	1973 Oct 2.91 10000 years	Spheroid 40?	1.0 long? 0.8 dia?	1973 Oct 3.0	74.00	115.15	7840	1438	1486	0.003	125
	Cosmos 591	1973-69D	1973 Oct 2.91 6000 years	Spheroid 40?	1.0 long? 0.8 dia?	1973 Oct 3.0	74.00	114.20	7797	1349	1488	0.009	108
	Cosmos 592	1973-69E	1973 Oct 2.91 6000 years	Spheroid 40?	1.0 long? 0.8 dia?	1973 Oct 3.0	74.00	114.01	7788	1333	1486	0.010	102
	Cosmos 593	1973-69F	1973 Oct 2.91 7000 years	Spheroid 40?	1.0 long? 0.8 dia?	1973 Oct 3.0	74.00	114.39	7805	1366	1487	0.008	108
	Cosmos 594	1973-69G	1973 Oct 2.91 8000 years	Spheroid 40?	1.0 long? 0.8 dia?	1973 Oct 3.0	74.01	114.57	7813	1382	1488	0.007	110
	Cosmos 595	1973-69H	1973 Oct 2.91 8000 years	Spheroid 40?	1.0 long? 0.8 dia?	1973 Oct 3.0	74.00	114.77	7822	1402	1486	0.005	111
	Cosmos 588 rocket	1973-69J	1973 Oct 2.91 20000 years	Cylinder 2200?	7.4 long 2.4 dia	1973 Oct 3.0	74.01	117.19	7933	1485	1625	0.009	260

	Name		Launch date, lifetime and descent date	Shape and weight (kg)	Size (m)	Date of orbital determination	Orbital inclination (deg)	Nodal period (min)	Semi major axis (km)	Perigee height (km)	Apogee height (km)	Orbital eccentricity	Argument of perigee (deg)
D R	Cosmos 596	1973-70A	1973 Oct 3.54 5.78 days 1973 Oct 9.32	Sphere-cylinder 5900?	5.9 long? 2.4 dia	1973 Oct 3.8	65.41	89.42	6625	206	287	0.006	66
D	Cosmos 596 rocket	1973-70B	1973 Oct 3.54 5.85 days 1973 Oct 9.39	Cylinder 2500?	7.5 long 2.6 dia	1973 Oct 4.3	65.41	89.24	6616	200	275	0.006	51
D	Capsule* ?	1973-70C	1973 Oct 3.54 13.93 days 1973 Oct 17.47	Ellipsoid ? 200?	0.9 long? 1.9 dia?	1973 Oct 9.4	65.40	89.25	6616	203	273	0.005	61
D R	Cosmos 597	1973-71A	1973 Oct 6.52 5.8 days 1973 Oct 12.3	Sphere-cylinder 6300?	6.5 long? 2.4 dia	1973 Oct 6.8 1973 Oct 8.6	65.42 65.42	89.45 89.07	6626 6607	206 207	290 251	0.006 0.003	64 79
D	Cosmos 597 rocket	1973-71B	1973 Oct 6.52 8.07 days 1973 Oct 14.59	Cylinder 2500?	7.5 long 2.6 dia	1973 Oct 6.9	65.40	89.35	6621	206	280	0.006	59
D	Cosmos 597 engine**	1973-71E	1973 Oct 6.52 12 days 1973 Oct 18	Cone 600? full	1.5 long? 2 dia?	1973 Oct 11.7	65.45	88.75	6591	200	226	0.002	83
D	Fragments	1973-71C,D											
D R	Cosmos 598	1973-72A	1973 Oct 10.45 5.8 days 1973 Oct 16.3	Sphere-cylinder 6300?	6.5 long? 2.4 dia	1973 Oct 10.8 1973 Oct 14.0	72.84 72.84	89.94 89.02	6649 6604	208 204	334 247	0.010 0.003	66 72
D	Cosmos 598 rocket	1973-72B	1973 Oct 10.45 10.58 days 1973 Oct 21.03	Cylinder 2500?	7.5 long 2.6 dia	1973 Oct 11.1	72.84	89.82	6643	203	327	0.009	60

* 1973-70C ejected from 1973-70A on 1973 Oct 9

** 1973-71E ejected from 1973-71A on 1973 Oct 11

1973-72 continued on page 343

	Name		Launch date, lifetime and descent date	Shape and weight (kg)	Size (m)	Date of orbital determination	Orbital inclination (deg)	Nodal period (min)	Semi major axis (km)	Perigee height (km)	Apogee height (km)	Orbital eccentricity	Argument of perigee (deg)
D	Cosmos 598 engine*	1973-72D	1973 Oct 10.45 10 days 1973 Oct 20	Cone 600? full	1.5 long? 2 dia?	1973 Oct 15.8	72.85	88.92	6599	199	242	0.003	67
D	Fragments	1973-72C,E											
D R	Cosmos 599	1973-73A	1973 Oct 15.37 12.9 days 1973 Oct 28.3	Sphere-cylinder 5700?	5.0 long 2.4 dia	1973 Oct 15.5	64.94	89.32	6619	202	280	0.006	38
D	Cosmos 599 rocket	1973-73B	1973 Oct 15.37 5.73 days 1973 Oct 21.10	Cylinder 2500?	7.5 long 2.6 dia	1973 Oct 16.3	64.94	89.18	6612	200	268	0.005	35
D	Fragments	1973-73C,D											
D R	Cosmos 600	1973-74A	1973 Oct 16.51 6.8 days 1973 Oct 23.3	Sphere-cylinder 6300?	6.5 long? 2.4 dia	1973 Oct 17.1 1973 Oct 18.7	72.83 72.82	89.97 88.90	6651 6597	205 201	340 237	0.010 0.003	71 105
D	Cosmos 600 rocket	1973-74B	1973 Oct 16.51 11.64 days 1973 Oct 28.15	Cylinder 2500?	7.5 long 2.6 dia	1973 Oct 17.0	72.83	89.90	6647	205	333	0.010	73
D	Cosmos 600 engine**	1973-74E	1973 Oct 16.51 13 days 1973 Oct 29	Cone 600? full	1.5 long? 2 dia?	1973 Oct 26.4	72.85	90.62	6683	256	354	0.007	355
D	Fragments	1973-74C,D											

* 1973-72D ejected from 1973-72A on 1973 Oct 15
** 1973-74E ejected from 1973-74A on 1973 Oct 22

	Name		Launch date, lifetime and descent date	Shape and weight (kg)	Size (m)	Date of orbital determination	Orbital inclination (deg)	Nodal period (min)	Semi major axis (km)	Perigee height (km)	Apogee height (km)	Orbital eccentricity	Argument of perigee (deg)
D	Cosmos 601	1973-75A	1973 Oct 16.59 303 days 1974 Aug 15	Ellipsoid 400?	1.8 long 1.2 dia	1973 Oct 18.5 1974 Mar 1.0	81.91 81.91	102.28 98.43	7244 7060	200 194	1531 1169	0.092 0.069	68 -
D	Cosmos 601 rocket	1973-75C	1973 Oct 16.59 162.98 days 1974 Mar 28.57	Cylinder 1500?	8 long 1.65 dia	1973 Oct 22.5 1974 Jan 1.0	81.92 81.92	102.05 98.30	7232 7053	210 204	1497 1146	0.089 0.067	58 -
D	Fragments	1973-75B,D-P											
	Molniya 2G	1973-76A	1973 Oct 19.44 9¾ years	Windmill + 6 vanes 1250?	4.2 long? 1.6 dia?	1973 Oct 25.6	62.84	717.93	26560	509	39855	0.741	289
D	Molniya 2G launcher rocket	1973-76B	1973 Oct 19.44 45.53 days 1973 Dec 3.97	Cylinder 2500?	7.5 long 2.6 dia	1973 Oct 21.6	62.81	91.94	6748	216	524	0.023	123
D	Molniya 2G launcher	1973-76C	1973 Oct 19.44 60.23 days 1973 Dec 18.67	Irregular	-	1973 Oct 21.6	62.82	92.62	6782	209	599	0.029	122
	Molniya 2G rocket	1973-76D	1973 Oct 19.44 9.7 years	Cylinder 440	2.0 long 2.0 dia	1973 Oct 29.2	62.87	733.19	26935	597	40517	0.741	288
D R	Cosmos 602	1973-77A	1973 Oct 20.43 8.8 days 1973 Oct 29.2	Sphere-cylinder 6300?	6.5 long? 2.4 dia	1973 Oct 20.7 1973 Oct 23.0	72.88 72.86	89.97 89.31	6651 6618	210 170	335 309	0.009 0.011	70 40
D	Cosmos 602 rocket	1973-77B	1973 Oct 20.43 10.90 days 1973 Oct 31.33	Cylinder 2500?	7.5 long 2.6 dia	1973 Oct 21.1	72.85	89.84	6644	201	331	0.010	60
D	Cosmos 602 engine*	1973-77D	1973 Oct 20.43 12 days 1973 Nov 1	Cone 600? full	1.5 long? 2 dia?	1973 Oct 28.7	72.87	88.71	6588	169	250	0.006	25
D	Fragment	1973-77C											

*1973-77D ejected from 1973-77A on 1973 Oct 28

	Name		Launch date, lifetime and descent date	Shape and weight (kg)	Size (m)	Date of orbital determination	Orbital inclination (deg)	Nodal period (min)	Semi major axis (km)	Perigee height (km)	Apogee height (km)	Orbital eccentricity	Argument of perigee (deg)
T	Explorer 50 (Imp 10)	1973-78A	1973 Oct 26.10 > million years	16-sided cylinder 371 full	1.58 long 1.35 dia	1973 Oct 26.1 1973 Oct 29.3	28.77 28.67	6971.0 17279	120881 221399	197 141185	228809 288857	0.946 0.333	120 52
D	Explorer 50 first stage	1973-78B	1973 Oct 26.10 3 days 1973 Oct 29	Cylinder 2750?	21.64 long 2.44 dia	1973 Oct 26.5	28.74	90.98	6708	147	513	0.027	76
	Explorer 50 second stage	1973-78C	1973 Oct 26.10 25 years	Cylinder 350?	4.9 long 1.43 dia	1973 Oct 27.2	28.85	112.30	7720	363	2321	0.127	172
D	Explorer 50 third stage	1973-78D	1973 Oct 26.10 3.4 years? 1977 Mar ?	Sphere - cone 66	1.32 long 0.94 dia	1973 Oct 26.1	28.77	6971.0	120881	197	228809	0.946	120
D	Fragment	1973-78E											
D R	Cosmos 603	1973-79A	1973 Oct 27.47 12.8 days 1973 Nov 9.3	Sphere-cylinder 6300?	6.5 long? 2.4 dia	1973 Oct 28.5 1973 Nov 5.4	72.86 72.84	90.15 89.12	6659 6608	205 172	357 288	0.011 0.009	67 41
D	Cosmos 603 rocket	1973-79B	1973 Oct 27.47 12.53 days 1973 Nov 9.00	Cylinder 2500?	7.5 long 2.6 dia	1973 Oct 28.7	72.84	90.00	6652	202	345	0.011	63
D	Cosmos 603 engine*	1973-79F	1973 Oct 27.47 17.26 days 1973 Nov 13.73	Cone 600? full	1.5 long? 2 dia?	1973 Nov 10.3	72.84	88.77	6591	169	257	0.007	32
D	Fragments	1973-79C-E											
	Cosmos 604	1973-80A	1973 Oct 29.59 60 years	Cylinder + 2 vanes ? 2500?	5 long? 1.5 dia?	1973 Oct 31.0	81.23	97.25	7004	615	636	0.002	290
	Cosmos 604 rocket	1973-80B	1973 Oct 29.59 60 years	Cylinder 1440	3.8 long 2.6 dia	1973 Oct 30.5	81.21	97.25	7004	583	668	0.005	203

* 1973-79F ejected from 1973-79A about 1973 Nov 8

	Name		Launch date, lifetime and descent date	Shape and weight (kg)	Size (m)	Date of orbital determination	Orbital inclination (deg)	Nodal period (min)	Semi major axis (km)	Perigee height (km)	Apogee height (km)	Orbital eccentricity	Argument of perigee (deg)
T	Navy Navigation Satellite 20	1973-81A	1973 Oct 30.03 900 years	Octagon + 4 vanes 58?	0.25 long? 0.46 dia?	1973 Oct 30.7	90.18	105.62	7400	895	1149	0.017	327
	Altair rocket	1973-81B	1973 Oct 30.03 600 years	Cylinder 24	1.50 long 0.46 dia	1973 Oct 31.0	90.19	105.59	7399	879	1162	0.019	325
D	Intercosmos 10	1973-82A	1973 Oct 30.79 1340 days 1977 Jul 1	Octagonal Ellipsoid 550?	1.8 long? 1.5 dia?	1973 Oct 31.6	74.03	102.10	7235	260	1454	0.083	120
D	Intercosmos 10 rocket	1973-82B	1973 Oct 30.79 1439 days 1977 Oct 8	Cylinder 2200?	7.4 long 2.4 dia	1973 Oct 31.5	74.03	102.00	7230	258	1446	0.082	120
D	Fragment	1973-82C											
D B R	Cosmos 605 *	1973-83A	1973 Oct 31.77 21.5 days 1973 Nov 22.3	Sphere-cylinder 5900?	5.9 long 2.4 dia	1973 Nov 1.8	62.80	90.66	6686	213	403	0.014	112
D	Cosmos 605 rocket	1973-83B	1973 Oct 31.77 27.75 days 1973 Nov 28.52	Cylinder 2500?	7.5 long 2.6 dia	1973 Nov 1.6	62.78	90.59	6683	213	396	0.014	110
D	Capsule**	1973-83C	1973 Oct 31.77 77 days 1974 Jan 16	Ellipsoid 200?	0.9 long 1.9 dia	1973 Dec 1.0	62.3	90.22	6664	193	379	0.014	-
D	Fragments	1973-83D-F											

* Biological satellite.

** 1973-83C ejected from 1973-83A on 1973 Nov 22.

	Name		Launch date, lifetime and descent date	Shape and weight (kg)	Size (m)	Date of orbital determination	Orbital inclination (deg)	Nodal period (min)	Semi major axis (km)	Perigee height (km)	Apogee height (km)	Orbital eccentricity	Argument of perigee (deg)
	Cosmos 606	1973-84A	1973 Nov 2.54 100 years?	Windmill + 6 vanes? 1250?	4.2 long? 1.6 dia?	1973 Nov 3.6 1973 Dec 1.0	62.91 62.79	709.92 717.51	26361 26550	657 635	39310 39708	0.733 0.736	318 -
D	Cosmos 606 launcher rocket	1973-84B	1973 Nov 2.54 64.88 days 1974 Jan 6.42	Cylinder 2500?	7.5 long 2.6 dia	1973 Nov 3.4	62.79	92.38	6770	218	566	0.026	118
D	Cosmos 606 launcher	1973-84C	1973 Nov 2.54 88 days 1974 Jan 29	Irregular	-	1973 Nov 3.7	62.82	92.90	6796	215	621	0.030	120
	Cosmos 606 rocket	1973-84D	1973 Nov 2.54 100 years?	Cylinder 440	2.0 long 2.0 dia	1973 Dec 2.5	62.75	706.54	26277	654	39144	0.732	318
	NOAA 3 (ITOS)	1973-86A	1973 Nov 6.71 10000 years	Box 306	1.25 long 1.02 square	1973 Nov 7.0	102.08	116.12	7883	1500	1509	0.0006	255
	NOAA 3 second stage*	1973-86B	1973 Nov 6.71 disintegrated	Cylinder 350?	4.9 long 1.43 dia	1973 Nov 8.8	102.06	116.18	7886	1503	1512	0.0006	230
12d	Fragments	1973-86C-GN											
D R	Cosmos 607	1973-87A	1973 Nov 10.53 11.8 days 1973 Nov 22.3	Sphere- cylinder 6300?	6.5 long? 2.4 dia	1973 Nov 11.5 1973 Nov 17.8	72.83 72.84	89.98 89.70	6651 6637	204 173	341 344	0.010 0.013	66 59
D	Cosmos 607 rocket	1973-87B	1973 Nov 10.53 12.76 days 1973 Nov 23.29	Cylinder 2500?	7.5 long 2.6 dia	1973 Nov 10.8	72.82	89.88	6646	204	332	0.010	69
D	Cosmos 607 engine**	1973-87D	1973 Nov 10.53 17.62 days 1973 Nov 28.15	Cone 600? full	1.5 long? 2 dia?	1973 Nov 21.7	72.82	89.56	6630	169	334	0.012	50
D	Fragments	1973-87C, E											

* NOAA 3 second stage disintegrated on 1973 Dec 28.38 near 37 deg South, 178 deg West
** 1973-87D ejected from 1973-87A on 1973 Nov 21.

Space Vehicle: Mariner 10 (1973-85A) and Centaur rocket (1973-85B).

	Name		Launch date, lifetime and descent date	Shape and weight (kg)	Size (m)	Date of orbital determination	Orbital inclination (deg)	Nodal period (min)	Semi major axis (km)	Perigee height (km)	Apogee height (km)	Orbital eccentricity	Argument of perigee (deg)
D	[Titan 3D]	1973-88A	1973 Nov 10.84 123 days 1974 Mar 13	Cylinder 13300? full	15 long 3.0 dia	1973 Nov 11.2	96.93	88.85	6595	159	275	0.009	151
D	Titan 3D rocket	1973-88C	1973 Nov 10.84 2.38 days 1973 Nov 13.22	Cylinder 1900	6 long 3.0 dia	1973 Nov 11.0	96.92	88.67	6586	159	257	0.008	150
D	Capsule	1973-88B	1973 Nov 10.84 1872 days 1978 Dec 26	Octagon 60?	0.3 long? 0.9 dia?	1973 Nov 11.9	96.33	94.59	6875	486	508	0.002	60
T	Capsule	1973-88D	1973 Nov 10.84 10000 years	Octagon 60?	0.3 long? 0.9 dia?	1973 Dec 1.0	96.93	114.64	7817	1419	1458	0.002	-
	Fragment	1973-88E											
D	Molniya 1AA	1973-89A	1973 Nov 14.86 2101 days? 1979 Aug 16?	Windmill + 6 vanes 1000?	3.4 long 1.6 dia	1973 Nov 19.3 1973 Nov 22.7	64.92 64.90	702.37 717.97	26204 26561	454 566	39197 39800	0.740 0.739	285 285
D	Molniya 1AA launcher rocket	1973-89B	1973 Nov 14.86 49.05 days 1974 Jan 2.91	Cylinder 2500?	7.5 long 2.6 dia	1973 Nov 15.6	64.96	91.33	6718	235	445	0.016	68
D	Molniya 1AA launcher	1973-89D	1973 Nov 14.86 72.68 days 1974 Jan 26.54	Irregular	-	1973 Nov 15.9	64.93	91.70	6737	241	476	0.017	70
D	Molniya 1AA rocket	1973-89E	1973 Nov 14.86 2105 days 1979 Aug 20	Cylinder 440	2.0 long 2.0 dia	1973 Nov 18.3	64.92	698.64	26082	412	38996	0.740	285
D	Fragment	1973-89C											

	Name		Launch date, lifetime and descent date	Shape and weight (kg)	Size (m)	Date of orbital determination	Orbital inclination (deg)	Nodal period (min)	Semi major axis (km)	Perigee height (km)	Apogee height (km)	Orbital eccentricity	Argument of perigee (deg)
D 3M R	Skylab 4* [Saturn 208]	1973-90A	1973 Nov 16.58 84.06 days 1974 Feb 8.64	Cone-cylinder 13980?	10.36 long 3.91 dia	1973 Nov 16.6 1973 Nov 17.0	50.04 50.04	88.22 93.11	6567 6808	154 422	224 437	0.005 0.001	- 322
D	Skylab 4 rocket	1973-90B	1973 Nov 16.58 $<\frac{1}{2}$ day 1973 Nov 16	Cylinder 13600?	18.7 long 6.6 dia	1973 Nov 16.8	50.06	88.16	6564	150	222	0.005	44
D	Cosmos 608	1973-91A	1973 Nov 20.52 231.64 days 1974 Jul 10.16	Ellipsoid 400?	1.8 long 1.2 dia	1973 Nov 21.2	70.97	92.29	6765	270	503	0.017	81
D	Cosmos 608 rocket	1973-91B	1973 Nov 20.52 122.59 days 1974 Mar 23.11	Cylinder 1500?	8 long 1.65 dia	1973 Nov 21.4	70.97	92.11	6756	274	481	0.015	82
D R	Cosmos 609	1973-92A	1973 Nov 21.42 12.9 days 1973 Dec 4.3	Sphere-cylinder 6300?	6.5 long? 2.4 dia	1973 Nov 22.4 1973 Nov 22.8	69.95 69.94	90.07 89.67	6656 6636	241 174	314 341	0.006 0.013	58 58
D	Cosmos 609 rocket	1973-92B	1973 Nov 21.42 12.90 days 1973 Dec 4.32	Cylinder 2500?	7.5 long 2.6 dia	1973 Nov 22.5	69.95	89.86	6645	207	327	0.009	40
D	Cosmos 609 engine**	1973-92C	1973 Nov 21.42 15.14 days 1973 Dec 6.56	Cone 600? full	1.5 long? 2 dia?	1973 Dec 4.0	69.96	88.94	6599	173	269	0.007	34
D	Fragment	1973-92D											
D	Cosmos 610	1973-93A	1973 Nov 27.01 2484 days 1980 Sep 15	Cylinder + paddles? 900?	2 long? 1 dia?	1973 Nov 27.4	74.04	95.27	6909	515	546	0.002	4
D	Cosmos 610 rocket	1973-93B	1973 Nov 27.01 2536 days 1980 Nov 6	Cylinder 2200?	7.4 long 2.4 dia	1973 Dec 1.6	74.04	95.09	6900	500	544	0.003	22
D	Fragment	1973-93C											

* Skylab 4 rendezvous with Skylab 1 on 1973 Nov 16.89; docked 1973 Nov 16.92; undocked 1974 Feb 8.44

** 1973-92C ejected from 1973-92A on 1973 Dec 3

	Name		Launch date, lifetime and descent date	Shape and weight (kg)	Size (m)	Date of orbital determination	Orbital inclination (deg)	Nodal period (min)	Semi major axis (km)	Perigee height (km)	Apogee height (km)	Orbital eccentricity	Argument of perigee (deg)
D	Cosmos 611	1973-94A	1973 Nov 28.42 203.35 days 1974 Jun 19.77	Ellipsoid 400?	1.8 long 1.2 dia	1973 Nov 30.3	70.97	92.06	6754	270	481	0.016	86
D	Cosmos 611 rocket	1973-94B	1973 Nov 28.42 118.31 days 1974 Mar 26.73	Cylinder 1500?	8 long 1.65 dia	1973 Nov 30.6	70.97	91.86	6744	271	460	0.014	88
D R	Cosmos 612	1973-95A	1973 Nov 28.49 12.8 days 1973 Dec 11.3	Sphere- cylinder 6300?	6.5 long? 2.4 dia	1973 Nov 28.8 1973 Dec 4.5	72.82 72.84	90.05 89.78	6654 6641	206 187	346 338	0.010 0.011	69 47
D	Cosmos 612 rocket	1973-95B	1973 Nov 28.49 14.17 days 1973 Dec 12.66	Cylinder 2500?	7.5 long 2.6 dia	1973 Nov 28.8	72.84	89.95	6649	205	337	0.010	67
D	Cosmos 612 engine*	1973-95C	1973 Nov 28.49 17.66 days 1973 Dec 16.15	Cone 600? full	1.5 long? 2 dia?	1973 Dec 10.4	72.84	89.16	6610	167	297	0.010	50
D	Fragments	1973-95D,E											
D R	Cosmos 613	1973-96A	1973 Nov 30.22 60.1 days 1974 Jan 29.3	Sphere- cylinder + 2 wings 6570?	7.5 long 2.2 dia	1973 Dec 1.1 1973 Dec 8.4	51.60 51.59	89.05 90.99	6609 6704	188 255	273 396	0.006 0.011	85 296
D	Cosmos 613 rocket	1973-96B	1973 Nov 30.22 4.29 days 1973 Dec 4.51	Cylinder 2500?	7.5 long 2.6 dia	1973 Dec 1.1	51.60	88.90	6601	184	262	0.006	86
D	Cosmos 613 orbital module	1973-96C	1973 Nov 30.22 99 days 1974 Mar 9	Spheroid 1200?	2.5 long 2.2 dia	1974 Jan 31.9	51.60	90.49	6679	239	363	0.009	-

* 1973-95C ejected from 1973-95A on 1973 Dec 10

	Name		Launch date, lifetime and descent date	Shape and weight (kg)	Size (m)	Date of orbital determination	Orbital inclination (deg)	Nodal period (min)	Semi major axis (km)	Perigee height (km)	Apogee height (km)	Orbital eccentricity	Argument of perigee (deg)
	Molniya 1AB	1973-97A	1973 Nov 30.55 11½ years	Windmill + 6 vanes 1000?	3.4 long 1.6 dia	1973 Dec 2.6 1974 Feb 1.0	62.89 62.7	740.03 717.71	27102 26555	619 484	40829 39869	0.742 0.742	284 -
D	Molniya 1AB launcher rocket	1973-97B	1973 Nov 30.55 29.92 days 1973 Dec 30.47	Cylinder 2500?	7.5 long 2.6 dia	1973 Dec 1.7	62.80	90.82	6694	216	415	0.015	128
D	Molniya 1AB launcher	1973-97C	1973 Nov 30.55 30.52 days 1973 Dec 31.07	Irregular	-	1973 Dec.1.7	62.80	90.98	6702	207	440	0.017	126
	Molniya 1AB rocket	1973-97D	1973 Nov 30.55 11½ years	Cylinder 440	2.0 long 2.0 dia	1974 Feb 19.2	62.79	734.90	26995	452	40782	0.747	281
	Cosmos 614	1973-98A	1973 Dec 4.63 120 years	Cylinder + paddles? 750?	2 long? 1 dia?	1973 Dec 4.8	74.06	100.66	7167	770	805	0.002	2
	Cosmos 614 rocket	1973-98B	1973 Dec 4.63 100 years	Cylinder 2200?	7.4 long 2.4 dia	1973 Dec 6.2	74.06	100.56	7162	765	803	0.003	356
	Fragment	1973-98C											
D	Cosmos 615	1973-99A	1973 Dec 13.47 734 days 1975 Dec 17	Ellipsoid 400?	1.8 long 1.2 dia	1973 Dec 13.8	71.02	95.70	6930	270	834	0.041	83
D	Cosmos 615 rocket	1973-99B	1973 Dec 13.47 406.98 days 1975 Jan 24.45	Cylinder 1500?	8 long 1.65 dia	1973 Dec 15.3	71.02	95.61	6926	271	824	0.040	83

	Name		Launch date, lifetime and descent date	Shape and weight (kg)	Size (m)	Date of orbital determination	Orbital inclination (deg)	Nodal period (min)	Semi major axis (km)	Perigee height (km)	Apogee height (km)	Orbital eccentricity	Argument of perigee (deg)
	DSCS 3 [Titan 3C]	1973-100A	1973 Dec 14.00 >million years	Cylinder + 2 dishes 565	1.83 long 2.74 dia	1974 Jan 1.0	2.5	1436.3	42169	35790	35791	0	-
T	DSCS 4	1973-100B	1973 Dec 14.00 >million years	Cylinder + 2 dishes 565	1.83 long 2.74 dia	1974 Jan 1.0	2.5	1436.7	42177	35797	35801	0	-
D	Titan 3C second stage	1973-100C	1973 Dec 14.00 3.80 days 1973 Dec 17.80	Cylinder 1900	6 long 3.0 dia	1973 Dec 15.4	28.60	89.79	6641	133	393	0.020	135
	Transtage	1973-100D	1973 Dec 14.00 >million years	Cylinder 1500?	6 long? 3.0 dia	1974 Jan 1.0	2.5	1445.5	42349	35806	36136	0.004	-
D	Explorer 51 (AE-C) *	1973-101A	1973 Dec 16.26 1822 days 1978 Dec 12	16-sided cylinder 659 (490 empty)	1.14 long 1.36 dia	1973 Dec 17.3	68.12	132.50	8609	158	4303	0.241	166
D	Explorer 51 second stage	1973-101B	1973 Dec 16.26 241.38 days 1974 Aug 14.64	Cylinder 600?	5.2 long 2.44 dia	1973 Dec 17.3	68.11	132.43	8607	159	4298	0.240	166
D R	Cosmos 616	1973-102A	1973 Dec 17.50 10.80 days 1973 Dec 28.30	Sphere-cylinder 5900?	5.9 long 2.4 dia	1973 Dec 17.7	72.86	89.90	6647	206	332	0.009	66
D	Cosmos 616 rocket	1973-102B	1973 Dec 17.50 13.67 days 1973 Dec 31.17	Cylinder 2500?	7.5 long 2.6 dia	1973 Dec 18.2	72.88	89.83	6644	204	327	0.009	60
D	Capsule **	1973-102E	1973 Dec 17.50 16 days 1974 Jan 2	Ellipsoid 200?	0.9 long 1.9 dia	1973 Dec 28.4	72.90	89.50	6627	197	301	0.008	-
D	Fragments	1973-102C,D,F											

* Atmospheric Explorer C. Manoeuvrable satellite. The orbit was changed many times.

** 1973-102E ejected from 1973-102A about 1973 Dec 28.

	Name		Launch date, lifetime and descent date	Shape and weight (kg)	Size (m)	Date of orbital determination	Orbital inclination (deg)	Nodal period (min)	Semi major axis (km)	Perigee height (km)	Apogee height (km)	Orbital eccentricity	Argument of perigee (deg)
D 2M R	Soyuz 13	1973-103A	1973 Dec 18.50 7.87 days 1973 Dec 26.37	Sphere-cylinder + 2 wings 6680?	7.5 long 2.2 dia	1973 Dec 18.6 1973 Dec 19.8	51.57 51.57	88.80 89.22	6596 6617	188 223	247 254	0.004 0.002	72 59
D	Soyuz 13 rocket	1973-103B	1973 Dec 18.50 3.60 days 1973 Dec 22.10	Cylinder 2500?	7.5 long 2.6 dia	1973 Dec 19.2	51.57	88.79	6595	187	247	0.004	84
D	Fragment	1973-103C											
	Cosmos 617	1973-104A	1973 Dec 19.40 5000 years	Spheroid 40?	1.0 long? 0.8 dia?	1973 Dec 26.6	74.03	114.04	7789	1336	1486	0.010	91
	Cosmos 618	1973-104B	1973 Dec 19.40 9000 years	Spheroid 40?	1.0 long? 0.8 dia?	1973 Dec 20.7	74.02	115.28	7846	1446	1489	0.003	120
	Cosmos 619	1973-104C	1973 Dec 19.40 9000 years	Spheroid 40?	1.0 long? 0.8 dia?	1973 Dec 20.4	74.02	115.06	7836	1423	1493	0.004	127
	Cosmos 620	1973-104D	1973 Dec 19.40 10000 years	Spheroid 40?	1.0 long? 0.8 dia?	1973 Dec 21.0	74.01	115.51	7856	1461	1495	0.002	156
	Cosmos 621	1973-104E	1973 Dec 19.40 8000 years	Spheroid 40?	1.0 long? 0.8 dia?	1973 Dec 20.7	74.03	114.84	7826	1410	1485	0.005	108
	Cosmos 622	1973-104F	1973 Dec 19.40 7000 years	Spheroid 40?	1.0 long? 0.8 dia?	1973 Dec 20.7	74.01	114.44	7807	1371	1487	0.007	105
	Cosmos 623	1973-104G	1973 Dec 19.40 7000 years	Spheroid 40?	1.0 long? 0.8 dia?	1973 Dec 20.4	74.02	114.63	7816	1389	1487	0.006	106

1973-104 continued on page 354

	Name		Launch date, lifetime and descent date	Shape and weight (kg)	Size (m)	Date of orbital determination	Orbital inclination (deg)	Nodal period (min)	Semi major axis (km)	Perigee height (km)	Apogee height (km)	Orbital eccentricity	Argument of perigee (deg)
	Cosmos 624	1973-104H	1973 Dec 19.40 6000 years	Spheroid 40?	1.0 long? 0.8 dia?	1973 Dec 20.7	74.02	114.24	7798	1366	1474	0.008	104
	Cosmos 617 rocket	1973-104J	1973 Dec 19.40 20000 years	Cylinder 2200?	7.4 long 2.4 dia	1973 Dec 20.8	74.03	117.13	7929	1476	1626	0.009	258
D R	Cosmos 625	1973-105A	1973 Dec 21.52 12.78 days 1974 Jan 3.30	Sphere-cylinder 6300?	6.5 long? 2.4 dia	1973 Dec 21.9 1973 Dec 27.4	72.83 72.83	89.77 89.80	6641 6642	204 188	321 340	0.009 0.012	70 51
D	Cosmos 625 rocket	1973-105B	1973 Dec 21.52 10.75 days 1974 Jan 1.27	Cylinder 2500?	7.5 long 2.6 dia	1973 Dec 22.3	72.83	89.65	6635	202	311	0.008	68
D	Cosmos 625 engine*	1973-105E	1973 Dec 21.52 15.61 days 1974 Jan 6.13	Core 600? full	1.5 long? 2 dia?	1973 Dec 30.8	72.81	89.37	6621	166	319	0.012	55
D	Fragments	1973-105C,D											
	Molniya 2H	1973-106A	1973 Dec 25.47 10½ years	Windmill + 6 vanes 1250?	4.2 long? 1.6 dia	1973 Dec 26.5 1974 Feb 1.0	62.89 62.90	736.95 718.01	27027 26562	488 434	40809 39934	0.746 0.744	281 -
D	Molniya 2H launcher	1973-106B	1973 Dec 25.47 30.44 days 1974 Jan 24.91	Irregular	-	1973 Dec 25.9	62.84	90.95	6700	194	450	0.019	121
D	Molniya 2H launcher rocket	1973-106C	1973 Dec 25.47 22.65 days 1974 Jan 17.12	Cylinder 2500?	7.5 long 2.6 dia	1973 Dec 28.4	62.83	90.79	6692	193	435	0.018	122
	Molniya 2H rocket	1973-106D	1973 Dec 25.47 10½ years	Cylinder 440	2.0 long 2.0 dia	1974 Jul 22.9	62.80	733.61	26945	611	40522	0.741	280

* 1973-105E ejected from 1973-105A on 1973 Dec 30

	Name		Launch date, lifetime and descent date	Shape and weight (kg)	Size (m)	Date of orbital determination	Orbital inclination (deg)	Nodal period (min)	Semi major axis (km)	Perigee height (km)	Apogee height (km)	Orbital eccentricity	Argument of perigee (deg)
	Aureole 2	1973-107A	1973 Dec 26.69 60 years	Octagonal ellipsoid 400?	1,8 long? 1.5 dia?	1973 Dec 28.0	74.01	109.16	7566	400	1975	0.184	118
	Aureole 2 rocket	1973-107B	1973 Dec 26.69 60 years	Cylinder 2200?	7.4 long 2.4 dia	1973 Dec 27.2	74.01	109.02	7559	396	1965	0.104	119
	Cosmos 626	1973-108A	1973 Dec 27.85 600 years	Cone-cylinder	6 long? 2 dia?	1973 Dec 29.0 1974 Feb 14.7	65.02 64.91	89.65 104.04	6636 7328	257 910	259 990	0.0001 0.005	321 180
D	Cosmos 626* rocket	1973-108C	1973 Dec 27.85 58 days 1974 Feb 23	Cylinder 1500?	8 long? 2.5 dia?	1974 Feb 11.7	65.02	89.79	6643	234	296	0.005	176
D	Cosmos 626* platform	1973-108D	1973 Dec 27.85 84.63 days 1974 Mar 22.48	Irregular	-	1974 Feb 19.4	65.01	89.43	6625	237	257	0.002	260
D	Fragments	1973-108B, E											
	Cosmos 627	1973-109A	1973 Dec 29.17 1200 years	Cylinder? 700?	1.3 long? 1.9 dia?	1973 Dec 30.6	82.95	105.08	7375	974	1019	0.003	274
	Cosmos 627 rocket	1973-109B	1973 Dec 29.17 600 years	Cylinder 2200?	7.4 long 2.4 dia	1973 Dec 29.5	82.95	104.93	7367	970	1008	0.003	268

* 1973-108C and 1973-108D attached to 1973-108A until orbit change about 1974 Feb 11

	Name		Launch date, lifetime and descent date	Shape and weight (kg)	Size (m)	Date of orbital determination	Orbital inclination (deg)	Nodal period (min)	Semi major axis (km)	Perigee height (km)	Apogee height (km)	Orbital eccentricity	Argument of perigee (deg)
	Cosmos 628	1974-01A	1974 Jan 17.42 1200 years	Cylinder? 700?	1.3 long? 1.9 dia?	1974 Jan 18.0	82.96	104.87	7365	958	1016	0.004	299
	Cosmos 628 rocket	1974-01B	1974 Jan 17.42 600 years	Cylinder 2200?	7.4 long 2.4 dia	1974 Jan 19.4	82.95	104.75	7359	959	1003	0.003	279
D	Skynet 2A	1974-02A	1974 Jan 19.07 6 days 1974 Jan 25	Cylinder 435 full 235 empty	1.33 long 1.90 dia	1974 Jan 20.7	37.60	121.48	8129	96	3406	0.204	171
D	Skynet 2A second stage	1974-02B	1974 Jan 19.07 1 day 1974 Jan 20	Cylinder + annulus 350?	6.4 long 1.52 and 2.44 dia	1974 Jan 19.7	28.70	88.05	6562	183	184	0	42
D	Fragments	1974-02C,D											
D R	Cosmos 629	1974-03A	1974 Jan 24.63 11.66 days 1974 Feb 5.29	Sphere-cylinder 5900?	5.9 long 2.4 dia	1974 Jan 25.1	62.81	89.35	6621	197	289	0.007	55
D	Cosmos 629 rocket	1974-03B	1974 Jan 24.63 8.20 days 1974 Feb 1.83	Cylinder 2500?	7.5 long 2.6 dia	1974 Jan 26.2	62.81	89.13	6610	191	273	0.006	52
D	Capsule	1974-03D	1974 Jan 24.63 16 days 1974 Feb 9	Ellipsoid 200?	0.9 long 1.9 dia	1974 Feb 5.3	62.81	89.21	6614	195	277	0.006	-
D	Fragment	1974-03C											

	Name		Launch date, lifetime and descent date	Shape and weight (kg)	Size (m)	Date of orbital determination	Orbital inclination (deg)	Nodal period (min)	Semi major axis (km)	Perigee height (km)	Apogee height (km)	Orbital eccentricity	Argument of perigee (deg)
D R	Cosmos 630 †	1974-04A	1974 Jan 30.46 13.7 days 1974 Feb 13.2	Sphere-cylinder 6300?	6.5 long? 2.4 dia	1974 Jan 31.6 1974 Feb 5.4	72.84 72.85	90.02 89.74	6653 6639	203 179	346 342	0.011 0.012	65 52
D	Cosmos 630 rocket	1974-04B	1974 Jan 30.46 15.45 days 1974 Feb 14.91	Cylinder 2500?	7.5 long 2.6 dia	1974 Jan 31.3	72.84	89.92	6648	200	339	0.010	66
D	Cosmos 630 engine*	1974-04F	1974 Jan 30.46 17.07 days 1974 Feb 16.53	Cone 600? full	1.5 long? 2 dia?	1974 Feb 13.1	72.82	88.87	6596	162	273	0.008	50
D	Fragments	1974-04C-E											
D	Cosmos 631	1974-05A	1974 Feb 6.03 2431 days 1980 Oct 3	Cylinder + paddles? 900?	2 long? 1 dia?	1974 Feb 9.5	74.04	95.31	6911	521	545	0.002	8
D	Cosmos 631 rocket	1974-05B	1974 Feb 6.03 2496 days 1980 Dec 7	Cylinder 2200?	7.4 long 2.4 dia	1974 Feb 10.2	74.04	95.19	6905	511	543	0.002	17
D	Fragments	1974-05C-E											
D R	Cosmos 632	1974-06A	1974 Feb 12.38 13.9 days 1974 Feb 26.3	Sphere-cylinder 6300?	6.5 long? 2.4 dia	1974 Feb 14.4	65.00	89.29	6618	176	303	0.010	60
D	Cosmos 632 rocket	1974-06B	1974 Feb 12.38 5 days 1974 Feb 17	Cylinder 2500?	7.5 long 2.6 dia	1974 Feb 14.4	65.01	88.86	6597	174	263	0.007	53
D	Cosmos 632 engine **	1974-06E	1974 Feb 12.38 26 days 1974 Mar 10	Cone 600? full	1.5 long? 2 dia?	1974 Feb 25.4	65.01	90.23	6665	192	382	0.014	-
D	Fragments	1974-06C,D											

* 1974-04F ejected from 1974-04A on 1974 Feb 12

** 1974-06E ejected from 1974-06A on 1974 Feb 25

	Name		Launch date, lifetime and descent date	Shape and weight (kg)	Size (m)	Date of orbital determination	Orbital inclination (deg)	Nodal period (min)	Semi major axis (km)	Perigee height (km)	Apogee height (km)	Orbital eccentricity	Argument of perigee (deg)
D	[Titan 3B Agena D]	1974-07A	1974 Feb 13.75 32 days 1974 Mar 17	Cylinder 3000?	8 long? 1.5 dia	1974 Feb 15.1	110.44	89.78	6642	134	393	0.020	149
	Tansei 2* [Mu 3C]	1974-08A	1974 Feb 16.21 3262 days	26-sided polyhedron? 56	0.75 long? 0.71 dia?	1974 Feb 17.5	31.23	121.60	8137	284	3233	0.181	102
	Tansei 2 rocket	1974-08B	1974 Feb 16.21 3260 days	Sphere-cone 230?	2.33 long 1.14 dia	1974 Feb 17.5	31.22	121.92	8151	281	3264	0.183	103
D	San Marco 4 **	1974-09A	1974 Feb 18.42 806 days 1976 May 4	Sphere 164	0.71 dia	1974 Feb 22.6 1975 May 1.0	2.92 2.92	95.89 93.55	6949 6835	231 235	910 678	0.049 0.032	342 -
D	San Marco 4 rocket	1974-09B	1974 Feb 18.42 111.53 days 1974 Jun 9.95	Cylinder 24	1.50 long 0.46 dia	1974 Feb 23.6	2.90	95.69	6939	233	889	0.047	357
D	Cosmos 633	1974-10A	1974 Feb 27.47 219.01 days 1974 Oct 4.48	Ellipsoid 400?	1.8 long 1.2 dia	1974 Mar 1.1	70.99	92.17	6759	271	491	0.016	74
D	Cosmos 633 rocket	1974-10B	1974 Feb 27.47 103.97 days 1974 Jun 11.44	Cylinder 1500?	8 long 1.65 dia	1974 Mar 1.1	70.99	92.06	6753	264	486	0.016	72

* Japanese satellite.

** Italian satellite,

	Name		Launch date, lifetime and descent date	Shape and weight (kg)	Size (m)	Date of orbital determination	Orbital inclination (deg)	Nodal period (min)	Semi major axis (km)	Perigee height (km)	Apogee height (km)	Orbital eccentricity	Argument of perigee (deg)
	Meteor 16	1974-11A	1974 Mar 5.49 500 years	Cylinder + 2 vanes 2200?	5 long? 1.5 dia?	1974 Mar 9.3	81.23	102.23	7241	832	894	0.004	249
	Meteor 16 rocket	1974-11B	1974 Mar 5.49 400 years	Cylinder 1440	3.8 long 2.6 dia	1974 Mar 9.5	81.24	102.26	7243	805	924	0.008	206
D	Cosmos 634	1974-12A	1974 Mar 5.67 217.79 days 1974 Oct 9.46	Ellipsoid 400?	1.8 long 1.2 dia	1974 Mar 7.3	70.92	92.18	6759	271	491	0.016	77
D	Cosmos 634 rocket	1974-12B	1974 Mar 5.67 103.64 days 1974 Jun 17.31	Cylinder 1500?	8 long 1.65 dia	1974 Mar 8.2	70.92	92.04	6752	272	476	0.015	78
D	Fragment	1974-12C											
T	Miranda*	1974-13A	1974 Mar 9.10 60 years	Box + 2 panels 93	0.82 long 0.66 square 2.56 span	1974 Mar 13.4	97.81	101.23	7193	714	916	0.014	194
	Miranda rocket	1974-13B	1974 Mar 9.10 60 years	Cylinder 24	1.50 long 0.46 dia	1974 Mar 25.7	97.82	101.23	7193	713	917	0.014	158
	Fragments	1974-13C,D											
D R	Cosmos 635	1974-14A	1974 Mar 14.44 11.79 days 1974 Mar 26.23	Sphere-cylinder 5900?	5.9 long 2.4 dia	1974 Mar 16.4	72.83	89.82	6643	204	326	0.009	62
D	Cosmos 635 rocket	1974-14B	1974 Mar 14.44 10.73 days 1974 Mar 25.17	Cylinder 2500?	7.5 long 2.6 dia	1974 Mar 16.9	72.83	89.53	6629	200	301	0.008	56
D	Capsule**	1974-14E	1974 Mar 14.44 23.69 days 1974 Apr 7.13	Ellipsoid 200?	0.9 long 1.9 dia	1974 Mar 19.2	72.83	89.74	6639	203	319	0.009	57
D	Fragments	1974-14C,D											

* UK technological satellite, known as X4 before launch.

** 1974-14E ejected from 1974-14A about 1974 Mar 18

	Name		Launch date, lifetime and descent date	Shape and weight (kg)	Size (m)	Date of orbital determination	Orbital inclination (deg)	Nodal period (min)	Semi major axis (km)	Perigee height (km)	Apogee height (km)	Orbital eccentricity	Argument of perigee (deg)
T	DMSP 5 [Thor Burner 2]	1974-15A	1974 Mar 16.34 80 years	12-sided frustum 195	1.64 long 1.31 to 1.10 dia	1974 Mar 16.9	98.94	101.54	7208	782	877	0.007	243
	Burner 2 rocket	1974-15B	1974 Mar 16.34 60 years	Sphere-cone 66	1.32 long 0.94 dia	1974 Mar 17.3	98.94	101.65	7213	784	886	0.007	238
D R	Cosmos 636	1974-16A	1974 Mar 20.36 13.9 days 1974 Apr 3.2	Sphere-cylinder 6300?	6.5 long? 2.4 dia	1974 Mar 21.3 1974 Mar 21.5	65.02 65.02	90.02 89.26	6654 6616	165 167	386 309	0.017 0.011	70 68
D	Cosmos 636 rocket	1974-16B	1974 Mar 20.36 4.34 days 1974 Mar 24.70	Cylinder 2500?	7.5 long 2.6 dia	1974 Mar 20.5	65.04	89.88	6647	168	370	0.015	70
D	Cosmos 636 engine*	1974-16F	1974 Mar 20.36 16 days 1974 Apr 5	Cone 600? full	1.5 long? 2 dia?	1974 Apr 3.5	65.03	90.57	6681	184	422	0.018	11
D	Fragments	1974-16C-E											
	Cosmos 637	1974-17A	1974 Mar 26.57 > million years	-	-	1974 Mar 26.6 1974 Mar 26.6 1974 Sep 1.0	51.54 49.73 0.25	88.52 647.52 1425.8	6582 24797 41963	178 226 35390	230 36611 35779	0.004 0.734 0.005	158 0 -
D	Cosmos 637 launcher	1974-17B	1974 Mar 26.57 1 day 1974 Mar 27	Irregular	-	1974 Mar 26.8	51.48	88.19	6566	181	194	0.001	211
D	Cosmos 637 launcher rocket	1974-17C	1974 Mar 26.57 1.90 days 1974 Mar 28.47	Cylinder 4000?	12 long? 4 dia	1974 Mar 27.0	51.52	88.02	6557	166	191	0.002	330
D	Cosmos 637 rocket	1974-17D	1974 Mar 26.57 966 days 1976 Nov 16	Cylinder 1900?	3.9 long? 3.9 dia	1974 Sep 15.1	47.29	630.18	24356	300	35655	0.726	46
	Cosmos 637 apogee motor	1974-17F	1974 Mar 26.57 > million years	Cylinder 440	2.0 long 2.0 dia		Synchronous orbit similar to 1974-17A						
D	Fragment	1974-17E											

* 1974-16F ejected from 1974-16A about 1974 Apr 2.

	Name		Launch date, lifetime and descent date	Shape and weight (kg)	Size (m)	Date of orbital determination	Orbital inclination (deg)	Nodal period (min)	Semi major axis (km)	Perigee height (km)	Apogee height (km)	Orbital eccentricity	Argument of perigee (deg)
D R	Cosmos 638	1974-18A	1974 Apr 3.32 9.9 days 1974 Apr 13.2	Sphere-cylinder + 2 wings 6680?	7.5 long 2.2 dia	1974 Apr 3.9 1974 Apr 7.7	51.78 51.78	89.41 89.77	6626 6644	187 258	309 274	0.009 0.001	90 265
D	Cosmos 638 rocket	1974-18B	1974 Apr 3.32 6.22 days 1974 Apr 9.54	Cylinder 2500?	7.5 long 2.6 dia	1974 Apr 4.0	51.78	89.36	6624	186	305	0.009	86
D	Fragments	1974-18C-E											
D R	Cosmos 639	1974-19A	1974 Apr 4.36 10.8 days 1974 Apr 15.2	Sphere-cylinder 6300?	6.5 long? 2.4 dia	1974 Apr 4.5 1974 Apr 6.8	81.31 81.31	88.85 88.84	6594 6594	206 181	226 250	0.001 0.005	317 25
D	Cosmos 639 rocket	1974-19B	1974 Apr 4.36 2.96 days 1974 Apr 7.32	Cylinder 2500?	7.5 long 2.6 dia	1974 Apr 5.5	81.31	88.52	6579	188	213	0.002	304
D	Cosmos 639 engine*	1974-19D	1974 Apr 4.36 13 days 1974 Apr 17	Cone 600? full	1.5 long? 2 dia?	1974 Apr 14.4	81.31	88.34	6570	163	220	0.004	343
D	Fragments	1974-19C,E,F											
D	[Titan 3D]	1974-20A	1974 Apr 10.85 109 days 1974 Jul 28	Cylinder 13300? full	15 long 3.0 dia	1974 Apr 12.2	94.52	88.91	6597	153	285	0.010	143
T	Capsule	1974-20B	1974 Apr 10.85 90 years	Octagon 60?	0.3 long? 0.9 dia?	1974 Apr 13.9	94.61	101.07	7186	786	830	0.003	127
D	Capsule	1974-20C	1974 Apr 10.85 2144 days 1980 Feb 22	Octagon 60?	0.3 long? 0.9 dia?	1974 Apr 12.6	94.00	95.01	6895	503	531	0.002	79
D	Titan 3D rocket	1974-20D	1974 Apr 10.85 1.79 days 1974 Apr 12.64	Cylinder 1900	6 long 3.0 dia	1974 Apr 11.7	94.50	88.43	6573	148	242	0.007	143

* 1974-19D ejected from 1974-19A on 1974 Apr 14

	Name		Launch date, lifetime and descent date	Shape and weight (kg)	Size (m)	Date of orbital determination	Orbital inclination (deg)	Nodal period (min)	Semi major axis (km)	Perigee height (km)	Apogee height (km)	Orbital eccentricity	Argument of perigee (deg)
D R	Cosmos 640	1974-21A	1974 Apr 11.52 11.83 days 1974 Apr 23.35	Sphere-cylinder 5700?	5.0 long 2.4 dia	1974 Apr 12.4	81.32	88.78	6591	201	225	0.002	307
D	Cosmos 640 rocket	1974-21B	1974 Apr 11.52 3.12 days 1974 Apr 14.64	Cylinder 2500?	7.5 long 2.6 dia	1974 Apr 12.7	81.32	88.48	6576	181	215	0.003	300
T	Westar 1	1974-22A	1974 Apr 13.98 >million years	Cylinder 574 full 300 empty	1.65 long 1.90 dia	1974 Sep 1.0	0.0	1435.4	42144	35761	35770	0.0001	-
D	Westar 1 second stage	1974-22B	1974 Apr 13.98 42 days 1974 May 25	Cylinder + annulus 350?	6.4 long 1.52 and 2.44 dia	1974 Apr 17.2	28.58	91.19	6719	227	454	0.017	225
D	Westar 1 third stage	1974-22C	1974 Apr 13.98 226 days 1974 Nov 25	Sphere-cone 66	1.32 long 0.94 dia	1974 Apr 14.2	24.75	637.64	24551	202	36143	0.732	178
	Molniya 1AC	1974-23A	1974 Apr 20.87 13 years	Windmill + 6 vanes 1000?	3.4 long 1.6 dia	1974 Apr 23.5 1974 Sep 1.0	62.86 63.0	737.63 717.81	27043 26557	624 606	40707 39752	0.741 0.737	288 -
D	Molniya 1AC launcher rocket	1974-23B	1974 Apr 20.87 60.29 days 1974 Jun 20.16	Cylinder 2500?	7.5 long 2.6 dia	1974 Apr 23.0	62.85	92.53	6778	219	580	0.027	120

1974-23 continued on page 363

	Name		Launch date, lifetime and descent date	Shape and weight (kg)	Size (m)	Date of orbital determination	Orbital inclination (deg)	Nodal period (min)	Semi major axis (km)	Perigee height (km)	Apogee height (km)	Orbital eccentricity	Argument of perigee (deg)
D	Molniya 1AC launcher	1974-23C	1974 Apr 20.87 74.36 days 1974 Jul 4.23	Irregular	-	1974 Apr 23.1	62.82	92.98	6799	217	625	0.030	120
	Molniya 1AC rocket	1974-23E	1974 Apr 20.87 13 years	Cylinder 440	2.0 long 2.0 dia	1974 Apr 23.5	62.88	734.50	26967	624	40553	0.740	288
D	Fragment	1974-23D											
	Cosmos 641	1974-24A	1974 Apr 23.59 7000 years	Spheroid 40?	1.0 long? 0.8 dia?	1974 Apr 24.6	74.01	114.60	7815	1389	1484	0.006	112
	Cosmos 642	1974-24B	1974 Apr 23.59 4000 years	Spheroid 40?	1.0 long? 0.8 dia?	1974 Apr 27.2	74.01	113.83	7780	1321	1483	0.010	99
	Cosmos 643	1974-24C	1974 Apr 23.59 6000 years	Spheroid 40?	1.0 long? 0.8 dia?	1974 Apr 26.2	74.01	114.22	7798	1355	1484	0.008	106
	Cosmos 644	1974-24D	1974 Apr 23.59 5000 years	Spheroid 40?	1.0 long? 0.8 dia?	1974 Apr 25.5	74.02	114.02	7788	1336	1484	0.009	107
	Cosmos 645	1974-24E	1974 Apr 23.59 7000 years	Spheroid 40?	1.0 long? 0.8 dia?	1974 Apr 28.5	74.02	114.40	7806	1370	1485	0.007	106
	Cosmos 646	1974-24F	1974 Apr 23.59 8000 years	Spheroid 40?	1.0 long? 0.8 dia?	1974 Apr 28.5	74.01	114.81	7824	1405	1487	0.005	117
	Cosmos 647	1974-24G	1974 Apr 23.59 9000 years	Spheroid 40?	1.0 long? 0.8 dia?	1974 Apr 25.9	74.01	115.00	7833	1424	1486	0.004	123

1974-24 continued on page 364

	Name		Launch date, lifetime and descent date	Shape and weight (kg)	Size (m)	Date of orbital determination	Orbital inclination (deg)	Nodal period (min)	Semi major axis (km)	Perigee height (km)	Apogee height (km)	Orbital eccentricity	Argument of perigee (deg)
	Cosmos 648	1974-24H	1974 Apr 23.59 10000 years	Spheroid 40?	1.0 long? 0.8 dia?	1974 Apr 25.9	74.01	115.23	7843	1440	1490	0.003	143
	Cosmos 641 rocket	1974-24J	1974 Apr 23.59 20000 years	Cylinder 2200?	7.4 long 2.4 dia	1974 Apr 25.5	74.02	116.99	7923	1479	1610	0.008	254
	Meteor 17	1974-25A	1974 Apr 24.50 500 years	Cylinder + 2 vanes 2200?	5 long? 1.5 dia?	1974 Apr 25.6	81.23	102.58	7258	865	894	0.002	278
	Meteor 17 rocket	1974-25B	1974 Apr 24.50 400 years	Cylinder 1440	3.8 long 2.6 dia	1974 Apr 26.1	81.24	102.67	7262	841	927	0.006	195
	Molniya 2J	1974-26A	1974 Apr 26.60 100 years?	Windmill + 6 vanes 1250?	4.2 long? 1.6 dia?	1974 Apr 27.7 1974 Sep 1.0	62.89 63.0	737.04 717.73	27029 26555	600 417	40702 39937	0.742 0.744	288 -
D	Molniya 2J launcher	1974-26B	1974 Apr 26.60 24.57 days 1974 May 21.17	Irregular	-	1974 Apr 26.7	62.85	90.96	6701	217	428	0.016	130
D	Molniya 2J launcher rocket	1974-26C	1974 Apr 26.60 21.13 days 1974 May 17.73	Cylinder 2500?	7.5 long 2.6 dia	1974 Apr 27.0	62.85	91.00	6703	203	446	0.018	118
	Molniya 2J rocket	1974-26E	1974 Apr 26.60 100 years?	Cylinder 440	2.0 long 2.0 dia	1974 Jul 18.6	62.91	732.75	26924	314	40778	0.751	279
D	Fragment	1974-26D											

	Name		Launch date, lifetime and descent date	Shape and weight (kg)	Size (m)	Date of orbital determination	Orbital inclination (deg)	Nodal period (min)	Semi major axis (km)	Perigee height (km)	Apogee height (km)	Orbital eccentricity	Argument of perigee (deg)
D R	Cosmos 649	1974-27A	1974 Apr 29.56 11.6 days 1974 May 11.2	Sphere-cylinder 6300?	6.5 long? 2.4 dia	1974 Apr 30.1	62.81	89.28	6618	181	299	0.009	67
D	Cosmos 649 rocket	1974-27B	1974 Apr 29.56 3.96 days 1974 May 3.52	Cylinder 2500?	7.5 long 2.6 dia	1974 Apr 29.8	62.79	89.20	6614	178	294	0.009	64
D	Cosmos 649 engine*	1974-27D	1974 Apr 29.56 16.86 days 1974 May 16.42	Cone 600? full	1.5 long? 2 dia?	1974 May 10.6	62.79	89.20	6614	177	295	0.009	69
D	Fragments	1974-27C,E-G											
	Cosmos 650	1974-28A	1974 Apr 29.71 6000 years	- 500?	-	1974 May 1.9	74.04	113.49	7764	1369	1402	0.002	242
	Cosmos 650 rocket	1974-28B	1974 Apr 29.71 4000 years	Cylinder 2200?	7.4 long 2.4 dia	1974 May 1.3	74.04	113.33	7756	1364	1392	0.002	219
	Cosmos 651**	1974-29A	1974 May 15.31 600 years	Cone-cylinder	6 long? 2 dia?	1974 May 16.1 1974 Sep 1.0	64.97 64.97	89.64 103.45	6635 7301	250 892	264 954	0.001 0.004	266 -
D	Cosmos 651 rocket	1974-29B	1974 May 15.31 75.92 days 1974 Jul 30.23	Cylinder 1500?	8 long? 2.5 dia?	1974 Jul 26.6	64.95	89.51	6629	243	258	0.001	281
D	Cosmos 651 platform	1974-29C	1974 May 15.31 112.75 days 1974 Sep 5.06	Irregular	-	1974 Jul 27.2	64.95	89.57	6632	245	262	0.001	286

* 1974-27D ejected from 1974-27A about 1974 May 10. ** 1974-29B and 29C attached to 1974-29A until orbit change between 1974 Jul 25.3 and 25.9.

	Name		Launch date, lifetime and descent date	Shape and weight (kg)	Size (m)	Date of orbital determination	Orbital inclination (deg)	Nodal period (min)	Semi major axis (km)	Perigee height (km)	Apogee height (km)	Orbital eccentricity	Argument of perigee (deg)
D R	Cosmos 652	1974-30A	1974 May 15.36 7.9 days 1974 May 23.3	Sphere-cylinder 6300?	6.5 long? 2.4 dia	1974 May 16.2	51.76	89.61	6636	173	343	0.013	80
D	Cosmos 652 rocket	1974-30B	1974 May 15.36 4.15 days 1974 May 19.51	Cylinder 2500?	7.5 long 2.6 dia	1974 May 16.2	51.76	89.38	6625	169	324	0.012	78
D	Cosmos 652 engine*	1974-30E	1974 May 15.36 21 days 1974 Jun 5	Cone 600? full	1.5 long? 2 dia?	1974 May 24.0	51.78	90.83	6696	218	417	0.015	50
D	Fragments	1974-30C,D,F,G											
D R	Cosmos 653	1974-31A	1974 May 15.52 11.65 days 1974 May 27.17	Sphere-cylinder 5700?	5.0 long 2.4 dia	1974 May 16.4	62.81	89.27	6618	192	287	0.007	47
D	Cosmos 653 rocket	1974-31B	1974 May 15.52 4.79 days 1974 May 20.31	Cylinder 2500?	7.5 long 2.6 dia	1974 May 16.7	62.80	89.04	6606	183	273	0.007	41
D	Fragment	1974-31C											
	Cosmos 654**	1974-32A	1974 May 17.29 600 years	Cone-cylinder	6 long? 2 dia?	1974 May 18.2 1974 Sep 1.0	64.99 64.99	89.63 104.44	6635 7347	248 913	265 1024	0.001 0.008	255 -
D	Cosmos 654 rocket	1974-32B	1974 May 17.29 79 days 1974 Aug 4	Cylinder 1500?	8 long? 2.5 dia?	1974 Jul 30.7	64.99	89.59	6633	248	261	0.001	272
D	Cosmos 654 platform	1974-32D	1974 May 17.29 113 days 1974 Sep 7	Irregular	-	1974 Sep 1.0	64.99	88.78	6593	203	226	0.002	-
D	Fragments	1974-32C,E											

* 1974-30E ejected from 1974-30A about 1974 May 22.

** 1974-32B and 32D attached to 1974-32A until orbit change on 1974 Jul 30.

	Name		Launch date, lifetime and descent date	Shape and weight (kg)	Size (m)	Date of orbital determination	Orbital inclination (deg)	Nodal period (min)	Semi major axis (km)	Perigee height (km)	Apogee height (km)	Orbital eccentricity	Argument of perigee (deg)
T	SMS 1*	1974-33A	1974 May 17.40 >million years	Cylinder + boom 627 full 243 empty	2.30 long 1.90 dia	1974 May 17.4 1974 May 23.3 1974 Sep 1.0	24.47 1.87 1.90	576.4 1340.4 1436.0	22944 40271 42164	182 32345 35741	32950 35440 35830	0.714 0.038 0.001	182 178 -
D	SMS 1 second stage	1974-33B	1974 May 17.40 3.03 days 1974 May 20.43	Cylinder + annulus 350?	6.4 long 1.52 and 2.44 dia	1974 May 19.9	28.32	88.02	6562	155	213	0.004	181
D	SMS 1 third stage	1974-33D	1974 May 17.40 179 days 1974 Nov 12	Sphere-cone 66	1.32 long 0.94 dia	1974 May 17.4	24.52	576.4	22944	182	32950	0.714	182
D	Fragments	1974-33C,E											
D	Intercosmos 11	1974-34A	1974 May 17.45 1938 days 1979 Sep 6	Octagonal ellipsoid 550?	1.8 long? 1.5 dia?	1974 May 19.5	50.64	94.50	6875	483	511	0.002	43
D	Intercosmos 11 rocket	1974-34B	1974 May 17.45 2061 days 1980 Jan 7	Cylinder 2200?	7.4 long 2.4 dia	1974 May 19.5	50.64	94.37	6870	472	511	0.003	46
D	Fragment	1974-34C											
D	Cosmos 655	1974-35A	1974 May 21.26 2374 days 1980 Nov 19	Cylinder + paddles? 900?	2 long? 1 dia?	1974 May 22.5	74.06	95.30	6911	523	542	0.001	16
D	Cosmos 655 rocket	1974-35B	1974 May 21.26 2329 days 1980 Oct 5	Cylinder 2200?	7.4 long 2.4 dia	1974 May 22.9	74.05	95.21	6906	514	542	0.002	10
D	Fragments	1974-35C-H											
D R	Cosmos 656	1974-36A	1974 May 27.31 2.0 days 1974 May 29.3	Sphere-cylinder 6570?	7.5 long 2.2 dia	1974 May 28.4	51.60	90.04	6658	195	364	0.013	71
D	Cosmos 656 rocket	1974-36B	1974 May 27.31 6.67 days 1974 Jun 2.98	Cylinder 2500?	7.5 long 2.6 dia	1974 May 29.9	51.57	89.17	6615	179	294	0.009	83

* Synchronous Meteorological Satellite. An apogee motor may have separated into a similar orbit.

	Name		Launch date, lifetime and descent date	Shape and weight (kg)	Size (m)	Date of orbital determination	Orbital inclination (deg)	Nodal period (min)	Semi major axis (km)	Perigee height (km)	Apogee height (km)	Orbital eccentricity	Argument of perigee (deg)
D	Luna 22 launcher rocket	1974-37B	1974 May 29.37 3.86 days 1974 Jun 2.23	Cylinder 4000?	12 long? 4 dia	1974 May 30.7	51.54	88.56	6585	187	226	0.003	4
D	Luna 22 launcher	1974-37C	1974 May 29.37 4 days 1974 Jun 2	-	-	1974 May 31.1	51.56	88.48	6581	178	227	0.004	2
D	Fragment	1974-37D											
D R	Cosmos 657	1974-38A	1974 May 30.53 13.64 days 1974 Jun 13.17	Sphere- cylinder 6300?	6.5 long? 2.4 dia	1974 May 31.6 1974 Jun 1.4	62.79 62.79	89.21 89.35	6615 6622	177 177	296 310	0.009 0.010	59 57
D	Cosmos 657 rocket	1974-38B	1974 May 30.53 3 days 1974 Jun 2	Cylinder 2500?	7.5 long 2.6 dia	1974 May 31.4	62.78	88.96	6602	171	277	0.008	53
D	Cosmos 657 engine*	1974-38D	1974 May 30.53 18.58 days 1974 Jun 18.11	Cone 600? full	1.5 long? 2 dia?	1974 Jun 12.9	62.79	89.05	6607	168	289	0.009	58
D	Fragments	1974-38C,E											
T	ATS 6 [Titan 3C]	1974-39A	1974 May 30.54 >million years	Box + dish + 2 paddles 1402**	4.0 high 9.15 dia 15.8 span	1974 Sep 1.0	1.6	1436.1	42164	35781	35791	0.0001	-
D	ATS 6 second stage	1974-39B	1974 May 30.54 5.31 days 1974 Jun 4.85	Cylinder 1900	6 long 3.0 dia	1974 May 30.5	28.60	91.80	6747	163	575	0.031	-
	ATS 6 rocket	1974-39C	1974 May 30.54 >million years	Cylinder 1500?	6 long? 3.0 dia	1974 Jul 1.0	1.8	1430.4	42053	35553	35797	0.003	-

Space Vehicle: Luna 22, 1974-37A

* 1974-38D ejected from 1974-38A on 1974 Jun 12.

** Payload weight 930kg.

	Name		Launch date, lifetime and descent date	Shape and weight (kg)	Size (m)	Date of orbital determination	Orbital inclination (deg)	Nodal period (min)	Semi major axis (km)	Perigee height (km)	Apogee height (km)	Orbital eccentricity	Argument of perigee (deg)
D	Explorer 52 (Hawkeye*)	1974-40A	1974 Jun 3.97 1427 days 1978 Apr 30	Cone-cylinder 27	0.75 long 0.25 to 0.75 dia	1974 Jun 7.0	89.80	3077.9	70082	513	126896	0.902	275
D	Explorer 52 rocket	1974-40B	1974 Jun 3.97 1226 days 1977 Oct 11	Cylinder 24	1.50 long 0.46 dia	1974 Jun 4.1 1975 Mar 1.0	89.70 89.70	96.00 95.19	6944 6905	337 330	794 723	0.033 0.028	210 -
D	Fragments	1974-40C,D											
D R	Cosmos 658	1974-41A	1974 Jun 6.27 11.85 days 1974 Jun 18.12	Sphere-cylinder 5700?	5.0 long 2.4 dia	1974 Jun 6.4	64.97	89.39	6623	204	286	0.006	37
D	Cosmos 658 rocket	1974-41B	1974 Jun 6.27 6.18 days 1974 Jun 12.45	Cylinder 2500?	7.5 long 2.6 dia	1974 Jun 6.3	64.97	89.29	6618	203	277	0.006	33
D	[Titan 3B Agena D]	1974-42A	1974 Jun 6.69 47 days 1974 Jul 23	Cylinder 3000?	8 long? 1.5 dia	1974 Jun 7.4	110.49	89.81	6643	136	394	0.019	153
D R	Cosmos 659	1974-43A	1974 Jun 13.52 12.66 days 1974 Jun 26.18	Sphere-cylinder 6300?	6.5 long? 2.4 dia	1974 Jun 15.4	62.81	89.30	6619	153	329	0.013	61
D	Cosmos 659 rocket	1974-43B	1974 Jun 13.52 5.20 days 1974 Jun 18.72	Cylinder 2500?	7.5 long 2.6 dia	1974 Jun 15.2	62.80	89.28	6618	173	307	0.010	65
D	Cosmos 659 engine**	1974-43D	1974 Jun 13.52 13.65 days 1974 Jun 27.17	Cone 600? full	1.5 long? 2 dia?	1974 Jun 25.5	62.81	89.10	6609	148	314	0.013	61
D	Fragment	1974-43C											

* A separated 5th-stage rocket may be in a similar orbit to Hawkeye.
** 1974-43D ejected from 1974-43A on 1974 Jun 25.

	Name		Launch date, lifetime and descent date	Shape and weight (kg)	Size (m)	Date of orbital determination	Orbital inclination (deg)	Nodal period (min)	Semi major axis (km)	Perigee height (km)	Apogee height (km)	Orbital eccentricity	Argument of perigee (deg)
	Cosmos 660	1974-44A	1974 Jun 18.54 35 years	-	-	1974 Jun 18.9	82.98	109.11	7563	397	1972	0.104	129
	Cosmos 660 rocket	1974-44B	1974 Jun 18.54 25 years	Cylinder 2200?	7.4 long 2.4 dia	1974 Jun 22.2	82.98	108.97	7556	395	1961	0.104	120
D	Cosmos 661	1974-45A	1974 Jun 21.38 2259 days 1980 Aug 27	Cylinder + paddles? 900?	2 long? 1 dia?	1974 Jun 22.2	74.04	95.24	6908	511	548	0.003	344
D	Cosmos 661 rocket	1974-45B	1974 Jun 21.38 2309 days 1980 Oct 16	Cylinder 2200?	7.4 long 2.4 dia	1974 Jun 23.2	74.04	95.04	6898	498	541	0.003	1
D	Fragment	1974-45C											
D r	Salyut 3 *	1974-46A	1974 Jun 24.95 214 days 1975 Jan 24	Cylinder + 3 wings 18500	14 long 4.15 to 2.0 dia	1974 Jun 25.7 1974 Jun 28.9 1974 Oct 26.4	51.58 51.58 51.57	89.10 89.80 89.93	6611 6646 6652	213 266 256	253 269 292	0.003 0.0002 0.003	81 91 295
D	Salyut 3 rocket	1974-46B	1974 Jun 24.95 8.68 days 1974 Jul 3.63	Cylinder 4000?	12 long? 4 dia	1974 Jun 26.5	51.60	88.86	6599	209	233	0.002	80
D	Fragments	1974-46C,D											
D	Cosmos 662	1974-47A	1974 Jun 26.52 794 days 1976 Aug 28	Ellipsoid 400?	1.8 long 1.2 dia	1974 Jun 29.2 1975 Jul 1.0	70.92 70.90	95.49 93.68	6920 6832	271 262	812 646	0.039 0.028	79 -
D	Cosmos 662 rocket	1974-47B	1974 Jun 26.52 487 days 1975 Oct 26	Cylinder 1500?	8 long 1.65 dia.	1974 Jun 29.2 1975 Jan 30.5	70.92 70.90	95.41 93.62	6916 6829	276 267	799 635	0.038 0.027	80 -

* Salyut 3 was de-orbited over the Pacific Ocean. Capsule ejected and recovered on 1974 Sep 23.4.

	Name		Launch date, lifetime and descent date	Shape and weight (kg)	Size (m)	Date of orbital determination	Orbital inclination (deg)	Nodal period (min)	Semi major axis (km)	Perigee height (km)	Apogee height (km)	Orbital eccentricity	Argument of perigee (deg)
	Cosmos 663	1974-48A	1974 Jun 27.65 1200 years	Cylinder? 700?	1.3 long? 1.9 dia?	1974 Jun 30.9	82.95	104.88	7368	972	1007	0.002	279
	Cosmos 663 rocket	1974-48B	1974 Jun 27.65 600 years	Cylinder 2200?	7.4 long 2.4 dia	1974 Jun 30.8	82.94	104.73	7360	972	992	0.001	271
D R	Cosmos 664	1974-49A	1974 Jun 29.54 11.80 days 1974 Jul 11.34	Sphere-cylinder 5900?	5.9 long 2.4 dia	1974 Jun 30.7	72.85	89.98	6651	205	341	0.010	61
D	Cosmos 664 rocket	1974-49B	1974 Jun 29.54 10.37 days 1974 Jul 9.91	Cylinder 2500?	7.5 long 2.6 dia	1974 Jul 1.9	72.85	89.71	6638	191	328	0.010	47
D	Capsule*	1974-49H	1974 Jun 29.54 14 days 1974 Jul 13	Ellipscid 200?	0.9 long 1.9 dia	1974 Jul 12.4	72.85	89.20	6612	190	277	0.007	30
D	Fragments	1974-49C-G											
	Cosmos 665	1974-50A	1974 Jun 29.67 $15\frac{3}{4}$ years	Windmill + 6 vanes 1250?	4.2 long? 1.6 dia?	1974 Jun 30.2 1974 Sep 1.0	62.82 62.82	710.65 717.91	26380 26560	625 703	39378 39660	0.734 0.733	318 -
D	Cosmos 665 launcher	1974-50B	1974 Jun 29.67 63.73 days 1974 Sep 1.40	Irregular	-	1974 Jul 1.4	62.82	92.82	6791	216	610	0.029	120
D	Cosmos 665 launcher rocket	1974-50D	1974 Jun 29.67 44.60 days 1974 Aug 13.27	Cylinder 2500?	7.5 long 2.6 dia	1974 Jul 3.6	62.84	92.69	6785	196	618	0.031	118
	Cosmos 665 rocket	1974-50C	1974 Jun 29.67 $15\frac{3}{4}$ years	Cylinder 440	2.0 long 2.0 dia	1974 Jun 30.2	62.82	707.45	26301	605	39241	0.734	318

* 1974-49H ejected from 1974-49A about 1974 Jul 10.

	Name		Launch date, lifetime and descent date	Shape and weight (kg)	Size (m)	Date of orbital determination	Orbital inclination (deg)	Nodal period (min)	Semi major axis (km)	Perigee height (km)	Apogee height (km)	Orbital eccentricity	Argument of perigee (deg)
D 2M R	Soyuz 14*	1974-51A	1974 Jul 3.79 15.72 days 1974 Jul 19.51	Sphere-cylinder 6570?	7.5 long 2.2 dia	1974 Jul 4.0 1974 Jul 5.1	51.58 51.58	88.55 89.84	6584 6648	195 268	217 271	0.002 0.0003	90 259
D	Soyuz 14 rocket	1974-51B	1974 Jul 3.79 2 days 1974 Jul 5	Cylinder 2500?	7.5 long 2.6 dia	1974 Jul 4.5	51.61	88.35	6574	182	210	0.002	69
D	Fragment	1974-51C											
	Meteor 18**	1974-52A	1974 Jul 9.61 500 years	Cylinder + 2 vanes 2200?	5 long? 1.5 dia?	1974 Jul 11.0 1974 Nov 1.0 1975 Nov 1.0	81.23 81.21 81.19	102.57 102.86 103.09	7257 7271 7282	865 879 890	893 906 918	0.002 0.002 0.002	256 - -
	Meteor 18 rocket	1974-52B	1974 Jul 9.61 400 years	Cylinder 1440	3.8 long 2.6 dia	1974 Jul 11.2	81.23	102.72	7264	853	919	0.005	178
D R	Cosmos 666	1974-53A	1974 Jul 12.54 12.7 days 1974 Jul 25.2	Sphere-cylinder 6300?	6.5 long? 2.4 dia	1974 Jul 13.0	62.81	89.59	6633	181	328	0.011	73
D	Cosmos 666 rocket	1974-53B	1974 Jul 12.54 5.93 days 1974 Jul 18.47	Cylinder 2500?	7.5 long 2.6 dia	1974 Jul 15.2	62.79	89.07	6608	170	289	0.009	65
D	Cosmos 666 engine	1974-53D	1974 Jul 12.54 17.69 days 1974 Jul 30.23	Cone 600? full	1.5 long? 2 dia?	1974 Jul 24.6	62.82	89.25	6616	168	308	0.011	66
D	Fragment	1974-53C											
T L	NTS 1† (Timation 3) [Atlas F]	1974-54A	1974 Jul 14.22 300 000 years	Octagon + 4 vanes 293 empty	0.56 long 1.22 dia	1974 Jul 22.0	125.08	468.40	19984	13445	13767	0.008	70

* Soyuz 14 docked with Salyut 3 about 1974 Jul 4.88; undocked 1974 Jul 19.38.

† Navigation Technology Satellite.

** Meteor 18 carried orbital adjustment motor.

1974-54 continued on page 373

	Name		Launch date, lifetime and descent date	Shape and weight (kg)	Size (m)	Date of orbital determination	Orbital inclination (deg)	Nodal period (min)	Semi major axis (km)	Perigee height (km)	Apogee height (km)	Orbital eccentricity	Argument of perigee (deg)
D	NTS 1 rocket	1974-54B	1974 Jul 14.22 2327 days 1980 Nov 26	Cone- cylinder 163	1.85 long 0.63 to 1.65 dia	1974 Jul 15.7	125.12	253.67	13277	193	13604	0.505	150
	NTS 1 apogee motor	1974-54C	1974 Jul 14.22 200000 years	Cylinder	0.88 long? 0.63 dia	1976 Mar 31.0	124.9	468.7	19988	13476	13744	0.007	-
D	Aeros 2*	1974-55A	1974 Jul 16.49 436 days 1975 Sep 25	Cylinder 127 Payload 28	0.74 long 0.91 dia	1974 Jul 17.6 1975 Feb 25.5	97.45 97.37	95.60 91.43	6925 6721	224 278	869 407	0.047 0.010	160 70
D	Aeros 2 rocket	1974-55B	1974 Jul 16.49 85.04 days 1974 Oct 9.53	Cylinder 24	1.50 long 0.46 dia	1974 Jul 18.5	97.44	95.59	6924	220	872	0.047	158
D	Fragment	1974-55C											
	Molniya 2K	1974-56A	1974 Jul 23.06 19 years	Windmill + 6 vanes 1250?	4.2 long? 1.6 dia	1974 Jul 24.1 1974 Jul 27.7	62.89 62.90	737.59 718.17	27043 26566	604 505	40726 39871	0.742 0.741	288 281
D	Molniya 2K launcher rocket	1974-56B	1974 Jul 23.06 34.02 days 1974 Aug 26.08	Cylinder 2500?	7.5 long 2.6 dia	1974 Jul 23.8	62.84	91.10	6707	217	441	0.017	130
D	Molniya 2K launcher	1974-56C	1974 Jul 23.06 36.80 days 1974 Aug 28.86	Irregular	-	1974 Jul 23.8	62.81	91.19	6712	216	451	0.018	130
	Molniya 2K rocket	1974-56D	1974 Jul 23.06 19 years	Cylinder 440	2.0 long 2.0 dia	1974 Jul 30.2	62.94	734.10	26957	456	40702	0.746	280

* FRG satellite.

	Name		Launch date, lifetime and descent date	Shape and weight (kg)	Size (m)	Date of orbital determination	Orbital inclination (deg)	Nodal period (min)	Semi major axis (km)	Perigee height (km)	Apogee height (km)	Orbital eccentricity	Argument of perigee (deg)
D R	Cosmos 667	1974-57A	1974 Jul 25.29 12.9 days 1974 Aug 7.2	Sphere-cylinder 6300?	6.5 long? 2.4 dia	1974 Jul 26.3	64.98	89.46	6626	176	320	0.011	61
D	Cosmos 667 rocket	1974-57B	1974 Jul 25.29 4.78 days 1974 Jul 30.07	Cylinder 2500?	7.5 long 2.6 dia	1974 Jul 26.3	64.97	89.19	6613	176	294	0.009	55
D	Cosmos 667 engine*	1974-57D	1974 Jul 25.29 24 days 1974 Aug 18	Cone 600? full	1.5 long? 2 dia?	1974 Aug 16.0	64.97	88.69	6588	165	255	0.007	-
D	Fragment	1974-57C											
D	Cosmos 668	1974-58A	1974 Jul 25.50 210.78 days 1975 Feb 21.28	Ellipsoid 400?	1.8 long 1.2 dia	1974 Jul 26.4	70.95	92.20	6760	270	494	0.017	83
D	Cosmos 668 rocket	1974-58B	1974 Jul 25.50 112.57 days 1974 Nov 15.07	Cylinder 1500?	8 long 1.65 dia	1974 Jul 26.4	70.95	92.01	6751	273	472	0.015	83
D R	Cosmos 669	1974-59A	1974 Jul 26.29 12.83 days 1974 Aug 8.12	Sphere-cylinder 5900?	5.9 long 2.4 dia	1974 Jul 27.1	81.32	88.91	6598	209	230	0.002	339
D	Cosmos 669 rocket	1974-59B	1974 Jul 26.29 3.83 days 1974 Jul 30.12	Cylinder 2500?	7.5 long 2.6 dia	1974 Jul 27.4	81.33	88.67	6586	198	217	0.001	322
D	Capsule**	1974-59G	1974 Jul 26.29 16.18 days 1974 Aug 11.47	Ellipsoid 200?	0.9 long 1.9 dia	1974 Aug 6.2	81.33	88.78	6591	201	225	0.002	283
D	Fragments	1974-59C-F											

* 1974-57D ejected from 1974-57A on 1974 Aug 6.

** 1974-59G ejected from 1974-59A about 1974 Aug 3.

	Name		Launch date, lifetime and descent date	Shape and weight (kg)	Size (m)	Date of orbital determination	Orbital inclination (deg)	Nodal period (min)	Semi major axis (km)	Perigee height (km)	Apogee height (km)	Orbital eccentricity	Argument of perigee (deg)
	Molniya S1*	1974-60A	1974 Jul 29.50 > million years	-	-	1974 Jul 29.5 1974 Sep 1.0	47.49 0.07	632.35 1436.2	24410 42167	340 35787	35724 35790	0.725 0.0	0 -
D	Molniya S1 launcher	1974-60B	1974 Jul 29.50 2.64 days 1974 Aug 1.14	Irregular	-	1974 Jul 29.9	51.49	88.21	6567	183	195	0.001	307
D	Molniya S1 launcher rocket	1974-60C	1974 Jul 29.50 2 days 1974 Jul 31	Cylinder 4000?	12 long? 4 dia	1974 Jul 29.9	51.47	88.28	6571	186	199	0.001	307
D	Molniya S1 rocket	1974-60D	1974 Jul 29.50 1642 days 1979 Jan 26	Cylinder 1900?	3.9 long? 3.9 dia	1974 Jul 30.9	47.49	634.14	24455	355	35799	0.725	1
D	Fragment	1974-60E											
D R	Cosmos 670	1974-61A	1974 Aug 6.01 3.0 days 1974 Aug 9.0	Sphere-cylinder 6570?	7.5 long 2.2 dia	1974 Aug 6.4	50.55	89.48	6631	211	294	0.006	95
D	Cosmos 670 rocket	1974-61B	1974 Aug 6.01 7.38 days 1974 Aug 13.39	Cylinder 2500?	7.5 long 2.6 dia	1974 Aug 6.3	50.56	89.30	6622	208	279	0.005	92
D R	Cosmos 671	1974-62A	1974 Aug 7.54 12.7 days 1974 Aug 20.2	Sphere-cylinder 6300?	6.5 long? 2.4 dia	1974 Aug 8.1 1974 Aug 9.1	62.82 62.82	89.84 89.30	6646 6619	182 169	353 312	0.013 0.011	75 60
D	Cosmos 671 rocket	1974-62B	1974 Aug 7.54 6.88 days 1974 Aug 14.42	Cylinder 2500?	7.5 long 2.6 dia	1974 Aug 8.1	62.80	89.68	6638	180	339	0.012	74

* First geostationary Molniya satellite. An apogee rocket may have separated from 1974-60A in equatorial orbit.

1974-62 continued on page 376

	Name		Launch date, lifetime and descent date	Shape and weight (kg)	Size (m)	Date of orbital determination	Orbital inclination (deg)	Nodal period (min)	Semi major axis (km)	Perigee height (km)	Apogee height (km)	Orbital eccentricity	Argument of perigee (deg)
D	Cosmos 671 engine*	1974-62C	1974 Aug 7.54 33 days 1974 Sep 9	Cone 600? full	1.5 long? 2 dia?	1974 Aug 19.4	62.81	89.14	6611	163	302	0.011	57
D	Fragment	1974-62D											
T	DMSP 6 [Thor Burner 2]	1974-63A	1974 Aug 9.14 80 years	12-sided frustum 195	1.64 long 1.31 to 1.10 dia	1974 Aug 10.0	98.86	101.76	7219	806	875	0.005	233
	Burner 2 rocket	1974-63B	1974 Aug 9.14 60 years	Sphere-cone 66	1.32 long 0.94 dia	1974 Aug 9.4	98.87	101.71	7217	805	872	0.005	243
D R	Cosmos 672	1974-64A	1974 Aug 12.27 5.9 days 1974 Aug 18.2	Sphere-cylinder + 2 wings 6680?	7.5 long 2.2 dia	1974 Aug 12.5 1974 Aug 14.5	51.76 51.76	88.59 89.09	6586 6611	195 227	221 238	0.002 0.001	76 61
D	Cosmos 672 rocket	1974-64B	1974 Aug 12.27 2.34 days 1974 Aug 14.61	Cylinder 2500?	7.5 long 2.6 dia	1974 Aug 12.6	51.76	88.52	6582	194	214	0.002	68
D	Fragment	1974-64C											
D	[Titan 3B Agena D]	1974-65A	1974 Aug 14.66 46 days 1974 Sep 29	Cylinder 3000?	8 long? 1.5 dia	1974 Aug 16.4	110.51	89.89	6647	135	402	0.020	150
	Cosmos 673	1974-66A	1974 Aug 16.16 25 years	Cylinder + 2 vanes? 2500?	5 long? 1.5 dia?	1974 Aug 18.6	81.21	97.17	7000	607	637	0.002	271
	Cosmos 673 rocket	1974-66B	1974 Aug 16.16 25 years	Cylinder 1440	3.8 long 2.6 dia	1974 Aug 18.2	81.22	97.29	7006	578	678	0.007	187
	Fragment	1974-66C											

* 1974-62C ejected from 1974-62A on 1974 Aug 19.

	Name		Launch date, lifetime and descent date	Shape and weight (kg)	Size (m)	Date of orbital determination	Orbital inclination (deg)	Nodal period (min)	Semi major axis (km)	Perigee height (km)	Apogee height (km)	Orbital eccentricity	Argument of perigee (deg)
D 2M R	Soyuz 15*	1974-67A	1974 Aug 26.83 2.01 days 1974 Aug 28.84	Sphere-cylinder 6570?	7.5 long 2.2 dia	1974 Aug 27.1 1974 Aug 27.5	51.62 51.60	88.52 89.67	6583 6639	173 251	236 271	0.005 0.002	26 336
D	Soyuz 15 rocket	1974-67B	1974 Aug 26.83 2.00 days 1974 Aug 28.83	Cylinder 2500?	7.5 long 2.6 dia	1974 Aug 27.1	51.60	88.43	6578	189	211	0.002	53
D R	Cosmos 674	1974-68A	1974 Aug 29.32 8.9 days 1974 Sep 7.2	Sphere-cylinder 6300?	6.5 long? 2.4 dia	1974 Aug 30.0	64.99	89.48	6627	175	323	0.011	56
D	Cosmos 674 rocket	1974-68B	1974 Aug 29.32 5.78 days 1974 Sep 4.10	Cylinder 2500?	7.5 long 2.6 dia	1974 Aug 29.6	65.00	89.40	6623	174	316	0.011	54
D	Cosmos 674 engine**	1974-68C	1974 Aug 29.32 16 days 1974 Sep 14	Cone 600? full	1.5 long? 2 dia?	1974 Sep 6.4	64.99	89.31	6619	174	307	0.010	53
D	Fragment	1974-68D											
	Cosmos 675	1974-69A	1974 Aug 29.62 5000 years	- 500?	-	1974 Aug 30.6	74.04	113.70	7774	1365	1426	0.004	203
	Cosmos 675 rocket	1974-69B	1974 Aug 29.62 4000 years	Cylinder 2200?	7.4 long 2.4 dia	1974 Sep 1.2	74.05	113.57	7768	1359	1421	0.004	189
D	ANS 1†	1974-70A	1974 Aug 30.59 1019 days 1977 Jun 14	Box + 2 panels 129	1.23 long 0.61 wide 0.73 deep	1974 Aug 30.8	98.03	99.13	7094	258	1173	0.064	211
D	ANS 1 rocket	1974-70B	1974 Aug 30.59 446 days 1975 Nov 19	Cylinder 24	1.50 long 0.46 dia	1974 Aug 31.7 1975 Mar 31.5	98.04 98.04	99.12 95.92	7093 6939	259 251	1171 871	0.064 0.045	208 -
D	Fragments	1974-70C-E											

* Soyuz 15 passed near to Salyut 3 about 1974 Aug 27.8. ** 1974-68C ejected from 1974-68A on 1974 Sep 6. †Astronomical Netherlands Satellite.

Name		Launch date, lifetime and descent date	Shape and weight (kg)	Size (m)	Date of orbital determination	Orbital inclination (deg)	Nodal period (min)	Semi major axis (km)	Perigee height (km)	Apogee height (km)	Orbital eccentricity	Argument of perigee (deg)
Cosmos 676	1974-71A	1974 Sep 11.74 120 years	Cylinder + paddles 750?	2 long? 1 dia?	1974 Sep 12.7	74.05	101.01	7184	796	816	0.001	58
Cosmos 676 rocket	1974-71B	1974 Sep 11.74 100 years	Cylinder 2200?	7.4 long 2.4 dia	1974 Sep 12.8	74.05	100.91	7180	787	816	0.002	54
Cosmos 677	1974-72A	1974 Sep 19.61 7000 years	Spheroid 40?	1.0 long? 0.8 dia?	1974 Sep 21.0	74.03	114.53	7812	1399	1469	0.004	86
Cosmos 678	1974-72B	1974 Sep 19.61 10000 years	Spheroid 40?	1.0 long? 0.8 dia?	1974 Sep 24.5	74.03	116.03	7880	1468	1535	0.004	260
Cosmos 679	1974-72C	1974 Sep 19.61 10000 years	Spheroid 40?	1.0 long? 0.8 dia?	1974 Sep 24.8	74.02	115.78	7869	1468	1513	0.003	265
Cosmos 680	1974-72D	1974 Sep 19.61 10000 years	Spheroid 40?	1.0 long? 0.8 dia?	1974 Sep 22.4	74.03	115.58	7859	1468	1494	0.002	258
Cosmos 681	1974-72E	1974 Sep 19.61 9000 years	Spheroid 40?	1.0 long? 0.8 dia?	1974 Sep 22.4	74.03	115.35	7849	1468	1474	0.0004	298
Cosmos 682	1974-72F	1974 Sep 19.61 9000 years	Spheroid 40?	1.0 long? 0.8 dia?	1974 Sep 25.9	74.03	115.15	7840	1455	1468	0.001	86
Cosmos 683	1974-72G	1974 Sep 19.61 9000 years	Spheroid 40?	1.0 long? 0.8 dia?	1974 Sep 28.3	74.03	114.95	7831	1436	1469	0.002	83

1974-72 continued on page 379

	Name		Launch date, lifetime and descent date	Shape and weight (kg)	Size (m)	Date of orbital determination	Orbital inclination (deg)	Nodal period (min)	Semi major axis (km)	Perigee height (km)	Apogee height (km)	Orbital eccentricity	Argument of perigee (deg)
	Cosmos 684	1974-72H	1974 Sep 19.61 8000 years	Spheroid 40?	1.0 long? 0.8 dia?	1974 Sep 25.6	74.02	114.74	7821	1418	1468	0.003	91
	Cosmos 677 rocket	1974-72J	1974 Sep 19.61 20000 years	Cylinder 2200?	7.4 long 2.4 dia	1974 Sep 21.0	74.02	117.82	7961	1471	1694	0.014	271
D R	Cosmos 685	1974-73A	1974 Sep 20.40 11.85 days 1974 Oct 2.25	Sphere- cylinder 5700?	5.0 long 2.4 dia	1974 Sep 21.3	64.98	89.39	6623	205	285	0.006	46
D	Cosmos 685 rocket	1974-73B	1974 Sep 20.40 6 days 1974 Sep 26	Cylinder 2500?	7.5 long 2.6 dia	1974 Sep 21.8	64.99	89.13	6610	198	266	0.005	31
D	Fragment	1974-73C											
D	Cosmos 686	1974-74A	1974 Sep 26.69 216.53 days 1975 May 1.22	Ellipsoid 400?	1.8 long 1.2 dia	1974 Sep 27.8	71.00	92.18	6759	273	489	0.016	74
D	Cosmos 686* rocket	1974-74B	1974 Sep 26.69 33.11 days 1974 Oct 29.80	Cylinder 1500?	8 long 1.65 dia	1974 Sep 27.3	70.93	91.69	6735	260	454	0.014	67
D	Fragments	1974-74C- V											
T	Westar 2	1974-75A	1974 Oct 10.95 > million years	Cylinder 574 full 300 empty	1.65 long 1.90 dia	1974 Nov 1.0 1975 Jan 1.0	0.4 0.0	1432.7 1435.9	42100 42166	35710 35780	35734 35795	0.0003 0.0002	- -
D	Westar 2 second stage	1974-75B	1974 Oct 10.95 1535 days 1978 Dec 23	Cylinder + annulus 350?	6.4 long 1.52 and 2.44 dia	1974 Oct 13.2	27.33	123.46	8220	230	3454	0.196	194

* Rocket disintegrated. The main piece may not be 1974-74B

1974-75 continued on page 380

	Name		Launch date, lifetime and descent date	Shape and weight (kg)	Size (m)	Date of orbital determination	Orbital inclination (deg)	Nodal period (min)	Semi major axis (km)	Perigee height (km)	Apogee height (km)	Orbital eccentricity	Argument of perigee (deg)
	Westar 2 third stage	1974-75C	1974 Oct 10.95 25 years	Sphere-cone 66	1.32 long 0.94 dia	1974 Oct 13.9	24.81	641.42	24648	227	36313	0.732	180
D	Cosmos 687	1974-76A	1974 Oct 11.48 1213 days 1978 Feb 5	Octagonal ellipsoid 550?	1.8 long? 1.5 dia?	1974 Oct 12.1	74.00	94.48	6870	286	698	0.030	133
D	Cosmos 687 rocket	1974-76B	1974 Oct 11.48 816 days 1977 Jan 4	Cylinder 2200?	7.4 long 2.4 dia	1974 Oct 12.3	73.99	94.40	6866	285	691	0.030	131
D	Ariel 5	1974-77A	1974 Oct 15.32 1977.45 days 1980 Mar 14.77	Cylinder 129	0.86 long 0.95 dia	1974 Oct 16.5	2.88	94.96	6905	504	549	0.003	268
D	Ariel 5 rocket	1974-77B	1974 Oct 15.32 1671 days 1979 May 13	Cylinder 24	1.50 long 0.46 dia	1974 Oct 21.9	2.88	94.97	6905	504	550	0.003	351
D R	Cosmos 688	1974-78A	1974 Oct 18.63 11.66 days 1974 Oct 30.29	Sphere-cylinder 6300?	6.5 long? 2.4 dia	1974 Oct 19.4	62.82	89.77	6642	179	349	0.013	74
D	Cosmos 688 rocket	1974-78B	1974 Oct 18.63 5.88 days 1974 Oct 24.51	Cylinder 2500?	7.5 long 2.6 dia	1974 Oct 20.9	62.81	89.27	6617	172	306	0.010	69
D	Cosmos 688 engine*	1974-78D	1974 Oct 18.63 12.58 days 1974 Oct 31.21	Cone 600? Full	1.5 long? 2 dia?	1974 Oct 29.7	62.81	88.88	6598	145	294	0.011	59
D	Fragment	1974-78C											

* 1974-78D ejected from 1974-78A on 1974 Oct 29.

	Name		Launch date, lifetime and descent date	Shape and weight (kg)	Size (m)	Date of orbital determination	Orbital inclination (deg)	Nodal period (min)	Semi major axis (km)	Perigee height (km)	Apogee height (km)	Orbital eccentricity	Argument of perigee (deg)
	Cosmos 689	1974-79A	1974 Oct 18.94 1200 years	Cylinder? 700?	1.3 long? 1.9 dia?	1974 Oct 19.4	82.94	105.12	7377	981	1017	0.002	255
	Cosmos 689 rocket	1974-79B	1974 Oct 18.94 600 years	Cylinder 2200?	7.4 long 2.4 dia	1974 Oct 19.4	82.94	105.00	7371	977	1009	0.002	242
D B R	Cosmos 690	1974-80A	1974 Oct 22.75 20.5 days 1974 Nov 12.2	Sphere-cylinder 5900?	5.9 long 2.4 dia	1974 Oct 27.6	62.81	90.29	6668	215	364	0.011	114
D	Cosmos 690 rocket	1974-80B	1974 Oct 22.75 18.64 days 1974 Nov 10.39	Cylinder 2500?	7.5 long 2.6 dia	1974 Oct 26.9	62.80	90.08	6658	214	345	0.010	110
D	Capsule	1974-80E	1974 Oct 22.75 53 days 1974 Dec 14	Ellipsoid 200?	0.9 long 1.9 dia	1974 Nov 12.5	62.80	90.02	6655	212	341	0.010	-
D	Fragments	1974-80C,D,F-H											
	Molniya 1AD	1974-81A	1974 Oct 24.53 14½ years*	Windmill + 6 vanes 1000?	3.4 long 1.6 dia	1974 Oct 25.6 1974 Oct 29.7	62.82 62.84	736.37 717.87	27013 26559	656 646	40614 39715	0.740 0.735	288 288
D	Molniya 1AD launcher rocket	1974-81B	1974 Oct 24.53 68.20 days 1974 Dec 31.73	Cylinder 2500?	7.5 long 2.6 dia	1974 Oct 26.6	62.84	92.95	6798	217	623	0.030	120
D	Molniya 1AD launcher	1974-81C	1974 Oct 24.53 70.97 days 1975 Jan 3.50	Irregular	-	1974 Oct 26.6	62.82	93.25	6813	211	658	0.033	119
	Molniya 1AD rocket	1974-81D	1974 Oct 24.53 14½ years*	Cylinder 440	2.0 long 2.0 dia	1974 Oct 25.6	62.81	731.93	26904	642	40410	0.739	288
D	Fragment	1974-81E											

* Possibility of decay in mid 1986 when perigee falls to 150km.

	Name		Launch date, lifetime and descent date	Shape and weight (kg)	Size (m)	Date of orbital determination	Orbital inclination (deg)	Nodal period (min)	Semi major axis (km)	Perigee height (km)	Apogee height (km)	Orbital eccentricity	Argument of perigee (deg)
D R	Cosmos 691	1974-82A	1974 Oct 25.40 11.86 days 1974 Nov 6.26	Sphere-cylinder 6300?	6.5 long? 2.4 dia	1974 Oct 27.1	65.04	89.50	6629	173	328	0.012	66
D	Cosmos 691 rocket	1974-82B	1974 Oct 25.40 4.14 days 1974 Oct 29.54	Cylinder 2500?	7.5 long 2.6 dia	1974 Oct 27.0	65.04	89.05	6606	169	287	0.009	63
D	Cosmos 691 engine*	1974-82C	1974 Oct 25.40 16 days 1974 Nov 10	Cone 600? full	1.5 long? 2 dia?	1974 Nov 4.6	65.03	89.23	6615	167	307	0.011	62
D	Fragment	1974-82D											
	Meteor 19	1974-83A	1974 Oct 28.43 500 years	Cylinder + 2 vanes 2200?	5 long? 1.5 dia?	1974 Oct 28.8	81.18	102.48	7253	843	907	0.004	302
	Meteor 19 rocket	1974-83B	1974 Oct 28.43 400 years	Cylinder 1440	3.8 long 2.6 dia	1974 Oct 28.7	81.18	102.62	7260	852	911	0.004	244
D	Luna 23 launcher	1974-84B	1974 Oct 28.60 3.43 days 1974 Nov 1.03	-	-	1974 Oct 29.1	51.54	88.72	6593	183	246	0.005	307
D	Luna 23 launcher rocket	1974-84C	1974 Oct 28.60 3.43 days 1974 Nov 1.03	Cylinder 4000?	12 long? 4 dia	1974 Oct 29.4	51.53	88.62	6588	179	240	0.005	316

* 1974-82C ejected from 1974-82A on 1974 Nov 4

Space Vehicle: Luna 23, 1974-84A

	Name		Launch date, lifetime and descent date	Shape and weight (kg)	Size (m)	Date of orbital determination	Orbital inclination (deg)	Nodal period (min)	Semi major axis (km)	Perigee height (km)	Apogee height (km)	Orbital eccentricity	Argument of perigee (deg)
D	[Titan 3D]	1974-85A	1974 Oct 29.81 141 days 1975 Mar 19	Cylinder 13300? full	15 long 3.0 dia	1974 Oct 30.0	96.69	88.86	6595	162	271	0.008	164
D	Titan 3D rocket	1974-85D	1974 Oct 29.81 2.02 days 1974 Oct 31.83	Cylinder 1900	6 long 3.0 dia	1974 Oct 30.2	96.69	88.73	6588	157	263	0.008	174
D	Capsule	1974-85B	1974 Oct 29.81 1912 days 1980 Jan 23	Octagon 60?	0.3 long? 0.9 dia?	1974 Nov 3.2	96.06	95.22	6906	520	535	0.001	160
D	SESP 73-5*	1974-85C	1974 Oct 29.81 208.70 days 1975 May 26.51	-	-	1974 Oct 31.9	96.98	126.59	8352	152	3795	0.218	133
D	Intercosmos 12	1974-86A	1974 Oct 31.42 253 days 1975 Jul 11	Octagonal ellipsoid 550?	1.8 long? 1.5 dia?	1974 Nov 2.4	74.02	94.11	6853	243	707	0.034	37
D	Intercosmos 12 rocket	1974-86B	1974 Oct 31.42 265 days 1975 Jul 23	Cylinder 2200?	7.4 long 2.4 dia	1974 Nov 3.2	74.00	94.01	6848	240	700	0.034	34
D	Fragments	1974-86C,D											
D R	Cosmos 692	1974-87A	1974 Nov 1.60 11.7 days 1974 Nov 13.3	Sphere-cylinder 5900?	5.9 long 2.4 dia	1974 Nov 1.6	62.82	89.41	6624	197	295	0.007	57
D	Cosmos 692 rocket	1974-87B	1974 Nov 1.60 5 days 1974 Nov 6	Cylinder 2500?	7.5 long 2.6 dia	1974 Nov 3.3	62.79	89.08	6608	184	275	0.007	48

* Space Experiments Support Programme - carried 8 atmospheric-density experiments.

1974-87 continued on page 384

	Name		Launch date, lifetime and descent date	Shape and weight (kg)	Size (m)	Date of orbital determination	Orbital inclination (deg)	Nodal period (min)	Semi major axis (km)	Perigee height (km)	Apogee height (km)	Orbital eccentricity	Argument of perigee (deg)
D	Capsule*	1974-87F	1974 Nov 1.60 22.32 days 1974 Nov 23.92	Ellipsoid 200?	0.9 long 1.9 dia	1974 Nov 11.6	62.81	89.24	6616	193	282	0.007	58
D	Fragments	1974-87C-E,G											
D R	Cosmos 693	1974-88A	1974 Nov 4.45 11.8 days 1974 Nov 16.3	Sphere-cylinder 5900?	5.9 long 2.4 dia	1974 Nov 5.4	81.33	89.14	6609	219	243	0.002	47
D	Cosmos 693 rocket	1974-88B	1974 Nov 4.45 4.59 days 1974 Nov 9.04	Cylinder 2500?	7.5 long 2.6 dia	1974 Nov 5.4	81.33	88.93	6599	212	229	0.001	23
D	Capsule**	1974-88E	1974 Nov 4.45 13 days 1974 Nov 17	Ellipsoid 200?	0.9 long 1.9 dia	1974 Nov 16.6	81.53	88.49	6577	182	215	0.002	85
D	Fragments	1974-88C,D											
T	NOAA 4 (ITOS)	1974-89A	1974 Nov 15.72 10000 years	Box 340	1.25 long 1.02 sq.	1974 Nov 18.4	101.75	115.00	7833	1447	1462	0.001	256
T	Oscar 7	1974-89B	1974 Nov 15.72 10000 years	8-sided cylinder 29	0.43 long 0.42 dia	1974 Nov 17.3	101.74	114.97	7831	1444	1462	0.001	214
	Intasat 1***	1974-89C	1974 Nov 15.72 10000 years	12-sided cylinder 20	0.45 long 0.44 dia	1974 Nov 18.4	101.73	114.95	7830	1442	1462	0.001	209
	NOAA 4† second stage	1974-89D	1974 Nov 15.72 disintegrated	Cylinder 350?	4.9 long 1.43 dia	1974 Nov 17.4	101.74	114.99	7832	1447	1461	0.001	210
8d	Fragments	1974-89E-EP											

* 1974-87F ejected from 1974-87A about 1974 Nov 11.
** 1974-88E ejected from 1974-88A about 1974 Nov 16.
*** First Spanish satellite.
† Disintegrated about 1975 Aug 20.50

	Name		Launch date, lifetime and descent date	Shape and weight (kg)	Size (m)	Date of orbital determination	Orbital inclination (deg)	Nodal period (min)	Semi major axis (km)	Perigee height (km)	Apogee height (km)	Orbital eccentricity	Argument of perigee (deg)
D R	Cosmos 694	1974-90A	1974 Nov 16.49 12.8 days 1974 Nov 29.3	Sphere-cylinder 6300?	6.5 long? 2.4 dia	1974 Nov 17.5 1974 Nov 22.8	72.83 72.83	89.37 89.59	6621 6632	173 172	313 336	0.011 0.012	56 46
D	Cosmos 694 rocket	1974-90B	1974 Nov 16.49 9.25 days 1974 Nov 25.74	Cylinder 2500?	7.5 long 2.6 dia	1974 Nov 18.5	72.82	89.50	6628	198	301	0.008	58
D	Cosmos 694 engine	1974-90E	1974 Nov 16.49 17 days 1974 Dec 3	Cone 600? full	1.5 long? 2 dia?	1974 Nov 29.5	72.83	89.36	6621	166	319	0.012	-
D	Fragments	1974-90C,D,F,G											
D	Cosmos 695	1974-91A	1974 Nov 20.50 237 days 1975 Jul 15	Ellipsoid 400?	1.8 long 1.2 dia	1974 Nov 21.3	71.00	91.96	6749	273	468	0.014	82
D	Cosmos 695 rocket	1974-91B	1974 Nov 20.50 114.77 days 1975 Mar 15.27	Cylinder 1500?	8 long 1.65 dia	1974 Nov 23.2	70.99	91.80	6741	272	453	0.013	79
	Molniya 3A	1974-92A	1974 Nov 21.44 11 years	Windmill + 6 vanes 1500?	4.2 long? 1.6 dia	1974 Nov 25.1 1974 Nov 26.6	62.82 62.82	737.26 717.50	27035 26549	628 503	40685 39839	0.741 0.741	288 288
D	Molniya 3A launcher rocket	1974-92B	1974 Nov 21.44 67.08 days 1975 Jan 27.52	Cylinder 2500?	7.5 long 2.6 dia	1974 Nov 22.5	62.83	92.70	6786	215	600	0.028	121
D	Molniya 3A launcher	1974-92C	1974 Nov 21.44 51.41 days 1975 Jan 11.85	Irregular	-	1974 Nov 22.5	62.84	92.82	6792	201	626	0.031	117
	Molniya 3A rocket	1974-92E	1974 Nov 21.44 11 years	Cylinder 440	2.0 long 2.0 dia	1974 Nov 24.0	62.73	733.95	26954	621	40530	0.740	288
D	Fragment	1974-92D											

	Name		Launch date, lifetime and descent date	Shape and weight (kg)	Size (m)	Date of orbital determination	Orbital inclination (deg)	Nodal period (min)	Semi major axis (km)	Perigee height (km)	Apogee height (km)	Orbital eccentricity	Argument of perigee (deg)
T	Intelsat 4 F-8	1974-93A	1974 Nov 21.99 >million years	Cylinder 1410 full 720 empty	2.82 long 2.39 dia	1974 Dec 31	1.77	1436.2	42167	35775	35801	0.0003	60
	Intelsat 4 F-8 rocket	1974-93B	1974 Nov 21.99 6000 years	Cylinder 1815	8.6 long 3.0 dia	1974 Nov 26.3	25.99	653.78	24963	557	36612	0.722	182
T	Skynet 2B	1974-94A	1974 Nov 23.02 >million years	Cylinder 435 full 235 empty	1.33 long 1.90 dia	1974 Dec 3.0 1975 May 1.0	2.30 1.9	1469.5 1436.2	42818 42167	36255 35784	36621 35794	0.0043 0.0001	181 -
D	Skynet 2B second stage	1974-94B	1974 Nov 23.02 45 days 1975 Jan 7	Cylinder + annulus 350?	6.4 long 1.52 and 2.44 dia	1974 Nov 24.5	28.17	97.95	7046	183	1152	0.069	187
D	Skynet 2B rocket	1974-94E	1974 Nov 23.02 718 days 1976 Nov 10	Sphere-cone 66	1.32 long 0.94 dia	1974 Nov 24.0	24.49	651.35	24893	176	36854	0.737	179
1d	Fragments	1974-94C,D											
D R	Cosmos 696	1974-95A	1974 Nov 27.49 11.8 days 1974 Dec 9.3	Sphere-cylinder 5700?	5.0 long 2.4 dia	1974 Nov 28.5	72.85	89.77	6641	205	321	0.009	63
D	Cosmos 696 rocket	1974-95B	1974 Nov 27.49 8.02 days 1974 Dec 5.51	Cylinder 2500?	7.5 long 2.6 dia	1974 Nov 28.1	72.86	89.69	6637	198	320	0.009	57
D	Fragments	1974-95C,D											
D 2M R	Soyuz 16	1974-96A	1974 Dec 2.40 5.94 days 1974 Dec 8.34	Sphere-cylinder + 2 wings 6680?	7.5 long 2.2 dia	1974 Dec 2.6 1974 Dec 3.1 1974 Dec 5.3	51.80 51.80 51.80	89.19 88.37 88.95	6616 6575 6604	184 183 225	291 210 226	0.008 0.002 0	91 94 -*

* Approximate orbit.

1974-96 continued on page 387

	Name		Launch date, lifetime and descent date	Shape and weight (kg)	Size (m)	Date of orbital determination	Orbital inclination (deg)	Nodal period (min)	Semi major axis (km)	Perigee height (km)	Apogee height (km)	Orbital eccentricity	Argument of perigee (deg)
D	Soyuz 16 rocket	1974-96B	1974 Dec 2.40 5.12 days 1974 Dec 7.52	Cylinder 2500?	7.5 long 2.6 dia	1974 Dec 3.2	51.80	89.11	6612	184	283	0.007	91
D	Fragments	1974-96C-F											
D	Helios 1 second stage	1974-97B	1974 Dec 10.30 5 years?	Cylinder 1815	8.6 long 3.0 dia	1974 Dec 10.5	31.77	4175	85880	1770	157235	0.905	216
D R	Cosmos 697	1974-98A	1974 Dec 13.57 11.6 days 1974 Dec 25.2	Sphere-cylinder 6300?	6.5 long? 2.4 dia	1974 Dec 14.4	62.80	90.16	6661	174	392	0.016	68
D	Cosmos 697 rocket	1974-98B	1974 Dec 13.57 7.70 days 1974 Dec 21.27	Cylinder 2500?	7.5 long 2.6 dia	1974 Dec 14.7	62.79	89.95	6651	171	375	0.015	66
D	Cosmos 697 engine?*	1974-98D	1974 Dec 13.57 13 days 1974 Dec 26	Cone 600? full	1.5 long? 2 dia?	1974 Dec 25.3	62.80	89.86	6647	177	360	0.014	61
D	Fragments	1974-98C,E											
	Meteor 20	1974-99A	1974 Dec 17.49 500 years	Cylinder + 2 vanes 2200?	5 long? 1.5 dia?	1974 Dec 24.0	81.24	102.38	7248	842	897	0.004	262
	Meteor 20 rocket	1974-99B	1974 Dec 17.49 400 years	Cylinder 1440	3.8 long 2.6 dia	1974 Dec 22.6	81.24	102.39	7248	820	920	0.007	207
D	Cosmos 698	1974-100A	1974 Dec 18.59 2183 days 1980 Dec 9	Cylinder + paddles? 900?	2 long? 1 dia?	1974 Dec 20.6	74.04	95.32	6912	515	552	0.003	7

Space Vehicle: Helios 1, 1974-97A; Helios 1 rocket, 1974-97C.

* 1974-98D ejected from 1974-98A about 1974 Dec 25.

1974-100 continued on page 388

	Name		Launch date, lifetime and descent date	Shape and weight (kg)	Size (m)	Date of orbital determination	Orbital inclination (deg)	Nodal period (min)	Semi major axis (km)	Perigee height (km)	Apogee height (km)	Orbital eccentricity	Argument of perigee (deg)
D	Cosmos 698 rocket	1974-100B	1974 Dec 18.59 2152 days 1980 Nov 8	Cylinder 2200?	7.4 long 2.4 dia	1974 Dec 21.9	74.04	95.22	6907	505	552	0.003	6
D	Fragments	1974-100C-J											
T	Symphonie 1*	1974-101A	1974 Dec 19.11 >million years	Octagon + 3 paddles 402 full 221 empty	0.58 long 1.85 dia	1974 Dec 19.1 1974 Dec 21.7 1975 May 1.0	13.23 1.18 0.2	688.4 1646.6 1436.1	25826 46190 42165	395 38705 35768	38500 40919 35806	0.738 0.024 0.0005	178 218 -
D	Symphonie 1 second stage	1974-101B	1974 Dec 19.11 1217 days 1978 Apr 19	Cylinder + annulus 350?	6.4 long 1.52 and 2.44 dia	1974 Dec 20.8	27.01	98.44	7070	283	1101	0.058	149
	Symphonie 1 third stage	1974-101G	1974 Dec 19.11 100 years	Sphere-cone 66	1.32 long 0.94 dia	1976 Nov 1.0	12.80	682.8	25685	409	38204	0.736	-
D	Fragments	1974-101C-F											
	Molniya 2L	1974-102A	1974 Dec 21.10 14 years	Windmill + 6 vanes 1250?	4.2 long? 1.6 dia	1974 Dec 23.2 1974 Dec 26.2	62.90 62.87	736.77 718.28	27022 26569	659 611	40629 39771	0.740 0.737	289 288
D	Molniya 2L launcher rocket	1974-102B	1974 Dec 21.10 81.53 days 1975 Mar 12.63	Cylinder 2500?	7.5 long 2.6 dia	1974 Dec 22.2	62.85	92.53	6778	217	582	0.027	121
D	Molniya 2L launcher	1974-102C	1974 Dec 21.10 77 days 1975 Mar 8	Irregular	-	1974 Dec 22.2	62.82	92.91	6797	211	626	0.031	122
	Molniya 2L rocket	1974-102D	1974 Dec 21.10 14 years	Cylinder 440	2.0 long 2.0 dia	1974 Dec 23.2	62.89	733.95	26954	616	40536	0.740	288

* Symphonie is a French-German satellite, launched by NASA.

	Name		Launch date, lifetime and descent date	Shape and weight (kg)	Size (m)	Date of orbital determination	Orbital inclination (deg)	Nodal period (min)	Semi major axis (km)	Perigee height (km)	Apogee height (km)	Orbital eccentricity	Argument of perigee (deg)
D	Cosmos 699*	1974-103A	1974 Dec 24.46	Cylinder?	-	1974 Dec 24.5	64.99	89.80	6644	114	418	0.023	52
			1027 days			1974 Dec 26.1	65.03	93.31	6812	428	440	0.001	265
			1977 Oct 16										
D	Cosmos 699 rocket	1974-103B	1974 Dec 24.46	Cylinder	8 long?	1974 Dec 24.7	65.06	89.22	6614	114	358	0.018	63
			1 day	1500?	2.5 dia?								
			1974 Dec 25										
D	Fragments	1974-103C-BC											
D	Salyut 4 †	1974-104A	1974 Dec 26.18	Cylinder +	14 long	1974 Dec 27.0	51.57	89.08	6610	212	251	0.003	81
			769.80 days	3 wings	4.15 to	1974 Dec 30.2	51.57	90.65	6687	276	341	0.005	328
			1977 Feb 2.98	18900	2.0 dia	1975 Jan 17.4	51.58	91.32	6721	336	349	0.001	304
D	Salyut 4 rocket	1974-104B	1974 Dec 26.18	Cylinder	12 long?	1974 Dec 27.0	51.58	88.83	6597	207	231	0.002	52
			6.81 days	4000?	4 dia								
			1975 Jan 1.99										
D	Fragments	1974-104C-U											
	Cosmos 700	1974-105A	1974 Dec 26.50	Cylinder?	1.3 long?	1974 Dec 26.8	82.96	104.80	7361	966	999	0.002	300
			1200 years	700?	1.9 dia?								
	Cosmos 700 rocket	1974-105B	1974 Dec 26.50	Cylinder	7.4 long	1974 Dec 27.5	82.95	104.68	7355	964	989	0.002	295
			600 years	2200?	2.4 dia								
D	Cosmos 701	1974-106A	1974 Dec 27.38	Sphere-	6.5 long?	1974 Dec 28.0	71.39	89.77	6640	205	319	0.009	47
R			12.88 days	cylinder	2.4 dia	1974 Dec 29.3	71.38	89.37	6620	170	314	0.011	56
			1975 Jan 9.26	6300?									
D	Cosmos 701 rocket	1974-106B	1974 Dec 27.38	Cylinder	7.5 long	1974 Dec 28.4	71.39	89.69	6636	202	314	0.008	45
			11.31 days	2500?	2.6 dia								
			1975 Jan 7.69										
D	Cosmos 701 engine**	1974-106F	1974 Dec 27.38	Cone	1.5 long?	1975 Jan 8.7	71.43	89.59	6631	175	331	0.012	34
			18.99 days	600?	2 dia?								
			1975 Jan 15.37	full									
D	Fragments	1974-106C-E,G											

* Partially disintegrated on 1975 Apr 17.91, near 3 deg North, 82 deg West.

** 1974-106F ejected from 1974-106A about 1975 Jan 8.

† De-orbited on command.

	Name		Launch date, lifetime and descent date	Shape and weight (kg)	Size (m)	Date of orbital determination	Orbital inclination (deg)	Nodal period (min)	Semi major axis (km)	Perigee height (km)	Apogee height (km)	Orbital eccentricity	Argument of perigee (deg)
D 2M R	Soyuz 17*	1975-01A	1975 Jan 10.90 29.56 days 1975 Feb 9.46	Sphere-cylinder 6570?	7.5 long 2.2 dia	1975 Jan 11.0 1975 Jan 11.6 1975 Jan 17.4	51.63 51.58 51.58	88.79 90.69 91.32	6595 6689 6721	185 274 336	249 347 349	0.005 0.005 0.001	88 266 304
D	Soyuz 17 rocket	1975-01B	1975 Jan 10.90 3.11 days 1975 Jan 14.01	Cylinder 2500?	7.5 long 2.6 dia	1975 Jan 11.7	51.63	88.59	6585	181	233	0.004	87
D	Fragments	1975-01C-J											
D R	Cosmos 702	1975-02A	1975 Jan 17.38 11.90 days 1975 Jan 29.28	Sphere-cylinder 5700?	5.0 long 2.4 dia	1975 Jan 18.6	71.33	89.70	6637	205	313	0.008	50
D	Cosmos 702 rocket	1975-02B	1975 Jan 17.38 11.99 days 1975 Jan 29.37	Cylinder 2500?	7.5 long 2.6 dia	1975 Jan 18.4	71.35	89.57	6631	202	304	0.008	39
D	Fragment	1975-02C											
D	Cosmos 703	1975-03A	1975 Jan 21.46 303 days 1975 Nov 20	Ellipsoid 400?	1.8 long 1.2 dia	1975 Jan 22.4	81.96	102.11	7236	197	1518	0.091	72
D	Cosmos 703 rocket	1975-03B	1975 Jan 21.46 213 days 1975 Aug 22	Cylinder 1500?	8 long 1.65 dia	1975 Jan 22.2	81.96	101.87	7225	200	1493	0.089	73
T	Landsat 2 (ERTS 2)	1975-04A	1975 Jan 22.75 100 years	Cone + 2 paddles 953	3.0 long 1.45 dia	1975 Jan 25.4	99.09	103.28	7291	907	918	0.001	266
	Landsat 2** second stage	1975-04B	1975 Jan 22.75 disintegrated	Cylinder + annulus 350?	6.4 long 1.52 and 2.44 dia	1975 Jan 25.3	97.83	101.51	7208	743	916	0.012	211
139d	Fragments	1975-04C-HF											

* Soyuz 17 docked with Salyut 4 about 1975 Jan 12.04; separated 1975 Feb 9.26.

** Disintegrated about 1976 Feb 5.43

	Name		Launch date, lifetime and descent date	Shape and weight (kg)	Size (m)	Date of orbital determination	Orbital inclination (deg)	Nodal period (min)	Semi major axis (km)	Perigee height (km)	Apogee height (km)	Orbital eccentricity	Argument of perigee (deg)
D R	Cosmos 704	1975-05A	1975 Jan 23.46 13.74 days 1975 Feb 6.20	Sphere-cylinder 6300?	6.5 long? 2.4 dia	1975 Jan 24.8 1975 Jan 27.4	72.86 72.87	89.62 89.25	6633 6615	205 169	305 304	0.008 0.010	68 60
D	Cosmos 704 rocket	1975-05B	1975 Jan 23.46 11.09 days 1975 Feb 3.55	Cylinder 2500?	7.5 long 2.6 dia	1975 Jan 24.8	72.86	89.47	6626	203	293	0.007	62
D	Cosmos 704 engine	1975-05F	1975 Jan 23.46 15 days 1975 Feb 7	Cone 600? full	1.5 long? 2 dia?	1975 Feb 2.9	72.83	89.39	6622	166	321	0.012	45
D	Fragments	1975-05C-E,G-J											
D	Cosmos 705	1975-06A	1975 Jan 28.50 294 days 1975 Nov 18	Ellipsoid 400?	1.8 long 1.2 dia	1975 Jan 28.7	70.97	92.29	6765	271	502	0.017	78
D	Cosmos 705 rocket	1975-06B	1975 Jan 28.50 148 days 1975 Jun 25	Cylinder 1500?	8 long 1.65 dia	1975 Jan 29.1	70.97	92.12	6756	273	483	0.016	78
	Cosmos 706	1975-07A	1975 Jan 30.63 30 years	Windmill + 6 vanes 1250?	4.2 long? 1.6 dia	1975 Jan 31.2	62.85	719.55	26602	623	39824	0.737	318
D	Cosmos 706 launcher rocket	1975-07B	1975 Jan 30.63 63.98 days 1975 Apr 4.61	Cylinder 2500?	7.5 long 2.6 dia	1975 Jan 31.7	62.82	92.53	6778	214	585	0.027	119
D	Cosmos 706 launcher	1975-07C	1975 Jan 30.63 30.29 days 1975 Mar 1.92	Irregular	-	1975 Jan 31.8	62.89	92.62	6782	179	629	0.033	118
	Cosmos 706 rocket	1975-07D	1975 Jan 30.63 30 years	Cylinder 440	2.0 long 2.0 dia	1975 Jan 31.7	62.87	716.77	26530	630	39674	0.736	318

	Name		Launch date, lifetime and descent date	Shape and weight (kg)	Size (m)	Date of orbital determination	Orbital inclination (deg)	Nodal period (min)	Semi major axis (km)	Perigee height (km)	Apogee height (km)	Orbital eccentricity	Argument of perigee (deg)
D	Cosmos 707	1975-08A	1975 Feb 5.55 2041 days 1980 Sep 7	Cylinder + paddles? 900?	2 long? 1 dia?	1975 Feb 6.7	74.03	95.14	6903	503	547	0.003	338
D	Cosmos 707 rocket	1975-08B	1975 Feb 5.55 2030 days 1980 Aug 27	Cylinder 2200?	7.4 long 2.4 dia	1975 Feb 6.5	74.03	95.04	6898	494	546	0.004	348
D	Fragments	1975-08C-E											
	Molniya 2M	1975-09A	1975 Feb 6.20 10¾ years	Windmill + 6 vanes 1250?	4.2 long? 1.6 dia	1975 Feb 6.8 1975 Feb 23.3	62.78 62.81	736.86 717.59	27025 26552	634 602	40660 39745	0.741 0.737	289 289
D	Molniya 2M launcher rocket	1975-09B	1975 Feb 6.20 78.57 days 1975 Apr 25.77	Cylinder 2500?	7.5 long 2.6 dia	1975 Feb 6.3	62.84	92.68	6784	216	596	0.028	119
D	Molniya 2M launcher	1975-09C	1975 Feb 6.20 76.19 days 1975 Apr 23.39	Irregular	-	1975 Feb 8.3	62.83	92.87	6794	212	620	0.030	121
	Molniya 2M rocket	1975-09D	1975 Feb 6.20 10¾ years	Cylinder 440	2.0 long 2.0 dia	1975 Feb 7.3	62.82	733.33	26939	612	40510	0.741	289
T L	Starlette* [Diamant B]	1975-10A	1975 Feb 6.69 2000 years	Quasi-sphere 47	0.26 dia	1975 Feb 20.6	49.82	104.13	7335	806	1108	0.021	75
	Starlette rocket	1975-10B	1975 Feb 6.69 200 years	Cylinder 68	1.60 long? 0.65 dia	1975 Feb 21.6	49.82	104.43	7349	804	1138	0.023	78
	Fragments	1975-10C-E											

* Satellite de Taille Adaptée avec Réflecteurs Laser pour les ETudes de la TErre.

	Name		Launch date, lifetime and descent date	Shape and weight (kg)	Size (m)	Date of orbital determination	Orbital inclination (deg)	Nodal period (min)	Semi major axis (km)	Perigee height (km)	Apogee height (km)	Orbital eccentricity	Argument of perigee (deg)
T	SMS 2*	1975-11A	1975 Feb 6.92 > million years	Cylinder 627 full 243 empty	2.30 long 1.90 dia	1975 Feb 13.6 1975 Apr 1.0	1.10 1.0	1456.4 1436.2	42561 42167	35680 35778	36685 35799	0.012 0.0003	- -
D	SMS 2 second stage	1975-11B	1975 Feb 6.92 2616 days 1982 Apr 6	Cylinder + annulus 350?	6.4 long 1.52 and 2.44 dia	1975 Feb 10.9	27.67	121.64	8142	278	3249	0.182	250
D	SMS 2 third stage	1975-11F	1975 Feb 6.92 4 years?	Sphere-cone 66	1.32 long 0.94 dia		Orbit similar to 1974-33D						
D	Fragments	1975-11C-E											
	Cosmos 708	1975-12A	1975 Feb 12.14 6000 years	Cylinder? 500?	-	1975 Feb 14.6	69.23	113.58	7769	1369	1413	0.003	271
	Cosmos 708 rocket	1975-12B	1975 Feb 12.14 4000 years	Cylinder 2200?	7.4 long 2.4 dia	1975 Feb 14.3	69.22	113.43	7762	1367	1400	0.002	260
D R	Cosmos 709	1975-13A	1975 Feb 12.61 12.65 days 1975 Feb 25.26	Sphere-cylinder 6300?	6.5 long? 2.4 dia	1975 Feb 14.5 1975 Feb 22.1	62.83 62.83	89.39 89.42	6624 6625	181 179	310 315	0.010 0.010	68 68
D	Cosmos 709 rocket	1975-13B	1975 Feb 12.61 5.17 days 1975 Feb 17.78	Cylinder 2500?	7.5 long 2.6 dia	1975 Feb 14.5	62.82	89.05	6607	177	280	0.008	64
D	Cosmos 709 engine **	1975-13F	1975 Feb 12.61 17.03 days 1975 Mar 1.64	Cone 600? full	1.5 long? 2 dia?	1975 Feb 24.3	62.82	88.96	6602	179	269	0.007	76
D	Fragments	1975-13C-E,G											

* An apogee motor may have separated into a similar orbit.

** 1975-13F ejected from 1975-13A about 1975 Feb 24.

	Name		Launch date, lifetime and descent date	Shape and weight (kg)	Size (m)	Date of orbital determination	Orbital inclination (deg)	Nodal period (min)	Semi major axis (km)	Perigee height (km)	Apogee height (km)	Orbital eccentricity	Argument of perigee (deg)
D	Taiyo (SRATS)* [Mu 3C]	1975-14A	1975 Feb 24.23 1952 days 1980 Jun 29	Octagonal cylinder 86	0.65 long 0.75 dia	1975 Feb 25.1	31.54	120.06	8067	249	3129	0.179	115
D	Taiyo rocket	1975-14B	1975 Feb 24.23 1909 days 1980 May 17	Sphere-cone 230?	2.33 long 1.14 dia	1975 Mar 5.6	31.55	120.17	8072	257	3131	0.178	167
D R	Cosmos 710	1975-15A	1975 Feb 26.38 13.83 days 1975 Mar 12.21	Sphere-cylinder 6300?	6.5 long? 2.4 dia	1975 Feb 27.1	64.99	89.61	6634	176	335	0.012	62
D	Cosmos 710 rocket	1975-15B	1975 Feb 26.38 5.48 days 1975 Mar 3.86	Cylinder 2500?	7.5 long 2.6 dia	1975 Feb 27.1	64.99	89.42	6624	174	318	0.011	59
D	Cosmos 710 engine	1975-15E	1975 Feb 26.38 18.01 days 1975 Mar 16.39	Cone 600? full	1.5 long? 2 dia?	1975 Mar 12.5	64.99	88.94	6601	168	278	0.008	-
D	Fragments	1975-15C,D											
	Cosmos 711	1975-16A	1975 Feb 28.58 10000 years	Spheroid 40?	1.0 long? 0.8 dia?	1975 Mar 2.9	74.00	115.53	7857	1462	1496	0.002	150
	Cosmos 712	1975-16B	1975 Feb 28.58 8000 years	Spheroid 40?	1.0 long? 0.8 dia?	1975 Mar 2.9	74.00	114.95	7831	1413	1492	0.005	117
	Cosmos 713	1975-16C	1975 Feb 28.58 7000 years	Spheroid 40?	1.0 long? 0.8 dia?	1975 Mar 1.6	74.00	114.75	7822	1398	1490	0.006	108

* Solar Radiation and Thermospheric Satellite. Japanese satellite.

1975-16 continued on page 395

	Name		Launch date, lifetime and descent date	Shape and weight (kg)	Size (m)	Date of orbital determination	Orbital inclination (deg)	Nodal period (min)	Semi major axis (km)	Perigee height (km)	Apogee height (km)	Orbital eccentricity	Argument of perigee (deg)
	Cosmos 714	1975-16D	1975 Feb 28.58 9000 years	Spheroid 40?	1.0 long? 0.8 dia?	1975 Mar 2.9	74.00	115.33	7848	1446	1494	0.003	131
	Cosmos 715	1975-16E	1975 Feb 28.58 10000 years	Spheroid 40?	1.0 long? 0.8 dia?	1975 Mar 4.2	74.00	115.75	7867	1470	1508	0.002	181
	Cosmos 716	1975-16F	1975 Feb 28.58 10000 years	Spheroid 40?	1.0 long? 0.8 dia?	1975 Mar 4.2	74.00	115.96	7877	1480	1517	0.002	214
	Cosmos 717	1975-16G	1975 Feb 28.58 10000 years	Spheroid 40?	1.0 long? 0.8 dia?	1975 Mar 4.2	74.00	116.21	7888	1481	1538	0.004	228
	Cosmos 718	1975-16H	1975 Feb 28.58 9000 years	Spheroid 40?	1.0 long? 0.8 dia?	1975 Mar 2.0	74.01	115.14	7839	1430	1492	0.004	122
	Cosmos 711 rocket	1975-16J	1975 Feb 28.58 20000 years	Cylinder 2200?	7.4 long 2.4 dia	1975 Mar 2.9	74.01	118.08	7972	1484	1704	0.014	261
T	SDS 1 [Titan 3B Agena D]	1975-17A	1975 Mar 10.20 100 years?	Cylinder?	-	1975 Mar 15.0	63.5	702.0	26194	295	39337	0.745	270
	SDS 1 rocket	1975-17B*	1975 Mar 10.20 100 years?	Cylinder 700	6 long? 1.5 dia	1975 Apr 1.0	63.5	708.0	26290	305	39518	0.746	-

* 1975-17B may be a second payload.

	Name		Launch date, lifetime and descent date	Shape and weight (kg)	Size (m)	Date of orbital determination	Orbital inclination (deg)	Nodal period (min)	Semi major axis (km)	Perigee height (km)	Apogee height (km)	Orbital eccentricity	Argument of perigee (deg)
D R	Cosmos 719	1975-18A	1975 Mar 12.37 12.86 days 1975 Mar 25.23	Sphere-cylinder 6300?	6.5 long? 2.4 dia	1975 Mar 13.0 1975 Mar 16.2	64.98 64.99	89.32 89.44	6619 6625	175 174	307 320	0.010 0.011	63 60
D	Cosmos 719 rocket	1975-18B	1975 Mar 12.37 4.08 days 1975 Mar 16.45	Cylinder 2500?	7.5 long 2.6 dia	1975 Mar 14.5	64.99	88.69	6589	164	257	0.007	58
D	Cosmos 719 engine*	1975-18E	1975 Mar 12.37 18.50 days 1975 Mar 30.87	Cone 600? full	1.5 long? 2 dia?	1975 Mar 24.7	64.98	89.16	6612	172	295	0.009	56
D	Fragments	1975-18C,D,F											
D R	Cosmos 720	1975-19A	1975 Mar 21.29 11.6 days 1975 Apr 1.9	Sphere-cylinder 5900?	5.9 long 2.4 dia	1975 Mar 24.2	62.81	89.33	6621	212	273	0.005	233
D	Cosmos 720 rocket	1975-19B	1975 Mar 21.29 9.38 days 1975 Mar 30.67	Cylinder 2500?	7.5 long 2.6 dia	1975 Mar 24.6	62.80	89.09	6609	209	252	0.003	257
D	Capsule**	1975-19F	1975 Mar 21.29 15 days 1975 Apr 5	Ellipsoid 200?	0.9 long 1.9 dia	1975 Apr 2.1	62.79	89.12	6610	206	258	0.004	240
D	Fragments	1975-19C-E,G-J											
D R	Cosmos 721	1975-20A	1975 Mar 26.37 11.85 days 1975 Apr 7.22	Sphere-cylinder 5900?	5.9 long 2.4 dia	1975 Mar 27.7	81.33	88.88	6596	208	228	0.002	327
D	Cosmos 721 rocket	1975-20B	1975 Mar 26.37 3.58 days 1975 Mar 29.95	Cylinder 2500?	7.5 long 2.6 dia	1975 Mar 27.0	81.33	88.78	6591	199	227	0.002	321

* 1975-18E ejected from 1975-18A about 1975 Mar 24

** 1975-19F ejected from 1975-19A about 1975 Apr 1

1975-20 continued on page 397

	Name		Launch date, lifetime and descent date	Shape and weight (kg)	Size (m)	Date of orbital determination	Orbital inclination (deg)	Nodal period (min)	Semi major axis (km)	Perigee height (km)	Apogee height (km)	Orbital eccentricity	Argument of perigee (deg)
D	Capsule	1975-20F	1975 Mar 26.37 15.27 days 1975 Apr 10.64	Ellipsoid 200?	0.9 long 1.9 dia	1975 Apr 8.7	81.33	88.26	6566	177	198	0.002	270
D	Fragments	1975-20C-E											
D R	Cosmos 722	1975-21A	1975 Mar 27.34 12.88 days 1975 Apr 9.22	Sphere-cylinder 6300?	6.5 long? 2.4 dia	1975 Mar 28.3 1975 Mar 28.8	71.35 71.35	89.94 89.62	6649 6633	204 173	337 336	0.010 0.012	54 53
D	Cosmos 722 rocket	1975-21B	1975 Mar 27.34 13.17 days 1975 Apr 9.51	Cylinder 2500?	7.5 long 2.6 dia	1975 Mar 28.3	71.35	89.82	6643	203	326	0.009	52
D	Cosmos 722 engine	1975-21G	1975 Mar 27.34 18.17 days 1975 Apr 14.51	Cone 600? full	1.5 long? 2 dia?	1975 Apr 9.9	71.35	89.12	6608	172	287	0.009	26
D	Fragments	1975-21C-F,H											
D	Intercosmos 13	1975-22A	1975 Mar 27.61 1986.14 days 1980 Sep 2.75	Octagonal ellipsoid 550?	1.8 long? 1.5 dia?	1975 Mar 28.8	82.95	104.88	7365	284	1689	0.095	64
D	Intercosmos 13 rocket	1975-22B	1975 Mar 27.61 1852 days 1980 Apr 21	Cylinder 2200?	7.4 long 2.4 dia	1975 Mar 28.9	82.95	104.74	7358	278	1681	0.095	64
	Meteor 21	1975-23A	1975 Apr 1.52 500 years	Cylinder + 2 vanes 2200?	5 long? 1.5 dia?	1975 Apr 1.9	81.21	102.59	7258	867	893	0.002	269
	Meteor 21 rocket	1975-23B	1975 Apr 1.52 400 years	Cylinder 1440	3.8 long 2.6 dia	1975 Apr 2.7	81.22	102.65	7261	845	920	0.005	201

	Name		Launch date, lifetime and descent date	Shape and weight (kg)	Size (m)	Date of orbital determination	Orbital inclination (deg)	Nodal period (min)	Semi major axis (km)	Perigee height (km)	Apogee height (km)	Orbital eccentricity	Argument of perigee (deg)
	Cosmos 723*	1975-24A	1975 Apr 2.46 600 years	Cone-cylinder	6 long? 2 dia?	1975 Apr 2.7 1975 May 16.3	65.02 64.7	89.64 103.74	6636 7313	249 917	266 952	0.001 0.002	258 -
D	Cosmos 723 rocket	1975-24B	1975 Apr 2.46 49 days 1975 May 21	Cylinder 1500?	8 long? 2.5 dia?	1975 May 16.0	65.00	89.59	6633	251	259	0.001	228
D	Cosmos 723 platform	1975-24D	1975 Apr 2.46 104 days 1975 Jul 15	Irregular	-	1975 May 15.9	65.01	89.66	6637	254	263	0.001	274
D	Fragments	1975-24C,E,F											
	Cosmos 724**	1975-25A	1975 Apr 7.46 600 years	Cone-cylinder	6 long? 2 dia?	1975 Apr 8.3 1975 Jun 12.5	64.97 65.5	89.63 103.04	6635 7281	248 869	266 937	0.001 0.005	255 -
D	Cosmos 724 platform	1975-25B	1975 Apr 7.46 122 days 1975 Aug 7	Irregular	-	1975 Jun 12.0	64.95	89.59	6633	248	262	0.001	319
D	Cosmos 724 rocket	1975-25C	1975 Apr 7.46 71 days 1975 Jun 17	Cylinder 1500?	8 long? 2.5 dia?	1975 Jun 12.5	64.96	89.49	6628	245	255	0.001	310
D	Cosmos 725	1975-26A	1975 Apr 8.77 273 days 1976 Jan 6	Ellipsoid 400?	1.8 long 1.2 dia	1975 Apr 9.4	70.99	92.08	6754	270	481	0.016	78
D	Cosmos 725 rocket	1975-26B	1975 Apr 8.77 154 days 1975 Sep 9	Cylinder 1500?	8 long 1.65 dia	1975 Apr 8.8	70.98	91.90	6745	272	461	0.014	81

* 1975-24B and 24D attached to 1975-24A until orbit change about 1975 May 16.31.

** 1975-25B and 25C attached to 1975-25A until orbit change about 1975 Jun 12.

	Name		Launch date, lifetime and descent date	Shape and weight (kg)	Size (m)	Date of orbital determination	Orbital inclination (deg)	Nodal period (min)	Semi major axis (km)	Perigee height (km)	Apogee height (km)	Orbital eccentricity	Argument of perigee (deg)
T L	GEOS 3*	1975-27A	1975 Apr 10.00 200 years	Octahedron + pyramid 241	1.11 high 1.22 wide	1975 Apr 10.3	114.96	101.82	7224	839	853	0.001	315
	GEOS 3 second stage	1975-27B	1975 Apr 10.00 80 years	Cylinder + annulus 350?	6.4 long 1.52 and 2.44 dia	1975 Apr 10.4	114.98	101.67	7216	833	843	0.0007	278
	Fragments	1975-27C-E											
	Cosmos 726	1975-28A	1975 Apr 11.33 1200 years	Cylinder? 700?	1.3 long? 1.9 dia	1975 Apr 11.6	82.99	104.65	7354	956	996	0.003	293
	Cosmos 726 rocket	1975-28B	1975 Apr 11.33 600 years	Cylinder 2200?	7.4 long 2.4 dia	1975 Apr 12.9	82.99	104.50	7347	956	981	0.002	278
	Molniya 3B	1975-29A	1975 Apr 14.75 14½ years	Windmill + 6 vanes 1500?	4.2 long? 1.6 dia	1975 Apr 16.8 1975 Apr 27.9	62.86 62.86	736.35 717.54	27013 26550	608 592	40661 39752	0.741 0.737	288 288
D	Molniya 3B launcher rocket	1975-29B	1975 Apr 14.75 67 days 1975 Jun 20	Cylinder 2500?	7.5 long 2.6 dia	1975 Apr 19.8	62.85	92.26	6764	217	555	0.025	123
D	Molniya 3B launcher	1975-29C	1975 Apr 14.75 51 days 1975 Jun 4	Irregular	-	1975 Apr 20.6	62.86	92.43	6773	196	593	0.029	117
	Molniya 3B rocket	1975-29D	1975 Apr 14.75 14½ years	Cylinder 440	2.0 long 2.0 dia	1975 Apr 22.4	62.85	733.16	26934	606	40506	0.741	288

* Geodynamic Experimental Ocean Satellite

	Name		Launch date, lifetime and descent date	Shape and weight (kg)	Size (m)	Date of orbital determination	Orbital inclination (deg)	Nodal period (min)	Semi major axis (km)	Perigee height (km)	Apogee height (km)	Orbital eccentricity	Argument of perigee (deg)
D R	Cosmos 727	1975-30A	1975 Apr 16.34 11.87 days 1975 Apr 28.21	Sphere-cylinder 6300?	6.5 long? 2.4 dia	1975 Apr 18.6	64.98	89.55	6631	172	334	0.012	68
D	Cosmos 727 rocket	1975-30B	1975 Apr 16.34 4.99 days 1975 Apr 21.33	Cylinder 2500?	7.5 long 2.6 dia	1975 Apr 18.4	64.99	89.07	6607	168	290	0.009	63
D	Cosmos 727 engine*	1975-30D	1975 Apr 16.34 19.01 days 1975 May 5.35	Cone 600? full	1.5 long? 2 dia?	1975 Apr 27.4	64.98	89.31	6619	168	314	0.001	64
D	Fragments	1975-30C,E											
D R	Cosmos 728	1975-31A	1975 Apr 18.42 10.79 days 1975 Apr 29.21	Sphere-cylinder 5900?	5.9 long 2.4 dia	1975 Apr 20.7	72.83	89.80	6642	205	323	0.009	65
D	Cosmos 728 rocket	1975-31B	1975 Apr 18.42 11.59 days 1975 Apr 30.01	Cylinder 2500?	7.5 long 2.6 dia	1975 Apr 25.2	72.82	89.03	6604	189	263	0.006	45
D	Capsule **	1975-31G	1975 Apr 18.42 27.90 days 1975 May 16.32	Ellipsoid 200?	0.9 long 1.9 dia	1975 Apr 28.8	72.83	89.67	6636	200	315	0.009	45
D	Fragments	1975-31C-F											
D	[Titan 3B Agena D]	1975-32A	1975 Apr 18.70 48 days 1975 Jun 5	Cylinder 3000?	8 long? 1.5 dia	1975 Apr 20.9	110.54	89.86	6646	134	401	0.020	155
	Aryabhata†	1975-33A	1975 Apr 19.32 25 years	Polyhedron 360	1.1 high 1.47 dia	1975 Apr 21.2	50.68	96.41	6968	569	610	0.003	36
	Aryabhata rocket	1975-33B	1975 Apr 19.32 15 years	Cylinder 2200?	7.4 long 2.4 dia	1975 Apr 25.1	50.68	96.31	6963	559	611	0.004	62
D	Fragments	1975-33C,D											

* 1975-30D ejected from 1975-30A about 1975 Apr 27.
** 1975-31G ejected from 1975-31A about 1975 Apr 28.

† First Indian satellite, launched by USSR.

	Name		Launch date, lifetime and descent date	Shape and weight (kg)	Size (m)	Date of orbital determination	Orbital inclination (deg)	Nodal period (min)	Semi major axis (km)	Perigee height (km)	Apogee height (km)	Orbital eccentricity	Argument of perigee (deg)
	Cosmos 729	1975-34A	1975 Apr 22.88 1200 years	Cylinder? 700?	1.3 long? 1.9 dia?	1975 Apr 24.9	82.97	105.05	7374	980	1011	0.002	267
	Cosmos 729 rocket	1975-34B	1975 Apr 22.88 600 years	Cylinder 2200?	7.4 long 2.4 dia	1975 Apr 24.9	82.96	104.93	7368	979	1001	0.001	258
D R	Cosmos 730	1975-35A	1975 Apr 24.34 11.85 days 1975 May 6.19	Sphere-cylinder 6300?	6.5 long? 2.4 dia	1975 Apr 25.4 1975 Apr 30.4	81.33 81.33	88.96 88.91	6600 6598	210 170	234 269	0.002 0.008	2 57
D	Cosmos 730 rocket	1975-35B	1975 Apr 24.34 5.36 days 1975 Apr 29.70	Cylinder 2500?	7.5 long 2.6 dia	1975 Apr 25.3	81.32	88.78	6591	201	225	0.002	339
D	Cosmos 730 engine*	1975-35E	1975 Apr 24.34 13 days 1975 May 7	Cone 600? full	1.5 long? 2 dia?	1975 May 5.5	81.33	88.66	6585	166	248	0.006	40
D	Fragments	1975-35C,D,F,G											
	Molniya 1AE	1975-36A	1975 Apr 29.44 100 years?	Windmill + 6 vanes 1000?	3.4 long 1.6 dia	1975 Apr 30.0 1975 Jul 1.0	62.83 62.9	736.47 717.69	27016 26554	430 446	40852 39906	0.748 0.743	280 -
D	Molniya 1AE launcher rocket	1975-36B	1975 Apr 29.44 29.61 days 1975 May 29.05	Cylinder 2500?	7.5 long 2.6 dia	1975 Apr 30.3	62.84	90.78	6691	213	413	0.015	128
D	Molniya 1AE launcher	1975-36C	1975 Apr 29.44 35 days 1975 Jun 3	Irregular	-	1975 Apr 30.6	62.81	91.21	6713	210	459	0.019	130
	Molniya 1AE rocket	1975-36D	1975 Apr 29.44 100 years?	Cylinder 440	2.0 long 2.0 dia	1975 May 3.6	62.89	732.85	26927	401	40696	0.748	280

* 1975-35E ejected from 1975-35A about 1975 May 5.

	Name		Launch date, lifetime and descent date	Shape and weight (kg)	Size (m)	Date of orbital determination	Orbital inclination (deg)	Nodal period (min)	Semi major axis (km)	Perigee height (km)	Apogee height (km)	Orbital eccentricity	Argument of perigee (deg)
D	Explorer 53 (SAS 3)	1975-37A	1975 May 7.95 1433 days 1979 Apr 9	Cylinder + 4 paddles 193	0.61 long 0.66 dia	1975 May 12.3	2.99	94.49	6882	499	508	0.0006	280
D	Explorer 53 rocket	1975-37B	1975 May 7.95 1586 days 1979 Sep 9	Cylinder 24	1.50 long 0.46 dia	1975 May 16.2	2.99	94.47	6881	498	507	0.0006	325
T	Telesat 3 (Anik 3)	1975-38A	1975 May 7.98 > million years	Cylinder 565 full 295 empty	1.52 long 1.83 dia	1975 May 8.0 1975 May 12.8 1975 Jul 1.0	24.8 0.05 0.05	634.3 1424.8 1436.2	24453 41956 42166	231 35222 35786	35919 35933 35789	0.730 0.008 0	- 129 -
D	Telesat 3 second stage	1975-38B	1975 May 7.98 545 days 1976 Nov 3	Cylinder + annulus 350?	6.4 long 1.52 and 2.44 dia	1975 May 8.0	28.36	109.33	7580	220	2185	0.130	230
	Telesat 3 third stage	1975-38D	1975 May 7.98 35 years	Sphere-cone 66	1.32 long 0.94 dia	1975 May 8.0	24.75	638.6	24576	239	36156	0.731	178
D	Fragment	1975-38C											
D	Pollux-D5A* (microrocket)	1975-39A	1975 May 17.44 80 days 1975 Aug 5	Double cone 35	0.56 long 0.61 dia	1975 May 19.2	29.96	100.24	7154	269	1283	0.071	44
D L	Castor-D5B* (accelerometer)	1975-39B	1975 May 17.44 1373 days 1979 Feb 18	Polyhedron 77	0.75 long 0.75 dia	1975 May 19.1	29.95	100.11	7148	272	1268	0.070	43
D	Diamant B rocket	1975-39C	1975 May 17.44 448 days 1976 Aug 7	Cylinder 68	1.60 long? 0.65 dia	1975 May 19.0	29.95	99.92	7139	271	1251	0.069	42
D	Fragments	1975-39D-G											

* French satellites.

	Name		Launch date, lifetime and descent date	Shape and weight (kg)	Size (m)	Date of orbital determination	Orbital inclination (deg)	Nodal period (min)	Semi major axis (km)	Perigee height (km)	Apogee height (km)	Orbital eccentricity	Argument of perigee (deg)
D	DSCS 5 [Titan 3C]	1975-40A	1975 May 20.59 6 days 1975 May 26	Cylinder + 2 dishes 565	1.83 long 2.74 dia	1975 May 21.2	28.58	88.34	6578	150	249	0.008	125
D	DSCS 6	1975-40B	1975 May 20.59 6 days 1975 May 26	Cylinder + 2 dishes 565	1.83 long 2.74 dia	1975 May 21.2	28.59	88.00	6561	143	222	0.006	111
D	Transtage	1975-40C	1975 May 20.59 1 day 1975 May 21	Cylinder 10000? full	6 long? 3.0 dia	1975 May 20.7	28.59	88.24	6573	150	239	0.007	117
D R	Cosmos 731	1975-41A	1975 May 21.29 11.9 days 1975 Jun 2.2	Sphere-cylinder 5900?	5.9 long 2.4 dia	1975 May 22.2	64.97	89.49	6628	203	296	0.007	44
D	Cosmos 731 rocket	1975-41B	1975 May 21.29 8.74 days 1975 May 30.03	Cylinder 2500?	7.5 long 2.6 dia	1975 May 22.3	64.97	89.33	6620	202	281	0.006	41
D	Capsule*	1975-41H	1975 May 21.29 29 days 1975 Jun 19	Ellipsoid 200?	0.9 long 1.9 dia	1975 Jun 1.0	64.97	89.35	6621	199	286	0.007	-
D	Fragments	**1975-41C-G**											
T	Intelsat 4 F-1	1975-42A	1975 May 22.92 > million years	Cylinder 1410 full 720 empty	2.82 long 2.39 dia	1975 Jul 1.0 1975 Oct 1.0	0.4 0.15	1446.3 1436.1	42363 42165	35787 35785	36182 35789	0.005 0	- -
	Intelsat 4 F-1 rocket	1975-42B	1975 May 22.92 6000 years	Cylinder 1815	8.6 long 3.0 dia	1975 Jun 4.0	26.10	654.71	24986	591	36625	0.721	186

* 1975-41H ejected from 1975-41A about 1975 May 31

	Name		Launch date, lifetime and descent date	Shape and weight (kg)	Size (m)	Date of orbital determination	Orbital inclination (deg)	Nodal period (min)	Semi major axis (km)	Perigee height (km)	Apogee height (km)	Orbital eccentricity	Argument of perigee (deg)
T	DMSP 7 [Thor Burner 2]	1975-43A	1975 May 24.14 80 years	12-sided frustum 195	1.64 long? 1.40 to 1.10 dia	1975 May 24.7	98.93	102.00	7231	813	892	0.005	230
	Burner 2 rocket	1975-43B	1975 May 24.14 60 years	Sphere -cone 66	1.32 long 0.94 dia	1975 May 24.4	98.89	101.94	7228	810	889	0.005	230
D 2M R	Soyuz 18*	1975-44A	1975 May 24.63 62.97 days 1975 Jul 26.60	Sphere- cylinder 6570?	7.5 long 2.2 dia	1975 May 24.7 1975 May 25.0 1975 May 26.3	51.69 51.60 51.59	88.60 89.45 91.34	6586 6628 6722	186 190 338	230 310 349	0.003 0.009 0.001	69 59 258
D	Soyuz 18 rocket	1975-44B	1975 May 24.63 2.41 days 1975 May 27.04	Cylinder 2500?	7.5 long 2.6 dia	1975 May 25.2	51.58	88.48	6580	186	218	0.002	89
D	Fragments	1975-44C-U											
	Cosmos 732	1975-45A	1975 May 28.02 7000 years	Spheroid 40?	1.0 long? 0.8 dia?	1975 Jun 1.0	74.02	114.65	7817	1405	1472	0.004	93
	Cosmos 733	1975-45B	1975 May 28.02 10000 years	Spheroid 40?	1.0 long? 0.8 dia?	1975 Jun 1.3	74.00	116.30	7892	1472	1555	0.005	256
	Cosmos 734	1975-45C	1975 May 28.02 9000 years	Spheroid 40?	1.0 long? 0.8 dia?	1975 May 29.6	74.01	115.10	7837	1445	1473	0.002	115
	Cosmos 735	1975-45D	1975 May 28.02 9000 years	Spheroid 40?	1.0 long? 0.8 dia?	1975 May 30.6	74.02	115.33	7848	1462	1477	0.001	159
	Cosmos 736	1975-45E	1975 May 28.02 10000 years	Spheroid 40?	1.0 long? 0.8 dia?	1975 Jun 1.0	74.02	115.55	7858	1471	1489	0.001	239

* Soyuz 18 docked with Salyut 4 from 1975 May 25.77 to 1975 Jul 26.46.

1975-45 continued on page 405

	Name		Launch date, lifetime and descent date	Shape and weight (kg)	Size (m)	Date of orbital determination	Orbital inclination (deg)	Nodal period (min)	Semi major axis (km)	Perigee height (km)	Apogee height (km)	Orbital eccentricity	Argument of perigee (deg)
	Cosmos 737	1975-45F	1975 May 28.02 10000 years	Spheroid 40?	1.0 long? 0.8 dia?	1975 May 29.5	74.02	116.04	7880	1471	1532	0.004	260
	Cosmos 738	1975-45G	1975 May 28.02 10000 years	Spheroid 40?	1.0 long? 0.8 dia?	1975 May 31.3	74.02	115.80	7869	1469	1512	0.003	242
	Cosmos 739	1975-45H	1975 May 28.02 8000 years	Spheroid 40?	1.0 long? 0.8 dia?	1975 Jun 1.3	74.01	114.88	7827	1425	1473	0.003	105
	Cosmos 732 rocket	1975-45J	1975 May 28.02 20000 years	Cylinder 2200?	7.4 long 2.4 dia	1975 Jun 1.6	73.97	118.04	7970	1480	1704	0.014	270
D R	Cosmos 740	1975-46A	1975 May 28.32 12.9 days 1975 Jun 10.2	Sphere-cylinder 6300?	6.5 long? 2.4 dia	1975 May 29.4	64.97	89.50	6628	173	327	0.012	60
D	Cosmos 740 rocket	1975-46B	1975 May 28.32 6 days 1975 Jun 3	Cylinder 2500?	7.5 long 2.6 dia	1975 May 28.7	64.98	89.44	6625	175	319	0.011	60
D	Cosmos 740 engine*	1975-46D	1975 May 28.32 21 days 1975 Jun 18	Cone 600? full	1.5 long? 2 dia?	1975 Jun 10.5	64.95	89.16	6611	167	299	0.010	55
D	Fragments	1975-46C,E											
D R	Cosmos 741	1975-47A	1975 May 30.28 11.86 days 1975 Jun 11.14	Sphere-cylinder 5700?	5.0 long 2.4 dia	1975 May 30.6	81.34	88.93	6599	210	231	0.002	347
D	Cosmos 741 rocket	1975-47B	1975 May 30.28 6 days 1975 Jun 5	Cylinder 2500?	7.5 long 2.6 dia	1975 May 31.0	81.37	88.79	6592	197	230	0.002	339

* 1975-46D ejected from 1975-46A about 1975 Jun 10.

	Name		Launch date, lifetime and descent date	Shape and weight (kg)	Size (m)	Date of orbital determination	Orbital inclination (deg)	Nodal period (min)	Semi major axis (km)	Perigee height (km)	Apogee height (km)	Orbital eccentricity	Argument of perigee (deg)
D R	Cosmos 742	1975-48A	1975 Jun 3.56 11.66 days 1975 Jun 15.22	Sphere-cylinder 6300?	6.5 long? 2.4 dia	1975 Jun 4.1 1975 Jun 5.7	62.85 62.82	89.82 89.25	6645 6617	178 148	355 329	0.013 0.014	72 57
D	Cosmos 742 rocket	1975-48B	1975 Jun 3.56 8 days 1975 Jun 11	Cylinder 2500?	7.5 long 2.6 dia	1975 Jun 4.3	62.83	89.67	6637	177	341	0.012	69
D	Cosmos 742 engine*	1975-48D	1975 Jun 3.56 13 days 1975 Jun 16	Cone 600? full	1.5 long? 2 dia?	1975 Jun 14.6	62.84	88.87	6598	143	296	0.012	68
D	Fragments	1975-48C,E											
	Molniya 1AF	1975-49A	1975 Jun 5.07 12 years	Windmill + 6 vanes 1000?	3.4 long 1.6 dia	1975 Jun 6.7 1975 Jul 1.0	62.82 62.83	736.82 717.79	27024 26557	435 435	40857 39922	0.748 0.744	280 -
	SRET 2	1975-49B	1975 Jun 5.07 12 years	Octahedron 29.6	0.56 dia	1975 Jun 5.6	62.83	737.77	27047	513	40825	0.745	280
D	Molniya 1AF launcher	1975-49C	1975 Jun 5.07 41 days 1975 Jul 16	Irregular	-	1975 Jun 5.9	62.84	90.89	6697	213	424	0.016	133
D	Molniya 1AF launcher rocket	1975-49D	1975 Jun 5.07 34 days 1975 Jul 9	Cylinder 2500?	7.5 long 2.6 dia	1975 Jun 5.9	62.83	90.86	6695	202	432	0.017	124
	Molniya 1AF rocket	1975-49F	1975 Jun 5.07 12 years	Cylinder 440	2.0 long 2.0 dia	1976 Feb 3.7	62.87	730.59	26870	742	40240	0.735	281
D	Fragment	1975-49E											

* 1975-48D ejected from 1975-48A about 1975 Jun 14

	Name		Launch date, lifetime and descent date	Shape and weight (kg)	Size (m)	Date of orbital determination	Orbital inclination (deg)	Nodal period (min)	Semi major axis (km)	Perigee height (km)	Apogee height (km)	Orbital eccentricity	Argument of perigee (deg)
D	Venus 9 launcher rocket	1975-50B	1975 Jun 8.11 1 day 1975 Jun 9	Cylinder 4000?	12 long? 4 dia	1975 Jun 8.4	51.50	88.14	6563	172	198	0.002	45
D	Venus 9 launcher	1975-50C	1975 Jun 8.11 1 day 1975 Jun 9	Irregular	-	1975 Jun 8.7	51.54	88.11	6562	171	196	0.002	46
D	[Titan 3D]	1975-51A	1975 Jun 8.77 150 days 1975 Nov 5	Cylinder 13300? full	15 long 3.0 dia	1975 Jun 9.5	96.38	88.77	6590	154	269	0.009	133
D	Titan 3D rocket	1975-51B	1975 Jun 8.77 3 days 1975 Jun 11	Cylinder 1900	6 long 3.0 dia	1975 Jun 8.9	96.37	88.67	6585	155	259	0.008	141
T?	SSU-A	1975-51C	1975 Jun 8.77 10000 years	Box + aerials?	0.3 x 0.9 x 2.4 ?	1975 Jun 19.4	95.09	113.68	7773	1389	1401	0.001	48
	Fragments	1975-51D,E											
T	Nimbus 6	1975-52A	1975 Jun 12.34 1600 years	Conical skeleton 827	3 long 1.45 dia	1975 Jun 27.0	99.96	107.30	7476	1092	1104	0.0007	212
	Nimbus 6 second stage	1975-52B	1975 Jun 12.34 1000 years	Cylinder + annulus 350?	6.4 long 1.52 and 2.44 dia	1975 Jun 23.9	99.96	107.32	7477	1096	1102	0.0004	206
D R	Cosmos 743	1975-53A	1975 Jun 12.52 12.66 days 1975 Jun 25.18	Sphere- cylinder 6300?	6.5 long? 2.4 dia	1975 Jun 13.6 1975 Jun 16.4	62.80 62.81	89.61 89.14	6634 6611	181 169	331 297	0.011 0.010	81 60

Space Vehicle: Venus 9, 1975-50A; Venus 9 lander, 1975-50D.

1975-53 continued on page 408

	Name		Launch date, lifetime and descent date	Shape and weight (kg)	Size (m)	Date of orbital determination	Orbital inclination (deg)	Nodal period (min)	Semi major axis (km)	Perigee height (km)	Apogee height (km)	Orbital eccentricity	Argument of perigee (deg)
D	Cosmos 743 rocket	1975-53B	1975 Jun 12.52 8 days 1975 Jun 20	Cylinder 2500?	7.5 long 2.6 dia	1975 Jun 13.3	62.79	89.47	6627	179	319	0.010	75
D	Cosmos 743 engine*	1975-53D	1975 Jun 12.52 17 days 1975 Jun 29	Cone 600? full	1.5 long? 2 dia?	1975 Jun 24.4	62.81	89.08	6608	164	296	0.010	62
D	Fragments	1975-53C,E,F											
D	Venus 10 launcher rocket	1975-54B	1975 Jun 14.13 1 day 1975 Jun 15	Cylinder 4000?	12 long? 4 dia	1975 Jun 14.4	51.54	88.09	6561	162	203	0.003	17
D	Venus 10 launcher	1975-54C	1975 Jun 14.13 1 day 1975 Jun 15	Irregular	-	1975 Jun 14.4	51.52	88.12	6562	162	206	0.003	8
T	[Titan 3C]	1975-55A	1975 Jun 18.42 >million years	Cylinder? 1400?	3.3 long? 2.5 dia?	1975 Jul 1.0	9.0	1422	41878	30200	40800	0.127	-
	Transtage	1975-55B	1975 Jun 18.42 >million years	Cylinder 1500?	6 long? 3.0 dia	1975 Jul 1.0	8.0	1401	41428	29700	40400	0.129	-
	Cosmos 744	1975-56A	1975 Jun 20.29 25 years	Cylinder + 2 vanes ? 2500?	5 long? 1.5 dia?	1975 Jun 26.4	81.25	97.11	6997	602	635	0.002	270
	Cosmos 744 rocket	1975-56B	1975 Jun 20.29 25 years	Cylinder 1440	3.8 long 2.6 dia	1975 Jun 27.2	81.27	97.29	7006	586	669	0.006	168

Space Vehicle: Venus 10, 1975-54A; Venus 10 lander, 1975-54D.

* 1975-53D ejected from 1975-53A about 1975 Jun 24

	Name		Launch date, lifetime and descent date	Shape and weight (kg)	Size (m)	Date of orbital determination	Orbital inclination (deg)	Nodal period (min)	Semi major axis (km)	Perigee height (km)	Apogee height (km)	Orbital eccentricity	Argument of perigee (deg)
T	OSO 8	1975-57A	1975 Jun 21.49 10 years	Cylinder + vane 1064	0.72 long 1.52 dia	1975 Jun 23.0	32.94	95.53	6930	544	560	0.001	6
D	OSO 8 second stage	1975-57B	1975 Jun 21.49 2119 days 1981 Apr 9	Cylinder + annulus 350?	6.4 long 1.52 and 2.44 dia	1975 Jun 24.0	32.93	95.46	6927	542	556	0.001	15
D	Cosmos 745	1975-58A	1975 Jun 24.51 262 days 1976 Mar 12	Ellipsoid 400?	1.8 long 1.2 dia	1975 Jun 25.5	71.00	92.35	6767	264	514	0.018	92
D	Cosmos 745 rocket	1975-58B	1975 Jun 24.51 143 days 1975 Nov 14	Cylinder 1500?	8 long 1.65 dia	1975 Jun 25.5	70.99	92.33	6766	264	512	0.018	93
D R	Cosmos 746	1975-59A	1975 Jun 25.54 12.66 days 1975 Jul 8.20	Sphere- cylinder 6300?	6.5 long? 2.4 dia	1975 Jun 26.7	62.80	89.54	6631	180	325	0.011	70
D	Cosmos 746 rocket	1975-59B	1975 Jun 25.54 6 days 1975 Jul 1	Cylinder 2500?	7.5 long 2.6 dia	1975 Jun 26.5	62.80	89.40	6624	175	316	0.011	64
D	Cosmos 746 engine	1975-59E	1975 Jun 25.54 17 days 1975 Jul 12	Cone 600? full	1.5 long? 2 dia?	1975 Jul 9.4	62.80	88.64	6586	160	256	0.007	65
D	Fragments	1975-59C,D,F											

	Name		Launch date, lifetime and descent date	Shape and weight (kg)	Size (m)	Date of orbital determination	Orbital inclination (deg)	Nodal period (min)	Semi major axis (km)	Perigee height (km)	Apogee height (km)	Orbital eccentricity	Argument of perigee (deg)
D R	Cosmos 747	1975-60A	1975 Jun 27.54 11.66 days 1975 Jul 9.20	Sphere-cylinder 5900?	5.9 long 2.4 dia	1975 Jun 27.8	62.83	89.32	6620	193	291	0.007	48
D	Cosmos 747 rocket	1975-60B	1975 Jun 27.54 6 days 1975 Jul 3	Cylinder 2500?	7.5 long 2.6 dia	1975 Jun 27.9	62.82	89.20	6614	189	283	0.007	44
D	Capsule	1975-60F	1975 Jun 27.54 20 days 1975 Jul 17	Ellipsoid 200?	0.9 long 1.9 dia	1975 Jul 10.2	62.83	88.89	6598	185	254	0.005	47
D	Fragments	1975-60C-E,G											
D R	Cosmos 748	1975-61A	1975 Jul 3.57 12.65 days 1975 Jul 16.22	Sphere-cylinder 6300?	6.5 long? 2.4 dia	1975 Jul 4.2 1975 Jul 4.6	62.81 62.82	89.44 89.83	6626 6645	178 178	317 356	0.011 0.013	65 64
D	Cosmos 748 rocket	1975-61B	1975 Jul 3.57 6 days 1975 Jul 9	Cylinder 2500?	7.5 long 2.6 dia	1975 Jul 4.1	62.81	89.34	6621	177	308	0.010	65
D	Cosmos 748 engine*	1975-61F	1975 Jul 3.57 14 days 1975 Jul 17	Cone 600? full	1.5 long? 2 dia?	1975 Jul 16.3	62.84	88.86	6596	164	272	0.008	51
D	Fragments	1975-61C-E											
D	Cosmos 749	1975-62A	1975 Jul 4.04 1911 days 1980 Sep 26	Cylinder + paddles? 900?	2 long? 1 dia?	1975 Jul 5.5	74.04	95.25	6908	509	550	0.003	346
D	Cosmos 749 rocket	1975-62B**	1975 Jul 4.04 2001.84 days 1980 Dec 25.88	Cylinder 2200?	7.4 long 2.4 dia	1975 Jul 6.2	74.04	95.14	6902	498	550	0.004	350
D	Fragment	1975-62C											

* Ejected from 1975-61A about 1975 Jul 16.

** Decay observed over Sussex, England.

	Name		Launch date, lifetime and descent date	Shape and weight (kg)	Size (m)	Date of orbital determination	Orbital inclination (deg)	Nodal period (min)	Semi major axis (km)	Perigee height (km)	Apogee height (km)	Orbital eccentricity	Argument of perigee (deg)
	Molniya 2N	1975-63A	1975 Jul 8.21 100 years?	Windmill + 6 vanes 1250?	4.2 long? 1.6 dia	1975 Jul 9.3 1975 Jul 17.8	62.87 62.89	736.87 719.03	27025 26587	432 460	40862 39958	0.748 0.743	280 280
D	Molniya 2N launcher rocket	1975-63B	1975 Jul 8.21 32 days 1975 Aug 9	Cylinder 2500?	7.5 long 2.6 dia	1975 Jul 8.3	62.83	90.71	6688	214	405	0.014	125
D	Molniya 2N launcher	1975-63C	1975 Jul 8.21 32 days 1975 Aug 9	Irregular	-	1975 Jul 8.8	62.83	91.02	6703	204	445	0.018	123
	Molniya 2N rocket	1975-63D	1975 Jul 8.21 100 years?	Cylinder 440	2.0 long 2.0 dia	1975 Jul 9.8	62.87	733.09	26933	436	40674	0.747	280
	Meteor 2-01	1975-64A	1975 Jul 11.18 500 years	Cylinder + 2 vanes 2750?	5 long? 1.5 dia?	1975 Jul 11.5	81.29	102.48	7253	858	891	0.002	247
	Meteor 2-01 rocket	1975-64B	1975 Jul 11.18 400 years	Cylinder 1440	3.8 long 2.6 dia	1975 Jul 20.0	81.29	102.60	7259	839	922	0.006	172
	Fragments	1975-64C,D											
D 2M R	Soyuz 19* (ASTP)	1975-65A	1975 Jul 15.51 5.94 days 1975 Jul 21.45	Sphere- cylinder + 2 wings 6680	7.48 long 2.30 dia	1975 Jul 15.7 1975 Jul 16.7	51.78 51.76	88.49 88.92	6581 6603	186 218	220 231	0.003 0.001	- 270
D	Soyuz 19 rocket	1975-65B	1975 Jul 15.51 2 days 1975 Jul 17	Cylinder 2500?	7.5 long 2.6 dia	1975 Jul 17.1	51.77	87.98	6556	165	190	0.002	21

* Soyuz 19 docked with Apollo 18 from 1975 Jul 17.67 to Jul 19.62.

	Name		Launch date, lifetime and descent date	Shape and weight (kg)	Size (m)	Date of orbital determination	Orbital inclination (deg)	Nodal period (min)	Semi major axis (km)	Perigee height (km)	Apogee height (km)	Orbital eccentricity	Argument of perigee (deg)
D 3M R	Apollo 18 (ASTP)	1975-66A	1975 Jul 15.83 9.06 days 1975 Jul 24.89	Cone-cylinder 12726	10.36 long 3.91 dia	1975 Jul 17.5 1975 Jul 18.0	51.76 51.75	88.41 88.91	6577 6602	170 217	228 231	0.004 0.001	- 277
D	Saturn IVB [Saturn 210]	1975-66B	1975 Jul 15.83 1 day 1975 Jul 16	Cylinder 13600?	18.7 long 6.6 dia	1975 Jul 15.8	51.76	87.56	6535	146	167	0.002	-
D	ASTP docking module*	1975-66C	1975 Jul 15.83 18 days 1975 Aug 2	Cylinder 2012	3.15 long 1.42 dia	1975 Jul 24.3	51.76	88.77	6595	210	224	0.001	-
D	Cosmos 750	1975-67A	1975 Jul 17.38 805.5 days 1977 Sep 29.9	Ellipsoid 400?	1.8 long 1.2 dia	1975 Jul 20.8	71.04	95.40	6916	272	803	0.038	77
D	Cosmos 750 rocket	1975-67B	1975 Jul 17.38 489 days 1976 Nov 17	Cylinder 1500?	8 long 1.65 dia	1975 Jul 20.7	71.04	95.15	6904	277	774	0.036	77
D	Fragment	1975-67C											
D R	Cosmos 751	1975-68A	1975 Jul 23.54 11.64 days 1975 Aug 4.18	Sphere-cylinder 5700?	5.0 long 2.4 dia	1975 Jul 27.3	62.82	89.58	6633	197	313	0.009	61
D	Cosmos 751 rocket	1975-68B	1975 Jul 23.54 7 days 1975 Jul 30	Cylinder 2500?	7.5 long 2.6 dia	1975 Jul 26.5	62.81	89.09	6609	173	288	0.009	55
D	Cosmos 752	1975-69A	1975 Jul 24.79 2046 days 1981 Feb 28	Cylinder?	4 long? 2 dia?	1975 Jul 25.4	65.85	94.56	6876	481	515	0.002	2
D	Cosmos 752 rocket	1975-69B	1975 Jul 24.79 1565 days 1979 Nov 5	Cylinder 2200?	7.4 long 2.4 dia	1975 Jul 28.0	65.85	94.44	6870	469	514	0.003	6

* Docking Module separated from Saturn 1VB on 1975 Jul 15.88, and attached to Apollo 18 until 1975 Jul 24.13.

	Name		Launch date, lifetime and descent date	Shape and weight (kg)	Size (m)	Date of orbital determination	Orbital inclination (deg)	Nodal period (min)	Semi major axis (km)	Perigee height (km)	Apogee height (km)	Orbital eccentricity	Argument of perigee (deg)
D	China 3	1975-70A	1975 Jul 26.56 50 days 1975 Sep 14	- 3500?	-	1975 Jul 27.4	69.02	90.98	6701	184	461	0.021	148
D	China 3 rocket	1975-70B	1975 Jul 26.56 30 days 1975 Aug 25	Cylinder	8 long 3 dia	1975 Jul 28.0	69.02	90.89	6696	183	453	0.020	146
D R	Cosmos 753	1975-71A	1975 Jul 31.54 12.66 days 1975 Aug 13.20	Sphere-cylinder 6300?	6.5 long? 2.4 dia	1975 Jul 31.6 1975 Aug 2.5	62.83 62.83	89.59 89.20	6634 6614	181 170	330 302	0.011 0.010	75 59
D	Cosmos 753 rocket	1975-71B	1975 Jul 31.54 7 days 1975 Aug 7	Cylinder 2500?	7.5 long 2.6 dia	1975 Jul 31.9	62.81	89.51	6630	181	322	0.011	70
D	Cosmos 753 engine*	1975-71D	1975 Jul 31.54 16 days 1975 Aug 16	Cone 600? full	1.5 long? 2 dia?	1975 Aug 12.4	62.82	89.06	6607	161	297	0.010	60
D	Fragments	1975-71C,E											
	COS-B**	1975-72A	1975 Aug 9.08 10½ years	Cylinder 275	1.21 long 1.40 dia	1975 Sep 1.0 1975 Dec 31	90.3 91.9	2203.9 2203.5	56100 56090	442 1800	99002 97630	0.878 0.854	- -
	COS-B second stage	1975-72B	1975 Aug 9.08 15 years	Cylinder + annulus 350?	6.4 long 1.52 and 2.44 dia	1975 Aug 21.5	89.23	139.63	8915	334	4740	0.247	313
	COS-B third stage	1975-72C	1975 Aug 9.08 10½ years	Sphere-cone 66	1.32 long 0.94 dia		Orbit similar to 1975-72A						

* 1975-71D ejected from 1975-71A about 1975 Aug 12.

** European Space Agency Celestial Observation Satellite, launched by NASA.
The transmitters were finally switched off on 1982 Apr 26.

	Name		Launch date, lifetime and descent date	Shape and weight (kg)	Size (m)	Date of orbital determination	Orbital inclination (deg)	Nodal period (min)	Semi major axis (km)	Perigee height (km)	Apogee height (km)	Orbital eccentricity	Argument of perigee (deg)
D R	Cosmos 754	1975-73A	1975 Aug 13.31 12.88 days 1975 Aug 26.19	Sphere-cylinder 6300?	6.5 long? 2.4 dia	1975 Aug 14.3 1975 Aug 15.4	71.37 71.37	89.83 89.41	6643 6622	204 172	326 316	0.009 0.011	48 59
D	Cosmos 754 rocket	1975-73B	1975 Aug 13.31 12 days 1975 Aug 25	Cylinder 2500?	7.5 long 2.6 dia	1975 Aug 14.4	71.38	89.69	6636	204	312	0.008	45
D	Cosmos 754 engine*	1975-73D	1975 Aug 13.31 18 days 1975 Aug 31	Cone 600? full	1.5 long? 2 dia?	1975 Aug 25.0	71.37	89.61	6632	172	336	0.012	40
D	Fragments	1975-73C,E,F											
	Cosmos 755	1975-74A	1975 Aug 14.56 1200 years	Cylinder? 700?	1.3 long? 1.9 dia?	1975 Aug 20.1	82.90	105.00	7372	974	1013	0.003	259
	Cosmos 755 rocket	1975-74B	1975 Aug 14.56 600 years	Cylinder 2200?	7.4 long 2.4 dia	1975 Aug 20.4	82.90	104.88	7366	973	1002	0.002	252
	Cosmos 756	1975-76A	1975 Aug 22.09 35 years	Cylinder + 2 vanes? 2500?	5 long? 1.5 dia?	1975 Aug 25.1	81.24	97.29	7006	622	634	0.001	267
	Cosmos 756 rocket	1975-76B	1975 Aug 22.09 35 years	Cylinder 1440	3.8 long 2.6 dia	1975 Aug 22.8	81.25	97.42	7013	604	665	0.004	184

Space Vehicle: Viking 1, 1975-75A; Viking 1 rocket, 1975-75B.

* 1975-73D ejected from 1975-73A about 1975 Aug 24.

	Name		Launch date, lifetime and descent date	Shape and weight (kg)	Size (m)	Date of orbital determination	Orbital inclination (deg)	Nodal period (min)	Semi major axis (km)	Perigee height (km)	Apogee height (km)	Orbital eccentricity	Argument of perigee (deg)
T	Symphonie 2*	1975-77A	1975 Aug 27.07 > million years	Octagon + 3 paddles 402 full 221 empty	0.58 long 1.85 dia	1975 Aug 27.1 1975 Sep 1.0 1976 Jan 1.0	13.16 0.0 0.1	678.3 1427.4 1436.1	25572 41995 42165	413 35364 35776	37974 35870 35797	0.734 0.006 0.0002	178 - -
	Symphonie 2 second stage	1975-77B	1975 Aug 27.07 25 years	Cylinder+ annulus 350?	6.4 long 1.52 and 2.44 dia	1975 Aug 29.3	25.34	109.52	7590	407	2016	0.106	188
	Symphonie 2 third stage	1975-77C	1975 Aug 27.07 100 years?	Sphere-cone 66	1.32 long 0.94 dia	1975 Aug 27.1	13.16	678.3	25572	413	37974	0.734	178
D	Fragment	1975-77D											
D R	Cosmos 757	1975-78A	1975 Aug 27.62 12.64 days 1975 Sep 9.26	Sphere-cylinder 6300?	6.5 long? 2.4 dia	1975 Aug 28.7 1975 Aug 29.8	62.82 62.82	89.46 89.24	6627 6616	182 168	316 308	0.010 0.011	72 64
D	Cosmos 757 rocket	1975-78B	1975 Aug 27.62 7 days 1975 Sep 3	Cylinder 2500?	7.5 long 2.6 dia	1975 Aug 28.7	62.82	89.24	6616	180	296	0.009	68
D	Cosmos 757 engine**	1975-78F	1975 Aug 27.62 15 days 1975 Sep 11	Cone 600? full	1.5 long? 2 dia?	1975 Sep 7.1	62.83	89.14	6611	166	300	0.010	63
D	Fragments	1975-78C-E											
	Molniya 1AG	1975-79A	1975 Sep 2.55 10 years	Windmill + 6 vanes 1000?	3.4 long 1.6 dia	1975 Sep 3.6 1975 Sep 23.2	62.90 62.87	736.78 717.75	27023 26556	623 606	40667 39749	0.741 0.737	288 288
D	Molniya 1AG launcher rocket	1975-79B	1975 Sep 2.55 69 days 1975 Nov 10	Cylinder 2500?	7.5 long 2.6 dia	1975 Sep 4.1	62.82	92.77	6789	217	604	0.028	120

* Symphonie is a French-German satellite, launched by NASA

** 1975-78F ejected from 1975-78A about 1975 Sep 6.

1975-79 continued on page 416

	Name		Launch date, lifetime and descent date	Shape and weight (kg)	Size (m)	Date of orbital determination	Orbital inclination (deg)	Nodal period (min)	Semi major axis (km)	Perigee height (km)	Apogee height (km)	Orbital eccentricity	Argument of perigee (deg)
D	Molniya 1AG launcher	1975-79C	1975 Sep 2.55 60 days 1975 Nov 1	Irregular	-	1975 Sep 4.1	62.83	92.75	6788	206	613	0.030	116
	Molniya 1AG rocket	1975-79E	1975 Sep 2.55 10 years	Cylinder 440	2.0 long 2.0 dia	1975 Sep 6.7	62.89	736.04	27005	656	40597	0.740	288
D	Fragment	1975-79D											
D	Cosmos 758*	1975-80A	1975 Sep 5.62 20 days 1975 Sep 25	Sphere-cylinder 6700?	7 long? 2.4 dia	1975 Sep 6.5 1975 Sep 13.2	67.14 67.31	89.50 92.29	6628 6765	174 195	326 579	0.011 0.028	67 31
D	Cosmos 758 rocket	1975-80B	1975 Sep 5.62 5 days 1975 Sep 10	Cylinder 2500?	7.5 long 2.6 dia	1975 Sep 6.2	67.15	89.37	6622	176	311	0.010	68
D	Fragments	1975-80C-CE											
	Molniya 2P	1975-81A	1975 Sep 9.02 100 years?	Windmill + 6 vanes 1250?	4.2 long? 1.6 dia	1975 Sep 13.2 1975 Sep 23.1	62.81 62.92	736.50 717.67	27016 26554	439 449	40837 39902	0.748 0.743	280 280
D	Molniya 2P launcher rocket	1975-81B	1975 Sep 9.02 30 days 1975 Oct 9	Cylinder 2500?	7.5 long 2.6 dia	1975 Sep 10.7	62.85	90.81	6693	213	417	0.015	130
D	Molniya 2P launcher	1975-81C	1975 Sep 9.02 39 days 1975 Oct 18	Irregular	-	1975 Sep 10.1	62.84	91.24	6714	213	459	0.018	129
	Molniya 2P rocket	1975-81D	1975 Sep 9.02 100 years?	Cylinder 440	2.0 long 2.0 dia	1975 Nov 1.0	62.6	733.81	26951	299	40846	0.752	-

* Partially disintegrated about 1975 Sep 6; probably first of 4th-generation observation satellites.

	Name		Launch date, lifetime and descent date	Shape and weight (kg)	Size (m)	Date of orbital determination	Orbital inclination (deg)	Nodal period (min)	Semi major axis (km)	Perigee height (km)	Apogee height (km)	Orbital eccentricity	Argument of perigee (deg)
	Kiku 1* (ETS 1) [Nu-1]	1975-82A	1975 Sep 9.23 1400 years	26-sided polyhedron 85	0.9 long? 0.86 dia	1975 Sep 15.5	46.99	105.88	7417	975	1103	0.009	229
	Kiku 1 rocket	1975-82B	1975 Sep 9.23 800 years	Cylinder 66?	1.74 long 1.65 dia	1975 Oct 3.3	46.98	105.88	7417	975	1103	0.009	296
D R	Cosmos 759	1975-84A	1975 Sep 12.23 11.63 days 1975 Sep 23.86	Sphere-cylinder 5900?	5.9 long 2.4 dia	1975 Sep 12.8	62.80	89.55	6632	231	276	0.003	214
D	Cosmos 759 rocket	1975-84B	1975 Sep 12.23 13 days 1975 Sep 25	Cylinder 2500?	7.5 long 2.6 dia	1975 Sep 13.3	62.79	89.44	6626	236	260	0.002	219
D	Capsule**	1975-84F	1975 Sep 12.23 14 days 1975 Sep 26	Ellipsoid 200?	0.9 long 1.9 dia	1975 Sep 24.3	62.81	89.33	6621	221	264	0.003	218
D	Fragments	1975-84C-E,G											
D R	Cosmos 760	1975-85A	1975 Sep 16.38 13.85 days 1975 Sep 30.23	Sphere-cylinder 6300?	6.5 long? 2.4 dia	1975 Sep 17.2 1975 Sep 18.2	64.96 64.96	89.59 89.38	6633 6623	174 172	335 317	0.012 0.011	59 56
D	Cosmos 760 rocket	1975-85B	1975 Sep 16.38 5 days 1975 Sep 21	Cylinder 2500?	7.5 long 2.6 dia	1975 Sep 17.2	64.97	89.44	6626	172	323	0.011	62
D	Cosmos 760 engine	1975-85D	1975 Sep 16.38 19 days 1975 Oct 5	Cone 600? full	1.5 long? 2 dia?	1975 Sep 29.7	64.96	89.47	6627	167	331	0.012	-
D	Fragments	1975-85C,E,F											

Space Vehicle: Viking 2, 1975-83A; Viking 2 rocket, 1975-83B

* Kiku is a Japanese Engineering Test Satellite.

** 1975-84F ejected from 1975-84A about 1975 Sep 23.

Name		Launch date, lifetime and descent date	Shape and weight (kg)	Size (m)	Date of orbital determination	Orbital inclination (deg)	Nodal period (min)	Semi major axis (km)	Perigee height (km)	Apogee height (km)	Orbital eccentricity	Argument of perigee (deg)
Cosmos 761	1975-86A	1975 Sep 17.30 7000 years	Spheroid 40?	1.0 long? 0.8 dia?	1975 Sep 20.5	73.99	114.74	7821	1402	1484	0.005	103
Cosmos 762	1975-86B	1975 Sep 17.30 9000 years	Spheroid 40?	1.0 long? 0.8 dia?	1975 Sep 23.6	74.00	115.19	7842	1440	1487	0.003	116
Cosmos 763	1975-86C	1975 Sep 17.30 10000 years	Spheroid 40?	1.0 long? 0.8 dia?	1975 Sep 23.6	74.00	115.86	7872	1476	1512	0.002	211
Cosmos 764	1975-86D	1975 Sep 17.30 10000 years	Spheroid 40?	1.0 long? 0.8 dia?	1975 Sep 21.6	74.00	116.09	7883	1481	1528	0.003	240
Cosmos 765	1975-86E	1975 Sep 17.30 10000 years	Spheroid 40?	1.0 long? 0.8 dia?	1975 Sep 20.5	74.00	116.36	7895	1480	1553	0.005	246
Cosmos 766	1975-86F	1975 Sep 17.30 8000 years	Spheroid 40?	1.0 long? 0.8 dia?	1975 Sep 20.6	74.00	114.97	7832	1421	1486	0.004	114
Cosmos 767	1975-86G	1975 Sep 17.30 9000 years	Spheroid 40?	1.0 long? 0.8 dia?	1975 Sep 20.6	74.00	115.41	7852	1457	1490	0.002	142
Cosmos 768	1975-86H	1975 Sep 17.30 10000 years	Spheroid 40?	1.0 long? 0.8 dia?	1975 Sep 19.9	74.00	115.63	7862	1474	1493	0.001	180
Cosmos 761 rocket	1975-86J	1975 Sep 17.30 20000 years	Cylinder 2200?	7.4 long 2.4 dia	1975 Sep 20.6	74.00	117.87	7963	1483	1687	0.013	261

	Name		Launch date, lifetime and descent date	Shape and weight (kg)	Size (m)	Date of orbital determination	Orbital inclination (deg)	Nodal period (min)	Semi major axis (km)	Perigee height (km)	Apogee height (km)	Orbital eccentricity	Argument of perigee (deg)
	Meteor 22	1975-87A	1975 Sep 18.02 500 years	Cylinder + 2 vanes 2200?	5 long? 1.5 dia?	1975 Sep 20.6	81.26	102.36	7248	838	901	0.004	271
	Meteor 22 rocket	1975-87B	1975 Sep 18.02 400 years	Cylinder 1440	3.8 long 2.6 dia	1975 Sep 20.6	81.27	102.50	7254	830	922	0.006	203
D R	Cosmos 769	1975-88A	1975 Sep 23.42 11.76 days 1975 Oct 5.18	Sphere-cylinder 5900?	5.9 long 2.4 dia	1975 Sep 24.3	72.83	89.62	6633	203	307	0.008	66
D	Cosmos 769 rocket	1975-88B	1975 Sep 23.42 8.8 days 1975 Oct 2.2	Cylinder 2500?	7.5 long 2.6 dia	1975 Sep 25.0	72.83	89.43	6624	200	292	0.007	57
D	Capsule*	1975-88C											
	Cosmos 770	1975-89A	1975 Sep 24.50 3000 years	Cylinder? 650?	-	1975 Sep 27.6	82.94	109.21	7568	1169	1210	0.003	259
	Cosmos 770 rocket	1975-89B	1975 Sep 24.50 2000 years	Cylinder 2200?	7.4 long 2.4 dia	1975 Sep 26.5	82.95	109.09	7562	1169	1198	0.002	252
D R	Cosmos 771	1975-90A	1975 Sep 25.41 12.85 days 1975 Oct 8.26	Sphere-cylinder 6300?	6.5 long? 2.4 dia	1975 Sep 25.5 1975 Sep 26.8	81.32 81.33	88.74 88.90	6589 6597	203 217	219 221	0.001 0.0003	178 142
D	Cosmos 771 rocket	1975-90B	1975 Sep 25.41 4 days 1975 Sep 29	Cylinder 2500?	7.5 long 2.6 dia	1975 Sep 26.1	81.34	88.76	6590	206	218	0.0008	329
D	Cosmos 771 engine**	1975-90C	1975 Sep 25.41 17 days 1975 Oct 12	Cone 600? full	1.5 long? 2 dia?	1975 Oct 6.8	81.33	88.86	6595	209	225	0.001	235
D	Fragment	1975-90D											

* 1.9m diameter capsule decayed 1975 Oct 13; life 20 days.

** 1975-90C ejected from 1975-90A about 1975 Oct 6.

	Name		Launch date, lifetime and descent date	Shape and weight (kg)	Size (m)	Date of orbital determination	Orbital inclination (deg)	Nodal period (min)	Semi major axis (km)	Perigee height (km)	Apogee height (km)	Orbital eccentricity	Argument of perigee (deg)
T	Intelsat 4A F-1	1975-91A	1975 Sep 26.01 > million years	Cylinder 1500 full 795 empty	2.82 long 2.39 dia	1975 Nov 1.0 1976 Jan 1.0	0.5 0.4	1426.1 1436.1	41969 42164	35358 35752	35823 35819	0.006 0.0008	- -
	Intelsat 4A F-1 rocket	1975-91B	1975 Sep 26.01 6000 years	Cylinder 1815	8.6 long 3.0 dia	1975 Sep 26.0	21.82	655.16	24982	566	36641	0.722	180
D	Aura* (D2-B) [Diamant BP-4]	1975-92A	**1975 Sep 27.36** **2560 days** **1982 Sep 30**	Cylinder + 4 vanes 110	0.80 long 0.70 dia	1975 Sep 28.1	37.13	96.78	6989	499	723	0.016	0
D	Aura rocket	1975-92B	**1975 Sep 27.36** **2376 days** **1982 Mar 30**	Cylinder 68	1.60 long? 0.65 dia	1975 Oct 8.7	37.16	96.90	6995	508	726	0.016	84
D	Fragments	1975-92C-G											
D R?	Cosmos 772	1975-93A	1975 Sep 29.18 3.0 days 1975 Oct 2.2	Sphere-cylinder 6570?	7.5 long 2.3 dia	1975 Sep 30.4	51.79	89.39	6625	195	299	0.008	87
D	Cosmos 772 rocket	1975-93B	1975 Sep 29.18 7 days 1975 Oct 6	Cylinder 2500?	7.5 long 2.6 dia	1975 Sep 30.0	51.81	89.35	6623	196	294	0.007	90
	Cosmos 773	1975-94A	1975 Sep 30.78 120 years	Cylinder + paddles? 750?	2 long? 1 dia?	1975 Oct 1.4	74.06	100.87	7178	791	808	0.001	20
	Cosmos 773 rocket	1975-94B	1975 Sep 30.78 100 years	Cylinder 2200?	7.4 long 2.4 dia	1975 Oct 2.5	74.06	100.77	7173	782	807	0.002	33
	Fragment	1975-94C											

* Aura, carrying solar experiments, is the final French national launch.

	Name		Launch date, lifetime and descent date	Shape and weight (kg)	Size (m)	Date of orbital determination	Orbital inclination (deg)	Nodal period (min)	Semi major axis (km)	Perigee height (km)	Apogee height (km)	Orbital eccentricity	Argument of perigee (deg)
D R	Cosmos 774	1975-95A	1975 Oct 1.36 13.86 days 1975 Oct 15.22	Sphere-cylinder 6300?	6.5 long? 2.4 dia	1975 Oct 2.4 1975 Oct 3.9	71.35 71.35	89.72 89.63	6638 6633	204 169	315 341	0.008 0.013	43 59
D	Cosmos 774 rocket	1975-95B	1975 Oct 1.36 9 days 1975 Oct 10	Cylinder 2500?	7.5 long 2.6 dia	1975 Oct 2.0	71.37	89.59	6631	201	305	0.008	38
D	Cosmos 774 engine*	1975-95F	1975 Oct 1.36 15 days 1975 Oct 16	Cone 600? full	1.5 long? 2 dia?	1975 Oct 11.3	71.35	89.14	6609	167	294	0.010	-
D	Fragments	1975-95C-E,G											
D	Explorer 54 (AE-D)	1975-96A	1975 Oct 6.38 158 days 1976 Mar 12	16-sided cylinder 659 full 490 empty	1.14 long 1.36 dia	1975 Oct 6.4	90.10	126.87	8364	155	3816	0.219	186
D	Explorer 54 second stage	1975-96B	1975 Oct 6.38 178 days 1976 Apr 1	Cylinder + annulus 350?	6.4 long 1.52 and 2.44 dia	1975 Oct 7.4	90.05	126.60	8350	146	3798	0.219	184
	Cosmos 775	1975-97A	1975 Oct 8.02 > million years	-	-	1975 Nov 1.0 1976 Jan 1.0	0.03 0.1	1445.9 1436.1	42357 42164	35737 35757	36220 35815	0.006 0.0007	84 -
D	Cosmos 775 launcher	1975-97B	1975 Oct 8.02 2 days 1975 Oct 10	Irregular	-	1975 Oct 8.3	51.49	88.26	6570	178	205	0.002	331
D	Cosmos 775 launcher rocket	1975-97C	1975 Oct 8.02 1 day 1975 Oct 9	Cylinder 4000?	12 long? 4 dia	1975 Oct 8.3	51.48	88.26	6570	178	205	0.002	331

* 1975-95F ejected from 1975-95A about 1975 Oct 11.

1975-97 continued on page 422

	Name		Launch date, lifetime and descent date	Shape and weight (kg)	Size (m)	Date of orbital determination	Orbital inclination (deg)	Nodal period (min)	Semi major axis (km)	Perigee height (km)	Apogee height (km)	Orbital eccentricity	Argument of perigee (deg)
D	Cosmos 775 rocket	1975-97D	1975 Oct 8.02 875 days 1978 Mar 1	Cylinder 1900?	3.9 long? 3.9 dia	1975 Nov 1.0	47.3	631.9	24393	281	35748	0.727	-
	Cosmos 775 apogee motor	1975-97F	1975 Oct 8.02 > million years	Cylinder 440	2.0 long 2.0 dia	Synchronous orbit similar to 1975-97A							
	Fragment	1975-97E											
D	[Titan 3B Agena D]	1975-98A	1975 Oct 9.80 52 days 1975 Nov 30	Cylinder 3000?	8 long? 1.5 dia	1975 Oct 12.0	96.41	89.34	6619	125	356	0.017	146
T	Triad 2 (TIP 2)*	1975-99A	1975 Oct 12.28 15 years	Dumb-bell 94	7.3 long 0.59 dia	1975 Oct 13.4 1978 Jul 8.1	90.74 90.4	95.34 98.83	6912 7079	362 582	705 820	0.025 0.017	176 288
D	Triad 2 rocket	1975-99B	1975 Oct 12.28 898 days 1978 Mar 28	Cylinder 24	1.50 long 0.46 dia	1975 Oct 13.3	90.74	95.32	6911	360	705	0.025	176
D	Fragments	1975-99C,D											
T	GOES 1** (SMS 3)	1975-100A	1975 Oct 16.94 > million years	Cylinder + boom 627 full 243 empty	2.30 long 1.90 dia	1975 Oct 24.1	1.00	1435.9	42161	35770	35796	0.0003	137
D	GOES 1 second stage	1975-100B	1975 Oct 16.94 81 days 1976 Jan 5	Cylinder + annulus 350?	6.4 long 1.52 and 2.44 dia	1975 Oct 22.5	28.26	95.68	6937	187	930	0.054	237
	GOES 1 third stage	1975-100C	1975 Oct 16.94 15 years	Sphere-cone 66	1.32 dia 0.94 dia	1975 Oct 17.0	23.76	650.96	24876	200	36796	0.736	180
D	Fragments	1975-100D,E											

* Transit Improvement Program.
** Geostationary Operational Environmental Satellite. An apogee motor may have separated and be in a similar orbit.

	Name		Launch date, lifetime and descent date	Shape and weight (kg)	Size (m)	Date of orbital determination	Orbital inclination (deg)	Nodal period (min)	Semi major axis (km)	Perigee height (km)	Apogee height (km)	Orbital eccentricity	Argument of perigee (deg)
D R	Cosmos 776	1975-101A	1975 Oct 17.61 11.66 days 1975 Oct 29.27	Sphere-cylinder 5900?	5.9 long 2.4 dia	1975 Oct 19.3	62.82	89.36	6622	200	288	0.007	62
D	Cosmos 776 rocket	1975-101B	1975 Oct 17.61 6 days 1975 Oct 23	Cylinder 2500?	7.5 long 2.6 dia	1975 Oct 19.7	62.82	89.02	6605	183	271	0.007	45
D	Capsule*	1975-101D	1975 Oct 17.61 15 days 1975 Nov 1	Ellipsoid 200?	0.9 long 1.9 dia	1975 Oct 27.6	62.83	89.17	6613	195	274	0.006	59
D	Fragments	1975-101C,E											
D	Cosmos 777**	1975-102A	1975 Oct 29.46 218 days 1976 Jun 3	Cylinder?	-	1975 Oct 29.5 1975 Nov 1.2	64.97 65.02	89.83 93.30	6646 6812	123 425	412 442	0.022 0.001	53 276
D	Cosmos 777 rocket	1975-102B	1975 Oct 29.46 1 day 1975 Oct 30	Cylinder 1500?	8 long? 2.5 dia?	1975 Oct 29.7	64.98	89.39	6624	118	373	0.019	56
D	Fragments	1975-102C-BQ											
	Cosmos 778	1975-103A	1975 Nov 4.42 1200 years	Cylinder? 700?	1.3 long? 1.9 dia?	1975 Nov 4.7	82.96	104.95	7369	978	1004	0.002	264
	Cosmos 778 rocket	1975-103B	1975 Nov 4.42 600 years	Cylinder 2200?	7.4 long 2.4 dia	1975 Nov 4.6	82.97	104.81	7362	973	995	0.002	243
D R	Cosmos 779	1975-104A	1975 Nov 4.64 13.64 days 1975 Nov 18.28	Sphere-cylinder 6300?	6.5 long? 2.4 dia	1975 Nov 6.6 1975 Nov 9.6	62.80 62.80	89.71 89.25	6640 6617	182 170	341 307	0.012 0.010	70 65

* 1975-101D ejected from 1975-101A about 1975 Oct 27.

** Cosmos 777 partially disintegrated in late January 1976.

1975-104 continued on page 424

	Name		Launch date, lifetime and descent date	Shape and weight (kg)	Size (m)	Date of orbital determination	Orbital inclination (deg)	Nodal period (min)	Semi major axis (km)	Perigee height (km)	Apogee height (km)	Orbital eccentricity	Argument of perigee (deg)
D	Cosmos 779 rocket	1975-104B	1975 Nov 4.64 6 days 1975 Nov 10	Cylinder 2500?	7.5 long 2.6 dia	1975 Nov 6.5	62.79	89.11	6610	172	291	0.009	65
D	Cosmos 779 engine*	1975-104C	1975 Nov 4.64 15 days 1975 Nov 19	Cone 600? full	1.5 long? 2 dia?	1975 Nov 16.1	62.80	88.95	6602	163	284	0.009	58
D	Fragment	1975-104D											
	Molniya 3C	1975-105A	1975 Nov 14.80 100 years?	Windmill + 6 vanes 1500?	4.2 long? 1.6 dia	1975 Nov 16.4 1975 Nov 24.9	62.90 62.80	737.26 717.21	27035 26542	523 483	40790 39844	0.745 0.742	281 280
D	Molniya 3C launcher rocket	1975-105B	1975 Nov 14.80 31 days 1975 Dec 15	Cylinder 2500?	7.5 long 2.6 dia	1975 Nov 16.4	62.78	90.92	6699	212	429	0.016	126
D	Molniya 3C launcher	1975-105C	1975 Nov 14.80 21 days 1975 Dec 5	Irregular	-	1975 Nov 16.4	62.79	90.93	6699	191	451	0.019	117
	Molniya 3C rocket	1975-105D	1975 Nov 14.80 100 years?	Cylinder 440	2.0 long 2.0 dia	1975 Dec 4.2	62.79	733.57	26940	492	40630	0.745	280
D B R	Soyuz 20**	1975-106A	1975 Nov 17.61 90.5 days 1976 Feb 16.1	Sphere- cylinder 6570?	7.5 long 2.3 dia	1975 Nov 17.8 1975 Nov 18.4 1975 Nov 20.2	51.62 51.62 51.59	88.72 89.74 91.39	6592 6643 6724	177 247 342	251 282 350	0.006 0.003 0.001	133 261 59
D	Soyuz 20 rocket	1975-106B	1975 Nov 17.61 3 days 1975 Nov 20	Cylinder 2500?	7.5 long 2.6 dia	1975 Nov 18.5	51.62	88.53	6583	185	224	0.003	89

* 1975-104C ejected from 1975-104A about 1975 Nov 15.

** Soyuz 20 (unmanned) docked with Salyut 4 about 1975 Nov 19.68.

	Name		Launch date, lifetime and descent date	Shape and weight (kg)	Size (m)	Date of orbital determination	Orbital inclination (deg)	Nodal period (min)	Semi major axis (km)	Perigee height (km)	Apogee height (km)	Orbital eccentricity	Argument of perigee (deg)
D	Explorer 55 (AE-E)	1975-107A	1975 Nov 20.09 2029.49 days 1981 Jun 10.58	16-sided cylinder 720 full 490 empty	1.14 long 1.36 dia	1975 Nov 25.0 1976 Nov 23.0	19.70 19.66	117.29 89.82	7948 6644	156 264	2983 267	0.178 0.0002	229 28
D	Explorer 55 second stage	1975-107B	1975 Nov 20.09 120 days 1976 Mar 19	Cylinder + annulus 350?	6.4 long 1.52 and 2.44 dia	1975 Nov 21.5	19.67	117.32	7949	157	2985	0.178	200
D R	Cosmos 780	1975-108A	1975 Nov 21.39 11.85 days 1975 Dec 3.24	Sphere-cylinder 5900?	5.9 long 2.4 dia	1975 Nov 24.2	65.01	89.28	6618	201	278	0.006	44
D	Cosmos 780 rocket	1975-108B	1975 Nov 21.39 6 days 1975 Nov 27	Cylinder 2500?	7.5 long 2.6 dia	1975 Nov 23.4	65.01	89.03	6605	199	255	0.004	36
D	Capsule*	1975-108D	1975 Nov 21.39 20 days 1975 Dec 11	Ellipsoid 200?	0.9 long 1.9 dia	1975 Dec 3.5	65.01	89.14	6611	197	268	0.005	-
D	Fragments	1975-108C,E											
D	Cosmos 781	1975-109A	1975 Nov 21.72 1832 days 1980 Nov 26	Cylinder + paddles 900?	2 long? 1 dia?	1975 Nov 22.0	74.03	95.21	6906	505	551	0.003	353
D	Cosmos 781 rocket	1975-109B	1975 Nov 21.72 1856 days 1980 Dec 20	Cylinder 2200?	7.4 long 2.4 dia	1975 Nov 23.0	74.04	95.11	6901	497	549	0.004	354
7d	Fragments	1975-109C-L											
D B R	Cosmos 782	1975-110A	1975 Nov 25.71 19.5 days 1975 Dec 15.2	Sphere-cylinder 5900?	5.9 long 2.4 dia	1975 Nov 27.2	62.81	90.52	6679	218	384	0.012	108
D	Cosmos 782 rocket	1975-110B	1975 Nov 25.71 28 days 1975 Dec 23	Cylinder 2500?	7.5 long 2.6 dia	1975 Nov 27.2	62.80	90.39	6673	217	372	0.012	103

1975-110 continued on page 426

* 1975-108D ejected from 1975-108A on 1975 Dec 2.

	Name		Launch date, lifetime and descent date	Shape and weight (kg)	Size (m)	Date of orbital determination	Orbital inclination (deg)	Nodal period (min)	Semi major axis (km)	Perigee height (km)	Apogee height (km)	Orbital eccentricity	Argument of perigee (deg)
D	Capsule*	1975-110D	1975 Nov 25.71 77 days 1976 Feb 10	Ellipsoid 200?	0.9 long 1.9 dia	1975 Dec 15.7	62.80	90.35	6671	216	369	0.011	112
D	Fragments	1975-110C,E											
D r	China 4 †	1975-111A	1975 Nov 26.15 33 days 1975 Dec 29	- 3500?	-	1975 Nov 28.0	62.95	91.09	6707	179	479	0.022	147
D	China 4 rocket	1975-111B	1975 Nov 26.15 26 days 1975 Dec 22	Cylinder	8 long 3 dia	1975 Nov 28.0	62.95	90.94	6700	177	466	0.022	147
D	Fragments	1975-111C-F											
	Cosmos 783	1975-112A	1975 Nov 28.01 120 years	Cylinder + paddles? 750?	2 long? 1 dia?	1975 Nov 29.9	74.06	100.99	7183	795	815	0.001	46
	Cosmos 783 rocket	1975-112B	1975 Nov 28.01 100 years	Cylinder 2200?	7.4 long 2.4 dia	1975 Nov 29.9	74.06	100.89	7178	785	815	0.002	47
	Fragment	1975-112C											
D R	Cosmos 784	1975-113A	1975 Dec 3.42 11.85 days 1975 Dec 15.27	Sphere- cylinder 5900?	5.9 long 2.4 dia	1975 Dec 3.7	81.33	88.99	6602	215	232	0.001	14
D	Cosmos 784 rocket	1975-113B	1975 Dec 3.42 4 days 1975 Dec 7	Cylinder 2500?	7.5 long 2.6 dia	1975 Dec 4.1	81.33	88.86	6595	213	221	0.001	325
D	Capsule**	1975-113G	1975 Dec 3.42 18 days 1975 Dec 21	Ellipsoid 200?	0.9 long 1.9 dia	1975 Dec 14.6	81.32	88.72	6588	203	217	0.001	320
D	Fragments	1975-113C-F											

* 1975-110D ejected from 1975-110A about 1975 Dec 15
** 1975-113G ejected from 1975-113A about 1975 Dec 14

† Capsule recovered on 1975 Dec 2

	Name		Launch date, lifetime and descent date	Shape and weight (kg)	Size (m)	Date of orbital determination	Orbital inclination (deg)	Nodal period (min)	Semi major axis (km)	Perigee height (km)	Apogee height (km)	Orbital eccentricity	Argument of perigee (deg)
D	[Titan 3D]	1975-114A	1975 Dec 4.86 119 days 1976 Apr 1	Cylinder 13300? full	15 long 3.0 dia	1975 Dec 6.5	96.27	88.44	6574	157	234	0.006	132
D	Capsule	1975-114B	1975 Dec 4.86 879 days 1978 May 1	-	-	1975 Dec 8.1	96.28	102.95	7275	236	1558	0.091	305
D	Titan 3D rocket	1975-114C	1975 Dec 4.86 2 days 1975 Dec 6	Cylinder 1900	6 long 3.0 dia	1975 Dec 5.4	96.27	88.12	6558	156	203	0.004	126
D	Fragment	1975-114D											
	Intercosmos 14	1975-115A	1975 Dec 11.71 7.2 years	Octagonal ellipsoid 550?	1.8 long? 1.5 dia?	1975 Dec 14.9	73.99	105.33	7388	335	1684	0.091	64
D	Intercosmos 14 rocket	1975-115B	1975 Dec 11.71 2544 days 1982 Nov 28	Cylinder 2200?	7.4 long 2.4 dia	1975 Dec 14.9	74.01	105.16	7379	325	1677	0.092	63
D	Fragments	1975-115C-F											
	Cosmos 785*	1975-116A	1975 Dec 12.53 600 years	Cone-cylinder	6 long? 2 dia?	1975 Dec 12.6 1975 Dec 15.2	64.96 65.07	89.61 104.26	6634 7339	251 898	261 1023	0.001 0.009	331 202
D	Cosmos 785 rocket	1975-116B	1975 Dec 12.53 2 days 1975 Dec 14	Cylinder 1500?	8 long? 2.5 dia?	1975 Dec 13.2	65.00	89.73	6640	241	283	0.003	-
D	Cosmos 785 platform	1975-116C	1975 Dec 12.53 55 days 1976 Feb 5	Irregular	-	1975 Dec 15.1	64.99	89.57	6632	248	260	0.001	270

* 1975-116B and 116C attached to 1975-116A until orbit change about 1975 Dec 13.21.

	Name		Launch date, lifetime and descent date	Shape and weight (kg)	Size (m)	Date of orbital determination	Orbital inclination (deg)	Nodal period (min)	Semi major axis (km)	Perigee height (km)	Apogee height (km)	Orbital eccentricity	Argument of perigee (deg)
T	RCA Satcom 1*	1975-117A	1975 Dec 13.08 > million years	Box + 2 panels 868 full 463 empty	1.62 high 1.25 wide 1.25 deep	1975 Dec 13.1 1976 Jan 1.0 1976 Mar 31.0	27.2 0.3 0.0	634.8 1439.7 1436.2	24467 42234 42165	185 35625 35783	35993 36086 35790	0.732 0.005 0.0001	- - -
D	RCA Satcom 1 second stage	1975-117B	1975 Dec 13.08 237 days 1976 Aug 7	Cylinder + annulus 350?	6.4 long 1.52 and 2.44 dia	1976 Jan 6.0	28.40	105.76	7413	190	1880	0.114	21
D	RCA Satcom 1 third stage	1975-117C	1975 Dec 13.08 2524 days 1982 Nov 10	Sphere-cone 66	1.32 long 0.94 dia	1975 Dec 13.1 1977 Jan 1.0	27.2 26.6	634.8 615.9	24467 23978	185 282	35993 34917	0.732 0.722	-
T	IMEWS 5 [Titan 3C]	1975-118A	1975 Dec 14.22 > million years	Cylinder +4 panels 820?	6 long? 2.5 dia?	1975 Dec 14.6 1976 Jan 1.0	26.3 3.0	633.0 1436	24445 42106	295 35671	35840 35785	0.727 0.001	180 ** -
D	Titan 3C second stage	1975-118B	1975 Dec 14.22 5 days 1975 Dec 19	Cylinder 1900	6 long 3.0 dia	1975 Dec 14.4	28.60	89.86	6653	151	398	0.019	115
	Transtage	1975-118C	1975 Dec 14.22 > million years	Cylinder 1500?	6 long? 3.0 dia	1980 Jun 30	3.4	1433.2	42108	35528	35931	0.005	-
	Fragment	1975-118D											
D	China 5	1975-119A	1975 Dec 16.39 42 days 1976 Jan 27	- 3500?	-	1975 Dec 17.7	69.00	90.26	6665	186	387	0.015	142
D	China 5 rocket	1975-119B	1975 Dec 16.39 25 days 1976 Jan 10	Cylinder	8 long 3 dia	1975 Dec 17.8	69.00	90.18	6661	185	380	0.015	142

* RCA is Radio Corporation of America.

** Approximate orbit

	Name		Launch date, lifetime and descent date	Shape and weight (kg)	Size (m)	Date of orbital determination	Orbital inclination (deg)	Nodal period (min)	Semi major axis (km)	Perigee height (km)	Apogee height (km)	Orbital eccentricity	Argument of perigee (deg)
D R	Cosmos 786	1975-120A	1975 Dec 16.41 12.81 days 1975 Dec 29.22	Sphere-cylinder 6300?	6.5 long? 2.4 dia	1975 Dec 17.7 1975 Dec 24.4	65.00 65.00	89.49 89.23	6628 6615	174 172	326 302	0.011 0.010	60 59
D	Cosmos 786 rocket	1975-120B	1975 Dec 16.41 5 days 1975 Dec 21	Cylinder 2500?	7.5 long 2.6 dia	1975 Dec 17.1	65.01	89.35	6621	174	312	0.010	60
D	Cosmos 786 engine*	1975-120C	1975 Dec 16.41 18 days 1976 Jan 3	Cone 600? full	1.5 long? 2 dia?	1976 Jan 1.6	65.00	88.45	6577	163	234	0.005	-
D	Fragments	1975-120D,E											
	Molniya 2Q	1975-121A	1975 Dec 17.47 9½ years	Windmill +6 vanes 1250?	4.2 long? 1.6 dia	1975 Dec 18.5 1975 Dec 25.6	62.81 62.86	736.01 717.99	27004 26562	431 436	40821 39931	0.748 0.744	280 281
D	Molniya 2Q launcher rocket	1975-121B	1975 Dec 17.47 28 days 1976 Jan 14	Cylinder 2500?	7.5 long 2.6 dia	1975 Dec 19.1	62.82	90.51	6679	211	390	0.013	134
D	Molniya 2Q launcher	1975-121C	1975 Dec 17.47 25 days 1976 Jan 11	Irregular	-	1975 Dec 19.1	62.82	90.80	6693	196	433	0.018	125
	Molniya 2Q rocket	1975-121D	1975 Dec 17.47 9½ years	Cylinder 440	2.0 long 2.0 dia	1976 Mar 1.0	62.81	732.38	26915	478	40596	0.745	-

* 1975-120C ejected from 1975-120A on 1975 Dec 29

	Name		Launch date, lifetime and descent date	Shape and weight (kg)	Size (m)	Date of orbital determination	Orbital inclination (deg)	Nodal period (min)	Semi major axis (km)	Perigee height (km)	Apogee height (km)	Orbital eccentricity	Argument of perigee (deg)
D	Prognoz 4	1975-122A	1975 Dec 22.09 $2\frac{1}{2}$ years?	Spheroid + 4 vanes 905	1.8 dia?	1975 Dec 22.1 1975 Dec 23.0	65.04 65.0	91.50 5740	6727 106195	232 634	465 199000	0.017 0.934	66 -
D	Prognoz 4 launcher rocket	1975-122B	1975 Dec 22.09 62 days 1976 Feb 22	Cylinder 2500?	7.5 long 2.6 dia	1975 Dec 23.5	64.98	91.50	6727	229	468	0.018	68
D	Prognoz 4 launcher	1975-122C	1975 Dec 22.09 49 days 1976 Feb 9	Irregular	-	1975 Dec 23.5	65.02	91.60	6732	209	498	0.021	66
D	Prognoz 4 rocket	1975-122D	1975 Dec 22.09 $2\frac{1}{2}$ years?	Cylinder 440	2.0 long 2.0 dia	Orbit similar to 1975-122A							
	Statsionar - Raduga 1**	1975-123A	1975 Dec 22.55 >million years	-	-	1975 Dec 23.0	0.1	1434	42178	35800	35800	0	-*
D	Raduga 1 launcher	1975-123B	1975 Dec 22.55 <0.5 day 1975 Dec 22	Irregular	-	1975 Dec 22.7	51.46	87.88	6550	146	197	0.004	151
D	Raduga 1 launcher rocket	1975-123C	1975 Dec 22.55 3 days 1975 Dec 25	Cylinder 4000?	12 long? 4 dia	1975 Dec 22.8	51.49	88.22	6567	182	195	0.001	309
	Raduga 1 rocket	1975-123D	1975 Dec 22.55 15 years	Cylinder 1900?	3.9 long? 3.9 dia	1976 Jan 18.3	47.14	630.35	24353	257	35694	0.728	7
	Raduga 1 apogee motor	1975-123F	1975 Dec 22.55 > million years	Cylinder 440	2.0 long 2.0 dia	Synchronous orbit similar to 1975-123A							
	Fragment	1975-123E											

* Approximate orbit.
** Raduga means Rainbow

	Name		Launch date, lifetime and descent date	Shape and weight (kg)	Size (m)	Date of orbital determination	Orbital inclination (deg)	Nodal period (min)	Semi major axis (km)	Perigee height (km)	Apogee height (km)	Orbital eccentricity	Argument of perigee (deg)
	Meteor 23	1975-124A	1975 Dec 25.79 500 years	Cylinder + 2 vanes 2200?	5 long? 1.5 dia?	1975 Dec 28.6	81.26	102.42	7250	842	902	0.004	287
	Meteor 23 rocket	1975-124B	1975 Dec 25.79 400 years	Cylinder 1440	3.8 long 2.6 dia	1975 Dec 27.1	81.26	102.49	7254	842	909	0.005	215
	Molniya 3D	1975-125A	1975 Dec 27.44 $10\frac{1}{2}$ years	Windmill + 6 vanes 1500?	4.2 long? 1.6 dia	1975 Dec 30.5 1976 Mar 1.0	62.81 62.9	735.10 717.69	26982 26554	443 507	40764 39845	0.747 0.741	280 -
D	Molniya 3D launcher rocket	1975-125B	1975 Dec 27.44 27 days 1976 Jan 23	Cylinder 2500?	7.5 long 2.6 dia	1975 Dec 28.6	62.77	90.78	6692	208	419	0.016	130
D	Molniya 3D launcher	1975-125C	1975 Dec 27.44 23 days 1976 Jan 19	Irregular	-	1975 Dec 28.6	62.83	90.94	6700	194	449	0.019	123
	Molniya 3D rocket	1975-125F	1975 Dec 27.44 $10\frac{1}{2}$ years	Cylinder 440	2.0 long 2.0 dia	1976 Jan 26.4	62.80	731.17	26886	453	40562	0.746	280
D	Fragments	1975-125D,E											

	Name		Launch date, lifetime and descent date	Shape and weight (kg)	Size (m)	Date of orbital determination	Orbital inclination (deg)	Nodal period (min)	Semi major axis (km)	Perigee height (km)	Apogee height (km)	Orbital eccentricity	Argument of perigee (deg)
D	Cosmos 787	1976-01A	1976 Jan 6.21 1802 days 1980 Dec 12	Cylinder + paddles? 900?	2 long? 1 dia?	1976 Jan 7.6	74.03	95.30	6911	518	547	0.002	16
D	Cosmos 787 rocket	1976-01B	**1976 Jan 6.21** **1895 days** **1981 Mar 15**	Cylinder 2200?	7.4 long 2.4 dia	1976 Jan 15.4	74.04	95.19	6905	507	547	0.003	3
D	Fragments	1976-01C -K											
D R	Cosmos 788	1976-02A	1976 Jan 7.65 12.60 days 1976 Jan 20.25	Sphere-cylinder 6300?	6.5 long? 2.4 dia	1976 Jan 8.4 1976 Jan 9.8	62.81 62.81	89.53 89.35	6630 6621	183 169	321 317	0.010 0.011	77 73
D	Cosmos 788 rocket	1976-02B	1976 Jan 7.65 7 days 1976 Jan 14	Cylinder 2500?	7.5 long 2.6 dia	1976 Jan 8.4	62.79	89.42	6625	180	313	0.010	75
D	Cosmos 788 engine*	1976-02D	1976 Jan 7.65 20 days 1976 Jan 27	Cone 600? full	1.5 long? 2 dia?	1976 Jan 20.4	62.81	89.45	6626	166	330	0.012	-
D	Fragments	1976-02C,E,F											
T	CTS 1**	1976-04A	1976 Jan 17.98 >million years	Cylinder + 2 wings 680 full?	1.88 long 1.83 dia 16.8 span	1976 Jan 20.0 1976 Jan 26.8 1976 Mar 1.0	27.29 0.67 0.55	635.88 1442.2 1436.1	24491 42285 42165	205 35786 35774	36021 36028 35800	0.731 0.003 0.0003	180 213 -
D	CTS 1 second stage	1976-04B	1976 Jan 17.98 28 days 1976 Feb 14	Cylinder + annulus 350?	6.4 long 1.52 and 2.44 dia	1976 Feb 11.0	28.66	89.02	6612	162	306	0.011	92

Space Vehicle: Helios 2, 1976-03A; Helios 2 rocket, 1976-03C; Helios 2 second stage, 1976-03B. * Jettisoned about 1976 Jan 19.
** Canadian Communications Technology Satellite, launched by NASA.

1976-04 continued on page 433

	Name		Launch date, lifetime and descent date	Shape and weight (kg)	Size (m)	Date of orbital determination	Orbital inclination (deg)	Nodal period (min)	Semi major axis (km)	Perigee height (km)	Apogee height (km)	Orbital eccentricity	Argument of perigee (deg)
D	CTS 1 third stage	1976-04D	1976 Jan 17.98 486 days 1977 May 17	Sphere- cone 66	1.32 long 0.94 dia	1976 Jan 21.3	27.18	637.64	24548	180	36160	0.733	181
D	Fragment	1976-04C											
	Cosmos 789	1976-05A	1976 Jan 20.71 1200 years	Cylinder? 700?	1.3 long? 1.9 dia?	1976 Jan 24.8	82.97	105.05	7374	975	1016	0.003	263
	Cosmos 789 rocket	1976-05B	1976 Jan 20.71 600 years	Cylinder 2200?	7.4 long 2.4 dia	1976 Jan 21.5	82.97	104.94	7368	974	1006	0.002	259
	Molniya 1AH	1976-06A	1976 Jan 22.49 100 years?	Windmill + 6 vanes 1000?	3.4 long 1.6 dia	1976 Jan 22.6 1976 Jan 27.4 1976 Feb 5.8	62.98 62.94 62.91	91.83 698.38 717.74	6743 26074 26556	233 465 476	497 38927 39879	0.020 0.738 0.742	63 280 280
D	Molniya 1AH launcher rocket	1976-06B	1976 Jan 22.49 71 days 1976 Apr 2	Cylinder 2500?	7.5 long 2.6 dia	1976 Jan 23.5	62.99	91.75	6739	234	488	0.019	61
D	Molniya 1AH launcher	1976-06C	1976 Jan 22.49 88 days 1976 Apr 19	Irregular	-	1976 Jan 23.5	62.98	91.86	6745	240	493	0.019	65
	Molniya 1AH rocket	1976-06D	1976 Jan 22.49 100 years?	Cylinder 440	2.0 long 2.0 dia	1976 Feb 17.1	62.96	695.47	26001	497	38749	0.735	280

	Name		Launch date, lifetime and descent date	Shape and weight (kg)	Size (m)	Date of orbital determination	Orbital inclination (deg)	Nodal period (min)	Semi major axis (km)	Perigee height (km)	Apogee height (km)	Orbital eccentricity	Argument of perigee (deg)
D	Cosmos 790	1976-07A	1976 Jan 22.94 1756 days 1980 Nov 12	Cylinder + paddles 900?	2 long? 1 dia?	1976 Jan 25.2	74.04	95.25	6908	511	549	0.003	353
D	Cosmos 790 rocket	1976-07B	1976 Jan 22.94 1783 days 1980 Dec 9	Cylinder 2200?	7.4 long 2.4 dia	1976 Jan 23.2	74.05	95.12	6902	499	548	0.004	3
	Cosmos 791	1976-08A	1976 Jan 28.44 8000 years	Spheroid 40?	1.0 long? 0.8 dia?	1976 Jan 28.7	74.05	114.81	7824	1402	1490	0.006	112
	Cosmos 792	1976-08B	1976 Jan 28.44 9000 years	Spheroid 40?	1.0 long? 0.8 dia?	1976 Jan 28.7	74.06	115.23	7843	1436	1494	0.004	133
	Cosmos 793	1976-08C	1976 Jan 28.44 8000 years	Spheroid 40?	1.0 long? 0.8 dia?	1976 Jan 28.7	74.06	115.02	7834	1418	1494	0.005	128
	Cosmos 794	1976-08D	1976 Jan 28.44 9000 years	Spheroid 40?	1.0 long? 0.8 dia?	1976 Jan 28.9	74.06	115.44	7853	1452	1497	0.003	148
	Cosmos 795	1976-08E	1976 Jan 28.44 10000 years	Spheroid 40?	1.0 long? 0.8 dia?	1976 Jan 28.9	74.05	115.66	7863	1467	1503	0.002	175
	Cosmos 796	1976-08F	1976 Jan 28.44 10000 years	Spheroid 40?	1.0 long? 0.8 dia?	1976 Jan 28.9	74.04	115.90	7874	1474	1518	0.003	202
	Cosmos 797	1976-08G	1976 Jan 28.44 10000 years	Spheroid 40?	1.0 long? 0.8 dia?	1976 Jan 28.9	74.05	116.13	7885	1480	1533	0.003	230

1976-08 continued on page 435

	Name		Launch date, lifetime and descent date	Shape and weight (kg)	Size (m)	Date of orbital determination	Orbital inclination (deg)	Nodal period (min)	Semi major axis (km)	Perigee height (km)	Apogee height (km)	Orbital eccentricity	Argument of perigee (deg)
	Cosmos 798	1976-08H	1976 Jan 28.44 10000 years	Spheroid 40?	1.0 long? 0.8 dia?	1976 Jan 28.9	74.05	116.40	7897	1481	1557	0.005	239
	Cosmos 791 rocket	1976-08J	1976 Jan 28.44 20000 years	Cylinder 2200?	7.4 long 2.4 dia	1976 Feb 1.7	74.06	118.03	7970	1486	1698	0.013	259
D R	Cosmos 799	1976-09A	1976 Jan 29.36 11.8 days 1976 Feb 10.2	Sphere- cylinder 5700?	5.0 long 2.4 dia	1976 Jan 30.5	71.40	89.64	6634	205	306	0.008	47
D	Cosmos 799 rocket	1976-09B	1976 Jan 29.36 13 days 1976 Feb 11	Cylinder 2500?	7.5 long 2.6 dia	1976 Jan 29.8	71.41	89.58	6631	208	297	0.007	47
D	Fragments	1976-09C,D											
T	Intelsat 4A F-2	1976-10A	1976 Jan 30.00 > million years	Cylinder 1500 full 795 empty	2.82 long 2.39 dia	1976 Feb 1.0 1976 May 1.0	0.1 0.1	1420.2 1436.2	41855 42167	35084 35784	35869 35794	0.009 0.0001	- -
	Intelsat 4A F-2 rocket	1976-10B	1976 Jan 30.00 6000 years	Cylinder 1815	8.6 long 3.0 dia	1976 May 1.0	21.5	655.0	24984	605	36606	0.721	-
	Cosmos 800	1976-11A	1976 Feb 3.34 1200 years	Cylinder? 700?	1.3 long? 1.9 dia?	1976 Feb 4.9	82.97	105.13	7378	984	1015	0.002	275
	Cosmos 800 rocket	1976-11B	1976 Feb 3.34 600 years	Cylinder 2200?	7.4 long 2.4 dia	1976 Feb 3.5	82.98	105.00	7371	983	1003	0.001	250

	Name		Launch date, lifetime and descent date	Shape and weight (kg)	Size (m)	Date of orbital determination	Orbital inclination (deg)	Nodal period (min)	Semi major axis (km)	Perigee height (km)	Apogee height (km)	Orbital eccentricity	Argument of perigee (deg)
D	Cosmos 801	1976-12A	1976 Feb 5.61 700 days 1978 Jan 5	Ellipsoid 400?	1.8 long 1.2 dia	1976 Feb 13.3	70.98	95.28	6910	268	796	0.038	72
D	Cosmos 801 rocket	1976-12B	1976 Feb 5.61 413 days 1977 Mar 24	Cylinder 1500?	8 long 1.65 dia	1976 Feb 9.6	71.03	95.20	6906	283	773	0.036	78
D	Fragments	1976-12C-S											
D R	Cosmos 802	1976-13A	1976 Feb 11.37 13.84 days 1976 Feb 25.21	Sphere-cylinder 6300?	6.5 long? 2.4 dia	1976 Feb 13.0 1976 Feb 13.7	64.99 64.98	89.56 89.33	6631 6620	172 173	334 311	0.012 0.010	62 63
D	Cosmos 802 rocket	1976-13B	1976 Feb 11.37 6 days 1976 Feb 17	Cylinder 2500?	7.5 long 2.6 dia	1976 Feb 12.4	64.98	89.35	6621	170	316	0.011	58
D	Cosmos 802 engine*	1976-13D	1976 Feb 11.37 18 days 1976 Feb 29	Cone 600? full	1.5 long? 2 dia?	1976 Feb 22.7	64.98	89.25	6616	170	306	0.010	57
D	Fragments	1976-13C,E											
	Cosmos 803	1976-14A	1976 Feb 12.54 40 years	Cylinder?	4 long? 2 dia?	1976 Feb 15.3	65.85	96.39	6964	554	618	0.005	2
	Cosmos 803 rocket	1976-14B	1976 Feb 12.54 8 years	Cylinder 2200?	7.4 long 2.4 dia	1976 Feb 15.5	65.86	96.31	6960	546	618	0.005	5
D	Fragment	1976-14C											
D	Cosmos 804**	1976-15A	1976 Feb 16.35 0.4 days? 1976 Feb 16.7?	Cylinder?	4 long? 2 dia?	1976 Feb 16.4 1976 Feb 16.7	65.15 65.86	93.08 96.38	6804 6964	149 556	703 615	0.041 0.004	55 3

* Jettisoned from 1976-13A about 1976 Feb 22.

** Cosmos 804 passed close to Cosmos 803 between 1976 Feb 16.41 and 16.68; then retrofired, with descent into Pacific Ocean?

	Name		Launch date, lifetime and descent date	Shape and weight (kg)	Size (m)	Date of orbital determination	Orbital inclination (deg)	Nodal period (min)	Semi major axis (km)	Perigee height (km)	Apogee height (km)	Orbital eccentricity	Argument of perigee (deg)
D	Cosmos 804 rocket	1976-15B	1976 Feb 16.35 10 days 1976 Feb 26	Cylinder 1500?	8 long? 2.5 dia?	1976 Feb 17.7	65.12	92.36	6769	142	640	0.037	52
D	DMSP 8 [Thor Burner 2]	1976-16A	1976 Feb 19.33 0.2 day? 1976 Feb 19	Cone-cylinder? 1355 full?	2.96 long? 0.94 to 1.31 dia?	1976 Feb 19.4	98.87	88.97	6601	90	355	0.020	328
T	Marisat 1	1976-17A	1976 Feb 19.94 >million years	Cylinder 655 full 362 empty	2.4 long? 1.9 dia?	1976 Feb 25.7	2.4	1436.1	42163	35703	35867	0.002	-
D	Marisat 1 second stage	1976-17B	1976 Feb 19.94 26 days 1976 Mar 16	Cylinder + annulus 350?	6.4 long 1.52 and 2.44 dia	1976 Feb 29.2	28.59	91.38	6729	175	527	0.026	219
	Marisat 1 third stage	1976-17C	**1976 Feb 19.94** **10 years**	Sphere-cone 66	1.32 long 0.94 dia	1976 Mar 1.0 1977 Mar 1.0	25.99 25.1	651.4 624.0	24895 24189	182 310	36851 35311	0.737 0.724	- -
	Marisat 1 apogee motor	1976-17F	1976 Feb 19.94 >million years	- 293 full	-		Orbit similar to 1976-17A						
D	Fragments	1976-17D,E											
D R	Cosmos 805*	1976-18A	1976 Feb 20.59 19.6 days 1976 Mar 11.2	Sphere-cylinder 6700?	7 long? 2.4 dia	1976 Feb 20.8 1976 Mar 6.8	67.13 67.13	89.72 89.61	6639 6634	171 168	351 343	0.014 0.013	80 55
D	Cosmos 805 rocket	1976-18B	1976 Feb 20.59 7 days 1976 Feb 27	Cylinder 2500?	7.5 long 2.6 dia	1976 Feb 24.1	67.13	89.00	6603	165	284	0.009	67
D	Fragment	1976-18C											

* No jettisoned engine was apparently tracked or designated.

	Name		Launch date, lifetime and descent date	Shape and weight (kg)	Size (m)	Date of orbital determination	Orbital inclination (deg)	Nodal period (min)	Semi major axis (km)	Perigee height (km)	Apogee height (km)	Orbital eccentricity	Argument of perigee (deg)
	Ume 1 (ISS 1)* [Nu]	1976-19A	1976 Feb 29.15 1200 years	Cylinder 139	0.82 long 0.94 dia	1976 Mar 15.5	69.67	105.20	7382	994	1013	0.001	215
	Ume 1 rocket	1976-19B	1976 Feb 29.15 600 years	Cylinder 66?	1.74 long 1.65 dia	1976 Mar 15.5	69.68	105.22	7383	994	1015	0.001	207
D R	Cosmos 806	1976-20A	1976 Mar 10.34 12.89 days 1976 Mar 23.23	Sphere-cylinder 6300?	6.5 long? 2.4 dia	1976 Mar 11.4 1976 Mar 21.3	71.37 71.37	89.65 89.82	6634 6643	178 177	334 352	0.012 0.013	41 20
D	Cosmos 806 rocket	1976-20B	1976 Mar 10.34 5 days 1976 Mar 15	Cylinder 2500?	7.5 long 2.6 dia	1976 Mar 11.3	71.37	89.39	6621	174	312	0.010	35
D	Cosmos 806 engine**	1976-20C	1976 Mar 10.34 20 days 1976 Mar 30	Cone 600? full	1.5 long? 2 dia?	1976 Mar 22.6	71.37	89.71	6637	173	345	0.013	17
D	Fragments	1976-20D,E											
	Molniya 1AJ	1976-21A	1976 Mar 11.83 100 years?	Windmill + 6 vanes 1000?	3.4 long 1.6 dia	1976 Mar 14.9 1976 Mar 24.0	62.84 62.86	734.41 717.93	26965 26560	491 487	40682 39877	0.745 0.742	280 280
D	Molniya 1AJ launcher	1976-21B	1976 Mar 11.83 33 days 1976 Apr 13	Irregular	-	1976 Mar 12.8	62.85	91.44	6724	203	488	0.021	129
D	Molniya 1AJ launcher rocket	1976-21C	1976 Mar 11.83 25 days 1976 Apr 5	Cylinder 2500?	7.5 long 2.6 dia	1976 Mar 13.9	62.83	91.41	6722	192	496	0.023	124
	Molniya 1AJ rocket	1976-21D	1976 Mar 11.83 100 years?	Cylinder 440	2.0 long 2.0 dia	1976 Sep 19.3	63.13	731.14	26884	451	40560	0.746	281

* Japanese Ionospheric Sounding Satellite.

** Jettisoned from 1976-20A about 1976 Mar 22.

	Name		Launch date, lifetime and descent date	Shape and weight (kg)	Size (m)	Date of orbital determination	Orbital inclination (deg)	Nodal period (min)	Semi major axis (km)	Perigee height (km)	Apogee height (km)	Orbital eccentricity	Argument of perigee (deg)
	Cosmos 807	1976-22A	1976 Mar 12.56 35 years	-	-	1976 Mar 17.6	82.97	109.13	7564	398	1973	0.104	117
	Cosmos 807 rocket	1976-22B	1976 Mar 12.56 25 years	Cylinder 2200?	7.4 long 2.4 dia	1976 Mar 20.3	82.97	109.00	7558	391	1968	0.104	109
T	LES 8	1976-23A	1976 Mar 15.06 >million years	Cylinder + box 454	3.0 long 1.6 dia?	1976 Apr 1.0	25.0	1436.1	42165	35787	35787	0	-*
T	LES 9	1976-23B	1976 Mar 15.06 >million years	Cylinder + box 454	3.0 long 1.6 dia?	1976 Apr 1.0	25.0	1436.1	42165	35787	35787	0	-*
T	Solrad 11A †	1976-23C	1976 Mar 15.06 >million years	Annulus + 4 vanes 181	0.41 deep 0.61 inner 1.47 outer diameter	1976 Jul 1.0	25.7	7344.3	125160	118383	119180	0.003	-
T	Solrad 11B	1976-23D	1976 Mar 15.06 >million years	Annulus + 4 vanes 181	size same as 1976-23C	1976 Jul 1.0	25.6	7116.7	122560	115720	116645	0.004	-
D	Titan 3C second stage	1976-23E	1976 Mar 15.06 < 1 day 1976 Mar 15	Cylinder 1900	6 long 3.0 dia	1976 Mar 15.2	28.60	87.42	6532	148	160	0.001	130
	Transtage	1976-23F	1976 Mar 15.06 >million years	Cylinder 1500?	6 long? 3.0 dia		Orbit similar to 1976-23A						
	Fragments**	1976-23G-K											

* Approximate orbits

** Two of these objects are probably Solrad apogee motors.

† Solar radiation

	Name		Launch date, lifetime and descent date	Shape and weight (kg)	Size (m)	Date of orbital determination	Orbital inclination (deg)	Nodal period (min)	Semi major axis (km)	Perigee height (km)	Apogee height (km)	Orbital eccentricity	Argument of perigee (deg)
	Cosmos 808	1976-24A	1976 Mar 16.73 25 years	Cylinder + 2 vanes? 2500?	5 long? 1.5 dia?	1976 Mar 24.8	81.25	97.10	6996	602	634	0.002	247
	Cosmos 808 rocket	1976-24B	1976 Mar 16.73 25 years	Cylinder 1440	3.8 long 2.6 dia	1976 Mar 22.6	81.26	97.18	7000	563	681	0.008	182
D R	Cosmos 809	1976-25A	1976 Mar 18.39 11.88 days 1976 Mar 30.27	Sphere-cylinder 5700?	5.0 long 2.4 dia	1976 Mar 19.4	65.03	89.55	6631	205	300	0.007	61
D	Cosmos 809 rocket	1976-25B	1976 Mar 18.39 8 days 1976 Mar 26	Cylinder 2500?	7.5 long 2.6 dia	1976 Mar 20.2	65.03	89.26	6617	196	281	0.006	47
D	Fragments	1976-25C,D											
	Molniya 1AK	1976-26A	1976 Mar 19.82 $8\frac{3}{4}$ years	Windmill + 6 vanes 1000?	3.4 long 1.6 dia	1976 Mar 23.3 1976 Mar 29.2	62.93 62.73	696.52 717.38	26027 26548	416 617	38882 39722	0.739 0.737	280 280
D	Molniya 1AK launcher rocket	1976-26B	1976 Mar 19.82 39 days 1976 Apr 27	Cylinder 2500?	7.5 long 2.6 dia	1976 Mar 20.4	63.01	91.03	6704	231	421	0.014	52
D	Molniya 1AK launcher	1976-26C	1976 Mar 19.82 49 days 1976 May 7	Irregular	-	1976 Mar 20.2	63.01	91.76	6740	220	503	0.021	64
D	Molniya 1AK rocket	1976-26D	1976 Mar 19.82 744 days 1978 Apr 2	Cylinder 440	2.0 long 2.0 dia	1976 Apr 4.4	63.01	696.66	26031	494	38812	0.736	280

	Name		Launch date, lifetime and descent date	Shape and weight (kg)	Size (m)	Date of orbital determination	Orbital inclination (deg)	Nodal period (min)	Semi major axis (km)	Perigee height (km)	Apogee height (km)	Orbital eccentricity	Argument of perigee (deg)
D	[Titan 3B Agena D]	1976-27A	1976 Mar 22.76 57 days 1976 May 18	Cylinder 3000?	8 long? 1.5 dia	1976 Mar 24.7	96.40	89.25	6614	125	347	0.017	136
D R	Cosmos 810	1976-28A	1976 Mar 26.63 12.7 days 1976 Apr 8.3	Sphere-cylinder 6300?	6.5 long? 2.4 dia	1976 Mar 27.4 1976 Mar 28.5	62.82 62.81	89.67 89.36	6637 6622	181 169	338 318	0.012 0.011	74 66
D	Cosmos 810 rocket	1976-28B	1976 Mar 26.63 6 days 1976 Apr 1	Cylinder 2500?	7.5 long 2.6 dia	1976 Mar 27.2	62.80	89.49	6628	179	321	0.011	73
D	Cosmos 810 engine*	1976-28D	1976 Mar 26.63 15 days 1976 Apr 10	Cone 600? full	1.5 long? 2 dia?	1976 Apr 7.0	62.81	89.02	6605	166	288	0.009	67
D	Fragments	1976-28C,E											
T	RCA Satcom 2	1976-29A	1976 Mar 26.95 million years	Box + vanes 868 full 463 empty	1.62 high 1.25 wide 1.25 deep	1976 Jul 1.0	0.0	1436.2	42166	35759	35817	0.0007	-
D	RCA Satcom 2 second stage	1976-29B	1976 Mar 26.95 251 days 1976 Dec 2	Cylinder + annulus 350?	6.4 long 1.52 and 2.44 dia	1976 Mar 28.4	28.40	106.67	7457	191	1966	0.119	187
D	RCA Satcom 2 third stage	1976-29C	1976 Mar 26.95 824 days 1978 Jun 28	Sphere-cone 66	1.32 long 0.94 dia	1976 Mar 27.0	27.2	635.6	24487	185	36032	0.732	-
D R	Cosmos 811	1976-30A	1976 Mar 31.54 11.8 days 1976 Apr 12.3	Sphere-cylinder 5900?	5.9 long 2.4 dia	1976 Apr 1.3	72.85	89.95	6650	206	338	0.010	61

*Jettisoned from 1976-28A about 1976 Apr 6

1976-30 continued on page 442

	Name		Launch date, lifetime and descent date	Shape and weight (kg)	Size (m)	Date of orbital determination	Orbital inclination (deg)	Nodal period (min)	Semi major axis (km)	Perigee height (km)	Apogee height (km)	Orbital eccentricity	Argument of perigee (deg)
D	Cosmos 811 rocket	1976-30B	1976 Mar 31.54 10 days 1976 Apr 10	Cylinder 2500?	7.5 long 2.6 dia	1976 Apr 1.3	72.85	89.80	6643	206	323	0.009	56
D	Capsule	1976-30E	1976 Mar 31.54 15 days 1976 Apr 15	Ellipsoid 200?	0.9 long 1.9 dia	1976 Apr 12.3	72.85	89.70	6638	201	318	0.009	-
D	Fragments	1976-30C,D,F											
D	Cosmos 812	1976-31A	1976 Apr 6.18 1668 days 1980 Oct 30	Cylinder + paddles? 900?	2 long? 1 dia?	1976 Apr 12.5	74.03	95.21	6906	508	548	0.003	332
D	Cosmos 812 rocket	1976-31B	1976 Apr 6.18 1735 days 1981 Jan 5	Cylinder 2200?	7.4 long 2.4 dia	1976 Apr 22.3	74.03	95.10	6901	499	546	0.003	332
D	Fragments	1976-31C,D											
	Meteor 24	1976-32A	1976 Apr 7.55 500 years	Cylinder + 2 vanes 2200?	5 long? 1.5 dia?	1976 Apr 21.6	81.26	102.33	7246	843	893	0.003	233
	Meteor 24 rocket	1976-32B	1976 Apr 7.55 400 years	Cylinder 1440	3.8 long 2.6 dia	1976 Apr 20.6	81.26	102.43	7251	827	918	0.006	172
D R	Cosmos 813	1976-33A	1976 Apr 9.36 11.85 days 1976 Apr 21.21	Sphere- cylinder 5700?	5.0 long 2.4 dia	1976 Apr 9.7	81.34	88.98	6601	210	236	0.002	358
D	Cosmos 813 rocket	1976-33B	1976 Apr 9.36 5 days 1976 Apr 14	Cylinder 2500?	7.5 long 2.6 dia	1976 Apr 9.7	81.34	88.93	6598	206	234	0.002	352

	Name		Launch date, lifetime and descent date	Shape and weight (kg)	Size (m)	Date of orbital determination	Orbital inclination (deg)	Nodal period (min)	Semi major axis (km)	Perigee height (km)	Apogee height (km)	Orbital eccentricity	Argument of perigee (deg)
D	Cosmos 814*	1976-34A	1976 Apr 13.72 < 0.28 day 1976 Apr 13	Cylinder?	4 long? 2 dia?	1976 Apr 13.7	65.07	90.48	6677	118	480	0.027	41
D	Cosmos 814 rocket	1976-34B	1976 Apr 13.72 3 days 1976 Apr 16	Cylinder 1500?	8 long? 2.5 dia?	1976 Apr 13.8	65.11	90.40	6673	141	449	0.023	46
T	NATO 3A	1976-35A	1976 Apr 22.87 > million years	Cylinder 720 full 310 empty	2.23 long 2.2 dia	1976 Apr 23.5 1976 May 1.0 1976 Oct 1.0	26.99 2.9 2.65	630.89 1423.4 1436.2	24418 41914 42166	177 35209 35778	35902 35863 35797	0.732 0.008 0.0002	184 - -
D	NATO 3A second stage	1976-35B	1976 Apr 22.87 38 days 1976 May 30	Cylinder + annulus 350?	6.4 long 1.52 and 2.44 dia	1976 Apr 23.4	28.23	93.39	6826	179	717	0.039	175
D	NATO 3A third stage	1976-35C	1976 Apr 22.87 544 days 1977 Oct 18	Sphere-cone 66	1.32 long 0.94 dia	1976 Apr 29.2	26.99	630.39	24404	201	35850	0.730	187
D R	Cosmos 815	1976-36A	1976 Apr 28.40 12.9 days 1976 May 11.3	Sphere-cylinder 5900?	5.9 long 2.4 dia	1976 Apr 29.1	81.33	89.01	6603	218	231	0.001	11
D	Cosmos 815 rocket	1976-36B	1976 Apr 28.40 4 days 1976 May 2	Cylinder 2500?	7.5 long 2.6 dia	1976 Apr 29.1	81.33	88.82	6593	207	222	0.001	337
D	Capsule**	1976-36F	1976 Apr 28.40 22 days 1976 May 20	Ellipsoid 200?	0.9 long 1.9 dia	1976 May 11.1	81.30	88.78	6591	207	219	0.001	297
D	Fragments	1976-36C-E,G,H											

* Cosmos 814 passed close to Cosmos 803 about 1976 Apr 13.75, then intentionally re-entered?

**Ejected from 1976-36A about 1976 May 10

	Name		Launch date, lifetime and descent date	Shape and weight (kg)	Size (m)	Date of orbital determination	Orbital inclination (deg)	Nodal period (min)	Semi major axis (km)	Perigee height (km)	Apogee height (km)	Orbital eccentricity	Argument of perigee (deg)
D	Cosmos 816	1976-37A	1976 Apr 28.57 1305 days 1979 Nov 24	Cylinder?	-	1976 Apr 29.0	65.83	94.56	6876	481	515	0.002	357
D	Cosmos 816 rocket	1976-37B	1976 Apr 28.57 1297 days 1979 Nov 16	Cylinder 2200?	7.4 long 2.4 dia	1976 Apr 29.7	65.83	94.46	6871	471	515	0.003	356
D	Fragments	1976-37C-AA											
T	NOSS 1* [Atlas]	1976-38A	1976 Apr 30.80 1600 years	Cylinder	-	1976 May 1.4	63.46	107.47	7488	1092	1128	0.002	277
	NOSS 1 rocket	1976-38B	1976 Apr 30.80 1000 years	-	-	1976 May 1.3	63.46	107.39	7485	1090	1124	0.002	289
T?	SSU 1	1976-38C	1976 Apr 30.80 1600 years	Box + aerials?	0.3 x 0.9 x 2.4?	1976 May 20.5	63.44	107.49	7489	1093	1129	0.002	88
T?	SSU 2	1976-38D	1976 Apr 30.80 1600 years	Box + aerials?	0.3 x 0.9 x 2.4?	1976 May 20.6	63.43	107.50	7490	1093	1130	0.002	73
T?	[Atlas]	1976-38E	1976 Apr 30.80 1600 years	-	-	1976 May 21.5	63.44	107.66	7498	1094	1145	0.003	89
T?	SSU 3	1976-38J	1976 Apr 30.80 1600 years	Box + aerials?	0.3 x 0.9 x 2.4?	1976 Jul 1.0	63.45	107.49	7489	1083	1139	0.004	-
	Fragments	1976-38F-H,K,L											

* Navy Ocean Surveillance Satellite: some of the objects listed as fragments may also be payloads.

	Name		Launch date, lifetime and descent date	Shape and weight (kg)	Size (m)	Date of orbital determination	Orbital inclination (deg)	Nodal period (min)	Semi major axis (km)	Perigee height (km)	Apogee height (km)	Orbital eccentricity	Argument of perigee (deg)
L	Lageos*	1976-39A	1976 May 4.33 > million years	Sphere 411	0.60 dia	1976 May 5.8	109.86	225.41	12269	5837	5945	0.004	258
	Lageos third stage	1976-39B	1976 May 4.33 15 years	Sphere- cone 77	1.32 long 0.94 dia	1976 May 6.0	109.69	153.42	9493	309	5920	0.296	176
T?	Lageos apogee motor	1976-39C	1976 May 4.33 50000 years	Sphere- cone 50?	0.98 long? 0.76 dia?	1976 May 7.5	109.84	225.42	12269	5837	5945	0.004	260
D R	Cosmos 817	1976-40A	1976 May 5.33 12.9 days 1976 May 18.2	Sphere- cylinder 5900?	5.9 long? 2.4 dia	1976 May 6.1	64.99	89.47	6627	173	324	0.011	63
D	Cosmos 817 rocket	1976-40B	1976 May 5.33 5 days 1976 May 10	Cylinder 2500?	7.5 long 2.6 dia	1976 May 6.1	65.00	89.30	6618	173	307	0.010	62
D	Capsule** ?	1976-40D	1976 May 5.33 19 days 1976 May 24	-	- 2 dia?	1976 May 17.5	64.98	89.11	6609	169	292	0.009	56
D	Fragments	1976-40C,E,F											
	Molyniya 3E	1976-41A	1976 May 12.75 13¾ years	Windmill + 6 vanes 1500?	4.2 long? 1.6 dia	1976 May 17.4 1976 May 18.9	62.81 62.87	736.64 717.90	27019 26559	625 629	40657 39733	0.741 0.736	288 288
D	Molniya 3E launcher rocket	1976-41B	1976 May 12.75 85 days 1976 Aug 5	Cylinder 2500?	7.5 long 2.6 dia	1976 May 14.5	62.81	92.74	6788	214	606	0.029	117

* Laser Geodynamic Satellite

** Ejected from 1976-40A about 1976 May 17; possibly an engine

1976-41 continued on page 446

	Name		Launch date, lifetime and descent date	Shape and weight (kg)	Size (m)	Date of orbital determination	Orbital inclination (deg)	Nodal period (min)	Semi major axis (km)	Perigee height (km)	Apogee height (km)	Orbital eccentricity	Argument of perigee (deg)
D	Molniya 3E launcher	1976-41C	1976 May 12.75 119 days 1976 Sep 8	Irregular	-	1976 May 16.3	62.80	93.04	6803	217	632	0.031	123
	Molniya 3E rocket	1976-41D	1976 May 12.75 $13\frac{3}{4}$ years	Cylinder 440	2.0 long 2.0 dia	1976 May 20.4	62.89	733.56	26944	621	40511	0.740	288
D	Fragment	1976-41E											
T	Comstar 1A	1976-42A	1976 May 13.94 > million years	Cylinder 1520 full 792 empty	2.82 long 2.36 dia	1976 May 14.0 1976 Aug 1.0	21.83 0.1	640.9 1436.0	24626 42163	550 35782	35945 35788	0.719 0	179 -
	Comstar 1A rocket	1976-42B	1976 May 13.94 6000 years	Cylinder 1815	8.6 long 3.0 dia	1976 May 14.0	21.82	649.2	24836	559	36357	0.721	179
	Meteor 25	1976-43A	1976 May 15.57 500 years	Cylinder + 2 vanes 2200?	5 long? 1.5 dia?	1976 May 16.8	81.24	102.39	7249	846	895	0.003	270
	Meteor 25 rocket	1976-43B	1976 May 15.57 400 years	Cylinder 1440	3.8 long 2.6 dia	1976 May 16.7	81.24	102.49	7254	846	905	0.004	218
D	Cosmos 818	1976-44A	1976 May 18.46 293 days 1977 Mar 7	Ellipsoid 400?	1.8 long 1.2 dia	1976 May 19.3	71.05	92.08	6754	271	481	0.016	80
D	Cosmos 818 rocket	1976-44B	1976 May 18.46 163 days 1976 Oct 28	Cylinder 1500?	8 long 1.65 dia	1976 May 21.5	71.05	91.82	6742	275	452	0.013	76

	Name		Launch date, lifetime and descent date	Shape and weight (kg)	Size (m)	Date of orbital determination	Orbital inclination (deg)	Nodal period (min)	Semi major axis (km)	Perigee height (km)	Apogee height (km)	Orbital eccentricity	Argument of perigee (deg)
D R	Cosmos 819	1976-45A	1976 May 20.29 11.9 days 1976 Jun 1.2	Sphere-cylinder 5700?	5.0 long 2.4 dia	1976 May 21.0	65.00	89.44	6626	202	293	0.007	36
D	Cosmos 819 rocket	1976-45B	1976 May 20.29 10 days 1976 May 30	Cylinder 2500?	7.5 long 2.6 dia	1976 May 21.5	65.01	89.27	6617	203	275	0.005	25
D	Fragment	1976-45C											
D R	Cosmos 820	1976-46A	1976 May 21.29 11.8 days 1976 Jun 2.1	Sphere-cylinder 6300?	6.5 long? 2.4 dia	1976 May 22.0 1976 May 22.3	81.36 81.36	88.78 88.97	6591 6601	209 209	217 236	0.001 0.002	251 273
D	Cosmos 820 rocket	1976-46B	1976 May 21.29 3 days 1976 May 24	Cylinder 2500?	7.5 long 2.6 dia	1976 May 21.8	81.36	88.63	6584	199	212	0.001	285
D	Cosmos 820* engine	1976-46E	1976 May 21.29 20 days 1976 Jun 10	Cone 600? full	1.5 long? 2 dia?	1976 Jun 1.6	81.36	88.74	6589	199	223	0.002	-
D	Fragments	1976-46C,D											
T	P76-5** [Scout]	1976-47A	1976 May 22.32 1400 years	Octagonal cylinder + vanes, 72	-	1976 May 24.8	99.68	105.73	7406	996	1060	0.004	159
	Altair rocket	1976-47B	1976 May 22.32 800 years	Cylinder 24	1.50 long 0.46 dia	1976 May 25.9	99.68	105.73	7406	996	1060	0.004	154
	Fragments	1976-47C,D											
D R	Cosmos 821	1976-48A	1976 May 26.38 12.8 days 1976 Jun 8.2	Sphere-cylinder 6300?	6.5 long? 2.4 dia	1976 May 26.6 1976 Jun 2.1	72.83 72.82	89.69 89.50	6637 6628	204 169	314 330	0.008 0.012	68 51

*Jettisoned from 1976-46A about 1976 Jun 1 **Transit navigation satellite modified for ionospheric experiments

1976-48 continued on page 448

	Name		Launch date, lifetime and descent date	Shape and weight (kg)	Size (m)	Date of orbital determination	Orbital inclination (deg)	Nodal period (min)	Semi major axis (km)	Perigee height (km)	Apogee height (km)	Orbital eccentricity	Argument of perigee (deg)
D	Cosmos 821 rocket	1976-48B	1976 May 26.38 12 days 1976 Jun 7	Cylinder 2500?	7.5 long 2.6 dia	1976 May 27.4	72.82	89.57	6631	204	302	0.007	69
D	Cosmos 821 engine	1976-48D	1976 May 26.38 19 days 1976 Jun 14	Cone 600? full	1.5 long? 2 dia?	1976 Jun 7.3	72.82	89.45	6625	167	326	0.012	41
D	Fragments	1976-48C,E,F											
D	Cosmos 822	1976-49A	1976 May 28.63 802 days 1978 Aug 8	Octagonal ellipsoid? 550?	1.8 long? 1.5 dia?	1976 Jun 1.9	74.05	94.54	6874	280	711	0.031	124
D	Cosmos 822 rocket	1976-49B	1976 May 28.63 574 days 1977 Dec 23	Cylinder 2200?	7.4 long 2.4 dia	1976 Jun 1.5	74.05	94.42	6868	276	704	0.031	123
T	SDS 2 [Titan 3B Agena D]	1976-50A	1976 Jun 2 100 years?	Cylinder?	-	1976 Jun 10	63.3	703.8	26225	380	39315	0.742	270*
	SDS 2 rocket	1976-50B	1976 Jun 2 100 years?	Cylinder 700	6 long? 1.5 dia	1976 Jun 18	63.3	700.2	26120	310	39175	0.744	270*
	Cosmos 823	1976-51A	1976 Jun 2.94 1200 years	Cylinder? 700?	1.3 long? 1.9 dia?	1976 Jun 4.0	82.96	105.04	7374	980	1011	0.002	265
	Cosmos 823 rocket	1976-51B	1976 Jun 2.94 600 years	Cylinder 2200?	7.4 long 2.4 dia	1976 Jun 3.8	82.96	104.92	7368	979	1000	0.001	252

*Approximate orbits.

	Name		Launch date, lifetime and descent date	Shape and weight (kg)	Size (m)	Date of orbital determination	Orbital inclination (deg)	Nodal period (min)	Semi major axis (km)	Perigee height (km)	Apogee height (km)	Orbital eccentricity	Argument of perigee (deg)
D R	Cosmos 824	1976-52A	1976 Jun 8.30 12.9 days 1976 Jun 21.2	Sphere- cylinder 6300?	6.5 long? 2.4 dia	1976 Jun 9.4 1976 Jun 10.2	71.37 71.37	89.82 89.39	6643 6621	204 169	325 317	0.009 0.011	48 58
D	Cosmos 824 rocket	1976-52B	1976 Jun 8.30 12 days 1976 Jun 20	Cylinder 2500?	7.5 long 2.6 dia	1976 Jun 9.8	71.38	89.65	6634	203	309	0.008	42
D	Cosmos 824 engine*	1976-52E	1976 Jun 8.30 19 days 1976 Jun 27	Cone 600? full	1.5 long? 2 dia?	1976 Jun 20.7	71.38	89.37	6620	166	318	0.011	37
D	Fragments	1976-52C,D,F,G											
T	Marisat 2	1976-53A	1976 Jun 10.01 > million years	Cylinder 655 full 362 empty	2.4 long? 1.9 dia?	1976 Jun 10.0 1976 Jun 17.6	26.00 2.5	653.01 1436.6	24933 42176	185 35788	36925 35807	0.737 0.0002	182 -
D	Marisat 2 second stage	1976-53B	1976 Jun 10.01 441 days 1977 Aug 25	Cylinder + annulus 350?	6.4 long 1.52 and 2.44 dia	1976 Jun 15.9	28.54	93.66	6839	273	649	0.028	221
	Marisat 2 third stage	1976-53F	1976 Jun 10.01 25 years	Sphere- cone 66	1.32 long 0.94 dia	1976 Jun 27.3	25.96	647.1	24780	190	36613	0.735	192
	Marisat 2 apogee motor	1976-53G	1976 Jun 10.01 > million years	- 293 full	-		Orbit similar to 1976-53A						
D	Fragments	1976-53C-E											

* Jettisoned from 1976-52A about 1976 Jun 20

Name		Launch date, lifetime and descent date	Shape and weight (kg)	Size (m)	Date of orbital determination	Orbital inclination (deg)	Nodal period (min)	Semi major axis (km)	Perigee height (km)	Apogee height (km)	Orbital eccentricity	Argument of perigee (deg)
Cosmos 825	1976-54A	1976 Jun 15.55 7000 years	Spheroid 40?	1.0 long? 0.8 dia?	1976 Jun 25.3	73.99	114.74	7821	1397	1489	0.006	91
Cosmos 826	1976-54B	1976 Jun 15.55 10000 years	Spheroid 40?	1.0 long? 0.8 dia?	1976 Jun 18.6	74.00	116.33	7893	1484	1546	0.004	238
Cosmos 827	1976-54C	1976 Jun 15.55 8000 years	Spheroid 40?	1.0 long? 0.8 dia?	1976 Jun 22.2	74.00	114.96	7831	1415	1491	0.005	106
Cosmos 828	1976-54D	1976 Jun 15.55 9000 years	Spheroid 40?	1.0 long? 0.8 dia?	1976 Jun 24.8	73.99	115.18	7841	1435	1491	0.004	106
Cosmos 829	1976-54E	1976 Jun 15.55 9000 years	Spheroid 40?	1.0 long? 0.8 dia?	1976 Jun 18.7	74.00	115.39	7851	1453	1492	0.003	127
Cosmos 830	1976-54F	1976 Jun 15.55 10000 years	Spheroid 40?	1.0 long? 0.8 dia?	1976 Jun 24.3	74.00	115.61	7861	1471	1495	0.001	141
Cosmos 831	1976-54G	1976 Jun 15.55 10000 years	Spheroid 40?	1.0 long? 0.8 dia?	1976 Jun 21.5	74.00	115.85	7872	1477	1510	0.002	193
Cosmos 832	1976-54H	1976 Jun 15.55 10000 years	Spheroid 40?	1.0 long? 0.8 dia?	1976 Jun 21.3	74.00	116.07	7882	1484	1523	0.002	228
Cosmos 825 rocket	1976-54J	1976 Jun 15.55 20000 years	Cylinder 2200?	7.4 long 2.4 dia	1976 Jun 19.3	73.99	117.99	7968	1488	1692	0.013	260

	Name		Launch date, lifetime and descent date	Shape and weight (kg)	Size (m)	Date of orbital determination	Orbital inclination (deg)	Nodal period (min)	Semi major axis (km)	Perigee height (km)	Apogee height (km)	Orbital eccentricity	Argument of perigee (deg)
D R	Cosmos 833	1976-55A	1976 Jun 16.55 12.6 days 1976 Jun 29.2	Sphere-cylinder 5900?	5.9 long? 2.4 dia	1976 Jun 17.5	62.82	89.44	6626	180	316	0.010	68
D	Cosmos 833 rocket	1976-55B	1976 Jun 16.55 7 days 1976 Jun 23	Cylinder 2500?	7.5 long 2.6 dia	1976 Jun 16.8	62.81	89.36	6622	177	311	0.010	64
D	Capsule? *	1976-55E	1976 Jun 16.55 22 days 1976 Jul 8	-	- 2 dia?	1976 Jun 29.3	62.82	89.16	6612	177	291	0.009	69
D	Fragments	1976-55C,D,F,G											
D	Intercosmos 15	1976-56A	1976 Jun 19.67 1247.2 days 1979 Nov 18.9	Octagonal ellipsoid? 550?	1.8 long? 1.5 dia?	1976 Jun 24.5	74.04	94.65	6879	484	518	0.002	351
D	Intercosmos 15 rocket	1976-56B	1976 Jun 19.67 1305 days 1980 Jan 15	Cylinder 2200?	7.4 long 2.4 dia	1976 Jun 23.9	74.04	94.57	6875	477	517	0.003	354
D r	Salyut 5 **	1976-57A	1976 Jun 22.76 412 days 1977 Aug 8	Cylinder + 3 wings 19000?	14 long 4.15 to 2.0 dia	1976 Jun 23.0 1976 Jun 25.4 1976 Jun 27.4	51.60 51.59 51.59	88.85 89.00 89.15	6599 6606 6614	208 212 214	233 244 257	0.002 0.002 0.003	130 111 100
D	Salyut 5 rocket	1976-57B	1976 Jun 22.76 8 days 1976 Jun 30	Cylinder 4000?	12 long? 4 dia	1976 Jun 23.1	51.61	88.80	6596	211	225	0.001	112
D	Fragments	1976-57C-K											

* Possibly an engine.

** Salyut 5 was de-orbited; a capsule was ejected and recovered on 1977 Feb 26.4.

	Name		Launch date, lifetime and descent date	Shape and weight (kg)	Size (m)	Date of orbital determination	Orbital inclination (deg)	Nodal period (min)	Semi major axis (km)	Perigee height (km)	Apogee height (km)	Orbital eccentricity	Argument of perigee (deg)
D R	Cosmos 834	1976-58A	1976 Jun 24.30 11.9 days 1976 Jul 6.2	Sphere-cylinder 5700?	5.0 long 2.4 dia	1976 Jun 25.4	81.37	89.05	6605	216	237	0.002	94
D	Cosmos 834 rocket	1976-58B	1976 Jun 24.30 6 days 1976 Jun 30	Cylinder 2500?	7.5 long 2.6 dia	1976 Jun 25.2	81.37	88.95	6600	216	227	0.001	93
T	IMEWS 6 [Titan 3C]	1976-59A	1976 Jun 26.13 > million years	-	-	1976 Jun 26.5 1976 Jul 1.0	26.3 0.5	633.0 1433.3	24445 42118	295 35620	35840 35860	0.727 0.003	180* -*
D	Titan 3C second stage	1976-59B	1976 Jun 26.13 4 days 1976 Jun 30	Cylinder 1900	6 long 3.0 dia	1976 Jun 26.2	28.60	90.41	6680	151	453	0.023	113
	Transtage	1976-59C	1976 Jun 26.13 > million years	Cylinder 1500?	6 long? 3.0 dia		Orbits similar to 1976-59A						
	Fragment	1976-59D											
D R	Cosmos 835	1976-60A	1976 Jun 29.31 12.84 days 1976 Jul 12.15	Sphere-cylinder 5900?	5.9 long? 2.4 dia	1976 Jun 30.6	64.96	89.41	6624	174	317	0.011	59
D	Cosmos 835 rocket	1976-60B	1976 Jun 29.31 5 days 1976 Jul 4	Cylinder 2500?	7.5 long 2.6 dia	1976 Jun 29.7	64.97	89.33	6620	175	308	0.010	58
D	Capsule** ?	1976-60C	1976 Jun 29.31 21 days 1976 Jul 20	-	- 2 dia?	1976 Jul 11.5	64.98	89.12	6610	172	291	0.009	51
D	Fragments	1976-60D,E											

*Unconfirmed orbits

** Ejected from 1976-60A on 1976 Jul 11; possibly an engine

	Name		Launch date, lifetime and descent date	Shape and weight (kg)	Size (m)	Date of orbital determination	Orbital inclination (deg)	Nodal period (min)	Semi major axis (km)	Perigee height (km)	Apogee height (km)	Orbital eccentricity	Argument of perigee (deg)
	Cosmos 836	1976-61A	1976 Jun 29.34 120 years	Cylinder + paddles? 750?	2 long? 1 dia?	1976 Jun 30.6	74.06	100.98	7183	791	818	0.002	106
	Cosmos 836 rocket	1976-61B	1976 Jun 29.34 100 years	Cylinder 2200?	7.4 long 2.4 dia	1976 Jun 30.0	74.07	100.88	7178	784	815	0.002	88
	Cosmos 837 *	1976-62A	1976 Jul 1.34 8 years	Windmill + 6 vanes? 1250?	4.2 long? 1.6 dia?	1976 Jul 2.3	62.75	98.51	7065	438	936	0.035	259
D	Cosmos 837 launcher rocket	1976-62B	1976 Jul 1.34 35 days 1976 Aug 5	Cylinder 2500?	7.5 long 2.6 dia	1976 Jul 2.3	62.80	90.73	6689	215	406	0.014	125
D	Cosmos 837 launcher	1976-62C	1976 Jul 1.34 20 days 1976 Jul 21	Irregular	-	1976 Jul 2.3	62.82	90.87	6696	189	446	0.019	123
D	Cosmos 837 rocket	1976-62E	**1976 Jul 1.34 2278 days 1982 Sep 26**	Cylinder 4800 full 440 empty	2.0 long 2.0 dia	1976 Jul 2.3	62.75	98.39	7059	440	922	0.034	260
D	Fragment	1976-62D											
D	Cosmos 838 **	1976-63A	1976 Jul 2.44 disintegrated	Cylinder?	-	1976 Jul 6.4	65.06	93.30	6812	428	440	0.0009	271
D	Cosmos 838 rocket	1976-63B	1976 Jul 2.44 1 day 1976 Jul 3	Cylinder 1500?	8 long? 2.5 dia?	1976 Jul 2.8	65.03	89.46	6628	117	382	0.020	48
D	Fragments	1976-63C-AS											

* Probable Molniya attempt; final stage shut down prematurely.

** Disintegrated during late June - early July 1977

	Name		Launch date, lifetime and descent date	Shape and weight (kg)	Size (m)	Date of orbital determination	Orbital inclination (deg)	Nodal period (min)	Semi major axis (km)	Perigee height (km)	Apogee height (km)	Orbital eccentricity	Argument of perigee (deg)
D 2M R	Soyuz 21*	1976-64A	1976 Jul 6.51 49.26 days 1976 Aug 24.77	Sphere-cylinder 6570?	7.5 long 2.2 dia	1976 Jul 7.2 1976 Jul 19.4	51.59 51.59	89.65 89.80	6638 6646	246 262	274 273	0.002 0.0008	249 324
D	Soyuz 21 rocket	1976-64B	1976 Jul 6.51 3 days 1976 Jul 9	Cylinder 2500?	7.5 long 2.6 dia	1976 Jul 7.2	51.60	88.46	6579	183	219	0.003	87
D	[Titan 3D]	1976-65A	1976 Jul 8.78 158 days 1976 Dec 13	Cylinder 13300? full	15 long 3.0 dia	1976 Jul 10.7	97.00	88.54	6579	159	242	0.006	131
T?	SESP 74-2 **	1976-65B	1976 Jul 8.78 15 years	-	-	1976 Jul 16.1	97.53	179.00	10520	236	8048	0.371	291
T	Capsule	1976-65C	1976 Jul 8.78 25 years	Octagon? 60?	0.3 long? 0.9 dia?	1976 Jul 18.7	96.38	97.34	7008	628	632	0.0003	219
D	Titan 3D rocket	1976-65D	1976 Jul 8.78 1.41 days 1976 Jul 10.19	Cylinder 1900	6 long 3.0 dia	1976 Jul 9.0	97.00	88.31	6567	150	228	0.006	130
T	Palapa 1†	1976-66A	1976 Jul 8.98 > million years	Cylinder 574 full 282 empty	1.56 long 1.90 dia	1976 Jul 9.0 1976 Sep 1.0	24.66 0.05	645.61 1436.1	24744 42165	231 35764	36501 35809	0.733 0.0005	179 -
D	Palapa 1 second stage	1976-66B	1976 Jul 8.98 2 days 1976 Jul 10	Cylinder + annulus 350?	6.4 long 1.52 and 2.44 dia	1976 Jul 9.1	28.73	87.97	6561	147	219	0.005	349

* Soyuz 21 docked with Salyut 5 on 1976 Jul 7.57; undocked 1976 Aug 24.63

** Small magnetospheric observatory

† First Indonesian satellite, launched by NASA

1976-66 continued on page 455

	Name		Launch date, lifetime and descent date	Shape and weight (kg)	Size (m)	Date of orbital determination	Orbital inclination (deg)	Nodal period (min)	Semi major axis (km)	Perigee height (km)	Apogee height (km)	Orbital eccentricity	Argument of perigee (deg)
	Palapa 1 third stage	1976-66C	1976 Jul 8.98 25 years	Sphere-cone 66	1.32 long 0.94 dia	1976 Jul 16.3	24.65	638.71	24570	239	36145	0.731	183
	Cosmos 839	1976-67A	1976 Jul 8.88 4000 years	Cylinder?	4 long? 2 dia?	1976 Jul 15.4	65.86	116.88	7919	984	2098	0.070	163
	Cosmos 839 rocket	1976-67B	1976 Jul 8.88 disintegrated	Cylinder 2200?	7.4 long 2.4 dia	1976 Jul 15.5	65.86	116.67	7910	972	2091	0.071	163
1d	Fragments	1976-67C-BD											
D R	Cosmos 840	1976-68A	1976 Jul 14.38 11.8 days 1976 Jul 26.2	Sphere-cylinder 5700?	5.0 long 2.4 dia	1976 Jul 15.6	72.87	89.73	6639	203	319	0.009	74
D	Cosmos 840 rocket	1976-68B	1976 Jul 14.38 11 days 1976 Jul 25	Cylinder 2500?	7.5 long 2.6 dia	1976 Jul 15.5	72.87	89.58	6632	198	309	0.008	66
D	Fragment	1976-68C											
	Cosmos 841	1976-69A	1976 Jul 15.55 120 years	Cylinder + paddles? 750?	2 long? 1 dia?	1976 Jul 17.8	74.05	100.83	7176	787	808	0.001	3
	Cosmos 841 rocket	1976-69B	1976 Jul 15.55 100 years	Cylinder 2200?	7.4 long 2.4 dia	1976 Jul 25.6	74.05	100.74	7171	779	807	0.002	5

	Name		Launch date, lifetime and descent date	Shape and weight (kg)	Size (m)	Date of orbital determination	Orbital inclination (deg)	Nodal period (min)	Semi major axis (km)	Perigee height (km)	Apogee height (km)	Orbital eccentricity	Argument of perigee (deg)
	Cosmos 842	1976-70A	1976 Jul 21.43 1200 years	Cylinder? 700?	1.3 long? 1.9 dia?	1976 Aug 1.5	82.98	104.96	7370	972	1011	0.003	253
	Cosmos 842 rocket	1976-70B	1976 Jul 21.43 600 years	Cylinder 2200?	7.4 long 2.4 dia	1976 Jul 24.5	82.98	104.84	7364	971	1000	0.002	269
D	Cosmos 843 *	1976-71A	1976 Jul 21.64 <0.36 day 1976 Jul 21	Cylinder?	4 long? 2 dia?	1976 Jul 21.7	65.08	89.27	6617	132	346	0.016	36
D	Cosmos 843 rocket	1976-71B	1976 Jul 21.64 2 days 1976 Jul 23	Cylinder 1500?	8 long? 2.5 dia?	1976 Jul 23.0	65.10	88.19	6564	127	244	0.009	37
D	Cosmos 844 **	1976-72A	1976 Jul 22.66 39 days 1976 Aug 30	Sphere-cylinder 6700?	7 long? 2.4 dia	1976 Jul 25.4	67.15	89.76	6641	172	353	0.014	70
D	Cosmos 844 rocket	1976-72B	1976 Jul 22.66 8 days 1976 Jul 30	Cylinder 2500?	7.5 long 2.6 dia	1976 Jul 24.0	67.14	89.67	6637	171	346	0.013	73
D	Fragments	1976-72C-KJ											
T	Comstar 1B	1976-73A	1976 Jul 22.92 > million years	Cylinder 1520 full 792 empty	2.82 long 2.36 dia	1976 Jul 22.9 1976 Nov 1.0	21.81 0.1	641.03 1436.2	24627 42166	552 35780	35946 35795	0.719 0.0002	179 -
	Comstar 1B rocket	1976-73B	1976 Jul 22.92 6000 years	Cylinder 1815	8.6 long 3.0 dia	1976 Jul 22.9	21.82	650.92	24880	561	36442	0.721	179

* Probably intended to pass close to Cosmos 839

** Exploded on 1976 Jul 25

	Name		Launch date, lifetime and descent date	Shape and weight (kg)	Size (m)	Date of orbital determination	Orbital inclination (deg)	Nodal period (min)	Semi major axis (km)	Perigee height (km)	Apogee height (km)	Orbital eccentricity	Argument of perigee (deg)
	Molniya 1AL	1976-74A	1976 Jul 23.66 10.9 years	Windmill + 6 vanes 1000?	3.4 long 1.6 dia	1976 Jul 30.0 1976 Sep 1.0	63.01 62.9	700.93 717.67	26139 26554	476 515	39045 39836	0.738 0.740	280 -
D	Molniya 1AL launcher rocket	1976-74B	1976 Jul 23.66 71 days 1976 Oct 2	Cylinder 2500?	7.5 long 2.6 dia	1976 Jul 23.8	62.90	91.54	6729	239	462	0.017	60
D	Molniya 1AL launcher	1976-74C	1976 Jul 23.66 57 days 1976 Sep 18	Irregular	-	1976 Jul 26.0	62.89	91.77	6740	215	509	0.022	60
	Molniya 1AL rocket	1976-74E	1976 Jul 23.66 10.9 years	Cylinder 440	2.0 long 2.0 dia	1976 Aug 5.2	62.90	698.49	26078	480	38919	0.737	280
D	Fragment	1976-74D											
D	Cosmos 845	1976-75A	1976 Jul 27.23 1572 days 1980 Nov 15	Cylinder + paddles? 900?	2 long? 1 dia?	1976 Jul 28.1	74.06	95.25	6908	514	546	0.002	1
D	Cosmos 845 rocket	1976-75B	1976 Jul 27.23 1677 days 1981 Feb 28	Cylinder 2200?	7.4 long 2.4 dia	1976 Jul 27.7	74.06	95.16	6904	504	547	0.003	356
D	Fragments	1976-75C-H											
D	Intercosmos 16*	1976-76A	1976 Jul 27.50 1078 days 1979 Jul 10	Octagonal ellipsoid? 550?	1.8 long? 1.5 dia?	1976 Aug 2.1	50.57	94.36	6869	464	517	0.004	38
D	Intercosmos 16 rocket	1976-76B	1976 Jul 27.50 1304 days 1980 Feb 21	Cylinder 2200?	7.4 long 2.4 dia	1976 Jul 28.1	50.57	94.23	6862	451	517	0.005	22

* Payload includes Swedish experiments.

	Name		Launch date, lifetime and descent date	Shape and weight (kg)	Size (m)	Date of orbital determination	Orbital inclination (deg)	Nodal period (min)	Semi major axis (km)	Perigee height (km)	Apogee height (km)	Orbital eccentricity	Argument of perigee (deg)
T	NOAA 5 (ITOS H)	1976-77A	1976 Jul 29.71 10000 years	Box 340	1.25 long 1.02 square	1976 Jul 31.5	102.10	116.34	7894	1509	1522	0.0008	242
	NOAA 5 second stage	1976-77B	1976 Jul 29.71 disintegrated	Cylinder + annulus? 350?	6.4 long? 1.52 and 2.44 dia?	1976 Aug 1.5	102.08	116.33	7893	1507	1523	0.001	227
1d	Fragments	1976-77C-EL											
	Cosmos 846	1976-78A	1976 Jul 29.83 1200 years	Cylinder? 700?	1.3 long? 1.9 dia?	1976 Aug 1.5	82.92	104.81	7363	954	1015	0.004	292
	Cosmos 846 rocket	1976-78B	1976 Jul 29.83 600 years	Cylinder 2200?	7.4 long 2.4 dia	1976 Jul 30.3	82.93	104.68	7356	953	1003	0.003	303
D R	Cosmos 847	1976-79A	1976 Aug 4.56 12.61 days 1976 Aug 17.17	Sphere-cylinder 5900?	5.9 long? 2.4 dia	1976 Aug 5.0	62.82	89.50	6629	181	321	0.011	78
D	Cosmos 847 rocket	1976-79B	1976 Aug 4.56 7 days 1976 Aug 11	Cylinder 2500?	7.5 long 2.6 dia	1976 Aug 5.1	62.78	89.39	6623	179	311	0.010	71
D	Capsule?*	1976-79D	1976 Aug 4.56 19 days 1976 Aug 23	-	- 2 dia?	1976 Aug 16.5	62.82	89.21	6614	165	307	0.011	-
D	Fragments	1976-79C,E											

* Possibly an engine

	Name		Launch date, lifetime and descent date	Shape and weight (kg)	Size (m)	Date of orbital determination	Orbital inclination (deg)	Nodal period (min)	Semi major axis (km)	Perigee height (km)	Apogee height (km)	Orbital eccentricity	Argument of perigee (deg)
T	SDS 3 [Titan 3B Agena D]	1976-80A	1976 Aug 6 100 years?	Cylinder?	-	1976 Aug 14	63.3	703.8	26225	380	39315	0.742	270*
	SDS 3 rocket	1976-80B	1976 Aug 6 100 years?	Cylinder 700	6 long? 1.5 dia	1976 Aug 22	63.3	700.2	26120	310	39175	0.744	270*
D	Luna 24 launcher	1976-81C	1976 Aug 9.63 6 days 1976 Aug 15	-	-	1976 Aug 10.0	51.54	88.74	6593	188	242	0.004	330
D	Luna 24 launcher rocket	1976-81D	1976 Aug 9.63 5 days 1976 Aug 14	Cylinder 4000?	12 long? 4 dia	1976 Aug 11.3	51.52	88.55	6584	186	225	0.003	339
D	Fragment	1976-81B											
D R	Cosmos 848	1976-82A	1976 Aug 12.57 12.62 days 1976 Aug 25.19	Sphere-cylinder 5700?	5.0 long 2.4 dia	1976 Aug 14.3	62.80	89.57	6633	206	303	0.007	87
D	Cosmos 848 rocket	1976-82B	1976 Aug 12.57 12 days 1976 Aug 24	Cylinder 2500?	7.5 long 2.6 dia	1976 Aug 13.5	62.79	89.46	6627	204	294	0.007	79
D	Fragments	1976-82C-E											
D	Cosmos 849	1976-83A	1976 Aug 18.40 614 days 1978 Apr 24	Ellipsoid 400?	1.8 long 1.2 dia	1976 Aug 20.4	70.97	95.95	6943	264	865	0.043	84
D	Cosmos 849 rocket	1976-83B	1976 Aug 18.40 441 days 1977 Nov 2	Cylinder 1500?	8 long 1.65 dia	1976 Aug 19.4	70.97	95.65	6928	268	831	0.040	85
D	**Fragment**	1976-83C											

Space Vehicle: Luna 24, 1976-81A; ascent stage, 1976-81E. *Approximate orbits.

	Name		Launch date, lifetime and descent date	Shape and weight (kg)	Size (m)	Date of orbital determination	Orbital inclination (deg)	Nodal period (min)	Semi major axis (km)	Perigee height (km)	Apogee height (km)	Orbital eccentricity	Argument of perigee (deg)
D	Cosmos 850	1976-84A	1976 Aug 26.46 263 days 1977 May 16	Ellipsoid 400?	1.8 long 1.2 dia	1976 Aug 26.6	70.94	92.20	6761	272	493	0.016	81
D	Cosmos 850 rocket	1976-84B	1976 Aug 26.46 141 days 1977 Jan 14	Cylinder 1500?	8 long 1.65 dia	1976 Aug 28.3	70.94	91.99	6750	275	469	0.014	82
	Cosmos 851	1976-85A	1976 Aug 27.61 25 years	Cylinder + 2 vanes? 2500?	5 long? 1.5 dia?	1976 Aug 28.7	81.20	96.78	6981	569	637	0.005	264
	Cosmos 851 rocket	1976-85B	1976 Aug 27.61 25 years	Cylinder 1440	3.8 long 2.6 dia	1976 Aug 28.7	81.21	96.80	6982	554	654	0.007	217
D R	Cosmos 852	1976-86A	1976 Aug 28.38 12.85 days 1976 Sep 10.23	Sphere-cylinder 5900?	5.9 long? 2.4 dia	1976 Aug 29.1	64.99	89.54	6631	173	332	0.012	66
D	Cosmos 852 rocket	1976-86B	1976 Aug 28.38 5 days 1976 Sep 2	Cylinder 2500?	7.5 long 2.6 dia	1976 Aug 29.1	64.99	89.36	6622	172	315	0.011	65
D	Capsule?*	1976-86D	1976 Aug 28.38 19 days 1976 Sep 16	-	- 2 dia?	1976 Sep 11.2	64.99	89.07	6607	167	291	0.009	-
D	Fragments	1976-86C,E,F											
D	China 6	1976-87A	1976 Aug 30.49 817 days 1978 Nov 25	Spheroid? 270?	1.25 dia?	1976 Aug 31.0	69.16	108.77	7548	195	2145	0.129	139
D	China 6 rocket	1976-87B	1976 Aug 30.49 523 days 1978 Feb 4	Cylinder	8 long 3 dia	1976 Aug 30.8	69.17	108.72	7546	190	2145	0.129	139

* Possibly an engine

	Name		Launch date, lifetime and descent date	Shape and weight (kg)	Size (m)	Date of orbital determination	Orbital inclination (deg)	Nodal period (min)	Semi major axis (km)	Perigee height (km)	Apogee height (km)	Orbital eccentricity	Argument of perigee (deg)
D	Cosmos 853*	1976-88A	1976 Sep 1.14 121 days 1976 Dec 31	Windmill + 6 vanes? 1250?	4.2 long? 1.6 dia	1976 Sep 3.3	62.82	91.57	6730	243	461	0.016	137
D	Cosmos 853 launcher rocket	1976-88B	1976 Sep 1.14 39 days 1976 Oct 10	Cylinder 2500?	7.5 long 2.6 dia	1976 Sep 3.3	62.83	91.20	6712	213	454	0.018	129
D	Cosmos 853 rocket	1976-88C	1976 Sep 1.14 261 days 1977 May 20	Cylinder 4800 full 440 empty	2.0 long 2.0 dia	1976 Sep 3.2	62.78	91.64	6734	240	471	0.017	130
D	Cosmos 853 launcher	1976-88D	1976 Sep 1.14 99 days 1976 Dec 9	Irregular	-	1976 Sep 3.4	62.81	91.68	6736	242	473	0.017	133
D	TIP 3 [Scout]	1976-89A	1976 Sep 1.88 1732 days 1981 May 30	Dumb-bell? 94?	7.3 long? 0.59 dia?	1976 Sep 4.3	90.31	96.02	6947	348	789	0.032	155
D	Altair rocket	1976-89B	1976 Sep 1.88 667 days 1978 Jun 29	Cylinder 24	1.50 long 0.46 dia	1976 Sep 4.3	90.32	95.99	6945	348	786	0.031	154
D	Fragments	1976-89C,D											
D R	Cosmos 854	1976-90A	1976 Sep 3.39 12.85 days 1976 Sep 16.24	Sphere- cylinder 6300?	6.5 long? 2.4 dia	1976 Sep 3.7 1976 Sep 5.1	81.35 81.35	89.27 89.02	6616 6603	167 168	308 282	0.011 0.009	83 80
D	Cosmos 854 rocket	1976-90B	1976 Sep 3.39 4 days 1976 Sep 7	Cylinder 2500?	7.5 long 2.6 dia	1976 Sep 3.9	81.35	89.12	6608	166	294	0.010	82

* Cosmos 853 may be a failed Molniya satellite; final stage failed to ignite.

1976-90 continued on page 462

	Name		Launch date, lifetime and descent date	Shape and weight (kg)	Size (m)	Date of orbital determination	Orbital inclination (deg)	Nodal period (min)	Semi major axis (km)	Perigee height (km)	Apogee height (km)	Orbital eccentricity	Argument of perigee (deg)
D	Cosmos 854 engine	1976-90C	1976 Sep 3.39 16 days 1976 Sep 19	Cone 600? full	1.5 long? 2 dia?	1976 Sep 17.2	81.35	88.77	6591	159	266	0.008	-
D	Fragment	1976-90D											
T	AMS 1* [Thor Burner 2]	1976-91A	1976 Sep 11.34 80 years	Irregular 450	6.40 long 1.68 dia	1976 Sep 14.3	98.70	101.60	7211	818	848	0.002	187
	Burner 2 rocket	1976-91B	1976 Sep 11.34 60 years	Sphere-cone 66	1.32 long 0.94 dia	1976 Sep 16.8	98.71	101.60	7211	817	849	0.002	166
1d	Fragments	1976-91C-G											
	Statsionar - Raduga 2	1976-92A	1976 Sep 11.77 >million years	-	-	1976 Sep 12.5	0.3	1440	42278	35900	35900	0	-**
D	Raduga 2 launcher rocket	1976-92B	1976 Sep 11.77 3 days 1976 Sep 14	Cylinder 4000?	12 long? 4 dia	1976 Sep 12.4	51.50	88.15	6564	177	195	0.001	316
D	Raduga 2 launcher	1976-92C	1976 Sep 11.77 1 day 1976 Sep 12	Irregular	-	1976 Sep 12.1	51.48	88.28	6571	189	196	0.001	27
D	Raduga 2 rocket	1976-92D	1976 Sep 11.77 311 days 1977 Jul 19	Cylinder 1900?	3.9 long? 3.9 dia	1976 Oct 2.1	47.33	633.94	24450	321	35823	0.726	5
D	Fragment	1976-92E											
D 2M R	Soyuz 22	1976-93A	1976 Sep 15.41 7.91 days 1976 Sep 23.32	Sphere-cylinder + 2 wings 6570?	7.5 long 2.3 dia	1976 Sep 15.5 1976 Sep 16.6 1976 Sep 22.5	64.75 64.75 64.75	89.31 89.58 89.42	6619 6632 6624	185 249 239	296 259 253	0.008 0.001 0.001	71 149 140

* USAF Advanced Meteorological Satellite (Block 5D).

** Approximate orbit. There may be a separated apogee motor in a similar orbit.

	Name		Launch date, lifetime and descent date	Shape and weight (kg)	Size (m)	Date of orbital determination	Orbital inclination (deg)	Nodal period (min)	Semi major axis (km)	Perigee height (km)	Apogee height (km)	Orbital eccentricity	Argument of perigee (deg)
D	Soyuz 22 rocket	1976-93B	1976 Sep 15.41 5 days 1976 Sep 20	Cylinder 2500?	7.5 long 2.6 dia	1976 Sep 16.8	64.76	89.03	6605	182	272	0.007	66
D	Fragment*	1976-93C											
D	[Titan 3B Agena D]	1976-94A	1976 Sep 15.79 51 days 1976 Nov 5	Cylinder 3000?	8 long? 1.5 dia	1976 Sep 16.4	96.39	89.18	6611	135	330	0.015	144
D	Fragment	1976-94B											
D R	Cosmos 855	1976-95A	1976 Sep 21.49 11.80 days 1976 Oct 3.29	Sphere-cylinder 5900?	5.9 long 2.4 dia	1976 Sep 22.4	72.88	89.96	6650	202	341	0.010	76
D	Cosmos 855 rocket	1976-95B	1976 Sep 21.49 11 days 1976 Oct 2	Cylinder 2500?	7.6 long 2.6 dia	1976 Sep 22.4	72.87	89.78	6641	197	328	0.010	69
D	Capsule	1976-95D	1976 Sep 21.49 17 days 1976 Oct 8	Ellipsoid 200?	0.9 long 1.9 dia	1976 Oct 2.6	72.88	89.82	6643	199	330	0.010	-
D	Fragment	1976-95C											
D R	Cosmos 856	1976-96A	1976 Sep 22.40 12.87 days 1976 Oct 5.27	Sphere-cylinder 5900?	5.9 long 2.4 dia	1976 Sep 23.4	65.01	89.53	6630	203	300	0.007	65
D	Cosmos 856 rocket	1976-96B	1976 Sep 22.40 8 days 1976 Sep 30	Cylinder 2500?	7.5 long 2.6 dia	1976 Sep 23.1	65.02	89.37	6622	202	286	0.006	55
D	Capsule**	1976-96E	1976 Sep 22.40 24 days 1976 Oct 16	Ellipsoid 200?	0.9 long 1.9 dia	1976 Oct 3.7	65.01	89.41	6624	201	291	0.007	61
D	Fragments	1976-96C,D											

* Probably the MKF-6 multizonal camera shroud (decayed 1976 Sep 16).

** Ejected from 1976-96A on 1976 Oct 3.

	Name		Launch date, lifetime and descent date	Shape and weight (kg)	Size (m)	Date of orbital determination	Orbital inclination (deg)	Nodal period (min)	Semi major axis (km)	Perigee height (km)	Apogee height (km)	Orbital eccentricity	Argument of perigee (deg)
D R	Cosmos 857	1976-97A	1976 Sep 24.63 12.61 days 1976 Oct 7.24	Sphere-cylinder 6300?	6.5 long? 2.4 dia	1976 Sep 25.7 1976 Oct 4.4	62.80 62.81	89.50 89.56	6629 6632	179 177	323 331	0.011 0.012	73 74
D	Cosmos 857 rocket	1976-97B	1976 Sep 24.63 4.66 days 1976 Sep 29.29	Cylinder 2500?	7.5 long 2.6 dia	1976 Sep 24.8	62.79	89.39	6624	176	315	0.010	68
D	Cosmos 857 engine*	1976-97C	1976 Sep 24.63 21 days 1976 Oct 15	Cone 600? full	1.5 long? 2 dia?	1976 Oct 6.4	62.80	89.49	6629	176	325	0.011	73
D	Fragments	1976-97D-F											
	Cosmos 858	1976-98A	1976 Sep 29.30 120 years	Cylinder + paddles? 750?	2 long? 1 dia?	1976 Sep 30.6	74.06	100.93	7181	792	813	0.001	19
	Cosmos 858 rocket	1976-98B	1976 Sep 29.30 100 years	Cylinder 2200?	7.4 long 2.4 dia	1976 Sep 30.5	74.06	100.82	7175	783	811	0.002	33
D R	Cosmos 859	1976-99A	1976 Oct 10.40 10.9 days 1976 Oct 21.3	Sphere-cylinder 6300?	6.5 long? 2.4 dia	1976 Oct 11.1 1976 Oct 13.4	65.00 64.99	89.60 89.36	6633 6621	173 172	337 314	0.012 0.011	67 66
D	Cosmos 859 rocket	1976-99B	1976 Oct 10.40 5.27 days 1976 Oct 15.67	Cylinder 2500?	7.5 long 2.6 dia	1976 Oct 10.6	65.00	89.52	6629	172	330	0.012	71
D	Cosmos 859 engine	1976-99E	1976 Oct 10.40 15 days 1976 Oct 25	Cone 600? full	1.5 long? 2 dia?	1976 Oct 20.6	64.99	89.01	6604	167	285	0.009	61
D	Fragments	1976-99C,D											

* Jettisoned from 1976-97A on 1976 Oct 6

	Name		Launch date, lifetime and descent date	Shape and weight (kg)	Size (m)	Date of orbital determination	Orbital inclination (deg)	Nodal period (min)	Semi major axis (km)	Perigee height (km)	Apogee height (km)	Orbital eccentricity	Argument of perigee (deg)
D 2M R	Soyuz 23*	1976-100A	1976 Oct 14.74 2.00 days 1976 Oct 16.74	Sphere- cylinder 6570?	7.5 long 2.3 dia	1976 Oct 15.0 1976 Oct 15.9	51.61 51.58	88.56 89.60	6584 6636	188 246	224 269	0.003 0.002	74 233
D	Soyuz 23 rocket	1976-100B	1976 Oct 14.74 2 days 1976 Oct 16	Cylinder 2500?	7.5 long 2.6 dia	1976 Oct 15.4	51.61	88.46	6580	184	219	0.003	83
T	Marisat 3	1976-101A	1976 Oct 14.95 >million years	Cylinder 655 full 362 empty	2.4 long? 1.9 dia?	1976 Oct 15.0 1976 Oct 22.6	26.00 2.6	652.92 1436.2	24931 42166	185 35051	36920 36525	0.737 0.017	174 -
D	Marisat 3 second stage	1976-101B	1976 Oct 14.95 32 days 1976 Nov 15	Cylinder + annulus 350?	6.4 long 1.52 and 2.44 dia	1976 Oct 16.4	28.58	92.40	6778	182	618	0.032	178
D	Marisat 3 third stage	1976-101E	1976 Oct 14.95 210 days 1977 May 12	Sphere- cone 66	1.32 long 0.94 dia	1976 Oct 15.0	25.9	644.7	24717	200	36478	0.734	-
	Marisat 3 apogee motor	1976-101F	1976 Oct 14.95 >million years	- 293 full	-		Orbit similar to 1976-101A						
D	Fragments	1976-101C,D											
	Meteor 26	1976-102A	1976 Oct 15.96 500 years	Cylinder + 2 vanes 2200?	5 long? 1.5 dia?	1976 Oct 22.9	81.27	102.48	7253	857	892	0.002	263
	Meteor 26 rocket	1976-102B	1976 Oct 15.96 400 years	Cylinder 1440	3.8 long 2.6 dia	1976 Oct 17.8	81.27	102.59	7258	836	924	0.006	192

* Soyuz 23 rendezvous with Salyut 5 about 1976 Oct 15.8, but failed to dock.

	Name		Launch date, lifetime and descent date	Shape and weight (kg)	Size (m)	Date of orbital determination	Orbital inclination (deg)	Nodal period (min)	Semi major axis (km)	Perigee height (km)	Apogee height (km)	Orbital eccentricity	Argument of perigee (deg)
	Cosmos 860*	1976-103A	1976 Oct 17.76 600 years	Cone-cylinder	6 long? 2 dia?	1976 Oct 23.4 1976 Nov 12.9	65.04 64.70	89.66 104.33	6637 7342	252 919	265 1008	0.001 0.006	274 350
D	Cosmos 860 rocket	1976-103D	1976 Oct 17.76 29 days 1976 Nov 15	Cylinder 1500?	8 long? 2.5 dia?	1976 Nov 12.0	65.03	89.49	6628	243	257	0.001	200
D	Cosmos 860 platform	1976-103B	1976 Oct 17.76 73 days 1976 Dec 29	Irregular	-	1976 Nov 12.1	65.04	89.54	6631	244	261	0.001	286
D	Fragments	1976-103C,E											
	Cosmos 861**	1976-104A	1976 Oct 21.71 600 years	Cone-cylinder	6 long? 2 dia?	1976 Oct 24.4 1977 Jan 24.5	64.96 64.86	89.65 104.31	6636 7341	251 921	265 1005	0.001 0.006	269 210
D	Cosmos 861 rocket	1976-104D	1976 Oct 21.71 65 days 1976 Dec 25	Cylinder 1500?	8 long? 2.5 dia?	1976 Dec 21.9	64.96	89.28	6618	226	253	0.002	264
D	Cosmos 861 platform	1976-104C	1976 Oct 21.71 107 days 1977 Feb 4	Irregular	-	1976 Dec 21.6	64.96	89.55	6631	244	262	0.001	260
D	Fragment	1976-104B											
	Cosmos 862 †	1976-105A	1976 Oct 22.39 disintegrated	Windmill + 6 vanes? 1250?	4.2 long? 1.6 dia?	1976 Oct 22.9 1976 Nov 1.0	62.81 62.81	712.32 718.11	26422 26565	571 598	39516 39775	0.737 0.738	318 -
D	Cosmos 862 launcher rocket	1976-105B	1976 Oct 22.39 66 days 1976 Dec 27	Cylinder 2500?	7.5 long 2.6 dia	1976 Oct 22.5	62.79	92.20	6762	222	545	0.024	116

* 1976-103B and 1976-103D attached to 1976-103A until orbit change on 1976 Nov 10.78
** 1976-104C and 1976-104D attached to 1976-104A until orbit change about 1976 Dec 20
† Disintegrated on 1977 Mar 15

1976-105 continued on page 467

	Name		Launch date, lifetime and descent date	Shape and weight (kg)	Size (m)	Date of orbital determination	Orbital inclination (deg)	Nodal period (min)	Semi major axis (km)	Perigee height (km)	Apogee height (km)	Orbital eccentricity	Argument of perigee (deg)
D	Cosmos 862 launcher	1976-105C	1976 Oct 22.39 49 days 1976 Dec 10	Irregular	-	1976 Oct 22.5	62.84	92.47	6775	198	596	0.029	114
	Cosmos rocket	1976-105D	1976 Oct 22.39 100 years?	Cylinder 440	2.0 long 2.0 dia	1976 Oct 29.8	62.70	711.90	26411	606	39460	0.736	318
	Fragments	1976-105E-P											
D R	Cosmos 863	1976-106A	1976 Oct 25.61 10.68 days 1976 Nov 5.29	Sphere-cylinder 6300?	6.5 long? 2.4 dia	1976 Oct 26.3 1976 Oct 26.7	62.81 62.81	89.74 89.40	6641 6624	178 170	348 322	0.013 0.011	85 75
D	Cosmos 863 rocket	1976-106B	1976 Oct 25.61 7 days 1976 Nov 1	Cylinder 2500?	7.5 long 2.6 dia	1976 Oct 26.3	62.80	89.60	6634	178	334	0.012	82
D	Cosmos 863 engine	1976-106E	1976 Oct 25.61 17 days 1976 Nov 11	Cone 600? full	1.5 long? 2 dia?	1976 Nov 6.0	62.80	89.28	6618	167	313	0.011	71
D	Fragments	1976-106C,D,F-K											
	Statsionar - Ekran 1	1976-107A	1976 Oct 26.62 >million years	Cylinder + plate	-	1976 Oct 27.5	0.2	1437	42228	35850	35850	0	- *
D	Ekran 1 launcher rocket	1976-107B	1976 Oct 26.62 3 days 1976 Oct 29	Cylinder 4000?	12 long? 4 dia	1976 Oct 27.0	51.48	88.18	6565	178	196	0.001	305
D	Ekran 1 launcher	1976-107C	1976 Oct 26.62 1 day 1976 Oct 27	Irregular	-	1976 Oct 27.0	51.47	88.08	6560	181	183	0.0002	16
D	Ekran 1 rocket	1976-107D	1976 Oct 26.62 243 days 1977 Jun 26	Cylinder 1900?	3.9 long? 3.9 dia	1976 Nov 20.9	47.21	626.78	24266	311	35464	0.724	7

* Approximate orbit. Ekran means screen.

1976-107 continued on page 468

	Name		Launch date, lifetime and descent date	Shape and weight (kg)	Size (m)	Date of orbital determination	Orbital inclination (deg)	Nodal period (min)	Semi major axis (km)	Perigee height (km)	Apogee height (km)	Orbital eccentricity	Argument of perigee (deg)
	Ekran 1 apogee motor	1976-107F	1976 Oct 26.62 > million years	Cylinder 440	2.0 long 2.0 dia	Synchronous orbit similar to 1976-107A							
D	Fragment	1976-107E											
	Cosmos 864	1976-108A	1976 Oct 29.53 1200 years	Cylinder 700?	1.4 long 2.0 dia	1976 Oct 30.2	82.94	104.90	7367	966	1011	0.003	292
	Cosmos 864 rocket	1976-108B	1976 Oct 29.53 600 years	Cylinder 2200?	7.4 long 2.4 dia	1976 Oct 30.1	82.94	104.79	7361	966	1000	0.002	291
D R	Cosmos 865	1976-109A	1976 Nov 1.48 11.78 days 1976 Nov 13.26	Sphere-cylinder 5700?	5.0 long 2.4 dia	1976 Nov 2.4	72.88	89.81	6643	203	326	0.009	67
D	Cosmos 865 rocket	1976-109B	1976 Nov 1.48 13 days 1976 Nov 14	Cylinder 2500?	7.5 long 2.6 dia	1976 Nov 1.7	72.87	89.69	6637	206	311	0.008	66
D	Fragments	1976-109C-H											
D R	Cosmos 866	1976-110A	1976 Nov 11.45 11.9 days 1976 Nov 23.3	Sphere-cylinder 6300?	6.5 long? 2.4 dia	1976 Nov 11.9 1976 Nov 12.7	64.98 64.98	89.16 89.45	6612 6626	180 177	287 318	0.008 0.011	50 52
D	Cosmos 866 rocket	1976-110B	1976 Nov 11.45 3 days 1976 Nov 14	Cylinder 2500?	7.5 long 2.6 dia	1976 Nov 12.1	64.98	88.93	6600	168	276	0.008	50
D	Cosmos 866 engine*	1976-110F	1976 Nov 11.45 13 days 1976 Nov 24	Cone 600? full	1.5 long? 2 dia?	1976 Nov 23.1	64.97	88.06	6557	124	233	0.008	63
D	Fragments	1976-110C-E,G											
D R	Cosmos 867	1976-111A	1976 Nov 23.69 12.64 days 1976 Dec 6.33	Sphere-cylinder 6300?	6.5 long? 2.4 dia	1976 Nov 23.8 1976 Nov 26.4	62.83 62.83	91.06 92.07	6704 6755	250 352	402 401	0.011 0.004	114 88

1976-111 continued on page 469

* Jettisoned from 1976-110A about 1976 Nov 22.

	Name		Launch date, lifetime and descent date	Shape and weight (kg)	Size (m)	Date of orbital determination	Orbital inclination (deg)	Nodal period (min)	Semi major axis (km)	Perigee height (km)	Apogee height (km)	Orbital eccentricity	Argument of perigee (deg)
D	Cosmos 867 rocket	1976-111B	1976 Nov 23.69 71 days 1977 Feb 1	Cylinder 2500?	7.5 long 2.6 dia	1976 Nov 30.6	62.80	90.83	6692	249	383	0.010	107
D	Cosmos 867 engine*	1976-111F	1976 Nov 23.69 410 days 1978 Jan 7	Cone 600? full	1.5 long? 2 dia?	1976 Dec 6.4	62.82	92.02	6752	351	397	0.003	95
D	Fragments	1976-111C-E,G-K											
D	Prognoz 5	1976-112A	1976 Nov 25.17 2½ years?	Spheroid + 4 vanes 930	1.8 dia?	1977 Jan 1.0	65.2	5728	106047	777	198560	0.933	-
D	Prognoz 5 launcher rocket	1976-112B	1976 Nov 25.17 59 days 1977 Jan 23	Cylinder 2500?	7.5 long 2.6 dia	1976 Nov 25.5	64.97	91.48	6726	231	464	0.017	66
D	Prognoz 5 launcher	1976-112C	1976 Nov 25.17 41 days 1977 Jan 5	Irregular	-	1976 Nov 26.5	65.00	91.50	6727	205	492	0.021	65
D	Prognoz 5 rocket	1976-112E	1976 Nov 25.17 2½ years?	Cylinder 440	2.0 long 2.0 dia	Orbit similar to 1976-112A							
D	Fragment	1976-112D											
D	Cosmos 868	1976-113A	1976 Nov 26.61 589 days 1978 Jul 8	Cylinder?	-	1976 Nov 26.6 1976 Nov 26.9	65.05 65.04	89.94 93.29	6651 6811	110 422	436 444	0.025 0.002	51 287
D	Cosmos 868 rocket	1976-113B	1976 Nov 26.61 1 day 1976 Nov 27	Cylinder 1500?	8 long? 2.5 dia?	1976 Nov 27.1	64.98	88.64	6586	113	303	0.014	53

* Jettisoned from 1976-111A about 1976 Dec 5.

	Name		Launch date, lifetime and descent date	Shape and weight (kg)	Size (m)	Date of orbital determination	Orbital inclination (deg)	Nodal period (min)	Semi major axis (km)	Perigee height (km)	Apogee height (km)	Orbital eccentricity	Argument of perigee (deg)
D R	Cosmos 869	1976-114A	1976 Nov 29.67 17.7 days 1976 Dec 17.4	Sphere-cylinder +2 wings 6570?	7.5 long 2.2 dia	1976 Nov 30.0 1976 Dec 10.3 1976 Dec 12.7	51.76 51.78 51.77	89.36 91.06 90.56	6624 6707 6682	198 268 299	293 390 309	0.007 0.009 0.001	89 127 254
D	Cosmos 869 rocket	1976-114B	1976 Nov 29.67 7 days 1976 Dec 6	Cylinder 2500?	7.5 long 2.6 dia	1976 Nov 29.9	51.78	89.24	6618	193	286	0.007	89
D	Cosmos 870	1976-115A	1976 Dec 2.01 1479 days 1980 Dec 20	Cylinder + paddles? 900?	2 long? 1 dia?	1976 Dec 6.6	74.00	95.26	6909	513	548	0.003	352
D	Cosmos 870 rocket	1976-115B	**1976 Dec 2.01** **1549 days** **1981 Feb 28**	Cylinder 2200?	7.4 long 2.4 dia	1976 Dec 4.6	74.00	95.14	6903	501	548	0.003	1
D	Fragments	1976-115C-F											
	Molniya 2R	1976-116A	1976 Dec 2.12 $14\frac{3}{4}$ years	Windmill + 6 vanes 1250?	4.2 long? 1.6 dia	1976 Dec 6.3 1976 Dec 10.3	62.88 62.57	735.70 717.93	26996 26560	637 575	40599 39789	0.740 0.738	288 288
D	Molniya 2R launcher	1976-116B	1976 Dec 2.12 92 days 1977 Mar 4	Irregular	-	1976 Dec 3.1	62.86	92.90	6796	211	624	0.030	119
D	Molniya 2R launcher rocket	1976-116C	1976 Dec 2.12 67 days 1977 Feb 7	Cylinder 2500?	7.5 long 2.6 dia	1976 Dec 3.6	62.86	92.89	6795	214	620	0.030	121
	Molniya 2R rocket	1976-116D	1976 Dec 2.12 $14\frac{3}{4}$ years	Cylinder 440	2.0 long 2.0 dia	1976 Dec 7.2	62.88	732.04	26907	629	40429	0.740	288
D	Fragments	1976-116E,F											

	Name		Launch date, lifetime and descent date	Shape and weight (kg)	Size (m)	Date of orbital determination	Orbital inclination (deg)	Nodal period (min)	Semi major axis (km)	Perigee height (km)	Apogee height (km)	Orbital eccentricity	Argument of perigee (deg)
D r	China 7*	1976-117A	1976 Dec 7.19 26 days 1977 Jan 2	Cylinder? 1st 3 days 3600, then 1200?	-	1976 Dec 10.3 1976 Dec 11.3	59.45 59.45	91.01 90.92	6704 6700	172 174	479 469	0.023 0.022	158 158
D	China 7 rocket	1976-117B	1976 Dec 7.19 22 days 1976 Dec 29	Cylinder	8 long 3 dia	1976 Dec 7.8	59.45	91.00	6703	169	481	0.023	154
D	Fragment	1976-117C											
	Cosmos 871	1976-118A	1976 Dec 7.43 8000 years	Spheroid 40?	1.0 long? 0.8 dia?	1976 Dec 12.4	74.03	114.74	7821	1420	1466	0.003	94
	Cosmos 872	1976-118B	1976 Dec 7.43 7000 years	Spheroid 40?	1.0 long? 0.8 dia?	1976 Dec 10.7	74.03	114.53	7812	1401	1466	0.004	84
	Cosmos 873	1976-118C	1976 Dec 7.43 10000 years	Spheroid 40?	1.0 long? 0.8 dia?	1976 Dec 12.5	74.03	115.60	7860	1466	1498	0.002	255
	Cosmos 874	1976-118D	1976 Dec 7.43 10000 years	Spheroid 40?	1.0 long? 0.8 dia?	1976 Dec 11.0	74.03	115.82	7870	1466	1518	0.003	270
	Cosmos 875	1976-118E	1976 Dec 7.43 9000 years	Spheroid 40?	1.0 long? 0.8 dia?	1976 Dec 12.4	74.03	114.95	7831	1439	1466	0.002	87
	Cosmos 876	1976-118F	1976 Dec 7.43 10000 years	Spheroid 40?	1.0 long? 0.8 dia?	1976 Dec 11.7	74.03	116.07	7882	1466	1541	0.005	261

* A Capsule returned to Earth - possibly about 1976 Dec 10.3,
although object 1976-117C (decayed Dec 9) might be the Capsule.

1976-118 continued on page 472

	Name		Launch date, lifetime and descent date	Shape and weight (kg)	Size (m)	Date of orbital determination	Orbital inclination (deg)	Nodal period (min)	Semi major axis (km)	Perigee height (km)	Apogee height (km)	Orbital eccentricity	Argument of perigee (deg)
	Cosmos 877	1976-118G	1976 Dec 7.43 9000 years	Spheroid 40?	1.0 long? 0.8 dia?	1976 Dec 11.7	74.03	115.15	7840	1457	1466	0.001	100
	Cosmos 878	1976-118H	1976 Dec 7.43 10000 years	Spheroid 40?	1.0 long? 0.8 dia?	1976 Dec 11.0	74.03	115.37	7850	1466	1477	0.001	274
	Cosmos 871 rocket	1976-118J	1976 Dec 7.43 20000 years	Cylinder 2200?	7.4 long 2.4 dia	1976 Dec 12.2	74.01	117.71	7956	1463	1692	0.014	265
D R	Cosmos 879	1976-119A	1976 Dec 9.42 12.83 days 1976 Dec 22.25	Sphere- cylinder 5700?	5.0 long 2.4 dia	1976 Dec 9.5	81.37	88.90	6597	213	225	0.001	336
D	Cosmos 879 rocket	1976-119B	1976 Dec 9.42 3 days 1976 Dec 12	Cylinder 2500?	7.5 long 2.6 dia	1976 Dec 10.3	81.38	88.66	6585	196	218	0.002	301
D	Fragments	1976-119C,D											
D	Cosmos 880	1976-120A *	1976 Dec 9.84 disintegrated	Cylinder?	4 long? 2 dia?	1976 Dec 12.4	65.85	96.44	6967	560	617	0.004	359
	Cosmos 880 rocket	1976-120B	1976 Dec 9.84 10 years	Cylinder 2200?	7.4 long 2.4 dia	1976 Dec 13.1	65.85	96.33	6961	551	615	0.005	14
36d	Fragments	1976-120C-BC											

* In the USA, the 1976-120A designation was subsequently transferred to a fragment which decayed on 1979 Oct 8

	Name		Launch date, lifetime and descent date	Shape and weight (kg)	Size (m)	Date of orbital determination	Orbital inclination (deg)	Nodal period (min)	Semi major axis (km)	Perigee height (km)	Apogee height (km)	Orbital eccentricity	Argument of perigee (deg)
D R?	Cosmos 881*	1976-121A	1976 Dec 15.07 <0.93 day 1976 Dec 15	-	-	1976 Dec 15.4	51.60	88.75	6594	198	233	0.003	151
D R?	Cosmos 882*	1976-121B	1976 Dec 15.07 <0.93 day 1976 Dec 15	-	-	1976 Dec 15.4	51.60	88.46	6579	189	213	0.002	204
D	Cosmos 881 rocket	1976-121C	1976 Dec 15.07 5 days 1976 Dec 20	Cylinder 4000?	12 long? 4 dia	1976 Dec 15.4	51.60	88.46	6579	189	213	0.002	204
D	Fragments	1976-121D-F											
	Cosmos 883	1976-122A	1976 Dec 15.58 1200 years	Cylinder 700?	1.3 long? 1.9 dia?	1976 Dec 16.2	82.95	104.86	7365	961	1012	0.003	291
	Cosmos 883 rocket	1976-122B	1976 Dec 15.58 600 years	Cylinder 2200?	7.4 long 2.4 dia	1976 Dec 17.0	82.95	104.74	7359	961	1000	0.003	287
D R	Cosmos 884	1976-123A	1976 Dec 17.40 11.87 days 1976 Dec 29.27	Sphere-cylinder 6300?	6.5 long? 2.4 dia	1976 Dec 17.7 1976 Dec 18.6	65.05 65.01	89.63 89.34	6635 6621	169 166	345 319	0.013 0.012	67 74
D	Cosmos 884 rocket	1976-123B	1976 Dec 17.40 6 days 1976 Dec 23	Cylinder 2500?	7.5 long 2.6 dia	1976 Dec 17.8	65.02	89.50	6629	171	330	0.012	67
D	Cosmos 884 engine	1976-123F	1976 Dec 17.40 17 days 1977 Jan 3	Cone 600? full	1.5 long? 2 dia?	1976 Dec 30.6	65.05	88.98	6603	159	291	0.010	65
D	Fragments	1976-123C-E											

* Probably manned-related

	Name		Launch date, lifetime and descent date	Shape and weight (kg)	Size (m)	Date of orbital determination	Orbital inclination (deg)	Nodal period (min)	Semi major axis (km)	Perigee height (km)	Apogee height (km)	Orbital eccentricity	Argument of perigee (deg)
D	Cosmos 885 †	1976-124A	1976 Dec 17.50 1031 days 1979 Oct 14	Cylinder?	4 long? 2 dia?	1976 Dec 20.6	65.84	94.40	6868	467	512	0.003	341
D	Cosmos 885 rocket	1976-124B	1976 Dec 17.50 992 days 1979 Sep 5	Cylinder 2200?	7.4 long 2.4 dia	1976 Dec 20.6	65.85	94.26	6861	457	508	0.004	353
D	Fragments	1976-124C-U											
D	[Titan 3D]*	1976-125A	1976 Dec 19.77 770 days 1979 Jan 28	Cylinder 13300? full	15 long 3.0 dia	1976 Dec 22.8 1976 Dec 23.8 1977 Mar 27.5	96.95 96.94 96.93	92.37 93.37 92.51	6768 6816 6775	247 341 264	533 535 530	0.021 0.014 0.020	159 149 164
D	Capsule?	1976-125B	1976 Dec 19.77 141 days 1977 May 9	-	-	1976 Dec 21.1	96.93	92.26	6763	248	521	0.020	165
D	Titan 3D rocket	1976-125C	1976 Dec 19.77 66 days 1977 Feb 23	Cylinder 1900	6 long 3.0 dia	1977 Jan 1.0	96.94	91.56	6728	254	445	0.014	-
D	Capsule?	1976-125D	1976 Dec 19.77 52 days 1977 Feb 9	-	-	1976 Dec 21.2	97.01	92.46	6773	256	533	0.021	161
	Cosmos 886**	1976-126A	1976 Dec 27.53 disintegrated	Cylinder?	4 long? 2 dia?	1976 Dec 27.6 1976 Dec 27.8	65.85 65.84	102.96 114.79	7276 7823	531 594	1265 2295	0.050 0.109	81 57
D	Cosmos 886 rocket	1976-126B	1976 Dec 27.53 3 days 1976 Dec 30	Cylinder 1500?	8 long? 2.5 dia?	1976 Dec 29.8	62.74	91.56	6730	132	571	0.033	50
8d	Fragments	1976-126C-BL											

† Cosmos 885 disintegrated during Mar-Apr 1977
* Titan 3D manoeuvred on 1976 Dec 23; 1977 Mar 25 and during 1978 June
** Cosmos 886 passed close to Cosmos 880 on 1976 Dec 27.6 and exploded.

	Name		Launch date, lifetime and descent date	Shape and weight (kg)	Size (m)	Date of orbital determination	Orbital inclination (deg)	Nodal period (min)	Semi major axis (km)	Perigee height (km)	Apogee height (km)	Orbital eccentricity	Argument of perigee (deg)
	Molniya 3F	1976-127A	1976 Dec 28.28 10 years	Windmill + 6 vanes 1500?	4.2 long? 1.6 dia	1977 Jan 3.9	62.81	716.97	26537	544	39773	0.739	288
D	Molniya 3F launcher	1976-127B	1976 Dec 28.28 78 days 1977 Mar 16	Irregular	-	1977 Jan 3.9	62.83	92.44	6773	217	573	0.026	121
D	Molniya 3F launcher rocket	1976-127C	1976 Dec 28.28 49 days 1977 Feb 15	Cylinder 2500?	7.5 long 2.6 dia	1977 Jan 4.9	62.84	92.30	6766	192	584	0.029	118
	Molniya 3F rocket	1976-127E	1976 Dec 28.28 10 years	Cylinder 440	2.0 long 2.0 dia	1977 Jan 3.4	62.90	732.48	26918	613	40466	0.740	288
D	Fragment	1976-127D											
	Cosmos 887	1976-128A	1976 Dec 28.32 1200 years	Cylinder 700?	1.3 long? 1.9 dia?	1977 Jan 5.0	82.94	104.84	7364	954	1018	0.004	277
	Cosmos 887 rocket	1976-128B	1976 Dec 28.32 600 years	Cylinder 2200?	7.4 long 2.4 dia	1977 Jan 1.1	82.94	104.72	7359	953	1008	0.004	286

	Name		Launch date, lifetime and descent date	Shape and weight (kg)	Size (m)	Date of orbital determination	Orbital inclination (deg)	Nodal period (min)	Semi major axis (km)	Perigee height (km)	Apogee height (km)	Orbital eccentricity	Argument of perigee (deg)
D R	Cosmos 888	1977-01A	1977 Jan 6.41 12.85 days 1977 Jan 19.26	Sphere-cylinder 6300?	6.5 long? 2.4 dia	1977 Jan 6.6 1977 Jan 16.7	64.97 64.97	89.45 89.40	6626 6623	170 168	325 322	0.012 0.012	68 63
D	Cosmos 888 rocket	1977-01B	1977 Jan 6.41 5 days 1977 Jan 11	Cylinder 2500?	7.5 long 2.6 dia	1977 Jan 6.8	64.99	89.30	6618	168	312	0.011	68
D	Cosmos 888 engine	1977-01D	1977 Jan 6.41 17 days 1977 Jan 23	Cone 600? full	1.5 long? 2 dia?	1977 Jan 19	65.0	89.48	6609	177	285	0.008	-
D	Fragments	1977-01C,E,F											
	Meteor 2-02	1977-02A	1977 Jan 6.97 500 years	Cylinder + 2 vanes 2750?	5 long? 1.5 dia?	1977 Jan 10.6	81.27	102.97	7276	890	906	0.001	56
	Meteor 2-02 rocket	1977-02B	1977 Jan 6.97 400 years	Cylinder 1440	3.8 long 2.6 dia	1977 Jan 13.5	81.28	103.05	7280	862	942	0.005	154
	Fragments	1977-02C, D											
D R	Cosmos 889	1977-03A	1977 Jan 20.36 11.92 days 1977 Feb 1.28	Sphere-cylinder 5700?	5.0 long 2.4 dia	1977 Jan 21.4	71.38	89.84	6644	202	329	0.010	60
D	Cosmos 889 rocket	1977-03B	1977 Jan 20.36 12 days 1977 Feb 1	Cylinder 2500?	7.5 long 2.6 dia	1977 Jan 21.5	71.38	89.56	6630	203	300	0.007	57

	Name		Launch date, lifetime and descent date	Shape and weight (kg)	Size (m)	Date of orbital determination	Orbital inclination (deg)	Nodal period (min)	Semi major axis (km)	Perigee height (km)	Apogee height (km)	Orbital eccentricity	Argument of perigee (deg)
	Cosmos 890	1977-04A	1977 Jan 20.84 1200 years	Cylinder 700?	1.3 long? 1.9 dia?	1977 Jan 21.1	82.96	105.17	7380	983	1020	0.003	265
	Cosmos 890 rocket	1977-04B	1977 Jan 20.84 600 years	Cylinder 2200?	7.4 long 2.4 dia	1977 Jan 23.2	82.96	105.05	7374	982	1009	0.002	249
T	NATO 3B	1977-05A	1977 Jan 28.03 > million years	Cylinder 701 full 340 empty	2.23 long 2.2 dia	1977 Jan 28.1 1977 Jul 1.0	27.00 2.6	633.41 1436.2	24431 42167	184 35784	35922 35794	0.731 0.0001	183 -
	NATO 3B second stage	1977-05B	1977 Jan 28.03 200 years	Cylinder + annulus 350?	6.4 long 1.52 and 2.44 dia	1977 Jan 30.8	28.01	104.09	7338	618	1301	0.047	85
D	NATO 3B third stage	1977-05C	1977 Jan 28.03 1470 days 1981 Feb 6	Sphere- cone 66	1.32 long 0.94 dia	1977 Feb 21.3	26.94	627.06	24269	145	35637	0.731	198*
	Fragments	1977-05D-F											
D	Cosmos 891	1977-06A	1977 Feb 2.52 1463 days 1981 Feb 4	Cylinder?	4 long? 2 dia?	1977 Feb 10.6	65.84	94.49	6873	473	516	0.003	348
D	Cosmos 891 rocket	1977-06B	1977 Feb 2.52 1049 days 1979 Dec 18	Cylinder 2200?	7.4 long 2.4 dia	1977 Feb 14.9	65.84	94.41	6868	464	516	0.004	354

* Approximate orbit

	Name		Launch date, lifetime and descent date	Shape and weight (kg)	Size (m)	Date of orbital determination	Orbital inclination (deg)	Nodal period (min)	Semi major axis (km)	Perigee height (km)	Apogee height (km)	Orbital eccentricity	Argument of perigee (deg)
T	IMEWS 7 [Titan 3C]	1977-07A	1977 Feb 6 > million years	-	-	1977 Feb 7.5 1977 Mar 1.0	26.3 0.5	633.0 1433.3	24445 42118	295 35620	35840 35860	0.727 0.003	180 † - †
D	Titan 3C second stage	1977-07B	1977 Feb 6 < 1 day 1977 Feb 6	Cylinder 1900	6 long 3.0 dia		Orbit similar to 1976-59B						
	Transtage	1977-07C	1977 Feb 6 > million years	Cylinder 1500?	6 long? 3.0 dia	1980 Jun 30	2.5	1436.1	42164	35705	35867	0.002	-
	Fragment	1977-07D											
D 2M R	Soyuz 24*	1977-08A	1977 Feb 7.68 17.7 days 1977 Feb 25.4	Sphere- cylinder 6570?	7.5 long 2.3 dia	1977 Feb 7.9 1977 Feb 8.4 1977 Feb 12.3	51.62 51.57 51.58	89.40 89.25 89.53	6626 6619 6633	173 217 251	323 264 258	0.011 0.004 0.001	99 60 89
D	Soyuz 24 rocket	1977-08B	1977 Feb 7.68 5 days 1977 Feb 12	Cylinder 2500?	7.5 long 2.6 dia	1977 Feb 8.7	51.62	89.24	6618	168	312	0.011	99
D	Fragment	1977-08C											
D R	Cosmos 892	1977-09A	1977 Feb 9.48 12.7 days 1977 Feb 22.2	Sphere- cylinder 6300?	6.5 long? 2.4 dia	1977 Feb 10.1 1977 Feb 11.2	72.86 72.86	90.40 89.66	6671 6635	159 171	427 343	0.020 0.013	89 72
D	Cosmos 892 rocket	1977-09B	1977 Feb 9.48 6 days 1977 Feb 15	Cylinder 2500?	7.5 long 2.6 dia	1977 Feb 10.1	72.86	90.22	6662	156	412	0.019	90
D	Cosmos 892 engine**	1977-09D	1977 Feb 9.48 17 days 1977 Feb 26	Cone 600? full	1.5 long? 2 dia?	1977 Feb 22.2	72.86	89.40	6622	166	322	0.012	41
D	Fragments	1977-09C,E-G											

† Approximate orbits.
* Soyuz 24 docked with Salyut 5 about 1977 Feb 8.80; undocked 1977 Feb 25.
** Jettisoned from 1977-09A about 1977 Feb 21.

	Name		Launch date, lifetime and descent date	Shape and weight (kg)	Size (m)	Date of orbital determination	Orbital inclination (deg)	Nodal period (min)	Semi major axis (km)	Perigee height (km)	Apogee height (km)	Orbital eccentricity	Argument of perigee (deg)
	Molniya 2S	1977-10A	1977 Feb 11.63 20 years	Windmill + 6 vanes 1250?	4.2 long? 1.6 dia	1977 Feb 14.2 1977 May 1.0	62.81 62.9	735.35 717.67	26988 26554	464 498	40756 39853	0.746 0.741	280 -
D	Molniya 2S launcher	1977-10B	1977 Feb 11.63 30 days 1977 Mar 13	Irregular	-	1977 Feb 14.0	62.84	91.02	6704	206	445	0.018	129
D	Molniya 2S launcher rocket	1977-10C	1977 Feb 11.63 19 days 1977 Mar 2	Cylinder 2500?	7.5 long 2.6 dia	1977 Feb 13.5	62.82	91.03	6704	184	468	0.021	123
	Molniya 2S rocket	1977-10E	1977 Feb 11.63 20 years	Cylinder 440	2.0 long 2.0 dia	1977 Feb 23.4	62.79	731.23	26887	493	40525	0.745	280
D	Fragment	1977-10D											
	Cosmos 893 (+ rocket)	1977-11A	1977 Feb 15.46 10 years	Cylinder + ellipsoid 2750?	9.2 long? 1.5 and 2.4 dia	1977 Feb 20.6	74.00	105.25	7384	332	1680	0.091	60
D	Fragments	1977-11B,C											
T	Tansei 3* [Mu-3H]	1977-12A	1977 Feb 19.22 2000 years	Polyhedral cylinder? 134	1 long? 1 dia?	1977 Feb 19.3	65.76	134.30	8687	796	3821	0.174	337
D	Tansei 3 rocket	1977-12B	1977 Feb 19.22 760 days 1979 Mar 21	Sphere-cone? 230?	2.33 long? 1.14 dia?	1977 Feb 21.5	65.50	95.87	6939	329	793	0.033	152
1d	Fragments	1977-12C-J											

* Japanese spacecraft launched by improved version of Mu-3C booster.

	Name		Launch date, lifetime and descent date	Shape and weight (kg)	Size (m)	Date of orbital determination	Orbital inclination (deg)	Nodal period (min)	Semi major axis (km)	Perigee height (km)	Apogee height (km)	Orbital eccentricity	Argument of perigee (deg)
	Cosmos 894	1977-13A	1977 Feb 21.72 1200 years	Cylinder 700?	1.3 long? 1.9 dia?	1977 Feb 24.5	82.94	105.00	7371	972	1014	0.003	273
	Cosmos 894 rocket	1977-13B	1977 Feb 21.72 600 years	Cylinder 2200?	7.4 long 2.4 dia	1977 Feb 26.7	82.93	104.89	7366	971	1004	0.002	260
T	Kiku 2* (ETS 2) [Nu]	1977-14A	1977 Feb 23.37 > million years	Polyhedral cylinder? 130 empty	1.4 long 1.8 dia	1977 Feb 23.6 1977 Mar 5.0	23.95 0.3	627 1435.9	24259 42161	186 35775	35576 35791	0.729 0.0002	176 -
	Kiku 2 rocket	1977-14B	1977 Feb 23.37 15 years	Cylinder 66?	1.74 long 1.65 dia	1977 May 1.0	23.5	627.5	24277	221	35577	0.728	-
	Cosmos 895	1977-15A	1977 Feb 26.89 30 years	Cylinder + 2 vanes? 2500?	5 long? 1.5 dia?	1977 Feb 27.5	81.19	97.19	7001	611	635	0.002	192
	Cosmos 895 rocket	1977-15B	1977 Feb 26.89 30 years	Cylinder 1440	3.8 long 2.6 dia	1977 Feb 27.6	81.20	97.27	7005	574	680	0.008	181
D R	Cosmos 896	1977-16A	1977 Mar 3.44 12.8 days 1977 Mar 16.2	Sphere-cylinder 6300?	6.5 long? 2.4 dia	1977 Mar 4.4 1977 Mar 6.9	72.87 72.89	88.56 89.72	6580 6638	195 177	209 343	0.001 0.012	307 226
D	Cosmos 896 rocket	1977-16B	1977 Mar 3.44 1 day 1977 Mar 4	Cylinder 2500?	7.5 long 2.6 dia	1977 Mar 3.7	72.88	88.24	6564	169	203	0.003	295

* Japanese Engineering Test Satellite; first orbit is approximate.

1977-16 continued on page 481

	Name		Launch date, lifetime and descent date	Shape and weight (kg)	Size (m)	Date of orbital determination	Orbital inclination (deg)	Nodal period (min)	Semi major axis (km)	Perigee height (km)	Apogee height (km)	Orbital eccentricity	Argument of perigee (deg)
D	Cosmos 896 engine	1977-16C	1977 Mar 3.44 22 days 1977 Mar 25	Cone 600? full	1.5 long? 2 dia?	1977 Mar 15.9	72.92	89.50	6627	160	338	0.013	197
D	Fragments	1977-16D-H											
D R	Cosmos 897	1977-17A	1977 Mar 10.46 12.79 days 1977 Mar 23.25	Sphere-cylinder 5900?	5.9 long? 2.4 dia	1977 Mar 11.3	72.85	89.63	6634	171	340	0.013	83
D	Cosmos 897 rocket	1977-17B	1977 Mar 10.46 4.71 days 1977 Mar 15.17	Cylinder 2500?	7.5 long 2.6 dia	1977 Mar 11.2	72.85	89.41	6623	169	320	0.011	80
D	Capsule?†	1977-17C	1977 Mar 10.46 21 days 1977 Mar 31	Ellipsoid 200?	0.9 long 1.9 dia	1977 Mar 23.3	72.84	90.17	6661	172	393	0.017	56
D	Fragments	1977-17D-G											
T	Palapa 2*	1977-18A	1977 Mar 10.97 > million years	Cylinder 574 full 293 empty	1.56 long 1.90 dia	1977 Mar 11.0 1977 May 1.0	24.66 0.1	645.57 1436.1	24743 42165	231 35764	36499 35809	0.733 0.0005	179 -
D	Palapa 2 second stage	1977-18C	1977 Mar 10.97 282 days 1977 Dec 17	Cylinder + annulus 350?	6.4 long 1.52 and 2.44 dia	1977 Mar 11.0	28.66	117.54	7948	188	2953	0.174	82
D	Palapa 2 third stage	1977-18B	1977 Mar 10.97 1872 days 1982 Apr 25	Sphere-cone 66	1.32 long 0.94 dia	1977 Mar 27.3	24.65	640.57	24630	247	36256	0.731	190
D	Fragments	1977-18D,E											

* Indonesian satellite launched by NASA.

† Possibly an engine

	Name		Launch date, lifetime and descent date	Shape and weight (kg)	Size (m)	Date of orbital determination	Orbital inclination (deg)	Nodal period (min)	Semi major axis (km)	Perigee height (km)	Apogee height (km)	Orbital eccentricity	Argument of perigee (deg)
D	[Titan 3B Agena D]	1977-19A	1977 Mar 13.78 74 days 1977 May 26	Cylinder 3000?	8 long? 1.5 dia	1977 Mar 15.5	96.40	89.25	6614	124	348	0.017	144
D	Fragment	1977-19B											
D R	Cosmos 898	1977-20A	1977 Mar 17.36 12.8 days 1977 Mar 30.2	Sphere-cylinder 5900?	5.9 long 2.4 dia	1977 Mar 18.9	81.35	88.99	6601	216	230	0.001	88
D	Cosmos 898 rocket	1977-20B	1977 Mar 17.36 5 days 1977 Mar 22	Cylinder 2500?	7.5 long 2.6 dia	1977 Mar 18.5	81.35	88.87	6595	214	220	0.0005	167
D	Capsule	1977-20D	1977 Mar 17.36 17 days 1977 Apr 3	Ellipsoid 200?	0.9 long 1.9 dia	1977 Mar 28.0	81.34	88.80	6592	210	217	0.0006	62
D	Fragment	1977-20C											
	Molniya 1AM	1977-21A	1977 Mar 24.50 20 years	Windmill + 6 vanes 1000?	3.4 long 1.6 dia	1977 Mar 24.5 1977 Apr 22.1	62.77 62.87	736.36 717.49	27013 26550	458 465	40812 39879	0.747 0.742	280 280
D	Molniya 1AM launcher	1977-21B	1977 Mar 24.50 26 days 1977 Apr 19	Irregular	-	1977 Mar 25.6	62.83	91.01	6703	211	439	0.017	131
D	Molniya 1AM launcher rocket	1977-21C	1977 Mar 24.50 19 days 1977 Apr 12	Cylinder 2500?	7.5 long 2.6 dia	1977 Mar 25.6	62.82	91.07	6706	192	464	0.020	121
	Molniya 1AM rocket	1977-21D	1977 Mar 24.50 20 years	Cylinder 440	2.0 long 2.0 dia	1977 Apr 17.5	62.83	732.89	26928	440	40659	0.747	280

	Name		Launch date, lifetime and descent date	Shape and weight (kg)	Size (m)	Date of orbital determination	Orbital inclination (deg)	Nodal period (min)	Semi major axis (km)	Perigee height (km)	Apogee height (km)	Orbital eccentricity	Argument of perigee (deg)
D	Cosmos 899	1977-22A	1977 Mar 24.93 1305 days 1980 Oct 19	Cylinder + paddles? 900?	2 long? 1 dia?	1977 Mar 25.6	74.05	95.15	6903	503	547	0.003	344
D	Cosmos 899 rocket	1977-22B	1977 Mar 24.93 1372 days 1980 Dec 25	Cylinder 2200?	7.4 long 2.4 dia	1977 Mar 27.6	74.05	95.03	6897	493	545	0.004	347
D	Fragment	1977-22C											
D	Cosmos 900*	1977-23A	1977 Mar 29.96 926 days 1979 Oct 11	- 900?	-	1977 Mar 30.3	82.95	94.43	6868	457	522	0.005	343
D	Cosmos 900 rocket	1977-23B	1977 Mar 29.96 891 days 1979 Sep 6	Cylinder 2200?	7.4 long 2.4 dia	1977 Apr 4.3	82.95	94.31	6862	448	519	0.005	331
	Meteor 27	1977-24A	1977 Apr 5.09 500 years	Cylinder + 2 vanes 2200?	5 long? 1.5 dia?	1977 Apr 6.5	81.25	102.50	7254	854	897	0.003	280
	Meteor 27 rocket	1977-24B	1977 Apr 5.09 400 years	Cylinder 1440	3.8 long 2.6 dia	1977 Apr 6.4	81.26	102.64	7261	842	923	0.006	200
	Fragment	1977-24C											
D	Cosmos 901	1977-25A	1977 Apr 5.44 449 days 1978 June 28	Ellipsoid 400?	1.8 long 1.2 dia	1977 Apr 6.4	70.99	95.54	6923	269	820	0.040	84
D	Cosmos 901 rocket	1977-25B	1977 Apr 5.44 354 days 1978 Mar 25	Cylinder 1500?	8 long 1.65 dia	1977 Apr 7.5	70.98	95.38	6915	273	800	0.038	84

* Included Czech and East German ionospheric and magnetospheric experiments.

	Name		Launch date, lifetime and descent date	Shape and weight (kg)	Size (m)	Date of orbital determination	Orbital inclination (deg)	Nodal period (min)	Semi major axis (km)	Perigee height (km)	Apogee height (km)	Orbital eccentricity	Argument of perigee (deg)
D R	Cosmos 902	1977-26A	1977 Apr 7.38 12.85 days 1977 Apr 20.23	Sphere-cylinder 6300?	6.5 long? 2.4 dia	1977 Apr 8.2 1977 Apr 17.2	81.39 81.39	89.00 89.12	6602 6608	168 172	279 287	0.008 0.009	77 43
D	Cosmos 902 rocket	1977-26B	1977 Apr 7.38 2 days 1977 Apr 9	Cylinder 2500?	7.5 long 2.6 dia	1977 Apr 7.9	81.39	88.78	6591	167	258	0.007	79
D	Cosmos 902 engine*	1977-26E	1977 Apr 7.38 16 days 1977 Apr 23	Cone 600? full	1.5 long? 2 dia?	1977 Apr 19.8	81.39	88.95	6599	159	283	0.009	32
D	Fragments	1977-26C,D											
	Cosmos 903	1977-27A	1977 Apr 11.07 100 years?	Windmill + 6 vanes? 1250?	4.2 long? 1.6 dia?	1977 Apr 12.6 1977 Apr 16.2	62.84 62.83	725.88 717.87	26756 26558	603 597	40153 39763	0.739 0.737	318 318
D	Cosmos 903 launcher	1977-27B	1977 Apr 11.07 56 days 1977 Jun 6	Irregular	-	1977 Apr 12.0	62.82	92.40	6771	213	573	0.027	121
D	Cosmos 903 launcher rocket	1977-27C	1977 Apr 11.07 20 days 1977 May 1	Cylinder 2500?	7.5 long 2.6 dia	1977 Apr 11.5	62.86	92.47	6775	176	617	0.033	116
	Cosmos 903 rocket	1977-27D	1977 Apr 11.07 100 years?	Cylinder 440	2.0 long 2.0 dia	1977 Apr 16.1	62.83	724.02	26710	610	40053	0.738	318
	Fragment	1977-27E											
D R	Cosmos 904	1977-28A	1977 Apr 20.38 13.9 days 1977 May 4.3	Sphere-cylinder 5700?	5.0 long 2.4 dia	1977 Apr 21.5	71.37	89.83	6644	203	328	0.009	60

* 1977-26E jettisoned from 1977-26A about 1977 Apr 19.

1977-28 continued on page 485

	Name		Launch date, lifetime and descent date	Shape and weight (kg)	Size (m)	Date of orbital determination	Orbital inclination (deg)	Nodal period (min)	Semi major axis (km)	Perigee height (km)	Apogee height (km)	Orbital eccentricity	Argument of perigee (deg)
D	Cosmos 904 rocket	1977-28B	1977 Apr 20.38 9.5 days 1977 Apr 29.9	Cylinder 2500?	7.5 long 2.6 dia	1977 Apr 21.9	71.38	89.66	6635	198	316	0.009	56
D	Fragment	1977-28C											
T	ESA-GEOS 1*	1977-29A	1977 Apr 20.43 100000 years	Cylinder 573 full 273 empty	1.10 long 1.62 dia	1977 Apr 21.6 1977 Apr 25.4	26.07 26.25	226.82 720.06	12340 26612	238 2110	11685 38357	0.464 0.681	186 179
D	ESA-GEOS 1 second stage	1977-29B	1977 Apr 20.43 <0.57 day 1977 Apr 20	Cylinder + annulus 350?	6.4 long 1.52 and 2.44 dia	1977 Apr 20.4	28.73	88.49	6578	165	234	0.005	5
	ESA-GEOS 1 third stage	1977-29C	1977 Apr 20.43 12 years	Sphere -cone 66	1.32 long 0.94 dia	1977 Apr 21.6	26.05	226.80	12339	238	11683	0.464	186
D	Fragment	1977-29D											
D R	Cosmos 905	1977-30A	1977 Apr 26.62 29.5 days 1977 May 26.1	Sphere- cylinder 6700?	7 long? 2.4 dia	1977 Apr 26.9 1977 May 4.3 1977 May 10.2	67.12 67.12 67.12	89.60 90.33 89.62	6633 6670 6634	171 170 178	339 413 334	0.013 0.018 0.012	63 61 62
D	Cosmos 905 rocket	1977-30B	1977 Apr 26.62 5 days 1977 May 1	Cylinder 2500?	7.5 long 2.6 dia	1977 Apr 26.9	67.12	89.54	6630	168	336	0.013	63
D	Cosmos 905 engine	1977-30D	1977 Apr 26.62 30 days 1977 May 26	Cone 600? full	1.5 long? 2 dia?	1977 May 26	67.2	88.50	6578	159	241	0.006	-
D	Fragment	1977-30C											

* Intended for 1436 min orbit, but third stage failed to reach nominal transfer orbit apogee. (European Space Agency GEOstationary Satellite.)

	Name		Launch date, lifetime and descent date	Shape and weight (kg)	Size (m)	Date of orbital determination	Orbital inclination (deg)	Nodal period (min)	Semi major axis (km)	Perigee height (km)	Apogee height (km)	Orbital eccentricity	Argument of perigee (deg)
D	Cosmos 906 †	1977-31A	1977 Apr 27.15 1061 days 1980 Mar 23	Cylinder + ellipsoid 2750?	9.2 long? 1.5 and 2.4 dia	1977 Apr 29.2	50.65	94.33	6867	463	515	0.004	33
	Molniya 3G	1977-32A	1977 Apr 28.39 20 years	Windmill + 6 vanes 1500?	4.2 long? 1.6 dia	1977 Apr 29.5 1977 Jul 1.0	62.79 62.8	736.03 717.59	27005 26552	436 506	40817 39841	0.748 0.741	280 -
D	Molniya 3G launcher	1977-32B	1977 Apr 28.39 17 days 1977 May 15	Irregular	-	1977 Apr 28.5	62.78	90.46	6675	210	384	0.013	132
D	Molniya 3G launcher rocket	1977-32C	1977 Apr 28.39 17 days 1977 May 15	Cylinder 2500?	7.5 long 2.6 dia	1977 Apr 29.5	62.81	90.84	6694	190	442	0.019	121
	Molniya 3G rocket	1977-32D	1977 Apr 28.39 20 years	Cylinder 440	2.0 long 2.0 dia		Orbit similar to 1977-32A first orbit						
D R	Cosmos 907	1977-33A	1977 May 5.59 10.6 days 1977 May 16.2	Sphere-cylinder 6300?	6.5 long? 2.4 dia	1977 May 6.5 1977 May 7.2	62.80 62.80	89.93 89.30	6651 6619	181 168	364 313	0.014 0.011	86 67
D	Cosmos 907 rocket	1977-33B	1977 May 5.59 8 days 1977 May 13	Cylinder 2500?	7.5 long 2.6 dia	1977 May 6.5	62.79	89.76	6642	179	349	0.013	83
D	Cosmos 907 engine*	1977-33G	1977 May 5.59 19 days 1977 May 24	Cone 600? full	1.5 long? 2 dia?	1977 May 21.0	62.80	88.80	6594	161	270	0.008	-
D	Fragments	1977-33C-F											

*Jettisoned from 1977-33A about 1977 May 15

† With rocket attached

	Name		Launch date, lifetime and descent date	Shape and weight (kg)	Size (m)	Date of orbital determination	Orbital inclination (deg)	Nodal period (min)	Semi major axis (km)	Perigee height (km)	Apogee height (km)	Orbital eccentricity	Argument of perigee (deg)
T	DSCS 7 [Titan 3C]	1977-34A	1977 May 12.61 > million years	Cylinder + 2 dishes 565	1.83 long 2.74 dia	1977 May 21.3	2.44	1426.7	41978	35438	35762	0.004	353
T	DSCS 8	1977-34B	1977 May 12.61 > million years	Cylinder + 2 dishes 565	1.83 long 2.74 dia	1977 May 24.1	2.43	1436.1	42165	35781	35792	0.0001	212
	Transtage	1977-34C	1977 May 12.61 > million years	Cylinder 1500?	6 long? 3.0 dia	1977 May 12.7 1977 May 13.0 1977 May 24.9	28.60 26.63 2.35	88.89 635.9 1507.1	6606 24492 43545	153 285 35762	302 35943 38572	0.011 0.728 0.032	117 - 188
D	Titan 3C second stage	1977-34D	1977 May 12.61 2 days 1977 May 14	Cylinder 1900	6 long 3.0 dia	1977 May 13.4	28.56	88.43	6583	149	260	0.008	124
D R	Cosmos 908	1977-35A	1977 May 17.43 13.83 days 1977 May 31.26	Sphere-cylinder 6300?	6.5 long? 2.4 dia	1977 May 18.5	51.79	89.06	6609	174	288	0.009	84
D	Cosmos 908 rocket	1977-35B	1977 May 17.43 4 days 1977 May 21	Cylinder 2500?	7.5 long 2.6 dia	1977 May 17.7	51.79	89.02	6607	174	284	0.008	88
D	Fragments *	1977-35C,D											
	Cosmos 909	1977-36A	1977 May 19.69 4000 years	Cylinder?	4 long? 2 dia?	1977 May 22.4	65.87	117.07	7928	990	2109	0.071	165
	Cosmos 909 rocket	1977-36B	1977 May 19.69 2000 years	Cylinder 2200?	7.4 long 2.4 dia	1977 May 28.8	65.87	116.94	7922	987	2100	0.070	162
	Fragment	1977-36C											

*Object 1977-35D is probably Cosmos 908 engine; it decayed 1977 June 3, life 17 days

	Name		Launch date, lifetime and descent date	Shape and weight (kg)	Size (m)	Date of orbital determination	Orbital inclination (deg)	Nodal period (min)	Semi major axis (km)	Perigee height (km)	Apogee height (km)	Orbital eccentricity	Argument of perigee (deg)
D	Cosmos 910	1977-37A	1977 May 23.52 0.05 day 1977 May 23.57	Cylinder?	4 long? 2 dia?	1977 May 23.6	65.86?	99.56?	7115?	[-300?]	1774?	0.146?	217?*
D	Cosmos 910 rocket	1977-37B	1977 May 23.52 4 days 1977 May 27	Cylinder 1500?	8 long? 2.5 dia?	1977 May 23.6	65.10	90.56	6681	141	465	0.024	46
D	Fragment	1977-37C											
T	[Atlas Agena D]	1977-38A	1977 May 23.6? >million years	Cylinder 700? full 350? empty	1.7 long? 1.4 dia?	1977 May 24.0 1977 Jun 1.0	28.2 0.2	733.2 1435.1	26975 42145	191 35679	41002 35855	0.756 0.002	-** -**
	Agena D rocket	1977-38B	1977 May 23.6? 10 years?	Cylinder 700	6 long? 1.5 dia	1977 May 24.0	28.2	733.2	26975	191	41002	0.756	-**
	Cosmos 911	1977-39A	1977 May 25.46 1200 years	Cylinder 700?	1.3 long? 1.9 dia?	1977 May 26.1	82.95	104.87	7365	970	1004	0.002	278
	Cosmos 911 rocket	1977-39B	1977 May 25.46 600 years	Cylinder 2200?	7.4 long 2.4 dia	1977 May 28.9	82.95	104.73	7358	966	994	0.002	245
D R	Cosmos 912	1977-40A	1977 May 26.30 12.8 days 1977 Jun 8.1	Sphere-cylinder 5900?	5.9 long 2.4 dia	1977 May 26.8	81.35	89.00	6602	217	231	0.001	48
D	Cosmos 912 rocket	1977-40B	1977 May 26.30 5 days 1977 May 31	Cylinder 2500?	7.5 long 2.6 dia	1977 May 26.9	81.37	88.89	6597	210	227	0.001	21
D	Capsule	1977-40E	1977 May 26.30 23 days 1977 Jun 18	Ellipsoid 200?	0.9 long 1.9 dia	1977 Jun 12.0	81.40	88.80	6592	207	221	0.001	-
D	Fragments	1977-40C,D											

*Orbit unconfirmed. Cosmos 910 was probably intended to pass close to Cosmos 909 about 1977 May 23.55; it then re-entered. Since it failed to complete one revolution, it is, strictly, not a satellite. **Approximate orbit.

	Name		Launch date, lifetime and descent date	Shape and weight (kg)	Size (m)	Date of orbital determination	Orbital inclination (deg)	Nodal period (min)	Semi major axis (km)	Perigee height (km)	Apogee height (km)	Orbital eccentricity	Argument of perigee (deg)
T	Intelsat 4A F-4	1977-41A	1977 May 26.91 >million years	Cylinder 1500 full 795 empty	2.82 long 2.39 dia	1977 May 26.9 1977 May 29.9 1977 Dec 31.9	21.84 0.28 0.1	640.23 1424.2 1436.2	24607 41929 42166	550 35346 35783	35907 35755 35793	0.718 0.005 0.0001	179 - -
	Intelsat 4A F-4 rocket	1977-41B	1977 May 26.91 6000 years	Cylinder 1815	8.6 long 3.0 dia	1977 Jul 1.0	21.82	649.9	24854	603	36348	0.719	-
D	Cosmos 913	1977-42A	1977 May 30.94 943 days 1979 Dec 29	Octagonal ellipsoid? 550?	1.8 long? 1.5 dia?	1977 May 31.5	74.04	94.55	6874	472	520	0.004	335
D	Cosmos 913 rocket	1977-42B	1977 May 30.94 886 days 1979 Nov 2	Cylinder 2200?	7.4 long 2.4 dia	1977 May 31.6	74.05	94.41	6867	462	516	0.004	345
D	Fragments	1977-42C-X											
D R	Cosmos 914	1977-43A	1977 May 31.32 12.9 days 1977 Jun 13.2	Sphere- cylinder 5900?	5.9 long 2.4 dia	1977 Jun 1.2	65.00	89.59	6633	203	306	0.008	64
D	Cosmos 914 rocket	1977-43B	1977 May 31.32 11 days 1977 Jun 11	Cylinder 2500?	7.5 long 2.6 dia	1977 May 31.4	65.00	89.59	6633	167	342	0.013	54
D	Capsule	1977-43E	1977 May 31.32 32 days 1977 Jul 2	Ellipsoid 200?	0.9 long 1.9 dia	1977 Jul 1.0	64.99	87.59	6534	150	162	0.001	-
D	Fragments	1977-43C,D											

	Name		Launch date, lifetime and descent date	Shape and weight (kg)	Size (m)	Date of orbital determination	Orbital inclination (deg)	Nodal period (min)	Semi major axis (km)	Perigee height (km)	Apogee height (km)	Orbital eccentricity	Argument of perigee (deg)
T	AMS 2 [Thor Burner 2]	1977-44A	1977 Jun 5.13 80 years	Irregular 450	6.40 long 1.68 dia	1977 Jun 5.2	99.20	101.74	7218	811	869	0.004	212
	Burner 2 rocket	1977-44B	1977 Jun 5.13 60 years	Sphere- cone 66	1.32 long 0.94 dia	1977 Jun 7.4	99.21	101.57	7209	799	863	0.004	207
	Fragments	1977-44C,D											
D R	Cosmos 915	1977-45A	1977 Jun 8.59 12.6 days 1977 Jun 21.2	Sphere- cylinder 6300?	6.5 long? 2.4 dia	1977 Jun 9.4 1977 Jun 9.6	62.80 62.81	89.10 89.32	6609 6620	173 177	289 307	0.009 0.010	58 63
D	Cosmos 915 rocket	1977-45B	1977 Jun 8.59 3 days 1977 Jun 11	Cylinder 2500?	7.5 long 2.6 dia	1977 Jun 8.8	62.79	89.02	6605	173	281	0.008	55
D	Cosmos 915 engine*	1977-45C	1977 Jun 8.59 17 days 1977 Jun 25	Cone 600? full	1.5 long? 2 dia?	1977 Jun 20.4	62.78	89.12	6610	172	292	0.009	61
D	Fragments	1977-45D,E											
D R	Cosmos 916	1977-46A	1977 Jun 10.34 11.61 days 1977 Jun 21.95	Sphere- cylinder 5900?	5.9 long 2.4 dia	1977 Jun 11.3	62.80	89.94	6650	246	298	0.004	168
D	Cosmos 916 rocket	1977-46B	1977 Jun 10.34 21 days 1977 Jul 1	Cylinder 2500?	7.5 long 2.6 dia	1977 Jun 11.9	62.79	89.86	6646	241	295	0.004	172
D	Capsule**	1977-46G	1977 Jun 10.34 21 days 1977 Jul 1	Ellipsoid 200?	0.9 long 1.9 dia	1977 Jun 22.7	62.81	89.83	6645	247	286	0.003	173
D	Fragments	1977-46C-F											

* Jettisoned from Cosmos 915 about 1977 Jun 20

** Ejected from Cosmos 916 about 1977 Jun 21

	Name		Launch date, lifetime and descent date	Shape and weight (kg)	Size (m)	Date of orbital determination	Orbital inclination (deg)	Nodal period (min)	Semi major axis (km)	Perigee height (km)	Apogee height (km)	Orbital eccentricity	Argument of perigee (deg)
	Cosmos 917	1977-47A	1977 Jun 16.09 100 years?	Windmill + 6 vanes 1250?	4.2 long? 1.6 dia	1977 Jun 17.1 1977 Jun 21.2	62.91 62.90	726.00 718.74	26759 26580	585 586	40176 39818	0.740 0.738	318 318
D	Cosmos 917 launcher	1977-47B	1977 Jun 16.09 62 days 1977 Aug 17	Irregular	-	1977 Jun 18.1	62.78	92.38	6771	216	569	0.026	119
D	Cosmos 917 launcher rocket	1977-47C	1977 Jun 16.09 28.80 days 1977 Jul 14.89	Cylinder 2500?	7.5 long 2.6 dia	1977 Jun 17.7	62.87	92.25	6764	184	588	0.030	119
	Cosmos 917 rocket	1977-47D	1977 Jun 16.09 100 years?	Cylinder 440	2.0 long 2.0 dia	1977 Jun 21.2	62.85	722.48	26672	587	40001	0.739	318
T	GOES 2	1977-48A	1977 Jun 16.45 > million years	Cylinder + boom. 627 (243 empty)	2.30 long 1.90 dia	1977 Jun 16.5 1977 Jun 21.2	23.85 0.88	652.07 1436.03	24909 42163	201 35266	36861 36304	0.736 0.012	178 144
	GOES 2 second stage	1977-48B	1977 Jun 16.45 200 years	Cylinder + annulus 350?	6.4 long 1.52 and 2.44 dia	1977 Jun 16.5 1977 Jun 18.1	28.40 28.41	94.02 108.83	6849 7557	179 572	763 1786	0.043 0.080	166 78
	GOES 2 third stage	1977-48F	1977 Jun 16.45 10 years?	Sphere- cone 66	1.32 long 0.94 dia	1977 Jun 16.5	23.80	655.93	25007	185	37073	0.738	178
	GOES 2 apogee motor	1977-48G	1977 Jun 16.45 > million years	- 384 full	-		Orbit similar to second 1977-48A orbit						
D	Fragments	1977-48C-E											

	Name		Launch date, lifetime and descent date	Shape and weight (kg)	Size (m)	Date of orbital determination	Orbital inclination (deg)	Nodal period (min)	Semi major axis (km)	Perigee height (km)	Apogee height (km)	Orbital eccentricity	Argument of perigee (deg)
D	Signe 3*	1977-49A	1977 Jun 17.15 733 days 1979 Jun 20	Cylinder 102	0.75 long? 0.7 dia?	1977 Jun 22.5	50.67	94.33	6867	459	519	0.004	35
D	Signe 3 rocket	1977-49B	1977 Jun 17.15 963 days 1980 Feb 5	Cylinder 2200?	7.4 long 2.4 dia	1977 Jun 22.0	50.66	94.23	6863	452	518	0.005	25
D	Cosmos 918**	1977-50A	1977 Jun 17.31 1 day 1977 Jun 18	Cylinder?	4 long? 2 dia?	1977 Jun 17.4	65.11	88.18	6564	128	243	0.009	14
D	Cosmos 918 rocket	1977-50B	1977 Jun 17.31 1 day 1977 Jun 18	Cylinder 1500?	8 long? 2.5 dia?	1977 Jun 17.7	65.07	87.70	6540	124	199	0.006	16
D	Cosmos 919	1977-51A	1977 Jun 18.44 436 days 1978 Aug 28	Ellipsoid 400?	1.8 long 1.2 dia	1977 Jun 19.6	71.02	95.56	6924	269	822	0.040	82
D	Cosmos 919 rocket	1977-51B	1977 Jun 18.44 307 days 1978 Apr 21	Cylinder 1500?	8 long 1.65 dia	1977 Jun 18.7	71.02	95.49	6920	270	814	0.039	84
D	Fragments	1977-51C,D											
D R	Cosmos 920	1977-52A	1977 Jun 22.34 12.84 days 1977 Jul 5.18	Sphere-cylinder 6300?	6.5 long? 2.4 dia	1977 Jun 23.0 1977 Jun 24.2	64.99 64.99	89.65 89.40	6636 6623	173 170	342 320	0.013 0.011	66 65
D	Cosmos 920 rocket	1977-52B	1977 Jun 22.34 5 days 1977 Jun 27	Cylinder 2500?	7.5 long 2.6 dia	1977 Jun 23.0	65.01	89.52	6629	167	335	0.013	67

* French satellite launched by the USSR (solar interplanetary gamma neutron experiment).

** May have passed close to Cosmos 909.

1977-52 concluded on page 493

	Name		Launch date, lifetime and descent date	Shape and weight (kg)	Size (m)	Date of orbital determination	Orbital inclination (deg)	Nodal period (min)	Semi major axis (km)	Perigee height (km)	Apogee height (km)	Orbital eccentricity	Argument of perigee (deg)
D	Cosmos 920 engine*	1977-52C	1977 Jun 22.34 18 days 1977 Jul 10	Cone 600? full	1.5 long? 2 dia?	1977 Jul 4.5	65.00	89.07	6607	168	290	0.009	61
D	Fragments	1977-52D,E											
T L	NTS 2 [Atlas F]	1977-53A	1977 Jun 23.4? 1 million years	Octagon + 2 vanes 431	0.79 long 1.65 dia	1977 Jun 23.4 1977 Jun 27.2 1977 Sep 1.0	63.18 63.28 63.28	351.87 705.18 717.91	16511 26244 26559	160 19545 20181	20106 20187 20181	0.604 0.012 0	158 198 -
	NTS 2 rocket	1977-53B	1977 Jun 23.4? 22 years	Cone-cylinder? 163?	1.85 long? 0.63 to 1.65 dia?	1977 Jun 26.0	63.19	347.67	16377	168	19830	0.600	158
	NTS 2 apogee motor	1977-53C	1977 Jun 23.4? 1 million years	Cylinder -	0.88 long? 0.63 dia?	1977 Jun 26.0	63.32	704.92	26237	19550	20168	0.012	198
	Molniya 1 AN	1977-54A	1977 Jun 24.24 16 years	Windmill+ 6 vanes 1000?	3.4 long 1.6 dia	1977 Jun 28.7 1977 Sep 1.0	62.93 63.1	699.66 717.73	26107 26555	447 457	39011 39896	0.738 0.743	280 -
D	Molniya 1 AN launcher	1977-54B	1977 Jun 24.24 61 days 1977 Aug 24	Irregular	-	1977 Jun 26.1	62.94	91.59	6732	235	472	0.018	62
D	Molniya 1 AN launcher rocket	1977-54C	1977 Jun 24.24 41 days 1977 Aug 4	Cylinder 2500?	7.5 long 2.6 dia	1977 Jun 26.0	62.95	91.48	6726	214	482	0.020	60
	Molniya 1 AN rocket	1977-54D	1977 Jun 24.24 16 years	Cylinder 440	2.0 long 2.0 dia	1977 Jul 18.9	63.00	695.50	26002	459	38789	0.737	280

	Name		Launch date, lifetime and descent date	Shape and weight (kg)	Size (m)	Date of orbital determination	Orbital inclination (deg)	Nodal period (min)	Semi major axis (km)	Perigee height (km)	Apogee height (km)	Orbital eccentricity	Argument of perigee (deg)
	Cosmos 921	1977-55A	1977 Jun 24.44 75 years	- 825?	-	1977 Jun 27.1	75.84	97.96	7038	620	700	0.006	270
	Cosmos 921 rocket	1977-55B	1977 Jun 24.44 70 years	Cylinder 2200?	7.4 long 2.4 dia	1977 Jun 26.5	75.84	97.93	7037	618	700	0.006	272
D	[Titan 3D]	1977-56A	1977 Jun 27.77 179 days 1977 Dec 23	Cylinder 13300? full	15 long 3.0 dia	1977 Jun 28.2	97.02	88.47	6575	155	239	0.006	126
D	Titan 3D rocket	1977-56B	1977 Jun 27.77 2 days 1977 Jun 29	Cylinder 1900	6 long 3.0 dia	1977 Jun 28.2	97.02	88.27	6565	153	221	0.005	118
	Meteor 28*	1977-57A	1977 Jun 29.78 60 years	Cylinder + 2 vanes 2200?	5 long? 1.5 dia	1977 Jul 1.1	97.91	97.46	7014	601	670	0.005	17
	Meteor 28 rocket	1977-57B	1977 Jun 29.78 60 years	Cylinder 1440	3.8 long 2.6 dia	1977 Jul 1.1	97.92	97.53	7018	627	652	0.002	284
D	Fragment	1977-57C											
D R	Cosmos 922	1977-58A	1977 Jun 30.59 12.61 days 1977 Jul 13.20	Sphere- cylinder 5700?	5.0 long 2.4 dia	1977 Jun 30.7	62.81	89.53	6630	205	299	0.007	79
D	Cosmos 922 rocket	1977-58B	1977 Jun 30.59 8 days 1977 Jul 8	Cylinder 2500?	7.5 long 2.6 dia	1977 Jun 30.8	62.80	89.46	6627	203	294	0.007	75
D	Fragments	1977-58C,D											

* First Meteor in Sun-synchronous orbit.

	Name		Launch date, lifetime and descent date	Shape and weight (kg)	Size (m)	Date of orbital determination	Orbital inclination (deg)	Nodal period (min)	Semi major axis (km)	Perigee height (km)	Apogee height (km)	Orbital eccentricity	Argument of perigee (deg)
	Cosmos 923	1977-59A	1977 Jul 1.50 120 years	Cylinder + paddles? 750?	2 long? 1 dia?	1977 Jul 4.4	74.05	101.05	7186	799	817	0.001	78
	Cosmos 923 rocket	1977-59B	1977 Jul 1.50 100 years	Cylinder 2200?	7.4 long 2.4 dia	1977 Jul 2.5	74.06	100.93	7180	788	816	0.002	73
D	Cosmos 924	1977-60A	**1977 Jul 4.93 1317 days 1981 Feb 10**	Cylinder + paddles? 900?	2 long? 1 dia?	1977 Jul 5.6	74.02	95.28	6910	513	550	0.003	353
D	Cosmos 924 rocket	1977-60B	**1977 Jul 4.93 1296 days 1981 Jan 20**	Cylinder 2200?	7.4 long 2.4 dia	1977 Jul 10.1	74.03	95.16	6904	502	549	0.003	350
D	Fragments*	1977-60C,D											
	Cosmos 925	1977-61A	1977 Jul 7.31 30 years	Cylinder + 2 vanes? 2500?	5 long? 1.5 dia?	1977 Jul 18.3	81.21	97.16	7000	609	634	0.002	234
	Cosmos 925 rocket	1977-61B	1977 Jul 7.31 30 years	Cylinder 1440	3.8 long 2.6 dia	1977 Jul 21.9	81.22	97.29	7006	578	677	0.007	145
	Cosmos 926	1977-62A	1977 Jul 8.73 1200 years	Cylinder 700?	1.3 long? 1.9 dia?	1977 Jul 14.6	82.94	105.13	7377	976	1022	0.003	260
	Cosmos 926 rocket	1977-62B	1977 Jul 8.73 600 years	Cylinder 2200?	7.4 long 2.4 dia	1977 Jul 18.3	82.94	105.01	7372	976	1011	0.002	242

* Fragment 1977-60C has probably decayed, and a different fragment (from a 100.7 min launch) now carries that designation in the USA.

	Name		Launch date, lifetime and descent date	Shape and weight (kg)	Size (m)	Date of orbital determination	Orbital inclination (deg)	Nodal period (min)	Semi major axis (km)	Perigee height (km)	Apogee height (km)	Orbital eccentricity	Argument of perigee (deg)
D R	Cosmos 927	1977-63A	1977 Jul 12.38 12.8 days 1977 Jul 25.2	Sphere-cylinder 6300?	6.5 long? 2.4 dia	1977 Jul 13.7 1977 Jul 19.3	72.87 72.89	89.65 89.88	6635 6647	153 151	361 386	0.016 0.018	77 59
D	Cosmos 927 rocket	1977-63B	1977 Jul 12.38 6 days 1977 Jul 18	Cylinder 2500?	7.5 long 2.6 dia	1977 Jul 13.1	72.88	89.81	6643	167	363	0.015	84
D	Cosmos 927 engine*	1977-63C	1977 Jul 12.38 15 days 1977 Jul 27	Cone 600? full	1.5 long? 2 dia?	1977 Jul 24.8	72.89	88.83	6594	116	316	0.015	45
D	Fragment	1977-63D											
	Cosmos 928	1977-64A	1977 Jul 13.21 1200 years	Cylinder 700?	1.3 long? 1.9 dia?	1977 Jul 16.9	82.96	104.79	7362	956	1011	0.004	287
	Cosmos 928 rocket	1977-64B	1977 Jul 13.21 600 years	Cylinder 2200?	7.4 long 2.4 dia	1977 Jul 15.5	82.96	104.70	7357	958	1000	0.003	284
T	**Himawari 1** (GMS 1)**	1977-65A	1977 Jul 14.44 > million years	Cylinder 670 full 315 empty	3.0 long† 2.16 dia?	1977 Jul 14.6 1977 Jul 17.6 1977 Nov 1.0	27.36 1.20 1.0	649.66 1429.43 1436.1	24844 42033 42165	187 35531 35775	36744 35779 35799	0.736 0.003 0.0003	180 98 -
	Himawari 1 second stage	1977-65B	1977 Jul 14.44 disintegrated	Cylinder + annulus 350?	6.4 long 1.52 and 2.44 dia	1977 Jul 14.4 1977 Jul 16.6	28.68 29.05	92.82 111.01	6798 7658	172 534	668 2025	0.037 0.097	165 67
D	**Himawari 1 third stage**	1977-65D	**1977 Jul 14.44 1576 days 1981 Nov 6**	Sphere-cone 66	1.32 long 0.94 dia	1977 Jul 14.5	27.09	658.10	25062	245	37123	0.736	180
54d	Fragments	1977-65C, E-FF											

* Jettisoned from Cosmos 927 about 1977 Jul 24.

** Japanese Geostationary Meteorological Satellite, launched by NASA, may have ejected an apogee motor.

† Length including antennae.

	Name		Launch date, lifetime and descent date	Shape and weight (kg)	Size (m)	Date of orbital determination	Orbital inclination (deg)	Nodal period (min)	Semi major axis (km)	Perigee height (km)	Apogee height (km)	Orbital eccentricity	Argument of perigee (deg)
D	Cosmos 929 †	1977-66A	1977 Jul 17.38 200 days 1978 Feb 2	Cylinder + vanes 19000?	14 long 4.15 to 2.0 dia	1977 Jul 18.6 1977 Aug 26 1977 Dec 19	51.59 51.58 51.58	89.36 90.78 93.38	6624 6693 6821	214 312 438	278 318 447	0.005 0.0005 0.001	88 - -
D	Cosmos 929 rocket	1977-66B	1977 Jul 17.38 12 days 1977 Jul 29	Cylinder 4000?	12 long? 4 dia	1977 Jul 18.6	51.59	89.15	6614	211	260	0.004	83
D	Fragment	1977-66C											
D	Cosmos 930*	1977-67A	1977 Jul 19.36 1028 days 1980 May 12	Cylinder + ellipsoid 2750?	9.2 long? 1.5 and 2.4 dia	1977 Jul 24.3	74.02	94.59	6876	481	514	0.002	354
	Cosmos 931	1977-68A	1977 Jul 20.20 100 years?	Windmill + 6 vanes? 1250?	4.2 long? 1.6 dia?	1977 Jul 22.3	62.96	724.12	26713	604	40065	0.739	318
D	Cosmos 931 launcher	1977-68B	1977 Jul 20.20 64 days 1977 Sep 22	Irregular	-	1977 Jul 23.3	62.81	92.57	6780	212	591	0.028	122
D	Cosmos 931 launcher rocket	1977-68C	1977 Jul 20.20 28 days 1977 Aug 17	Cylinder 2500?	7.5 long 2.6 dia	1977 Jul 23.6	62.85	92.20	6762	182	585	0.030	119
	Cosmos 931 rocket	1977-68D	1977 Jul 20.20 100 years?	Cylinder 440	2.0 long 2.0 dia	1977 Jul 26.8	62.90	720.90	26633	605	39905	0.738	318
	Fragments	**1977-68E,F**											

* With rocket attached
† De-orbited over Pacific Ocean. Probably a test of engine and airlock modifications for Salyut 6.

	Name		Launch date, lifetime and descent date	Shape and weight (kg)	Size (m)	Date of orbital determination	Orbital inclination (deg)	Nodal period (min)	Semi major axis (km)	Perigee height (km)	Apogee height (km)	Orbital eccentricity	Argument of perigee (deg)
D R	Cosmos 932	1977-69A	1977 Jul 20.32 12.9 days 1977 Aug 2.2	Sphere-cylinder 6300?	6.5 long? 2.4 dia	1977 Jul 22.3 1977 Jul 25.1	65.02 65.02	89.09 89.57	6608 6632	149 150	311 358	0.012 0.016	53 59
D	Cosmos 932 rocket	1977-69B	1977 Jul 20.32 4 days 1977 Jul 24	Cylinder 2500?	7.5 long 2.6 dia	1977 Jul 20.5	65.03	89.33	6620	173	311	0.010	64
D	Cosmos 932 engine*	1977-69D	1977 Jul 20.32 14 days 1977 Aug 3	Cone 600? full	1.5 long? 2 dia?	1977 Aug 1.6	65.05	88.61	6584	151	261	0.008	65
D	Fragment	1977-69C											
D	Cosmos 933	1977-70A	1977 Jul 22.42 467 days 1978 Nov 1	Cylinder	4 long? 2 dia?	1977 Jul 23.6	65.84	92.46	6774	384	408	0.002	343
D	Cosmos 933 rocket	1977-70B	1977 Jul 22.42 320 days 1978 Jun 7	Cylinder 2200?	7.4 long 2.4 dia	1977 Jul 23.9	65.85	92.38	6770	376	408	0.002	343
	Statsionar-Raduga 3	1977-71A	1977 Jul 23.89 > million years	-	-	1977 Aug 4.6	0.21	1436.3	42170	35730	35854	0.001	313
D	Raduga 3 launcher rocket	1977-71B	1977 Jul 23.89 3 days 1977 Jul 26	Cylinder 4000?	12 long? 4 dia	1977 Jul 24.1	51.47	88.21	6567	179	198	0.001	286
D	Raduga 3 launcher	1977-71C	1977 Jul 23.89 3 days 1977 Jul 26	Irregular	-	1977 Jul 24.1	51.46	88.30	6571	191	195	0.0003	32
D	Raduga 3 rocket	1977-71D	1977 Jul 23.89 317 days 1978 Jun 5	Cylinder 1900?	3.9 long? 3.9 dia	1977 Aug 16.5	47.2	632.15	24402	290	35758	0.727	6

* Jettisoned from Cosmos 932 about 1977 Aug 1.

1977-71 continued on page 499

	Name		Launch date, lifetime and descent date	Shape and weight (kg)	Size (m)	Date of orbital determination	Orbital inclination (deg)	Nodal period (min)	Semi major axis (km)	Perigee height (km)	Apogee height (km)	Orbital eccentricity	Argument of perigee (deg)
	Raduga 3 apogee motor	1977-71F	1977 Jul 23.89 > million years	Cylinder 440	2.0 long 2.0 dia	Synchronous orbit similar to 1977-71A							
D	Fragment	1977-71E											
D R	Cosmos 934	1977-72A	1977 Jul 27.76 12.62 days 1977 Aug 9.38	Sphere-cylinder 6300?	6.5 long? 2.4 dia	1977 Jul 28.3 1977 Jul 28.9	62.81 62.82	89.35 89.60	6621 6634	231 167	255 344	0.002 0.013	289 205
D	Cosmos 934 rocket	1977-72B	1977 Jul 27.76 9 days 1977 Aug 5	Cylinder 2500?	7.5 long 2.6 dia	1977 Jul 28.2	62.79	89.26	6617	223	254	0.002	300
D	Cosmos 934 engine	1977-72D	1977 Jul 27.76 21 days 1977 Aug 17	Cone 600? full	1.5 long? 2 dia?	1977 Aug 8.7	62.81	89.50	6629	170	331	0.012	213
D	Fragments	1977-72C,E,F											
D R	Cosmos 935	1977-73A	1977 Jul 29.34 12.88 days 1977 Aug 11.22	Sphere-cylinder 5700?	5.0 long 2.4 dia	1977 Jul 30.6	81.33	89.20	6612	217	251	0.003	71
D	Cosmos 935 rocket	1977-73B	1977 Jul 29.34 5 days 1977 Aug 3	Cylinder 2500?	7.5 long 2.6 dia	1977 Jul 30.4	81.33	88.92	6598	205	235	0.002	23
D R B	Cosmos 936*	1977-74A	1977 Aug 3.59 18.55 days 1977 Aug 22.14	Sphere-cylinder 5900?	5.9 long 2.4 dia	1977 Aug 7.4	62.80	90.63	6686	219	396	0.013	112
D	Cosmos 936 rocket	1977-74B	1977 Aug 3.59 25 days 1977 Aug 28	Cylinder 2500?	7.5 long 2.6 dia	1977 Aug 8.2	62.79	90.35	6671	214	371	0.012	104

* Satéllite with international biological experiments (1977-74 continued on page 500).

	Name		Launch date, lifetime and descent date	Shape and weight (kg)	Size (m)	Date of orbital determination	Orbital inclination (deg)	Nodal period (min)	Semi major axis (km)	Perigee height (km)	Apogee height (km)	Orbital eccentricity	Argument of perigee (deg)
D	Capsule*	1977-74D	1977 Aug 3.59 66 days 1977 Oct 8	Ellipsoid 200?	0.9 long 1.9 dia	1977 Aug 24.5	62.80	90.40	6673	215	375	0.012	113
D	Fragment	1977-74C											
D	HEAO 1 **	1977-75A	1977 Aug 12.27 580 days 1979 Mar 15	Hexagonal Cylinder 2720	5.8 long 2.4 dia	1977 Aug 12.5	22.76	93.16	6816	428	447	0.001	295
D	HEAO 1 rocket	1977-75B	1977 Aug 12.27 105 days 1977 Nov 25	Cylinder 1815	8.6 long 3.0 dia	1977 Aug 14.5	22.81	91.49	6734	329	383	0.004	240
D	Cosmos 937	1977-77A	1977 Aug 24.30 421 days 1978 Oct 19	Cylinder?	-	1977 Aug 24.4 1977 Aug 29.9	65.05 65.04	92.07 93.31	6751 6812	149 424	597 444	0.033 0.001	64 273
D	Cosmos 937 rocket	1977-77B	1977 Aug 24.30 1 day 1977 Aug 25	Cylinder 1500?	8 long? 2.5 dia?	1977 Aug 24.5	65.01	89.28	6618	100	379	0.021	45
D	Fragment	1977-77C											
D R	Cosmos 938	1977-78A	1977 Aug 24.61 12.63 days 1977 Sep 6.24	Sphere-cylinder 6300?	6.5 long? 2.4 dia	1977 Aug 25.3 1977 Aug 25.5	62.81 62.81	89.70 89.37	6639 6622	181 156	340 332	0.012 0.013	83 72
D	Cosmos 938 rocket	1977-78B	1977 Aug 24.61 6 days 1977 Aug 30	Cylinder 2500?	7.5 long 2.6 dia	1977 Aug 25.2	62.80	89.49	6628	178	322	0.011	75
D	Cosmos 938 † engine	1977-78E	1977 Aug 24.61 14 days 1977 Sep 7	Cone 600? full	1.5 long? 2 dia?	1977 Sep 6	62.8	88.85	6596	169	267	0.007	-
D	Fragments	1977-78C,D,F											

Space Vehicle: Voyager 2 and rockets (1977-76A, 76B and 76C).

** High-Energy Astronomy Observatory.

* Jettisoned from Cosmos 936 about 1977 Aug 22.

† Jettisoned from Cosmos 938 about 1977 Sep 5.

Name		Launch date, lifetime and descent date	Shape and weight (kg)	Size (m)	Date of orbital determination	Orbital inclination (deg)	Nodal period (min)	Semi major axis (km)	Perigee height (km)	Apogee height (km)	Orbital eccentricity	Argument of perigee (deg)
Cosmos 939	1977-79A	1977 Aug 24.76 8000 years	Spheroid 40?	1.0 long? 0.8 dia?	1977 Aug 28.8	74.02	114.88	7828	1435	1464	0.002	88
Cosmos 940	1977-79B	1977 Aug 24.76 6000 years	Spheroid 40?	1.0 long? 0.8 dia?	1977 Aug 28.8	74.02	114.46	7809	1397	1464	0.004	82
Cosmos 941	1977-79C	1977 Aug 24.76 7000 years	Spheroid 40?	1.0 long? 0.8 dia?	1977 Aug 29.8	74.02	114.67	7818	1416	1464	0.003	92
Cosmos 942	1977-79D	1977 Aug 24.76 10,000 years	Spheroid 40?	1.0 long? 0.8 dia?	1977 Aug 29.8	74.02	115.98	7878	1464	1535	0.004	261
Cosmos 943	1977-79E	1977 Aug 24.76 9000 years	Spheroid 40?	1.0 long? 0.8 dia?	1977 Aug 29.5	74.02	115.08	7837	1453	1464	0.001	92
Cosmos 944	1977-79F	1977 Aug 24.76 9000 years	Spheroid 40?	1.0 long? 0.8 dia?	1977 Aug 29.6	74.02	115.30	7847	1464	1473	0.001	280
Cosmos 945	1977-79G	1977 Aug 24.76 10,000 years	Spheroid 40?	1.0 long? 0.8 dia?	1977 Aug 28.8	74.02	115.52	7857	1464	1493	0.002	257
Cosmos 946	1977-79H	1977 Aug 24.76 10,000 years	Spheroid 40?	1.0 long? 0.8 dia?	1977 Aug 29.6	74.02	115.73	7866	1464	1512	0.003	269
Cosmos 939 rocket	1977-79J	1977 Aug 24.76 20,000 years	Cylinder 2200?	7.4 long 2.4 dia	1977 Aug 28.0	74.02	117.60	7951	1462	1683	0.014	269

	Name		Launch date, lifetime and descent date	Shape and weight (kg)	Size (m)	Date of orbital determination	Orbital inclination (deg)	Nodal period (min)	Semi major axis (km)	Perigee height (km)	Apogee height (km)	Orbital eccentricity	Argument of perigee (deg)
T	Sirio 1*	1977-80A	1977 Aug 25.99 >million years	Cylinder + nozzle 220	2 long 1.4 dia	1977 Aug 26.2 1977 Aug 28.6	22.96 0.24	659.92 1417.95	25108 41809	245 33653	37215 37208	0.736 0.043	179 292
	Sirio 1 second stage	1977-80B	1977 Aug 25.99 1600 years	Cylinder + annulus 350?	6.4 long 1.52 and 2.44 dia	1977 Aug 26.0 1977 Aug 26.3	28.10 27.10	99.08 115.30	7093 7853	210 870	1220 2080	0.071 0.077	162 76
D	Sirio 1 third stage	1977-80C	**1977 Aug 25.99** **1903 days** **1982 Nov 10**	Sphere- cone 66	1.32 long 0.94 dia	1977 Oct 31.0	22.92	659.40	25094	259	37172	0.736	222
D	Fragments	1977-80D,E											
D R	Cosmos 947	1977-81A	1977 Aug 27.43 12.77 days 1977 Sep 9.20	Sphere- cylinder 5700?	5.0 long 2.4 dia	1977 Aug 29.7	72.85	89.75	6640	203	321	0.009	63
D	Cosmos 947 rocket	1977-81B	1977 Aug 27.43 11 days 1977 Sep 7	Cylinder 2500?	7.5 long 2.6 dia	1977 Aug 27.7	72.85	89.67	6636	200	316	0.009	68
D	Fragment	1977-81C											
	Molniya 1AP	1977-82A	1977 Aug 30.76 $10\frac{3}{4}$ years	Windmill + 6 vanes 1000?	3.4 long 1.6 dia	1977 Aug 31.8 1977 Sep 13.9	62.83 62.85	735.58 717.77	26993 26556	445 483	40785 39873	0.747 0.742	280 280
D	Molniya 1AP launcher rocket	1977-82B	1977 Aug 30.76 27 days 1977 Sep 26	Cylinder 2500?	7.5 long 2.6 dia	1977 Aug 31.9	62.85	90.94	6700	209	434	0.017	131

* Italian communications satellite, launched by NASA.

1977-82 continued on page 503

	Name		Launch date, lifetime and descent date	Shape and weight (kg)	Size (m)	Date of orbital determination	Orbital inclination (deg)	Nodal period (min)	Semi major axis (km)	Perigee height (km)	Apogee height (km)	Orbital eccentricity	Argument of perigee (deg)
D	Molniya 1AP launcher	1977-82C	1977 Aug 30.76 24 days 1977 Sep 23	Irregular	-	1977 Aug 31.4	62.84	91.14	6709	198	464	0.020	121
	Molniya 1AP rocket	1977-82E	1977 Aug 30.76 $10\frac{3}{4}$ years	Cylinder 440	2.0 long 2.0 dia	1977 Sep 22.2	62.84	732.08	26908	482	40578	0.745	280
D	Fragment	1977-82D											
D R	Cosmos 948	1977-83A	1977 Sep 2.38 12.84 days 1977 Sep 15.22	Sphere-cylinder 5900?	5.9 long 2.4 dia	1977 Sep 2.9	81.36	89 04	6604	217	235	0.001	62
D	Cosmos 948 rocket	1977-83B	1977 Sep 2.38 5 days 1977 Sep 7	Cylinder 2500?	7.5 long 2.6 dia	1977 Sep 2.6	81.35	88.95	6600	214	229	0.001	46
D	Capsule	1977-83C	1977 Sep 2.38 18 days 1977 Sep 20	Ellipsoid 200?	0.9 long 1.9 dia	1977 Sep 15	81.3	88.71	6588	204	216	0.001	-
D	Fragment	1977-83D											
D R	Cosmos 949*	1977-85A	1977 Sep 6.73 29.5 days 1977 Oct 6.2	Sphere-cylinder 6700?	7 long? 2.4 dia	1977 Sep 7.3 1977 Sep 20.5 1977 Sep 23.5	62.80 62.80 62.80	89.50 89.61 89.89	6629 6635 6649	177 149 177	325 364 364	0.011 0.016 0.014	65 59 83
D	Cosmos 949 rocket	1977-85B	1977 Sep 6.73 5 days 1977 Sep 11	Cylinder 2500?	7.5 long 2.6 dia	1977 Sep 7.3	62.79	89.41	6625	172	321	0.011	60
D	Fragment	1977-85C											

Space Vehicle: Voyager 1 and rockets (1977-84A, 84B and 84C).

* Included manoeuvring engine which was not separately tracked or designated.

	Name		Launch date, lifetime and descent date	Shape and weight (kg)	Size (m)	Date of orbital determination	Orbital inclination (deg)	Nodal period (min)	Semi major axis (km)	Perigee height (km)	Apogee height (km)	Orbital eccentricity	Argument of perigee (deg)
D R	Cosmos 950	1977-86A	1977 Sep 13.64 13.6 days 1977 Sep 27.2	Sphere-cylinder 5700?	5.0 long 2.4 dia	1977 Sep 14.6	62.81	89.36	6622	205	282	0.006	72
D	Cosmos 950 rocket	1977-86B	1977 Sep 13.64 5 days 1977 Sep 18	Cylinder 2500?	7.5 long 2.6 dia	1977 Sep 14.5	62.80	89.07	6607	197	261	0.005	57
D	Fragment	1977-86C											
	Cosmos 951	1977-87A	1977 Sep 13.83 1200 years	Cylinder 700?	1.3 long? 1.9 dia?	1977 Sep 16.5	82.97	104.98	7371	968	1017	0.003	276
	Cosmos 951 rocket	1977-87B	1977 Sep 13.83 600 years	Cylinder 2200?	7.4 long 2.4 dia	1977 Sep 17.5	82.97	104.88	7366	968	1007	0.003	264
	Cosmos 952*	1977-88A	1977 Sep 16.60 600 years	Cone-cylinder	6 long? 2 dia?	1977 Sep 18.6 1977 Oct 8.7	64.97 64.94	89.65 104.13	6636 7332	251 910	265 998	0.001 0.006	272 267
D	Cosmos 952 rocket	1977-88C	1977 Sep 16.60 25 days 1977 Oct 11	Cylinder 1500?	8 long? 2.5 dia?	1977 Oct 9.6	64.97	89.08	6607	224	233	0.0007	269
D	Cosmos 952 platform	1977-88B	1977 Sep 16.60 52 days 1977 Nov 7	Irregular	-	1977 Oct 14.7	64.97	89.37	6622	235	253	0.001	278

* 1977-88B and 1977-88C attached to 1977-88A until orbit raised on 1977 Oct 8.5

	Name		Launch date, lifetime and descent date	Shape and weight (kg)	Size (m)	Date of orbital determination	Orbital inclination (deg)	Nodal period (min)	Semi major axis (km)	Perigee height (km)	Apogee height (km)	Orbital eccentricity	Argument of perigee (deg)
D R	Cosmos 953	1977-89A	1977 Sep 16.61 12.6 days 1977 Sep 29.2	Sphere-cylinder 6300?	6.5 long? 2.4 dia	1977 Sep 17.2 1977 Sep 22.2	62.80 62.81	89.58 89.00	6633 6604	180 151	330 300	0.011 0.011	77 65
D	Cosmos 953 rocket	1977-89B	1977 Sep 16.61 5 days 1977 Sep 21	Cylinder 2500?	7.5 long 2.6 dia	1977 Sep 17.2	62.79	89.43	6626	176	319	0.011	71
D	Cosmos 953* engine	1977-89C	1977 Sep 16.61 14 days 1977 Sep 30	Cone 600? full	1.5 long? 2 dia?	1977 Sep 28.4	62.81	89.07	6607	149	309	0.012	86
D	Fragment	1977-89D											
D P	Cosmos 954**	1977-90A	1977 Sep 18.58 127.92 days 1978 Jan 24.50	Cone-cylinder	14 long? 2.5 dia?	1977 Sep 20.9 1978 Jan 6.2	64.98 64.98	89.65 89.27	6636 6617	251 233	265 245	0.001 0.001	270 -
	Cosmos 955	1977-91A	1977 Sep 20.05 30 years	Cylinder + 2 vanes? 2500?	5 long? 1.5 dia?	1977 Sep 25.5	81.24	97.46	7014	630	641	0.001	355
	Cosmos 955 rocket	1977-91B	1977 Sep 20.05 30 years	Cylinder 1440	3.8 long 2.6 dia	1977 Sep 25.0	81.23	97.55	7019	592	689	0.007	172
D	Fragment	1977-91C											

* Jettisoned from 1977-89A about 1977 Sep 28.

** Manoeuvred until 1977 Nov 1; attitude stabilized until 1978 Jan 6; then lost pressurization and started to tumble, with great increase in drag and rapid decay. Fragments picked up in Canada.

	Name		Launch date, lifetime and descent date	Shape and weight (kg)	Size (m)	Date of orbital determination	Orbital inclination (deg)	Nodal period (min)	Semi major axis (km)	Perigee height (km)	Apogee height (km)	Orbital eccentricity	Argument of perigee (deg)
	Statsionar Ekran 2	1977-92A	1977 Sep 20.73 > million years	Cylinder + plate	-	1977 Sep 21.5	0.40	1426.55	41979	35580	35622	0.0005	28
D	Ekran 2 launcher rocket	1977-92B	1977 Sep 20.73 1 day 1977 Sep 21	Cylinder 4000?	12 long? 4 dia	1977 Sep 21.0	51.45	88.21	6567	184	193	0.0007	291
D	Ekran 2 launcher	1977-92C	1977 Sep 20.73 1 day 1977 Sep 21	Irregular	-	1977 Sep 21.0	51.44	88.22	6567	186	192	0.0005	81
D	Ekran 2 rocket	1977-92E	1977 Sep 20.73 222 days 1978 Apr 30	Cylinder 1900?	3.9 long? 3.9 dia	1977 Oct 26.9	47.18	626.34	24254	242	35510	0.727	10
	Ekran 2 apogee motor	1977-92G	1977 Sep 20.73 > million years	Cylinder 440	2.0 long 2.0 dia	Synchronous orbit similar to 1977-92A							
D	Fragments	1977-92D,F											
	Prognoz 6	1977-93A	1977 Sep 22.04 22 years	Spheroid + 4 vanes 910	1.8 dia?	1977 Sep 22.0 1977 Sep 22.1	65.04 65.00	91.43 5688	6723 105556	226 488	464 197867	0.018 0.935	70 290
D	Prognoz 6 launcher rocket	1977-93B	1977 Sep 22.04 37 days 1977 Oct 29	Cylinder 2500?	7.5 long 2.6 dia	1977 Sep 23.2	65.04	91.39	6721	226	460	0.017	69
D	Prognoz 6 launcher	1977-93C	1977 Sep 22.04 31 days 1977 Oct 23	Irregular	-	1977 Sep 26.3	65.07	91.31	6717	207	471	0.020	65
	Prognoz 6 rocket	1977-93E	1977 Sep 22.04 22 years	Cylinder 440	2.0 long 2.0 dia	Orbit similar to 1977-93A							
D	Fragment	1977-93D											
D	[Titan 3B Agena D]	1977-94A	1977 Sep 23.78 76 days 1977 Dec 8	Cylinder 3000?	8 long? 1.5 dia	1977 Sep 26.2	96.49	89.30	6617	125	352	0.017	145

	Name		Launch date, lifetime and descent date	Shape and weight (kg)	Size (m)	Date of orbital determination	Orbital inclination (deg)	Nodal period (min)	Semi major axis (km)	Perigee height (km)	Apogee height (km)	Orbital eccentricity	Argument of perigee (deg)
D	Cosmos 956	1977-95A	1977 Sep 24.43 1737 days 1982 Jun 27	- 825?	-	1977 Sep 25.2	75.83	96.89	6987	355	863	0.036	342
D	Cosmos 956 rocket	1977-95B	1977 Sep 24.43 1864 days 1982 Nov 1	Cylinder 2200?	7.4 long 2.4 dia	1977 Sep 24.8	75.84	96.84	6985	351	862	0.037	343
D L	Intercosmos 17	1977-96A	1977 Sep 24.69 775 days 1979 Nov 8	Octagonal ellipsoid 550?	1.8 long? 1.5 dia?	1977 Sep 26.4	82.96	94.44	6868	466	514	0.003	344
D	Intercosmos 17 rocket	1977-96B	1977 Sep 24.69 717 days 1979 Sep 11	Cylinder 2200?	7.4 long 2.4 dia	1977 Sep 26.4	82.96	94.36	6864	458	514	0.004	346
D	Fragment	1977-96C											
D	Salyut 6	1977-97A	1977 Sep 29.29 1764 days 1982 Jul 29	Cylinder + 3 vanes 18900	14 long 4.15 to 2.0 dia	1977 Oct 1.5	51.59	89.14	6613	214	256	0.003	102
D	Salyut 6 rocket	1977-97B	1977 Sep 29.29 6 days 1977 Oct 5	Cylinder 4000?	12 long? 4 dia	1977 Sep 30.1	51.63	88.78	6595	209	225	0.001	79
D	Fragments*	1977-97C-DK											
D R	Cosmos 957	1977-98A	1977 Sep 30.41 12.9 days 1977 Oct 13.3	Sphere-cylinder 6300?	6.5 long? 2.4 dia	1977 Oct 1.3 1977 Oct 2.3	64.97 64.98	89.82 89.51	6644 6629	171 150	361 351	0.014 0.015	66 58
D	Cosmos 957 rocket	1977-98B	1977 Sep 30.41 5.5 days 1977 Oct 5.9	Cylinder 2500?	7.5 long 2.6 dia	1977 Oct 1.3	64.98	89.59	6633	171	338	0.013	62

* See also footnote to 1979-103D and 1981-23C.

1977-98 continued on page 508

	Name		Launch date, lifetime and descent date	Shape and weight (kg)	Size (m)	Date of orbital determination	Orbital inclination (deg)	Nodal period (min)	Semi major axis (km)	Perigee height (km)	Apogee height (km)	Orbital eccentricity	Argument of perigee (deg)
D	Cosmos 957 engine	1977-98E	1977 Sep 30.41 15 days 1977 Oct 15	Cone 600? full	1.5 long? 2 dia?	1977 Oct 13.9	64.96	88.46	6576	152	244	0.007	51
D	Fragments	1977-98C,D,F											
D 2M R	Soyuz 25*	1977-99A	1977 Oct 9.11 2.03 days 1977 Oct 11.14	Sphere-cylinder 6570?	7.5 long 2.3 dia	1977 Oct 9.3 1977 Oct 9.4 1977 Oct 10.2	51.64 51.60 51.62	88.78 90.22 91.29	6595 6665 6719	194 265 329	240 309 353	0.004 0.003 0.002	73 282 295
D	Soyuz 25 rocket	1977-99B	1977 Oct 9.11 2 days 1977 Oct 11	Cylinder 2500?	7.5 long 2.6 dia	1977 Oct 9.5	51.64	88.63	6588	189	230	0.003	79
D R	Cosmos 958	1977-100A	1977 Oct 11.64 12.64 days 1977 Oct 24.28	Sphere-cylinder 6300?	6.5 long? 2.4 dia	1977 Oct 11.7 1977 Oct 15.5	62.81 62.81	90.59 91.96	6682 6750	257 323	351 420	0.007 0.007	105 202
D	Cosmos 958 rocket	1977-100B	1977 Oct 11.64 29 days 1977 Nov 9	Cylinder 2500?	7.5 long 2.6 dia	1977 Oct 12.1	62.79	90.48	6677	255	342	0.006	99
D	Cosmos 958 engine	1977-100E	1977 Oct 11.64 339 days 1978 Sep 15	Cone 600? full	1.5 long? 2 dia?	1977 Oct 25.1 1977 Nov 1.0	62.82 62.81	92.28 92.23	6765 6763	352 352	422 417	0.005 0.005	218
D	Fragments	1977-100C,D,F,G											
D	Cosmos 959	1977-101A	1977 Oct 21.42 40 days 1977 Nov 30	Cylinder?	4 long? 2 dia?	1977 Oct 24.0	65.84	94.57	6876	146	850	0.051	63
D	Cosmos 959 rocket	1977-101B	1977 Oct 21.42 18 days 1977 Nov 8	Cylinder 2200?	7.4 long? 2.4 dia?	1977 Oct 24.1	65.84	94.17	6856	144	812	0.049	63
D	Fragment	1977-101C											

* Soyuz 25 rendezvous with Salyut 6 on 1977 Oct 10.17, but failed to dock

	Name		Launch date, lifetime and descent date	Shape and weight (kg)	Size (m)	Date of orbital determination	Orbital inclination (deg)	Nodal period (min)	Semi major axis (km)	Perigee height (km)	Apogee height (km)	Orbital eccentricity	Argument of perigee (deg)
T	ISEE 1*	1977-102A	1977 Oct 22.58 10 years	16-sided Cylinder 340	1.61 long 1.73 dia	1977 Oct 28.0	28.95	3440.9	75499	337	137904	0.911	311
T	ISEE 2	1977-102B	1977 Oct 22.58 10 years	Cylinder + 3 booms 166	1.14 long 1.27 dia	1977 Oct 28.0	28.96	3439.1	75472	341	137847	0.911	311
D	ISEE second stage	1977-102C	1977 Oct 22.58 366 days 1978 Oct 23	Cylinder + annulus 350?	6.4 long 1.52 and 2.44 dia	1977 Oct 24.4	28.74	95.68	6937	277	840	0.041	326
	ISEE third stage	1977-102D	1977 Oct 22.58 10 years	Sphere -cone 66	1.32 long 0.94 dia	1977 Oct 22.6	28.76	3552.3	77119	278	141204	0.914	310
D	Cosmos 960	1977-103A	1977 Oct 25.23 1093.3 days 1980 Oct 22.5	Cylinder + paddles? 900?	2 long? 1 dia?	1977 Oct 28.0	74.04	95.13	6902	502	546	0.003	337
D	Cosmos 960 rocket	1977-103B	1977 Oct 25.23 1132 days 1980 Nov 30	Cylinder 2200?	7.4 long 2.4 dia	1977 Oct 29.4	74.04	95.01	6896	494	542	0.004	348
D	Fragment	1977-103C											
D	Cosmos 961**	1977-104A	1977 Oct 26.22 0.14 day 1977 Oct 26.36	Cylinder?	4 long? 2 dia?	1977 Oct 26.2 1977 Oct 26.3	66 66.4	88.76 101.8	6592 7223	125 269	302 1421	0.013 0.080	- -
D	Cosmos 961 rocket	1977-104B	1977 Oct 26.22 1 day 1977 Oct 27	Cylinder 1500?	8 long? 2.5 dia?	1977 Oct 26.3	65.09	88.24	6566	129	247	0.009	22

* International Sun-Earth Explorer, launched for ESA by NASA: ISEE 1 and 2 are sometimes called 'Mother' and 'Daughter' respectively.

** Cosmos 961 probably passed close to Cosmos 959 about 1977 Oct 26.3 (both orbits unconfirmed). De-orbited into Pacific Ocean and burnt up near 36 deg north, 143 deg east. Re-entry observed from Japan.

	Name		Launch date, lifetime and descent date	Shape and weight (kg)	Size (m)	Date of orbital determination	Orbital inclination (deg)	Nodal period (min)	Semi major axis (km)	Perigee height (km)	Apogee height (km)	Orbital eccentricity	Argument of perigee (deg)
	Molniya 3H	1977-105A	1977 Oct 28.07 100 years?	Windmill + 6 vanes 1500?	4.2 long? 1.6 dia	1977 Nov 3.2	62.80	734.89	26977	428	40769	0.748	280
D	Molniya 3H launcher rocket	1977-105B	1977 Oct 28.07 21.59 days 1977 Nov 18.66	Cylinder 2500?	7.5 long 2.6 dia	1977 Oct 29.9	62.82	90.97	6701	206	440	0.017	133
D	Molniya 3H launcher	1977-105C	1977 Oct 28.07 29 days 1977 Nov 26	Irregular	-	1977 Oct 29.5	62.80	91.14	6709	208	454	0.018	120
	Molniya 3H rocket	1977-105E	1977 Oct 28.07 100 years?	Cylinder 440	2.0 long 2.0 dia	1977 Nov 21.0	62.86	731.66	26897	430	40608	0.747	280
D	Fragment	1977-105D											
T	Transat [Scout]	1977-106A	1977 Oct 28.20 2000 years	Octagonal cylinder 94	0.75 long 0.45 dia	1977 Oct 30.1	89.92	107.03	7466	1069	1107	0.003	19
	Altair rocket	1977-106B	1977 Oct 28.20 2000 years	Cylinder 24	1.50 long 0.46 dia	1977 Oct 31.1	89.91	106.98	7463	1065	1104	0.003	20
	Fragment	1977-106C											
	Cosmos 962	1977-107A	1977 Oct 28.66 1200 years	Cylinder 700?	1.3 long? 1.9 dia?	1977 Oct 28.9	82.96	104.93	7368	968	1012	0.003	289
	Cosmos 962 rocket	1977-107B	1977 Oct 28.66 600 years	Cylinder 2200?	7.4 long 2.4 dia	1977 Oct 29.4	82.93	104.81	7362	968	1000	0.002	282

	Name		Launch date, lifetime and descent date	Shape and weight (kg)	Size (m)	Date of orbital determination	Orbital inclination (deg)	Nodal period (min)	Semi major axis (km)	Perigee height (km)	Apogee height (km)	Orbital eccentricity	Argument of perigee (deg)
	Meteosat 1*	1977-108A	1977 Nov 23.07 >million years	Cylinder 697 full 295 empty	3.20 long 2.10 dia	1977 Nov 23.1 1977 Nov 25.4	27.48 0.73	654.78 1411.5	24978 41681	198 34913	37002 35692	0.737 0.009	180 60
	Meteosat 1 second stage	1977-108B	1977 Nov 23.07 200 years	Cylinder + annulus 350?	6.4 long 1.52 and 2.44 dia	1977 Nov 23.1 1977 Nov 23.2	28.70 28.31	92.55 117.08	6778 7928	172 487	627 2612	0.034 0.134	160 48
	Meteosat 1 third stage	1977-108C	1977 Nov 23.07 10 years	Sphere- cone 66	1.32 long 0.94 dia	1977 Nov 23.1	27.51	656.29	25016	185	37091	0.738	180
	Cosmos 963	1977-109A	1977 Nov 24.60 3000 years	Spheroid + 2 paddles? 650?	1.6 dia?	1977 Nov 25.6	82.93	109.35	7574	1182	1210	0.002	238
	Cosmos 963 rocket	1977-109B	1977 Nov 24.60 2000 years	Cylinder 2200?	7.4 long 2.4 dia	1977 Nov 25.7	82.93	109.23	7568	1179	1201	0.001	216
D R	Cosmos 964	1977-110A	1977 Dec 4.50 12.76 days 1977 Dec 17.26	Sphere- cylinder 6300?	6.5 long? 2.4 dia	1977 Dec 5.1 1977 Dec 9.2	72.88 72.88	89.85 89.24	6645 6614	171 169	362 303	0.014 0.010	81 74
D	Cosmos 964 rocket	1977-110B	1977 Dec 4.50 5 days 1977 Dec 9	Cylinder 2500?	7.5 long 2.4 dia	1977 Dec 5.1	72.88	89.66	6635	168	346	0.013	78
D	Cosmos 964 engine**	1977-110D	1977 Dec 4.50 16 days 1977 Dec 20	Cone 600? full	1.5 long? 2 dia?	1977 Dec 16.7	72.87	89.59	6632	164	343	0.013	54
D	Fragments	1977-110C,E											

* Meteorological Satellite launched for ESA by NASA.

** Ejected from Cosmos 964 about 1977 Dec 16.

	Name		Launch date, lifetime and descent date	Shape and weight (kg)	Size (m)	Date of orbital determination	Orbital inclination (deg)	Nodal period (min)	Semi major axis (km)	Perigee height (km)	Apogee height (km)	Orbital eccentricity	Argument of perigee (deg)
D	Cosmos 965	1977-111A	1977 Dec 8.46 738 days 1979 Dec 16	Octagonal ellipsoid? 550?	1.8 long? 1.5 dia?	1977 Dec 11.3	74.03	94.44	6869	465	516	0.004	326
D	Cosmos 965 rocket	1977-111B	1977 Dec 8.46 706 days 1979 Nov 14	Cylinder 2200?	7.4 long 2.4 dia	1977 Dec 9.0	74.03	94.34	6864	457	514	0.004	336
D	Fragments	1977-111C-AC											
T	NOSS 2 [Atlas]	1977-112A	1977 Dec 8.74 1600 years	Cylinder	-	1977 Dec 19.0	63.43	107.50	7490	1054	1169	0.008	157
	NOSS 2 rocket	1977-112B	1977 Dec 8.74 1000 years	-	-	1977 Dec 9.0	63.39	107.41	7485	1101	1113	0.001	118
T?	SSU 4	1977-112D	1977 Dec 8.74 1600 years	Box + aerials	0.3 x 0.9 x 2.4?	1977 Dec 29.5	63.44	107.50	7490	1054	1169	0.008	157
T?	SSU 5	1977-112E	1977 Dec 8.74 1600 years	Box + aerials	0.3 x 0.9 x 2.4?	1977 Dec 29.7	63.44	107.50	7490	1055	1168	0.008	156
T?	SSU 6	1977-112F	1977 Dec 8.74 1600 years	Box + aerials	0.3 x 0.9 x 2.4?	1977 Dec 29.7	63.44	107.50	7490	1055	1168	0.008	156
	Fragments	1977-112C,G,H											
D 2M R	Soyuz 26*	1977-113A	1977 Dec 10.05 37.43 days 1978 Jan 16.48	Sphere-cylinder 6570?	7.5 long 2.3 dia	1977 Dec 10.1 1977 Dec 10.7 1977 Dec 12.6	51.64 51.62 51.59	88.74 90.20 91.39	6593 6664 6724	195 251 337	235 321 354	0.003 0.005 0.001	64 256 347
D	Soyuz 26 rocket	1977-113B	1977 Dec 10.05 2 days 1977 Dec 12	Cylinder 2500?	7.5 long 2.6 dia	1977 Dec 10.2	51.64	88.53	6583	187	222	0.003	71

* Soyuz 26 docked with Salyut 6 (second airlock) on 1977 Dec 11.13. Undocked from Salyut 6 on 1978 Jan 16 but landed with the Soyuz 27 cosmonauts (see page 516). The Soyuz 26 cosmonauts returned to Earth in Soyuz 27 craft on 1978 Mar 16.

	Name		Launch date, lifetime and descent date	Shape and weight (kg)	Size (m)	Date of orbital determination	Orbital inclination (deg)	Nodal period (min)	Semi major axis (km)	Perigee height (km)	Apogee height (km)	Orbital eccentricity	Argument of perigee (deg)
T	[Atlas Agena D]	1977-114A	1977 Dec 11.54? >million years	Cylinder 700 full? 350 empty?	1.7 long? 1.4 dia?	1977 Dec 11 1977 Dec 12.0 1978 Jan 1.0	29.9 28.2 0.2	87.68 733.2 1435.1	6545 26975 42145	146 191 35679	188 41002 35855	0.003 0.756 0.002	- -* -*
	Agena D rocket	1977-114B	1977 Dec 11.54? 10 years ?	Cylinder 700	6 long? 1.5 dia	1977 Dec 12.0	28.2	733.2	26975	191	41002	0.756	-*
D R	Cosmos 966	1977-115A	1977 Dec 12.41 11.87 days 1977 Dec 24.28	Sphere-cylinder 5900?	5.9 long 2.4 dia	1977 Dec 12.5	65.03	89.50	6628	204	296	0.007	60
D	Cosmos 966 rocket	1977-115B	1977 Dec 12.41 6 days 1977 Dec 18	Cylinder 2500?	7.5 long 2.6 dia	1977 Dec 12.6	65.04	89.36	6621	202	284	0.006	52
D	Capsule	1977-115D	1977 Dec 12.41 26 days 1978 Jan 7	Ellipsoid 200 ?	0.9 long 1.9 dia	1977 Dec 23	65.0	89.28	6617	211	266	0.004	-
D	Fragment	1977-115C											
	Cosmos 967	1977-116A	1977 Dec 13.66 1200 years	Cylinder?	4 long? 2 dia?	1977 Dec 19.0	65.84	104.77	7362	963	1005	0.003	301
	Cosmos 967 rocket	1977-116B	1977 Dec 13.66 600 years	Cylinder 2200?	7.4 long 2.4 dia	1977 Dec 14.2	65.85	104.80	7364	963	1008	0.003	308
	Fragments	1977-116C,D											
	Meteor 2-03	1977-117A	1977 Dec 14.40 500 years	Cylinder + 2 vanes 2750?	5 long? 1.5 dia?	1977 Dec 15.7	81.22	102.48	7253	856	894	0.003	273
	Meteor 2-03 rocket	1977-117B	1977 Dec 14.40 400 years	Cylinder 1440	3.8 long 2.6 dia	1977 Dec 15.6	81.22	102.50	7254	842	910	0.005	209

*Unconfirmed orbits.

	Name		Launch date, lifetime and descent date	Shape and weight (kg)	Size (m)	Date of orbital determination	Orbital inclination (deg)	Nodal period (min)	Semi major axis (km)	Perigee height (km)	Apogee height (km)	Orbital eccentricity	Argument of perigee (deg)
T	Sakura 1* (CS 1)	1977-118A	1977 Dec 15.03 >million years	Cylinder 676 full 340 empty	3.51 long 2.18 dia	1977 Dec 16.0 1977 Dec 16.1	28.70 0.06	629.28 1440.0	24321 42241	155 35568	35732 36157	0.731 0.007	180 72
	Sakura 1 second stage	1977-118B	1977 Dec 15.03 70 years	Cylinder + annulus 350?	6.4 long 1.52 and 2.44 dia	1977 Dec 15.0 1977 Dec 18.5	28.76 28.65	92.05 111.14	6753 7664	165 482	585 2089	0.031 0.105	174 79
D	Sakura 1 third stage	1977-118D	1977 Dec 15.03 1394 days 1981 Oct 9	Sphere-cone 66	1.32 long 0.94 dia	1977 Dec 15.0	28.78	634.69	24464	166	36006	0.733	179
	Fragment	1977-118C											
	Cosmos 968	1977-119A	1977 Dec 16.19 120 years	Cylinder + paddles? 750?	2 long? 1 dia?	1977 Dec 16.9	74.03	100.80	7174	782	810	0.002	351
	Cosmos 968 rocket	1977-119B	1977 Dec 16.19 100 years	Cylinder 2200?	7.4 long 2.4 dia	1977 Dec 16.9	74.03	100.66	7167	774	804	0.002	14
	Fragments	1977-119C,D											
D R	Cosmos 969	1977-120A	1977 Dec 20.66 13.59 days 1978 Jan 3.25	Sphere-cylinder 6300?	6.5 long? 2.4 dia	1977 Dec 21.5 1977 Dec 23.3	62.81 62.81	89.45 89.20	6627 6614	180 166	317 306	0.010 0.011	74 65
D	Cosmos 969 rocket	1977-120B	1977 Dec 20.66 6 days 1977 Dec 26	Cylinder 2500?	7.5 long 2.6 dia	1977 Dec 21.0	62.80	89.37	6623	177	312	0.010	70
D	Cosmos 969 engine	1977-120C	1977 Dec 20.66 16 days 1978 Jan 5	Cone 600? full	1.5 long? 2 dia?	1978 Jan 3.1	62.81	89.37	6623	160	329	0.013	65
D	Fragments	1977-120D-F											
	Cosmos 970**	1977-121A	1977 Dec 21.44 disintegrated	Cylinder?	4 long? 2 dia?	1977 Dec 21.5 1977 Dec 21.6	65.16 65.85	94.67 106.04	6681 7423	144 949	861 1141	0.052 0.013	55 116

* Japanese Communications Satellite launched by NASA. ** Passed close to Cosmos 967, then exploded; continued on page 515

	Name		Launch date, lifetime and descent date	Shape and weight (kg)	Size (m)	Date of orbital determination	Orbital inclination (deg)	Nodal period (min)	Semi major axis (km)	Perigee height (km)	Apogee height (km)	Orbital eccentricity	Argument of perigee (deg)
D	Cosmos 970 rocket	1977-121B	1977 Dec 21.44 4 days 1977 Dec 25	Cylinder 1500?	8 long? 2.5 dia?	1977 Dec 21.4	65.16	94.78	6886	141	875	0.053	55
	Fragments	1977-121C-BC											
	Cosmos 971	1977-122A	1977 Dec 23.68 1200 years	Cylinder 700?	1.3 long? 1.9 dia?	1977 Dec 26.6	82.93	105.04	7373	980	1010	0.002	247
	Cosmos 971 rocket	1977-122B	1977 Dec 23.68 600 years	Cylinder 2200?	7.4 long 2.4 dia	1977 Dec 27.4	82.93	104.93	7368	977	1002	0.002	222
	Cosmos 972	1977-123A	1977 Dec 27.33 200 years	- 825?	-	1977 Dec 27.6	75.85	103.92	7322	716	1172	0.031	52
	Cosmos 972 rocket	1977-123B	1977 Dec 27.33 150 years	Cylinder 2200?	7.4 long 2.4 dia	1977 Dec 27.7	75.83	103.88	7320	716	1168	0.031	52
D	Fragment	1977-123C											
D R	Cosmos 973	1977-124A	1977 Dec 27.39 12.91 days 1978 Jan 9.30	Sphere-cylinder 5900?	5.9 long 2.4 dia	1977 Dec 28.2	71.45	89.81	6642	203	325	0.009	58
D	Cosmos 973 rocket	1977-124B	1977 Dec 27.39 9 days 1978 Jan 5	Cylinder 2500?	7.5 long 2.6 dia	1977 Dec 29.1	71.43	89.46	6625	209	284	0.006	45
D	Capsule	1977-124D	1977 Dec 27.39 29 days 1978 Jan 25	Ellipsoid 200?	0.9 long 1.9 dia	1978 Jan 8.8	71.43	89.64	6634	199	312	0.009	32
D	Fragments	1977-124C,E											

	Name		Launch date, lifetime and descent date	Shape and weight (kg)	Size (m)	Date of orbital determination	Orbital inclination (deg)	Nodal period (min)	Semi major axis (km)	Perigee height (km)	Apogee height (km)	Orbital eccentricity	Argument of perigee (deg)
D R	Cosmos 974	1978-01A	1978 Jan 6.66 12.60 days 1978 Jan 19.26	Sphere-cylinder 6300?	6.5 long? 2.4 dia	1978 Jan 7.5 1978 Jan 14.2	62.81 62.81	89.61 89.49	6634 6628	178 175	334 325	0.012 0.011	78 74
D	Cosmos 974 rocket	1978-01B	1978 Jan 6.66 5 days 1978 Jan 11	Cylinder 2500?	7.5 long 2.6 dia	1978 Jan 8.7	62.80	89.09	6609	168	293	0.009	70
D	Cosmos 974 engine	1978-01D	1978 Jan 6.66 20 days 1978 Jan 26	Cone 600? full	1.5 long? 2 dia?	1977 Jan 19	62.8	89.47	6627	189	308	0.009	-
D	Fragments	1978-01C,E											
T	Intelsat 4A F-3	1978-02A	1978 Jan 7.01 > million years	Cylinder 1500 full 795 empty	2.82 long 2.39 dia	1978 Jan 7.0 1978 Mar 15.5	21.82 0.37	640.44 1436.12	24612 42165	549 35783	35918 35790	0.719 0.0001	179 -
	Intelsat 4A F-3 rocket	1978-02B	1978 Jan 7.01 6000 years	Cylinder 1815	8.6 long 3.0 dia	1978 Jan 20.3	21.60	648.98	24830	612	36292	0.719	188
D 2M R	Soyuz 27*	1978-03A	1978 Jan 10.52 64.95 days 1978 Mar 16.47	Sphere-cylinder 6570?	7.5 long 2.3 dia	1978 Jan 10.6 1978 Jan 10.8 1978 Jan 11.6	51.71 51.58 51.60	88.71 89.90 91.28	6592 6651 6718	190 241 330	237 304 350	0.004 0.005 0.002	90 227 101
D	Soyuz 27 rocket	1978-03B	1978 Jan 10.52 3 days 1978 Jan 13	Cylinder 2500?	7.5 long 2.6 dia	1978 Jan 10.8	51.64	88.68	6590	193	231	0.003	79

* Soyuz 27 docked with Salyut 6 (first airlock) on 1978 Jan 11.59; undocked from Salyut 6 on 1978 Mar 16.33, but landed with the Soyuz 26 crew. The Soyuz 27 cosmonauts returned to Earth in the Soyuz 26 craft, which undocked from Salyut 6 (second airlock) on 1978 Jan 16. See page 512.

Name		Launch date, lifetime and descent date	Shape and weight (kg)	Size (m)	Date of orbital determination	Orbital inclination (deg)	Nodal period (min)	Semi major axis (km)	Perigee height (km)	Apogee height (km)	Orbital eccentricity	Argument of perigee (deg)
Cosmos 975	1978-04A	1978 Jan 10.56 60 years	Cylinder + 2 vanes? 2500?	5 long? 1.5 dia?	1978 Jan 12.4	81.22	97.62	7022	634	653	0.001	57
Cosmos 975 rocket	1978-04B	1978 Jan 10.56 60 years	Cylinder 1440	3.8 long 2.6 dia	1978 Jan 19.6	81.23	97.63	7022	594	694	0.007	146
Cosmos 976	1978-05A	1978 Jan 10.87 9000 years	Spheroid 40?	1.0 long? 0.8 dia?	1978 Jan 12.9	74.03	115.14	7839	1457	1465	0.0005	77
Cosmos 977	1978-05B	1978 Jan 10.87 7000 years	Spheroid 40?	1.0 long? 0.8 dia?	1978 Jan 14.9	74.03	114.54	7812	1403	1465	0.004	81
Cosmos 978	1978-05C	1978 Jan 10.87 8000 years	Spheroid 40?	1.0 long? 0.8 dia?	1978 Jan 18.6	74.03	114.74	7821	1421	1465	0.003	87
Cosmos 979	1978-05D	1978 Jan 10.87 9000 years	Spheroid 40?	1.0 long? 0.8 dia?	1978 Jan 14.5	74.03	114.95	7831	1440	1465	0.002	86
Cosmos 980	1978-05E	1978 Jan 10.87 10000 years	Spheroid 40?	1.0 long? 0.8 dia?	1978 Jan 13.9	74.03	115.36	7850	1465	1478	0.0008	287
Cosmos 981	1978-05F	1978 Jan 10.87 10000 years	Spheroid 40?	1.0 long? 0.8 dia?	1978 Jan 12.7	74.03	115.59	7860	1465	1498	0.002	264

1978-05 continued on page 518

	Name		Launch date, lifetime and descent date	Shape and weight (kg)	Size (m)	Date of orbital determination	Orbital inclination (deg)	Nodal period (min)	Semi major axis (km)	Perigee height (km)	Apogee height (km)	Orbital eccentricity	Argument of perigee (deg)
	Cosmos 982	1978-05G	1978 Jan 10.87 10000 years	Spheroid 40?	1.0 long? 0.8 dia?	1978 Jan 13.9	74.03	115.81	7870	1465	1518	0.003	273
	Cosmos 983	1978-05H	1978 Jan 10.87 10000 years	Spheroid 40?	1.0 long? 0.8 dia?	1978 Jan 18.6	74.03	116.05	7881	1465	1540	0.005	258
	Cosmos 976 rocket	1978-05J	1978 Jan 10.87 20000 years	Cylinder 2200?	7.4 long 2.4 dia	1978 Jan 14.0	74.03	117.74	7957	1465	1693	0.014	268
D R	Cosmos 984	1978-06A	1978 Jan 13.64 12.7 days 1978 Jan 26.3	Sphere-cylinder 5700?	5.0 long 2.4 dia	1978 Jan 14.6	62.81	89.45	6627	206	291	0.007	80
D	Cosmos 984 rocket	1978-06B	1978 Jan 13.64 9 days 1978 Jan 22	Cylinder 2500?	7.5 long 2.6 dia	1978 Jan 14.7	62.79	89.31	6620	207	276	0.005	74
	Cosmos 985*	1978-07A	1978 Jan 17.14 1200 years	Cylinder 700?	1.3 long? 1.9 dia?	1978 Jan 17.6	82.94	104.79	7362	945	1022	0.005	298
	Cosmos 985 rocket	1978-07B	1978 Jan 17.14 600 years	Cylinder 2200?	7.4 long 2.4 dia	1978 Jan 23.7	82.94	104.66	7356	943	1012	0.005	279

* Navigation satellite

	Name		Launch date, lifetime and descent date	Shape and weight (kg)	Size (m)	Date of orbital determination	Orbital inclination (deg)	Nodal period (min)	Semi major axis (km)	Perigee height (km)	Apogee height (km)	Orbital eccentricity	Argument of perigee (deg)
D	Progress 1*	1978-08A	1978 Jan 20.35 18.81 days 1978 Feb 8.16	Sphere- cylinder 7020	7.9 long 2.3 dia	1978 Jan 20.5 1978 Jan 21.7 1978 Jan 22.4	51.61 51.66 51.60	88.73 90.29 91.25	6593 6670 6717	173 250 329	256 334 348	0.005 0.006 0.002	86 199 -
D	Progress 1 rocket	1978-08B	1978 Jan 20.35 3 days 1978 Jan 23	Cylinder 2500?	7.5 long 2.6 dia	1978 Jan 20.7	51.65	88.66	6589	188	234	0.004	79
	Molniya 3J	1978-09A	1978 Jan 24.29 12 years	Windmill + 6 vanes 1500?	4.2 long? 1.6 dia	1978 Jan 26.4 1978 Feb 1.0	62.81 62.78	736.26 718.01	27010 26563	646 652	40618 39718	0.740 0.735	288 288
D	Molniya 3J launcher rocket	1978-09B	1978 Jan 24.29 47 days 1978 Mar 12	Cylinder 2500?	7.5 long 2.6 dia	1978 Jan 24.4	62.78	92.95	6798	215	624	0.030	117
D	Molniya 3J launcher	1978-09C	1978 Jan 24.29 38 days 1978 Mar 3	Irregular	-	1978 Jan 24.5	62.81	93.18	6809	219	643	0.031	120
	Molniya 3J rocket	1978-09E	1978 Jan 24.29 12 years	Cylinder 440	2.0 long 2.0 dia	1979 Jun 30.0	63.2	732.6	26920	813	40272	0.733	-
D	Fragment	1978-09D											
D R	Cosmos 986**	1978-10A	1978 Jan 24.41 13.8 days 1978 Feb 7.2	Sphere- cylinder 6300?	6.5 long? 2.4 dia	1978 Jan 25.1 1978 Jan 25.7	65.01 65.02	89.39 89.64	6623 6636	172 171	318 344	0.011 0.013	70 68
D	Cosmos 986 rocket	1978-10B	1978 Jan 24.41 4 days 1978 Jan 28	Cylinder 2500?	7.5 long 2.6 dia	1978 Jan 24.5	65.01	89.29	6618	172	308	0.010	67

* Unmanned fuel-and-supplies ferry, without Soyuz descent capability. Docked with Salyut 6 (second airlock) on 1978 Jan 22.42. Separated Feb 6.25. De-orbited over Pacific Ocean two days later. ** Cosmos 986 manoeuvred, but no jettisoned engine was apparently tracked or designated.

	Name		Launch date, lifetime and descent date	Shape and weight (kg)	Size (m)	Date of orbital determination	Orbital inclination (deg)	Nodal period (min)	Semi major axis (km)	Perigee height (km)	Apogee height (km)	Orbital eccentricity	Argument of perigee (deg)
D r	China 8*	1978-11A	1978 Jan 26.21 12 days 1978 Feb 7	Cylinder? 3600, then 1200?	-	1978 Jan 26.7	57.03	90.90	6698	161	479	0.024	160
D	China 8 rocket	1978-11B	1978 Jan 26.21 11 days 1978 Feb 6	Cylinder	8 long 3 dia	1978 Jan 26.7	57.02	90.79	6693	160	469	0.023	159
T	IUE 1**	1978-12A	1978 Jan 26.73 >million years	Octagonal- cylinder 669 full	4.3 long 1.3 dia	1978 Jan 26.8 1978 Jan 28.3	28.71 28.63	840.64 1435.7	29505 42157	173 25669	46081 45888	0.778 0.240	257 257
D	IUE 1 second stage	1978-12B	1978 Jan 26.73 26 days 1978 Feb 21	Cylinder + annulus 350?	6.4 long 1.52 and 2.44 dia	1978 Jan 26.8	28.74	96.56	6972	164	1024	0.062	254
	IUE 1 third stage	1978-12C	1978 Jan 26.73 7 years	Sphere- cone 66	1.32 long 0.94 dia	1978 Jan 27.0	28.71	846.69	29648	177	46367	0.779	257
D R	Cosmos 987	1978-13A	1978 Jan 31.62 13.6 days 1978 Feb 14.2	Sphere- cylinder 6300?	6.5 long? 2.4 dia	1978 Feb 1.5 1978 Feb 6.1	62.80 62.80	89.44 89.72	6626 6640	175 173	321 351	0.011 0.013	71 71
D	Cosmos 987 rocket	1978-13B	1978 Jan 31.62 4 days 1978 Feb 4	Cylinder 2500?	7.5 long 2.6 dia	1978 Feb 2.2	62.80	89.18	6613	169	301	0.010	67

*Capsule returned to Earth about 1978 Jan 30. **IUE is International Ultraviolet Explorer, launched by NASA.

1978-13 continued on page 521

	Name		Launch date, lifetime and descent date	Shape and weight (kg)	Size (m)	Date of orbital determination	Orbital inclination (deg)	Nodal period (min)	Semi major axis (km)	Perigee height (km)	Apogee height (km)	Orbital eccentricity	Argument of perigee (deg)
D	Cosmos 987 engine*	1978-13E	1978 Jan 31.62 20 days 1978 Feb 20	Cone 600? full	1.5 long? 2.0 dia?	1978 Feb 13	62.8	88.56	6582	166	241	0.006	-
D	Fragments	1978-13C,D											
T	Kyokko** (Exos A) [Mu-3H]	1978-14A	1978 Feb 4.29 300 years	Cylinder 103	0.8 long 0.95 dia	1978 Feb 7.8	65.37	134.27	8687	642	3975	0.192	334
D	Kyokko rocket	1978-14B	1978 Feb 4.29 373 days 1979 Feb 12	Cylinder?	-	1978 Feb 5.9	65.09	94.26	6862	331	636	0.022	160
	Fragments	1978-14C-F											
D R	Cosmos 988	1978-15A	1978 Feb 8.51 11.80 days 1978 Feb 20.31	Sphere-cylinder 5900?	5.9 long 2.4 dia	1978 Feb 11.8	72.84	89.87	6646	201	335	0.010	66
D	Cosmos 988 rocket	1978-15B	1978 Feb 8.51 9 days 1978 Feb 17	Cylinder 2500?	7.5 long 2.6 dia	1978 Feb 12.8	72.84	89.35	6620	195	289	0.007	52
D	Capsule	1978-15F	1978 Feb 8.51 31 days 1978 Mar 11	Ellipsoid 200?	0.9 long 1.9 dia	1978 Feb 20	72.8	89.26	6615	203	271	0.005	-
D	Fragments	1978-15C-E,G											
T	Fleetsatcom 1	1978-16A	1978 Feb 9.89 > million years	Hexagonal cylinder 1884 full	1.27 long 2.44 dia	1978 Feb 9.9 1979 Jun 30.0	26.46 1.7	634.16 1436.0	24451 42163	167 35755	35978 35816	0.732 0.001	182 -
D	Fleetsatcom 1 rocket	1978-16B	1978 Feb 9.89 1664 days 1982 Aug 31	Cylinder 1815	8.6 long 3.0 dia	1978 Feb 9.9	26.40	620.40	24096	172	35263	0.728	182
	Fragment	1978-16C											

* Jettisoned from Cosmos 987 about 1978 Feb 13.

** Japanese contribution to International Magnetospheric Study; Kyokko means aurora.

	Name		Launch date, lifetime and descent date	Shape and weight (kg)	Size (m)	Date of orbital determination	Orbital inclination (deg)	Nodal period (min)	Semi major axis (km)	Perigee height (km)	Apogee height (km)	Orbital eccentricity	Argument of perigee (deg)
D R	Cosmos 989	1978-17A	1978 Feb 14.40 13.8 days 1978 Feb 28.2	Sphere-cylinder 5900?	5.9 long? 2.4 dia	1978 Feb 20.1	65.05	89.36	6622	169	318	0.011	67
D	Cosmos 989 rocket	1978-17B	1978 Feb 14.40 3 days 1978 Feb 17	Cylinder 2500?	7.5 long 2.6 dia	1978 Feb 15.4	65.06	89.28	6618	165	314	0.011	72
D	Capsule ?†	1978-17C	1978 Feb 14.40 15 days 1978 Mar 1	Ellipsoid 200?	0.9 long 1.9 dia	1978 Feb 28	65.0	88.86	6598	175	264	0.007	-
D	Fragment	1978-17D											
T	Ume 2* (ISS 2) [Nu]	1978-18A	1978 Feb 16.17 1400 years	Cylinder 140	0.82 long 0.94 dia	1978 Feb 18.7	69.37	107.25	7478	975	1224	0.017	202
	Ume 2 rocket	1978-18B	1978 Feb 16.17 700 years	Cylinder 66?	1.74 long 1.65 dia	1978 Feb 18.7	69.36	107.24	7477	975	1223	0.017	202
	Fragment	**1978-18C**											
	Cosmos 990	1978-19A	1978 Feb 17.69 120 years	Cylinder + paddles? 750?	2 long? 1 dia?	1978 Feb 18.2	74.05	100.80	7174	783	809	0.002	359
	Cosmos 990 rocket	1978-19B	1978 Feb 17.69 100 years	Cylinder 2200?	7.4 long 2.4 dia	1978 Feb 19.9	74.04	100.67	7168	774	805	0.002	21

*Japanese Ionospheric Sounding Satellite. † Possibly an engine.

	Name		Launch date, lifetime and descent date	Shape and weight (kg)	Size (m)	Date of orbital determination	Orbital inclination (deg)	Nodal period (min)	Semi major axis (km)	Perigee height (km)	Apogee height (km)	Orbital eccentricity	Argument of perigee (deg)
T	Navstar 1 (GPS)* [Atlas F]	1978-20A	1978 Feb 22.99 1 million years	Cylinder + 4 vanes 433	-	1978 Mar 7.0	63.27	718.67	26580	20095	20308	0.004	348
	Navstar 1 rocket	1978-20B	1978 Feb 22.99 40 years	Cone-cylinder? 163?	1.85 long? 0.63 to 1.65 dia?	1978 Feb 23.2	63.00	354.81	16605	161	20292	0.606	158
T	SDS 4 [Titan 3B Agena D]	1978-21A	1978 Feb 25.2? 100 years?	Cylinder	-	1978 Feb 25	63.15	703.7	26222	311	39377	0.745	270**
	SDS 4 rocket	1978-21B	1978 Feb 25.2? 100 years?	Cylinder 700?	6 long? 1.5 dia	Orbit similar to 1978-21A							
	Cosmos 991	1978-22A	1978 Feb 28.28 1200 years	Cylinder 700?	1.3 long? 1.9 dia?	1978 Mar 1.1	82.98	104.84	7364	963	1009	0.003	301
	Cosmos 991 rocket	1978-22B	1978 Feb 28.28 600 years	Cylinder 2200?	7.4 long 2.4 dia	1978 Mar 2.2	82.98	104.74	7359	960	1002	0.003	287
D 2M R	Soyuz 28†	1978-23A	1978 Mar 2.64 7.93 days 1978 Mar 10.57	Sphere-cylinder 6570?	7.5 long 2.3 dia	1978 Mar 2.9 1978 Mar 3.3 1978 Mar 3.9	51.63 51.62 51.62	88.82 90.02 91.35	6597 6657 6722	192 251 334	246 306 353	0.004 0.004 0.001	86 283 233
D	Soyuz 28 rocket	1978-23B	1978 Mar 2.64 3 days 1978 Mar 5	Cylinder 2500?	7.5 long 2.6 dia	1978 Mar 3.2	51.63	88.80	6596	191	245	0.004	87

* Global Positioning System
** Approximate orbit

† Soyuz 28 docked with Salyut 6 (2nd airlock) on 1978 Mar 3 at 17:10 UT, with one Russian and one Czechoslovak cosmonaut; undocked 1978 Mar 10.43

	Name		Launch date, lifetime and descent date	Shape and weight (kg)	Size (m)	Date of orbital determination	Orbital inclination (deg)	Nodal period (min)	Semi major axis (km)	Perigee height (km)	Apogee height (km)	Orbital eccentricity	Argument of perigee (deg)
	Molniya 1AQ	1978-24A	1978 Mar 2.92 $14\frac{1}{4}$ years	Windmill + 6 vanes 1000?	3.4 long 1.6 dia	1978 Mar 5.0 1978 Mar 13.1	62.82 62.83	738.14 717.79	27056 26557	617 615	40739 39743	0.741 0.737	288 288
D	Molniya 1AQ launcher	1978-24B	1978 Mar 2.92 39 days 1978 Apr 10	Irregular	-	1978 Mar 4.5	62.83	92.70	6786	210	605	0.029	116
D	Molniya 1AQ launcher rocket	1978-24C	1978 Mar 2.92 38 days 1978 Apr 9	Cylinder 2500?	7.5 long 2.6 dia	1978 Mar 4.2	62.76	92.48	6775	219	575	0.026	118
	Molniya 1AQ rocket	1978-24D	1978 Mar 2.92 $14\frac{1}{4}$ years	Cylinder 440	2.0 long 2.0 dia	1979 Jun 30.0	63.9	729.4	26843	671	40259	0.737	-
D R	Cosmos 992	1978-25A	1978 Mar 4.32 12.9 days 1978 Mar 17.2	Sphere-cylinder 5700?	5.0 long 2.4 dia	1978 Mar 4.8	71.34	89.79	6641	203	323	0.009	59
D	Cosmos 992 rocket	1978-25B	1978 Mar 4.32 6 days 1978 Mar 10	Cylinder 2500?	7.5 long 2.6 dia	1978 Mar 4.8	71.34	89.63	6633	201	309	0.008	50
D	Fragment	1978-25C											
T	Landsat 3 (ERTS 3)	1978-26A	1978 Mar 5.75 100 years	Cone + 2 paddles 960	3.0 long 1.45 dia 3.96 span	1978 Mar 5.8	99.14	103.21	7287	900	918	0.001	307
T	Oscar 8	1978-26B	1978 Mar 5.75 100 years	Rectangular box 27	0.43 x 0.30 x 0.15?	1978 Mar 5.9	98.99	103.23	7288	903	917	0.001	221
	Landsat 3 second stage*	1978-26C	**1978 Mar 5.75** **disintegrated**	Cylinder + annulus 350?	6.4 long 1.52 and 2.44 dia	1978 Mar 5.9	98.95	103.22	7288	906	913	0.001	306
16d	**Fragments**	**1978-26D-FT**											

* Carried PIX — a 34 kg Plasma Interaction Experiment. Landsat 3 second stage disintegrated about 1981 Jan 27.19 over Antarctica near 80 deg south, 61 deg west.

	Name		Launch date, lifetime and descent date	Shape and weight (kg)	Size (m)	Date of orbital determination	Orbital inclination (deg)	Nodal period (min)	Semi major axis (km)	Perigee height (km)	Apogee height (km)	Orbital eccentricity	Argument of perigee (deg)
D R	Cosmos 993	1978-27A	1978 Mar 10.45 12.7 days 1978 Mar 23.2	Sphere-cylinder 6300?	6.5 long? 2.4 dia	1978 Mar 11.8 1978 Mar 18.1	72.86 72.85	89.63 90.15	6634 6660	171 170	340 393	0.013 0.017	79 64
D	Cosmos 993 rocket	1978-27B	1978 Mar 10.45 4 days 1978 Mar 14	Cylinder 2500?	7.5 long 2.6 dia	1978 Mar 11.3	72.86	89.32	6618	170	310	0.011	76
D	Cosmos 993 engine	1978-27D	1978 Mar 10.45 14 days 1978 Mar 24	Cone 600? full	1.5 long? 2 dia?	1978 Mar 23	72.8	89.04	6604	180	272	0.007	-
D	Fragment	1978-27C											
	Cosmos 994	1978-28A	1978 Mar 15.66 1200 years	Cylinder 700?	1.3 long? 1.9 dia?	1978 Mar 18.4	82.93	105.05	7374	980	1011	0.002	261
	Cosmos 994 rocket	1978-28B	1978 Mar 15.66 600 years	Cylinder 2200?	7.4 long 2.4 dia	1978 Mar 18.4	82.93	104.93	7368	977	1003	0.002	245
D	[Titan 3D]	1978-29A	1978 Mar 16.78 179 days 1978 Sep 11	Cylinder 13300? full	15 long 3.0 dia	1978 Mar 23.2	96.43	88.52	6578	160	240	0.006	125
T	Capsule	1978-29B	1978 Mar 16.78 60 years	Octagon? 60?	0.3 long? 0.9 dia	1978 Mar 19.8	95.83	97.59	7020	639	645	0.0004	290
D	Titan 3D rocket	1978-29C	1978 Mar 16.78 1 day 1978 Mar 17	Cylinder 1900	6 long 3.0 dia	1978 Mar 17	96.4	88.50	6577	166	231	0.005	-

	Name		Launch date, lifetime and descent date	Shape and weight (kg)	Size (m)	Date of orbital determination	Orbital inclination (deg)	Nodal period (min)	Semi major axis (km)	Perigee height (km)	Apogee height (km)	Orbital eccentricity	Argument of perigee (deg)
D R	Cosmos 995	1978-30A	1978 Mar 17.45 12.85 days 1978 Mar 30.30	Sphere-cylinder 5700?	5.0 long 2.4 dia	1978 Mar 18.5	81.34	89.05	6604	217	235	0.001	52
D	Cosmos 995 rocket	1978-30B	1978 Mar 17.45 3 days 1978 Mar 20	Cylinder 2500?	7.5 long 2.6 dia	1978 Mar 17.8	81.34	88.81	6592	208	220	0.001	8
	Cosmos 996	1978-31A	1978 Mar 28.06 1200 years	Cylinder 700?	1.3 long? 1.9 dia?	1978 Mar 28.3	82.93	104.80	7362	957	1010	0.004	299
	Cosmos 996 rocket	1978-31B	1978 Mar 28.06 600 years	Cylinder 2200?	7.4 long 2.4 dia	1978 Mar 31.5	82.92	104.70	7357	957	1000	0.003	288
D R?	Cosmos 997	1978-32A	1978 Mar 30.00 < 1 day 1978 Mar 30	-	-		Orbit similar to 1978-32C						
D R?	Cosmos 998	1978-32B	1978 Mar 30.00 <1 day 1978 Mar 30	-	-		Orbit similar to 1978-32C						
D	Cosmos 997 rocket	1978-32C	1978 Mar 30.00 3 days 1978 Apr 2	Cylinder 4000?	12 long? 4 dia	1978 Mar 30.1	51.60	88.45	6579	188	213	0.002	237
D	Fragment	1978-32D											

	Name		Launch date, lifetime and descent date	Shape and weight (kg)	Size (m)	Date of orbital determination	Orbital inclination (deg)	Nodal period (min)	Semi major axis (km)	Perigee height (km)	Apogee height (km)	Orbital eccentricity	Argument of perigee (deg)
D R	Cosmos 999	1978-33A	1978 Mar 30.33 12.89 days 1978 Apr 12.22	Sphere-cylinder 5900?	5.9 long? 2.4 dia	1978 Mar 31.3	71.39	89.79	6641	174	352	0.013	55
D	Cosmos 999 rocket	1978-33B	1978 Mar 30.33 4 days 1978 Apr 3	Cylinder 2500?	7.5 long 2.6 dia	1978 Mar 31.4	71.40	89.43	6623	174	316	0.011	50
D	Capsule?†	1978-33E	1978 Mar 30.33 25 days 1978 Apr 24	Ellipsoid? 200?	0.9 long? 1.9 dia?	1978 Apr 12	71.3	89.39	6621	189	297	0.008	-
D	Fragments	1978-33C,D											
	Cosmos 1000*	1978-34A	1978 Mar 31.58 1200 years	Cylinder 700?	1.3 long? 1.9 dia?	1978 Mar 31.9	82.93	104.90	7367	965	1012	0.003	290
	Cosmos 1000 rocket	1978-34B	1978 Mar 31.58 600 years	Cylinder 2200?	7.4 long 2.4 dia	1978 Mar 31.9	82.94	104.78	7361	964	1001	0.003	294
T	Intelsat 4A F-6 ††	1978-35A	1978 Mar 31.98 > million years	Cylinder 1500 full 795 empty	2.82 long 2.39 dia	1978 Apr 1.0 1978 Jul 15.0	21.85 0.3	641.03 1436.1	24627 42165	549 35768	35949 35806	0.719 0.0005	179 -
	Intelsat 4A F-6 rocket	1978-35B	1978 Mar 31.98 6000 years	Cylinder 1815	8.6 long 3.0 dia	1978 Apr 25.9	21.90	647.40	24790	596	36227	0.719	195

* Navigational beacon. † Possibly an engine. †† Intelsat 4A F-5 failed to reach orbit on 1977 Sep 29.

	Name		Launch date, lifetime and descent date	Shape and weight (kg)	Size (m)	Date of orbital determination	Orbital inclination (deg)	Nodal period (min)	Semi major axis (km)	Perigee height (km)	Apogee height (km)	Orbital eccentricity	Argument of perigee (deg)
D R?	Cosmos 1001	1978-36A	1978 Apr 4.63 10.87 days 1978 Apr 15.50	-	-	1978 Apr 4.8 1978 Apr 6.2 1978 Apr 11.6	51.62 51.63 51.61	88.71 89.28 90.73	6592 6620 6691	199 196 307	228 288 318	0.002 0.007 0.001	84 66 302
D	Cosmos 1001 rocket	1978-36B	1978 Apr 4.63 2 days 1978 Apr 6	-	-	1978 Apr 4.8	51.63	88.59	6586	198	217	0.001	90
D	Fragments	1978-36C-F											
D R	Cosmos 1002	1978-37A	1978 Apr 6.39 12.84 days 1978 Apr 19.23	Sphere-cylinder 5700?	5.0 long 2.4 dia	1978 Apr 8.8	65.05	89.37	6622	205	283	0.006	52
D	Cosmos 1002 rocket	1978-37B	1978 Apr 6.39 4 days 1978 Apr 10	Cylinder 2500?	7.5 long 2.6 dia	1978 Apr 7.3	65.05	89.13	6610	197	267	0.005	28
D	Fragments	1978-37C,D											
T	[Atlas Agena D]*	1978-38A	1978 Apr 8.03 > million years	Cylinder 700 full? 350 empty?	1.7 long? 1.4 dia?	1978 Apr 8.0 1978 Apr 8.4	29.9 28.4	87.72 615.5	6547 23970	149 150	189 35033	0.003 0.728	- -
						Probably entered synchronous orbit similar to 1977-114A							
	Agena D rocket	1978-38B	1978 Apr 8.03 10 years?	Cylinder 700	6 long? 1.5 dia	Orbit similar to 1978-38A second orbit							
T	Yuri (BSE 1)**	1978-39A	1978 Apr 7.92 > million years	Irregular 678 full 327 empty	3.09 long 1.32 wide 1.2 deep	1978 Apr 8.7 1978 Apr 9.5 1978 Apr 26.5	27.23 0.10 0.08	627.64 1415.75 1436.0	24287 41766 42163	164 35115 35784	35653 35662 35786	0.731 0.007 0	180 266 247
	Yuri second stage	1978-39B	1978 Apr 7.92 250 years	Cylinder+ annulus 350?	6.4 long 1.52 and 2.44 dia	1978 Apr 7.9 1978 Apr 20.2	28.64 28.23	92.93 111.35	6796 7673	165 569	671 2021	0.037 0.095	173 143

* Early warning satellite development.

** Broadcasting Satellite Experiment launched for Japan by NASA.

1978-39 continued on page 529

	Name		Launch date, lifetime and descent date	Shape and weight (kg)	Size (m)	Date of orbital determination	Orbital inclination (deg)	Nodal period (min)	Semi major axis (km)	Perigee height (km)	Apogee height (km)	Orbital eccentricity	Argument of perigee (deg)
	Yuri third stage	1978-39C	1978 Apr 7.92 10 years	Sphere - cone 66	1.32 long 0.94 dia	1978 Dec 19.6	26.92	392.40	17750	225	22520	0.628	206
D R	Cosmos 1003	1978-40A	1978 Apr 20.65 13.6 days 1978 May 4.3	Sphere-cylinder 6300?	6.5 long? 2.4 dia	1978 Apr 20.9 1978 Apr 23.9	62.81 62.81	89.54 88.89	6631 6599	178 162	328 279	0.011 0.009	71 52
D	Cosmos 1003 rocket	1978-40B	1978 Apr 20.65 3 days 1978 Apr 23	Cylinder 2500?	7.5 long 2.6 dia	1978 Apr 21.2	62.80	89.32	6620	169	315	0.011	68
D	Cosmos 1003 engine	1978-40D	1978 Apr 20.65 14 days 1978 May 4	Cone 600? full	1.5 long? 2 dia?	1978 May 4	62.8	89.00	6604	167	284	0.009	-
D	Fragments	1978-40C,E											
D	HCMM (AEM 1)* [Scout]	1978-41A	1978 Apr 26.43 1336 days 1981 Dec 22	Hexagonal prism 134	0.64 long 0.7 wide	1978 May 1.4 1980 Aug 30	97.60 97.66	96.72 95.47	6979 6919	560 537	641 546	0.006 0.001	245 257
D	HCMM rocket	1978-41B	1978 Apr 26.43 943 days 1980 Nov 24	Cylinder 24	1.50 long 0.46 dia	1978 Apr 30.1	97.60	96.89	6987	564	653	0.006	250
T	AMS 3 [Thor Burner 2]	1978-42A	1978 May 1.13 80 years	Irregular 513	6.40 long 1.68 dia	1978 May 1.9	98.71	101.47	7206	820	835	0.001	212
	Burner 2 rocket	1978-42B	1978 May 1.13 60 years	Sphere - cone 66	1.32 long 0.94 dia	1980 Jun 30	98.70	101.41	7203	820	829	0.0006	174
	Fragments	1978-42C-E											

* Heat Capacity Mapping Mission (Applications Explorer Mission).

	Name		Launch date, lifetime and descent date	Shape and weight (kg)	Size (m)	Date of orbital determination	Orbital inclination (deg)	Nodal period (min)	Semi major axis (km)	Perigee height (km)	Apogee height (km)	Orbital eccentricity	Argument of perigee (deg)
D R	Cosmos 1004	1978-43A	1978 May 5.65 12.6 days 1978 May 18.3	Sphere - cylinder 5700?	5.0 long 2.4 dia	1978 May 6.3	62.81	89.43	6626	205	290	0.006	78
D	Cosmos 1004 rocket	1978-43B	1978 May 5.65 4.24 days 1978 May 9.89	Cylinder 2500?	7.5 long 2.6 dia	1978 May 5.9	62.80	89.27	6618	200	279	0.006	65
D	Fragments	1978-43C-G											
T	OTS 2 *	1978-44A	1978 May 11.96 > million years	Hexagonal box 865 full 444 empty	2.13 long 1.68 wide 2.39 high	1978 May 12.0 1980 Jun 30	27.32 0.0	633.80 1436.2	24442 42168	184 35756	35943 35819	0.732 0.0007	179 -
	OTS 2 second stage	1978-44B	1978 May 11.96 60000 years	Cylinder + annulus 350?	6.4 long 1.52 and 2.44 dia	1978 May 11.9 1978 May 15.3	28.45 27.93	107.09 139.69	7470 8924	180 1568	2004 3524	0.122 0.110	173 86
	OTS 2 third stage	1978-44C	1978 May 11.96 20 years	Sphere - cone 66	1.32 long 0.94 dia	1978 May 12.0	27.32	633.17	24436	182	35933	0.732	179
	Cosmos 1005	1978-45A	1978 May 12.17 60 years	Cylinder + 2 vanes? 2500?	5 long? 1.5 dia?	1978 May 16.6	81.24	97.54	7018	627	653	0.002	0
	Cosmos 1005 rocket	1978-45B	1978 May 12.17 60 years	Cylinder 1440	3.8 long 2.6 dia	1978 May 15.2	81.25	97.69	7025	603	691	0.006	167

* Orbital Test Satellite launched for ESA by NASA (OTS 1 failed to enter orbit on 1977 Sep 13). Entered synchronous orbit on 1978 May 13.

	Name		Launch date, lifetime and descent date	Shape and weight (kg)	Size (m)	Date of orbital determination	Orbital inclination (deg)	Nodal period (min)	Semi major axis (km)	Perigee height (km)	Apogee height (km)	Orbital eccentricity	Argument of perigee (deg)
D	Cosmos 1006	1978-46A	1978 May 12.46 306 days 1979 Mar 14	Cylinder?	4 long? 2 dia?	1978 May 14.8	65.85	92.45	6773	382	408	0.002	342
D	Cosmos 1006 rocket	1978-46B	1978 May 12.46 208 days 1978 Dec 6	Cylinder 2200?	7.4 long 2.4 dia	1978 May 13.9	65.85	92.33	6767	373	405	0.002	7
T	Navstar 2 (GPS) [Atlas F]	1978-47A	1978 May 13.44 1 million years	Cylinder + 4 vanes 433	-	1978 May 22.0	63.13	711.30	26396	19952	20084	0.003	98
	Navstar 2 rocket	1978-47B	1978 May 13.44 30 years	Cone - cylinder? 163?	1.85 long? 0.63 to 1.65 dia?	1978 May 16.1	63.07	350.45	16466	162	20014	0.603	158
D R	Cosmos 1007	1978-48A	1978 May 16.45 12.7 days 1978 May 29.2	Sphere-cylinder 6300?	6.5 long? 2.4 dia	1978 May 16.7 1978 May 16.9	72.83 72.84	89.69 89.80	6637 6643	168 170	350 359	0.014 0.014	82 83
D	Cosmos 1007 rocket	1978-48B	1978 May 16.45 4 days 1978 May 20	Cylinder 2500?	7.5 long 2.6 dia	1978 May 17.1	72.83	89.57	6631	170	336	0.013	80
D	Cosmos 1007 engine	1978-48D	1978 May 16.45 15 days 1978 May 31	Cone 600? full	1.5 long? 2 dia?	1978 May 28.7	72.80	89.27	6616	158	317	0.012	53
D	Fragments	1978-48C,E											

	Name		Launch date, lifetime and descent date	Shape and weight (kg)	Size (m)	Date of orbital determination	Orbital inclina-tion (deg)	Nodal period (min)	Semi major axis (km)	Perigee height (km)	Apogee height (km)	Orbital eccen-tricity	Argument of perigee (deg)
D	Cosmos 1008	1978-49A	1978 May 17.61 967 days 1981 Jan 8	Cylinder + paddles? 900?	2 long? 1 dia?	1978 May 21.5	74.04	95.12	6902	499	549	0.004	332
D	Cosmos 1008 rocket	1978-49B	1978 May 17.61 996 days 1981 Feb 6	Cylinder 2200?	7.4 long 2.4 dia	1978 May 21.3	74.05	95.00	6896	488	548	0.004	334
D	Fragments**	1978-49C-F											
D	Cosmos 1009 *	1978-50A	1978 May 19.02 0.17 day? 1978 May 19.19?	Cylinder?	4 long? 2 dia?		Initial transfer orbit similar to 1978-50B						
						1978 May 19.1	65.86	108.64	7543	966	1364	0.026	41
D	Cosmos 1009 rocket	1978-50B	1978 May 19.02 17 days 1978 Jun 5	Cylinder 1500?	8 long? 2.5 dia?	1978 May 19.9	65.14	97.41	7014	147	1125	0.070	57
D	Fragments	1978-50C,D											
D R	Cosmos 1010	1978-52A	1978 May 23.32 12.83 days 1978 Jun 5.15	Sphere - cylinder 5900?	5.9 long 2.4 dia	1978 May 24.2	81.37	88.99	6602	217	230	0.001	55
D	Cosmos 1010 rocket	1978-52B	1978 May 23.32 3 days 1978 May 26	Cylinder 2500?	7.5 long 2.6 dia	1978 May 23.4	81.37	88.91	6598	212	227	0.001	16
D	Capsule	1978-52C	1978 May 23.32 15 days 1978 Jun 7	Ellipsoid 200?	0.9 long 1.9 dia	1978 Jun 5	81.3	88.41	6573	191	198	0.0005	-

Space Vehicle : Pioneer Venus 1 (1978-51A), and Centaur rocket (1978-51B)

* May have passed close to Cosmos 967 (1977-116A). Probably re-entered near 10°N, 147°E?

** Fragment 1978-49F has probably decayed and a different fragment (from a 100.9 min launch) now carries that designation in the USA.

	Name		Launch date, lifetime and descent date	Shape and weight (kg)	Size (m)	Date of orbital determination	Orbital inclination (deg)	Nodal period (min)	Semi major axis (km)	Perigee height (km)	Apogee height (km)	Orbital eccentricity	Argument of perigee (deg)
	Cosmos 1011	1978-53A	1978 May 23.70 1200 years	Cylinder 700?	1.3 long? 1.9 dia?	1978 May 24.9	82.91	104.90	7367	961	1016	0.004	286
	Cosmos 1011 rocket	1978-53B	1978 May 23.70 600 years	Cylinder 2200?	7.4 long 2.4 dia	1978 May 24.2	82.92	104.78	7361	961	1004	0.003	283
D R	Cosmos 1012	1978-54A	1978 May 25.61 12.6 days 1978 Jun 7.2	Sphere - cylinder 5700?	5.0 long 2.4 dia	1978 May 26.6	62.80	89.15	6612	202	265	0.005	51
D	Cosmos 1012 rocket	1978-54B	1978 May 25.61 3 days 1978 May 28	Cylinder 2500?	7.5 long 2.6 dia	1978 May 26.1	62.80	88.90	6599	189	253	0.005	25
	Molniya 1AR	1978-55A	1978 Jun 2.51 100 years?	Windmill + 6 vanes 1000?	3.4 long 1.6 dia	1978 Jun 3.1 1978 Jun 9.7	62.85 62.81	736.26 718.15	27010 26567	422 423	40842 39954	0.748 0.744	280 280
D	Molniya 1AR launcher	1978-55B	1978 Jun 2.51 19 days 1978 Jun 21	Irregular	-	1978 Jun 3.0	62.85	90.85	6695	211	422	0.016	130
D	Molniya 1AR launcher rocket	1978-55C	1978 Jun 2.51 15 days 1978 Jun 17	Cylinder 2500?	7.5 long 2.6 dia	1978 Jun 3.1	62.84	90.83	6694	195	436	0.018	120
	Molniya 1AR rocket	1978-55E	1978 Jun 2.51 100 years?	Cylinder 440	2.0 long 2.0 dia	1979 Jun 30.0	63.2	732.5	26917	619	40460	0.740	-
D	Fragment	1978-55D											

Name		Launch date, lifetime and descent date	Shape and weight (kg)	Size (m)	Date of orbital determination	Orbital inclination (deg)	Nodal period (min)	Semi major axis (km)	Perigee height (km)	Apogee height (km)	Orbital eccentricity	Argument of perigee (deg)
Cosmos 1013	1978-56A	1978 Jun 7.91 10000 years	Spheroid 40?	1.0 long? 0.8 dia?	1978 Jun 11.5	74.02	116.40	7897	1480	1557	0.005	239
Cosmos 1014	1978-56B	1978 Jun 7.91 10000 years	Spheroid 40?	1.0 long? 0.8 dia?	1978 Jun 11.9	74.02	116.15	7885	1480	1534	0.003	234
Cosmos 1015	1978-56C	1978 Jun 7.91 10000 years	Spheroid 40?	1.0 long? 0.8 dia?	1978 Jun 11.4	74.02	115.93	7875	1475	1519	0.003	212
Cosmos 1016	1978-56D	1978 Jun 7.91 10000 years	Spheroid 40?	1.0 long? 0.8 dia?	1978 Jun 11.4	74.02	115.70	7865	1473	1501	0.002	186
Cosmos 1017	1978-56E	1978 Jun 7.91 9000 years	Spheroid 40?	1.0 long? 0.8 dia?	1978 Jun 11.5	74.02	115.49	7856	1460	1495	0.002	153
Cosmos 1018	1978-56F	1978 Jun 7.91 9000 years	Spheroid 40?	1.0 long? 0.8 dia?	1978 Jun 10.9	74.02	115.27	7846	1444	1491	0.003	131
Cosmos 1019	1978-56G	1978 Jun 7.91 8000 years	Spheroid 40?	1.0 long? 0.8 dia?	1978 Jun 11.5	74.02	115.06	7836	1425	1491	0.004	122
Cosmos 1020	1978-56H	1978 Jun 7.91 8000 years	Spheroid 40?	1.0 long? 0.8 dia?	1978 Jun 10.9	74.02	114.85	7827	1410	1487	0.005	108
Cosmos 1013 rocket	1978-56J	1978 Jun 7.91 20000 years	Cylinder 2200?	7.4 long 2.4 dia	1978 Jun 10.5	74.02	117.95	7967	1486	1691	0.013	262

	Name		Launch date, lifetime and descent date	Shape and weight (kg)	Size (m)	Date of orbital determination	Orbital inclination (deg)	Nodal period (min)	Semi major axis (km)	Perigee height (km)	Apogee height (km)	Orbital eccentricity	Argument of perigee (deg)
D R	Cosmos 1021	1978-57A	1978 Jun 10.36 12.85 days 1978 Jun 23.21	Sphere-cylinder 6300?	6.5 long? 2.4 dia	1978 Jun 11.8 1978 Jun 18.4	65.03 65.03	89.35 89.46	6621 6627	173 171	313 326	0.011 0.012	59 56
D	Cosmos 1021 rocket	1978-57B	1978 Jun 10.36 3 days 1978 Jun 13	Cylinder 2500?	7.5 long 2.6 dia	1978 Jun 10.5	65.04	89.29	6618	172	308	0.010	58
D	Cosmos 1021 engine	1978-57D	1978 Jun 10.36 16 days 1978 Jun 26	Cone 600? full	1.5 long? 2 dia?	1978 Jun 24.3	65.02	88.72	6590	167	256	0.007	52
D	Fragment	1978-57C											
T	IMEWS 8 [Titan 3C]	1978-58A	1978 Jun 10.80 > million years	-	-	1978 Jun 11.0 1978 Jul 1.0	26.3 12.0	633.0 1446.3	24445 42362	295 29929	35840 42039	0.727 0.143	180 -
	Transtage	1978-58B	1978 Jun 10.80 > million years	Cylinder 1500?	6 long? 3.0 dia		Orbits similar to 1978-58A						
D	Titan 3C second stage	1978-58C	1978 Jun 10.80 4 days 1978 Jun 14	Cylinder 1900	6 long 3.0 dia	1978 Jun 10.8	29.98	91.14	6716	152	524	0.028	125

	Name		Launch date, lifetime and descent date	Shape and weight (kg)	Size (m)	Date of orbital determination	Orbital inclination (deg)	Nodal period (min)	Semi major axis (km)	Perigee height (km)	Apogee height (km)	Orbital eccentricity	Argument of perigee (deg)
D R	Cosmos 1022	1978-59A	1978 Jun 12.44 12.76 days 1978 Jun 25.20	Sphere- cylinder 6300?	6.5 long? 2.4 dia	1978 Jun 14.4 1978 Jun 22.6	72.84 72.84	89.67 89.69	6636 6637	171 167	344 350	0.013 0.014	77 60
D	Cosmos 1022 rocket	1978-59B	1978 Jun 12.44 4 days 1978 Jun 16	Cylinder 2500?	7.5 long 2.6 dia	1978 Jun 12.8	72.85	89.54	6629	167	335	0.013	80
D	Cosmos 1022 engine	1978-59C	1978 Jun 12.44 16 days 1978 Jun 28	Cone 600? full	1.5 long? 2 dia?	1978 Jun 24.7	72.86	89.48	6626	166	330	0.012	51
D	Fragment	1978-59D											
D	[Titan 3D]	1978-60A	1978 Jun 14.77 1166 days 1981 Aug 23	Cylinder 13300? full	15 long 3.0 dia	1978 Jun 15.7 1978 Jun 16.5	96.96 96.82	91.90 92.42	6744 6771	223 276	509 509	0.021 0.017	159 155
D	Titan 3D rocket	1978-60B	1978 Jun 14.77 40 days 1978 Jul 24	Cylinder 1900	6 long 3.0 dia	1978 Jun 15.3	96.96	91.74	6736	221	495	0.020	158
D	Fragments	1978-60C-F											
D 2M R	Soyuz 29*	1978-61A	1978 Jun 15.85 79.64 days 1978 Sep 3.49	Sphere- cylinder 6570?	7.5 long 2.3 dia	1978 Jun 16.0 1978 Jun 16.2 1978 Jun 22.6	51.63 51.64 51.63	88.85 90.07 91.39	6599 6659 6724	193 253 338	248 309 353	0.004 0.004 0.001	74 262 32
D	Soyuz 29 rocket	1978-61B	1978 Jun 15.85 3 days 1978 Jun 18	Cylinder 2500?	7.5 long 2.6 dia	1978 Jun 16.4	51.62	88.63	6588	186	233	0.004	76

* Soyuz 29 docked with Salyut 6 (first airlock) 1978 Jun 16.92; undocked from Salyut 6 on 1978 Sep 3, but landed with the Soyuz 31 crew (see page 543).

	Name		Launch date, lifetime and descent date	Shape and weight (kg)	Size (m)	Date of orbital determination	Orbital inclination (deg)	Nodal period (min)	Semi major axis (km)	Perigee height (km)	Apogee height (km)	Orbital eccentricity	Argument of perigee (deg)
T	GOES 3	1978-62A	1978 Jun 16.46 >million years	Cylinder + boom 627 full 243 empty	2.30 long 1.90 dia	1978 Jun 16.5 1978 Jun 17.1 1979 Jun 30.0	23.90 1.78 0.3	649.03 1446.85 1436.1	24832 42375 42165	198 35473 35781	36709 36521 35794	0.735 0.012 0	178 164 -
	GOES 3 second stage	1978-62B	1978 Jun 16.46 200 years	Cylinder + annulus 350?	6.4 long 1.52 and 2.44 dia	1978 Jun 16.5 1978 Jun 20.4	28.42 28.43	93.83 108.01	6840 7520	178 553	746 1730	0.042 0.078	165 97
	GOES 3 third stage	1978-62C	1978 Jun 16.46 10 years?	Sphere-cone 66	1.32 long 0.94 dia	1978 Jun 16.5	23.79	655.06	24988	167	37053	0.738	177
	GOES 3 apogee motor	1978-62D	1978 Jun 16.46 >million years	- 384 full	-	Orbit similar to second 1978-62A orbit							
	Cosmos 1023	1978-63A	1978 Jun 21.40 120 years	Cylinder + paddles? 750?	2 long? 1 dia?	1978 Jun 24.7	74.08	100.76	7172	783	805	0.002	356
	Cosmos 1023 rocket	1978-63B	1978 Jun 21.40 100 years	Cylinder 2200?	7.4 long 2.4 dia	1978 Jun 24.7	74.08	100.63	7166	773	803	0.002	18
L	Seasat 1 [Atlas Agena D]	1978-64A	1978 Jun 27.05 200 years	Cylinder + 4 wings 2300	21 long 1.5 dia 11 span	1978 Jun 27.4	108.02	100.63	7166	776	800	0.002	263

	Name		Launch date, lifetime and descent date	Shape and weight (kg)	Size (m)	Date of orbital determination	Orbital inclina-tion (deg)	Nodal period (min)	Semi major axis (km)	Perigee height (km)	Apogee height (km)	Orbital eccen-tricity	Argument of perigee (deg)
D 2M R	Soyuz 30*	1978-65A	1978 Jun 27.65 7.91 days 1978 Jul 5.56	Sphere-cylinder 6570?	7.5 long 2.3 dia	1978 Jun 27.8	51.64	88.82	6597	194	244	0.004	64
D	Soyuz 30 rocket	1978-65B	1978 Jun 27.65 3 days 1978 Jun 30	Cylinder 2500?	7.5 long 2.6 dia	1978 Jun 28.4	51.64	88.53	6583	184	226	0.003	52
	Cosmos 1024	1978-66A	1978 Jun 28.13 100 years?	Windmill + 6 vanes? 1250?	4.2 long? 1.6 dia	1978 Jun 30.2 1978 Jul 18.2	62.83 62.76	724.73 717.41	26728 26547	605 617	40094 39721	0.739 0.736	318 318
D	Cosmos 1024 launcher rocket	1978-66B	1978 Jun 28.13 33 days 1978 Jul 31	Cylinder 2500?	7.5 long 2.6 dia	1978 Jun 28.2	62.82	92.12	6758	212	547	0.025	124
D	Cosmos 1024 launcher	1978-66C	1978 Jun 28.13 14 days 1978 Jul 12	Irregular	-	1978 Jun 28	62.9	92.42	6773	185	605	0.031	-
	Cosmos 1024 rocket	1978-66D	1978 Jun 28.13 100 years?	Cylinder 440	2.0 long 2.0 dia	1980 Jun 30	63.2	720.2	26615	1643	38831	0.699	-
	Cosmos 1025	1978-67A	1978 Jun 28.73 60 years	-	-	1978 Jul 4.4	82.49	97.84	7032	640	668	0.002	276
	Cosmos 1025 rocket	1978-67B	1978 Jun 28.73 60 years	Cylinder 2200?	7.4 long 2.4 dia	1978 Jun 30.1	82.49	97.81	7031	638	667	0.002	297

* Soyuz 30 docked with Salyut 6 (second airlock) and Soyuz 29 on 1978 Jun 28.71, with one Russian and one Polish cosmonaut: Soyuz 30 undocked 1978 Jul 5.43

	Name		Launch date, lifetime and descent date	Shape and weight (kg)	Size (m)	Date of orbital determination	Orbital inclina-tion (deg)	Nodal period (min)	Semi major axis (km)	Perigee height (km)	Apogee height (km)	Orbital eccen-tricity	Argument of perigee (deg)
T	Comstar 1C	1978-68A	1978 Jun 29.93? >million years	Cylinder 1520 full 790 empty	6.32 long 2.36 dia	1978 Jun 30.0 1978 Jul 13.0 1979 Jun 30.0	21.80 0.08 0.0	639.16 1428.15 1436.0	24579 42008 42166	550 35470 35783	35852 35780 35792	0.718 0.004 0	179 - -
	Comstar 1C rocket	1978-68B	1978 Jun 29.93? 6000 years	Cylinder 1815	8.6 long 3.0 dia	1980 Jun 30	21.1	649.7	24849	569	36372	0.720	-
D R	Cosmos 1026	1978-69A	1978 Jul 2.40 4.02 days 1978 Jul 6.42	-	-	1978 Jul 2.7	51.78	88.99	6606	207	248	0.003	40
D	Cosmos 1026 rocket	1978-69B	1978 Jul 2.40 5 days 1978 Jul 7	Cylinder 2500?	7.5 long? 2.6 dia?	1978 Jul 2.7	51.78	88.94	6603	206	244	0.003	35
D	Progress 2*	1978-70A	1978 Jul 7.48 27.63 days 1978 Aug 4.11	Sphere-cylinder 7020	7.9 long 2.3 dia	1978 Jul 7.7 1978 Jul 9.2	51.62 51.63	88.60 89.98	6586 6655	182 245	234 308	0.004 0.005	109 210
D	Progress 2 rocket	1978-70B	1978 Jul 7.48 3 days 1978 Jul 10	Cylinder 2500?	7.5 long 2.6 dia	1978 Jul 8.1	51.63	88.62	6587	182	236	0.004	88
T	ESA-GEOS 2	1978-71A	1978 Jul 14.45 >million years	Cylinder 573 full 273 empty	1.10 long 1.62 dia	1978 Jul 14.7 1978 Jul 18.0 1979 Jun 30.0	25.85 0.80 0.2	626.60 1421.17 1436.0	24256 41874 42162	214 35377 35754	35542 35614 35815	0.728 0.003 0.001	179 313 -
D	ESA-GEOS 2 second stage	1978-71B	1978 Jul 14.45 150 days 1978 Dec 11	Cylinder + annulus 350?	6.4 long 1.52 and 2.44 dia	1978 Jul 14.5 1978 Jul 14.7	28.73 28.09	88.41 123.80	6574 8232	158 165	233 3543	0.006 0.205	3 40
	ESA-GEOS 2 third stage	1978-71C	1978 Jul 14.45 30 years	Sphere-cone 66	1.32 long 0.94 dia	1978 Aug 31.6	25.42	626.82	24262	197	35571	0.729	211

*Progress 2 docked with Salyut 6 (2nd airlock) - Soyuz 29 complex on 1978 Jul 9.54; undocked 1978 Aug 2.21. De-orbited over Pacific Ocean two days later.

	Name		Launch date, lifetime and descent date	Shape and weight (kg)	Size (m)	Date of orbital determination	Orbital inclination (deg)	Nodal period (min)	Semi major axis (km)	Perigee height (km)	Apogee height (km)	Orbital eccentricity	Argument of perigee (deg)
	Molniya 1AS	1978-72A	1978 Jul 14.63 14 years	Windmill + 6 vanes 1000?	3.4 long 1.6 dia	1978 Jul 15.7 1978 Jul 24.8	62.83 62.83	736.44 718.15	27015 26566	607 606	40666 39769	0.741 0.737	288 288
D	Molniya 1AS launcher rocket	1978-72B	1978 Jul 14.63 47 days 1978 Aug 30	Cylinder 2500?	7.5 long 2.6 dia	1978 Jul 16.2	62.84	92.66	6784	215	597	0.028	121
D	Molniya 1AS launcher	1978-72C	1978 Jul 14.63 40 days 1978 Aug 23	Irregular	-	1978 Jul 16.2	62.84	92.74	6788	204	616	0.030	116
	Molniya 1AS rocket	1978-72D	1978 Jul 14.63 14 years	Cylinder 440	2.0 long 2.0 dia	1980 Jun 30	64.3	732.7	26923	646	40443	0.739	-
	Statsionar-Raduga 4	1978-73A	1978 Jul 18.92 >million years	-	-	1978 Jul 19.2	0.50	1477.84	42980	36473	36730	0.003	62
D	Raduga 4 launcher rocket	1978-73B	1978 Jul 18.92 2 days 1978 Jul 20	Cylinder 4000?	12 long? 4 dia	1978 Jul 19.2	51.63	88.06	6559	170	192	0.002	281
D	Raduga 4 launcher	1978-73C	1978 Jul 18.92 1 day 1978 Jul 19	Irregular	-	1978 Jul 19.2	51.62	88.00	6556	161	195	0.003	183
	Raduga 4 rocket	1978-73D	1978 Jul 18.92 4½ years	Cylinder 1900?	3.9 long? 3.9 dia	1978 Jul 20.0	47.3	649.05	24802	349	36499	0.729	5
	Raduga 4 apogee motor	1978-73E	1978 Jul 18.92 > million years	Cylinder 440	2.0 long 2.0 dia	1980 Sep 30.7	1.23	1475.86	42941	36497	36628	0.0015	166

	Name		Launch date, lifetime and descent date	Shape and weight (kg)	Size (m)	Date of orbital determination	Orbital inclination (deg)	Nodal period (min)	Semi major axis (km)	Perigee height (km)	Apogee height (km)	Orbital eccentricity	Argument of perigee (deg)
	Cosmos 1027	1978-74A	1978 Jul 27.20 1200 years	Cylinder 700?	1.3 long? 1.9 dia?	1978 Aug 17.4	82.94	104.82	7363	966	1004	0.003	226
	Cosmos 1027 rocket	1978-74B	1978 Jul 27.20 600 years	Cylinder 2200?	7.4 long 2.4 dia	1978 Aug 2.9	82.93	104.71	7358	968	991	0.002	273
T	SDS 5 [Titan 3B Agena D]	1978-75A	1978 Aug 5.2? 100 years?	Cylinder	-	1978 Aug 31	62.5	697.1	26062	315	39053	0.743	-
	SDS 5 rocket	1978-75B	1978 Aug 5.2? 100 years?	Cylinder 700?	6 long? 1.5 dia	1978 Aug 21	63.3	700.2	26120	310	39175	0.744	270*
D R	Cosmos 1028	1978-76A	1978 Aug 5.63 29.5 days 1978 Sep 4.1	Sphere-cylinder 6700?	7 long? 2.4 dia	1978 Aug 6.2 1978 Aug 7.8	67.14 67.14	88.66 89.54	6587 6631	170 168	247 337	0.006 0.013	78 65
D	Cosmos 1028 rocket	1978-76B	1978 Aug 5.63 2 days 1978 Aug 7	Cylinder 2500?	7.5 long 2.6 dia	1978 Aug 6.2	67.14	88.52	6580	169	234	0.005	87
D	Cosmos 1028 engine	1978-76D	1978 Aug 5.63 30 days 1978 Sep 4	Cone 600? full	1.5 long? 2 dia?	1978 Sep 4.2	67.13	88.88	6598	149	290	0.011	69
D	Fragment	1978-76C											

* USAF payload launched from Vandenberg, California. Orbit and launch time unconfirmed.

	Name		Launch date, lifetime and descent date	Shape and weight (kg)	Size (m)	Date of orbital determination	Orbital inclination (deg)	Nodal period (min)	Semi major axis (km)	Perigee height (km)	Apogee height (km)	Orbital eccentricity	Argument of perigee (deg)
D	Progress 3*	1978-77A	1978 Aug 7.94 15.84 days 1978 Aug 23.78	Sphere-cylinder 7020	7.9 long 2.3 dia	1978 Aug 8.2 1978 Aug 9.6 1978 Aug 21.7	51.64 51.63 51.62	88.66 89.36 91.35	6589 6624 6722	190 243 335	232 249 352	0.003 0.0004 0.001	77 334 256
D	Progress 3 rocket	1978-77B	1978 Aug 7.94 2 days 1978 Aug 9	Cylinder 2500?	7.5 long 2.6 dia	1978 Aug 8.1	51.64	88.55	6584	185	226	0.003	77
D	Pioneer Venus 2 Atlas stage	1978-78B	1978 Aug 8.31 < 0.7 day 1978 Aug 8	Cylinder 3400	20 long 3.0 dia	1978 Aug 8.3	28.6?	87.5?	6536?	150?	165?	0.001?	-
T	ISEE 3	1978-79A	1978 Aug 12.63 Indefinite	Cylinder 469 full	1.61 long 1.73 dia	1978 Aug 12.7	28.89	73702	582300	180	1151664	0.989	320**
D	ISEE 3 second stage	1978-79B	1978 Aug 12.63 72 days 1978 Oct 23	Cylinder + annulus 350?	6.4 long 1.52 and 2.44 dia	1978 Aug 13.9	28.73	100.86	7184	176	1436	0.088	327
	ISEE 3 third stage	1978-79D	1978 Aug 12.63 Indefinite	Sphere - cone 66	1.32 long 0.94 dia		Orbit similar to 1978-79A transfer orbit						
	Fragment	1978-79C											
	Molniya 1 AT	1978-80A	1978 Aug 22.99 18½ years	Windmill +6 vanes 1000?	3.4 long 1.6 dia	1978 Aug 23.6 1978 Aug 29.7	62.87 62.83	735.68 718.23	26996 26568	443 464	40793 39915	0.747 0.743	280 280

* Progress 3 docked with Salyut 6 (2nd airlock) - Soyuz 29 complex on 1978 Aug 10.00; undocked 1978 Aug 21 and later de-orbited over Pacific Ocean.

** Entered heliocentric orbit - a 'halo' orbit around the Sun-Earth/Moon libration point, at distance of 1.6 million km from Earth on Earth-Sun line.

Space Vehicle: Pioneer Venus 2 (1978-78A), and Centaur rocket (1978-78C).

1978-80 continued on page 543

	Name		Launch date, lifetime and descent date	Shape and weight (kg)	Size (m)	Date of orbital determination	Orbital inclination (deg)	Nodal period (min)	Semi major axis (km)	Perigee height (km)	Apogee height (km)	Orbital eccentricity	Argument of perigee (deg)
D	Molniya 1 AT launcher rocket	1978-80B	1978 Aug 22.99 24 days 1978 Sep 15	Cylinder 2500?	7.5 long 2.6 dia	1978 Aug 23.9	62.81	90.95	6700	190	454	0.020	114
D	Molniya 1 AT launcher	1978-80C	1978 Aug 22.99 15 days 1978 Sep 6	Irregular	-	1978 Aug 24.5	62.82	91.01	6703	211	439	0.017	124
	Molniya 1 AT rocket	1978-80D	1978 Aug 22.99 18½ years	Cylinder 440	2.0 long 2.0 dia	1980 Jun 30	63.1	732.4	26913	1321	39749	0.714	-
D 2M R	Soyuz 31*	1978-81A	1978 Aug 26.62 67.84 days 1978 Nov 2.46	Sphere-cylinder 6570?	7.5 long 2.3 dia	1978 Aug 26.8 1978 Aug 27.1 1978 Aug 28.4	51.62 51.63 51.63	88.80 90.23 91.40	6596 6667 6725	193 256 339	243 322 354	0.004 0.005 0.001	62 281 306
D	Soyuz 31 rocket	1978-81B	1978 Aug 26.62 2 days 1978 Aug 28	Cylinder 2500?	7.5 long 2.6 dia	1978 Aug 27.1	51.62	88.64	6588	191	229	0.003	57
D R	Cosmos 1029	1978-82A	1978 Aug 29.63 9.68 days 1978 Sep 8.31	Sphere-cylinder 6300?	6.5 long? 2.4 dia	1978 Aug 30.7 1978 Aug 31.2	62.81 62.81	89.57 89.33	6633 6621	179 171	330 314	0.011 0.011	76 63
D	Cosmos 1029 rocket	1978-82B	1978 Aug 29.63 5 days 1978 Sep 3	Cylinder 2500?	7.5 long 2.6 dia	1978 Aug 31.0	62.80	89.21	6615	173	300	0.010	72
D P	Cosmos 1029 engine†	1978-82C	1978 Aug 29.63 13.07 days 1978 Sep 11.70	Cone 600? full	1.5 long? 2 dia?	1978 Sep 8.1	62.81	89.46	6628	168	332	0.012	70
D	Fragment	1978-82D											

† A 20kg, 0.6m piece was picked up near Garnat-sur-Engievre (Allier), France

* Soyuz 31 docked with Salyut 6 (2nd airlock) and Soyuz 29 on 1978 Aug 27.69, with one Russian and one East German cosmonaut. The crew returned to Earth in Soyuz 29, undocking 1978 Sep 3 (see page 536). Soyuz 31 (piloted by Soyuz 29 crew) undocked from 2nd airlock and re-docked with 1st airlock 1978 Sep 7. Soyuz 31 finally undocked from Salyut 6 on 1978 Nov 2.32, landing the Soyuz 29 crew with a duration record for manned space flight of 139.61 days.

	Name		Launch date, lifetime and descent date	Shape and weight (kg)	Size (m)	Date of orbital determination	Orbital inclination (deg)	Nodal period (min)	Semi major axis (km)	Perigee height (km)	Apogee height (km)	Orbital eccentricity	Argument of perigee (deg)
	Cosmos 1030	1978-83A	1978 Sep 6.13 25 years	Windmill + 6 vanes? 1250?	4.2 long? 1.6 dia	1978 Sep 9.2 1978 Sep 19.2	62.80 62.80	725.64 719.16	26749 26591	613 654	40129 39771	0.739 0.735	318 318
D	Cosmos 1030 launcher rocket	1978-83B	1978 Sep 6.13 32 days 1978 Oct 8	Cylinder 2500?	7.5 long 2.6 dia	1978 Sep 6.2	62.79	92.51	6777	213	585	0.027	121
D	Cosmos 1030 launcher	1978-83C	1978 Sep 6.13 15 days 1978 Sep 21	Irregular	-	1978 Sep 6.4	62.87	92.48	6776	173	622	0.033	119
	Cosmos 1030 rocket	1978-83D	1978 Sep 6.13 25 years	Cylinder 440	2.0 long 2.0 dia	1978 Oct 1.3	62.97	723.36	26693	605	40025	0.738	318
	Fragments	1978-83E,F											
D	Venus 11* launcher	1978-84C	1978 Sep 9.15 1 day 1978 Sep 10	Irregular	-	1978 Sep 9.3	51.55	88.19	6566	170	205	0.003	0
D R	Cosmos 1031	1978-85A	1978 Sep 9.63 12.60 days 1978 Sep 22.23	Sphere- cylinder 6300?	6.5 long? 2.4 dia	1978 Sep 10.2 1978 Sep 11.2	62.82 62.82	89.59 89.33	6634 6621	182 171	329 314	0.011 0.011	82 83
D	Cosmos 1031 rocket	1978-85B	1978 Sep 9.63 5 days 1978 Sep 14	Cylinder 2500?	7.5 long 2.6 dia	1978 Sep 10.1	62.81	89.47	6628	178	321	0.011	80
D	Cosmos 1031 engine	1978-85D	1978 Sep 9.63 15 days 1978 Sep 24	Cone 600? full	1.5 long? 2 dia?	1978 Sep 22.1	62.82	89.12	6610	161	303	0.011	74
D	Fragment	1978-85C											

Space Vehicle: Venus 11 (1978-84A)

* Venus 11 launcher rocket, similar to 1976-81D, was designated 1978-84B

	Name		Launch date, lifetime and descent date	Shape and weight (kg)	Size (m)	Date of orbital determination	Orbital inclination (deg)	Nodal period (min)	Semi major axis (km)	Perigee height (km)	Apogee height (km)	Orbital eccentricity	Argument of perigee (deg)
D	Venus 12* launcher	1978-86B	1978 Sep 14.11 1 day 1978 Sep 15	Irregular	-	1978 Sep 14.3	51.51	88.15	6564	164	207	0.003	0
T	Jikiken (Exos B) [Mu-3H]	1978-87A	1978 Sep 16.21 20 years	12-sided polygon 70	-	1978 Sep 16.4	31.09	532.85	21772	230	30558	0.697	-
	Jikiken rocket	1978-87B	1978 Sep 16.21 20 years	Cylinder?	-	1979 Jan 1.0	31.1	517.9	21363	215	29754	0.691	-
D R	Cosmos 1032	1978-88A	1978 Sep 19.34 12.83 days 1978 Oct 2.17	Sphere-cylinder 5900?	5.9 long 2.4 dia	1978 Sep 19.9	81.34	88.93	6599	215	226	0.001	59
D	Cosmos 1032 rocket	1978-88B	1978 Sep 19.34 3 days 1978 Sep 22	Cylinder 2500?	7.5 long 2.6 dia	1978 Sep 19.9	81.35	88.73	6589	204	217	0.001	329
D	Fragments**	1978-88C,D											
D R	Cosmos 1033	1978-89A	1978 Oct 3.46 12.84 days 1978 Oct 16.30	Sphere-cylinder 5900?	5.9 long 2.4 dia	1978 Oct 6.4	81.37	88.95	6600	212	231	0.001	102
D	Cosmos 1033 rocket	1978-89B	1978 Oct 3.46 3 days 1978 Oct 6	Cylinder 2500?	7.5 long 2.6 dia	1978 Oct 3.8	81.37	88.92	6599	206	235	0.002	118
D	Capsule	1978-89C	1978 Oct 3.46 14 days 1978 Oct 17	Ellipsoid 200?	0.9 long 1.9 dia	1978 Oct 15.3	81.36	88.43	6574	189	202	0.001	64

Space Vehicle: Venus 12 (1978-86A)

* Venus 12 launcher rocket, similar to 1976-81D, was apparently not tracked or designated.

** Object 1978-88D was a capsule; it decayed 1978 Oct 3, life 14 days.

	Name		Launch date, lifetime and descent date	Shape and weight (kg)	Size (m)	Date of orbital determination	Orbital inclination (deg)	Nodal period (min)	Semi major axis (km)	Perigee height (km)	Apogee height (km)	Orbital eccentricity	Argument of perigee (deg)
D	Progress 4*	1978-90A	1978 Oct 3.96 22.74 days 1978 Oct 26.70	Sphere-cylinder 7020	7.9 long 2.3 dia	1978 Oct 4.2 1978 Oct 7.5 1978 Oct 21.8	51.65 51.64 51.64	88.75 91.19 91.68	6594 6714 6739	185 325 359	247 347 362	0.005 0.002 0.0002	98 81 338
D	Progress 4 rocket	1978-90B	1978 Oct 3.96 2 days 1978 Oct 5	Cylinder 2500?	7.5 long 2.6 dia	1978 Oct 4.1	51.65	88.66	6590	184	239	0.004	97
	Cosmos 1034	1978-91A	1978 Oct 4.16 8000 years	Spheroid 40?	1.0 long? 0.8 dia?	1978 Oct 4.7	74.03	114.97	7832	1423	1484	0.004	110
	Cosmos 1035	1978-91B	1978 Oct 4.16 7000 years	Spheroid 40?	1.0 long? 0.8 dia?	1978 Oct 8.8	74.03	114.74	7822	1405	1482	0.005	90
	Cosmos 1036	1978-91C	1978 Oct 4.16 9000 years	Spheroid 40?	1.0 long? 0.8 dia?	1978 Oct 4.7	74.04	115.19	7842	1443	1484	0.003	114
	Cosmos 1037	1978-91D	1978 Oct 4.16 9000 years	Spheroid 40?	1.0 long? 0.8 dia?	1978 Oct 8.9	74.03	115.41	7852	1463	1484	0.001	123
	Cosmos 1038	1978-91E	1978 Oct 4.16 10000 years	Spheroid 40?	1.0 long? 0.8 dia?	1978 Oct 8.9	74.03	115.64	7862	1480	1488	0.001	202
	Cosmos 1039	1978-91F	1978 Oct 4.16 10000 years	Spheroid 40?	1.0 long? 0.8 dia?	1978 Oct 8.9	74.03	116.38	7896	1481	1554	0.005	252

*Progress 4 docked with Salyut 6 (2nd airlock) - Soyuz 31 complex on 1978 Oct 6.04; undocked 1978 Oct 24.55 and later de-orbited over Pacific Ocean.

1978-91 continued on page 547

	Name		Launch date, lifetime and descent date	Shape and weight (kg)	Size (m)	Date of orbital determination	Orbital inclination (deg)	Nodal period (min)	Semi major axis (km)	Perigee height (km)	Apogee height (km)	Orbital eccentricity	Argument of perigee (deg)
	Cosmos 1040	1978-91G	1978 Oct 4.16 10000 years	Spheroid 40?	1.0 long? 0.8 dia?	1978 Oct 8.9	74.03	116.11	7883	1481	1529	0.003	257
	Cosmos 1041	1978-91H	1978 Oct 4.16 10000 years	Spheroid 40?	1.0 long? 0.8 dia?	1978 Oct 7.8	74.03	115.88	7873	1480	1510	0.002	235
	Cosmos 1034 rocket	1978-91J	1978 Oct 4.16 20000 years	Cylinder 2200?	7.4 long 2.4 dia	1978 Oct 8.9	74.04	118.06	7972	1484	1703	0.014	266
D R	Cosmos 1042	1978-92A	1978 Oct 6.65 12.64 days 1978 Oct 19.29	Sphere- cylinder 6300?	6.5 long? 2.4 dia	1978 Oct 8.9	62.80	89.26	6617	179	299	0.009	72
D	Cosmos 1042 rocket	1978-92B	1978 Oct 6.65 4 days 1978 Oct 10	Cylinder 2500?	7.5 long 2.6 dia	1978 Oct 7.7	62.79	88.99	6604	174	277	0.008	67
D	Cosmos 1042 engine?	1978-92C	1978 Oct 6.65 15 days 1978 Oct 21	Cone 600? full	1.5 long? 2 dia?	1978 Oct 19	62.8	89.24	6616	171	304	0.010	-
D	Fragment	1978-92D											
T	Navstar 3 (GPS) [Atlas F]	1978-93A	1978 Oct 7.02 1 million years	Cylinder + 4 vanes 433	-	1978 Oct 17.1	62.81	722.61	26677	20285	20312	0.0005	127
	Navstar 3 rocket	1978-93B	1978 Oct 7.02 30 years	Cone- cylinder? 163?	1.85 long? 0.63 to 1.65 dia?	1978 Oct 7.4	62.95	350.79	16477	158	20040	0.603	159

	Name		Launch date, lifetime and descent date	Shape and weight (kg)	Size (m)	Date of orbital determination	Orbital inclination (deg)	Nodal period (min)	Semi major axis (km)	Perigee height (km)	Apogee height (km)	Orbital eccentricity	Argument of perigee (deg)
	Cosmos 1043	1978-94A	1978 Oct 10.82 60 years	Cylinder + 2 vanes? 2500?	5 long? 1.5 dia?	1978 Oct 12.9	81.20	97.31	7007	622	635	0.001	301
	Cosmos 1043 rocket	1978-94B	1978 Oct 10.82 60 years	Cylinder 1440	3.8 long 2.6 dia	1978 Oct 14.5	81.21	97.38	7010	579	685	0.008	178
	Molniya 3K	1978-95A	1978 Oct 13.22 25 years *	Windmill + 6 vanes 1500?	4.2 long? 1.6 dia	1978 Oct 13.8 1978 Oct 24.9	62.79 62.82	736.21 717.66	27009 26554	432 424	40829 39928	0.748 0.744	280 280
D	Molniya 3K launcher rocket	1978-95B	1978 Oct 13.22 15 days 1978 Oct 28	Cylinder 2500?	7.5 long 2.6 dia	1978 Oct 14.5	62.82	90.97	6701	213	433	0.016	127
D	Molniya 3K launcher	1978-95C	1978 Oct 13.22 7 days 1978 Oct 20	Irregular	-	1978 Oct 14.5	62.79	90.47	6677	175	422	0.019	114
	Molniya 3K rocket	1978-95E	1978 Oct 13.22 25 years *	Cylinder 440	2.0 long 2.0 dia	1979 Jun 30.0	63.2	734.5	26967	397	40781	0.749	-
D	Fragment	1978-95D											
	Tiros 11 [Atlas Burner 2]	1978-96A	1978 Oct 13.47 500 years	Cylinder 734	3.71 long 1.88 dia	1978 Oct 16.7	98.91	102.12	7236	850	866	0.001	256
	Tiros 11 rocket	1978-96B	1978 Oct 13.47 400 years	Sphere-cone 682 full? 66 empty	1.32 long 0.94 dia	1978 Oct 14.4	98.91	102.11	7236	855	860	0.0003	292
	Fragment	1978-96C											

* Decay possible in 1994 when perigee falls below 200 km

	Name		Launch date, lifetime and descent date	Shape and weight (kg)	Size (m)	Date of orbital determination	Orbital inclination (deg)	Nodal period (min)	Semi major axis (km)	Perigee height (km)	Apogee height (km)	Orbital eccentricity	Argument of perigee (deg)
D R	Cosmos 1044	1978-97A	1978 Oct 17.63 12.65 days 1978 Oct 30.28	Sphere-cylinder 5700?	5.0 long 2.4 dia	1978 Oct 18.4	62.82	89.46	6627	203	295	0.007	69
D	Cosmos 1044 rocket	1978-97B	1978 Oct 17.63 5 days 1978 Oct 22	Cylinder 2500?	7.5 long 2.6 dia	1978 Oct 18.4	62.82	89.30	6619	211	271	0.005	78
D	Fragments	1978-97C-E											
T	Nimbus 7	1978-98A	1978 Oct 24.34 1000 years	Conical skeleton? 832?	3 long 2 dia	1978 Oct 26.0	99.29	104.08	7327	943	953	0.0007	240
	Nimbus 7 second stage*	1978-98B	1978 Oct 24.34 500 years	Cylinder + annulus 350?	6.4 long 1.52 and 2.44 dia	1978 Oct 26.3	99.28	104.08	7327	943	953	0.0007	252
D	Intercosmos 18 (Magic)**	1978-99A	1978 Oct 24.79 875 days 1981 Mar 17	Octagonal ellipsoid 550?	1.8 long? 1.5 dia?	1978 Oct 25.5	82.97	96.40	6963	406	764	0.026	166
D	Intercosmos 18 rocket	1978-99B	1978 Oct 24.79 932 days 1981 May 13	Cylinder 2200?	7.4 long 2.4 dia	1978 Oct 29.1	82.96	96.30	6958	403	757	0.025	153
D	Magion 1 †	1978-99C	1978 Oct 24.79 1053 days 1981 Sep 11	Prism 15	0.30 x 0.30 x 0.15	1978 Nov 17.5	82.95	96.36	6961	404	762	0.026	89

* Carried CAMEO - Chemically active materials ejected in orbit (Barium released 1978 Oct 29, and Lithium released 1978 Nov 6). Weight 89 kg.

† Czechoslovak MAGnetospheric and IONospheric satellite, ejected from Intercosmos 18 on 1978 Nov 14.74 ** Magnetospheric Intercosmos.

	Name		Launch date, lifetime and descent date	Shape and weight (kg)	Size (m)	Date of orbital determination	Orbital inclination (deg)	Nodal period (min)	Semi major axis (km)	Perigee height (km)	Apogee height (km)	Orbital eccentricity	Argument of perigee (deg)
	Cosmos 1045	1978-100A	1978 Oct 26.29 15000 years	-	-	1978 Oct 29.1	82.55	120.41	8078	1689	1710	0.001	173
	Radio 1	1978-100B	1978 Oct 26.29 15000 years	-	-	1978 Oct 29.1	82.55	120.39	8077	1688	1709	0.001	183
	Radio 2	1978-100C	1978 Oct 26.29 15000 years	Cylinder + bar 40	0.39 long 0.42 dia	1978 Oct 28.6	82.55	120.40	8077	1689	1709	0.001	181
	Cosmos 1045 rocket	1978-100D	1978 Oct 26.29 15000 years	Cylinder 2200?	7.4 long 2.4 dia	1978 Oct 30.1	82.55	120.35	8075	1688	1705	0.001	175
	Fragment	1978-100E											
D	Prognoz 7	1978-101A	1978 Oct 30.23 693 days 1980 Sep 22	Spheroid + 4 vanes 915?	1.8 dia?	1978 Oct 30.2	64.91	5881.1	107928	472	202627	0.937	290
D	Prognoz 7 launcher rocket	1978-101B	1978 Oct 30.23 18 days 1978 Nov 17	Cylinder 2500?	7.5 long 2.6 dia	1978 Nov 4.9	65.02	90.79	6692	206	422	0.016	59
D	Prognoz 7 launcher	1978-101C	1978 Oct 30.23 26 days 1978 Nov 25	Irregular	-	1978 Nov 4.9	64.99	91.13	6709	233	429	0.015	63
D	Prognoz 7 rocket	1978-101D	1978 Oct 30.23 747 days? 1980 Nov 15?	Cylinder 440	2.0 long 2.0 dia	Orbit similar to 1978-101A							

	Name		Launch date, lifetime and descent date	Shape and weight (kg)	Size (m)	Date of orbital determination	Orbital inclination (deg)	Nodal period (min)	Semi major axis (km)	Perigee height (km)	Apogee height (km)	Orbital eccentricity	Argument of perigee (deg)
D R	Cosmos 1046	1978-102A	1978 Nov 1.50 11.78 days 1978 Nov 13.28	Sphere-cylinder 5900?	5.9 long 2.4 dia	1978 Nov 5.2	72.86	89.77	6641	202	324	0.009	62
D	Cosmos 1046 rocket	1978-102B	1978 Nov 1.50 6 days 1978 Nov 7	Cylinder 2500?	7.5 long 2.6 dia	1978 Nov 5.4	72.86	88.83	6595	180	253	0.006	50
D	Capsule	1978-102D	1978 Nov 1.50 14 days 1978 Nov 15	Ellipsoid 200?	0.9 long 1.9 dia	1978 Nov 13	72.9	91.43	6724	209	482	0.020	-
D	Fragments	1978-102C,E-G											
D	HEAO 2	1978-103A	**1978 Nov 13.23** **1228 days** **1982 Mar 25**	Hexagonal cylinder 2720	5.8 long 2.4 dia	1978 Nov 17.2	23.51	95.07	6909	520	541	0.002	345
D	HEAO 2 rocket	1978-103B	1978 Nov 13.23 120 days 1979 Mar 13	Cylinder 1815	8.6 long 3.0 dia	1978 Nov 18.0	23.51	92.86	6801	379	467	0.006	10
D R	Cosmos 1047	1978-104A	1978 Nov 15.49 12.77 days 1978 Nov 28.26	Sphere-cylinder 6300?	6.5 long? 2.4 dia	1978 Nov 15.5	72.86	89.77	6641	171	354	0.014	80
D	Cosmos 1047 rocket	1978-104B	1978 Nov 15.49 4 days 1978 Nov 19	Cylinder 2500?	7.5 long 2.6 dia	1978 Nov 18.0	72.86	88.85	6595	160	274	0.009	69
D	Cosmos 1047 engine	1978-104C	1978 Nov 15.49 17 days 1978 Dec 2	Cone 600? full	1.5 long? 2 dia?	1978 Nov 30.3	72.86	88.74	6590	159	265	0.008	45

	Name		Launch date, lifetime and descent date	Shape and weight (kg)	Size (m)	Date of orbital determination	Orbital inclination (deg)	Nodal period (min)	Semi major axis (km)	Perigee height (km)	Apogee height (km)	Orbital eccentricity	Argument of perigee (deg)
	Cosmos 1048	1978-105A	1978 Nov 16.91 120 years	Cylinder + paddles? 750?	2 long? 1 dia?	1978 Nov 17.1	74.03	100.89	7179	785	816	0.002	350
	Cosmos 1048 rocket	1978-105B	1978 Nov 16.91 100 years	Cylinder 2200?	7.4 long 2.4 dia	1978 Nov 17.1	74.04	100.81	7175	777	816	0.003	358
	Fragments	1978-105C,D											
T	NATO 3C	1978-106A	1978 Nov 19.04 > million years	Cylinder 706 full 346 empty	2.23 long 2.18 dia	1978 Nov 19.1 1978 Nov 30.2	27.20 4.41	632.80 1428.6	24416 42020	185 35516	35890 35768	0.731 0.003	171 217
D	NATO 3C second stage	1978-106B	1978 Nov 19.04 23 days 1978 Dec 12	Cylinder + annulus 350?	6.4 long 1.52 and 2.44 dia	1978 Nov 21.1	28.24	93.50	6831	174	732	0.041	179
D	NATO 3C third stage	1978-106C	1978 Nov 19.04 1470 days 1982 Nov 28	Sphere - cone 66	1.32 long 0.94 dia	1978 Nov 19	27.2	632.76	24414	183	35889	0.731	171
D R	Cosmos 1049	1978-107A	1978 Nov 21.50 12.76 days 1978 Dec 4.26	Sphere-cylinder 6300?	6.5 long? 2.4 dia	1978 Nov 22.8 1978 Nov 26.2	72.86 72.86	89.59 89.65	6632 6635	169 169	338 344	0.013 0.013	80 71
D	Cosmos 1049 rocket	1978-107B	1978 Nov 21.50 5 days 1978 Nov 26	Cylinder 2500?	7.5 long 2.6 dia	1978 Nov 21.7	72.86	89.56	6630	169	335	0.012	82
D	Cosmos 1049 engine	1978-107E	1978 Nov 21.50 20 days 1978 Dec 11	Cone 600? full	1.5 long? 2 dia?	1978 Dec 4	72.8	89.58	6630	181	322	0.011	-
D	Fragments	1978-107C,D											

	Name		Launch date, lifetime and descent date	Shape and weight (kg)	Size (m)	Date of orbital determination	Orbital inclination (deg)	Nodal period (min)	Semi major axis (km)	Perigee height (km)	Apogee height (km)	Orbital eccentricity	Argument of perigee (deg)
D R	Cosmos 1050	1978-108A	1978 Nov 28.68 13.62 days 1978 Dec 12.30	Sphere-cylinder 6300?	6.5 long? 2.4 dia	1978 Nov 30.1 1978 Dec 3.2	62.80 62.80	89.81 89.43	6644 6625	254 224	278 270	0.002 0.003	126 77
D	Cosmos 1050 rocket	1978-108B	1978 Nov 28.68 10 days 1978 Dec 8	Cylinder 2500?	7.5 long 2.6 dia	1978 Dec 2.7	62.79	89.41	6624	240	252	0.001	73
D	Cosmos 1050 engine	1978-108E	1978 Nov 28.68 17 days 1978 Dec 15	Cone 600? full	1.5 long? 2 dia?	1978 Dec 12.2	62.80	89.07	6607	160	298	0.011	230
D	Fragments	1978-108C, D											
	Cosmos 1051	1978-109A	1978 Dec 5.76 7000 years	Spheroid 40?	1.0 long? 0.8 dia?	1978 Dec 8.4	74.02	114.72	7820	1397	1487	0.006	102
	Cosmos 1052	1978-109B	1978 Dec 5.76 8000 years	Spheroid 40?	1.0 long? 0.8 dia?	1978 Dec 6.1	74.02	114.92	7829	1412	1490	0.005	117
	Cosmos 1053	1978-109C	1978 Dec 5.76 9000 years	Spheroid 40?	1.0 long 0.8 dia?	1978 Dec 8.3	74.02	115.12	7839	1433	1488	0.003	119
	Cosmos 1054	1978-109D	1978 Dec 5.76 9000 years	Spheroid 40?	1.0 long? 0.8 dia?	1978 Dec 9.2	74.02	115.33	7848	1449	1491	0.003	129
	Cosmos 1055	1978-109E	1978 Dec 5.76 10000 years	Spheroid 40?	1.0 long? 0.8 dia?	1978 Dec 8.2	74.02	115.5	7858	1460	1500	0.002	160
	Cosmos 1056	1978-109F	1978 Dec 5.76 10000 years	Spheroid 40?	1.0 long? 0.8 dia?	1978 Dec 8.3	74.02	115.77	7868	1472	1508	0.002	197

1978-109 continued on page 554

	Name		Launch date, lifetime and descent date	Shape and weight (kg)	Size (m)	Date of orbital determination	Orbital inclination (deg)	Nodal period (min)	Semi major axis (km)	Perigee height (km)	Apogee height (km)	Orbital eccentricity	Argument of perigee (deg)
	Cosmos 1057	1978-109G	1978 Dec 5.76 9000 years	Spheroid 40?	1.0 long? 0.8 dia?	1978 Dec 8.3	74.02	115.99	7878	1482	1518	0.002	231
	Cosmos 1058	1978-109H	1978 Dec 5.76 10000 years	Spheroid 40?	1.0 long? 0.8 dia?	1978 Dec 9.2	74.02	116.24	7889	1481	1541	0.004	236
	Cosmos 1051 rocket	1978-109J	1978 Dec 5.76 20000 years	Cylinder 2200?	7.4 long 2.4 dia	1980 Jun 30	74.0	118.06	7973	1484	1706	0.014	-
D R	Cosmos 1059	1978-110A	1978 Dec 7.65 12.63 days 1978 Dec 20.28	Sphere-cylinder 6300?	6.5 long? 2.4 dia	1978 Dec 7.8 1978 Dec 9.4	62.81 62.81	89.67 89.35	6637 6621	180 172	338 314	0.012 0.011	80 72
D	Cosmos 1059 rocket	1978-110B	1978 Dec 7.65 4 days 1978 Dec 11	Cylinder 2500?	7.5 long 2.6 dia	1978 Dec 7.8	62.80	89.53	6630	177	327	0.011	77
D	Cosmos 1059 engine	1978-110C	1978 Dec 7.65 16 days 1978 Dec 23	Cone 600? full	1.5 long? 2 dia?	1978 Dec 19.4	62.84	89.76	6642	168	359	0.014	66
D	Fragment	1978-110D											
D R	Cosmos 1060	1978-111A	1978 Dec 8.40 12.8 days 1978 Dec 21.2	Sphere-cylinder 5700?	5.0 long 2.4 dia	1978 Dec 8.5	65.03	89.47	6627	206	292	0.006	53
D	Cosmos 1060 rocket	1978-111B	1978 Dec 8.40 5 days 1978 Dec 13	Cylinder 2500?	7.5 long 2.6 dia	1978 Dec 8.5	65.03	89.32	6620	201	282	0.006	36

	Name		Launch date, lifetime and descent date	Shape and weight (kg)	Size (m)	Date of orbital determination	Orbital inclination (deg)	Nodal period (min)	Semi major axis (km)	Perigee height (km)	Apogee height (km)	Orbital eccentricity	Argument of perigee (deg)
T	Navstar 4 (GPS) [Atlas F]	1978-112A	1978 Dec 11.18 1 million years	Cylinder + 4 vanes 433	-	1978 Dec 14.2	63.27	722.38	26670	20267	20316	0.001	88
	Navstar 4 rocket	1978-112B	1978 Dec 11.18 30 years	Cone-cylinder? 163?	1.85 long? 0.63 to 1.65 dia?	1978 Dec 13.8	63.09	349.31	16431	120	19986	0.605	162
T	DSCS 11*	1978-113A	1978 Dec 14.03 > million years	Cylinder + 2 dishes 590	1.83 long 2.74 dia	1978 Dec 14.8 1979 Mar 7.7	2.49 2.27	1452.2 1436.2	42482 42168	35796 35784	36412 35796	0.007 0.0001	2 204
T	DSCS 12	1978-113B	1978 Dec 14.03 > million years	Cylinder + 2 dishes 590	1.83 long 2.74 dia	1978 Dec 14.8 1979 Feb 28.8	2.50 2.30	1464.3 1436.0	42715 42164	36261 35776	36413 35796	0.002 0.0002	352 268
D	Titan 3C second stage	1978-113C	1978 Dec 14.03 <1 day 1978 Dec 14	Cylinder 1900	6 long 3.0 dia	1978 Dec 14.6	28.59	87.90	6556	150	206	0.004	121
	Transtage	1978-113D	1978 Dec 14.03 > million years	Cylinder 1500?	6 long? 3.0 dia	1978 Dec 14.8	2.50	1452.0	42478	35788	36412	0.007	2
D R	Cosmos 1061	1978-114A	1978 Dec 14.64 12.63 days 1978 Dec 27.27	Sphere-cylinder 5900?	5.9 long 2.4 dia	1978 Dec 15.6	62.82	89.62	6635	203	310	0.008	78
D	Cosmos 1061 rocket	1978-114B	1978 Dec 14.64 6 days 1978 Dec 20	Cylinder 2500?	7.5 long 2.6 dia	1978 Dec 15.1	62.81	89.42	6625	200	293	0.007	73

* DSCS 9 and 10 failed to reach orbit in 1978 March.

1978-114 continued on page 556

	Name		Launch date, lifetime and descent date	Shape and weight (kg)	Size (m)	Date of orbital determination	Orbital inclination (deg)	Nodal period (min)	Semi major axis (km)	Perigee height (km)	Apogee height (km)	Orbital eccentricity	Argument of perigee (deg)
D	Capsule	1978-114C	1978 Dec 14.64 21 days 1979 Jan 4	Ellipsoid 200?	0.9 long 1.9 dia	1978 Dec 27	62.8	89.56	6632	212	295	0.006	-
D	Cosmos 1062	1978-115A	1978 Dec 15.56 857 days 1981 Apr 20	Cylinder + paddles? 900?	2 long? 1 dia?	1978 Dec 16.9	74.04	95.18	6905	504	550	0.003	350
D	Cosmos 1062 rocket	1978-115B	1978 Dec 15.56 918 days 1981 Jun 20	Cylinder 2200?	7.4 long 2.4 dia	1978 Dec 16.9	74.04	95.09	6901	494	551	0.004	357
D	Fragment	1978-115C											
T	Telesat 4 (Anik)	1978-116A	1978 Dec 16.01 > million years	Box + vanes 922 full 474 empty	2.14 high 2.17 square	1978 Dec 16.0 1979 Jan 31.6	27.25 0.03	632.91 1435.9	24419 42164	185 35781	35896 35790	0.731 0.0001	178 58
D	Telesat 4 second stage (DRIMS)*	1978-116B	1978 Dec 16.01 128 days 1979 Apr 23	Cylinder + annulus 350?	6.4 long 1.52 and 2.44 dia	1978 Dec 16.7	28.42	107.68	7504	183	2069	0.126	181
	Telesat 4 third stage	1978-116C	1978 Dec 16.01 20 years?	Sphere- cone 66	1.8 long 0.94 dia		Orbit similar to 1978-116A transfer orbit						
	Cosmos 1063	1978-117A	1978 Dec 19.07 60 years	Cylinder + 2 vanes? 2500?	5 long? 1.5 dia?	1978 Dec 21.8	81.23	97.38	7011	631	634	0.0002	55
	Cosmos 1063 rocket	1978-117B	1978 Dec 19.07 60 years	Cylinder 1440	3.8 long 2.6 dia	1978 Dec 22.0	81.24	97.44	7014	581	690	0.008	170

*Delta Redundant Inertial Measurement Systems

	Name		Launch date, lifetime and descent date	Shape and weight (kg)	Size (m)	Date of orbital determination	Orbital inclination (deg)	Nodal period (min)	Semi major axis (km)	Perigee height (km)	Apogee height (km)	Orbital eccentricity	Argument of perigee (deg)
	Gorizont 1	1978-118A	1978 Dec 19.52 > million years	-	-	1978 Dec 20.4 1979 Jun 30.0	11.3 11.3	1420 1436.1	41850 42166	22580 22553	48365 49023	0.308 0.314	70 -
D	Gorizont 1 launcher rocket*	1978-118B	1978 Dec 19.52 2 days 1978 Dec 21	Cylinder 4000?	12 long? 4 dia	1978 Dec 20.1	51.61	88.08	6560	174	190	0.001	206
	Gorizont 1 apogee motor	1978-118C	1978 Dec 19.52 > million years	Cylinder 400	2.0 long? 2.0 dia?	1980 Jul 24.7	10.75	1417.3	41798	22124	48715	0.318	28
D	Gorizont 1 rocket	1978-118D	1978 Dec 19.52 1 year?	Cylinder 1900?	3.9 long? 3.9 dia			Orbit similar to 1978-73D					
	Cosmos 1064**	1978-119A	1978 Dec 20.87 8 years	Cylinder 700?	1.3 long? 1.9 dia?	1978 Dec 21.9	82.95	98.69	7073	424	965	0.038	99
	Cosmos 1064 rocket	1978-119B	1978 Dec 20.87 7 years	Cylinder 2200?	7.4 long 2.4 dia	1978 Dec 21.2	82.95	98.58	7067	416	962	0.039	102
D	Cosmos 1065	1978-120A	1978 Dec 22.92 222 days 1979 Aug 1	Octagonal ellipsoid? 550?	1.8 long? 1.5 dia?	1978 Dec 24.3	50.68	93.45	6824	344	548	0.015	151
D	Cosmos 1065 rocket	1978-120B	1978 Dec 22.92 239 days 1979 Aug 18	Cylinder 2200?	7.4 long 2.4 dia	1978 Dec 23.0	50.68	93.41	6822	344	544	0.015	144
D	Fragments	1978-120C-H											
	Cosmos 1066†	1978-121A	1978 Dec 23.37 500 years	Cylinder + 2 vanes 2750?	5 long? 1.5 dia?	1978 Dec 23.5	81.24	102.05	7233	818	891	0.005	272
	Cosmos 1066 rocket	1978-121B	1978 Dec 23.37 400 years	Cylinder 1440	3.8 long 2.6 dia	1978 Dec 23.6	81.25	102.10	7235	816	898	0.006	246

* There may have been a launch platform in a similar orbit.

† Probably an attempted Meteor 2 satellite

** Possibly a navigation satellite where the rocket failed to circularize the orbit at apogee

	Name		Launch date, lifetime and descent date	Shape and weight (kg)	Size (m)	Date of orbital determination	Orbital inclination (deg)	Nodal period (min)	Semi major axis (km)	Perigee height (km)	Apogee height (km)	Orbital eccentricity	Argument of perigee (deg)
	Cosmos 1067	1978-122A	1978 Dec 26.56 3000 years	Spheroid + paddles? 650?	1.6 dia?	1978 Dec 26.8	82.97	109.07	7561	1158	1208	0.003	267
	Cosmos 1067 rocket	1978-122B	1978 Dec 26.56 2000 years	Cylinder 2200?	7.4 long 2.4 dia	1978 Dec 26.7	82.97	108.95	7555	1157	1197	0.003	256
D R	Cosmos 1068	1978-123A	1978 Dec 26.65 12.6 days 1979 Jan 8.3	Sphere-cylinder 6300?	6.5 long? 2.4 dia	1978 Dec 26.9 1978 Dec 27.6	62.80 62.80	90.17 89.40	6662 6624	177 173	391 318	0.016 0.011	74 64
D P	Cosmos 1068 rocket *	1978-123B	1978 Dec 26.65 5.15 days 1978 Dec 31.80	Cylinder 2500?	7.5 long 2.6 dia	1978 Dec 27.1	62.78	89.92	6650	174	369	0.015	71
D	Cosmos 1068 engine	1978-123D	1978 Dec 26.65 14 days 1979 Jan 9	Cone 600? full	1.5 long? 2 dia?	1979 Jan 6.7	62.80	89.38	6623	166	324	0.012	62
D	Fragments	1978-123C,E											
D R	Cosmos 1069	1978-124A	1978 Dec 28.69 12.64 days 1979 Jan 10.33	Sphere-cylinder 5900?	5.9 long 2.4 dia	1978 Dec 29.5	62.82	89.75	6641	241	285	0.003	188
D	Cosmos 1069 rocket	1978-124B	1978 Dec 28.69 8 days 1979 Jan 5	Cylinder 2500?	7.5 long 2.6 dia	1978 Dec 29.5	62.82	89.62	6635	241	272	0.002	193
D	Capsule	1978-124F	1978 Dec 28.69 25 days 1979 Jan 22	Ellipsoid 200?	0.9 long 1.9 dia	1979 Jan 14.2	62.81	90.44	6676	242	354	0.008	319
D	Fragments	1978-124C-E											

* Decay was observed over Europe and pieces were picked up near Hannover

	Name		Launch date, lifetime and descent date	Shape and weight (kg)	Size (m)	Date of orbital determination	Orbital inclination (deg)	Nodal period (min)	Semi major axis (km)	Perigee height (km)	Apogee height (km)	Orbital eccentricity	Argument of perigee (deg)
D R	Cosmos 1070	1979-01A	1979 Jan 11.63 8.7 days 1979 Jan 20.3	Sphere-cylinder 5900?	5.9 long 2.4 dia	1979 Jan 12.8	62.81	89.47	6627	205	293	0.007	88
D	Cosmos 1070 rocket	1979-01B	1979 Jan 11.63 5 days 1979 Jan 16	Cylinder 2500?	7.5 long 2.6 dia	1979 Jan 12.9	62.80	89.13	6610	202	262	0.005	78
D	Capsule	1979-01F	1979 Jan 11.63 16 days 1979 Jan 27	Ellipsoid 200?	0.9 long 1.9 dia	1979 Jan 19.7	62.81	89.30	6619	201	281	0.006	85
D	Fragments	1979-01C-E,G,H											
D R	Cosmos 1071	1979-02A	1979 Jan 13.65 12.63 days 1979 Jan 26.28	Sphere-cylinder 6300?	6.5 long? 2.4 dia	1979 Jan 13.9 1979 Jan 16.2	62.80 62.80	89.67 89.34	6637 6621	179 159	339 326	0.012 0.013	80 72
D	Cosmos 1071 rocket	1979-02B	1979 Jan 13.65 4 days 1979 Jan 17	Cylinder 2500?	7.5 long 2.6 dia	1979 Jan 13.9	62.79	89.45	6626	173	323	0.011	86
D	Cosmos 1071 engine	1979-02F	1979 Jan 13.65 16 days 1979 Jan 29	Cone 600? full	1.5 long? 2 dia?	1979 Jan 26.6	62.88	91.47	6725	268	426	0.012	35
D	Fragments	1979-02C-E											
	Cosmos 1072	1979-03A	1979 Jan 16.73 1200 years	Cylinder 700?	1.3 long? 1.9 dia?	1979 Jan 17.2	82.93	104.98	7371	965	1020	0.004	285
	Cosmos 1072 rocket	1979-03B	1979 Jan 16.73 600 years	Cylinder 2200?	7.4 long 2.4 dia	1979 Jan 21.5	82.93	104.87	7365	966	1008	0.003	264

	Name		Launch date, lifetime and descent date	Shape and weight (kg)	Size (m)	Date of orbital determination	Orbital inclination (deg)	Nodal period (min)	Semi major axis (km)	Perigee height (km)	Apogee height (km)	Orbital eccentricity	Argument of perigee (deg)
	Molniya 3L	1979-04A	1979 Jan 18.66 19½ years	Windmill + 6 vanes 1500?	4.2 long? 1.6 dia	1979 Jan 21.3 1979 Sep 1.0	62.82 62.9	735.98 717.73	27003 26555	433 720	40817 39634	0.748 0.733	280 -
D	Molniya 3L launcher rocket	1979-04B	1979 Jan 18.66 13 days 1979 Jan 31	Cylinder 2500?	7.5 long 2.6 dia	1979 Jan 21.7	62.77	90.51	6679	208	393	0.014	128
D	Molniya 3L launcher	1979-04C	1979 Jan 18.66 11 days 1979 Jan 29	Irregular	-	1979 Jan 22.8	62.81	90.40	6673	183	407	0.017	122
	Molniya 3L rocket	1979-04D	1979 Jan 18.66 19½ years	Cylinder 440	2.0 long 2.0 dia	1979 Nov 1.0	62.9	732.91	26928	817	40283	0.733	-
	Meteor 29	1979-05A	1979 Jan 25.24 60 years	Cylinder + 2 vanes 2200?	5 long? 1.5 dia?	1979 Jan 29.0	98.00	97.42	7012	622	645	0.002	288
	Meteor 29 rocket	1979-05B	1979 Jan 25.24 60 years	Cylinder 1440	3.8 long 2.6 dia	1979 Jan 27.9	98.02	97.56	7019	606	675	0.005	240
D R	Cosmos 1073	1979-06A	1979 Jan 30.64 12.6 days 1979 Feb 12.2	Sphere- cylinder 6300?	6.5 long? 2.4 dia	1979 Jan 30.7 1979 Feb 3.1	62.81 62.80	89.59 89.31	6633 6619	182 162	328 320	0.011 0.012	75 75
D	Cosmos 1073 rocket	1979-06B	1979 Jan 30.54 4 days 1979 Feb 3	Cylinder 2500?	7.5 long 2.6 dia	1979 Jan 30.7	62.79	89.50	6629	178	323	0.011	74

	Name		Launch date, lifetime and descent date	Shape and weight (kg)	Size (m)	Date of orbital determination	Orbital inclination (deg)	Nodal period (min)	Semi major axis (km)	Perigee height (km)	Apogee height (km)	Orbital eccentricity	Argument of perigee (deg)
D	Cosmos 1073 engine	1979-06C	1979 Jan 30.64 15 days 1979 Feb 14	Cone 600? full	1.5 long? 2 dia?	1979 Feb 11.3	62.81	89.75	6641	154	372	0.016	66
D	Fragment	1979-06D											
T	Scatha* [Thor Delta]	1979-07A	1979 Jan 30.91 > million years	Cylinder 357	1.8 long 1.8 dia	1979 Jan 30.9 1979 Apr 1.9	27.39 7.81	794.8 1416.2	28423 41775	185 27543	43905 43251	0.769 0.188	179 186
D	Scatha second stage	1979-07B	1979 Jan 30.91 26 days 1979 Feb 25	Cylinder + annulus 350?	6.4 long 1.52 and 2.44 dia	1979 Jan 31.0	28.62	95.07	6907	189	869	0.049	168
	Scatha third stage	1979-07C	1979 Jan 30.91 5 years ?	Sphere-cone 66	1.32 long 0.94 dia	Orbit similar to 1979-07A transfer orbit							
D R	Cosmos 1074	1979-08A	1979 Jan 31.38 60.04 days 1979 Apr 1.42	Sphere-cylinder + 2 wings 6850	7.5 long 2.3 to 2.72 dia	1979 Feb 1.0 1979 Feb 4.5 1979 Feb 8.4	51.60 51.66 51.65	88.77 90.76 91.96	6595 6693 6752	195 310 364	238 319 383	0.003 0.001 0.001	96 296 243
D P?	Cosmos 1074 rocket**	1979-08B	1979 Jan 31.38 2 days 1979 Feb 2	-	-	1979 Feb 1.0	51.60	88.61	6587	189	228	0.003	106
D	Cosmos orbital† module	1979-08C	1979 Jan 31.38 191 days 1979 Aug 10	Spheroid 1260?	2.3 dia	1979 May 1.0	51.6	91.19	6714	330	341	0.001	
D	Fragments	1979-08D-G											
	Ayame 1 (ECS 1)†† [Nu]	1979-09A	1979 Feb 6.37 > million years	Cylinder 260 full 130 empty	0.95 long 1.40 dia	1979 Feb 6.4 1979 Mar 1.0	24.09 0.4	606.80 1380.6	23769 41072	191 33966	34590 35421	0.724 0.018	179 -
	Ayame 1 rocket	1979-09B	1979 Feb 6.37 5 years?	Cylinder 66?	1.74 long 1.65 dia	1979 Feb 6.2	24.41	625.20	24275	190	35603	0.729	179

* Spacecraft charging at high altitudes.

** Fragments found in February 1979 near Eastbourne, England were probably from this rocket.

† Part of 1979-08A until about 1979 Apr 1

†† Japanese Experimental Communications Satellite (Ayame means Sweet Flower); contact lost when final stage collided with satellite.

	Name		Launch date, lifetime and descent date	Shape and weight (kg)	Size (m)	Date of orbital determination	Orbital inclination (deg)	Nodal period (min)	Semi major axis (km)	Perigee height (km)	Apogee height (km)	Orbital eccentricity	Argument of perigee (deg)
D	Cosmos 1075	1979-10A	1979 Feb 8.42 984 days 1981 Oct 19	Cylinder?	4 long? 2 dia?	1979 Feb 11.0	65.83	94.50	6873	473	516	0.003	350
D	Cosmos 1075 rocket	1979-10B	1979 Feb 8.42 568 days 1980 Aug 29	Cylinder 2200?	7.4 long 2.4 dia	1979 Feb 10.6	65.83	94.40	6868	465	515	0.004	6
	Cosmos 1076*	1979-11A	1979 Feb 12.54 60 years	-	-	1979 Feb 13.5	82.53	97.78	7030	637	666	0.002	281
	Cosmos 1076 rocket	1979-11B	1979 Feb 12.54 60 years	Cylinder 2200?	7.4 long 2.4 dia	1979 Feb 18.6	82.54	97.75	7028	635	664	0.002	274
	Cosmos 1077	1979-12A	1979 Feb 13.91 30 years	Cylinder + 2 vanes? 2500?	5 long? 1.5 dia?	1979 Feb 18.8	81.23	97.30	7006	625	631	0.0005	244
	Cosmos 1077 rocket	1979-12B	1979 Feb 13.91 30 years	Cylinder 1440	3.8 long 2.6 dia	1979 Feb 18.2	81.24	97.37	7010	576	687	0.008	171
T	Sage** (AEM 2) [Scout]	1979-13A	1979 Feb 18.68 40 years	Hexagonal prism 147	0.64 long	1979 Feb 18.8	54.93	96.74	6983	549	661	0.008	249
D	Sage rocket	1979-13B	1979 Feb 18.68 944 days 1981 Sep 19	Cylinder 24	1.50 long 0.46 dia	1979 Feb 19.1	54.93	96.76	6984	550	662	0.008	249
D	Fragment	1979-13C											

* Oceanographic satellite.

** Stratospheric Aerosol and Gas Experiment (Applications Explorer Mission).

	Name		Launch date, lifetime and descent date	Shape and weight (kg)	Size (m)	Date of orbital determination	Orbital inclination (deg)	Nodal period (min)	Semi major axis (km)	Perigee height (km)	Apogee height (km)	Orbital eccentricity	Argument of perigee (deg)
T	Hakucho* (Corsa B) [Mu-3C]	1979-14A	1979 Feb 21.21 10 years	Octagonal prism 96	0.65 long 0.80 dia	1979 Feb 24.0	29.90	95.61	6935	541	572	0.002	121
D	Hakucho rocket	1979-14B	1979 Feb 21.21 1083 days 1982 Feb 8	Sphere-cone 230?	2.33 long 1.14 dia	1979 Feb 24.6	29.90	95.63	6936	542	573	0.002	133
	Statsionar-Ekran 3	1979-15A	1979 Feb 21.33 > million years	Cylinder + plate	-	1979 Nov 1.0	0.1	1436.1	42165	35656	35917	0.003	-
D	Ekran 3 launcher rocket	1979-15B	1979 Feb 21.33 1 day 1979 Feb 22	Cylinder 4000?	12 long? 4 dia	1979 Feb 21.7	51.60	88.13	6563	181	188	0.001	228
D	Ekran 3 launcher	1979-15C	1979 Feb 21.33 1 day 1979 Feb 22	Irregular	-	1979 Feb 21.9	51.61	88.06	6559	174	188	0.001	210
	Ekran 3 apogee motor	1979-15D	1979 Feb 21.33 > million years	Cylinder 440?	2.0 long? 2.0 dia?		Orbit similar to 1979-15A						
D	Ekran 3 rocket	1979-15E	1979 Feb 21.33 1 year?	Cylinder 1900?	3.9 long? 3.9 dia		Transfer orbit similar to 1979-35D						
D R	Cosmos 1078	1979-16A	1979 Feb 22.51 7.8 days 1979 Mar 2.3	Sphere-cylinder 6300?	6.5 long? 2.4 dia	1979 Feb 23.2	72.86	88.99	6602	168	280	0.008	88
D	Cosmos 1078 rocket	1979-16B	1979 Feb 22.51 2 days 1979 Feb 24	Cylinder 2500?	7.5 long 2.6 dia	1979 Feb 22.6	72.88	88.85	6595	167	267	0.008	85
D	Cosmos 1078 engine	1979-16C	1979 Feb 22.51 8 days 1979 Mar 2	Cone 600? full	1.5 long? 2 dia?	1979 Mar 1.5	72.86	88.01	6553	147	203	0.004	-

* Japanese Cosmic Radiation Satellite (Hakucho means Swan). Re-launch of Corsa A, which failed on 1976 Feb 4.

	Name		Launch date, lifetime and descent date	Shape and weight (kg)	Size (m)	Date of orbital determination	Orbital inclination (deg)	Nodal period (min)	Semi major axis (km)	Perigee height (km)	Apogee height (km)	Orbital eccentricity	Argument of perigee (deg)
T	Solwind (P78-1) [Atlas Burner 2]	1979-17A	1979 Feb 24.35 40 years	- 850	-	1979 Feb 25.3	97.65	96.36	6961	563	602	0.003	304
	Solwind rocket	1979-17B	1979 Feb 24.35 5 years	Sphere-cone 66?	1.32 long 0.94 dia	1979 Feb 24.6	97.66	96.34	6960	562	601	0.003	306
D	Fragments	1979-17C,D											
D 2M R	Soyuz 32*	1979-18A	1979 Feb 25.50 108.18 days 1979 Jun 13.68	Sphere-cylinder 6570?	7.5 long 2.3 dia	1979 Feb 25.7 1979 Feb 26.0	51.59 51.63	88.93 89.65	6603 6639	193 240	256 281	0.005 0.003	79 218
D	Soyuz 32 rocket**	1979-18B	1979 Feb 25.50 2.35 days 1979 Feb 27.85	Cylinder 2500?	7.5 long 2.6 dia	1979 Feb 25.8	51.60	88.86	6599	191	251	0.005	83
D R	Cosmos 1079	1979-19A	1979 Feb 27.63 11.8 days 1979 Mar 11.4	Sphere-cylinder 6700?	7 long? 2.4 dia	1979 Feb 27.9	67.14	89.60	6634	174	337	0.012	64
D	Cosmos 1079 rocket	1979-19B	1979 Feb 27.63 3 days 1979 Mar 2	Cylinder 2500?	7.5 long 2.6 dia	1979 Feb 27.7	67.14	89.49	6628	167	333	0.013	66

* Soyuz 32 docked with Salyut 6 (1st airlock) on 1979 Feb 26.56. Soyuz 32 returned to Earth unmanned; the Soyuz 32 cosmonauts remaining aboard Salyut until recovery by Soyuz 34 (see p.574).

** Decay observed from ship at 15.5°S, 52.2°E

	Name		Launch date, lifetime and descent date	Shape and weight (kg)	Size (m)	Date of orbital determination	Orbital inclination (deg)	Nodal period (min)	Semi major axis (km)	Perigee height (km)	Apogee height (km)	Orbital eccentricity	Argument of perigee (deg)
	Intercosmos 19	1979-20A	1979 Feb 27.71 30 years	Octagonal ellipsoid 550?	1.8 long? 1.5 dia?	1979 Mar 1.0	73.98	99.75	7124	501	991	0.034	162
	Intercosmos 19 rocket	1979-20B	1979 Feb 27.71 30 years	Cylinder 2200?	7.4 long 2.4 dia	1979 Mar 5.0	73.99	99.65	7120	495	988	0.035	153
D	Fragment	1979-20C											
	Meteor 2-04	1979-21A	1979 Mar 1.78 500 years	Cylinder + 2 vanes 2750?	5 long? 1.5 dia?	1979 Mar 5.0	81.22	102.33	7246	839	897	0.004	274
	Meteor 2-04 rocket	1979-21B	1979 Mar 1.78 400 years	Cylinder 1440	3.8 long 2.6 dia	1979 Mar 5.5	81.23	102.31	7245	825	909	0.006	234
	Fragment	1979-21C											
D	Progress 5*	1979-22A	1979 Mar 12.24 23.74 days 1979 Apr 4.98	Sphere-cylinder 7020	7.9 long 2.3 dia	1979 Mar 12.4 1979 Mar 14.5	51.65 51.62	88.83 90.64	6598 6687	183 293	256 325	0.006 0.002	100 4
D	Progress 5 rocket	1979-22B	1979 Mar 12.24 2 days 1979 Mar 14	Cylinder 2500?	7.5 long 2.6 dia	1979 Mar 12.5	51.62	88.68	6591	183	242	0.004	80
D R	Cosmos 1080	1979-23A	1979 Mar 14.45 13.8 days 1979 Mar 28.2	Sphere-cylinder 6300?	6.5 long? 2.4 dia	1979 Mar 15.1 1979 Mar 15.6	72.85 72.86	89.14 88.64	6610 6585	169 168	294 245	0.009 0.006	87 87
D	Cosmos 1080 rocket	1979-23B	1979 Mar 14.45 2 days 1979 Mar 16	Cylinder 2500?	7.5 long 2.6 dia	1979 Mar 14.7	72.85	88.96	6601	168	277	0.008	86

* Progress 5 docked with Salyut 6 (second airlock)-Soyuz 32 on 1979 Mar 14.31. Undocked 1979 Apr 3.67, then de-orbited over Pacific Ocean next day.

1979-23 continued on p.566

	Name		Launch date, lifetime and descent date	Shape and weight (kg)	Size (m)	Date of orbital determination	Orbital inclination (deg)	Nodal period (min)	Semi major axis (km)	Perigee height (km)	Apogee height (km)	Orbital eccentricity	Argument of perigee (deg)
D	Cosmos 1080 engine	1979-23E	1979 Mar 14.45 14 days 1979 Mar 28	Cone 600? full	1.5 long? 2 dia?	1979 Mar 28	72.8	89.81	6643	173	356	0.014	-
D	Fragments	1979-23C,D											
	Cosmos 1081	1979-24A	1979 Mar 15.12 7000 years	Spheroid 40?	1.0 long? 0.8 dia?	1979 Mar 19.4	74.02	114.60	7815	1406	1467	0.004	84
	Cosmos 1082	1979-24B	1979 Mar 15.12 8000 years	Spheroid 40?	1.0 long? 0.8 dia?	1979 Mar 24.8	74.03	114.78	7823	1424	1466	0.003	89
	Cosmos 1083	1979-24C	1979 Mar 15.12 9000 years	Spheroid 40?	1.0 long? 0.8 dia?	1979 Mar 19.4	74.02	114.99	7832	1443	1465	0.001	97
	Cosmos 1084	1979-24D	1979 Mar 15.12 9000 years	Spheroid 40?	1.0 long? 0.8 dia?	1979 Mar 19.5	74.02	115.24	7844	1463	1469	0	132
	Cosmos 1085	1979-24E	1979 Mar 15.12 10000 years	Spheroid 40?	1.0 long? 0.8 dia?	1979 Mar 19.1	74.03	115.70	7865	1467	1507	0.003	255
	Cosmos 1086	1979-24F	1979 Mar 15.12 10000 years	Spheroid 40?	1.0 long? 0.8 dia?	1979 Mar 19.7	74.03	115.46	7854	1468	1484	0.001	269
	Cosmos 1087	1979-24G	1979 Mar 15.12 10000 years	Spheroid 40?	1.0 long? 0.8 dia?	1979 Mar 19.6	74.03	115.92	7875	1468	1526	0.004	265
	Cosmos 1088	1979-24H	1979 Mar 15.12 10000 years	Spheroid 40?	1.0 long? 0.8 dia?	1979 Mar 23.4	74.02	116.14	7885	1466	1548	0.005	255

1979-24 continued on p.567

	Name		Launch date, lifetime and descent date	Shape and weight (kg)	Size (m)	Date of orbital determination	Orbital inclination (deg)	Nodal period (min)	Semi major axis (km)	Perigee height (km)	Apogee height (km)	Orbital eccentricity	Argument of perigee (deg)
	Cosmos 1081 rocket	1979-24J	1979 Mar 15.12 20000 years	Cylinder 2200?	7.4 long 2.4 dia	1979 Mar 15.4	74.03	117.74	7957	1463	1695	0.015	273
D	[Titan 3D]	1979-25A	1979 Mar 16.77 190 days 1979 Sep 22	Cylinder 13300? full	15 long 3.0 dia	1979 Mar 19.0	96.39	88.80	6592	170	258	0.007	149
T	Capsule	1979-25B	1979 Mar 16.77 60 years	Octagon? 60?	0.3 long? 0.9 dia?	1979 Mar 18.4	95.78	97.23	7003	621	628	0.0005	27
D	Titan 3D rocket	1979-25C	1979 Mar 16.77 2 days 1979 Mar 18	Cylinder 1900	6 long 3.0 dia	1979 Mar 16.9	96.40	88.75	6590	161	262	0.008	141
	Cosmos 1089*	1979-26A	1979 Mar 21.17 1200 years	Cylinder 700?	1.3 long? 1.9 dia?	1979 Mar 25.5	82.97	104.90	7367	973	1005	0.002	266
	Cosmos 1089 rocket	1979-26B	1979 Mar 21.17 600 years	Cylinder 2200?	7.4 long 2.4 dia	1979 Mar 23.9	82.97	104.79	7362	972	995	0.002	259
D R	Cosmos 1090	1979-27A	1979 Mar 31.45 12.8 days 1979 Apr 13.2	Sphere-cylinder 5700?	5.0 long 2.4 dia	1979 Apr 2.5	72.85	89.79	6642	202	326	0.009	68
D	Cosmos 1090 rocket	1979-27B	1979 Mar 31.45 5 days 1979 Apr 5	Cylinder 2500?	7.5 long 2.6 dia	1979 Apr 2.5	72.85	89.35	6620	192	292	0.008	57

*Navigation satellite.

	Name		Launch date, lifetime and descent date	Shape and weight (kg)	Size (m)	Date of orbital determination	Orbital inclination (deg)	Nodal period (min)	Semi major axis (km)	Perigee height (km)	Apogee height (km)	Orbital eccentricity	Argument of perigee (deg)
	Cosmos 1091	1979-28A	1979 Apr 7.26 1200 years	Cylinder 700?	1.3 long? 1.9 dia?	1979 Apr 9.6	82.92	104.94	7369	969	1012	0.003	273
	Cosmos 1091 rocket	1979-28B	1979 Apr 7.26 600 years	Cylinder 2200?	7.4 long 2.4 dia	1979 Apr 7.5	82.92	104.84	7364	969	1003	0.002	277
D 2M R	Soyuz 33*	1979-29A	1979 Apr 10.73 1.96 days 1979 Apr 12.69	Sphere-cylinder 6800?	7.5 long 2.3 to 2.72 dia	1979 Apr 10.8 1979 Apr 11.5 1979 Apr 12.4	51.61 51.64 51.64	88.98 90.02 90.92	6606 6657 6701	194 245 292	261 312 353	0.005 0.005 0.005	78 256 224
D	Soyuz 33 rocket	1979-29B	1979 Apr 10.73 3 days 1979 Apr 13	Cylinder 2500?	7.5 long 2.6 dia	1979 Apr 12.8	51.61	87.60	6537	156	161	0.0004	79
	Cosmos 1092	1979-30A	1979 Apr 11.91 1200 years	Cylinder 700?	1.3 long? 1.9 dia?	1979 Apr 16.4	82.95	104.90	7367	968	1009	0.003	269
	Cosmos 1092 rocket	1979-30B	1979 Apr 11.91 600 years	Cylinder 2200?	7.4 long 2.4 dia	1979 Apr 15.0	82.95	104.80	7362	968	999	0.002	272
	Molniya 1 AU	1979-31A	1979 Apr 12.02 11 years	Windmill + 6 vanes 1000?	3.4 long 1.6 dia	1979 Apr 16.2 1979 Apr 20.2	62.89 62.92	735.31 718.03	26987 26563	623 630	40595 39739	0.741 0.736	288 288
D	Molniya 1 AU launcher rocket	1979-31B	1979 Apr 12.02 26 days 1979 May 8	Cylinder 2500?	7.5 long 2.6 dia	1979 Apr 16.6	62.85	92.25	6764	211	561	0.026	124

*Soyuz 33 approached Salyut 6 - Soyuz 32 on 1979 Apr 11.8, but failed to dock. Carried one Russian and first Bulgarian cosmonaut.

1979-31 continued on page 569

	Name		Launch date, lifetime and descent date	Shape and weight (kg)	Size (m)	Date of orbital determination	Orbital inclination (deg)	Nodal period (min)	Semi major axis (km)	Perigee height (km)	Apogee height (km)	Orbital eccentricity	Argument of perigee (deg)
D	Molniya 1 AU launcher	1979-31C	1979 Apr 12.02 23 days 1979 May 5	Irregular	-	1979 Apr 16.6	62.87	92.42	6772	196	592	0.029	118
	Molniya 1 AU rocket	1979-31D	1979 Apr 12.02 11 years	Cylinder 440	2.0 long 2.0 dia	1979 Apr 18.2	62.97	732.46	26917	622	40456	0.740	288
	Cosmos 1093	1979-32A	1979 Apr 14.23 60 years	Cylinder + 2 vanes? 2500?	5 long? 1.5 dia?	1979 Apr 17.5	81.25	97.30	7006	621	635	0.001	293
	Cosmos 1093 rocket	1979-32B	1979 Apr 14.23 60 years	Cylinder 1440	3.8 long 2.6 dia	1979 Apr 16.5	81.25	97.53	7018	584	695	0.008	145
D	Cosmos 1094	1979-33A	1979 Apr 18.50 203 days 1979 Nov 7	Cylinder?	-	1979 Apr 23.9	65.04	93.31	6812	426	442	0.001	277
D	Cosmos 1094 rocket*	1979-33B	1979 Apr 18.50 0.45 day 1979 Apr 18.95	Cylinder 1500?	8 long? 2.5 dia?	1979 Apr 18.7	65.04	89.36	6622	105	382	0.021	47
D	Fragment	1979-33C											
D R	Cosmos 1095	1979-34A	1979 Apr 20.48 13.8 days 1979 May 4.2	Sphere-cylinder 6300?	6.5 long? 2.4 dia	1979 Apr 21.0 1979 Apr 22.1	72.84 72.85	90.30 92.30	6667 6765	199 354	379 420	0.014 0.005	85 172
D	Cosmos 1095 rocket	1979-34B	1979 Apr 20.48 7 days 1979 Apr 27	Cylinder 2500?	7.5 long 2.6 dia	1979 Apr 23.3	72.84	89.67	6636	190	326	0.010	74
D	Cosmos 1095 engine	1979-34D	1979 Apr 20.48 212 days 1979 Nov 18	Cone 600? full	1.5 long? 2 dia?	1979 Sep 1.0	72.85	91.37	6719	317	365	0.004	-
D	Fragments	1979-34C,E-G											

* Decay observed from Denmark and Holland.

	Name		Launch date, lifetime and descent date	Shape and weight (kg)	Size (m)	Date of orbital determination	Orbital inclination (deg)	Nodal period (min)	Semi major axis (km)	Perigee height (km)	Apogee height (km)	Orbital eccentricity	Argument of perigee (deg)
	Statsionar - Raduga 5*	1979-35A	1979 Apr 25.16 >million years	-	-	1979 May 3.9	0.41	1436.2	42167	35789	35789	0	300
D	Raduga 5 launcher rocket	1979-35B	1979 Apr 25.16 1 day 1979 Apr 26	Cylinder 4000?	12 long? 4 dia	1979 Apr 25.8	51.62	88.04	6557	175	183	0.0006	236
D	Raduga 5 launcher	1979-35C	1979 Apr 25.16 <1 day 1979 Apr 25	Irregular	-	1979 Apr 25.7	51.59	88.20	6565	100	273	0.013	181
D	Raduga 5 rocket	1979-35D	1979 Apr 25.16 9 months?	Cylinder 1900?	3.9 long? 3.9 dia	1979 May 8.4	47.42	635.48	24484	318	35893	0.726	4
D	Cosmos 1096	1979-36A	1979 Apr 25.42 213 days 1979 Nov 24	Cylinder?	-	1979 Apr 25.9	65.06	93.32	6813	428	442	0.001	282
D	Cosmos 1096 rocket	1979-36B	1979 Apr 25.42 <0.6 day 1979 Apr 25	Cylinder 1500?	8 long? 2.5 dia?	1979 Apr 25.8	65.02	89.32	6620	113	370	0.019	54
D R	Cosmos 1097	1979-37A	1979 Apr 27.72 29.5 days 1979 May 27.2	Sphere-cylinder 6700?	7 long? 2.4 dia	1979 Apr 28.5	62.79	89.53	6630	173	331	0.012	65
D	Cosmos 1097 rocket	1979-37B	1979 Apr 27.72 3 days 1979 Apr 30	Cylinder 2500?	7.5 long 2.6 dia	1979 Apr 28.0	62.78	89.46	6627	171	326	0.012	62

* There is probably a separated apogee motor in a similar orbit.

	Name		Launch date, lifetime and descent date	Shape and weight (kg)	Size (m)	Date of orbital determination	Orbital inclina-tion (deg)	Nodal period (min)	Semi major axis (km)	Perigee height (km)	Apogee height (km)	Orbital eccen-tricity	Argument of perigee (deg)
T	Fleetsatcom 2	1979-38A	1979 May 4.79 > million years	Hexagonal cylinder 1884 full	1.27 long 2.44 dia	1979 May 4.8 1980 Nov 1.0	26.37 1.8	634.29 1436.1	24454 42165	166 35691	35986 35883	0.732 0.002	181 -
D	Fleetsatcom 2 rocket	1979-38B	1979 May 4.79 1 year?	Cylinder 1815	8.6 long 3.0 dia	1979 May 4.8	26.31	622.02	24138	172	35347	0.729	182
D	Progress 6*	1979-39A	1979 May 13.18 27.62 days 1979 Jun 9.80	Sphere-cylinder 7020	7.9 long 2.3 dia	1979 May 13.3 1979 May 14.6	51.62 51.62	88.80 90.94	6597 6702	190 313	247 334	0.004 0.002	80 287
D	Progress 6 rocket	1979-39B	1979 May 13.18 2 days 1979 May 15	Cylinder 2500?	7.5 long 2.6 dia	1979 May 14.2	51.61	88.35	6574	174	218	0.003	65
D R	Cosmos 1098	1979-40A	1979 May 15.49 12.75 days 1979 May 28.24	Sphere-cylinder 6300?	6.5 long? 2.4 dia	1979 May 16.8	72.87	89.75	6640	170	354	0.014	77
D	Cosmos 1098 rocket	1979-40B	1979 May 15.49 4 days 1979 May 19	Cylinder 2500?	7.5 long 2.6 dia	1979 May 16.1	72.86	89.52	6629	169	332	0.012	76
D	Cosmos 1098 engine	1979-40C	1979 May 15.49 15 days 1979 May 30	Cone 600? Full	1.5 long? 2 dia?	1979 May 29.5	72.87	88.46	6572	148	239	0.007	45
D R	Cosmos 1099	1979-41A	1979 May 17.30 12.8 days 1979 May 30.1	Sphere-cylinder 5900?	5.9 long 2.4 dia	1979 May 18.3	81.35	89.14	6609	215	247	0.002	85
D	Cosmos 1099 rocket	1979-41B	1979 May 17.30 4 days 1979 May 21	Cylinder 2500?	7.5 long 2.6 dia	1979 May 17.8	81.36	89.03	6604	213	238	0.002	91
D	Fragments**	1979-41C-E											

* Progress 6 docked with Salyut 6 - Soyuz 32 complex (2nd airlock) on 1979 May 15.26. Undocked 1979 Jun 8.33, then de-orbited over Pacific Ocean next day.

** 1979-41C (decayed 1979 Jun 22) was a Capsule.

	Name		Launch date, lifetime and descent date	Shape and weight (kg)	Size (m)	Date of orbital determination	Orbital inclination (deg)	Nodal period (min)	Semi major axis (km)	Perigee height (km)	Apogee height (km)	Orbital eccentricity	Argument of perigee (deg)
D R?	Cosmos 1100	1979-42A	1979 May 22.96 <0.2 day 1979 May 23	- 12000?	-	1979 May 23.1	51.58	88.59	6586	193	222	0.002	190
D R?	Cosmos 1101	1979-42B	1979 May 22.96 <0.2 day 1979 May 23	- 12000?	-		Orbit similar to 1979-42A						
D	Cosmos 1100 rocket	1979-42C	1979 May 22.96 4 days 1979 May 26	Cylinder 4000?	12 long? 4 dia	1979 May 23	51.5	88.55	6584	182	229	0.004	-
D	Fragments	1979-42D,E											
D R	Cosmos 1102	1979-43A	1979 May 25.29 12.8 days 1979 Jun 7.1	Sphere-cylinder 5900?	5.9 long 2.4 dia	1979 May 25.4	81.34	89.24	6614	212	260	0.004	87
D	Cosmos 1102 rocket	1979-43B	1979 May 25.29 4 days 1979 May 29	Cylinder 2500?	7.5 long 2.6 dia	1979 May 26.4	81.37	88.93	6598	207	233	0.002	76
D	Capsule	1979-43D	1979 May 25.29 49 days 1979 Jul 13	Ellipsoid 200?	0.9 long 1.9 dia	1979 Jun 6.6	81.34	88.74	6589	191	230	0.003	13
D	Fragments	1979-43C,E											
D	[Titan 3B Agena D]	1979-44A	1979 May 28.76 90 days 1979 Aug 26	Cylinder 3000?	8 long? 1.5 dia	1979 May 29.5	96.41	88.67	6586	131	285	0.012	136

	Name		Launch date, lifetime and descent date	Shape and weight (kg)	Size (m)	Date of orbital determination	Orbital inclination (deg)	Nodal period (min)	Semi major axis (km)	Perigee height (km)	Apogee height (km)	Orbital eccentricity	Argument of perigee (deg)
D R	Cosmos 1103	1979-45A	1979 May 31.69 13.59 days 1979 Jun 14.28	Sphere-cylinder 6300?	6.5 long? 2.4 dia	1979 Jun 1.4 1979 Jun 1.9	62.82 62.82	90.82 91.95	6694 6749	257 325	375 417	0.009 0.007	103 206
D	Cosmos 1103 rocket	1979-45B	1979 May 31.69 24 days 1979 Jun 24	Cylinder 2500?	7.5 long 2.6 dia	1979 Jun 4.8	62.80	90.55	6681	248	357	0.008	95
D	Cosmos 1103 engine	1979-45G	1979 May 31.69 168 days 1979 Nov 15	Cone 600? full	1.5 long? 2 dia?	1979 Sep 1.0	62.82	91.30	6717	309	369	0.004	-
D	Fragments	1979-45C-F											
	Cosmos 1104	1979-46A	1979 May 31.75 1200 years	Cylinder 700?	1.3 long? 1.9 dia?	1979 Jun 4.5	82.95	104.85	7364	962	1010	0.003	274
	Cosmos 1104 rocket	1979-46B	1979 May 31.75 600 years	Cylinder 2200?	7.4 long 2.4 dia	1979 Jun 5.9	82.94	104.75	7359	962	1000	0.003	267
T	Ariel 6*	1979-47A	1979 Jun 2.98 10 years	Cylinder + 4 panels 154	1.31 long 0.70 dia	1979 Jun 5.0	55.03	97.21	7005	600	654	0.004	223
	Ariel 6 rocket	1979-47B	1979 Jun 2.98 8 years	Cylinder 24	1.50 long 0.46 dia	1979 Jun 9.6	55.03	97.21	7005	601	653	0.004	233

* Final British satellite in the series launched by NASA; known as UK 6 before launch

	Name		Launch date, lifetime and descent date	Shape and weight (kg)	Size (m)	Date of orbital determination	Orbital inclination (deg)	Nodal period (min)	Semi major axis (km)	Perigee height (km)	Apogee height (km)	Orbital eccentricity	Argument of perigee (deg)
	Molniya 3M	1979-48A	1979 Jun 5.98 12 years	Windmill + 6 vanes 1500?	4.2 long? 1.6 dia	1979 Jun 7.6 1979 Jun 12.2	62.84 62.83	735.19 718.00	26984 26562	439 449	40773 39918	0.747 0.743	280 280
D	Molniya 3M launcher rocket	1979-48B	1979 Jun 5.98 19 days 1979 Jun 24	Cylinder 2500?	7.5 long 2.6 dia	1979 Jun 10.5	62.85	90.87	6697	206	431	0.017	128
D	Molniya 3M launcher	1979-48C	1979 Jun 5.98 16 days 1979 Jun 21	Irregular	-	1979 Jun 10.4	62.82	90.60	6683	196	414	0.016	119
	Molniya 3M rocket	1979-48D	1979 Jun 5.98 10 years	Cylinder 440	2.0 long 2.0 dia	1979 Jun 17.2	62.88	731.12	26883	450	40560	0.746	280
D R	Soyuz 34*	1979-49A	1979 Jun 6.76 73.76 days 1979 Aug 19.52	Sphere-cylinder 6800?	7.5 long 2.3 to 2.72 dia	1979 Jun 6.9 1979 Jun 8.3	51.61 51.61	88.90 91.00	6601 6705	192 289	254 364	0.004 0.006	81 227
D	Soyuz 34 rocket	1979-49B	1979 Jun 6.76 3 days 1979 Jun 9	Cylinder 2500?	7.5 long 2.6 dia	1979 Jun 7.4	51.61	88.63	6588	189	230	0.003	75
T	AMS 4 [Thor Burner 2]	1979-50A	1979 Jun 6.77 80 years	Irregular 513	6.40 long 1.68 dia	1979 Jun 13.4	98.77	101.50	7207	819	838	0.001	198
	Burner 2 rocket	1979-50B	1979 Jun 6.77 60 years	Sphere-cone 66	1.32 long 0.94 dia	1979 Jun 16.2	98.78	101.48	7206	819	836	0.001	179
2d	Fragments	1979-50C-G											

* Unmanned Soyuz 34 docked with Soyuz 32 - Salyut 6 (2nd airlock) on 1979 Jun 8.83 (Progress 6 had vacated the 2nd airlock on Jun 8.33, for de-orbiting the next day). On 1979 Jun 14.68 Soyuz 34 was detached from 2nd airlock and re-attached to 1st airlock. (Soyuz 32 had vacated the 1st airlock on Jun 13, prior to unmanned recovery.) Soyuz 34 landed carrying the Soyuz 32 cosmonauts, with a manned duration record of 175.02 days.

	Name		Launch date, lifetime and descent date	Shape and weight (kg)	Size (m)	Date of orbital determination	Orbital inclination (deg)	Nodal period (min)	Semi major axis (km)	Perigee height (km)	Apogee height (km)	Orbital eccentricity	Argument of perigee (deg)
T	Bhaskara 1*	1979-51A	1979 Jun 7.44 9 years	26-faced Polyhedron 444	1.19 high 1.55 dia	1979 Jun 9.4	50.67	95.17	6908	519	541	0.002	61
D	Bhaskara 1 rocket	1979-51B	1979 Jun 7.44 986 days 1982 Feb 17	Cylinder 2200?	7.4 long 2.4 dia	1979 Jun 16.7	50.67	95.09	6904	510	542	0.002	78
D R	Cosmos 1105**	1979-52A	1979 Jun 8.30 12.85 days 1979 Jun 21.15	Sphere-cylinder 5900?	5.9 long 2.4 dia	1979 Jun 9.3	81.35	89.18	6611	212	254	0.003	93
D	Cosmos 1105 rocket	1979-52B	1979 Jun 8.30 4 days 1979 Jun 12	Cylinder 2500?	7.5 long 2.6 dia	1979 Jun 8.6	81.35	89.08	6606	212	244	0.002	78
D	Capsule	1979-52D	1979 Jun 8.30 50 days 1979 Jul 28	Ellipsoid 200?	0.9 long 1.9 dia	1979 Jul 22.5	81.35	88.61	6583	202	207	0.0004	-
D	Fragment	1979-52C											
T	IMEWS 9 [Titan 3C]	1979-53A	1979 Jun 10.57 >million years	-	-	1979 Jun 12.3 1979 Sep 1.0	1.95 1.8	1448.5 1435.9	42409 42161	35801 35712	36261 35854	0.005 0.002	359 -
D	Titan 3C second stage	1979-53B	1979 Jun 10.57 1 day 1979 Jun 11	Cylinder 1900	6 long 3.0 dia	1979 Jun 11.0	28.60	88.23	6573	152	237	0.006	121
	Titan 3C transtage	1979-53C	1979 Jun 10.57 >million years	Cylinder 1500?	6 long? 3.0 dia	1979 Nov 1.0	1.6	1448.3	42404	35798	36253	0.005	-

* Second Indian satellite launched by USSR (Bhaskar Acharya was a 7th-century Indian astronomer).

** Earth resources satellite.

	Name		Launch date, lifetime and descent date	Shape and weight (kg)	Size (m)	Date of orbital determination	Orbital inclination (deg)	Nodal period (min)	Semi major axis (km)	Perigee height (km)	Apogee height (km)	Orbital eccentricity	Argument of perigee (deg)
D R	Cosmos 1106*	1979-54A	1979 Jun 12.29 12.85 days 1979 Jun 25.14	Sphere-cylinder 5900?	5.9 long 2.4 dia	1979 Jun 12.3	81.36	89.05	6605	216	237	0.002	79
D	Cosmos 1106 rocket	1979-54B	1979 Jun 12.29 3 days 1979 Jun 15	Cylinder 2500?	7.5 long 2.6 dia	1979 Jun 14.3	81.36	88.55	6580	191	212	0.002	354
D	Capsule	1979-54D	1979 Jun 12.29 18 days 1979 Jun 30	Ellipsoid 200?	0.9 long 1.9 dia	1979 Jun 24.1	81.36	88.66	6585	201	213	0.0009	41
D	Fragment	1979-54C											
D R	Cosmos 1107	1979-55A	1979 Jun 15.46 13.78 days 1979 Jun 29.24	Sphere-cylinder 5900?	5.9 long? 2.4 dia	1979 Jun 16.2	72.86	89.50	6628	198	301	0.008	85
D	Cosmos 1107 rocket	1979-55B	1979 Jun 15.46 5 days 1979 Jun 20	Cylinder 2500?	7.5 long 2.6 dia	1979 Jun 15.9	72.89	89.35	6620	194	290	0.007	80
D	Capsule?**	1979-55D	1979 Jun 15.46 18 days 1979 Jul 3	-	- 2 dia?	1979 Jun 29	72.9	89.71	6638	165	355	0.014	-
D	Fragment	1979-55C											
D R	Cosmos 1108	1979-56A	1979 Jun 22.29 12.85 days 1979 Jul 5.14	Sphere-cylinder 5900?	5.9 long 2.4 dia	1979 Jun 23.2	81.33	89.11	6608	214	245	0.002	86
D	Cosmos 1108 rocket	1979-56B	1979 Jun 22.29 3 days 1979 Jun 25	Cylinder 2500?	7.5 long 2.6 dia	1979 Jun 23.2	81.33	88.87	6596	211	224	0.001	60
D	Capsule	1979-56E	1979 Jun 22.29 29 days 1979 Jul 21	Ellipsoid 200?	0.9 long 1.9 dia	1979 Jul 17.7	81.29	88.29	6567	167	210	0.003	-
D	Fragments	1979-56C,D											

* Earth resources satellite
** Possibly an engine

	Name		Launch date, lifetime and descent date	Shape and weight (kg)	Size (m)	Date of orbital determination	Orbital inclination (deg)	Nodal period (min)	Semi major axis (km)	Perigee height (km)	Apogee height (km)	Orbital eccentricity	Argument of perigee (deg)
T	NOAA 6 [Atlas Burner 2]	1979-57A	1979 Jun 27.66 120 years	Box 340?	1.25 long? 1.02 square?	1979 Jun 29.8	98.75	101.31	7198	812	828	0.001	286
	NOAA 6 rocket	1979-57B	1979 Jun 27.66 100 years	Sphere-cone 66	1.32 long 0.94 dia	1979 Jun 28.1	98.75	101.30	7198	804	835	0.002	268
	Fragment	1979-57C											
	Cosmos 1109	1979-58A	1979 Jun 27.76 17 years	Windmill + 6 vanes? 1250?	4.2 long? 1.6 dia?	1979 Jun 30.0 1979 Jul 4.8	62.89 62.89	724.20 718.01	26715 26562	613 628	40060 39740	0.738 0.736	318 318
D	Cosmos 1109 launcher rocket	1979-58B	1979 Jun 27.76 29 days 1979 Jul 26	Cylinder 2500?	7.5 long 2.6 dia	1979 Jul 2.2	62.83	91.99	6751	209	537	0.024	123
D	Cosmos 1109 launcher	1979-58C	1979 Jun 27.76 16 days 1979 Jul 13	Irregular	-	1979 Jun 30.0	62.81	92.15	6759	176	586	0.030	120
	Cosmos 1109 rocket	1979-58D	1979 Jun 27.76 100 years?	Cylinder 440	2.0 long 2.0 dia	1979 Jul 3.8	62.88	721.75	26654	630	39922	0.737	318
	Fragments	1979-58E-H											
D	Progress 7*	1979-59A	1979 Jun 28.39 21.71 days 1979 Jul 20.10	Sphere-cylinder 7020	7.9 long 2.3 dia	1979 Jun 28.6 1979 Jun 29.9 1979 Jul 2.0	51.63 51.64 51.62	88.82 90.66 91.60	6597 6688 6735	186 275 353	251 345 360	0.005 0.005 0.001	87 215 27
D	Progress 7 rocket	1979-59B	1979 Jun 28.39 2 days 1979 Jun 30	Cylinder 2500?	7.5 long 2.6 dia	1979 Jun 29.1	51.64	88.49	6581	182	224	0.003	69

* Docked with Salyut 6 - Soyuz 34 complex (2nd airlock) on 1979 Jun 30.47; undocked Jul 18. De-orbited over Pacific Ocean.

	Name		Launch date, lifetime and descent date	Shape and weight (kg)	Size (m)	Date of orbital determination	Orbital inclination (deg)	Nodal period (min)	Semi major axis (km)	Perigee height (km)	Apogee height (km)	Orbital eccentricity	Argument of perigee (deg)
	Cosmos 1110	1979-60A	1979 Jun 28.84 120 years	Cylinder + paddles? 750?	2 long? 1 dia?	1979 Jun 30.9	74.02	100.94	7181	791	814	0.002	17
	Cosmos 1110 rocket	1979-60B	1979 Jun 28.84 100 years	Cylinder 2200?	7.4 long 2.4 dia	1979 Jul 1.2	74.03	100.82	7175	781	813	0.002	32
D R	Cosmos 1111	1979-61A	1979 Jun 29.67 14.6 days 1979 Jul 14.3	Sphere- cylinder 6300?	6.5 long? 2.4 dia	1979 Jun 30.2 1979 Jul 2.1	62.80 62.81	90.60 91.96	6683 6750	255 328	354 415	0.007 0.006	110 201
D	Cosmos 1111 rocket	1979-61B	1979 Jun 29.67 21 days 1979 Jul 20	Cylinder 2500?	7.5 long 2.6 dia	1979 Jun 30.5	62.80	90.45	6675	251	343	0.007	-
D	Cosmos 1111 engine	1979-61C	1979 Jun 29.67 159 days 1979 Dec 5	Cone 600? full	1.5 long? 2 dia?	1979 Sep 1.0	62.81	91.58	6731	322	383	0.005	-
D	Fragments	1979-61D-F											
	Gorizont 2	1979-62A	1979 Jul 5.98 > million years	-	-	1979 Jul 7.0 1979 Sep 1.0	0.8 0.7	1477 1436.1	42930 42165	36550 35763	36550 35810	0 0.0006	- -
D	Gorizont 2 launcher rocket	1979-62B	1979 Jul 5.98 2 days 1979 Jul 7	Cylinder 4000?	12 long? 4 dia	1979 Jul 7.3	51.62	87.52	6533	147	162	0.001	260
D	Gorizont 2 launcher	1979-62C	1979 Jul 5.98 2 days 1979 Jul 7	Irregular	-	1979 Jul 6.0	51.62	88.34	6573	188	202	0.001	-
	Gorizont 2 apogee motor	1979-62D	1979 Jul 5.98 > million years	Cylinder 440?	2.0 long? 2.0 dia?		Orbit similar to 1979-62A						
D	Gorizont 2 rocket	1979-62E	1979 Jul 5.98 1 year?	Cylinder 1900?	3.9 long? 3.9 dia		Transfer orbit similar to 1979-35D						

	Name		Launch date, lifetime and descent date	Shape and weight (kg)	Size (m)	Date of orbital determination	Orbital inclination (deg)	Nodal period (min)	Semi major axis (km)	Perigee height (km)	Apogee height (km)	Orbital eccentricity	Argument of perigee (deg)
D	Cosmos 1112	1979-63A	1979 Jul 6.35 199 days 1980 Jan 21	Octagonal ellipsoid? 550?	1.8 long? 1.5 dia?	1979 Jul 9.1	50.69	93.38	6821	344	542	0.015	157
D	Cosmos 1112 rocket*	1979-63B	1979 Jul 6.35 186 days 1980 Jan 8	Cylinder 2200?	7.4 long 2.4 dia	1979 Jul 9.9	50.69	93.25	6815	338	536	0.014	157
D	Fragments	1979-63C-AB											
D R	Cosmos 1113	1979-64A	1979 Jul 10.38 12.86 days 1979 Jul 23.24	Sphere-cylinder 5900?	5.9 long 2.4 dia	1979 Jul 10.6	65.00	89.52	6630	173	330	0.012	64
D	Cosmos 1113 rocket	1979-64B	1979 Jul 10.38 5 days 1979 Jul 15	Cylinder 2500?	7.5 long 2.6 dia	1979 Jul 14.1	65.00	87.95	6552	149	198	0.004	56
D	Capsule?	1979-64D	1979 Jul 10.38 17 days 1979 Jul 27	Ellipsoid 200?	0.9 long 1.9 dia	1979 Jul 23	65.0	88.98	6603	155	294	0.011	-
D	Fragments	1979-64C,E											
D	Cosmos 1114	1979-65A	1979 Jul 11.66 899 days 1981 Dec 26	Cylinder + paddles? 900?	2 long? 1 dia?	1979 Jul 14.2	74.05	95.22	6907	506	552	0.003	346
D	Cosmos 1114 rocket	1979-65B	1979 Jul 11.66 940 days 1982 Feb 5	Cylinder 2200?	7.4 long 2.4 dia	1979 Jul 15.9	74.05	95.11	6902	496	551	0.004	346

* Disintegrated

	Name		Launch date, lifetime and descent date	Shape and weight (kg)	Size (m)	Date of orbital determination	Orbital inclination (deg)	Nodal period (min)	Semi major axis (km)	Perigee height (km)	Apogee height (km)	Orbital eccentricity	Argument of perigee (deg)
D R	Cosmos 1115*	1979-66A	1979 Jul 13.35 12.85 days 1979 Jul 26.20	Sphere-cylinder 5900?	5.9 long 2.4 dia	1979 Jul 14.8	81.35	89.04	6604	217	235	0.001	75
D	Cosmos 1115 rocket	1979-66B	1979 Jul 13.35 4 days 1979 Jul 17	Cylinder 2500?	7.5 long 2.6 dia	1979 Jul 15.6	81.34	88.44	6574	187	205	0.001	7
D	Capsule	1979-66C	1979 Jul 13.35 38 days 1979 Aug 20	Ellipsoid 200?	0.9 long 1.9 dia	1979 Jul 26	81.3	88.74	6589	210	211	0	-
	Cosmos 1116	1979-67A	1979 Jul 20.50 30 years	Cylinder + 2 vanes? 2500?	5.0 long? 1.5 dia?	1979 Jul 24.0	81.19	97.07	6995	590	643	0.004	286
	Cosmos 1116 rocket	1979-67B	1979 Jul 20.50 30 years	Cylinder 1440	3.8 long 2.6 dia	1979 Jul 20.9	81.21	97.20	7001	596	650	0.004	243
D R	Cosmos 1117	1979-68A	1979 Jul 25.64 12.6 days 1979 Aug 7.2	Sphere-cylinder 6300?	6.5 long? 2.4 dia	1979 Jul 26.7	62.80	89.53	6630	179	325	0.011	75
D	Cosmos 1117 rocket	1979-68B	1979 Jul 25.64 4 days 1979 Jul 29	Cylinder 2500?	7.5 long 2.6 dia	1979 Jul 25.9	62.78	89.42	6625	177	316	0.011	72
D	Cosmos 1117 engine	1979-68C	1979 Jul 25.64 17 days 1979 Aug 11	Cone 600? full	1.5 long? 2.0 dia?	1979 Aug 6.5	62.80	89.08	6608	168	291	0.009	-
D	Fragments	1979-68D,E											

* 2000th launch.

	Name		Launch date, lifetime and descent date	Shape and weight (kg)	Size (m)	Date of orbital determination	Orbital inclination (deg)	Nodal period (min)	Semi major axis (km)	Perigee height (km)	Apogee height (km)	Orbital eccentricity	Argument of perigee (deg)
D R	Cosmos 1118	1979-69A	1979 Jul 27.30 12.8 days 1979 Aug 9.1	Sphere-cylinder 5700?	5.0 long 2.4 dia	1979 Jul 27.6	81.35	89.12	6608	217	243	0.002	73
D	Cosmos 1118 rocket	1979-69B	1979 Jul 27.30 4 days 1979 Jul 31	Cylinder 2500?	7.5 long 2.6 dia	1979 Jul 29.6	81.35	88.68	6586	197	219	0.002	24
D	Fragment	1979-69C											
	Molniya 1AV	1979-70A	1979 Jul 31.17 19½ years	Windmill + 6 vanes 1000?	3.4 long 1.6 dia	1979 Aug 23.3	62.84	717.73	26555	452	39902	0.743	280
D	Molniya 1AV launcher rocket	1979-70B	1979 Jul 31.17 20 days 1979 Aug 20	Cylinder 2500?	7.5 long 2.6 dia	1979 Aug 2.2	62.84	90.71	6689	216	405	0.014	126
D	Molniya 1AV launcher	1979-70C	1979 Jul 31.17 14 days 1979 Aug 14	Irregular	-	1979 Aug 1.7	62.80	90.81	6694	194	437	0.018	119
	Molniya 1AV rocket	1979-70D	1979 Jul 31.17 19½ years	Cylinder 440	2.0 long 2.0 dia	1979 Nov 1.0	62.9	733.10	26933	517	40592	0.744	-
D R	Cosmos 1119	1979-71A	1979 Aug 3.45 11.9 days 1979 Aug 15.3	Sphere-cylinder 5700?	5.0 long 2.4 dia	1979 Aug 4.2	81.35	89.10	6607	216	242	0.002	83
D	Cosmos 1119 rocket	1979-71B	1979 Aug 3.45 4 days 1979 Aug 7	Cylinder 2500?	7.5 long 2.6 dia	1979 Aug 5.7	81.35	88.46	6575	183	211	0.002	297

	Name		Launch date, lifetime and descent date	Shape and weight (kg)	Size (m)	Date of orbital determination	Orbital inclination (deg)	Nodal period (min)	Semi major axis (km)	Perigee height (km)	Apogee height (km)	Orbital eccentricity	Argument of perigee (deg)
T	Westar 3	1979-72A	1979 Aug 10.01 >million years	Cylinder 574 full?	1.65 long? 1.90 dia?	1979 Aug 10.0 1980 Jan 1.0	24.64 0.0	649.14 1436.2	24834 42165	231 35780	36681 35794	0.734 0.0002	179 -
D	Westar 3 second stage	1979-72B	1979 Aug 10.01 <1 day 1979 Aug 10	Cylinder + annulus 350?	6.4 long 1.52 and 2.44 dia	1979 Aug 10.1	31.47	88.18	6562	130	239	0.008	294
	Westar 3 third stage	1979-72C	1979 Aug 10.01 10 years?	Sphere- cone 66	1.32 long 0.94 dia	1979 Aug 10.0	24.64	649.14	24834	231	36681	0.734	179
D R	Cosmos 1120	1979-73A	1979 Aug 11.39 12.9 days 1979 Aug 24.3	Sphere- cylinder 6300?	6.5 long? 2.4 dia	1979 Aug 11.5 1979 Aug 21.8	70.56 70.42	89.84 89.76	6644 6640	170 173	362 351	0.014 0.013	60 42
D	Cosmos 1120 rocket	1979-73B	1979 Aug 11.39 4 days 1979 Aug 15	Cylinder 2500?	7.5 long 2.6 dia	1979 Aug 11.7	70.42	89.66	6635	173	341	0.013	50
D	Cosmos 1120 engine	1979-73D	1979 Aug 11.39 20 days 1979 Aug 31	Cone 600? full	1.5 long? 2 dia?	1979 Aug 28.0	70.42	89.09	6608	163	296	0.010	-
D	Fragment	1979-73C											
D R	Cosmos 1121*	1979-74A	1979 Aug 14.65 29.5 days 1979 Sep 13.1	Sphere- cylinder 6700?	7 long? 2.4 dia	1979 Aug 15.5	67.16	89.69	6638	171	348	0.013	71
D	Cosmos 1121 rocket	1979-74B	1979 Aug 14.65 4 days 1979 Aug 18	Cylinder 2500?	7.5 long 2.6 dia	1979 Aug 14.9	67.15	89.58	6632	169	339	0.013	69

* Manoeuvrable, but no jettisoned engine apparently tracked or designated.

	Name		Launch date, lifetime and descent date	Shape and weight (kg)	Size (m)	Date of orbital determination	Orbital inclination (deg)	Nodal period (min)	Semi major axis (km)	Perigee height (km)	Apogee height (km)	Orbital eccentricity	Argument of perigee (deg)
D R	Cosmos 1122*	1979-75A	1979 Aug 17.32 12.9 days 1979 Aug 30.2	Sphere-cylinder 5700?	5.0 long? 2.4 dia	1979 Aug 17.9	81.34	89.06	6605	214	239	0.002	39
D	Cosmos 1122 rocket	1979-75B	1979 Aug 17.32 3 days 1979 Aug 20	Cylinder 2500?	7.5 long 2.6 dia	1979 Aug 17.7	81.36	88.91	6597	207	231	0.002	16
D	Fragments	1979-75C-E											
D R	Cosmos 1123	1979-76A	1979 Aug 21.47 12.8 days 1979 Sep 3.3	Sphere-cylinder 5900?	5.9 long 2.4 dia	1979 Aug 26.8	81.36	88.99	6601	212	234	0.002	42
D	Cosmos 1123 rocket	1979-76B	1979 Aug 21.47 3 days 1979 Aug 24	Cylinder 2500?	7.5 long 2.6 dia	1979 Aug 21.8	81.35	88.95	6599	211	231	0.001	28
D	Capsule	1979-76E	1979 Aug 21.47 17 days? 1979 Sep 7	Ellipsoid 200?	0.9 long 1.9 dia	1979 Sep 3.1	81.36	88.71	6588	202	218	0.001	4
D	Fragments	1979-76C,D											
	Cosmos 1124	1979-77A	1979 Aug 28.01 100 years?	Windmill + 6 vanes? 1250?	4.2 long? 1.6 dia?	1979 Aug 29.1 1979 Sep 13.6	62.98 62.97	727.27 716.52	26792 26525	561 586	40267 39708	0.741 0.737	318 318
D	Cosmos 1124 launcher rocket	1979-77B	1979 Aug 28.01 31 days 1979 Sep 28	Cylinder 2500?	7.5 long 2.6 dia	1979 Sep 5.1	62.91	92.11	6760	215	549	0.025	120
D	Cosmos 1124 launcher	1979-77C	1979 Aug 28.01 14 days 1979 Sep 11	Irregular	-	1979 Sep 2.8	62.91	91.35	6722	174	514	0.025	119
	Cosmos 1124 rocket	1979-77D	1979 Aug 28.01 100 years?	Cylinder 440	2.0 long 2.0 dia	1979 Nov 1.0	62.98	723.82	26705	622	40031	0.738	-
	Fragments	1979-77E-H											

* Earth resources

	Name		Launch date, lifetime and descent date	Shape and weight (kg)	Size (m)	Date of orbital determination	Orbital inclination (deg)	Nodal period (min)	Semi major axis (km)	Perigee height (km)	Apogee height (km)	Orbital eccentricity	Argument of perigee (deg)
	Cosmos 1125	1979-78A	1979 Aug 28.04 120 years	Cylinder + paddles? 750?	2 long? 1 dia?	1979 Aug 28.2	74.05	100.84	7176	784	812	0.002	38
	Cosmos 1125 rocket	1979-78B	1979 Aug 28.04 100 years	Cylinder 2200?	7.4 long 2.4 dia	1979 Nov 1.0	74.06	100.74	7171	780	806	0.002	-
D R	Cosmos 1126	1979-79A	1979 Aug 31.47 13.8 days 1979 Sep 14.3	Sphere- cylinder 6300?	6.5 long? 2.4 dia	1979 Sep 1.2 1979 Sep 1.6	72.85 72.87	90.42 92.26	6674 6763	197 377	395 393	0.015 0.001	83 88
D	Cosmos 1126 rocket	1979-79B	1979 Aug 31.47 9 days 1979 Sep 9	Cylinder 2500?	7.5 long 2.6 dia	1979 Sep 3.0	72.86	90.03	6654	191	361	0.013	74
D	Cosmos 1126 engine	1979-79D	1979 Aug 31.47 169 days 1980 Feb 16	Cone 600? full	1.5 long? 2 dia?	1979 Nov 1.0	72.87	91.76	6738	349	371	0.002	-
D	Fragments	1979-79C,E-G											
D R	Cosmos 1127*	1979-80A	1979 Sep 5.43 12.9 days 1979 Sep 18.3	Sphere- cylinder 6300?	6.5 long? 2.4 dia	1979 Sep 6.3 1979 Sep 7.4	81.35 81.35	89.40 89.83	6622 6643	215 259	272 271	0.004 0.001	87 278
D	Cosmos 1127 rocket	1979-80B	1979 Sep 5.43 5 days 1979 Sep 10	Cylinder 2500?	7.5 long 2.6 dia	1979 Sep 6.3	81.36	89.19	6611	211	255	0.003	72
D	Cosmos 1127 engine	1979-80D	1979 Sep 5.43 25 days 1979 Sep 30	Cone 600? full	1.5 long? 2 dia?	1979 Sep 17.6	81.36	89.82	6643	257	272	0.001	261
D	Fragment	1979-80C											

* Earth resources

	Name		Launch date, lifetime and descent date	Shape and weight (kg)	Size (m)	Date of orbital determination	Orbital inclination (deg)	Nodal period (min)	Semi major axis (km)	Perigee height (km)	Apogee height (km)	Orbital eccentricity	Argument of perigee (deg)
D R	Cosmos 1128	1979-81A	1979 Sep 14.65 12.6 days 1979 Sep 27.3	Sphere-cylinder 6300?	6.5 long? 2.4 dia	1979 Sep 15.2 1979 Sep 17.1	62.80 62.80	89.54 89.40	6631 6624	177 154	328 337	0.011 0.014	72 72
D	Cosmos 1128 rocket	1979-81B	1979 Sep 14.65 3 days 1979 Sep 17	Cylinder 2500?	7.5 long 2.6 dia	1979 Sep 15.0	62.78	89.38	6623	173	316	0.011	68
D	Cosmos 1128 engine	1979-81C	1979 Sep 14.65 13 days 1979 Sep 27	Cone 600? full	1.5 long? 2 dia?	1979 Sep 26.4	62.80	89.50	6629	147	354	0.016	69
D	HEAO 3	1979-82A	1979 Sep 20.23 809 days 1981 Dec 7	Hexagonal cylinder 2720	5.8 long 2.4 dia	1979 Sep 20.6	43.61	94.38	6871	485	501	0.001	295
D	HEAO 3 rocket	1979-82B	1979 Sep 20.23 54 days 1979 Nov 13	Cylinder 1815	8.6 long 3.0 dia	1979 Sep 23.2	43.67	92.21	6766	358	417	0.004	306
D R B	Cosmos 1129*	1979-83A	1979 Sep 25.65 18.5 days 1979 Oct 14.1	Sphere-cylinder 5900?	5.9 long 2.4 dia	1979 Sep 26.8	62.82	90.45	6676	218	377	0.012	109
D	Cosmos 1129 rocket	1979-83B	1979 Sep 25.65 11 days 1979 Oct 6	Cylinder 2500?	7.5 long 2.6 dia	1979 Sep 27.2	62.81	90.26	6666	214	362	0.011	106
D	Capsule	1979-83E	1979 Sep 25.65 173 days 1980 Mar 16	Ellipsoid 200?	0.9 long 1.9 dia	1979 Nov 1.0	62.82	90.50	6678	209	391	0.014	-
D	Fragments	1979-83C,D											

* International biological satellite carrying rats, quails, etc.

Name		Launch date, lifetime and descent date	Shape and weight (kg)	Size (m)	Date of orbital determination	Orbital inclination (deg)	Nodal period (min)	Semi major axis (km)	Perigee height (km)	Apogee height (km)	Orbital eccentricity	Argument of perigee (deg)
Cosmos 1130	1979-84A	1979 Sep 25.87 7000 years	Spheroid 40?	1.0 long? 0.8 dia?	1979 Sep 26.9	74.03	114.69	7819	1400	1482	0.005	104
Cosmos 1131	1979-84B	1979 Sep 25.87 8000 years	Spheroid 40?	1.0 long? 0.8 dia?	1979 Sep 26.9	74.04	114.85	7826	1410	1486	0.005	121
Cosmos 1132	1979-84C	1979 Sep 25.87 9000 years	Spheroid 40?	1.0 long? 0.8 dia?	1979 Sep 27.0	74.02	115.02	7834	1429	1483	0.004	115
Cosmos 1133	1979-84D	1979 Sep 25.87 9000 years	Spheroid 40?	1.0 long? 0.8 dia?	1979 Sep 26.8	74.05	115.18	7841	1441	1485	0.003	133
Cosmos 1134	1979-84E	1979 Sep 25.87 10000 years	Spheroid 40?	1.0 long? 0.8 dia?	1979 Sep 26.8	74.01	115.35	7849	1455	1486	0.002	136
Cosmos 1135	1979-84F	1979 Sep 25.87 10000 years	Spheroid 40?	1.0 long? 0.8 dia?	1979 Sep 28.5	74.02	115.53	7873	1465	1493	0.002	170
Cosmos 1136	1979-84G	1979 Sep 25.87 10000 years	Spheroid 40?	1.0 long? 0.8 dia?	1979 Sep 26.8	74.03	115.70	7865	1472	1501	0.002	202
Cosmos 1137	1979-84H	1979 Sep 25.87 10000 years	Spheroid 40?	1.0 long? 0.8 dia?	1979 Sep 26.8	74.03	115.90	7874	1472	1519	0.003	216
Cosmos 1130 rocket	1979-84J	1979 Sep 25.87 20000 years	Cylinder 2200?	7.4 long 2.4 dia	1979 Sep 27.1	74.03	117.88	7963	1480	1690	0.013	264

	Name		Launch date, lifetime and descent date	Shape and weight (kg)	Size (m)	Date of orbital determination	Orbital inclination (deg)	Nodal period (min)	Semi major axis (km)	Perigee height (km)	Apogee height (km)	Orbital eccentricity	Argument of perigee (deg)
D R	Cosmos 1138	1979-85A	1979 Sep 28.52 13.8 days 1979 Oct 12.3	Sphere-cylinder 6300?	6.5 long? 2.4 dia	1979 Sep 29.3 1979 Sep 29.9	72.86 72.87	90.24 92.31	6664 6766	199 364	373 411	0.013 0.003	82 203
D	Cosmos 1138 rocket	1979-85B	1979 Sep 28.52 7 days 1979 Oct 5	Cylinder 2500?	7.5 long 2.6 dia	1979 Sep 28.6	72.87	90.18	6661	199	367	0.013	83
D	Cosmos 1138 engine	1979-85F	1979 Sep 28.52 151 days 1980 Feb 26	Cone 600? full	1.5 long? 2 dia?	1979 Oct 11.9	72.86	92.30	6765	362	412	0.004	-
D	Fragments	1979-85C-E											
T	IMEWS 10 [Titan 3C]	1979-86A	1979 Oct 1.48 > million years	-	-	1979 Oct 30	7.5	1445.5	42348	30443	41497	0.131	-
D	Titan 3C second stage	1979-86B	1979 Oct 1.48 6 days 1979 Oct 7	Cylinder 1900	6 long 3.0 dia	1979 Oct 1.5	29.05	91.14	6716	151	524	0.028	117
	Titan 3C transtage	1979-86C	1979 Oct 1.48 > million years	Cylinder 1500?	6 long? 3.0 dia	Synchronous orbit similar to 1979-86A							
	Statsionar-Ekran 4*	1979-87A	1979 Oct 3.72 > million years	Cylinder + plate	-	1979 Oct 4.5 1979 Nov 1.0	0.45 0.4	1424 1436.1	41935 42165	35557 35707	35557 35866	0 0.002	- -
D	Ekran 4 launcher rocket	1979-87B	1979 Oct 3.72 2 days 1979 Oct 5	Cylinder 4000?	12 long? 4 dia	1979 Oct 4.0	51.62	88.13	6563	177	193	0.001	233

* There is probably a separated apogee motor in a similar orbit.

1979-87 continued on page 588

	Name		Launch date, lifetime and descent date	Shape and weight (kg)	Size (m)	Date of orbital determination	Orbital inclination (deg)	Nodal period (min)	Semi major axis (km)	Perigee height (km)	Apogee height (km)	Orbital eccentricity	Argument of perigee (deg)
D	Ekran 4 launcher	1979-87C	1979 Oct 3.72 2 days? 1979 Oct 5?	Irregular	-		Orbit similar to 1979-87B						
D	Ekran 4 rocket	1979-87D	1979 Oct 3.72 1 year?	Cylinder 1900?	3.9 long? 3.9 dia		Transfer orbit similar to 1979-35D						
D R	Cosmos 1139	1979-88A	1979 Oct 5.48 12.8 days 1979 Oct 18.3	Sphere-cylinder 5700?	5.0 long 2.4 dia	1979 Oct 8.4	72.85	89.81	6643	202	327	0.009	65
D	Cosmos 1139 rocket	1979-88B	1979 Oct 5.48 6 days 1979 Oct 11	Cylinder 2500?	7.5 long 2.6 dia	1979 Oct 8.0	72.85	89.33	6619	199	282	0.006	75
D	Fragment	1979-88C											
	Cosmos 1140	1979-89A	1979 Oct 11.69 120 years	Cylinder + paddles? 750?	2 long? 1 dia?	1979 Oct 13.4	74.07	100.73	7171	780	805	0.002	346
	Cosmos 1140 rocket	1979-89B	1979 Oct 11.69 100 years	Cylinder 2200?	7.4 long 2.4 dia	1979 Oct 13.7	74.08	100.63	7166	774	802	0.002	9
	Cosmos 1141	1979-90A	1979 Oct 16.51 1200 years	Cylinder 700?	1.3 long? 1.9 dia?	1979 Oct 18.4	82.95	104.76	7360	961	1003	0.003	285
	Cosmos 1141 rocket	1979-90B	1979 Oct 16.51 600 years	Cylinder 2200?	7.4 long 2.4 dia	1979 Oct 16.7	82.88	104.66	7355	961	993	0.002	288
1d	Fragments	1979-90C,D											

	Name		Launch date, lifetime and descent date	Shape and weight (kg)	Size (m)	Date of orbital determination	Orbital inclination (deg)	Nodal period (min)	Semi major axis (km)	Perigee height (km)	Apogee height (km)	Orbital eccentricity	Argument of perigee (deg)
	Molniya 1AW	1979-91A	1979 Oct 20.30 15 years	Windmill + 6 vanes 1000?	3.4 long 1.6 dia	1979 Oct 21.4 1980 Jan 1.0	62.83 63.0	735.87 717.63	27001 26553	618 593	40627 39756	0.741 0.738	288 -
D	Molniya 1AW launcher rocket	1979-91B	1979 Oct 20.30 22 days 1979 Nov 11	Cylinder 2500?	7.5 long 2.6 dia	1979 Oct 21.8	62.85	92.29	6766	216	560	0.025	122
D	Molniya 1AW launcher	1979-91C	1979 Oct 20.30 21 days 1979 Nov 10	Irregular	-	1979 Oct 21.8	62.85	92.58	6781	201	604	0.030	117
	Molniya 1AW rocket	1979-91D	1979 Oct 20.30 15 years	Cylinder 440	2.0 long 2.0 dia	1979 Oct 27.0	62.89	732.80	26924	627	40465	0.740	288
D R	Cosmos 1142	1979-92A	1979 Oct 22.53 12.8 days 1979 Nov 4.3	Sphere-cylinder 6300?	6.5 long? 2.4 dia	1979 Oct 23.3 1979 Oct 29.4	72.86 72.86	90.32 92.27	6668 6764	198 353	382 418	0.014 0.005	83 184
D	Cosmos 1142 rocket	1979-92B	1979 Oct 22.53 6 days 1979 Oct 28	Cylinder 2500?	7.5 long 2.6 dia	1979 Oct 23.5	72.84	90.07	6656	194	361	0.013	79
D	Cosmos 1142 engine	1979-92C	1979 Oct 22.53 155 days 1980 Mar 25	Cone 600? full	1.5 long? 2 dia?	1979 Nov 6.4	72.85	92.21	6761	353	412	0.004	167
D	Fragments	1979-92D-F											
	Cosmos 1143	1979-93A	1979 Oct 26.76 60 years	Cylinder + 2 vanes? 2500?	5 long? 1.5 dia?	1979 Oct 27.4	81.24	97.44	7013	624	646	0.002	8
	Cosmos 1143 rocket	1979-93B	1979 Oct 26.76 60 years	Cylinder 1440	3.8 long 2.6 dia	1979 Oct 27.4	81.24	97.54	7018	614	666	0.004	156

	Name		Launch date, lifetime and descent date	Shape and weight (kg)	Size (m)	Date of orbital determination	Orbital inclination (deg)	Nodal period (min)	Semi major axis (km)	Perigee height (km)	Apogee height (km)	Orbital eccentricity	Argument of perigee (deg)
D	Magsat* [Scout]	1979-94A	1979 Oct 30.59 225 days 1980 Jun 11	Box + 4 paddles 181	1.64 long 0.77 square	1979 Nov 1.3	96.80	93.82	6837	355	562	0.015	129
D	Magsat rocket	1979-94B	1979 Oct 30.59 89 days 1980 Jan 27	Cylinder 24	1.50 long 0.46 dia	1979 Nov 1.2	96.80	93.76	6834	355	557	0.015	129
D	Fragment	1979-94C											
	Meteor 2-05	1979-95A	1979 Oct 31.40 500 years	Cylinder + 2 vanes 2750?	5 long? 1.5 dia?	1979 Nov 2.1	81.21	102.62	7260	873	890	0.001	214
	Meteor 2-05 rocket	1979-95B	1979 Oct 31.40 400 years	Cylinder 1440	3.8 long 2.6 dia	1979 Nov 1.2	81.26	102.69	7263	848	922	0.005	174
D	Intercosmos 20	1979-96A	1979 Nov 1.34 488 days 1981 Mar 3	Octagonal ellipsoid 550?	1.8 long? 1.5 dia?	1979 Nov 2.0	74.05	94.42	6869	462	519	0.004	331
D	Intercosmos 20 rocket	1979-96B	1979 Nov 1.34 511 days 1981 Mar 26	Cylinder 2200?	7.4 long 2.4 dia	1979 Nov 1.9	74.04	94.35	6865	455	519	0.005	339
D R	Cosmos 1144	1979-97A	1979 Nov 2.67 31.5 days 1979 Dec 4.2	Sphere- cylinder 6700?	7 long? 2.4 dia	1979 Nov 4.7	67.16	89.44	6626	158	337	0.013	67
D	Cosmos 1144 rocket	1979-97B	1979 Nov 2.67 3.28 days 1979 Nov 5.95	Cylinder 2500?	7.5 long 2.6 dia	1979 Nov 4.4	67.16	89.11	6609	158	304	0.011	65

* Magnetic field satellite.

	Name		Launch date, lifetime and descent date	Shape and weight (kg)	Size (m)	Date of orbital determination	Orbital inclination (deg)	Nodal period (min)	Semi major axis (km)	Perigee height (km)	Apogee height (km)	Orbital eccentricity	Argument of perigee (deg)
T	DSCS 13	1979-98A	1979 Nov 21.09? > million years	Cylinder + 2 dishes 590	1.83 long 2.74 dia	1979 Nov 23.8	2.42	1431.14	42070	35594	35789	0.002	15
T	DSCS 14	1979-98B	1979 Nov 21.09? > million years	Cylinder + 2 dishes 590	1.83 long 2.74 dia	1979 Nov 23.9	2.41	1450.73	42452	35792	36357	0.007	175
	Transtage	1979-98C	1979 Nov 21.09? > million years	Cylinder 1500?	6 long? 3.0 dia	1979 Nov 24.0	2.44	1510.35	43608	35818	38641	0.032	179
D	Titan 3C second stage	1979-98D	1979 Nov 21.09? <1 day 1979 Nov 21	Cylinder 1900	6 long 3.0 dia	1979 Nov 21.1	28.6	87.72	6547	152	186	0.003	-
	Cosmos 1145	1979-99A	1979 Nov 27.42 60 years	Cylinder + 2 vanes? 2500?	5 long? 1.5 dia?	1979 Nov 28.2	81.22	97.33	7008	624	635	0.001	314
	Cosmos 1145 rocket	1979-99B	1979 Nov 27.42 60 years	Cylinder 1440	3.8 long 2.6 dia	1979 Nov 28.2	81.22	97.39	7011	588	677	0.006	181
D	Cosmos 1146	1979-100A	1979 Dec 5.38 721 days 1981 Nov 25	Cylinder?	4 long? 2 dia?	1979 Dec 6.0	65.85	93.98	6847	444	494	0.004	337
D	Cosmos 1146 rocket	1979-100B	1979 Dec 5.38 423 days 1981 Jan 31	Cylinder 2200?	7.4 long 2.4 dia	1979 Dec 6.0	65.85	93.91	6843	436	493	0.004	20

	Name		Launch date, lifetime and descent date	Shape and weight (kg)	Size (m)	Date of orbital determination	Orbital inclination (deg)	Nodal period (min)	Semi major axis (km)	Perigee height (km)	Apogee height (km)	Orbital eccentricity	Argument of perigee (deg)
	RCA Satcom 3*	1979-101A	1979 Dec 7.07 Indefinite	Box + vanes 868 full 463 empty	1.62 high 1.25 wide 1.25 deep	1979 Dec 7.1 1979 Dec 7.2	28.73 23.82	88.21 636.99	6564 24523	153 166	219 36124	0.005 0.733	122 178
							Intended to reach synchronous orbit*						
D	RCA Satcom 3 second stage	1979-101B	1979 Dec 7.07 1 month? 1980 Jan?	Cylinder + annulus 350?	6.4 long 1.52 and 2.44 dia	1979 Dec 8.1	27.35	111.38	7675	157	2437	0.149	181
D	RCA Satcom 3 third stage	1979-101D	1979 Dec 7.07 3 years 1982 Dec?	Sphere-cone 66	1.32 long 0.94 dia		Orbit similar to 1979-101A transfer orbit						
D	Fragment	1979-101C											
D R	Cosmos 1147	1979-102A	1979 Dec 12.52 13.75 days 1979 Dec 26.27	Sphere-cylinder 6300?	6.5 long? 2.4 dia	1979 Dec 12.5 1979 Dec 13.5	72.86 72.86	90.30 92.29	6667 6765	196 354	382 419	0.014 0.005	88 200
D	Cosmos 1147 rocket	1979-102B	1979 Dec 12.52 7 days 1979 Dec 19	Cylinder 2500?	7.5 long 2.6 dia	1979 Dec 12.5	72.86	90.18	6661	194	372	0.013	-
D	Cosmos 1147 engine	1979-102D	1979 Dec 12.52 88 days 1980 Mar 9	Cone 600? full	1.5 long? 2 dia?	1979 Dec 31.9	72.86	91.48	6725	290	403	0.008	-
D	Fragment	1979-102C											
D R	Soyuz Transport 1**	1979-103A	1979 Dec 16.52 100.39 days 1980 Mar 25.91	Sphere-cylinder + 2 wings 6850	7.5 long 2.3 to 2.72 dia	1979 Dec 16.8 1979 Dec 17.5 1979 Dec 31.9	51.63 51.63 51.63	88.42 89.34 91.88	6578 6623 6718	194 231 363	205 259 376	0.001 0.002 0.001	162 166 -
D	Soyuz T-1 rocket	1979-103B	1979 Dec 16.52 2 days 1979 Dec 18	Cylinder 2500?	7.5 long 2.6 dia	1979 Dec 17.2	51.59	88.26	6570	179	204	0.002	99
D	Soyuz T-1 orbital† module	1979-103D	1979 Dec 16.52 164 days 1980 May 28?	Spheroid 1260?	2.3 dia	1980 May 1.0	51.63	90.46	6678	296	303	0.001	
D	Fragment	1979-103C											

* Contact lost during apogee motor firing. Present orbit unknown. Possibly disintegrated.

** Unmanned cargo craft docked with Salyut 6 (1st airlock) on 1979 Dec 19.59. Undocked on 1980 Mar 23.88.

† Part of 1979-103A until about 1980 Mar 25. This orbital module was probably designated as Salyut 6 fragment 1977-97BR in the United States.

Year of launch 1979 continued

	Name		Launch date, lifetime and descent date	Shape and weight (kg)	Size (m)	Date of orbital determination	Orbital inclination (deg)	Nodal period (min)	Semi major axis (km)	Perigee height (km)	Apogee height (km)	Orbital eccentricity	Argument of perigee (deg)
	CAT 1 (+ ballast) [Ariane 1-01]*	1979-104A	1979 Dec 24.72 10 years	Sphere 1602**	-	1979 Dec 24.7	17.55	635.30	24480	201	36003	0.731	190
D	Ariane 1-01 third stage	1979-104B	1979 Dec 24.72 1056 days 1982 Nov 14	Cylinder 1634 empty	8.6 long 2.6 dia	1980 Mar 1.0	18.1	604.8	23691	204	34421	0.722	-
	Gorizont 3	1979-105A	1979 Dec 28 > million years	-	-	1980 Jan 1.0 1980 Mar 1.0	0.8 0.6	1461.9 1436.1	42668 42164	36240 35689	36339 35883	0.001 0.002	- -
D	Gorizont 3 launcher rocket	1979-105B	1979 Dec 28 2 days 1979 Dec 30	Cylinder 4000?	12 long? 4 dia	1979 Dec 29.8	51.63	87.43	6528	142	157	0.001	230
D	Gorizont 3 rocket	1979-105C	1979 Dec 28 2.3 years 1982 Apr?	Cylinder 1900?	3.9 long? 3.9 dia	1980 Mar 1.0	47.1	639.9	24597	245	36192	0.731	-
	Gorizont 3 apogee motor	1979-105E	1979 Dec 28 > million years	Cylinder 440?	2.0 long? 2.0 dia?	1980 Mar 1.0	0.7	1459.3	42616	36168	36308	0.002	-
D	Fragment	1979-105D											
D R	Cosmos 1148	1979-106A	1979 Dec 28.6 12.7 days 1980 Jan 10.3	Sphere-cylinder 6700?	7 long? 2.4 dia	1979 Dec 29.7	67.14	89.64	6636	173	343	0.013	66
D	Cosmos 1148 rocket	1979-106B	1979 Dec 28.6 3 days 1979 Dec 31	Cylinder 2500?	7.5 long 2.6 dia	1979 Dec 31.1	67.13	88.36	6572	145	242	0.007	58
D	Cosmos 1148 engine	1979-106D	1979 Dec 28.6 18 days 1980 Jan 15	Cone 600? full	1.5 long? 2 dia?	1980 Jan 10.8	67.14	89.32	6620	161	323	0.012	-
D	Fragments	1979-106C,E,F											

* Capsule Ariane Technologique. First ESA launch, from Kourou, French Guiana.
** 217 kg capsule plus 1385 kg ballast.

	Name		Launch date, lifetime and descent date	Shape and weight (kg)	Size (m)	Date of orbital determination	Orbital inclination (deg)	Nodal period (min)	Semi major axis (km)	Perigee height (km)	Apogee height (km)	Orbital eccentricity	Argument of perigee (deg)
D R	Cosmos 1149	1980-01A	1980 Jan 9.51 13.78 days 1980 Jan 23.29	Sphere-cylinder 6300?	6.5 long? 2.4 dia	1980 Jan 10.2 1980 Jan 15.4	72.87 72.87	90.32 92.30	6668 6765	188 353	392 420	0.015 0.005	84 185
D	Cosmos 1149 rocket	1980-01B	1980 Jan 9.51 8 days 1980 Jan 17	Cylinder 2500?	7.5 long 2.6 dia	1980 Jan 10.3	72.87	90.24	6664	196	376	0.013	82
D	Cosmos 1149 engine	1980-01D	1980 Jan 9.51 154 days 1980 Jun 11	Cone 600? full	1.5 long? 2 dia?	1980 Jan 23.4	72.88	92.27	6764	353	418	0.005	167
D	Fragments	1980-01C,E-G											
	Molniya 1AX	1980-02A	1980 Jan 11.52 16½ years*	Windmill + 6 vanes 1000?	3.4 long 1.6 dia	1980 Jan 13.6 1980 Jan 22.7	62.84 62.88	736.53 717.87	27017 26559	435 442	40842 39920	0.748 0.743	280 280
D	Molniya 1AX launcher rocket	1980-02B	1980 Jan 11.52 17 days 1980 Jan 28	Cylinder 2500?	7.5 long 2.6 dia	1980 Jan 11.6	62.84	90.99	6702	210	438	0.017	129
D	Molniya 1AX launcher	1980-02C	1980 Jan 11.52 13 days 1980 Jan 24	Irregular	-	1980 Jan 11.6	62.85	91.59	6731	215	491	0.020	109
	Molniya 1AX rocket	1980-02F	1980 Jan 11.52 16½ years*	Cylinder 440	2.0 long 2.0 dia	1980 Jan 22.3	62.88	732.82	26926	448	40648	0.746	280
D	Fragments	1980-02D,E											

* Decay possible in 1992.

	Name		Launch date, lifetime and descent date	Shape and weight (kg)	Size (m)	Date of orbital determination	Orbital inclination (deg)	Nodal period (min)	Semi major axis (km)	Perigee height (km)	Apogee height (km)	Orbital eccentricity	Argument of perigee (deg)
	Cosmos 1150	1980-03A	1980 Jan 14.82 1200 years	Cylinder 700?	1.3 long? 1.9 dia?	1980 Jan 15.1	82.95	105.01	7372	971	1017	0.003	277
	Cosmos 1150 rocket	1980-03B	1980 Jan 14.82 600 years	Cylinder 2200?	7.4 long 2.4 dia	1980 Jan 15.1	82.95	104.88	7366	969	1006	0.003	278
T	Fleetsatcom 3	1980-04A	1980 Jan 18.06 > million years	Hexagonal cylinder 1884 full	1.27 long 2.44 dia	1980 Jan 18.1 1980 Jan 20.1	26.36 2.40	634.56 1423.06	24461 41911	167 35405	35999 35661	0.732 0.003	181 183
	Fleetsatcom 3 rocket	1980-04B	1980 Jan 18.06 7 years	Cylinder 1815	8.6 long 3.0 dia	1980 Jan 18.1	26.30	619.94	24084	171	35241	0.728	182
	Cosmos 1151*	1980-05A	1980 Jan 23.29 60 years	-	-	1980 Jan 24.1	82.52	97.78	7030	637	666	0.002	256
	Cosmos 1151 rocket	1980-05B	1980 Jan 23.29 60 years	Cylinder 2200?	7.4 long 2.4 dia	1980 Jan 24.6	82.52	97.76	7029	637	664	0.002	259
D R	Cosmos 1152	1980-06A	1980 Jan 24.66 12.8 days 1980 Feb 6.5	Sphere-cylinder 6700?	7 long? 2.4 dia	1980 Jan 25.6	67.14	89.66	6637	173	345	0.013	70
D	Cosmos 1152 rocket	1980-06B	1980 Jan 24.66 4 days 1980 Jan 28	Cylinder 2500?	7.5 long 2.6 dia	1980 Jan 25.4	67.14	89.51	6630	171	332	0.012	70

* Oceanographic satellite

	Name		Launch date, lifetime and descent date	Shape and weight (kg)	Size (m)	Date of orbital determination	Orbital inclination (deg)	Nodal period (min)	Semi major axis (km)	Perigee height (km)	Apogee height (km)	Orbital eccentricity	Argument of perigee (deg)
	Cosmos 1153	1980-07A	1980 Jan 25.93 1200 years	Cylinder 700?	1.3 long? 1.9 dia?	1980 Jan 26.7	82.93	105.00	7372	967	1020	0.004	285
	Cosmos 1153 rocket	1980-07B	1980 Jan 25.93 600 years	Cylinder 2200?	7.4 long 2.4 dia	1980 Jan 27.3	82.93	104.89	7366	967	1009	0.003	274
	Cosmos 1154	1980-08A	1980 Jan 30.54 60 years	Cylinder + 2 vanes? 2500?	5 long? 1.5 dia?	1980 Jan 30.7	81.23	97.48	7015	630	644	0.001	66
	Cosmos 1154 rocket	1980-08B	1980 Jan 30.54 60 years	Cylinder 1440	3.8 long 2.6 dia	1980 Jan 31.0	81.24	97.57	7020	608	675	0.005	165
D R	Cosmos 1155	1980-09A	1980 Feb 7.46 13.7 days 1980 Feb 21.2	Sphere- cylinder 6300?	6.5 long? 2.4 dia	1980 Feb 7.5 1980 Feb 9.2 1980 Feb 20.0	72.86 72.87 72.86	90.45 92.29 91.35	6674 6765 6718	195 357 296	397 416 384	0.015 0.004 0.007	88 208 109
D	Cosmos 1155 rocket	1980-09B	1980 Feb 7.46 9 days 1980 Feb 16	Cylinder 2500?	7.5 long 2.6 dia	1980 Feb 7.8	72.88	90.38	6671	194	391	0.015	89
D	Cosmos 1155 engine	1980-09E	1980 Feb 7.46 63 days 1980 Apr 10	Cone 600? full	1.5 long? 2 dia?	1980 Feb 24.6	72.87	91.27	6714	293	379	0.006	99
D	Fragments	1980-09C,D											

	Name		Launch date, lifetime and descent date	Shape and weight (kg)	Size (m)	Date of orbital determination	Orbital inclination (deg)	Nodal period (min)	Semi major axis (km)	Perigee height (km)	Apogee height (km)	Orbital eccentricity	Argument of perigee (deg)
D	[Titan 3D]	1980-10A	1980 Feb 7.88 996 days 1982 Oct 30	Cylinder 13300? full	15 long 3.0 dia	1980 Feb 9.7 1980 Feb 9.9	96.98 97.05	91.86 92.69	6743 6783	229 309	500 501	0.020 0.014	175 179
D	Titan 3D rocket	1980-10B	1980 Feb 7.88 35 days 1980 Mar 13	Cylinder 1900	6 long 3.0 dia	1980 Feb 8.3	96.98	91.73	6736	230	486	0.019	179
D	Fragment	1980-10C											
T	Navstar 5 [Atlas F]	1980-11A	1980 Feb 9.97 1 million years	Cylinder + 4 vanes 433	-	1980 Feb 10.1 1980 Feb 13.6	63.10 63.72	352.54 715.23	16532 26493	163 20083	20144 20147	0.604 0.001	159 285
	Navstar 5 rocket	1980-11B	1980 Feb 9.97 32 years	Cone-cylinder? 163?	1.85 long? 0.63 to 1.65 dia?	1980 Feb 15.3	63.10	350.45	16467	166	20011	0.603	159
	Cosmos 1156	1980-12A	1980 Feb 11.98 8000 years	Spheroid 40?	1.0 long? 0.8 dia?	1980 Feb 15.2	74.02	114.64	7816	1400	1475	0.005	100
	Cosmos 1157	1980-12B	1980 Feb 11.98 9000 years	Spheroid 40?	1.0 long? 0.8 dia?	1980 Feb 15.2	74.02	114.85	7825	1417	1477	0.004	114
	Cosmos 1158	1980-12C	1980 Feb 11.98 9000 years	Spheroid 40?	1.0 long? 0.8 dia?	1980 Feb 15.2	74.02	115.05	7835	1435	1478	0.003	120
	Cosmos 1159	1980-12D	1980 Feb 11.98 9000 years	Spheroid 40?	1.0 long? 0.8 dia?	1980 Feb 15.2	74.02	115.26	7845	1453	1481	0.002	141

1980-12 continued on page 598

	Name		Launch date, lifetime and descent date	Shape and weight (kg)	Size (m)	Date of orbital determination	Orbital inclination (deg)	Nodal period (min)	Semi major axis (km)	Perigee height (km)	Apogee height (km)	Orbital eccentricity	Argument of perigee (deg)
	Cosmos 1160	1980-12E	1980 Feb 11.98 9000 years	Spheroid 40?	1.0 long? 0.8 dia?	1980 Feb 17.7	74.02	115.47	7855	1467	1486	0.001	181
	Cosmos 1161	1980-12F	1980 Feb 11.98 10000 years	Spheroid 40?	1.0 long? 0.8 dia?	1980 Feb 15.3	74.02	115.71	7865	1469	1505	0.002	221
	Cosmos 1162	1980-12G	1980 Feb 11.98 10000 years	Spheroid 40?	1.0 long? 0.8 dia?	1980 Feb 13.6	74.02	115.94	7876	1472	1523	0.003	245
	Cosmos 1163	1980-12H	1980 Feb 11.98 10000 years	Speroid 40?	1.0 long? 0.8 dia?	1980 Feb 20.4	74.02	116.20	7887	1472	1545	0.005	239
	Cosmos 1156 rocket	1980-12J	1980 Feb 11.98 20000 years	Cylinder 2200?	7.4 long 2.4 dia	1980 Feb 12.3	74.02	117.86	7963	1473	1696	0.014	266
D	Fragment	1980-12K											
D	Cosmos 1164*	1980-13A	1980 Feb 12.04 <1 day 1980 Feb 12	Windmill + cylinder 6050? full	6.2 long? 2.0 dia		Initial parking orbit similar to 1980-13B						
D	Cosmos 1164 launcher rocket	1980-13B	1980 Feb 12.04 27 days 1980 Mar 10	Cylinder 2500?	7.5 long 2.6 dia	1980 Feb 13.4	62.81	92.36	6769	212	570	0.027	119
D	Cosmos 1164 launcher	1980-13C	1980 Feb 12.04 51 days 1980 Apr 3	Irregular	-	1980 Feb 13.4	62.87	93.00	6801	235	611	0.028	125
D	Fragment	1980-13D											

* Probably an intended Molniya satellite; final stage rocket probably malfunctioned.

	Name		Launch date, lifetime and descent date	Shape and weight (kg)	Size (m)	Date of orbital determination	Orbital inclination (deg)	Nodal period (min)	Semi major axis (km)	Perigee height (km)	Apogee height (km)	Orbital eccentricity	Argument of perigee (deg)
T	SMM*	1980-14A	1980 Feb 14.66 8 years**	Cylinder + 2 panels 2315	4.0 long 2.3 dia	1980 Feb 21.0	28.51	95.86	6946	566	569	0.0002	224
	SMM second stage	1980-14B	1980 Feb 14.66 8 years	Cylinder + annulus 350?	6.4 long 1.52 and 2.44 dia	1980 Feb 14.8	28.51	95.78	6942	549	578	0.002	294
T L	Tansei 4† [Mu 3S]	1980-15A	1980 Feb 17.03 3¼ years	Octagonal cylinder + 4 panels 185	0.82 long 1.00 dia	1980 Feb 18.6	38.71	95.73	6938	517	602	0.006	170
	Tansei 4 rocket	1980-15B	1980 Feb 17.03 4 years	Sphere- cone? 230?	2.33 long? 1.14 dia?	1980 Feb 17.6	38.70	95.74	6938	517	603	0.006	165
D	Fragments	1980-15C,D											
	Statsionar- Raduga 6	1980-16A	1980 Feb 20.34 > million years	-	-	1980 Feb 23.4	0.60	1486.15	43141	36087	37438	0.016	186
D	Raduga 6 launcher rocket	1980-16B	1980 Feb 20.34 2 days 1980 Feb 22	Cylinder 4000?	12 long? 4 dia	1980 Feb 20.6	51.60	88.12	6562	173	195	0.002	273
D	Raduga 6 launcher	1980-16C	1980 Feb 20.34 2 days 1980 Feb 22	Irregular	-	1980 Feb 20.6	51.63	88.13	6563	178	191	0.001	278
	Raduga 6 apogee motor	1980-16D	1980 Feb 20.34 > million years	Cylinder 440	2.0 long 2.0 dia	1980 Mar 6.7	0.40	1475.06	42927	36508	36590	0.001	24
D	Raduga 6 rocket	1980-16E	1980 Feb 20.34 327 days 1981 Jan 12	Cylinder 1900?	3.9 long? 3.9 dia	1980 May 31.3	47.03	647.60	24797	243	36595	0.733	26
D	Fragment	1980-16F											

* Solar maximum mission.
** If not retrieved by Space Shuttle.
† Japanese satellite.

	Name		Launch date, lifetime and descent date	Shape and weight (kg)	Size (m)	Date of orbital determination	Orbital inclination (deg)	Nodal period (min)	Semi major axis (km)	Perigee height (km)	Apogee height (km)	Orbital eccentricity	Argument of perigee (deg)
D R	Cosmos 1165	1980-17A	1980 Feb 21.50 12.76 days 1980 Mar 5.26	Sphere-cylinder 6300?	6.5 long? 2.4 dia	1980 Feb 23.5	72.87	89.72	6638	170	350	0.014	77
D	Cosmos 1165 rocket	1980-17B	1980 Feb 21.50 4 days 1980 Feb 25	Cylinder 2500?	7.5 long 2.6 dia	1980 Feb 21.9	72.87	89.58	6631	169	337	0.013	79
D	Cosmos 1165 engine	1980-17D	1980 Feb 21.50 14 days 1980 Mar 6	Cone 600? full	1.5 long? 2 dia?	1980 Mar 6.7	72.90	89.00	6602	145	302	0.012	50
D	Fragment	1980-17C											
	Ayame 2 (ECS 2) [Nu]	1980-18A	1980 Feb 22.36 > million years	Cylinder 260 full 130 empty	0.95 long 1.40 dia	1980 Feb 23.5 1980 Dec 31	24.57 0.1	625.52 1441.4	24228 42268	190 35196	35510 36584	0.729 0.016	179 -
D	Ayame 2 second stage	1980-18B	1980 Feb 22.36 6 days 1980 Feb 28	Cylinder 550?	5.4 long 1.4 dia	1980 Feb 25.6	30.45	88.80	6601	190	255	0.005	178
	Ayame 2 rocket	1980-18C	1980 Feb 22.36 20 years	Cylinder 66?	1.74 long 1.65 dia	1980 Mar 2.7	24.53	627.29	24276	223	35574	0.728	185
D	Fragment	1980-18D											
T	NOSS 3 [Atlas]	1980-19A	1980 Mar 3.47 1600 years	Cylinder 64?	-	1980 Mar 3.5	63.03	107.12	7471	1035	1150	0.008	308
	NOSS 3 rocket	1980-19B	1980 Mar 3.47 1000 years	-	-	1980 Mar 3.7	63.52	107.42	7486	1071	1144	0.005	300

1980-19 continued on page 601

	Name		Launch date, lifetime and descent date	Shape and weight (kg)	Size (m)	Date of orbital determination	Orbital inclination (deg)	Nodal period (min)	Semi major axis (km)	Perigee height (km)	Apogee height (km)	Orbital eccentricity	Argument of perigee (deg)
T?	SSU	1980-19C	1980 Mar 3.47 1600 years	Box + aerials	0.3 x 0.9 x 2.4?	1980 Mar 24.6	63.49	107.40	7485	1048	1166	0.008	163
T?	SSU	1980-19E	1980 Mar 3.47 1600 years	Box + aerials	0.3 x 0.9 x 2.4?	1980 Mar 24.6	63.49	107.40	7485	1048	1166	0.008	163
T?	SSU	1980-19G	1980 Mar 3.47 1600 years	Box + aerials	0.3 x 0.9 x 2.4?	1980 Mar 24.6	63.49	107.40	7485	1048	1166	0.008	163
	Fragments	1980-19D,F,H											
D R	Cosmos 1166	1980-20A	1980 Mar 4.44 13.77 days 1980 Mar 18.21	Sphere- cylinder 6300?	6.5 long? 2.4 dia	1980 Mar 4.8 1980 Mar 6.3	72.85 72.85	90.32 92.29	6668 6764	198 356	382 416	0.014 0.004	84 209
D	Cosmos 1166 rocket	1980-20B	1980 Mar 4.44 9 days 1980 Mar 13	Cylinder 2500?	7.5 long 2.6 dia	1980 Mar 4.9	72.85	90.23	6664	195	376	0.014	81
D	Cosmos 1166 engine	1980-20D	1980 Mar 4.44 184 days 1980 Sep 4	Cone 600? full	1.5 long? 2 dia?	1980 Mar 21.3	72.85	92.25	6762	355	413	0.004	176
D	Fragments	1980-20C,E,F											
D	Cosmos 1167	1980-21A	1980 Mar 14.45 566 days 1981 Oct 1	Cylinder?	-	1980 Mar 14.7	65.03	93.31	6812	426	442	0.001	270
D	Cosmos 1167 rocket	1980-21B	1980 Mar 14.45 <0.6 day 1980 Mar 14	Cylinder 1500?	8 long? 2.5 dia?	1980 Mar 14.9	65.01	88.25	6567	98	280	0.014	63
11d	Fragments	1980-21C-N											

	Name		Launch date, lifetime and descent date	Shape and weight (kg)	Size (m)	Date of orbital determination	Orbital inclination (deg)	Nodal period (min)	Semi major axis (km)	Perigee height (km)	Apogee height (km)	Orbital eccentricity	Argument of perigee (deg)
	Cosmos 1168	1980-22A	1980 Mar 17.90 1200 years	Cylinder 700?	1.3 long? 1.9 dia?	1980 Mar 20.2	82.95	104.92	7368	964	1015	0.003	280
	Cosmos 1168 rocket	1980-22B	1980 Mar 17.90 600 years	Cylinder 2200?	7.4 long 2.4 dia	1980 Mar 20.7	82.95	104.79	7362	964	1003	0.003	273
	Fragment	1980-22C											
	Cosmos 1169	1980-23A	1980 Mar 27.32 2.9 years	Cylinder?	4 long? 2 dia?	1980 Mar 27.5	65.84	94.52	6874	477	515	0.003	352
D	Cosmos 1169 rocket	1980-23B	1980 Mar 27.32 587 days 1981 Nov 4	Cylinder 2200?	7.4 long 2.4 dia	1980 Mar 27.5	65.84	94.44	6870	469	515	0.003	356
D	Progress 8*	1980-24A	1980 Mar 27.79 29.50 days 1980 Apr 26.29	Sphere-cylinder 7020	7.9 long 2.3 dia	1980 Mar 29.0	51.61	89.16	6614	230	242	0.001	173
D	Progress 8 rocket	1980-24B	1980 Mar 27.79 2 days 1980 Mar 29	Cylinder 2500?	7.5 long 2.6 dia	1980 Mar 29.4	51.61	87.83	6548	154	186	0.002	78
D R	Cosmos 1170	1980-25A	1980 Apr 1.34 10.9 days 1980 Apr 12.2	Sphere-cylinder 6300?	6.5 long? 2.4 dia	1980 Apr 2.1	70.37	89.92	6648	174	366	0.014	54
D	Cosmos 1170 rocket	1980-25B	1980 Apr 1.34 4 days 1980 Apr 5	Cylinder 2500?	7.5 long 2.6 dia	1980 Apr 2.4	70.37	89.54	6629	169	333	0.012	48

1980-25 continued on p. 603

* Docked with Salyut 6 (2nd airlock) on 1980 Mar 29.83, separated 1980 Apr 25.34 and de-orbited over Pacific Ocean.

	Name		Launch date, lifetime and descent date	Shape and weight (kg)	Size (m)	Date of orbital determination	Orbital inclination (deg)	Nodal period (min)	Semi major axis (km)	Perigee height (km)	Apogee height (km)	Orbital eccentricity	Argument of perigee (deg)
D	Cosmos 1170 engine	1980-25D	1980 Apr 1.34 13 days 1980 Apr 14	Cone 600? full	1.5 long? 2 dia?	1980 Apr 12.3	70.37	89.17	6610	153	311	0.012	26
D	Fragments	1980-25C,E,F											
	Cosmos 1171	1980-26A	1980 Apr 3.32 1200 years	Cylinder?	4 long? 2 dia?	1980 Apr 4.2	65.84	104.89	7368	947	1033	0.006	297
	Cosmos 1171 rocket	1980-26B	1980 Apr 3.32 600 years	Cylinder 2200?	7.4 long 2.4 dia	1980 Apr 4.8	65.84	104.75	7361	967	999	0.002	304
	Fragment	1980-26C											
D 2M R	Soyuz 35*	1980-27A	1980 Apr 9.57 55.06 days 1980 Jun 3.63	Sphere-cylinder 6800?	7.5 long 2.3 to 2.72 dia	1980 Apr 9.7	51.62	88.87	6600	197	247	0.004	69
D	Soyuz 35 rocket	1980-27B	1980 Apr 9.57 2 days 1980 Apr 11	Cylinder 2500?	7.5 long 2.6 dia	1980 Apr 10.1	51.65	88.69	6591	194	232	0.003	66
	Cosmos 1172	1980-28A	1980 Apr 12.85 16 years	Windmill + 6 vanes? 1250?	4.2 long? 1.6 dia	1980 Apr 12.9 1980 Apr 20.9	62.77 62.82	726.03 719.47	26760 26600	608 621	40155 39822	0.739 0.737	320 320
D	Cosmos 1172 launcher rocket	1980-28B	1980 Apr 12.85 14 days 1980 Apr 26	Cylinder 2500?	7.5 long 2.6 dia	1980 Apr 13.7	62.86	92.39	6771	180	605	0.031	117

* Soyuz 35 docked with Progress 8 - Salyut 6 (1st airlock) on 1980 Apr 10.64. The Soyuz 35 cosmonauts remained aboard Salyut and the Soyuz 35 spacecraft returned with Soyuz 36 crew (see page 608).

1980-28 continued on p.604

	Name		Launch date, lifetime and descent date	Shape and weight (kg)	Size (m)	Date of orbital determination	Orbital inclination (deg)	Nodal period (min)	Semi major axis (km)	Perigee height (km)	Apogee height (km)	Orbital eccentricity	Argument of perigee (deg)
D	Cosmos 1172 launcher	1980-28C	1980 Apr 12.85 26 days 1980 May 8	Irregular	-	1980 Apr 13.1	62.79	92.55	6778	213	587	0.028	122
	Cosmos 1172 rocket	1980-28E	1980 Apr 12.85 16 years	Cylinder 440	2.0 long 2.0 dia	1980 Apr 17.9	62.76	722.24	26666	615	39961	0.738	320
D	Fragment	1980-28D											
D R	Cosmos 1173	1980-29A	1980 Apr 17.36 10.88 days 1980 Apr 28.24	Sphere-cylinder 6300?	6.5 long? 2.4 dia	1980 Apr 18.3 1980 Apr 23.3	70.30 70.29	89.59 89.49	6633 6628	155 162	354 337	0.015 0.013	60 60
D	Cosmos 1173 rocket	1980-29B	1980 Apr 17.36 4 days 1980 Apr 21	Cylinder 2500?	7.5 long 2.6 dia	1980 Apr 18.4	70.31	89.50	6628	169	330	0.012	52
D	Cosmos 1173 engine	1980-29C	1980 Apr 17.36 13 days 1980 Apr 30	Cone 600? full	1.5 long? 2 dia?	1980 Apr 28.4	70.29	89.33	6619	154	328	0.013	47
D	Fragment	1980-29D											
	Cosmos 1174*	1980-30A	1980 Apr 18.04 disintegrated	Cylinder?	4 long? 2 dia?	1980 Apr 18.0 1980 Apr 18.1 1980 Apr 19.2	65.15 65.84 66.13	89.13 98.63 105.54	6610 7072 7399	124 362 381	340 1025 1660	0.016 0.047 0.086	36 272 248
D	Cosmos 1174 rocket	1980-30B	1980 Apr 18.04 2 days 1980 Apr 20	Cylinder 1500?	8 long? 2.5 dia?	1980 Apr 19.0	65.12	89.69	6638	133	386	0.019	46
16d	Fragments	1980-30C-AG											

* Passed within 8 km of Cosmos 1171 on 1980 Apr 20, then exploded (see Cosmos 967 and 970).

	Name		Launch date, lifetime and descent date	Shape and weight (kg)	Size (m)	Date of orbital determination	Orbital inclination (deg)	Nodal period (min)	Semi major axis (km)	Perigee height (km)	Apogee height (km)	Orbital eccentricity	Argument of perigee (deg)
D	Cosmos 1175 launcher	1980-31A	1980 Apr 18.73 40 days 1980 May 28	Irregular	-	1980 Apr 20.1	62.83	91.56	6730	253	451	0.015	132
D	Cosmos 1175 launcher rocket	1980-31B	1980 Apr 18.73 14 days 1980 May 2	Cylinder 2500?	7.5 long 2.6 dia	1980 Apr 19.8	62.84	90.92	6699	210	431	0.016	131
D	Cosmos 1175*	1980-31C	1980 Apr 18.73 164 days 1980 Sep 29	Windmill + 6 vanes? 1250?	4.2 long? 1.6 dia?	1980 Apr 19.9	62.81	92.30	6766	313	463	0.011	121
D	Cosmos 1175 rocket	1980-31D	1980 Apr 18.73 42 days 1980 May 30	Cylinder 4800 full 440 empty	2.0 long 2.0 dia	1980 Apr 19.9	62.81	91.77	6741	267	458	0.014	134
T	Navstar 6 [Atlas F]	1980-32A	1980 Apr 26.91 1 million years	Cylinder + 4 vanes 433	-	1980 Apr 26.9 1980 Apr 29.3	63.02 62.88	355.05 707.73	16610 26308	157 19628	20307 20232	0.607 0.011	159 141
	Navstar 6 rocket	1980-32B	1980 Apr 26.91 34½ years	Cone-cylinder? 163?	1.85 long? 0.63 to 1.65 dia?	1980 May 2.0	63.05	352.54	16532	165	20142	0.604	159
D	Progress 9**	1980-33A	1980 Apr 27.27 24.7 days 1980 May 22.0	Sphere-cylinder 7020	7.9 long 2.3 dia	1980 Apr 27.3 1980 Apr 28.3	51.65 51.65	88.84 90.39	6598 6675	185 244	255 350	0.005 0.008	89 206
D	Progress 9 rocket	1980-33B	1980 Apr 27.27 2 days 1980 Apr 29	Cylinder 2500?	7.5 long 2.6 dia	1980 Apr 27.4	51.64	88.72	6592	183	245	0.005	89

* Probably an intended Molniya or early warning satellite.

** Progress 9 docked with Soyuz 35 - Salyut 6 (2nd airlock) on 1980 Apr 29.34, undocked on 1980 May 20.79, then de-orbited over Pacific Ocean.

	Name		Launch date, lifetime and descent date	Shape and weight (kg)	Size (m)	Date of orbital determination	Orbital inclination (deg)	Nodal period (min)	Semi major axis (km)	Perigee height (km)	Apogee height (km)	Orbital eccentricity	Argument of perigee (deg)
	Cosmos 1176*	1980-34A	1980 Apr 29.49 600 years	Cone-cylinder	6 long? 2.5 dia?	1980 Apr 30.3 1980 Sep 10	65.02 64.84	89.65 103.44	6636 7296	250 870	266 966	0.001 0.007	268 252
D	Cosmos 1176 platform	1980-34B	1980 Apr 29.49 158 days 1980 Oct 4	Irregular	- -	1980 Sep 10.5	65.01	89.57	6632	244	264	0.001	286
D	Cosmos 1176 rocket	1980-34C	1980 Apr 29.49 136.01 days 1980 Sep 12.50	Cylinder 1500?	8 long? 2.5 dia?	1980 Sep 10.8	65.02	89.30	6619	229	252	0.002	281
	Fragment**	1980-34D											
D R	Cosmos 1177	1980-35A	1980 Apr 29.57 44.2 days? 1980 Jun 12.8?	Sphere-cylinder 6700?	7 long? 2.4 dia	1980 Apr 29.8	67.14	89.68	6638	174	346	0.013	68
D	Cosmos 1177 rocket	1980-35B	1980 Apr 29.57 3 days 1980 May 2	Cylinder 2500?	7.5 long 2.6 dia	1980 Apr 29.8	67.13	89.56	6632	174	334	0.012	70
D	Cosmos 1177 engine	1980-35C	1980 Apr 29.57 48 days 1980 Jun 16	Cone 600?	1.5 long? 2.0 dia?	1980 Jun 13.0	67.13	88.13	6561	149	216	0.005	-
D R	Cosmos 1178	1980-36A	1980 May 7.54 14.8 days 1980 May 22.3	Sphere-cylinder 6300?	6.5 long? 2.4 dia	1980 May 7.9 1980 May 8.5	72.85 72.83	90.38 92.30	6671 6765	200 358	386 415	0.014 0.004	86 213
D	Cosmos 1178 rocket	1980-36B	1980 May 7.54 7 days 1980 May 14	Cylinder 2500?	7.5 long 2.6 dia	1980 May 7.9	72.82	90.30	6667	196	382	0.014	85

* Cosmos 1176 jettisoned rocket and platform, then manoeuvred into 103 min orbit about 1980 Sep 10.2.

** Probably the uranium fuel core ejected from nuclear reactor of 1980-34A when in 103 min orbit.

1980-36 continued on page 607

	Name		Launch date, lifetime and descent date	Shape and weight (kg)	Size (m)	Date of orbital determination	Orbital inclination (deg)	Nodal period (min)	Semi major axis (km)	Perigee height (km)	Apogee height (km)	Orbital eccentricity	Argument of perigee (deg)
D	Cosmos 1178 engine	1980-36E	1980 May 7.54 193 days 1980 Nov 16	Cone 600? full	1.5 long? 2 dia?	1980 May 21.9	72.82	92.27	6764	355	416	0.004	187
D	Fragments	1980-36C,D,F											
	Cosmos 1179	1980-37A	1980 May 14.54 10 years	Octagonal ellipsoid 550?	1.8 long? 1.5 dia?	1980 May 14.9	82.97	103.60	7306	303	1552	0.086	112
	Cosmos 1179 rocket	1980-37B	1980 May 14.54 972 days	Cylinder 2200?	7.4 long 2.4 dia	1980 May 14.9	82.96	103.50	7301	300	1546	0.085	111
D R	Cosmos 1180	1980-38A	1980 May 15.24 11.65 days 1980 May 26.89	Sphere-cylinder 5900?	5.9 long 2.4 dia	1980 May 15.8	62.81	89.79	6643	239	291	0.004	184
D	Cosmos 1180 rocket	1980-38B	1980 May 15.24 8 days 1980 May 23	Cylinder 2500?	7.5 long 2.6 dia	1980 May 15.5	62.79	89.74	6641	243	282	0.003	186
D	Capsule	1980-38E	1980 May 15.24 16 days 1980 May 31	Ellipsoid 200?	0.9 long 1.9 dia	1980 May 28.3	62.80	91.14	6710	264	399	0.010	326
D	Fragments	1980-38C,D,F											
	Cosmos 1181	1980-39A	1980 May 20.39 1200 years	Cylinder 700?	1.3 long? 1.9 dia?	1980 May 23.2	82.95	104.97	7370	976	1008	0.002	262
	Cosmos 1181 rocket	1980-39B	1980 May 20.39 600 years	Cylinder 2200?	7.4 long 2.4 dia	1980 May 22.6	82.95	104.85	7364	975	997	0.002	247

	Name		Launch date, lifetime and descent date	Shape and weight (kg)	Size (m)	Date of orbital determination	Orbital inclination (deg)	Nodal period (min)	Semi major axis (km)	Perigee height (km)	Apogee height (km)	Orbital eccentricity	Argument of perigee (deg)
D R	Cosmos 1182	1980-40A	1980 May 23.30 12.9 days 1980 Jun 5.2	Sphere-cylinder 5700?	5.0 long? 2.4 dia	1980 May 23.4	82.34	89.14	6609	211	251	0.003	100
D	Cosmos 1182 rocket	1980-40B	1980 May 23.30 3 days 1980 May 26	Cylinder 2500?	7.5 long 2.6 dia	1980 May 23.7	82.33	89.01	6603	207	242	0.003	94
D	Fragments	1980-40C,D											
D 2M R	Soyuz 36*	1980-41A	1980 May 26.76 65.88 days 1980 Jul 31.64	Sphere-cylinder 6800?	7.5 long 2.3 to 2.72 dia	1980 May 26.8 1980 May 27.1	51.60 51.62	89.00 90.10	6606 6661	191 251	265 314	0.006 0.005	84 249
D	Soyuz 36 rocket	1980-41B	1980 May 26.76 3 days 1980 May 29	Cylinder 2500?	7.5 long 2.6 dia	1980 May 27.0	51.60	88.84	6598	189	251	0.005	79
D R	Cosmos 1183	1980-42A	1980 May 28.44 13.84 days 1980 Jun 11.28	Sphere-cylinder 6300?	6.5 long? 2.4 dia	1980 May 28.6 1980 Jun 3.2	72.89 72.89	90.42 92.26	6673 6765	201 358	389 416	0.014 0.004	80 200
D	Cosmos 1183 rocket	1980-42B	1980 May 28.44 8 days 1980 Jun 5	Cylinder 2500?	7.5 long 2.6 dia	1980 May 29.1	72.90	90.24	6664	198	374	0.013	77
D	Cosmos 1183 engine	1980-42G	1980 May 28.44 190 days 1980 Dec 4	Cone 600?	1.5 long? 2.0 dia ?	1980 Sep 1.0	72.89	91.61	6731	331	375	0.003	-
D	Fragments	1980-42C-F											
D	NOAA-B† [Atlas F]	1980-43A	1980 May 29.45 339 days 1981 May 3	Rect box 1405 full 723 empty	3.71 long 1.88 dia	1980 May 30.1	92.23	102.05	7233	264	1445	0.082	73
D	Fragments	1980-43B,C											

* Soyuz 36 docked with Soyuz 35-Salyut 6 (2nd airlock) on 1980 May 27.83, with one Russian and first Hungarian cosmonaut. They returned to Earth in Soyuz 35, undocking 1980 June 3 (see page 603). Soyuz 36 landed with Soyuz 37 crew.

† Intended orbit not achieved.

	Name		Launch date, lifetime and descent date	Shape and weight (kg)	Size (m)	Date of orbital determination	Orbital inclination (deg)	Nodal period (min)	Semi major axis (km)	Perigee height (km)	Apogee height (km)	Orbital eccentricity	Argument of perigee (deg)
	Cosmos 1184	1980-44A	1980 Jun 4.32 60 years	Cylinder + 2 vanes? 2500?	5 long? 1.5 dia?	1980 Jun 4.4	81.25	97.43	7013	623	647	0.002	348
	Cosmos 1184 rocket	1980-44B	1980 Jun 4.32 60 years	Cylinder 1440	3.8 long 2.6 dia	1980 Jun 18.6	81.26	97.60	7021	599	686	0.006	132
D 2M R	Soyuz Transport 2*	1980-45A	1980 Jun 5.60 3.9 days 1980 Jun 9.5	Sphere- cylinder + 2 wings 6850	7.5 long 2.3 to 2.72 dia	1980 Jun 5.7 1980 Jun 6.2	51.63 51.62	88.70 90.26	6591 6668	195 262	231 318	0.003 0.004	97 204
D	Soyuz T-2 rocket	1980-45B	1980 Jun 5.60 2 days 1980 Jun 7	Cylinder 2500?	7.5 long 2.6 dia	1980 Jun 6.4	51.63	88.39	6576	187	208	0.002	85
D	Soyuz T-2 orbital module**	1980-45C	1980 Jun 5.60 92 days 1980 Sep 5	Spheroid 1260?	2.3 dia	1980 Jun 9.8	51.62	91.23	6716	330	346	0.001	313
D R	Cosmos 1185	1980-46A	1980 Jun 6.29 13.86 days 1980 Jun 20.15	Sphere- cylinder 6300?	6.5 long? 2.4 dia	1980 Jun 7.2 1980 Jun 7.8	82.34 82.34	89.49 89.89	6626 6646	214 261	282 275	0.005 0.001	85 320
D	Cosmos 1185 rocket	1980-46B	1980 Jun 6.29 5 days 1980 Jun 11	Cylinder 2500?	7.5 long 2.6 dia	1980 Jun 6.5	82.35	89.37	6620	214	270	0.004	76
D	Cosmos 1185 engine	1980-46C	1980 Jun 6.29 28 days 1980 Jul 4	Cone 600? full	1.5 long? 2 dia?	1980 Jun 20.2	82.33	89.73	6638	251	269	0.001	248
D	Fragment	1980-46D											

* Soyuz T-2 docked with Soyuz 36-Salyut 6 (2nd airlock) Jun 6.67. ** Was part of 1980-45A until jettisoned prior to retrofire 1980 Jun 9.
(Soyuz 36 had undocked from 2nd airlock and re-attached to 1st airlock on 1980 Jun 4.69)

	Name		Launch date, lifetime and descent date	Shape and weight (kg)	Size (m)	Date of orbital determination	Orbital inclination (deg)	Nodal period (min)	Semi major axis (km)	Perigee height (km)	Apogee height (km)	Orbital eccentricity	Argument of perigee (deg)
D	Cosmos 1186	1980-47A	1980 Jun 6.46 574 days 1982 Jan 1	Octagonal ellipsoid? 550?	1.8 long? 1.5 dia?	1980 Jun 7.2	74.02	94.54	6874	473	519	0.003	340
D	Cosmos 1186 rocket	1980-47B	1980 Jun 6.46 558 days 1981 Dec 16	Cylinder 2200?	7.4 long 2.4 dia	1980 Jun 6.6	74.03	94.46	6870	465	519	0.004	347
D	Fragments	1980-47C-AC											
D R	Cosmos 1187	1980-48A	1980 Jun 12.52 13.77 days 1980 Jun 26.29	Sphere-cylinder 6300?	6.5 long? 2.4 dia	1980 Jun 12.6 1980 Jun 13.6	72.85 72.85	89.57 89.69	6631 6637	199 227	307 291	0.008 0.005	91 70
D	Cosmos 1187 rocket	1980-48B	1980 Jun 12.52 5 days 1980 Jun 17	Cylinder 2500?	7.5 long 2.6 dia	1980 Jun 12.6	72.85	89.50	6628	198	301	0.008	94
D	Cosmos 1187 engine	1980-48E	1980 Jun 12.52 36 days 1980 Jul 18	Cone 600? full	1.5 long? 2 dia?	1980 Jun 27.2	72.83	89.27	6616	214	263	0.004	45
D	Fragments	1980-48C,D											
	Gorizont 4	1980-49A	1980 Jun 14.04 > million years	-	-	1980 Jun 14.1 1980 Jun 27.3	47.50 0.81	632.00 1436.1	24400 42164	314 35744	35730 35828	0.726 0.001	0 254
D	Gorizont 4 launcher rocket	1980-49B	1980 Jun 14.04 1 day 1980 Jun 15	Cylinder 4000?	12 long? 4 dia	1980 Jun 14.3	51.64	88.08	6561	172	193	0.002	261
D	Gorizont 4 launcher	1980-49C	1980 Jun 14.04 < 1 day 1980 Jun 14	Irregular	-	1980 Jun 14.3	51.62	87.98	6556	172	184	0.001	162

1980-49 continued on page 611

	Name		Launch date, lifetime and descent date	Shape and weight (kg)	Size (m)	Date of orbital determination	Orbital inclination (deg)	Nodal period (min)	Semi major axis (km)	Perigee height (km)	Apogee height (km)	Orbital eccentricity	Argument of perigee (deg)
D	Gorizont 4 rocket	1980-49D	1980 Jun 14.04 0.6 year 1981 Jan?	Cylinder 1900?	3.9 long? 3.9 dia	1980 Jun 27.6	47.26	646.14	24760	280	36490	0.731	3
	Gorizont 4 apogee motor	1980-49F	1980 Jun 14.04 > million years	Cylinder 440?	2.0 long? 2.0 dia?	1980 Jun 14.2	0.45	1472.8	42883	36468	36541	0.001	0
D	Fragment	1980-49E											
	Cosmos 1188	1980-50A	1980 Jun 14.87 100 years?	Windmill + 6 vanes? 1250?	4.2 long? 1.6 dia	1980 Jun 14.9	62.92	725.53	26747	609	40129	0.739	318
	Cosmos 1188 rocket	1980-50B	1980 Jun 14.87 100 years?	Cylinder 440	2.0 long 2.0 dia	1980 Jun 16.9	62.91	722.92	26683	613	39997	0.738	318
D	Cosmos 1188 launcher	1980-50C	1980 Jun 14.87 33 days 1980 Jul 17	Irregular	-	1980 Jun 14.9	62.81	92.56	6779	221	581	0.026	122
D	Cosmos 1188 launcher rocket	1980-50D	1980 Jun 14.87 22 days 1980 Jul 6	Cylinder 2500?	7.5 long 2.6 dia	1980 Jun 14.9	62.88	92.98	6800	191	652	0.034	117
	Meteor 30	1980-51A	1980 Jun 18.26 60 years	Cylinder + 2 vanes 2200?	5 long? 1.5 dia?	1980 Jun 18.4	97.94	97.24	7004	584	667	0.006	12
	Meteor 30 rocket	1980-51B	1980 Jun 18.26 60 years	Cylinder 1440	3.8 long 2.6 dia	1980 Jun 19.0	97.94	97.47	7015	614	659	0.003	298

	Name		Launch date, lifetime and descent date	Shape and weight (kg)	Size (m)	Date of orbital determination	Orbital inclination (deg)	Nodal period (min)	Semi major axis (km)	Perigee height (km)	Apogee height (km)	Orbital eccentricity	Argument of perigee (deg)
D	[Titan 3D]	1980-52A	1980 Jun 18.77 261 days 1981 Mar 6	Cylinder 13300? full	15 long 3.0 dia	1980 Jun 19.2	96.46	88.87	6595	169	265	0.007	145
D	Titan 3D rocket	1980-52B	1980 Jun 18.77 2 days 1980 Jun 20	Cylinder 1900	6 long 3.0 dia	1980 Jun 19.2	96.46	88.60	6582	164	244	0.006	146
T	Capsule	1980-52C	1980 Jun 18.77 5000 years	Octagon? 60?	0.3 long? 0.9 dia?	1980 Jun 19.4	96.62	112.31	7710	1331	1333	0.0001	113
	Molniya 1AY	1980-53A	1980 Jun 21.84 11½ years	Windmill + 6 vanes 1000?	3.4 long 1.6 dia	1980 Jun 21.9	62.83	737.68	27045	631	40703	0.741	288
D	Molniya 1AY launcher rocket	1980-53B	1980 Jun 21.84 37 days 1980 Jul 28	Cylinder 2500?	7.5 long 2.6 dia	1980 Jun 21.8	62.86	93.02	6801	214	632	0.031	119
D	Molniya 1AY launcher	1980-53C	1980 Jun 21.84 34 days 1980 Jul 25	Irregular	-	1980 Jun 22.0	62.83	92.51	6776	199	597	0.029	113
	Molniya 1AY rocket	1980-53D	1980 Jun 21.84 11½ years	Cylinder 440	2.0 long 2.0 dia	1980 Jun 24.4	62.86	733.40	26940	637	40487	0.740	288
D R	Cosmos 1189	1980-54A	1980 Jun 26.52 13.78 days 1980 Jul 10.30	Sphere-cylinder 6300?	6.5 long? 2.4 dia	1980 Jun 26.6	72.88	89.55	6630	198	305	0.008	85
D	Cosmos 1189 rocket	1980-54B	1980 Jun 26.52 5 days 1980 Jul 1	Cylinder 2500?	7.5 long 2.6 dia	1980 Jun 26.6	72.86	89.38	6622	196	291	0.007	83

1980-54 continued on page 613

	Name		Launch date, lifetime and descent date	Shape and weight (kg)	Size (m)	Date of orbital determination	Orbital inclination (deg)	Nodal period (min)	Semi major axis (km)	Perigee height (km)	Apogee height (km)	Orbital eccentricity	Argument of perigee (deg)
D	Cosmos 1189 engine	1980-54D	1980 Jun 26.52 39 days 1980 Aug 4	Cone 600? full	1.5 long? 2 dia?	1980 Jul 11.1	72.86	90.02	6653	226	324	0.007	49
D	Fragments	1980-54C,E											
D	Progress 10*	1980-55A	1980 Jun 29.20 19.89 days 1980 Jul 19.09	Sphere-cylinder 7020	7.9 long 2.3 dia	1980 Jun 30.2	51.62	89.04	6608	200	260	0.005	92
D	Progress 10 rocket	1980-55B	1980 Jun 29.20 3 days 1980 Jul 2	Cylinder 2500?	7.5 long 2.6 dia	1980 Jun 29.8	51.62	88.79	6596	183	252	0.005	77
T	Cosmos 1190	1980-56A	1980 Jul 1.30 120 years	Cylinder + paddles? 750?	2 long? 1 dia?	1980 Jul 1.4	74.05	100.86	7177	792	806	0.001	39
	Cosmos 1190 rocket	1980-56B	1980 Jul 1.30 100 years	Cylinder 2200?	7.4 long 2.4 dia	1980 Jul 1.4	74.05	100.79	7174	783	808	0.002	20
T	Cosmos 1191	1980-57A	1980 Jul 2.04 100 years?	Windmill+ 6 vanes? 1250?	4.2 long? 1.6 dia	1980 Jul 5.1 1980 Jul 15.1	62.67 62.69	725.41 718.45	26744 26573	605 617	40127 39773	0.739 0.737	316 316
D	Cosmos 1191 launcher rocket	1980-57B	1980 Jul 2.04 31 days 1980 Aug 2	Cylinder 2500?	7.5 long 2.6 dia	1980 Jul 2.2	62.71	92.63	6783	212	597	0.028	119
D	Cosmos 1191 launcher	1980-57C	1980 Jul 2.04 29 days 1980 Jul 31	Irregular	-	1980 Jul 2.7	62.80	92.77	6790	198	625	0.031	117
	Cosmos 1191 rocket	1980-57D	1980 Jul 2.04 100 years?	Cylinder 440	2.0 long 2.0 dia	1980 Jul 17.1	62.73	722.16	26664	620	39952	0.738	316

* Progress 10 docked with Soyuz 36 - Salyut 6 (2nd airlock) on 1980 Jul 1.25, and undocked on 1980 Jul 17.93. De-orbited over Pacific Ocean.

	Name		Launch date, lifetime and descent date	Shape and weight (kg)	Size (m)	Date of orbital determination	Orbital inclination (deg)	Nodal period (min)	Semi major axis (km)	Perigee height (km)	Apogee height (km)	Orbital eccentricity	Argument of perigee (deg)
T	Cosmos 1192	1980-58A	1980 Jul 9.03 7000 years	Spheroid 40?	1.0 long? 0.8 dia?	1980 Jul 10.4	74.02	114.61	7815	1398	1476	0.005	94
T	Cosmos 1193	1980-58B	1980 Jul 9.03 8000 years	Spheroid 40?	1.0 long? 0.8 dia?	1980 Jul 10.4	74.01	114.82	7825	1414	1479	0.004	106
T	Cosmos 1194	1980-58C	1980 Jul 9.03 9000 years	Spheroid 40?	1.0 long? 0.8 dia?	1980 Jul 10.4	74.01	115.01	7834	1433	1478	0.003	106
T	Cosmos 1195	1980-58D	1980 Jul 9.03 10,000 years	Spheroid 40?	1.0 long? 0.8 dia?	1980 Jul 11.3	74.02	115.21	7843	1452	1477	0.002	114
T	Cosmos 1196	1980-58E	1980 Jul 9.03 10,000 years	Spheroid 40?	1.0 long? 0.8 dia?	1980 Jul 10.4	74.02	115.41	7852	1470	1477	0.0005	147
T	Cosmos 1197	1980-58F	1980 Jul 9.03 10,000 years	Spheroid 40?	1.0 long? 0.8 dia?	1980 Jul 10.4	74.02	115.63	7862	1473	1494	0.001	228
T	Cosmos 1198	1980-58G	1980 Jul 9.03 10,000 years	Spheroid 40?	1.0 long? 0.8 dia?	1980 Jul 10.4	74.02	115.83	7871	1475	1510	0.002	260
T	Cosmos 1199	1980-58H	1980 Jul 9.03 9000 years	Spheroid 40?	1.0 long? 0.8 dia?	1980 Jul 10.5	74.02	116.08	7882	1475	1533	0.004	258
	Cosmos 1192 rocket	1980-58J	1980 Jul 9.03 20,000 years	Cylinder 2200?	7.4 long 2.4 dia	1980 Jul 12.4	74.02	117.73	7957	1473	1684	0.013	266

	Name		Launch date, lifetime and descent date	Shape and weight (kg)	Size (m)	Date of orbital determination	Orbital inclination (deg)	Nodal period (min)	Semi major axis (km)	Perigee height (km)	Apogee height (km)	Orbital eccentricity	Argument of perigee (deg)
D R	Cosmos 1200	1980-59A	1980 Jul 9.53 13.78 days 1980 Jul 23.31	Sphere-cylinder 6300?	6.5 long? 2.4 dia	1980 Jul 10.1 1980 Jul 10.3	72.86 72.86	89.57 89.70	6631 6637	198 227	307 291	0.008 0.005	85 70
D	Cosmos 1200 rocket	1980-59B	1980 Jul 9.53 6 days 1980 Jul 15	Cylinder 2500?	7.5 long 2.6 dia	1980 Jul 10.5	72.86	89.39	6622	194	293	0.007	84
D	Cosmos 1200 engine	1980-59D	1980 Jul 9.53 39 days 1980 Aug 17	Cone 600? full	1.5 long? 2 dia?	1980 Jul 23.4	72.88	89.73	6639	172	350	0.013	88
D	Fragments	1980-59C,E,F											
T	Statsionar - Ekran 5	1980-60A	1980 Jul 15.13 > million years	Cylinder + plate	-	1980 Aug 6.6	1.98	1419.98	41850	35263	35681	0.005	7
D	Ekran 5 launcher rocket	1980-60B	1980 Jul 15.13 1 day 1980 Jul 16	Cylinder 4000?	12 long? 4 dia	1980 Jul 15.4	51.64	88.10	6562	177	190	0.001	244
D	Ekran 5 launcher	1980-60C	1980 Jul 15.13 < 1 day 1980 Jul 15	Irregular	-	1980 Jul 15.4	51.62	87.46	6530	138	165	0.002	3
D	Ekran 5 rocket	1980-60D	1980 Jul 15.13 $1\frac{3}{4}$ years 1982 Apr?	Cylinder 1900?	3.9 long? 3.9 dia	1980 Jul 29.4	47.35	628.00	24297	334	35504	0.724	3
	Ekran 5 apogee motor	1980-60F	1980 Jul 15.13 > million years	Cylinder 440?	2.0 long? 2.0 dia?	Synchronous orbit similar to 1979-87A							
D	Fragment	1980-60E											

	Name		Launch date, lifetime and descent date	Shape and weight (kg)	Size (m)	Date of orbital determination	Orbital inclination (deg)	Nodal period (min)	Semi major axis (km)	Perigee height (km)	Apogee height (km)	Orbital eccentricity	Argument of perigee (deg)
D R	Cosmos 1201†	1980-61A	1980 Jul 15.32 12.85 days 1980 Jul 28.17	Sphere-cylinder 5700?	5.0 long? 2.4 dia	1980 Jul 16.2	82.33	89.12	6608	213	247	0.003	102
D	Cosmos 1201 rocket	1980-61B	1980 Jul 15.32 3 days 1980 Jul 18	Cylinder 2500?	7.5 long 2.6 dia	1980 Jul 15.4	82.34	89.03	6604	212	239	0.002	108
D	Fragments	1980-61C,D**											
D	Rohini 1B* [SLV-3]	1980-62A	1980 Jul 18.11 306 days 1981 May 20	Spheroid? 35	0.6 dia?	1980 Jul 18.8	44.74	96.85	6990	305	919	0.044	169
D	Rohini 1B rocket	1980-62B	1980 Jul 18.11 370 days 1981 Jul 23	Cylinder?	-	1980 Jul 20.9	44.75	96.82	6989	302	919	0.044	180
T	Molniya 3N	1980-63A	1980 Jul 18.45 100 years?	Windmill + 6 vanes 1500?	4.2 long? 1.6 dia	1980 Jul 18.5	62.81	736.48	27016	457	40818	0.747	280
D	Molniya 3N launcher rocket	1980-63B	1980 Jul 18.45 15 days 1980 Aug 2	Cylinder 2500?	7.5 long 2.6 dia	1980 Jul 18.6	62.79	90.79	6693	201	428	0.017	138
D	Molniya 3N launcher	1980-63C	1980 Jul 18.45 17 days 1980 Aug 4	Irregular	-	1980 Jul 18.5	62.80	91.23	6714	208	463	0.019	131
	Molniya 3N rocket	1980-63D	1980 Jul 18.45 100 years?	Cylinder 440	2.0 long 2.0 dia	1980 Aug 4.8	62.88	733.00	26930	446	40658	0.747	280

† Cooperative Earth resources satellite.

** Object 1980-61D (decayed 1980 Aug 2) may be a Capsule.

* First successful Indian national launch from Sriharikota, Bengal Bay (Rohini 1A failed in autumn 1979).

	Name		Launch date, lifetime and descent date	Shape and weight (kg)	Size (m)	Date of orbital determination	Orbital inclination (deg)	Nodal period (min)	Semi major axis (km)	Perigee height (km)	Apogee height (km)	Orbital eccentricity	Argument of perigee (deg)
D 2M R	Soyuz 37*	1980-64A	1980 Jul 23.77 79.64 days 1980 Oct 11.41	Sphere-cylinder 6800?	7.5 long 2.3 to 2.72 dia	1980 Jul 23.8 1980 Jul 24.0	51.58 51.58	89.06 89.97	6609 6654	190 258	272 294	0.006 0.003	84 258
D	Soyuz 37 rocket	1980-64B	1980 Jul 23.77 4 days 1980 Jul 27	Cylinder 2500?	7.5 long 2.6 dia	1980 Jul 24.3	51.58	89.02	6607	189	269	0.006	84
D R	Cosmos 1202	1980-65A	1980 Jul 24.53 13.76 days 1980 Aug 7.29	Sphere-cylinder 6300?	6.5 long? 2.4 dia	1980 Jul 25.3	72.85	89.57	6631	198	307	0.008	87
D	Cosmos 1202 rocket	1980-65B	1980 Jul 24.53 5 days 1980 Jul 29	Cylinder 2500?	7.5 long 2.6 dia	1980 Jul 24.7	72.85	89.46	6625	199	295	0.007	81
D	Cosmos 1202 engine	1980-65C	1980 Jul 24.53 30 days 1980 Aug 23	Cone 600? full	1.5 long? 2 dia?	1980 Aug 8.0	72.85	89.21	6613	216	254	0.003	27
D	Fragments	1980-65D,E											
D R	Cosmos 1203	1980-66A	1980 Jul 31.33 13.85 days 1980 Aug 14.18	Sphere-cylinder 6300?	6.5 long? 2.4 dia	1980 Jul 31.4 1980 Aug 2.2	82.32 82.32	89.36 89.88	6620 6646	213 261	270 274	0.004 0.001	91 308
D	Cosmos 1203 rocket	1980-66B	1980 Jul 31.33 6 days 1980 Aug 6	Cylinder 2500?	7.5 long 2.6 dia	1980 Jul 31.8	82.32	89.31	6617	213	265	0.004	82
D	Cosmos 1203 engine	1980-66D	1980 Jul 31.33 30 days 1980 Aug 30	Cone 600? full	1.5 long? 2 dia?	1980 Aug 14.9	82.32	89.78	6641	255	270	0.001	248
D	Fragments	1980-66C,E,F											

* Soyuz 37 docked with Soyuz 36 - Salyut 6 (2nd airlock) on 1980 July 24 at 20h 02m, with one Russian and first Vietnamese cosmonaut. They returned to Earth in Soyuz 36, undocking 1980 Jul 31 at 11h 55m (see page 608). Soyuz 37 (piloted by Soyuz 35 crew) then undocked from 2nd airlock on 1980 Aug 1.70 and re-attached to 1st airlock. Soyuz 37 landed with Soyuz 35 crew, with a manned duration record of 184.8 days.

	Name		Launch date, lifetime and descent date	Shape and weight (kg)	Size (m)	Date of orbital determination	Orbital inclination (deg)	Nodal period (min)	Semi major axis (km)	Perigee height (km)	Apogee height (km)	Orbital eccentricity	Argument of perigee (deg)
D	Cosmos 1204	1980-67A	1980 Jul 31.43 207 days 1981 Feb 23	Octagonal ellipsoid? 550?	1.8 long? 1.5 dia?	1980 Jul 31.6	50.66	93.35	6820	345	538	0.014	147
D	Cosmos 1204 rocket	1980-67B	1980 Jul 31.43 197 days 1981 Feb 13	Cylinder 2200?	7.4 long 2.4 dia	1980 Jul 31.6	50.66	93.24	6815	341	532	0.014	144
D	Fragments	1980-67C-Z											
D R	Cosmos 1205	1980-68A	1980 Aug 12.50 13.75 days 1980 Aug 26.25	Sphere-cylinder 6300?	6.5 long? 2.4 dia	1980 Aug 12.7 1980 Aug 14.4	72.82 72.81	89.56 89.68	6631 6637	199 229	306 288	0.008 0.004	82 72
D	Cosmos 1205 rocket	1980-68B	1980 Aug 12.50 5 days 1980 Aug 17	Cylinder 2500?	7.5 long 2.6 dia	1980 Aug 12.7	72.81	89.45	6625	198	296	0.007	79
D	Cosmos 1205 engine	1980-68C	1980 Aug 12.50 29 days 1980 Sep 10	Cone 600? full	1.5 long? 2 dia?	1980 Aug 26.3	72.81	89.52	6629	223	278	0.004	42
D	Fragment	1980-68D											
T	Cosmos 1206	1980-69A	1980 Aug 15.23 60 years	Cylinder + 2 vanes? 2500?	5 long? 1.5 dia?	1980 Aug 17.3	81.21	97.37	7010	630	634	0.0003	0
	Cosmos 1206 rocket	1980-69B	1980 Aug 15.23 60 years	Cylinder 1440	3.8 long 2.6 dia	1980 Aug 17.3	81.22	97.53	7017	592	686	0.007	165

	Name		Launch date, lifetime and descent date	Shape and weight (kg)	Size (m)	Date of orbital determination	Orbital inclination (deg)	Nodal period (min)	Semi major axis (km)	Perigee height (km)	Apogee height (km)	Orbital eccentricity	Argument of perigee (deg)
D R	Cosmos 1207	1980-70A	1980 Aug 22.42 12.86 days 1980 Sep 4.28	Sphere-cylinder 5900?	5.9 long 2.4 dia	1980 Aug 23.2	82.32	89.19	6612	211	256	0.003	105
D	Cosmos 1207 rocket	1980-70B	1980 Aug 22.42 4 days 1980 Aug 26	Cylinder 2500?	7.5 long 2.6 dia	1980 Aug 23.0	82.32	89.02	6603	209	241	0.002	100
D	Capsule	1980-70C	1980 Aug 22.42 17 days 1980 Sep 8	Ellipsoid 200?	0.9 long 1.9 dia	1980 Sep 4.9	82.32	88.75	6589	209	213	0.0003	57
D	Fragment	1980-70D											
D R	Cosmos 1208	1980-71A	1980 Aug 26.65 28.5 days 1980 Sep 24.2	Sphere-cylinder 6700?	7 long? 2.4 dia	1980 Aug 27.3	67.14	89.60	6634	173	339	0.012	67
D	Cosmos 1208 rocket	1980-71B	1980 Aug 26.65 4 days 1980 Aug 30	Cylinder 2500?	7.5 long 2.6 dia	1980 Aug 27.1	67.14	89.49	6629	172	329	0.012	67
D R	Cosmos 1209	1980-72A	1980 Sep 3.43 13.85 days 1980 Sep 17.28	Sphere-cylinder 6300?	6.5 long? 2.4 dia	1980 Sep 3.7 1980 Sep 4.5	82.34 82.34	89.44 89.88	6624 6646	211 261	280 274	0.005 0.001	100 312
D	Cosmos 1209 rocket	1980-72B	1980 Sep 3.43 5 days 1980 Sep 8	Cylinder 2500?	7.5 long 2.6 dia	1980 Sep 4.4	82.35	89.22	6613	204	265	0.005	109
D	Cosmos 1209 engine	1980-72C	1980 Sep 3.43 27 days 1980 Sep 30	Cone 600? full	1.5 long? 2 dia?	1980 Sep 20.3	82.34	89.52	6628	241	258	0.001	227

	Name		Launch date, lifetime and descent date	Shape and weight (kg)	Size (m)	Date of orbital determination	Orbital inclination (deg)	Nodal period (min)	Semi major axis (km)	Perigee height (km)	Apogee height (km)	Orbital eccentricity	Argument of perigee (deg)
T	Meteor 2-06	1980-73A	1980 Sep 9.46 500 years	Cylinder + 2 vanes 2750?	5 long? 1.5 dia?	1980 Sep 10.0	81.25	102.39	7249	848	894	0.003	271
	Meteor 2-06 rocket	1980-73B	1980 Sep 9.46 400 years	Cylinder 1440	3.8 long 2.6 dia	1980 Sep 10.0	81.25	102.41	7250	837	907	0.005	221
T	GOES 4	1980-74A	1980 Sep 9.94 > million years	Cylinder + boom 627 full 243 empty	2.30 long 1.90 dia	1980 Sep 10.0 1980 Sep 11.6 1980 Sep 28.0	26.50 0.26 0.19	917.03 1759.9 1436.2	31267 48287 42166	167 34170 35776	49610 49647 35800	0.791 0.160 0.0003	180 187 324
D	GOES 4 second stage	1980-74B	1980 Sep 9.94 83 days 1980 Dec 1	Cylinder + annulus 350?	6.4 long 1.52 and 2.44 dia	1980 Sep 9.9 1980 Sep 10.5	28.73 28.10	88.14 112.78	6561 7740	157 164	208 2559	0.004 0.155	120 179
	GOES 4 apogee motor	1980-74C	1980 Sep 9.94 > million years	- 384 full	-	1980 Oct 3.0	0.09	1766.29	48410	34330	49734	0.159	40
	GOES 4 third stage	1980-74D	1980 Sep 9.94 10 years?	Sphere- cone 66	1.32 long 0.94 dia		In 917 min transfer orbit similar to 1980-74A						
D 2M R	Soyuz 38*	1980-75A	1980 Sep 18.80 7.86 days 1980 Sep 26.66	Sphere- cylinder 6800?	7.5 long 2.3 to 2.72 dia	1980 Sep 18.8 1980 Sep 19.0	51.61 51.62	88.95 90.22	6604 6666	195 263	257 313	0.005 0.004	78 258
D	Soyuz 38 rocket	1980-75B	1980 Sep 18.80 3 days 1980 Sep 21	Cylinder 2500?	7.5 long 2.6 dia	1980 Sep 19.2	51.62	88.71	6592	191	237	0.004	79
D	Fragment	1980-75C											

* Soyuz 38 docked with Soyuz 37 - Salyut 6 (2nd airlock) on 1980 Sep 19.87, with one Russian and first Cuban cosmonaut; undocked 1980 Sep 26.

	Name		Launch date, lifetime and descent date	Shape and weight (kg)	Size (m)	Date of orbital determination	Orbital inclination (deg)	Nodal period (min)	Semi major axis (km)	Perigee height (km)	Apogee height (km)	Orbital eccentricity	Argument of perigee (deg)
D R	Cosmos 1210	1980-76A	1980 Sep 19.43 13.83 days 1980 Oct 3.26	Sphere-cylinder 6300?	6.5 long? 2.4 dia	1980 Sep 19.5 1980 Sep 20.3	82.33 82.33	88.76 89.60	6590 6632	180 187	244 320	0.005 0.010	84 67
D	Cosmos 1210 rocket	1980-76B	1980 Sep 19.43 2 days 1980 Sep 21	Cylinder 2500?	7.5 long 2.6 dia	1980 Sep 20.2	82.33	88.36	6570	172	212	0.003	75
D	Cosmos 1210 engine	1980-76E	1980 Sep 19.43 40 days 1980 Oct 29	Cone 600? full	1.5 long? 2 dia?	1980 Oct 5.7	82.33	90.25	6665	230	343	0.008	356
D	Fragments	1980-76C,D,F,G											
D R	Cosmos 1211	1980-77A	1980 Sep 23.44 10.87 days 1980 Oct 4.31	Sphere-cylinder 5700?	5.0 long 2.4 dia	1980 Sep 23.6	82.35	89.11	6607	216	242	0.002	88
D	Cosmos 1211 rocket	1980-77B	1980 Sep 23.44 3 days 1980 Sep 26	Cylinder 2500?	7.5 long 2.6 dia	1980 Sep 23.7	82.36	88.99	6601	213	233	0.001	74
D R	Cosmos 1212*	1980-78A	1980 Sep 26.43 12.84 days 1980 Oct 9.27	Sphere-cylinder 5900?	5.9 long 2.4 dia	1980 Sep 26.9	82.34	89.11	6607	209	249	0.003	109
D	Cosmos 1212 rocket	1980-78B	1980 Sep 26.43 3 days 1980 Sep 29	Cylinder 2500?	7.5 long 2.6 dia	1980 Sep 26.9	82.36	88.97	6600	204	240	0.003	127
D	Capsule	1980-78D	1980 Sep 26.43 15 days 1980 Oct 11	Ellipsoid 200?	0.9 long 1.9 dia	1980 Oct 9.3	82.34	88.33	6568	182	198	0.001	48
D	Fragment	1980-78C											

* Earth resources satellite.

	Name		Launch date, lifetime and descent date	Shape and weight (kg)	Size (m)	Date of orbital determination	Orbital inclination (deg)	Nodal period (min)	Semi major axis (km)	Perigee height (km)	Apogee height (km)	Orbital eccentricity	Argument of perigee (deg)
D	Progress 11*	1980-79A	1980 Sep 28.63 73.95 days 1980 Dec 11.58	Sphere-cylinder 7020	7.9 long 2.3 dia	1980 Sep 28.8 1980 Sep 28.9	51.62 51.62	88.79 90.45	6596 6678	189 260	247 340	0.004 0.006	73 197
D	Progress 11 rocket	1980-79B	1980 Sep 28.63 2 days 1980 Sep 30	Cylinder 2500?	7.5 long 2.6 dia	1980 Sep 28.9	51.61	88.62	6588	184	235	0.004	82
D R	Cosmos 1213	1980-80A	1980 Oct 3.50 13.79 days 1980 Oct 17.29	Sphere-cylinder 6300?	6.5 long? 2.4 dia	1980 Oct 4.5 1980 Oct 14.2	72.87 72.87	89.69 90.23	6637 6664	229 235	289 336	0.005 0.008	73 50
D	Cosmos 1213 rocket	1980-80B	1980 Oct 3.50 4 days 1980 Oct 7	Cylinder 2500?	7.5 long 2.6 dia	1980 Oct 4.5	72.87	89.34	6619	191	291	0.008	81
D	Cosmos 1213 engine	1980-80C	1980 Oct 3.50 34 days 1980 Nov 6	Cone 600? full	1.5 long? 2 dia?	1980 Oct 17.7	72.86	92.22	6761	259	507	0.018	351
D	Fragments	1980-80D-F											
T	Statsionar-Raduga 7	1980-81A	1980 Oct 5.72 > million years	-	-	1980 Oct 18.7	0.33	1436.0	42162	35730	35840	0.001	303
D	Raduga 7 launcher rocket	1980-81B	1980 Oct 5.72 2 days 1980 Oct 7	Cylinder 4000?	12 long? 4 dia	1980 Oct 6.1	51.68	88.04	6559	174	187	0.001	223
D	Raduga 7 launcher	1980-81C	1980 Oct 5.72 1 day 1980 Oct 6	Irregular	-	1980 Oct 6.0	51.63	87.78	6546	135	200	0.005	144
D	Raduga 7 rocket	1980-81D	1980 Oct 5.72 0.5 years 1981 Apr?	Cylinder 1900?	3.9 long? 3.9 dia	1980 Oct 19.0	47.35	634.05	24450	249	35896	0.729	3

* Progress 11 docked with Soyuz 37 - Salyut 6 (2nd airlock) on 1980 Sep 30.71. Progress 11 undocked from Soyuz T3 - Salyut 6 on 1980 Dec 9.43 and was de-orbited into the Pacific Ocean.

	Name		Launch date, lifetime and descent date	Shape and weight (kg)	Size (m)	Date of orbital determination	Orbital inclination (deg)	Nodal period (min)	Semi major axis (km)	Perigee height (km)	Apogeè height (km)	Orbital eccentricity	Argument of perigee (deg)
	Raduga 7 apogee motor	1980-81F	1980 Oct 5.72 > million years	Cylinder 440?	2.0 long? 2.0 dia?	1981 Jun 5.5	0.59	1474.9	42925	36451	36642	0.002	239
D	Fragment	1980-81E											
D R	Cosmos 1214	1980-82A	1980 Oct 10.55 12.75 days 1980 Oct 23.30	Sphere-cylinder 6700?	7 long? 2.4 dia	1980 Oct 11.2	67.15	89.67	6638	174	345	0.013	67
D	Cosmos 1214 rocket	1980-82B	1980 Oct 10.55 3 days 1980 Oct 13	Cylinder 2500?	7.5 long 2.6 dia	1980 Oct 10.7	67.15	89.61	6635	173	340	0.013	67
D	Fragments	1980-82C-E											
T	Cosmos 1215	1980-83A	1980 Oct 14.86 2.6 years	Cylinder + paddles? 900?	2 long? 1 dia?	1980 Oct 16.2	74.04	95.12	6902	498	550	0.004	338
D	Cosmos 1215 rocket	1980-83B	1980 Oct 14.86 682 days 1982 Aug 27	Cylinder 2200?	7.4 long 2. 4 dia	1980 Oct 19.1	74.06	95.06	6899	485	557	0.005	327
D	Fragments	1980-83C,D											
D R	Cosmos 1216	1980-84A	1980 Oct 16.52 13.77 days 1980 Oct 30.29	Sphere-cylinder 6300?	6.5 long? 2.4 dia	1980 Oct 16.8 1980 Oct 17.1	72.87 72.87	90.29 92.27	6667 6764	198 367	379 404	0.014 0.003	83 218
D	Cosmos 1216 rocket	1980-84B	1980 Oct 16.52 7 days 1980 Oct 23	Cylinder 2500?	7.5 long 2.6 dia	1980 Oct 16.8	72.87	90.18	6661	197	369	0.013	83
D	Cosmos 1216 engine	1980-84F	1980 Oct 16.52 179 days 1981 Apr 13	Cone 600? full	1.5 long? 2 dia?	1980 Oct 30.5	72.47	92.48	6775	362	432	0.005	149
D	Fragments	1980-84C-E,G,H											

	Name		Launch date, lifetime and descent date	Shape and weight (kg)	Size (m)	Date of orbital determination	Orbital inclination (deg)	Nodal period (min)	Semi major axis (km)	Perigee height (km)	Apogee height (km)	Orbital eccentricity	Argument of perigee (deg)
T	Cosmos 1217	1980-85A	1980 Oct 24.46 100 years?	Windmill + 6 vanes? 1250?	4.2 long? 1.6 dia	1980 Oct 25.5 1980 Nov 4.0	62.92 62.93	725.33 717.39	26742 26547	596 612	40131 39725	0.739 0.737	318 318
D	Cosmos 1217 launcher rocket	1980-85B	1980 Oct 24.46 24 days 1980 Nov 17	Cylinder 2500?	7.5 long 2.6 dia	1980 Oct 25.4	62.93	92.45	6774	178	613	0.032	118
D	Cosmos 1217 launcher	1980-85C	1980 Oct 24.46 14 days 1980 Nov 7	Irregular	-	1980 Oct 25.4	62.80	92.59	6781	211	594	0.028	115
	Cosmos 1217 rocket	1980-85D	1980 Oct 24.46 100 years?	Cylinder 440	2.0 long 2.0 dia	1980 Nov 4.5	62.91	722.00	26662	613	39955	0.738	318
D R	Cosmos 1218	1980-86A	1980 Oct 30.42 43.2 days? 1980 Dec 12.6?	Sphere-cylinder 6700?	7 long? 2.4 dia	1980 Oct 30.7	64.89	89.73	6640	171	353	0.014	65
D	Cosmos 1218 rocket	1980-86B	1980 Oct 30.42 3 days 1980 Nov 2	Cylinder 2500?	7.5 long 2.6 dia	1980 Oct 30.7	64.89	89.63	6635	172	342	0.013	65
T	Fleetsatcom 4	1980-87A	1980 Oct 31.16 > million years	Hexagonal cylinder 1884 full 1005 empty	1.27 long 2.44 dia	1980 Oct 31.2 1980 Nov 2.1	26.29 2.40	620.14 1418.72	24089 41825	173 34903	35249 35991	0.728 0.013	182 248
	Fleetsatcom 4 rocket	1980-87B	1980 Oct 31.16 15 years	Cylinder 1815	8.6 long 3.0 dia	1980 Nov 18.5	26.49	607.57	23756	193	34563	0.723	194

	Name		Launch date, lifetime and descent date	Shape and weight (kg)	Size (m)	Date of orbital determination	Orbital inclination (deg)	Nodal period (min)	Semi major axis (km)	Perigee height (km)	Apogee height (km)	Orbital eccentricity	Argument of perigee (deg)
D R	Cosmos 1219	1980-88A	1980 Oct 31.50 12.78 days 1980 Nov 13.28	Sphere-cylinder 6300?	6.5 long? 2.4 dia	1980 Nov 1.3	72.85	89.70	6638	228	291	0.005	71
D	Cosmos 1219 rocket	1980-88B	1980 Oct 31.50 4 days 1980 Nov 4	Cylinder 2500?	7.5 long 2.6 dia	1980 Nov 1.5	72.85	89.48	6627	188	309	0.009	96
D	Cosmos 1219 engine	1980-88E	1980 Oct 31.50 26 days 1980 Nov 26	Cone 600? full	1.5 long? 2 dia?	1980 Nov 13.4	72.83	89.67	6636	216	300	0.006	53
D	Fragments	1980-88C,D,F,G											
	Cosmos 1220	1980-89A	1980 Nov 4.63 manoeuvrable	Cylinder?	-	1980 Nov 7.3	65.04	93.30	6812	427	440	0.001	271
D	Cosmos 1220 rocket	1980-89B	1980 Nov 4.63 1 day 1980 Nov 5	Cylinder 1500?	8 long? 2.5 dia?	1980 Nov 4.9	64.90	88.74	6592	111	316	0.016	8
6d	Fragments	1980-89C-BG											
D R	Cosmos 1221	1980-90A	1980 Nov 12.52 13.78 days 1980 Nov 26.30	Sphere-cylinder 6300?	6.5 long? 2.4 dia	1980 Nov 12.6 1980 Nov 14.0	72.90 72.89	90.49 92.29	6676 6765	196 356	399 417	0.015 0.004	85 210
D	Cosmos 1221 rocket	1980-90B	1980 Nov 12.52 8 days 1980 Nov 20	Cylinder 2500?	7.5 long 2.6 dia	1980 Nov 13.2	72.90	90.35	6669	195	387	0.014	82
D	Cosmos 1221 engine	1980-90C	1980 Nov 12.52 153 days 1981 Apr 14	Cone 600? full	1.5 long? 2 dia?	1980 Nov 27.1	72.89	92.21	6761	353	413	0.004	181
D	Fragments	1980-90D,E											

	Name		Launch date, lifetime and descent date	Shape and weight (kg)	Size (m)	Date of orbital determination	Orbital inclination (deg)	Nodal period (min)	Semi major axis (km)	Perigee height (km)	Apogee height (km)	Orbital eccentricity	Argument of perigee (deg)
T	SBS 1* [Thor Delta]	1980-91A	1980 Nov 15.95 > million years	Cylinder + panel 550	2.82 long 2.16 dia	1980 Nov 16.0 1980 Nov 19.3 1980 Dec 9.1	27.70 0.42 0.18	655.93 1393.51 1436.08	25007 41320 42164	166 34060 35769	37092 35830 35803	0.738 0.021 0.0004	179 133 165
D	SBS 1 third stage (PAM)†	1980-91B	1980 Nov 15.95 169 days 1981 May 3	Cylinder 2154 full	2.29 long 1.5 dia	1981 Mar 25.0	27.29	460.3	19747	137	26600	0.670	271
T	Molniya 1AZ	1980-92A	1980 Nov 16.19 15½ years	Windmill + 6 vanes 1000?	3.4 long 1.6 dia	1980 Nov 17.3 1980 Dec 12.3	62.78 62.88	736.23 717.81	27010 26557	601 593	40662 39764	0.742 0.737	288 288
D	Molniya 1AZ launcher rocket**	1980-92B	1980 Nov 16.19 27.54 days 1980 Dec 13.73	Cylinder 2500?	7.5 long 2.6 dia	1980 Nov 16.4	62.85	92.71	6786	195	621	0.031	118
D	Molniya 1AZ launcher	1980-92C	1980 Nov 16.19 26 days 1980 Dec 12	Irregular	-	1980 Nov 16.4	62.80	92.68	6785	212	601	0.029	122
	Molniya 1AZ rocket	1980-92D	1980 Nov 16.19 15½ years	Cylinder 440	2.0 long 2.0 dia	1980 Nov 30.5	62.76	733.61	26945	582	40552	0.742	288
T	Cosmos 1222	1980-93A	1980 Nov 21.50 60 years	Cylinder + 2 vanes? 2500?	5 long? 1.5 dia?	1980 Nov 22.4	81.23	97.38	7010	631	633	0.0001	2
	Cosmos 1222 rocket	1980-93B	1980 Nov 21.50 60 years	Cylinder 1440	3.8 long 2.6 dia	1980 Nov 21.9	81.27	97.48	7015	607	667	0.004	168

* Satellite Business Systems. Solar panel is hollow cylinder. Second stage entered sub-orbital trajectory.
† Payload Assist Module.
** Decay observed over Japan.

	Name		Launch date, lifetime and descent date	Shape and weight (kg)	Size (m)	Date of orbital determination	Orbital inclination (deg)	Nodal period (min)	Semi major axis (km)	Perigee height (km)	Apogee height (km)	Orbital eccentricity	Argument of perigee (deg)
D 3M R	Soyuz-Transport 3*	1980-94A	1980 Nov 27.60 12.79 days 1980 Dec 10.39	Sphere-cylinder +2 wings 6850	7.5 long 2.3 to 2.72 dia	1980 Nov 28.0	51.62	89.60	6636	255	260	0.0004	241
D	Soyuz T-3 rocket	1980-94B	1980 Nov 27.60 2 days 1980 Nov 29	Cylinder 2500?	7.5 long 2.6 dia	1980 Nov 28.0	51.62	88.57	6585	190	224	0.003	104
D	Soyuz T-3** orbital module	1980-94C	1980 Nov 27.60 60 days 1981 Jan 26	Spheroid 1260?	2.3 dia	1980 Dec 10.7	51.62	90.90	6701	289	357	0.005	100
T	Cosmos 1223	1980-95A	1980 Nov 27.97 100 years?	Windmill + 6 vanes? 1250?	4.2 long? 1.6 dia	1980 Dec 2.0	62.87	717.73	26555	605	39749	0.737	316
D	Cosmos 1223 launcher rocket	1980-95B	1980 Nov 27.97 25 days 1980 Dec 22	Cylinder 2500?	7.5 long 2.6 dia	1980 Nov 28.0	62.80	92.95	6798	201	639	0.032	93
D	Cosmos 1223 launcher	1980-95C	1980 Nov 27.97 24 days 1980 Dec 21	Irregular	-	1980 Nov 28.0	62.79	92.51	6776	213	583	0.027	119
	Cosmos 1223 rocket	1980-95E	1980 Nov 27.97 100 years?	Cylinder 440	2.0 long 2.0 dia	1980 Dec 22.1	62.91	723.49	26700	623	40020	0.738	316
D	Fragment	1980-95D											

* First 3-man Soyuz since June 1971. Soyuz T-3 docked with Progress 11 - Salyut 6 (1st airlock) on 1980 Nov 28.66.
** Designated 1977-97CN in the USA.

	Name		Launch date, lifetime and descent date	Shape and weight (kg)	Size (m)	Date of orbital determination	Orbital inclination (deg)	Nodal period (min)	Semi major axis (km)	Perigee height (km)	Apogee height (km)	Orbital eccentricity	Argument of perigee (deg)
D R	Cosmos 1224	1980-96A	1980 Dec 1.51 13.8 days 1980 Dec 15.3	Sphere-cylinder 6300?	6.5 long? 2.4 dia	1980 Dec 1.5 1980 Dec 2.1	72.87 72.85	90.29 92.15	6666 6758	198 351	378 408	0.013 0.004	83 225
D	Cosmos 1224 rocket	1980-96B	1980 Dec 1.51 8 days 1980 Dec 9	Cylinder 2500?	7.5 long 2.6 dia	1980 Dec 1.9	72.85	90.20	6662	196	371	0.013	81
D	Cosmos 1224 engine	1980-96C	1980 Dec 1.51 154 days 1981 May 4	Cone 600? full	1.5 long? 2 dia?	1980 Dec 17.6	72.86	92.19	6760	337	426	0.007	167
D	Fragments	1980-96D,E											
T	Cosmos 1225	1980-97A	1980 Dec 5.18 1200 years	Cylinder 700?	1.3 long? 1.9 dia?	1980 Dec 6.3	82.94	104.93	7369	950	1031	0.005	291
	Cosmos 1225 rocket	1980-97B	1980 Dec 5.18 600 years	Cylinder 2200?	7.4 long 2.4 dia	1980 Dec 5.8	82.94	104.81	7363	949	1020	0.005	286
T	Intelsat 5 F-2 [Atlas Centaur]	1980-98A	1980 Dec 6.98 > million years	Box + 2 vanes + dishes 1928 full	15.7 span	1980 Dec 7.1 1980 Dec 7.8	23.76 0.91	633.82 1417.67	24442 41803	167 35143	35961 35707	0.732 0.007	179 163
	Intelsat 5 F-2 rocket	1980-98B	1980 Dec 6.98 3 years?	Cylinder 1815	8.6 long 3.0 dia	1980 Dec 7.0 1980 Dec 7.1	28.31 23.75	89.57 614.96	6631 23955	148 170	358 34983	0.016 0.727	146 179
T	Cosmos 1226	1980-99A	1980 Dec 10.87 1200 years	Cylinder 700?	1.3 long? 1.9 dia?	1980 Dec 11.1	82.94	104.92	7368	966	1014	0.003	285
	Cosmos 1226 rocket	1980-99B	1980 Dec 10.87 600 years	Cylinder 2200?	7.4 long 2.4 dia	1980 Dec 12.0	82.94	104.82	7363	966	1004	0.003	278

	Name		Launch date, lifetime and descent date	Shape and weight (kg)	Size (m)	Date of orbital determination	Orbital inclination (deg)	Nodal period (min)	Semi major axis (km)	Perigee height (km)	Apogee height (km)	Orbital eccentricity	Argument of perigee (deg)
T	SDS 6 [Titan 3B Agena D]	1980-100A	1980 Dec 13.67 100 years?	Cylinder?	-	1980 Dec 13.7	63.8	697.4	26070	250	39130	0.746	-
	SDS 6 rocket	1980-100B	1980 Dec 13.67 100 years?	Cylinder 700?	6 long 1.5 dia	Orbit similar to 1980-100A							
D R	Cosmos 1227	1980-101A	1980 Dec 16.51 11.8 days 1980 Dec 28.3	Sphere-cylinder 6300?	6.5 long? 2.4 dia	1980 Dec 16.5 1980 Dec 17.1	72.85 72.84	89.50 89.81	6628 6643	199 229	300 300	0.008 0.005	86 77
D	Cosmos 1227 rocket	1980-101B	1980 Dec 16.51 4 days 1980 Dec 20	Cylinder 2500?	7.5 long 2.6 dia	1980 Dec 17.1	72.83	89.40	6623	193	296	0.008	86
D	Cosmos 1227 engine	1980-101C	1980 Dec 16.51 30 days 1981 Jan 15	Cone 600? full	1.5 long? 2 dia?	1980 Dec 28.8	72.84	89.55	6630	217	287	0.005	59
D	Fragments	1980-101D,E											
T	Cosmos 1228	1980-102A	1980 Dec 23.95 7000 years	Spheroid 40?	1.0 long? 0.8 dia?	1980 Dec 30.5	74.00	114.44	7807	1394	1464	0.004	73
T	Cosmos 1229	1980-102B	1980 Dec 23.95 9000 years	Spheroid 40?	1.0 long? 0.8 dia?	1980 Dec 29.6	74.00	114.67	7818	1416	1464	0.003	84
T	Cosmos 1230	1980-102C	1980 Dec 23.95 8000 years	Spheroid 40?	1.0 long? 0.8 dia?	1980 Dec 28.9	74.00	114.50	7809	1399	1462	0.004	85
T	Cosmos 1231	1980-102D	1980 Dec 23.95 8000 years	Spheroid 40?	1.0 long? 0.8 dia?	1980 Dec 30.9	74.00	114.57	7813	1406	1463	0.004	78

1980-102 continued on page 630

	Name		Launch date, lifetime and descent date	Shape and weight (kg)	Size (m)	Date of orbital determination	Orbital inclination (deg)	Nodal period (min)	Semi major axis (km)	Perigee height (km)	Apogee height (km)	Orbital eccentricity	Argument of perigee (deg)
T	Cosmos 1232	1980-102E	1980 Dec 23.95 8000 years	Spheroid 40?	1.0 long? 0.8 dia?	1980 Dec 29.9	74.00	114.65	7817	1414	1464	0.003	77
T	Cosmos 1233	1980-102F	1980 Dec 23.95 6000 years	Spheroid 40?	1.0 long? 0.8 dia?	1980 Dec 28.8	74.00	114.71	7820	1420	1464	0.003	75
T	Cosmos 1234	1980-102G	1980 Dec 23.95 8000 years	Spheroid 40?	1.0 long? 0.8 dia?	1980 Dec 26.3	74.00	114.61	7815	1411	1462	0.003	79
T	Cosmos 1235	1980-102H	1980 Dec 23.95 7000 years	Spheroid 40?	1.0 long? 0.8 dia?	1980 Dec 29.9	74.00	114.66	7818	1415	1464	0.003	78
T	Cosmos 1228 rocket	1980-102J	1980 Dec 23.95 20000 years	Cylinder 2200?	7.4 long 2.4 dia	1980 Dec 27.7	73.99	114.97	7831	1441	1464	0.001	118
T	Prognoz 8	1980-103A	1980 Dec 25.17 3.4 years	Spheroid + 4 vanes 985?	1.8 dia?	1980 Dec 25.3	65.83	5689	105563	980	197390	0.930	291
D	Prognoz 8 launcher	1980-103B	1980 Dec 25.17 26 days 1981 Jan 20	Irregular	-	1980 Dec 25.9	65.06	91.86	6745	234	499	0.020	67
D	Prognoz 8 launcher rocket	1980-103C	1980 Dec 25.17 20 days 1981 Jan 14	Cylinder 2500?	7.5 long 2.6 dia	1980 Dec 26.2	65.10	91.55	6729	202	500	0.022	69
	Prognoz 8 rocket	1980-103E	1980 Dec 25.17 3.4 years	Cylinder 440	2.0 long 2.0 dia	Orbit similar to 1980-103A							
D	Fragment	1980-103D											

	Name		Launch date, lifetime and descent date	Shape and weight (kg)	Size (m)	Date of orbital determination	Orbital inclination (deg)	Nodal period (min)	Semi major axis (km)	Perigee height (km)	Apogee height (km)	Orbital eccentricity	Argument of perigee (deg)
T	Statsionar-Ekran 6	1980-104A	1980 Dec 26.50 > million years	Cylinder + plate	-	1980 Dec 27.1	0.07	1439.85	42237	35859	35859	0	142
D	Ekran 6 launcher rocket	1980-104B	1980 Dec 26.50 2 days 1980 Dec 28	Cylinder 4000?	12 long? 4 dia	1980 Dec 26.9	51.60	88.16	6564	176	195	0.001	212
D	Ekran 6 launcher	1980-104C	1980 Dec 26.50 1 day 1980 Dec 27	Irregular	-	1980 Dec 27.0	51.60	87.42	6527	141	156	0.001	21
D	Ekran 6 rocket	1980-104D	1980 Dec 26.50 1.3 years 1982 Apr?	Cylinder 1900?	3.9 long? 3.9 dia	1980 Dec 31.0	47.24	626.01	24270	296	35488	0.725	1
	Ekran 6 apogee motor	1980-104E	1980 Dec 26.50 > million years	Cylinder 440?	2.0 long? 2.0 dia	1981 Jun 7.3	0.08	1420.8	41867	35351	35627	0.003	284
D R	Cosmos 1236	1980-105A	1980 Dec 26.68 25.6 days 1981 Jan 21.3	Sphere-cylinder 6700?	7 long? 2.4 dia	1980 Dec 27.2	67.13	89.80	6644	169	363	0.015	74
D	Cosmos 1236 rocket	1980-105B	1980 Dec 26.68 4 days 1980 Dec 30	Cylinder 2500?	7.5 long 2.6 dia	1980 Dec 27.7	67.13	89.53	6631	169	336	0.013	73
	Fragments	1980-105C,D											

	Name		Launch date, lifetime and descent date	Shape and weight (kg)	Size (m)	Date of orbital determination	Orbital inclination (deg)	Nodal period (min)	Semi major axis (km)	Perigee height (km)	Apogee height (km)	Orbital eccentricity	Argument of perigee (deg)
D R	Cosmos 1237	1981-01A	1981 Jan 6.51 13.80 days 1981 Jan 20.31	Sphere-cylinder 6300?	6.5 long? 2.4 dia	1981 Jan 6.6 1981 Jan 7.2	72.88 72.86	90.35 92.29	6669 6765	195 356	387 417	0.014 0.004	87 210
D	Cosmos 1237 rocket	1981-01B	1981 Jan 6.51 9 days 1981 Jan 15	Cylinder 2500?	7.5 long 2.6 dia	1981 Jan 7.2	72.86	90.15	6659	194	368	0.013	84
D	Cosmos 1237 engine	1981-01D	1981 Jan 6.51 150 days 1981 Jun 5	Cone 600? full	1.5 long? 2 dia?	1981 Jan 23.9	72.86	92.21	6761	353	413	0.004	173
D	Fragments	1981-01C,E-G											
T	Molniya 3P	1981-02A	1981 Jan 9.63 18½ years	Windmill + 6 vanes? 1500?	4.2 long? 1.6 dia	1981 Jan 11.7 1981 Jan 16.3	62.80 62.79	735.74 718.11	26998 26565	439 448	40800 39925	0.747 0.743	280 280
	Molniya 3P rocket	1981-02B	1981 Jan 9.63 18¾ years	Cylinder 440	2.0 long 2.0 dia	1981 Jan 10.7	62.80	732.06	26908	432	40627	0.747	280
D	Molniya 3P launcher rocket	1981-02C	1981 Jan 9.63 18 days 1981 Jan 27	Cylinder 2500?	7.5 long 2.6 dia	1981 Jan 10.9	62.83	91.11	6708	206	454	0.018	131
D	Molniya 3P launcher	1981-02D	1981 Jan 9.63 13 days 1981 Jan 22	Irregular	-	1981 Jan 10.5	62.82	91.01	6703	189	461	0.020	120

	Name		Launch date, lifetime and descent date	Shape and weight (kg)	Size (m)	Date of orbital determination	Orbital inclination (deg)	Nodal period (min)	Semi major axis (km)	Perigee height (km)	Apogee height (km)	Orbital eccentricity	Argument of perigee (deg)
T	Cosmos 1238	1981-03A	1981 Jan 16.38 30 years	Octagonal ellipsoid? 550?	1.8 long? 1.5 dia?	1981 Jan 17.2	82.98	109.05	7560	406	1958	0.103	128
	Cosmos 1238 rocket	1981-03B	1981 Jan 16.38 30 years	Cylinder 2200?	7.4 long 2.4 dia	1981 Jan 17.5	82.97	108.94	7555	405	1948	0.102	127
D R	Cosmos 1239	1981-04A	1981 Jan 16.50 11.85 days 1981 Jan 28.35	Sphere-cylinder 5700?	5.0 long 2.4 dia	1981 Jan 16.9	82.33	89.03	6603	216	234	0.001	98
D	Cosmos 1239 rocket	1981-04B	1981 Jan 16.50 4 days 1981 Jan 20	Cylinder 2500?	7.5 long 2.6 dia	1981 Jan 16.9	82.34	88.93	6598	210	230	0.001	85
D R	Cosmos 1240	1981-05A	1981 Jan 20.46 28 days 1981 Feb 17	Sphere-cylinder 6700?	7 long? 2.4 dia	1981 Jan 20.8	64.88	89.77	6642	171	357	0.014	63
D	Cosmos 1240 rocket	1981-05B	1981 Jan 20.46 4 days 1981 Jan 24	Cylinder 2500?	7.5 long 2.6 dia	1981 Jan 20.8	64.88	89.66	6637	168	349	0.014	61
	Cosmos 1241	1981-06A	1981 Jan 21.35 1200 years	Cylinder?	4 long? 2 dia?	1981 Jan 22.3	65.82	104.97	7372	977	1011	0.002	287
	Cosmos 1241 rocket	1981-06B	1981 Jan 21.35 600 years	Cylinder 2200?	7.4 long 2.4 dia	1981 Jan 21.7	65.82	104.83	7365	975	999	0.002	272
	Fragment	1981-06C											

	Name		Launch date, lifetime and descent date	Shape and weight (kg)	Size (m)	Date of orbital determination	Orbital inclination (deg)	Nodal period (min)	Semi major axis (km)	Perigee height (km)	Apogee height (km)	Orbital eccentricity	Argument of perigee (deg)
D	Progress 12*	1981-07A	1981 Jan 24.60 55.11 days 1981 Mar 20.71	Sphere-cylinder 7020	7.9 long 2.3 dia	1981 Jan 24.8 1981 Jan 25.6	51.67 51.67	89.06 90.02	6610 6657	181 247	282 310	0.008 0.005	95 221
D	Progress 12 rocket	1981-07B	1981 Jan 24.60 3 days 1981 Jan 27	Cylinder 2500?	7.5 long 2.6 dia	1981 Jan 24.8	51.67	89.00	6607	181	276	0.007	95
T	Cosmos 1242	1981-08A	1981 Jan 27.63 60 years	Cylinder + 2 vanes? 2500?	5 long? 1.5 dia?	1981 Jan 27.8	81.17	97.58	7020	626	658	0.002	95
	Cosmos 1242 rocket	1981-08B	1981 Jan 27.63 60 years	Cylinder 1440	3.8 long 2.6 dia	1981 Jan 27.8	81.21	97.72	7027	636	661	0.002	117
T	Molniya 1BA	1981-09A	1981 Jan 30.75 100 years?**	Windmill + 6 vanes 1000?	3.4 long 1.6 dia	1981 Feb 2.8	62.83	735.65	26996	430	40805	0.748	280
D	Molniya 1BA launcher rocket	1981-09B	1981 Jan 30.75 15 days 1981 Feb 14	Cylinder 2500?	7.5 long 2.6 dia	1981 Jan 31.3	62.84	91.06	6706	208	447	0.018	132
D	Molniya 1BA launcher	1981-09C	1981 Jan 30.75 10 days 1981 Feb 9	Irregular	-	1981 Jan 31.3	62.83	90.84	6695	190	443	0.019	121
	Molniya 1BA rocket	1981-09D	1981 Jan 30.75 100 years?**	Cylinder 440	2.0 long 2.0 dia	1981 Jan 30.7	62.79	733.25	26937	459	40659	0.746	280

* Progress 12 docked with Salyut 6 (2nd airlock) on 1981 Jan 26.66 and undocked on 1981 Mar 19.76

** Decay possible in 2003

	Name		Launch date, lifetime and descent date	Shape and weight (kg)	Size (m)	Date of orbital determination	Orbital inclination (deg)	Nodal period (min)	Semi major axis (km)	Perigee height (km)	Apogee height (km)	Orbital eccentricity	Argument of perigee (deg)
D	Cosmos 1243*	1981-10A	1981 Feb 2.10 <1 day 1981 Feb 2	Cylinder?	4 long? 2 dia?	1981 Feb 2.2	65.82	97.85	7035	297	1017	0.051	250
D	Cosmos 1243 rocket	1981-10B	1981 Feb 2.10 2 days 1981 Feb 4	Cylinder 1500?	8 long? 2.5 dia?	1981 Feb 2.5	65.08	89.06	6607	131	326	0.015	38
D	Intercosmos 21	1981-11A	1981 Feb 6.33 516 days 1982 Jul 7	Octagonal ellipsoid 550?	1.8 long? 1.5 dia?	1981 Feb 7.4	74.04	94.52	6873	475	515	0.003	336
D	Intercosmos 21 rocket	1981-11B	1981 Feb 6.33 501 days 1982 Jun 22	Cylinder 2200?	7.4 long 2.4 dia	1981 Feb 7.4	74.04	94.44	6869	468	514	0.003	344
T	Kiku 3** (ETS 4) [Nu-2]	1981-12A	1981 Feb 11.35 30 years	Cylinder? 640	2.8 long 2.1 dia	1981 Feb 13.8	28.60	636.26	24515	248	36025	0.730	182
D	Kiku 3 second stage	1981-12B	1981 Feb 11.35 17 days 1981 Feb 28	Cylinder 550?	5.4 long? 1.4 dia?	1981 Feb 14.0	30.36	90.33	6676	221	374	0.011	211
	Kiku 3 third stage	1981-12C	1981 Feb 11.35 30 years	Cylinder 66?	1.74 long 1.65 dia	1981 Sep 5.0	28.18	626.5	24254	308	35443	0.724	304

* Probably passed within 8 km of Cosmos 1241 on 1981 Feb 2.23, then de-orbited.
** Japanese Engineering Test Satellite launched by uprated Nu rocket.

	Name		Launch date, lifetime and descent date	Shape and weight (kg)	Size (m)	Date of orbital determination	Orbital inclination (deg)	Nodal period (min)	Semi major axis (km)	Perigee height (km)	Apogee height (km)	Orbital eccentricity	Argument of perigee (deg)
T	Cosmos 1244	1981-13A	1981 Feb 12.76 1200 years	Cylinder 700?	1.3 long? 1.9 dia?	1981 Feb 13.0	82.95	104.90	7367	963	1014	0.003	283
	Cosmos 1244 rocket	1981-13B	1981 Feb 12.76 600 years	Cylinder 2200?	7.4 long 2.4 dia	1981 Feb 13.3	82.95	104.79	7362	968	999	0.002	274
D R	Cosmos 1245	1981-14A	1981 Feb 13.47 13.8 days 1981 Feb 27.3	Sphere-cylinder 6300?	6.5 long? 2.4 dia	1981 Feb 14.1 1981 Feb 16.5	72.85 72.84	90.00 92.28	6652 6764	190 356	357 416	0.013 0.004	85 205
D	Cosmos 1245 rocket	1981-14B	1981 Feb 13.47 8 days 1981 Feb 21	Cylinder 2500?	7.5 long 2.6 dia	1981 Feb 13.5	72.83	90.21	6662	198	370	0.013	84
D	Cosmos 1245 engine	1981-14F	1981 Feb 13.47 159 days 1981 Jul 22	Cone 600? full	1.5 long? 2 dia?	1981 Mar 4.7	72.83	92.21	6761	354	412	0.004	171
D	Fragments	1981-14C-E											
D R	Cosmos 1246	1981-15A	1981 Feb 18.38 23 days 1981 Mar 13	Sphere-cylinder 6700?	7 long? 2.4 dia	1981 Feb 18.6	64.90	89.19	6613	198	272	0.006	58
D	Cosmos 1246 rocket	1981-15B	1981 Feb 18.38 3 days 1981 Feb 21	Cylinder 2500?	7.5 long 2.6 dia	1981 Feb 18.6	64.91	89.09	6608	196	264	0.005	52

	Name		Launch date, lifetime and descent date	Shape and weight (kg)	Size (m)	Date of orbital determination	Orbital inclination (deg)	Nodal period (min)	Semi major axis (km)	Perigee height (km)	Apogee height (km)	Orbital eccentricity	Argument of perigee (deg)
T	Cosmos 1247	1981-16A	1981 Feb 19.48 100 years?	Windmill + 6 vanes? 1250?	4.2 long? 1.6 dia	1981 Feb 20.0	62.93	707.33	26298	608	39232	0.734	316
D	Cosmos 1247 launcher rocket	1981-16B	1981 Feb 19.48 22 days 1981 Mar 13	Cylinder 2500?	7.5 long 2.6 dia	1981 Feb 20.4	62.85	91.94	6749	146	595	0.033	120
D	Cosmos 1247 launcher	1981-16C	1981 Feb 19.48 6 days 1981 Feb 25	Irregular	-	1981 Feb 20.4	62.84	92.53	6777	207	591	0.028	124
	Cosmos 1247 rocket	1981-16E	1981 Feb 19.48 100 years?	Cylinder 440	2.0 long 2.0 dia	1981 Feb 27.3	62.89	703.58	26205	616	39037	0.733	316
1d	Fragment	1981-16D,F-H											
T	Hinotori* (ASTRO 1) [Mu 3S]	1981-17A	1981 Feb 21.40 30 years	Polyhedral cylinder? 185?	0.8 long? 1.0 dia?	1981 Feb 22.3	31.34	96.64	6983	571	638	0.005	111
	Hinotori rocket	1981-17B	1981 Feb 21.40 30 years	Sphere- cone? 230?	2.33 long? 1.14 dia?	1981 Feb 22.3	31.33	96.63	6982	571	637	0.005	110
D	Fragment	1981-17C											
T	Comstar 1D	1981-18A	1981 Feb 21.97 > million years	Cylinder 1520 full 792 empty	2.82 long 2.36 dia	1981 Feb 22.0 1981 Mar 4.0	20.67 0.21	652.43 1425.91	24918 41968	554 35388	36526 35791	0.722 0.005	179 212
	Comstar 1D rocket	1981-18B	1981 Feb 21.97 6000 years	Cylinder 1815	8.6 long 3.0 dia	1981 Mar 27.5	20.54	650.5	24869	636	36345	0.718	200

* Hinotori means Firebird (solar studies)

	Name		Launch date, lifetime and descent date	Shape and weight (kg)	Size (m)	Date of orbital determination	Orbital inclination (deg)	Nodal period (min)	Semi major axis (km)	Perigee height (km)	Apogee height (km)	Orbital eccentricity	Argument of perigee (deg)
D	[Titan 3B-Agena D]?	1981-19A	1981 Feb 28.80 112 days 1981 Jun 20	Cylinder 3000?	8 long? 1.5 dia	1981 Mar 1.8	96.38	89.25	6615	138	336	0.015	158
D R	Cosmos 1248	1981-20A	1981 Mar 5.63 29.5 days 1981 Apr 4.1	Sphere-cylinder 6700?	7 long? 2.4 dia	1981 Mar 6.4	67.14	89.68	6637	173	345	0.013	71
D	Cosmos 1248 rocket	1981-20B	1981 Mar 5.63 3 days 1981 Mar 8	Cylinder 2500?	7.5 long 2.6 dia	1981 Mar 6.1	67.13	89.52	6629	166	336	0.013	74
	Cosmos 1249*	1981-21A	1981 Mar 5.76 600 years	Cone-cylinder	6 long? 2.5 dia?	1981 Mar 6.4 1981 Jul 1.0	64.99 65.00	89.66 103.89	6637 7320	252 898	265 985	0.001 0.006	271 255
D	Cosmos 1249 platform	1981-21D	1981 Mar 5.76 136 days 1981 Jul 19	Irregular	-	1981 Jun 21.1	64.99	89.53	6630	243	261	0.001	260
D	Cosmos 1249 rocket	1981-21B	1981 Mar 5.76 109 days 1981 Jun 22	Cylinder 1500?	8 long? 2.5 dia?	1981 Jun 20.4	64.99	89.31	6620	233	251	0.001	244
	Fragment**	1981-21C											
T	Cosmos 1250	1981-22A	1981 Mar 6.48 7000 years	Spheroid 40?	1.0 long? 0.8 dia?	1981 Mar 6.9	74.03	114.51	7811	1399	1467	0.004	92
T	Cosmos 1251	1981-22B	1981 Mar 6.48 7000 years	Spheroid 40?	1.0 long? 0.8 dia?	1981 Mar 6.9	74.02	114.67	7818	1406	1474	0.004	108

* Cosmos 1249 jettisoned rocket and platform before manoeuvring into 104 min orbit about 1981 Jun 19.

1981-22 continued on page 639

** Probably the uranium fuel core ejected from nuclear reactor of 1981-21A when in 104 min orbit.

	Name		Launch date, lifetime and descent date	Shape and weight (kg)	Size (m)	Date of orbital determination	Orbital inclination (deg)	Nodal period (min)	Semi major axis (km)	Perigee height (km)	Apogee height (km)	Orbital eccentricity	Argument of perigee (deg)
T	Cosmos 1252	1981-22C	1981 Mar 6.48 8000 years	Spheroid 40?	1.0 long? 0.8 dia?	1981 Mar 6.9	74.02	114.82	7825	1420	1474	0.003	107
T	Cosmos 1253	1981-22D	1981 Mar 6.48 9000 years	Spheroid 40?	1.0 long? 0.8 dia?	1981 Mar 12.0	74.03	115.18	7842	1442	1485	0.003	46
T	Cosmos 1254	1981-22E	1981 Mar 6.48 9000 years	Spheroid 40?	1.0 long? 0.8 dia?	1981 Mar 11.7	74.03	114.98	7832	1434	1474	0.003	105
T	Cosmos 1255	1981-22F	1981 Mar 6.48 9000 years	Spheroid 40?	1.0 long? 0.8 dia?	1981 Mar 12.0	74.03	115.13	7839	1448	1474	0.002	109
T	Cosmos 1256	1981-22G	1981 Mar 6.48 10000 years	Spheroid 40?	1.0 long? 0.8 dia?	1981 Mar 12.0	74.04	115.30	7847	1459	1479	0.001	156
T	Cosmos 1257	1981-22H	1981 Mar 6.48 10000 years	Spheroid 40?	1.0 long? 0.8 dia?	1981 Mar 11.2	74.03	115.46	7854	1470	1482	0.001	207
	Cosmos 1250 rocket	1981-22J	1981 Mar 6.48 20000 years	Cylinder 2200?	7.4 long 2.4 dia	1981 Mar 11.9	74.03	117.71	7956	1468	1687	0.014	260
D 2M R	Soyuz - Transport 4*	1981-23A	1981 Mar 12.79 74.78 days 1981 May 26.57	Sphere- cylinder + 2 wings 6850	7.5 long 2.3 to 2.72 dia	1981 Mar 13.1	51.61	90.06	6658	245	315	0.005	48
D	Soyuz T-4 rocket	1981-23B	1981 Mar 12.79 2 days 1981 Mar 14	Cylinder 2500?	7.5 long 2.6 dia	1981 Mar 13.1	51.62	88.49	6581	189	217	0.002	111

* Soyuz T-4 docked with Progress 12 - Salyut 6 (1st airlock) on 1981 Mar 14.86.

1981-23 continued on page 640

	Name		Launch date, lifetime and descent date	Shape and weight (kg)	Size (m)	Date of orbital determination	Orbital inclination (deg)	Nodal period (min)	Semi major axis (km)	Perigee height (km)	Apogee height (km)	Orbital eccentricity	Argument of perigee (deg)
D	Soyuz T-4 orbital module	1981-23C*	1981 Mar 12.79 181 days 1981 Sep 9	Spheroid 1260?	2.3 dia	1981 Jun 2.2	51.64	91.52	6730	338	365	0.002	71
D	Cosmos 1258**	1981-24A	1981 Mar 14.71 <1 day 1981 Mar 14	Cylinder?	4 long? 2 dia?	1981 Mar 14.8	65.83	98.00	7043	303	1026	0.051	246
D	Cosmos 1258 rocket	1981-24B	1981 Mar 14.71 1 day 1981 Mar 15	Cylinder 1500?	8 long? 2.5 dia?	1981 Mar 15.1	65.09	89.11	6609	136	326	0.014	40
T	IMEWS 11 [Titan 3C]	1981-25A	1981 Mar 16.88 > million years	-	-	1981 Mar 17.6	1.99	1421.15	41873	35463	35527	0.001	69
D	Titan 3C second stage	1981-25B	1981 Mar 16.88 2 days 1981 Mar 18	Cylinder 1900	6 long 3.0 dia	1981 Mar 17.8	28.60	88.31	6576	136	260	0.009	129
	Titan 3C transtage	1981-25C	1981 Mar 16.88 > million years	Cylinder 1500?	6 long? 3.0 dia	1981 Apr 13.2	1.91	1420.9	41867	35454	35523	0.001	62
D R	Cosmos 1259	1981-26A	1981 Mar 17.36 13.91 days 1981 Mar 31.27	Sphere-cylinder 6300?	6.5 long? 2.4 dia	1981 Mar 17.4	70.35	90.43	6674	208	383	0.013	66
D	Cosmos 1259 rocket	1981-26B	1981 Mar 17.36 8 days 1981 Mar 25	Cylinder 2500?	7.5 long 2.6 dia	1981 Mar 18.0	70.36	90.25	6665	205	368	0.012	63
D	Cosmos 1259 engine	1981-26F	1981 Mar 17.36 166 days 1981 Aug 30	Cone 600? full	1.5 long? 2 dia?	1981 Mar 31.5	70.36	92.18	6759	347	415	0.005	189
D	Fragments	1981-26C-E,G											

* Designated 1977-97DH in the USA. Part of 1981-23A until about 1981 May 26

** Cosmos 1258 probably passed within 8 km of Cosmos 1241, then de-orbited.

	Name		Launch date, lifetime and descent date	Shape and weight (kg)	Size (m)	Date of orbital determination	Orbital inclination (deg)	Nodal period (min)	Semi major axis (km)	Perigee height (km)	Apogee height (km)	Orbital eccentricity	Argument of perigee (deg)
T	Statsionar-Raduga 8	1981-27A	1981 Mar 18.20 > million years	-	-	1981 Mar 18.3	0.75	1475.17	42929	36551	36551	0	180
D	Raduga 8 launcher	1981-27B	1981 Mar 18.20 < 1 day 1981 Mar 18	Irregular	-	1981 Mar 18.2	51.63	88.36	6575	189	204	0.001	143
D	Raduga 8 launcher rocket	1981-27C	1981 Mar 18.20 2 days 1981 Mar 20	Cylinder 4000?	12 long? 4 dia	1981 Mar 19.0	51.63	88.10	6562	181	186	0.0003	113
D	Raduga 8 rocket	1981-27D	1981 Mar 18.20 196 days? 1981 Sep 30?	Cylinder 1900?	3.9 long? 3.9 dia	1981 Apr 21.5	47.30	647.6	24794	245	36586	0.733	9
	Raduga 8 apogee motor	1981-27F	1981 Mar 18.20 >million years	Cylinder 440?	2.0 long? 2.0 dia?		Synchronous orbit similar to 1981-27A						
D	Fragment	1981-27E											
D	Cosmos 1260*	1981-28A	1981 Mar 20.99? 428 days 1982 May 22	Cylinder?	-	1981 Mar 21.4 1982 Jan 14.3	65.03 65.03	93.32 96.84	6813 6988	425 463	444 757	0.001 0.021	271 345
D	Cosmos 1260 rocket	1981-28B	1981 Mar 20.99? 1 day 1981 Mar 21	Cylinder 1500?	8 long? 2.5 dia?	1981 Mar 21.2	65.00	89.31	6619	111	371	0.020	54
25d	Fragments	1981-28C-BH											
D 2M R	Soyuz 39†	1981-29A	1981 Mar 22.62 7.87 days 1981 Mar 30.49	Sphere-cylinder 6800?	7.5 long 2.3 to 2.72 dia	1981 Mar 22.8	51.68	89.00	6606	195	261	0.005	77
D	Soyuz 39 rocket	1981-29B	1981 Mar 22.62 3 days 1981 Mar 25	Cylinder 2500?	7.5 long 2.6 dia	1981 Mar 23.9	51.64	88.44	6578	177	223	0.004	74

† Soyuz 39 docked with Soyuz T-4 - Salyut 6 (2nd airlock) on 1981 Mar 23.69, with one Russian and first Mongolian cosmonaut. (Progress 12 had vacated 2nd airlock on 1981 Mar 19.76). Undocked 1981 Mar 30.

* Cosmos 1260 disintegrated during 1982 May.

	Name		Launch date, lifetime and descent date	Shape and weight (kg)	Size (m)	Date of orbital determination	Orbital inclination (deg)	Nodal period (min)	Semi major axis (km)	Perigee height (km)	Apogee height (km)	Orbital eccentricity	Argument of perigee (deg)
T	Molniya 3Q	1981-30A	1981 Mar 24.15 12¼ years	Windmill + 6 vanes 1500?	4.2 long? 1.6 dia	1981 Mar 24.2	62.73	736.00	27004	609	40643	0.741	288
D	Molniya 3Q launcher rocket	1981-30B	1981 Mar 24.15 20 days 1981 Apr 13	Cylinder 2500?	7.5 long 2.6 dia	1981 Mar 24.3	62.80	92.34	6768	214	566	0.026	121
D	Molniya 3Q launcher	1981-30C	1981 Mar 24.15 19 days 1981 Apr 12	Irregular	-	1981 Mar 24.2	62.82	92.72	6787	196	622	0.031	118
	Molniya 3Q rocket	1981-30D	1981 Mar 24.15 12¼ years	Cylinder 440	2.0 long 2.0 dia	1981 Apr 11.5	62.78	732.75	26925	671	40423	0.738	288
T	Cosmos 1261	1981-31A	1981 Mar 31.41 100 years?	Windmill + 6 vanes 1250?	4.2 long? 1.6 dia	1981 Mar 31.5	62.95	710.47	26376	589	39406	0.736	316
D	Cosmos 1261 launcher	1981-31B	1981 Mar 31.41 17 days 1981 Apr 17	Irregular	-	1981 Mar 31.6	62.91	92.60	6781	192	613	0.031	119
D	Cosmos 1261 launcher rocket	1981-31C	1981 Mar 31.41 15 days 1981 Apr 15	Cylinder 2500?	7.5 long 2.6 dia	1981 Mar 31.6	62.81	92.17	6760	213	550	0.025	121
	Cosmos 1261 rocket	1981-31D	1981 Mar 31.41 100 years?	Cylinder 440	2.0 long 2.0 dia	1981 Apr 25.5	62.91	707.5	26300	615	39228	0.734	316
	Fragments	1981-31E-G											

	Name		Launch date, lifetime and descent date	Shape and weight (kg)	Size (m)	Date of orbital determination	Orbital inclination (deg)	Nodal period (min)	Semi major axis (km)	Perigee height (km)	Apogee height (km)	Orbital eccentricity	Argument of perigee (deg)
D R	Cosmos 1262	1981-32A	1981 Apr 7.45 13.8 days 1981 Apr 21.2	Sphere-cylinder 6300?	6.5 long? 2.4 dia	1981 Apr 8.1	72.87	90.42	6673	197	393	0.015	84
D	Cosmos 1262 rocket	1981-32B	1981 Apr 7.45 6 days 1981 Apr 13	Cylinder 2500?	7.5 long 2.6 dia	1981 Apr 7.8	72.88	89.63	6634	181	330	0.011	98
D	Cosmos 1262 engine	1981-32E	1981 Apr 7.45 21 days 1981 Apr 28	Cone 600? full	1.5 long? 2 dia?	1981 Apr 21.6	72.88	89.54	6629	204	298	0.007	93
D	Fragments	1981-32C,D,F											
T	Cosmos 1263	1981-33A	1981 Apr 9.50 30 years	Octagonal ellipsoid 550?	1.8 long? 1.5 dia?	1981 Apr 10.1	82.98	109.09	7562	397	1970	0.104	129
	Cosmos 1263 rocket	1981-33B	1981 Apr 9.50 30 years	Cylinder 2200?	7.4 long 2.4 dia	1981 Apr 9.9	82.98	108.95	7555	392	1962	0.104	129
D 2M R	STS 1* (Columbia F1)†	1981-34A	1981 Apr 12.50 2.27 days 1981 Apr 14.77	Cylinder + delta wing 68800 empty	37.1 long 17.4 high 23.8 span	1981 Apr 12.7	40.35	89.23	6619	237	245	0.001	263
D R	Cosmos 1264	1981-35A	1981 Apr 15.44 13.85 days 1981 Apr 29.29	Sphere-cylinder 6300?	6.5 long? 2.4 dia	1981 Apr 15.9	70.37	90.48	6676	208	388	0.014	66
D	Cosmos 1264 rocket	1981-35B	1981 Apr 15.44 8 days 1981 Apr 23	Cylinder 2500?	7.5 long 2.6 dia	1981 Apr 16.1	70.38	90.39	6672	207	380	0.013	68
D	Cosmos 1264 engine	1981-35C	1981 Apr 15.44 192 days 1981 Oct 24	Cone 600? full	1.5 long? 2 dia	1981 Apr 30.9	70.37	92.29	6765	360	413	0.004	184
D	Fragments	1981-35D-F											

*Space Transportation System (Space Shuttle), launched from Cape Canaveral, landed at Edwards AFB, California.

†First flight (F1) of the Columbia spacecraft

	Name		Launch date, lifetime and descent date	Shape and weight (kg)	Size (m)	Date of orbital determination	Orbital inclination (deg)	Nodal period (min)	Semi major axis (km)	Perigee height (km)	Apogee height (km)	Orbital eccentricity	Argument of perigee (deg)
D R	Cosmos 1265	1981-36A	1981 Apr 16.48 11.81 days 1981 Apr 28.29	Sphere-cylinder 6300?	6.5 long? 2.4 dia	1981 Apr 17.2	72.85	89.65	6635	226	288	0.005	69
D	Cosmos 1265 rocket	1981-36B	1981 Apr 16.48 3 days 1981 Apr 19	Cylinder 2500?	7.5 long 2.6 dia	1981 Apr 16.5	72.84	89.32	6619	200	281	0.006	84
D	Cosmos 1265 engine	1981-36E	1981 Apr 16.48 5 months 1981 Sep?	Cone 600? full	1.5 long? 2 dia?	1981 Apr 28.9	72.74	92.56	6780	253	551	0.022	7
D	Fragments	1981-36C,D,F											
	Cosmos 1266*	1981-37A	1981 Apr 21.16 600 years	Cone-cylinder	6 long? 2.5 dia?	1981 Apr 21.4 1981 May 1.2	64.97 64.76	89.66 103.64	6637 7306	249 891	268 965	0.001 0.005	251 281
D	Cosmos 1266 platform	1981-37B	1981 Apr 21.16 29 days 1981 May 20	Irregular	-	1981 Apr 29.0	64.97	89.56	6632	247	260	0.001	267
D	Cosmos 1266 rocket	1981-37C	1981 Apr 21.16 10 days 1981 May 1	Cylinder 1500?	8 long? 2.5 dia?	1981 Apr 30.5	64.96	88.93	6600	212	232	0.002	282
	Fragment**	1981-37D											
T	SDS 7? [Titan 3B Agena D]?	1981-38A	1981 Apr 24 100 years?	Cylinder?	-		Orbit similar to 1980-100A?						
	SDS 7 rocket?	1981-38B	1981 Apr 24 100 years?	Cylinder 700?	6 long 1.5 dia		Orbit similar to 1980-100A?						

*Cosmos 1266 jettisoned rocket and platform, then manoeuvred into 104 min orbit about 1981 Apr 29.0.

** Probably the uranium fuel core ejected from nuclear reactor of 1981-37A when in 104 min orbit.

	Name		Launch date, lifetime and descent date	Shape and weight (kg)	Size (m)	Date of orbital determination	Orbital inclination (deg)	Nodal period (min)	Semi major axis (km)	Perigee height (km)	Apogee height (km)	Orbital eccentricity	Argument of perigee (deg)
D	Cosmos 1267*	1981-39A	1981 Apr 25.15 460 days 1982 Jul 29	Cylinder + 3 vanes? 15100	11 long? 4.15 to 2.0 dia?	1981 Apr 25.6	51.58	88.94	6604	192	259	0.005	87
D	Cosmos 1267 rocket	1981-39B	1981 Apr 25.15 3 days 1981 Apr 28	Cylinder 4000?	12 long? 4 dia?	1981 Apr 25.6	51.60	88.64	6589	188	233	0.003	79
D	Fragments	1981-39C,D											
D R	Cosmos 1268	1981-40A	1981 Apr 28.38 13.88 days 1981 May 12.26	Sphere- cylinder 6300?	6.5 long? 2.4 dia	1981 Apr 29.0	70.38	90.30	6667	210	368	0.012	64
D	Cosmos 1268 rocket	1981-40B	1981 Apr 28.38 8 days 1981 May 6	Cylinder 2500?	7.5 long 2.6 dia	1981 Apr 30.4	70.38	89.96	6650	203	341	0.010	53
D	Cosmos 1268 engine	1981-40C	1981 Apr 28.38 22 days 1981 May 20	Cone 600? full	1.5 long? 2 dia?	1981 May 12.4	70.38	89.42	6624	220	271	0.004	46
D	Fragments	1981-40D-G											
T	Cosmos 1269	1981-41A	1981 May 7.56 120 years	Cylinder + paddles? 750?	2 long? 1 dia?	1981 May 7.9	74.06	100.94	7181	796	810	0.001	40
	Cosmos 1269 rocket	1981-41B	1981 May 7.56 100 years	Cylinder 2200?	7.4 long 2.4 dia	1981 May 8.7	74.06	100.85	7176	787	810	0.002	40

* Cosmos 1267 docked with Salyut 6 on 1981 Jun 19.33

	Name		Launch date, lifetime and descent date	Shape and weight (kg)	Size (m)	Date of orbital determination	Orbital inclination (deg)	Nodal period (min)	Semi major axis (km)	Perigee height (km)	Apogee height (km)	Orbital eccentricity	Argument of perigee (deg)
D 2M R	Soyuz 40*	1981-42A	1981 May 14.76 7.86 days 1981 May 22.62	Sphere-cylinder 6800?	7.5 long 2.3 to 2.72 dia	1981 May 14.9	51.62	89.05	6608	191	269	0.006	83
D	Soyuz 40 rocket	1981-42B	1981 May 14.76 2 days 1981 May 16	Cylinder 2500?	7.5 long 2.6 dia	1981 May 14.9	51.63	88.86	6599	192	249	0.004	73
T	Meteor 2-07	1981-43A	1981 May 14.91 500 years	Cylinder + 2 vanes 2750?	5 long? 1.5 dia?	1981 May 15.1	81.27	102.46	7252	855	893	0.003	251
	Meteor 2-07 rocket	1981-43B	1981 May 14.91 400 years	Cylinder 1440	3.8 long 2.6 dia	1981 May 15.8	81.28	102.57	7257	836	922	0.006	202
T	Nova 1** [Scout]	1981-44A	1981 May 15.26 3000 years	Cylinder + 4 vanes	-	1981 May 16.2 1981 May 27.2 1981 Jun 3.0	90.16 89.99 89.96	97.67 107.26 108.99	7024 7476 7557	354 997 1170	937 1199 1187	0.041 0.013 0.001	168 152 247
D	Nova 1 rocket	1981-44B	1981 May 15.26 337 days 1982 Apr 17	Cylinder 24	1.50 long 0.46 dia	1981 May 18.8	90.16	97.62	7022	353	934	0.041	159
D	Fragment	1981-44C											
D R	Cosmos 1270	1981-45A	1981 May 18.50 30 days 1981 Jun 17	Sphere-cylinder 6700?	7 long? 2.4 dia	1981 May 18.7	64.86	89.71	6639	173	349	0.013	64
D	Cosmos 1270 rocket	1981-45B	1981 May 18.50 3 days 1981 May 21	Cylinder 2500?	7.5 long 2.6 dia	1981 May 18.7	64.89	89.65	6636	177	339	0.012	68
D	Fragment	1981-45C											

* Soyuz 40 docked with Soyuz T4 - Salyut 6 (2nd airlock) on 1981 May 15.83 with one Russian and first Romanian cosmonaut. Undocked 1981 May 22.48

** Improved US Navy navigation satellite.

	Name		Launch date, lifetime and descent date	Shape and weight (kg)	Size (m)	Date of orbital determination	Orbital inclination (deg)	Nodal period (min)	Semi major axis (km)	Perigee height (km)	Apogee height (km)	Orbital eccentricity	Argument of perigee (deg)
T	Cosmos 1271	1981-46A	1981 May 19.16 60 years	Cylinder + 2 vanes? 2500?	5 long? 1.5 dia?	1981 May 22.2	81.22	97.52	7017	628	650	0.002	8
	Cosmos 1271 rocket	1981-46B	1981 May 19.16 60 years	Cylinder 1440	3.8 long 2.6 dia	1981 May 24.2	81.24	97.70	7026	603	693	0.006	150
D R	Cosmos 1272	1981-47A	1981 May 21.39 13.9 days 1981 Jun 4.3	Sphere- cylinder 6300?	6.5 long? 2.4 dia	1981 May 21.6 1981 May 25.3	70.38 70.39	90.42 92.35	6673 6768	209 362	380 417	0.013 0.004	66 204
D	Cosmos 1272 rocket	1981-47B	1981 May 21.39 10 days 1981 May 31	Cylinder 2500?	7.5 long 2.6 dia	1981 May 22.1	70.38	90.24	6664	207	364	0.012	66
D	Cosmos 1272 engine	1981-47D	1981 May 21.39 187 days 1981 Nov 24	Cone 600? full	1.5 long? 2 dia?	1981 Jun 7.9	70.39	92.30	6765	360	414	0.004	182
D	Fragments	1981-47C,E-H											
D R	Cosmos 1273	1981-48A	1981 May 22.30 12.8 days 1981 Jun 4.1	Sphere- cylinder 5900?	5.9 long 2.4 dia	1981 May 22.4	82.30	89.27	6615	210	264	0.004	85
D	Cosmos 1273 rocket	1981-48B	1981 May 22.30 3 days 1981 May 25	Cylinder 2500?	7.5 long 2.6 dia	1981 May 23.3	82.33	88.86	6595	204	229	0.002	58
D	Capsule	1981-48F	1981 May 22.30 15 days 1981 June 6	Ellipsoid 200?	0.9 long 1.9 dia	1981 Jun 4.7	82.35	90.73	6688	213	407	0.014	335
D	Fragments	1981-48C-E											

	Name		Launch date, lifetime and descent date	Shape and weight (kg)	Size (m)	Date of orbital determination	Orbital inclination (deg)	Nodal period (min)	Semi major axis (km)	Perigee height (km)	Apogee height (km)	Orbital eccentricity	Argument of perigee (deg)
T	GOES 5	1981-49A	1981 May 22.94 > million years	Cylinder + antennae 836 full 397 empty	3.12 long 2.15 dia	1981 May 22.9 1981 May 27.0 1981 Jun 4.2	26.50 0.45 0.51	918.91 1743.7 1430.1	31309 47989 42048	162 33763 35453	49700 49459 35887	0.791 0.164 0.005	180 290 306
D	GOES 5 second stage	1981-49B	1981 May 22.94 109 days 1981 Sep 8	Cylinder 350? empty	7 long 1.4 dia	1981 May 22.9 1981 May 23.0	28.73 28.13	88.10 113.57	6559 7769	157 163	204 2618	0.004 0.158	118 176
D	GOES 5 third stage	1981-49C	1981 May 22.94 9 months? 1982 Feb?	Sphere -cone 66	1.8 long 0.94 dia	1981 May 23.0	26.50	917.79	31284	167	49645	0.791	180
	GOES 5 apogee motor	1981-49D	1981 May 22.94 > million years	- 439 full	-	1981 May 24.5	0.45	1751.5	48133	33841	49668	0.164	290
T	Intelsat 5 F-1 [Atlas Centaur]	1981-50A	1981 May 23.95 > million years	Box + 2 vanes + dishes 1928 full	15.7 span	1981 May 24.0 1981 May 24.4	24.43 0.47	632.79 1404.9	24416 41553	167 33615	35908 36735	0.732 0.038	179 268
	Intelsat 5 F-1 rocket	1981-50B	1981 May 23.95 3 years	Cylinder 1815	9.1 long 3.0 dia	1981 May 23.9 1981 May 24.0	28.31 24.42	89.56 611.37	6631 23861	148 167	357 34799	0.016 0.726	146 179
D	Rohini 2* [SLV-3]	1981-51A	1981 May 31.21 8 days 1981 June 8	Spheroid? 38	0.6 dia?	1981 May 31.4	46.27	90.48	6680	186	418	0.017	81
D	Rohini 2 rocket	1981-51B	1981 May 31.21 6 days 1981 June 6	Cylinder?	-	1981 May 31.4	46.26	90.43	6678	186	414	0.017	81

* Second Indian launch from Sriharikota, Bengal Bay.

	Name		Launch date, lifetime and descent date	Shape and weight (kg)	Size (m)	Date of orbital determination	Orbital inclination (deg)	Nodal period (min)	Semi major axis (km)	Perigee height (km)	Apogee height (km)	Orbital eccentricity	Argument of perigee (deg)
D R	Cosmos 1274	1981-52A	1981 Jun 3.59 29.5 days 1981 Jul 3.1	Sphere-cylinder 6700?	7 long ? 2.4 dia	1981 Jun 4.3	67.15	89.77	6642	172	355	0.014	75
D	Cosmos 1274 rocket	1981-52B	1981 Jun 3.59 4 days 1981 Jun 7	Cylinder 2500?	7.5 long 2.6 dia	1981 Jun 3.9	67.18	89.56	6631	178	328	0.011	76
D	Fragment*	1981-52C											
T	Cosmos 1275	1981-53A	1981 Jun 4.65 1200 years	Cylinder 700?	1.3 long? 1.9 dia?	1981 Jun 5.6	82.96	104.91	7367	964	1014	0.003	279
	Cosmos 1275 rocket	1981-53B	1981 Jun 4.65 600 years	Cylinder 2200?	7.4 long 2.4 dia	1981 Jun 5.3	82.97	104.81	7362	963	1005	0.003	270
3d	Fragments	1981-53C-GQ											
T	Molniya 3R	1981-54A	1981 Jun 9.15 17 years**	Windmill + 6 vanes 1500?	4.2 long? 1.6 dia	1981 Jun 10.2	62.81	736.56	27017	434	40844	0.748	280
D	Molniya 3R launcher rocket	1981-54B	1981 Jun 9.15 18 days 1981 Jun 27	Cylinder 2500?	7.5 long 2.6 dia	1981 Jun 9.5	62.84	90.99	6702	195	453	0.019	120
D	Molniya 3R launcher	1981-54C	1981 Jun 9.15 14 days 1981 Jun 23	Irregular	-	1981 Jun 9.3	62.84	90.93	6699	213	429	0.016	130
	Molniya 3R rocket	1981-54E	1981 Jun 9.15 17 years**	Cylinder 440	2.0 long 2.0 dia	1981 Jun 15.3	62.84	733.76	26949	444	40698	0.747	280
D	Fragment	1981-54D											

* Possibly a jettisoned engine?

** Decay possible in 1994

	Name		Launch date, lifetime and descent date	Shape and weight (kg)	Size (m)	Date of orbital determination	Orbital inclination (deg)	Nodal period (min)	Semi major axis (km)	Perigee height (km)	Apogee height (km)	Orbital eccentricity	Argument of perigee (deg)
D R	Cosmos 1276	1981-55A	1981 Jun 16.29 12.86 days 1981 Jun 29.15	Sphere-cylinder 5900?	5.9 long 2.4 dia	1981 Jun 16.6	82.37	89.07	6606	216	239	0.002	87
D	Cosmos 1276 rocket	1981-55B	1981 Jun 16.29 3 days 1981 Jun 19	Cylinder 2500?	7.5 long 2.6 dia	1981 Jun 16.5	82.36	88.96	6600	216	228	0.001	38
D	Capsule	1981-55C	1981 Jun 16.29 20 days 1981 July 6	Ellipsoid 200?	0.9 long 1.9 dia	1981 Jun 29.3	82.37	88.90	6597	208	230	0.002	84
D	Fragments	1981-55D-F											
D R	Cosmos 1277	1981-56A	1981 Jun 17.40 13.9 days 1981 Jul 1.3	Sphere-cylinder 6300?	6.5 long? 2.4 dia	1981 Jun 17.5	70.41	90.39	6672	208	379	0.013	67
D	Cosmos 1277 rocket	1981-56B	1981 Jun 17.40 11 days 1981 Jun 28	Cylinder 2500?	7.5 long 2.6 dia	1981 Jun 17.5	70.39	90.31	6668	208	371	0.012	65
D	Cosmos 1277 engine	1981-56J	1981 Jun 17.40 178 days 1981 Dec 12	Cone 600? full	1.5 long? 2 dia?	1981 Jul 1.5	70.39	92.31	6766	358	417	0.004	187
D	Fragments	1981-56C-H,K											
T	Meteosat 2* [Ariane 1-03]	1981-57A	1981 Jun 19.52 > million years	Cylinder 697 full 295 empty	3.20 long 2.10 dia	1981 Jun 19.8 1981 Jul 27.9	10.48 1.01	631.68 1442.09	24386 42288	210 35847	35806 35973	0.730 0.002	180 58
T	Apple**	1981-57B	1981 Jun 19.52 > million years	Cylinder + 2 panels 670 full	1.6 long? 1.7 dia?	1981 Jun 20.2 1981 Dec 31	10.50 1.1	632.80 1436.7	24415 42175	201 35582	35873 36012	0.731 0.005	181 -

* European weather satellite. Entered synchronous orbit on 1981 Jun 20.18 and drift stopped at 0° longitude on 1981 Jul 21. (Ariane 1-02 failed in May 1980)

1981-57 continued on page 651

** Indian communications test satellite (Ariane Passenger Payload Experiment); one panel did not deploy. Entered synchronous orbit on 1981 Jun 21.95.

	Name		Launch date, lifetime and descent date	Shape and weight (kg)	Size (m)	Date of orbital determination	Orbital inclina-tion (deg)	Nodal period (min)	Semi major axis (km)	Perigee height (km)	Apogee height (km)	Orbital eccen-tricity	Argument of perigee (deg)
	CAT 3*	1981-57C	1981 Jun 19.52 20 years	Cylinder 217	1.1 long? 1.6 dia?	1981 Jun 21.1	10.45	632.12	24398	202	35838	0.730	181
	Ariane third stage	1981-57D	1981 Jun 19.52 20 years	Cylinder 1634 empty	8.6 long 2.6 dia	1981 Jul 7.0	10.39	633.76	24439	222	35900	0.730	194
	MAGE-1**	1981-57F	1981 Jun 19.52 > million years	Cylinder + nozzle	1.12 long 0.77 dia		Entered synchronous orbit similar to 1981-57A on 1981 Jun 20						
	Fragment†	1981-57E											
T	Cosmos 1278	1981-58A	1981 June 19.82 100 years?	Windmill + 6 vanes 1250?	4.2 long? 1.6 dia	1981 Jun 21.9	62.84	727.43	26796	623	40213	0.739	318
D	Cosmos 1278 launcher rocket	1981-58B	1981 Jun 19.82 39 days 1981 Jul 28	Cylinder 2500?	7.5 long 2.6 dia	1981 Jun 20.6	62.84	92.60	6781	215	590	0.028	120
D	Cosmos 1278 launcher	1981-58C	1981 Jun 19.82 40 days 1981 Jul 29	Irregular	-	1981 Jun 20.2	62.82	92.80	6791	213	612	0.029	121
	Cosmos 1278 rocket	1981-58D	1981 Jun 19.82 100 years?	Cylinder 440	2.0 long 2.0 dia	1981 Jun 24.4	62.84	724.06	26711	622	40044	0.738	318
T	NOAA 7 [Atlas F]	1981-59A	1981 Jun 23.45 350 years	Box + panel 1405 full 723 empty	3.71 long 1.88 dia	1981 Jun 24.5	98.90	102.04	7232	845	863	0.001	293
	Fragments	1981-59B,C											

* Capsule Ariane Technologique.

** Moteur d'Apogée Geostationaire Européen

† Conical adaptor separating MAGE and APPLE

	Name		Launch date, lifetime and descent date	Shape and weight (kg)	Size (m)	Date of orbital determination	Orbital inclination (deg)	Nodal period (min)	Semi major axis (km)	Perigee height (km)	Apogee height (km)	Orbital eccentricity	Argument of perigee (deg)
T	Molniya 1BB	1981-60A	1981 Jun 24.81 11½ years	Windmill + 6 vanes 1000?	3.4 long 1.6 dia	1981 Jun 25.3	62.79	736.13	27007	617	40641	0.741	288
D	Molniya 1BB launcher rocket	1981-60B	1981 Jun 24.81 42 days 1981 Aug 5	Cylinder 2500?	7.5 long 2.6 dia	1981 Jun 24.9	62.81	92.90	6796	216	619	0.030	119
D	Molniya 1BB launcher	1981-60C	1981 Jun 24.81 28 days 1981 Jul 22	Irregular	-	1981 Jun 24.9	62.84	92.74	6788	193	626	0.032	115
	Molniya 1BB rocket	1981-60D	1981 Jun 24.81 11½ years	Cylinder 440	2.0 long 2.0 dia	1981 Jun 26.8	62.81	732.17	26910	618	40446	0.740	288
T	Statsionar- Ekran 7	1981-61A	1981 Jun 26.00 > million years	Cylinder + plate	-	1981 Jun 26.0	0.07	1426.46	41978	35599	35600	0	28
D	Ekran 7 launcher	1981-61B	1981 Jun 26.00 < 1 day 1981 Jun 26	Irregular	-	1981 Jun 26.4	51.62	87.96	6555	164	189	0.002	272
D	Ekran 7 launcher rocket	1981-61C	1981 Jun 26.00 2 days 1981 Jun 28	Cylinder 4000?	12 long? 4 dia	1981 Jun 26.3	51.62	88.24	6569	185	196	0.001	208
D	Ekran 7 rocket	1981-61D	1981 Jun 26.00 10 months 1982 Apr?	Cylinder 1900?	3.9 long? 3.9 dia	1981 Sep 17.3	47.06	625.79	24246	227	35509	0.728	23
	Ekran 7 apogee motor	1981-61F	1981 Jun 26.00 > million years	Cylinder 440?	2.0 long? 2.0 dia?	1981 Sep 27.0	0.25	1425.66	41966	35580	35596	0	231
D	Fragment	1981-61E											

	Name		Launch date, lifetime and descent date	Shape and weight (kg)	Size (m)	Date of orbital determination	Orbital inclination (deg)	Nodal period (min)	Semi major axis (km)	Perigee height (km)	Apogee height (km)	Orbital eccentricity	Argument of perigee (deg)
D R	Cosmos 1279	1981-62A	1981 Jul 1.40 13.90 days 1981 Jul 15.30	Sphere-cylinder 6300?	6.5 long? 2.4 dia	1981 Jul 1.8	70.39	90.27	6666	212	363	0.011	61
D	Cosmos 1279 rocket	1981-62B	1981 Jul 1.40 13 days 1981 Jul 14	Cylinder 2500?	7.5 long 2.6 dia	1981 Jul 1.8	70.39	90.25	6665	210	363	0.011	61
D	Cosmos 1279 engine	1981-62D	1981 Jul 1.40 172 days 1981 Dec 20	Cone 600? full	1.5 long? 2 dia?	1981 Jul 22.0	70.40	92.29	6765	357	416	0.004	176
D	Fragments	1981-62C,E-G											
D R	Cosmos 1280	1981-63A	1981 Jul 2.30 12.8 days 1981 Jul 15.1	Sphere-cylinder 6300?	6.5 long? 2.4 dia	1981 Jul 2.8	82.31	89.50	6627	211	286	0.006	96
D	Cosmos 1280 rocket	1981-63B	1981 Jul 2.30 6 days 1981 Jul 8	Cylinder 2500?	7.5 long 2.6 dia	1981 Jul 2.8	82.31	89.38	6621	212	273	0.005	90
D	Cosmos 1280 engine	1981-63D	1981 Jul 2.30 39 days 1981 Aug 10	Cone 600? full	1.5 long? 2 dia?	1981 Jul 16.3	82.30	89.60	6631	242	264	0.002	218
D	Fragments	1981-63C,E,F											
D R	Cosmos 1281	1981-64A	1981 Jul 7.52 13.78 days 1981 Jul 21.30	Sphere-cylinder 6300?	6.5 long? 2.4 dia	1981 Jul 7.6	72.84	90.44	6674	197	394	0.015	85
D	Cosmos 1281 rocket	1981-64B	1981 Jul 7.52 10 days 1981 Jul 17	Cylinder 2500?	7.5 long 2.6 dia	1981 Jul 8.0	72.84	90.31	6667	197	381	0.014	82

1981-64 continued on page 654

	Name		Launch date, lifetime and descent date	Shape and weight (kg)	Size (m)	Date of orbital determination	Orbital inclination (deg)	Nodal period (min)	Semi major axis (km)	Perigee height (km)	Apogee height (km)	Orbital eccentricity	Argument of perigee (deg)
D	Cosmos 1281 engine	1981-64E	1981 Jul 7.52 155 days 1981 Dec 9	Cone 600? full	1.5 long? 2 dia?	1981 Jul 24.0	72.84	92.28	6764	358	414	0.004	177
D	Fragments	1981-64C,D											
T	Meteor 31	1981-65A	1981 Jul 10.22 60 years	Cylinder + 2 vanes 2200?	5 long? 1.5 dia?	1981 Jul 10.9	97.94	97.56	7019	610	671	0.004	25
	Meteor 31 rocket	1981-65B	1981 Jul 10.22 60 years	Cylinder 1440	3.8 long 2.6 dia	1981 Jul 11.0	97.95	97.76	7029	634	667	0.002	277
D	Iskra 1*	1981-65C	1981 Jul 10.22 3 months? †	- 28	0.6 dia?	1981 Jul 11.1	97.98	97.76	7029	638	663	0.002	269
D R	Cosmos 1282	1981-66A	1981 Jul 15.55 29.5 days 1981 Aug 14.1	Sphere-cylinder 6700?	7 long? 2.4 dia	1981 Jul 15.8	64.92	89.59	6633	173	337	0.012	62
D	Cosmos 1282 rocket	1981-66B	1981 Jul 15.55 4 days 1981 Jul 19	Cylinder 2500?	7.5 long 2.6 dia	1981 Jul 15.7	64.91	89.43	6625	173	321	0.011	65
D R	Cosmos 1283	1981-67A	1981 Jul 17.34 13.83 days 1981 Jul 31.17	Sphere-cylinder 6300?	6.5 long? 2.4 dia	1981 Jul 18.2	82.34	88.84	6594	182	250	0.005	94
D	Cosmos 1283 rocket	1981-67B	1981 Jul 17.34 2 days 1981 Jul 19	Cylinder 2500?	7.5 long 2.6 dia	1981 Jul 17.7	82.33	88.67	6586	177	238	0.005	95

* Development satellite built by Moscow Aviation Institute.
† Possibly decayed on 7 October 1981

1981-67 continued on page 655

	Name		Launch date, lifetime and descent date	Shape and weight (kg)	Size (m)	Date of orbital determination	Orbital inclination (deg)	Nodal period (min)	Semi major axis (km)	Perigee height (km)	Apogee height (km)	Orbital eccentricity	Argument of perigee (deg)
D	Cosmos 1283 engine	1981-67D	1981 Jul 17.34 83 days 1981 Oct 8	Cone 600? full	1.5 long? 2 dia?	1981 Aug 2.6	82.33	91.45	6724	319	373	0.004	189
D	Fragments	1981-67C,E											
D R	Cosmos 1284	1981-68A	1981 Jul 29.56 13.80 days 1981 Aug 12.36	Sphere-cylinder 6300?	6.5 long? 2.4 dia	1981 Jul 30.3	82.33	88.76	6590	183	241	0.004	92
D	Cosmos 1284 rocket	1981-68B	1981 Jul 29.56 2 days 1981 Jul 31	Cylinder 2500?	7.5 long 2.6 dia	1981 Jul 29.9	82.33	88.59	6582	179	228	0.004	85
D	Cosmos 1284 engine	1981-68C	1981 Jul 29.56 79 days 1981 Oct 16	Cone 600? full	1.5 long? 2 dia?	1981 Aug 18.0	82.33	91.39	6721	318	368	0.004	182
D	Fragments	1981-68D-F											
T	Statsionar-Raduga 9	1981-69A	1981 Jul 30.90 > million years	-	-	1981 Jul 30.9	0.40	1476.8	42961	36582	36583	0	28
D	Raduga 9 launcher rocket	1981-69B	1981 Jul 30.90 2 days 1981 Aug 1	Cylinder 4000?	12 long? 4 dia	1981 Jul 31.2	51.62	88.18	6566	183	192	0.001	201
D	Raduga 9 launcher	1981-69C	1981 Jul 30.90 1 day 1981 Jul 31	Irregular	-	1981 Jul 31.1	51.61	88.26	6570	182	201	0.001	124
D	Raduga 9 rocket	1981-69D	1981 Jul 30.90 9 months 1982 Apr?	Cylinder 1900?	3.9 long? 3.9 dia	1981 Aug 15.3	47.31	648.88	24828	348	36552	0.729	4
	Raduga 9 apogee motor	1981-69F	1981 Jul 30.90 > million years	Cylinder 440?	2.0 long? 2.0 dia?	1981 Sep 20.1	0.36	1473.90	42906	36434	36622	0.002	293
D	Fragment	1981-69E											

	Name		Launch date, lifetime and descent date	Shape and weight (kg)	Size (m)	Date of orbital determination	Orbital inclination (deg)	Nodal period (min)	Semi major axis (km)	Perigee height (km)	Apogee height (km)	Orbital eccentricity	Argument of perigee (deg)
T	Dynamics Explorer 1	1981-70A	1981 Aug 3.41 5000 years	Polygonal cylinder 403	1.14 long 1.35 dia	1981 Aug 27.0	89.91	410.92	18305	559	23295	0.621	274
T	Dynamics Explorer 2	1981-70B	1981 Aug 3.41 18 months	Polygonal cylinder 415	1.14 long 1.35 dia	1981 Aug 22.0	89.99	97.66	7025	298	996	0.050	141
	Dynamics Explorer second stage	1981-70C	1981 Aug 3.41 5 years	Cylinder + annulus 350?	6.4 long 1.52 and 2.44 dia	1981 Aug 9.7	89.98	98.24	7054	382	969	0.042	179
	Dynamics Explorer third stage	1981-70E	1981 Aug 3.41 5000 years	Sphere-cone 66	1.32 long 0.94 dia	1981 Aug 25.5	89.90	412.98	18371	564	23422	0.622	275
D	Fragments	1981-70D,F-H											
T	Cosmos 1285	1981-71A	1981 Aug 4.01 100 years?	Windmill + 6 vanes 1250?	4.2 long? 1.6 dia	1981 Aug 7.1	62.96	727.59	26800	594	40250	0.740	316
D	Cosmos 1285 launcher	1981-71B	1981 Aug 4.01 22 days 1981 Aug 26	Irregular	-	1981 Aug 4.5	62.92	92.55	6779	190	612	0.031	118
D	Cosmos 1285 launcher rocket	1981-71C	1981 Aug 4.01 28 days 1981 Sep 1	Cylinder 2500?	7.5 long 2.6 dia	1981 Aug 4.5	62.83	92.31	6767	216	562	0.026	118
	Cosmos 1285 rocket	1981-71D	1981 Aug 4.01 100 years?	Cylinder 440	2.0 long 2.0 dia	1981 Aug 20.1	62.94	722.90	26682	602	40006	0.739	316
	Fragment	1981-71E											

	Name		Launch date, lifetime and descent date	Shape and weight (kg)	Size (m)	Date of orbital determination	Orbital inclination (deg)	Nodal period (min)	Semi major axis (km)	Perigee height (km)	Apogee height (km)	Orbital eccentricity	Argument of perigee (deg)
D	Cosmos 1286	1981-72A	1981 Aug 4.35 438 days 1982 Oct 16	Cylinder?	-	1981 Aug 4.8 1982 Jan 13.9	65.04 65.02	93.33 93.18	6817 6811	432 432	445 434	0.001 0	268 103
D	Cosmos 1286 rocket	1981-72B	1981 Aug 4.35 <0.7 day 1981 Aug 4	Cylinder 1500?	8 long? 2.5 dia?	1981 Aug 4.7	65.01	89.10	6609	103	358	0.019	61
D	Fragment	1981-72C											
T	Fleetsatcom 5	1981-73A	1981 Aug 6.34 >million years	Cylinder 1884 full 1005 empty	1.27 long 2.44 dia	1981 Aug 10.0	6.33	1558.08	44522	35102	41185	0.068	329
	Fleetsatcom 5 rocket	1981-73B	1981 Aug 6.34 1.7 years	Cylinder 1815	8.6 long 3.0 dia	1981 Aug 18.0	26.60	606.35	23741	160	34566	0.725	189
T	Cosmos 1287	1981-74A	1981 Aug 6.49 10000 years	Spheroid 40?	1.0 long? 0.8 dia?	1981 Aug 8.3	74.03	115.79	7869	1466	1515	0.003	260
T	Cosmos 1288	1981-74B	1981 Aug 6.49 10000 years	Spheroid 40?	1.0 long? 0.8 dia?	1981 Aug 8.0	74.03	115.58	7859	1468	1494	0.002	268
T	Cosmos 1289	1981-74C	1981 Aug 6.49 10000 years	Spheroid 40?	1.0 long? 0.8 dia?	1981 Aug 8.0	74.02	115.37	7850	1462	1481	0.001	278
T	Cosmos 1290	1981-74D	1981 Aug 6.49 10000 years	Spheroid 40?	1.0 long? 0.8 dia?	1981 Aug 8.2	74.03	115.18	7841	1460	1466	0	90

1981-74 continued on page 658

	Name		Launch date, lifetime and descent date	Shape and weight (kg)	Size (m)	Date of orbital determination	Orbital inclination (deg)	Nodal period (min)	Semi major axis (km)	Perigee height (km)	Apogee height (km)	Orbital eccentricity	Argument of perigee (deg)
T	Cosmos 1291	1981-74E	1981 Aug 6.49 10000 years	Spheroid 40?	1.0 long? 0.8 dia?	1981 Aug 8.1	74.03	115.18	7841	1460	1466	0	88
T	Cosmos 1292	1981-74F	1981 Aug 6.49 8000 years	Spheroid 40?	1.0 long? 0.8 dia?	1981 Aug 8.1	74.03	114.83	7825	1428	1466	0.002	95
T	Cosmos 1293	1981-74G	1981 Aug 6.49 7000 years	Spheroid 40?	1.0 long? 0.8 dia?	1981 Aug 8.3	74.03	114.65	7817	1411	1467	0.004	99
T	Cosmos 1294	1981-74H	1981 Aug 6.49 6000 years	Spheroid 40?	1.0 long? 0.8 dia?	1981 Aug 7.6	74.04	114.46	7809	1395	1466	0.005	89
	Cosmos 1287 rocket	1981-74J	1981 Aug 6.49 20000 years	Cylinder 2200?	7.4 long 2.4 dia	1981 Aug 24.1	74.04	117.52	7947	1469	1669	0.013	247
T L	Intercosmos 22 (Bulgaria 1300)	1981-75A	1981 Aug 7.57 500 years	Cylinder + 2 vanes 1500	17 span	1981 Aug 12.0	81.22	101.91	7226	800	895	0.007	267
	Intercosmos 22 rocket	1981-75B	1981 Aug 7.57 400 years	Cylinder 1440	3.8 long 2.6 dia	1981 Aug 8.1	81.23	101.97	7229	801	901	0.007	241
T	Himawari 2 (GMS 2) [Nu-2]	1981-76A	1981 Aug 10.84 > million years	Cylinder 670 full? 281 empty?	3.0 long? 2.1 dia?	1981 Aug 12.0 1981 Aug 26.0	28.99 0.2?	656.19 1435.98	25014 42162	181 35776	37090 35792	0.738 0.0002	180 212
D	Himawari 2 second stage	1981-76B	1981 Aug 10.84 14 days 1981 Aug 24	Cylinder 550?	5.4 long 1.4 dia	1981 Aug 11.3	30.33	91.43	6730	170	533	0.027	191

1981-76 continued on page 659

	Name		Launch date, lifetime and descent date	Shape and weight (kg)	Size (m)	Date of orbital determination	Orbital inclination (deg)	Nodal period (min)	Semi major axis (km)	Perigee height (km)	Apogee height (km)	Orbital eccentricity	Argument of perigee (deg)
	Himawari 2 rocket	1981-76C	1981 Aug 10.84 9 years	Cylinder 66?	1.74 long 1.65 dia	1981 Sep 13.2	28.7	648.78	24825	176	36718	0.736	199
T	Cosmos 1295	1981-77A	1981 Aug 12.24 1200 years	Cylinder 700?	1.3 long? 1.9 dia?	1981 Aug 16.7	82.92	104.79	7362	952	1015	0.004	284
	Cosmos 1295 rocket	1981-77B	1981 Aug 12.24 600 years	Cylinder 2200?	7.4 long 2.4 dia	1981 Aug 21.7	82.92	104.67	7356	950	1005	0.004	269
D R	Cosmos 1296	1981-78A	1981 Aug 13.68 30.5 days 1981 Sep 13.2	Sphere-cylinder 6700?	7 long? 2.4 dia	1981 Aug 14.0	67.14	89.77	6641	172	354	0.014	73
D	Cosmos 1296 rocket	1981-78B	1981 Aug 13.68 4 days 1981 Aug 17	Cylinder 2500?	7.5 long 2.6 dia	1981 Aug 14.0	67.13	89.63	6634	171	341	0.013	72
D R	Cosmos 1297	1981-79A	1981 Aug 18.40 11.8 days 1981 Aug 30.2	Sphere-cylinder 6300?	6.5 long? 2.4 dia	1981 Aug 18.5	72.86	90.15	6660	199	364	0.012	83
D	Cosmos 1297 rocket	1981-79B	1981 Aug 18.40 7 days 1981 Aug 25	Cylinder 2500?	7.5 long 2.6 dia	1981 Aug 18.5	72.86	90.06	6655	198	356	0.012	81
D	Cosmos 1297 engine	1981-79D	1981 Aug 18.40 18 days 1981 Sep 5	Cone 600? full	1.5 long? 2 dia?	1981 Aug 31.9	72.86	88.92	6598	212	227	0.001	91
D	Fragments	1981-79C,E-G											

	Name		Launch date, lifetime and descent date	Shape and weight (kg)	Size (m)	Date of orbital determination	Orbital inclination (deg)	Nodal period (min)	Semi major axis (km)	Perigee height (km)	Apogee height (km)	Orbital eccentricity	Argument of perigee (deg)
D R	Cosmos 1298	1981-80A	1981 Aug 21.43 42 days 1981 Oct 2	Sphere-cylinder 6700?	7 long? 2.4 dia	1981 Aug 21.5	64.89	89.54	6631	174	331	0.012	58
D	Cosmos 1298 rocket	1981-80B	1981 Aug 21.43 3 days 1981 Aug 24	Cylinder 2500?	7.5 long 2.6 dia	1981 Aug 21.5	64.90	89.45	6626	173	323	0.011	58
	Cosmos 1299*	1981-81A	1981 Aug 24.69 600 years	Cone-cylinder	6 long? 2.5 dia?	1981 Aug 24.9 1981 Sep 6.1	65.00 65.12	89.65 104.00	6636 7325	248 910	267 984	0.001 0.005	266 240
D	Cosmos 1299 rocket	1981-81C	1981 Aug 24.69 14 days 1981 Sep 7	Cylinder 1500?	8 long? 2.5 dia?	1981 Sep 6.1	65.00	89.38	6622	234	254	0.001	260
D	Cosmos 1299 platform	1981-81E	1981 Aug 24.69 34 days 1981 Sep 27	Irregular	-	1981 Sep 6.1	65.00	89.56	6631	245	261	0.001	251
1d	Fragments**	1981-81B,D											
T	Cosmos 1300	1981-82A	1981 Aug 24.90 60 years	-	-	1981 Aug 25.6	82.50	97.79	7030	638	666	0.002	283
	Cosmos 1300 rocket	1981-82B	1981 Aug 24.90 60 years	Cylinder 2200?	7.4 long 2.4 dia	1981 Aug 25.6	82.50	97.76	7029	636	665	0.002	287
D R	Cosmos 1301	1981-83A	1981 Aug 27.44 13.86 days 1981 Sep 10.30	Sphere-cylinder 6300?	6.5 long? 2.4 dia	1981 Aug 28.2	82.31	89.38	6621	213	272	0.004	95
D	Cosmos 1301 rocket	1981-83B	1981 Aug 27.44 4 days 1981 Aug 31	Cylinder 2500?	7.5 long 2.6 dia	1981 Aug 28.1	82.32	89.22	6613	210	259	0.004	94

* Cosmos 1299 jettisoned rocket and platform, before moving to 104 min orbit on 1981 Sep 5.

** Fragment 1981-81D is probably the uranium fuel core ejected from nuclear reactor 1981-81A when in 104 min orbit.

	Name		Launch date, lifetime and descent date	Shape and weight (kg)	Size (m)	Date of orbital determination	Orbital inclination (deg)	Nodal period (min)	Semi major axis (km)	Perigee height (km)	Apogee height (km)	Orbital eccentricity	Argument of perigee (deg)
D	Cosmos 1301 engine	1981-83D	1981 Aug 27.44 28 days 1981 Sep 24	Cone 600? full	1.5 long? 2 dia?	1981 Sep 12.9	82.31	89.67	6635	250	264	0.001	299
D	Fragments	1981-83C,E,F											
T	Cosmos 1302	1981-84A	1981 Aug 28.68 120 years	Cylinder + paddles? 750?	2 long? 1 dia?	1981 Aug 30.2	74.03	100.83	7176	783	812	0.002	353
	Cosmos 1302 rocket	1981-84B	1981 Aug 28.68 100 years	Cylinder 2200?	7.4 long 2.4 dia	1981 Aug 29.4	74.03	100.69	7169	774	807	0.002	17
	Fragment	1981-84C											
T	[Titan 3D]	1981-85A	1981 Sep 3.77 manoeuvrable	Cylinder 13300? full	15 long 3.0 dia	1981 Sep 4.1	96.99	92.27	6763	244	526	0.021	163
D	Titan 3D rocket	1981-85B	1981 Sep 3.77 39 days 1981 Oct 12	Cylinder 1900	6 long 3.0 dia	1981 Sep 4.1	96.97	92.13	6756	243	513	0.020	163
D R	Cosmos 1303	1981-86A	1981 Sep 4.34 13.90 days 1981 Sep 18.24	Sphere- cylinder 6300?	6.5 long? 2.4 dia	1981 Sep 4.7 1981 Sep 7.3	70.40 70.41	90.36 92.33	6670 6767	208 361	376 416	0.013 0.004	67 206
D	Cosmos 1303 rocket	1981-86B	1981 Sep 4.34 8 days 1981 Sep 12	Cylinder 2500?	7.5 long 2.6 dia	1981 Sep 4.4	70.40	90.23	6664	207	364	0.012	66
D	Cosmos 1303 engine	1981-86G	1981 Sep 4.34 155 days 1982 Feb 6	Cone 600? full	1.5 long? 2 dia?	1981 Oct 10.0	70.40	92.04	6753	348	402	0.004	150
D	Fragments	1981-86C-F											

	Name		Launch date, lifetime and descent date	Shape and weight (kg)	Size (m)	Date of orbital determination	Orbital inclination (deg)	Nodal period (min)	Semi major axis (km)	Perigee height (km)	Apogee height (km)	Orbital eccentricity	Argument of perigee (deg)
T	Cosmos 1304	1981-87A	1981 Sep 4.46 1200 years	Cylinder 700?	1.3 long? 1.9 dia?	1981 Sep 5.2	82.94	103.99	7324	912	980	0.005	327
	Cosmos 1304 rocket	1981-87B	1981 Sep 4.46 600 years	Cylinder 2200?	7.4 long 2.4 dia	1981 Sep 4.7	82.93	103.89	7320	909	974	0.004	332
T?	Cosmos 1305†	1981-88A	1981 Sep 11.37 100 years?	Windmill + 6 vanes? 1000?	3.4 long? 1.6 dia?	1981 Sep 12.5	62.83	263.73	13624	626	13865	0.486	287
D	Cosmos 1305 launcher rocket	1981-88B	1981 Sep 11.37 24 days 1981 Oct 5	Cylinder 2500?	7.5 long 2.6 dia	1981 Sep 12.1	62.77	92.66	6784	215	597	0.028	119
D	Cosmos 1305 launcher	1981-88C	1981 Sep 11.37 22 days 1981 Oct 3	Irregular	-	1981 Sep 13.0	62.81	92.76	6789	200	622	0.031	123
	Cosmos 1305 rocket	1981-88F	1981 Sep 11.37 100 years?	Cylinder 4800 full 440 empty*	2.0 long 2.0 dia	1981 Sep 22.5	62.82	262.44	13577	611	13787	0.485	287
D	Fragments	1981-88D,E											
D	Cosmos 1306	1981-89A	1981 Sep 14.86 305 days 1982 Jul 16	Cylinder?	-	1981 Sep 18.3	64.96	93.28	6814	409	462	0.004	23
D	Cosmos 1306 rocket	1981-89B	1981 Sep 14.86 1 day 1981 Sep 15	Cylinder 1500?	8 long? 2.5 dia?	1981 Sep 15.2	64.94	88.15	6562	102	265	0.012	49
D	Fragments	1981-89C-F											

† Probably an attempted Molniya launch.

* Probably shut down prematurely; approximate weight with residual fuel 3300 kg.

	Name		Launch date, lifetime and descent date	Shape and weight (kg)	Size (m)	Date of orbital determination	Orbital inclination (deg)	Nodal period (min)	Semi major axis (km)	Perigee height (km)	Apogee height (km)	Orbital eccentricity	Argument of perigee (deg)
D R	Cosmos 1307	1981-90A	1981 Sep 15.48 13.75 days 1981 Sep 29.23	Sphere-cylinder 6300?	6.5 long? 2.4 dia	1981 Sep 15.7 1981 Sep 17.6	72.86 72.87	90.44 92.30	6674 6765	198 356	394 418	0.015 0.005	85 207
D	Cosmos 1307 rocket	1981-90B	1981 Sep 15.48 8 days 1981 Sep 23	Cylinder 2500?	7.5 long 2.6 dia	1981 Sep 16.2	72.87	90.30	6667	197	381	0.014	83
D	Cosmos 1307 engine	1981-90G	1981 Sep 15.48 142 days 1982 Feb 4	Cone 600? full	1.5 long? 2 dia?	1981 Oct 10.0	72.86	92.11	6757	345	413	0.005	158
D	Fragments	1981-90C-F,H											
T	Cosmos 1308	1981-91A	1981 Sep 18.15 1200 years	Cylinder 700?	1.3 long? 1.9 dia?	1981 Sep 18.6	82.92	104.86	7365	970	1004	0.002	299
	Cosmos 1308 rocket	1981-91B	1981 Sep 18.15 600 years	Cylinder 2200?	7.4 long 2.4 dia	1981 Sep 18.8	82.92	104.75	7360	967	996	0.002	293
D R	Cosmos 1309	1981-92A	1981 Sep 18.40 12.86 days 1981 Oct 1.26	Sphere-cylinder 5700?	5.0 long 2.4 dia	1981 Sep 18.8	82.30	89.22	6613	212	257	0.003	93
D	Cosmos 1309 rocket	1981-92B	1981 Sep 18.40 3 days 1981 Sep 21	Cylinder 2500?	7.5 long 2.6 dia	1981 Sep 18.5	82.31	89.22	6613	213	256	0.003	81

	Name		Launch date, lifetime and descent date	Shape and weight (kg)	Size (m)	Date of orbital determination	Orbital inclination (deg)	Nodal period (min)	Semi major axis (km)	Perigee height (km)	Apogee height (km)	Orbital eccentricity	Argument of perigee (deg)
D	China 9A	1981-93A	1981 Sep 19.90 7 days 1981 Sep 26	Balloon + sphere	1.2 dia? 0.75 dia?	1981 Sep 20.0	59.47	103.27	7293	232	1598	0.093	148
D	China 9B*	1981-93B	1981 Sep 19.90 382 days 1982 Oct 6	Truncated cone	1.65 high? 0.65 to 1.65 dia?	1981 Sep 20.4	59.47	103.49	7303	235	1615	0.094	149
D	China 9C*	1981-93D	1981 Sep 19.90 332 days 1982 Aug 17	Octagonal prism + 4 vanes	1.1 high 1.2 max dia	1981 Sep 20.2	59.46	103.42	7300	234	1610	0.094	148
	China 9 rocket	1981-93E	1981 Sep 19.90 19 months	Cylinder?	8 long 3 dia	1981 Sep 24.9	59.47	103.23	7292	236	1592	0.093	153
D	Fragments	1981-93C,F-K											
T	Aureole 3**	1981-94A	1981 Sep 21.55 60 years	Octagonal cylinder 1000	1.8 long? 1.5 dia?	1981 Sep 23.3	82.50	109.52	7582	406	2001	0.105	141
	Aureole 3 rocket	1981-94B	1981 Sep 21.55 50 years	Cylinder 2200?	7.4 long 2.4 dia	1981 Sep 22.3	82.50	109.47	7579	406	1996	0.105	143
T	Cosmos 1310	1981-95A	1981 Sep 23.34 5 years	Cylinder?	4 long? 2 dia?	1981 Sep 24.1	65.84	94.55	6876	477	518	0.003	351
	Cosmos 1310 rocket	1981-95B	1981 Sep 23.34 2 years	Cylinder 2200?	7.4 long 2.4 dia	1981 Sep 23.9	65.84	94.41	6869	466	515	0.004	1

* Chinese payload for Earth resources.

** Included French magnetosphere, ionosphere, and Arctic auroral density (Arcad) experiments; launched by USSR.

	Name		Launch date, lifetime and descent date	Shape and weight (kg)	Size (m)	Date of orbital determination	Orbital inclination (deg)	Nodal period (min)	Semi major axis (km)	Perigee height (km)	Apogee height (km)	Orbital eccentricity	Argument of perigee (deg)
T	SBS 2 [Thor Delta]	1981-96A	1981 Sep 24.96 >million years	Cylinder + panel 550	2.82 long 2.16 dia	1981 Sep 25.0 1981 Sep 27.0	27.70 0.33	650.32 1401.68	24865 41488	166 34288	36807 35932	0.737 0.020	179 71
D	SBS 2 third stage (PAM)	1981-96B	1981 Sep 24.96 177 days? 1982 Mar 20?	Cylinder 2154 full	2.29 long 1.5 dia	1981 Sep 25.0	27.70	650.32	24865	166	36807	0.737	179
T	Cosmos 1311	1981-97A	1981 Sep 28.88 2 years	-	-	1981 Sep 28.9	82.99	94.46	6869	463	519	0.004	345
	Cosmos 1311 rocket	1981-97B	1981 Sep 28.88 506 days	Cylinder 2200?	7.4 long 2.4 dia	1981 Sep 28.9	82.96	94.44	6868	465	515	0.004	4
D	Fragments	1981-97C-T											
T	Cosmos 1312	1981-98A	1981 Sep 30.33 10,000 years	-	-	1981 Sep 30.8	82.59	115.99	7877	1493	1505	0.001	122
	Cosmos 1312 rocket	1981-98B	1981 Sep 30.33 10,000 years	Cylinder 2200?	7.4 long 2.4 dia	1981 Sep 30.8	82.60	115.93	7875	1490	1503	0.001	99
D R	Cosmos 1313	1981-99A	1981 Oct 1.38 13.87 days 1981 Oct 15.25	Sphere-cylinder 6300?	6.5 long? 2.4 dia	1981 Oct 1.6	70.36	89.48	6627	206	291	0.006	69
D	Cosmos 1313 rocket	1981-99B	1981 Oct 1.38 4 days 1981 Oct 5	Cylinder 2500?	7.5 long 2.6 dia	1981 Oct 1.7	70.32	89.35	6621	203	282	0.006	60
D	Cosmos 1313 engine	1981-99F	1981 Oct 1.38 25 days 1981 Oct 26	Cone 600? full	1.5 long? 2 dia?	1981 Oct 16.7	70.35	89.32	6619	220	262	0.003	311
D	Fragments	1981-99C-E											

	Name		Launch date, lifetime and descent date	Shape and weight (kg)	Size (m)	Date of orbital determination	Orbital inclination (deg)	Nodal period (min)	Semi major axis (km)	Perigee height (km)	Apogee height (km)	Orbital eccentricity	Argument of perigee (deg)
T	SME*	1981-100A	1981 Oct 6.48 12 years	Cylinder + disc 437	1.7 high 1.25 dia	1981 Oct 6.5	97.47	95.47	6918	538	542	0.0003	317
T	UOSAT**	1981-100B	1981 Oct 6.48 10 years	Box 52?	1.0 × 0.5 ×0.5 approx	1981 Oct 7.9	97.46	95.46	6918	538	541	0.0002	312
	Thor Delta second stage	1981-100C	1981 Oct 6.48 100 years	Cylinder + annulus 350?	6.4 long 1.52 and 2.44 dia	1981 Oct 6.5 1981 Oct 6.6 1981 Oct 9.3	97.48 97.46 99.89	91.87 95.49 119.40	6743 6919 8031	185 536 554	544 546 2752	0.027 0.001 0.137	163 59 101
D	Fragment	1981-100D,E											
D R	Cosmos 1314	1981-101A	1981 Oct 9.45 12.8 days 1981 Oct 22.3	Sphere- cylinder 6300?	6.5 long? 2.4 dia	1981 Oct 9.7	82.34	89.03	6604	214	237	0.002	101
D	Cosmos 1314 rocket	1981-101B	1981 Oct 9.45 2 days 1981 Oct 11	Cylinder 2500?	7.5 long 2.6 dia	1981 Oct 9.5	82.33	88.94	6599	215	227	0.001	104
D	Cosmos 1314 engine	1981-101D	1981 Oct 9.45 16 days 1981 Oct 25	Cone 600? full	1.5 long? 2 dia?	1981 Oct 22.6	82.33	88.79	6592	206	221	0.001	30
D	Fragments	1981-101C,E,F											
T	Statsionar-Raduga 10	1981-102A	1981 Oct 9.71 >million years	-	-	1981 Oct 9.7	0.07	1443.45	42310	35932	35932	0	28
D	Raduga 10 launcher rocket	1981-102B	1981 Oct 9.71 2 days 1981 Oct 11	Cylinder 4000?	12 long? 4 dia	1981 Oct 9.9	51.62	88.22	6568	186	193	0.001	210

* Solar Mesosphere Explorer. ** University of Surrey satellite.

1981-102 continued on page 667

	Name		Launch date, lifetime and descent date	Shape and weight (kg)	Size (m)	Date of orbital determination	Orbital inclina-tion (deg)	Nodal period (min)	Semi major axis (km)	Perigee height (km)	Apogee height (km)	Orbital eccen-tricity	Argument of perigee (deg)
D	Raduga 10 launcher	1981-102C	1981 Oct 9.71 1 day 1981 Oct 10	Irregular	-	1981 Oct 9.9	51.61	88.05	6559	175	187	0.001	127
D	Raduga 10 rocket	1981-102D	1981 Oct 9.71 265 days 1982 Jul 1	Cylinder 1900?	3.9 long? 3.9 dia	1981 Oct 25.2	47.39	635.50	24486	309	35907	0.727	4
	Raduga 10 apogee motor	1981-102F	1981 Oct 9.71 >million years	Cylinder 440?	2.0 long? 2.0 dia?	Synchronous orbit similar to 1981-102A							
D	Fragment	1981-102E											
T	Cosmos 1315	1981-103A	1981 Oct 13.96 60 years	Cylinder + 2 vanes? 2500?	5 long? 1.5 dia?	1981 Oct 14.2	81.19	97.69	7025	627	667	0.003	21
	Cosmos 1315 rocket	1981-103B	1981 Oct 13.96 60 years	Cylinder 1440	3.8 long 2.6 dia	1981 Oct 21.4	81.20	97.77	7029	613	689	0.005	144
D R	Cosmos 1316	1981-104A	1981 Oct 15.39 13.88 days 1981 Oct 29.27	Sphere-cylinder 6300?	6.5 long? 2.4 dia	1981 Oct 15.4	70.36	90.46	6675	209	385	0.013	63
D	Cosmos 1316 rocket	1981-104B	1981 Oct 15.39 8 days 1981 Oct 23	Cylinder 2500?	7.5 long 2.6 dia	1981 Oct 16.2	70.35	90.28	6666	207	368	0.012	65
D	Cosmos 1316 engine	1981-104C	1981 Oct 15.39 24 days 1981 Nov 8	Cone 600? full	1.5 long? 2 dia?	1981 Oct 29.4	70.33	89.49	6627	227	271	0.003	57
D	Fragments	1981-104D,E											

	Name		Launch date, lifetime and descent date	Shape and weight (kg)	Size (m)	Date of orbital determination	Orbital inclination (deg)	Nodal period (min)	Semi major axis (km)	Perigee height (km)	Apogee height (km)	Orbital eccentricity	Argument of perigee (deg)
T	Molniya 3S	1981-105A	1981 Oct 17.25 16 years	Windmill + 6 vanes 1500?	4.2 long? 1.6 dia	1981 Oct 18.3	62.82	736.30	27011	618	40648	0.741	288
D	Molniya 3S launcher	1981-105B	1981 Oct 17.25 18 days 1981 Nov 4	Irregular	-	1981 Oct 17.8	62.86	92.72	6787	194	623	0.032	116
D	Molniya 3S launcher rocket	1981-105C	1981 Oct 17.25 20 days 1981 Nov 6	Cylinder 2500?	7.5 long 2.6 dia	1981 Oct 17.5	62.81	92.41	6771	213	573	0.027	118
	Molniya 3S rocket	1981-105E	1981 Oct 17.25 16 years	Cylinder 440	2.0 long 2.0 dia	1981 Oct 21.4	62.88	733.36	26939	616	40506	0.740	288
D	Fragment	1981-105D											
D	Venus 13 launcher rocket	1981-106B	1981 Oct 30.26 1 day 1981 Oct 31	Cylinder 4000?	12 long? 4 dia	1981 Oct 30.8	51.56	87.85	6549	139	203	0.005	349
D	Venus 13 launcher	1981-106C	1981 Oct 30.26 1 day 1981 Oct 31	Irregular	-	1981 Oct 31.0	51.52	87.64	6539	146	175	0.002	341
T	IMEWS 12* [Titan 3C]	1981-107A	1981 Oct 31.40 >million years	-	-	Synchronous orbit similar to 1981-25A							
D	Titan 3C second stage	1981-107B	1981 Oct 31.40 2 days 1981 Nov 2	Cylinder 1900	6 long 3.0 dia	1981 Nov 1.2	29.32	89.19	6619	146	336	0.014	134
	Titan 3C transtage	1981-107C	1981 Oct 31.40 >million years	Cylinder 1500?	6 long? 3.0 long	Synchronous orbit similar to 1981-25C							

Space Vehicle: Venus 13, 1981-106A.

* All data on this USAF launch are provisional.

	Name		Launch date, lifetime and descent date	Shape and weight (kg)	Size (m)	Date of orbital determination	Orbital inclination (deg)	Nodal period (min)	Semi major axis (km)	Perigee height (km)	Apogee height (km)	Orbital eccentricity	Argument of perigee (deg)
T	Cosmos 1317	1981-108A	1981 Oct 31.96 100 years?	Windmill + 6 vanes 1250?	4.2 long? 1.6 dia	1981 Nov 1.0	62.87	725.73	26752	584	40163	0.740	316
D	Cosmos 1317 launcher rocket	1981-108B	1981 Oct 31.96 23 days 1981 Nov 23	Cylinder 2500?	7.5 long 2.6 dia	1981 Nov 1.5	62.84	92.40	6771	213	573	0.027	120
D	Cosmos 1317 launcher	1981-108C	1981 Oct 31.96 14 days 1981 Nov 14	Irregular	-	1981 Nov 1.5	62.91	92.46	6774	182	610	0.032	119
	Cosmos 1317 rocket	1981-108D	1981 Oct 31.96 100 years?	Cylinder 440	2.0 long 2.0 dia	1981 Nov 10.0	62.87	723.38	26694	588	40044	0.739	316
D R	Cosmos 1318	1981-109A	1981 Nov 3.55 30.5 days 1981 Dec 4.1	Sphere- cylinder 6700?	7 long? 2.4 dia	1981 Nov 3.7	67.14	89.75	6641	172	353	0.014	70
D	Cosmos 1318 rocket	1981-109B	1981 Nov 3.55 3 days 1981 Nov 6	Cylinder 2500?	7.5 long 2.6 dia	1981 Nov 3.7	67.04	89.68	6637	175	343	0.013	67
D	Venus 14 launcher	1981-110B	1981 Nov 4.23 1 day 1981 Nov 5	Irregular	-	1981 Nov 4.9	51.56	87.79	6546	153	183	0.002	345
D	Venus 14 launcher rocket	1981-110C	1981 Nov 4.23 1 day 1981 Nov 5	Cylinder 4000?	12 long? 4 dia	1981 Nov 5.0	51.57	87.67	6540	151	173	0.002	291

Space Vehicle: Venus 14, 1981-110A.

	Name		Launch date, lifetime and descent date	Shape and weight (kg)	Size (m)	Date of orbital determination	Orbital inclination (deg)	Nodal period (min)	Semi major axis (km)	Perigee height (km)	Apogee height (km)	Orbital eccentricity	Argument of perigee (deg)
D 2M R	STS 2* (Columbia F2)	1981-111A	1981 Nov 12.63 2.26 days 1981 Nov 14.89	Cylinder + delta wing 68800 empty	37.1 long 17.4 high 23.8 span	1981 Nov 13.6	38.03	89.56	6636	253	262	0.001	264
D R	Cosmos 1319	1981-112A	1981 Nov 13.40 13.89 days 1981 Nov 27.29	Sphere-cylinder 6300?	6.5 long? 2.4 dia	1981 Nov 13.6	70.36	90.38	6671	209	377	0.013	67
D	Cosmos 1319 rocket	1981-112B	1981 Nov 13.40 9 days 1981 Nov 22	Cylinder 2500?	7.5 long 2.6 dia	1981 Nov 13.6	70.36	90.30	6667	208	370	0.012	66
D	Cosmos 1319 engine	1981-112F	1981 Nov 13.40 154 days 1982 Apr 15	Cone 600? full	1.5 long? 2 dia?	1981 Dec 11.1	70.36	92.01	6751	341	404	0.005	194
D	Fragments	1981-112C-E,G-J											
T	Molniya 1BC	1981-113A	1981 Nov 17.65 15½ years **	Windmill + 6 vanes 1000?	3.4 long 1.6 dia	1981 Nov 17.7	62.81	702.03	26167	441	39136	0.739	280
D	Molniya 1BC launcher rocket	1981-113B	1981 Nov 17.65 10 days 1981 Nov 27	Cylinder 2500?	7.5 long 2.6 dia	1981 Nov 17.9	62.78	90.62	6684	207	404	0.015	135
D	Molniya 1BC launcher	1981-113C	1981 Nov 17.65 12 days 1981 Nov 29	Irregular	-	1981 Nov 17.8	62.83	91.09	6707	197	461	0.020	120
	Molniya 1BC rocket	1981-113D	1981 Nov 17.65 15½ years **	Cylinder 440	2.0 long 2.0 dia	1981 Nov 30.8	62.83	699.03	26094	460	38972	0.738	280

* Launched from Cape Canaveral; landed at Edwards AFB, California. Cargo bay carried OSTA-1 pallet (not separated). OSTA is Office of Space and Terrestrial Applications.

** Decay possible in 1994

	Name		Launch date, lifetime and descent date	Shape and weight (kg)	Size (m)	Date of orbital determination	Orbital inclination (deg)	Nodal period (min)	Semi major axis (km)	Perigee height (km)	Apogee height (km)	Orbital eccentricity	Argument of perigee (deg)
T	RCA Satcom 3R	1981-114A	1981 Nov 20.07 >million years	Box + vanes 1078 full	1.62 high 1.25 wide 1.25 deep	1981 Nov 20.8 1981 Nov 22.9	27.51 0.79	627.13 1418.8	24280 41826	193 35206	35611 35690	0.729 0.007	179 230
	RCA Satcom 3R third stage (PAM)	1981-114B	1981 Nov 20.07 20 years?	Cylinder 2154 full	2.29 long 1.5 dia	1981 Nov 20.1	27.40	633.57	24436	185	35930	0.731	178
T	Bhaskara 2*	1981-115A	1981 Nov 20.36 9 years	26-faced polyhedron 444	1.19 high 1.55 dia	1981 Nov 20.8	50.64	95.20	6909	520	542	0.002	64
	Bhaskara 2 rocket	1981-115B	1981 Nov 20.36 5 years	Cylinder 2200?	7.4 long 2.4 dia	1981 Nov 20.7	50.64	95.10	6904	512	540	0.002	54
T	Cosmos 1320	1981-116A	1981 Nov 28.75 10000 years	Spheroid 40?	1.0 long? 0.8 dia?	1981 Nov 29.6	73.97	117.32	7938	1482	1638	0.010	260
T	Cosmos 1321	1981-116B	1981 Nov 28.75 10000 years	Spheroid 40?	1.0 long? 0.8 dia?	1981 Nov 29.6	73.99	117.29	7937	1482	1635	0.009	262
T	Cosmos 1322	1981-116C	1981 Nov 28.75 10000 years	Spheroid 40?	1.0 long? 0.8 dia?	1981 Nov 29.5	73.98	117.26	7935	1483	1631	0.009	260
T	Cosmos 1323	1981-116D	1981 Nov 28.75 10000 years	Spheroid 40?	1.0 long? 0.8 dia?	1981 Nov 30.3	73.98	117.21	7933	1483	1627	0.009	261
T	Cosmos 1324	1981-116E	1981 Nov 28.75 10000 years	Spheroid 40?	1.0 long? 0.8 dia?	1981 Nov 30.2	73.99	117.15	7931	1482	1623	0.009	259

* Indian satellite launched by USSR.

1981-116 continued on page 672

	Name		Launch date, lifetime and descent date	Shape and weight (kg)	Size (m)	Date of orbital determination	Orbital inclination (deg)	Nodal period (min)	Semi major axis (km)	Perigee height (km)	Apogee height (km)	Orbital eccentricity	Argument of perigee (deg)
T	Cosmos 1325	1981-116F	1981 Nov 28.75 10000 years	Spheroid 40?	1.0 long? 0.8 dia?	1981 Nov 30.3	73.98	117.12	7929	1483	1619	0.009	258
T	Cosmos 1326	1981-116G	1981 Nov 28.75 10000 years	Spheroid 40?	1.0 long? 0.8 dia?	1981 Nov 29.5	73.98	117.12	7929	1485	1617	0.008	257
T	Cosmos 1327	1981-116H	1981 Nov 28.75 10000 years	Spheroid 40?	1.0 long? 0.8 dia?	1981 Nov 29.6	73.99	117.05	7926	1486	1609	0.008	261
	Cosmos 1320 rocket	1981-116J	1981 Nov 28.75 20000 years	Cylinder 2200?	7.4 long 2.4 dia	1981 Nov 29.4	73.98	117.69	7955	1483	1670	0.012	264
T	Cosmos 1328	1981-117A	1981 Dec 3.49 60 years	-	-	1981 Dec 3.8	82.52	97.77	7029	637	665	0.002	290
	Cosmos 1328 rocket	1981-117B	1981 Dec 3.49 60 years	Cylinder 2200?	7.4 long 2.4 dia	1981 Dec 3.9	82.52	97.75	7028	634	666	0.002	300
D R	Cosmos 1329	1981-118A	1981 Dec 4.41 13.86 days 1981 Dec 18.27	Sphere-cylinder 6300?	6.5 long? 2.4 dia	1981 Dec 4.4	65.02	89.45	6626	232	264	0.002	55
D	Cosmos 1329 rocket	1981-118B	1981 Dec 4.41 4 days 1981 Dec 8	Cylinder 2500?	7.5 long 2.6 dia	1981 Dec 5.1	65.03	89.25	6616	226	250	0.002	17

1981-118 continued on page 673

	Name		Launch date, lifetime and descent date	Shape and weight (kg)	Size (m)	Date of orbital determination	Orbital inclination (deg)	Nodal period (min)	Semi major axis (km)	Perigee height (km)	Apogee height (km)	Orbital eccentricity	Argument of perigee (deg)
D	Cosmos 1329 engine	1981-118D	1981 Dec 4.41 32 days 1982 Jan 5	Cone 600? full	1.5 long? 2 dia?	1981 Dec 18.4	65.02	89.66	6636	230	286	0.004	49
D	Fragments	1981-118C,E,F											
T	Intelsat 5 F-3 [Atlas Centaur]	1981-119A	1981 Dec 15.98 >million years	Box +2 vanes + dishes 1870 full	15.9 span 6.4 wide	1981 Dec 16.0 1981 Dec 23.7	23.69 0.33	633.54 1439.09	24435 42223	166 35676	35947 36014	0.732 0.004	179 331
	Intelsat 5 F-3 rocket	1981-119B	1981 Dec 15.98 10 years	Cylinder 1815	8.6 long 3.0 dia	1981 Dec 16.0 1981 Dec 16.1	28.31 23.68	89.55 614.14	6630 23933	148 169	356 34941	0.016 0.726	146 179
T	Radio 3	1981-120A	1981 Dec 17.45 15000 years	-	-	1981 Dec 18.0	82.95	118.52	7992	1564	1663	0.006	97
T	Radio 8	1981-120B	1981 Dec 17.45 15000 years	-	-	1981 Dec 17.9	82.96	119.79	8049	1656	1686	0.002	215
T	Radio 5	1981-120C	1981 Dec 17.45 15000 years	-	-	1981 Dec 17.9	82.96	119.56	8039	1653	1668	0.001	174
T	Radio 4	1981-120D	1981 Dec 17.45 15000 years	-	-	1981 Dec 17.9	82.96	119.40	8032	1639	1668	0.002	145
T	Radio 7	1981-120E	1981 Dec 17.45 15000 years	-	-	1981 Dec 18.0	82.96	119.20	8022	1626	1662	0.002	116

1981-120 continued on page 674

	Name		Launch date, lifetime and descent date	Shape and weight (kg)	Size (m)	Date of orbital determination	Orbital inclination (deg)	Nodal period (min)	Semi major axis (km)	Perigee height (km)	Apogee height (km)	Orbital eccentricity	Argument of perigee (deg)
T	Radio 6	1981-120F	1981 Dec 17.45 15000 years	-	-	1981 Dec 18.5	82.95	118.72	8001	1579	1666	0.005	107
	Radio 3 rocket	1981-120G	1981 Dec 17.45 15000 years	Cylinder 2200?	7.4 long 2.4 dia	1981 Dec 18.4	82.96	120.95	8102	1663	1785	0.007	258
D R	Cosmos 1330	1981-121A	1981 Dec 19.50 31 days 1982 Jan 19	Sphere-cylinder 6700?	7 long? 2.4 dia	1981 Dec 19.8	70.36	89.99	6652	168	379	0.016	70
D	Cosmos 1330 rocket	1981-121B	1981 Dec 19.50 4 days 1981 Dec 23	Cylinder 2500?	7.5 long 2.6 dia	1981 Dec 19.7	70.38	89.84	6644	166	366	0.015	70
T	MARECS 1* [Ariane 1-04]	1981-122A	1981 Dec 20.06 >million years	Octagonal cylinder 582 full 497 empty	2.5 long 2.0 dia	1981 Dec 20.1 1981 Dec 23.7	10.55 2.32	635.90 1430.71	24495 42060	199 35640	36035 35724	0.731 0.001	174 77
	CAT 4	1981-122B	1981 Dec 20.06 20 years	Cylinder? 217?	1.1 long? 1.6 dia?	1981 Dec 21.2	10.54	632.34	24406	236	35820	0.729	176
	Ariane third stage	1981-122C	1981 Dec 20.06 8 years	Cylinder 1634 empty	8.6 long 2.6 dia	1982 Jan 7.8	10.45	616.96	24007	200	35058	0.726	189

* MARitime European Communications Satellite.

	Name		Launch date, lifetime and descent date	Shape and weight (kg)	Size (m)	Date of orbital determination	Orbital inclination (deg)	Nodal period (min)	Semi major axis (km)	Perigee height (km)	Apogee height (km)	Orbital eccentricity	Argument of perigee (deg)
T	Molniya 1BD	1981-123A	1981 Dec 23.56 100 years?*	Windmill + 6 vanes 1000?	3.4 long 1.6 dia	1981 Dec 27.0	62.95	699.39	26100	484	38960	0.737	280
D	Molniya 1BD launcher rocket	1981-123B	1981 Dec 23.56 24 days 1982 Jan 16	Cylinder 2500?	7.5 long 2.6 dia	1981 Dec 23.6	62.98	91.45	6725	234	459	0.017	62
D	Molniya 1BD launcher	1981-123C	1981 Dec 23.56 18 days 1982 Jan 10	Irregular	-	1981 Dec 23.7	62.99	91.55	6730	208	495	0.021	61
	Molniya 1BD rocket	1981-123D	1981 Dec 23.56 100 years?*	Cylinder 440	2.0 long 2.0 dia	1981 Dec 28.9	62.97	695.32	25998	476	38763	0.736	280

* Decay possible in 2003

	Name		Launch date, lifetime and descent date	Shape and weight (kg)	Size (m)	Date of orbital determination	Orbital inclination (deg)	Nodal period (min)	Semi major axis (km)	Perigee height (km)	Apogee height (km)	Orbital eccentricity	Argument of perigee (deg)
T	Cosmos 1331	1982-01A	1982 Jan 7.65 120 years	Cylinder + paddles? 750?	2 long? 1 dia?	1982 Jan 7.9	74.05	100.73	7171	774	812	0.003	349
	Cosmos 1331 rocket	1982-01B	1982 Jan 7.65 100 years	Cylinder 2200?	7.4 long 2.4 dia	1982 Jan 7.9	74.06	100.63	7166	766	810	0.003	0
	Fragments	1982-01C,D											
D R	Cosmos 1332	1982-02A	1982 Jan 12.52 12.86 days 1982 Jan 25.38	Sphere-cylinder 5700?	5.0 long 2.4 dia	1982 Jan 12.9	82.32	89.14	6609	211	250	0.003	106
D	Cosmos 1332 rocket	1982-02B	1982 Jan 12.52 4 days 1982 Jan 16	Cylinder 2500?	7.5 long 2.6 dia	1982 Jan 12.9	82.31	88.98	6601	211	235	0.002	90
T	Cosmos 1333	1982-03A	1982 Jan 14.33 1200 years	Cylinder 700?	1.3 long? 1.9 dia?	1982 Jan 16.4	82.94	105.02	7372	971	1017	0.003	275
	Cosmos 1333 rocket	1982-03B	1982 Jan 14.33 600 years	Cylinder 2200?	7.4 long 2.4 dia	1982 Jan 15.3	82.94	104.91	7367	971	1007	0.003	269
T	RCA Satcom 4	1982-04A	1982 Jan 16.08 > million years	Box + vanes 1082 full	1.62 high 1.27 wide 1.29 deep	1982 Jan 19.0 1982 Jan 25.9	27.44 0.21	632.75 1422.05	24425 41891	194 35293	35900 35733	0.731 0.005	180 181
	RCA Satcom 4 third stage (PAM)	1982-04B	1982 Jan 16.08 20 years	Cylinder 2154 full	2.29 long 1.5 dia	1982 Feb 27.4	27.33	604.50	23696	166	34470	0.724	205

	Name		Launch date, lifetime and descent date	Shape and weight (kg)	Size (m)	Date of orbital determination	Orbital inclination (deg)	Nodal period (min)	Semi major axis (km)	Perigee height (km)	Apogee height (km)	Orbital eccentricity	Argument of perigee (deg)
D R	Cosmos 1334	1982-05A	1982 Jan 20.48 13.8 days 1982 Feb 3.3	Sphere-cylinder 6300?	6.5 long? 2.4 dia	1982 Jan 21.0 1982 Jan 25.3	72.86 72.85	89.37 89.75	6621 6640	196 230	290 294	0.007 0.005	95 69
D	Cosmos 1334 rocket	1982-05B	1982 Jan 20.48 4 days 1982 Jan 24	Cylinder 2500?	7.5 long 2.6 dia	1982 Jan 20.9	72.86	89.21	6613	195	275	0.006	90
D	Cosmos 1334 engine	1982-05G	1982 Jan 20.48 28 days 1982 Feb 17	Cone 600? full	1.5 long? 2 dia?	1982 Feb 4.3	72.85	89.56	6631	226	279	0.004	44
D	Fragments	1982-05C-F,H,J											
D	[Titan 3D?]	1982-06A	1982 Jan 21.81 122 days 1982 May 23	-	-	1982 Jan 22.4 1982 Jan 25.3	97.32 97.25	91.84 96.73	6742 6978	177 553	550 646	0.028 0.007	165 327
D	Titan 3D rocket?	1982-06B	1982 Jan 21.81 1 day 1982 Jan 22	-	-	1982 Jan 21.9	97.34	91.26	6713	158	511	0.026	165
	Fragments	1982-06C-F											
T	Cosmos 1335	1982-07A	1982 Jan 29.46 4 years	Octagonal ellipsoid 550?	1.8 long? 1.5 dia?	1982 Jan 30.0	74.05	94.63	6878	482	518	0.003	352
	Cosmos 1335 rocket	1982-07B	1982 Jan 29.46 2½ years	Cylinder 2200?	7.4 long 2.4 dia	1982 Jan 30.0	74.06	94.51	6872	473	515	0.003	0
D	Fragments	1982-07C-U											

	Name		Launch date, lifetime and descent date	Shape and weight (kg)	Size (m)	Date of orbital determination	Orbital inclination (deg)	Nodal period (min)	Semi major axis (km)	Perigee height (km)	Apogee height (km)	Orbital eccentricity	Argument of perigee (deg)
D R	Cosmos 1336	1982-08A	1982 Jan 30.48 27 days 1982 Feb 26	Sphere-cylinder 6700?	7 long? 2.4 dia	1982 Jan 31.2	70.34	89.73	6639	170	352	0.014	69
D	Cosmos 1336 rocket	1982-08B	1982 Jan 30.48 3 days 1982 Feb 2	Cylinder 2500?	7.5 long 2.6 dia	1982 Jan 30.6	70.34	89.55	6630	177	327	0.011	56
D	Fragment	1982-08C											
T	Statsionar - Ekran 8	1982-09A	1982 Feb 5.39 > million years	Cylinder + plate	-	1982 Feb 5.6	0.07	1426.95	41987	35609	35609	0	28
D	Ekran 8 launcher rocket	1982-09B	1982 Feb 5.39 2 days 1982 Feb 7	Cylinder 4000?	12 long? 4 dia	1982 Feb 5.6	51.62	88.22	6568	185	194	0.001	205
D	Ekran 8 launcher	1982-09C	1982 Feb 5.39 1 day 1982 Feb 6	Irregular	-	1982 Feb 5.7	51.61	88.10	6562	176	191	0.001	122
D	Ekran 8 rocket	1982-09D	1982 Feb 5.39 10 months 1982 Dec?	Cylinder 1900?	3.9 long? 3.9 dia	1982 Feb 17.7	47.28	628.76	24320	300	35583	0.725	3
	Ekran 8 apogee motor	1982-09F	1982 Feb 5.39 > million years	Cylinder 440?	2.0 long? 2.0 dia?	Synchronous orbit similar to 1982-09A							
	Fragment	1982-09E											

	Name		Launch date, lifetime and descent date	Shape and weight (kg)	Size (m)	Date of orbital determination	Orbital inclina-tion (deg)	Nodal period (min)	Semi major axis (km)	Perigee height (km)	Apogee height (km)	Orbital eccen-tricity	Argument of perigee (deg)
D	Cosmos 1337	1982-10A	1982 Feb 11.05 164 days 1982 Jul 25	Cylinder?	-	1982 Feb 12.3	65.04	93.32	6816	429	447	0.001	272
D	Cosmos 1337 rocket	1982-10B	1982 Feb 11.05 < 1 day 1982 Feb 11	Cylinder 1500?	8 long? 2.5 dia?	1982 Feb 11.4	65.00	88.46	6577	102	296	0.015	55
D R	Cosmos 1338	1982-11A	1982 Feb 16.47 13.8 days 1982 Mar 2.3	Sphere-cylinder 6300?	6.5 long? 2.4 dia	1982 Feb 16.7 1982 Feb 22.3	72.84 72.85	90.14 92.30	6659 6765	186 358	376 416	0.014 0.004	93 199
D	Cosmos 1338 rocket†	1982-11B	1982 Feb 16.47 8 days 1982 Feb 24	Cylinder 2500?	7.5 long 2.6 dia	1982 Feb 17.0	72.84	90.09	6657	197	360	0.012	85
D	Cosmos 1338 engine	1982-11D	1982 Feb 16.47 223 days 1982 Sep 27	Cone 600? full	1.5 long? 2 dia?	1982 Mar 4.0	72.85	92.22	6764	357	415	0.004	179
D	Fragments	1982-11C,E,F											
T	Cosmos 1339	1982-12A	1982 Feb 17.91 1200 years	Cylinder 700?	1.3 long? 1.9 dia?	1982 Feb 18.5	82.91	104.85	7364	955	1018	0.004	282
	Cosmos 1339 rocket	1982-12B	1982 Feb 17.91 600 years	Cylinder 2200?	7.4 long 2.4 dia	1982 Feb 18.6	82.90	104.74	7359	955	1007	0.004	284

† Decay observed over western USA

	Name		Launch date, lifetime and descent date	Shape and weight (kg)	Size (m)	Date of orbital determination	Orbital inclination (deg)	Nodal period (min)	Semi major axis (km)	Perigee height (km)	Apogee height (km)	Orbital eccentricity	Argument of perigee (deg)
T	Cosmos 1340	1982-13A	1982 Feb 19.07 60 years	Cylinder + 2 vanes? 2500?	5 long? 1.5 dia?	1982 Feb 19.3	81.21	97.54	7018	626	654	0.002	120
	Cosmos 1340 rocket	1982-13B	1982 Feb 19.07 60 years	Cylinder 1440	3.8 long 2.6 dia	1982 Feb 19.5	81.20	97.61	7021	614	672	0.004	151
T	Westar 4	1982-14A	1982 Feb 26.00? > million years	Cylinder 1100 full 585 empty	2.74 to 6.84 long* 2.16 dia	1982 Feb 26.0 1982 Mar 1.8	27.50 0.23	648.39 1421.70	24815 41885	166 35005	36708 36008	0.736 0.012	179 183
	Westar 4 third stage (PAM)	1982-14B	1982 Feb 26.00? 8 years	Cylinder 2154 full	2.29 long 1.5 dia	1982 Apr 16.3	27.26	611.5	23867	187	34790	0.725	210
T	Molniya 1BE	1982-15A	1982 Feb 26.84 100 years?	Windmill + 6 vanes 1000?	3.4 long 1.6 dia	1982 Feb 27.9	62.86	735.33	26988	476	40743	0.746	280
D	Molniya 1BE launcher	1982-15B	1982 Feb 26.84 10 days 1982 Mar 8	Irregular	-	1982 Feb 27.4	62.85	91.04	6704	184	468	0.021	121
D	Molniya 1BE launcher rocket	1982-15C	1982 Feb 26.84 15 days 1982 Mar 13	Cylinder 2500?	7.5 long 2.6 dia	1982 Feb 27.4	62.84	91.15	6710	206	457	0.019	131
	Molniya 1BE rocket	1982-15D	1982 Feb 26.84 100 years?	Cylinder 440	2.0 long 2.0 dia	1982 Mar 1.9	62.86	731.16	26886	467	40548	0.745	280

* With solar panel extended

	Name		Launch date, lifetime and descent date	Shape and weight (kg)	Size (m)	Date of orbital determination	Orbital inclination (deg)	Nodal period (min)	Semi major axis (km)	Perigee height (km)	Apogee height (km)	Orbital eccentricity	Argument of perigee (deg)
T	Cosmos 1341	1982-16A	1982 Mar 3.24 100 years?	Windmill + 6 vanes 1250?	4.2 long? 1.6 dia	1982 Mar 5.3	62.90	708.18	26319	631	39251	0.734	318
D	Cosmos 1341 launcher rocket	1982-16B	1982 Mar 3.24 24 days 1982 Mar 27	Cylinder 2500?	7.5 long 2.6 dia	1982 Mar 3.9	62.82	92.61	6782	211	596	0.028	120
D	Cosmos 1341 launcher	1982-16C	1982 Mar 3.24 13 days 1982 Mar 16	Irregular	-	1982 Mar 4.1	62.90	92.41	6772	177	610	0.032	122
	Cosmos 1341 rocket	1982-16D	1982 Mar 3.24 100 years?	Cylinder 440	2.0 long 2.0 dia	1982 Mar 22.5	62.92	708.98	26339	708	39214	0.731	318
T	Intelsat 5 F-4 [Atlas Centaur]	1982-17A	1982 Mar 5.02 > million years	Box + dishes + 2 vanes 1928 full	15.7 span 6.4 wide	1982 Mar 5.7 1982 Mar 25.4	23.83 0.35	630.83 1436.13	24376 42165	245 35737	35751 35836	0.728 0.001	179 88
	Intelsat 5 F-4 rocket	1982-17B	1982 Mar 5.02 1.2 years	Cylinder 1815	8.6 long 3.0 dia	1982 Mar 5.0 1982 Mar 5.1	28.32 24.12	89.56 611.55	6631 23866	149 169	357 34807	0.016 0.726	146 178
D R	Cosmos 1342	1982-18A	1982 Mar 5.45 13.8 days 1982 Mar 19.2	Sphere- cylinder 6300?	6.5 long? 2.4 dia	1982 Mar 6.0	72.86	89.86	6645	230	303	0.005	86
D	Cosmos 1342 rocket	1982-18B	1982 Mar 5.45 4 days 1982 Mar 9	Cylinder 2500?	7.5 long 2.6 dia	1982 Mar 6.1	72.86	89.26	6615	193	281	0.007	90

1982-18 continued on page 682

	Name		Launch date, lifetime and descent date	Shape and weight (kg)	Size (m)	Date of orbital determination	Orbital inclination (deg)	Nodal period (min)	Semi major axis (km)	Perigee height (km)	Apogee height (km)	Orbital eccentricity	Argument of perigee (deg)
D	Cosmos 1342 engine	1982-18E	1982 Mar 5.45 27 days 1982 Apr 1	Cone 600? full	1.5 long? 2 dia?	1982 Mar 21.3	72.85	89.47	6626	222	273	0.004	47
D	Fragments	1982-18C,D											
T	IMEWS 13 [Titan 3C]	1982-19A	1982 Mar 6 > million years	-	-	1982 Mar 8.6	1.97	1424.40	41937	35520	35598	0.001	156
	Titan 3C transtage	1982-19B	1982 Mar 6 > million years	Cylinder 1500?	6 long? 3.0 dia	1982 Mar 17.5	1.96	1424.35	41936	35517	35599	0.001	141
T	Gorizont 5	1982-20A	1982 Mar 15.20 > million years	-	-	1982 Mar 15.5	0.70	1463.27	42697	36319	36319	0	28
D	Gorizont 5 launcher rocket	1982-20B	1982 Mar 15.20 1 day 1982 Mar 16	Cylinder 4000?	12 long? 4 dia	1982 Mar 15.2	51.61	88.38	6576	193	202	0.001	101
D	Gorizont 5 launcher	1982-20C	1982 Mar 15.20 1 day 1982 Mar 16	Irregular	-	1982 Mar 15.5	51.64	88.13	6563	178	192	0.001	243
D	Gorizont 5 rocket	1982-20E	1982 Mar 15.20 7 months 1982 Oct?	Cylinder 1900?	3.9 long? 3.9 dia	1982 Mar 21.5	47.42	643.01	24679	266	36337	0.731	2
	Gorizont 5 apogee motor	1982-20F	1982 Mar 15.20 > million years	Cylinder 440?	2.0 long? 2.0 dia?	Synchronous orbit similar to 1982-20A							
D	Fragment	1982-20D											

	Name		Launch date, lifetime and descent date	Shape and weight (kg)	Size (m)	Date of orbital determination	Orbital inclination (deg)	Nodal period (min)	Semi major axis (km)	Perigee height (km)	Apogee height (km)	Orbital eccentricity	Argument of perigee (deg)
D R	Cosmos 1343	1982-21A	1982 Mar 17.44 13.78 days 1982 Mar 31.22	Sphere-cylinder 6300?	6.5 long? 2.4 dia	1982 Mar 18.4	72.84	89.69	6637	229	288	0.005	78
D	Cosmos 1343 rocket	1982-21B	1982 Mar 17.44 4 days 1982 Mar 21	Cylinder 2500?	7.5 long 2.6 dia	1982 Mar 17.9	72.84	89.19	6612	193	274	0.006	84
D	Cosmos 1343 engine	1982-21C	1982 Mar 17.44 19 days 1982 Apr 5	Cone 600? full	1.5 long? 2 dia?	1982 Mar 31.6	72.83	89.06	6605	215	239	0.002	112
D	Fragments	1982-21D,E											
D 2M R	STS 3* (Columbia F3)	1982-22A	1982 Mar 22.67 8.00 days 1982 Mar 30.67	Cylinder + delta wing 68800 empty	37.1 long 17.4 high 23.8 span	1982 Mar 22.7	38.02	89.35	6626	242	254	0.001	5
T	Molniya 3T	1982-23A	1982 Mar 24.01 $10\frac{3}{4}$ years	Windmill + 6 vanes 1500?	4.2 long? 1.6 dia	1982 Mar 26.1	62.87	735.83	27000	624	40619	0.741	288
D	Molniya 3T launcher	1982-23B	1982 Mar 24.01 19 days 1982 Apr 12	Irregular	-	1982 Mar 24.3	62.85	92.83	6793	193	636	0.033	115
D	Molniya 3T launcher rocket	1982-23C	1982 Mar 24.01 26 days 1982 Apr 19	Cylinder 2500?	7.5 long 2.6 dia	1982 Mar 24.3	62.83	92.70	6786	214	602	0.029	122
	Molniya 3T rocket	1982-23D	1982 Mar 24.01 $10\frac{3}{4}$ years	Cylinder 440	2.0 long 2.0 dia	1982 Mar 25.1	62.92	732.51	26919	629	40452	0.740	288

* Launched from Cape Canaveral; landed at White Sands, New Mexico. Cargo bay carried OSS 1 and 160 kg Plasma Diagnostics Package (not separated). OSS is Office of Space Science.

	Name		Launch date, lifetime and descent date	Shape and weight (kg)	Size (m)	Date of orbital determination	Orbital inclination (deg)	Nodal period (min)	Semi major axis (km)	Perigee height (km)	Apogee height (km)	Orbital eccentricity	Argument of perigee (deg)
T	Cosmos 1344	1982-24A	1982 Mar 24.82 1200 years	Cylinder 700?	1.3 long? 1.9 dia?	1982 Mar 26.3	82.92	104.96	7370	971	1012	0.003	278
	Cosmos 1344 rocket	1982-24B	1982 Mar 24.82 600 years	Cylinder 2200?	7.4 long 2.4 dia	1982 Mar 26.3	82.92	104.88	7366	973	1002	0.002	272
T	Meteor 2-08	1982-25A	1982 Mar 25.41 1200 years	Cylinder + 2 vanes 2750?	5 long? 1.5 dia?	1982 Mar 26.3	82.54	104.13	7331	942	964	0.001	266
	Meteor 2-08 rocket	1982-25B	1982 Mar 25.41 600 years	Cylinder 2200?	7.4 long 2.4 dia	1982 Mar 26.1	82.54	104.10	7330	941	962	0.001	274
T	Cosmos 1345	1982-26A	1982 Mar 31.38 5 years	Cylinder + paddles? 900?	2 long? 1 dia?	1982 Mar 31.8	74.04	95.15	6904	504	547	0.003	340
	Cosmos 1345 rocket	1982-26B	1982 Mar 31.38 5 years	Cylinder 2200?	7.4 long 2.4 dia	1982 Mar 31.6	74.04	95.07	6900	496	547	0.004	347
T	Cosmos 1346	1982-27A	1982 Mar 31.69 60 years	Cylinder + 2 vanes? 2500?	5 long? 1.5 dia?	1982 Apr 3.3	81.17	97.58	7020	622	661	0.003	357
	Cosmos 1346 rocket	1982-27B	1982 Mar 31.69 60 years	Cylinder 1440	3.8 long 2.6 dia	1982 Mar 31.8	81.16	97.71	7026	639	657	0.001	8

	Name		Launch date, lifetime and descent date	Shape and weight (kg)	Size (m)	Date of orbital determination	Orbital inclination (deg)	Nodal period (min)	Semi major axis (km)	Perigee height (km)	Apogee height (km)	Orbital eccentricity	Argument of perigee (deg)
D R	Cosmos 1347	1982-28A	1982 Apr 2.43 50 days 1982 May 22	Sphere-cylinder 6700?	7 long? 2.4 dia	1982 Apr 2.7	70.35	89.65	6635	173	340	0.013	65
D	Cosmos 1347 rocket	1982-28B	1982 Apr 2.43 3 days 1982 Apr 5	Cylinder 2500?	7.5 long 2.6 dia	1982 Apr 2.7	70.35	89.52	6629	177	324	0.012	66
T	Cosmos 1348	1982-29A	1982 Apr 7.57 100 years?	Windmill + 6 vanes 1250?	4.2 long? 1.6 dia	1982 Apr 8.1	62.85	708.72	26333	593	39316	0.735	318
D	Cosmos 1348 launcher rocket	1982-29B	1982 Apr 7.57 28 days 1982 May 5	Cylinder 2500?	7.5 long 2.6 dia	1982 Apr 8.0	62.86	92.49	6776	212	584	0.027	122
D	Cosmos 1348 launcher	1982-29C	1982 Apr 7.57 20 days 1982 Apr 27	Irregular	-	1982 Apr 8.3	62.90	92.69	6786	192	624	0.032	118
	Cosmos 1348 rocket	1982-29D	1982 Apr 7.57 100 years?	Cylinder 440	2.0 long 2.0 dia	1982 May 7.5	62.88	705.47	26250	604	39140	0.734	318
T	Cosmos 1349	1982-30A	1982 Apr 8.01 1200 years	Cylinder 700?	1.3 long? 1.9 dia?	1982 Apr 8.7	82.93	104.97	7370	970	1014	0.003	287
	Cosmos 1349 rocket	1982-30B	1982 Apr 8.01 600 years	Cylinder 2200?	7.4 long 2.4 dia	1982 Apr 9.3	82.93	104.84	7364	969	1003	0.002	276
T	Insat 1A*	1982-31A	1982 Apr 10.28 > million years	Box + sail + antennae 1152 full	1.55 high 1.42 deep 2.18 wide	1982 Apr 15.7	0.07	1435.39	42153	35760	35789	0.0004	248

* Indian communications satellite (launched by NASA).

1982-31 continued on page 686

	Name		Launch date, lifetime and descent date	Shape and weight (kg)	Size (m)	Date of orbital determination	Orbital inclination (deg)	Nodal period (min)	Semi major axis (km)	Perigee height (km)	Apogee height (km)	Orbital eccentricity	Argument of perigee (deg)
	Insat 1A third stage (PAM)	1982-31B	1982 Apr 10.28 2 years	Cylinder 2154 full	2.29 long 1.5 dia	1982 May 2.8	28.54	597.07	23497	154	34084	0.722	193
D R	Cosmos 1350	1982-32A	1982 Apr 15.61 31 days 1982 May 16	Sphere-cylinder 6700?	7 long? 2.4 dia	1982 Apr 16.0	67.14	89.78	6642	172	355	0.014	74
D	Cosmos 1350 rocket	1982-32B	1982 Apr 15.61 3 days 1982 Apr 18	Cylinder 2500?	7.5 long 2.6 dia	1982 Apr 16.0	67.13	89.62	6634	168	343	0.013	71
T	Salyut 7	1982-33A	1982 Apr 19.83 manoeuvrable	Cylinder + 3 vanes 20100?	14 long 4.15 to 2.0 dia	1982 Apr 20.1	51.59	89.16	6614	212	260	0.004	89
D	Salyut 7 rocket	1982-33B	1982 Apr 19.83 5 days 1982 Apr 24	Cylinder 4000?	12 long? 4 dia	1982 Apr 20.2	51.60	88.89	6601	211	234	0.002	98
D	Iskra 2*	1982-33C	1982 Apr 19.83 81 days 1982 Jul 9	- 28	0.6 dia?	1982 May 17.8	51.59	91.27	6718	335	345	0.001	273
D	Iskra 3†	1982-33AD	1982 Apr 19.83 241 days 1982 Dec 16	- 28	0.6 dia?	1982 Nov 22.3	51.63	91.38	6724	345	346	0	207
26d	Fragments**	1982-33D-AF											
T	Cosmos 1351	1982-34A	1982 Apr 21.07 11 months	Octagonal ellipsoid? 550?	1.8 long? 1.5 dia?	1982 Apr 22.3	50.69	93.47	6825	348	546	0.014	151
	Cosmos 1351 rocket	1982-34B	1982 Apr 21.07 10 months	Cylinder 2200?	7.4 long 2.4 dia	1982 Apr 22.3	50.70	93.38	6821	346	540	0.014	149

* Amateur radio relay satellite ejected from Salyut 7 disposal hatch on 1982 May 17.

** See also footnotes to 1982-42C, 1982-63C and 1982-80C.

† Amateur radio relay satellite ejected from Salyut 7 disposal hatch on 1982 Nov 18.

	Name		Launch date, lifetime and descent date	Shape and weight (kg)	Size (m)	Date of orbital determination	Orbital inclina-tion (deg)	Nodal period (min)	Semi major axis (km)	Perigee height (km)	Apogee height (km)	Orbital eccen-tricity	Argument of perigee (deg)
D R	Cosmos 1352	1982-35A	1982 Apr 21.39 13.9 days 1982 May 5.3	Sphere-cylinder 6300?	6.5 long? 2.4 dia	1982 Apr 21.8	70.37	90.22	6663	209	361	0.011	65
D	Cosmos 1352 rocket	1982-35B	1982 Apr 21.39 8 days 1982 Apr 29	Cylinder 2500?	7.5 long 2.6 dia	1982 Apr 22.3	70.38	90.00	6652	206	342	0.010	60
D	Cosmos 1352 engine	1982-35D	1982 Apr 21.39 223 days 1982 Nov 30	Cone 600? full	1.5 long? 2 dia?	1982 Jun 5.6	70.39	91.88	6747	349	389	0.003	70
D	Fragments	1982-35C,E,F											
D R	Cosmos 1353	1982-36A	1982 Apr 23.41 12.86 days 1982 May 6.27	Sphere-cylinder 5900?	5.9 long 2.4 dia	1982 Apr 24.1	82.34	89.07	6605	212	242	0.002	112
D	Cosmos 1353 rocket	1982-36B	1982 Apr 23.41 3 days 1982 Apr 26	Cylinder 2500?	7.5 long 2.6 dia	1982 Apr 23.8	82.32	88.91	6597	206	232	0.002	113
D	Capsule	1982-36C	1982 Apr 23.41 17 days 1982 May 10	Ellipsoid 200?	0.9 long 1.9 dia	1982 May 7.7	82.35	88.51	6580	195	209	0.001	96
T	Cosmos 1354	1982-37A	1982 Apr 28.12 120 years	Cylinder + paddles? 750?	2 long? 1 dia?	1982 Apr 28.2	74.03	100.97	7183	794	815	0.001	14
	Cosmos 1354 rocket	1982-37B	1982 Apr 28.12 100 years	Cylinder 2200?	7.4 long 2.4 dia	1982 Apr 28.4	74.05	100.86	7177	787	811	0.002	19

	Name		Launch date, lifetime and descent date	Shape and weight (kg)	Size (m)	Date of orbital determination	Orbital inclination (deg)	Nodal period (min)	Semi major axis (km)	Perigee height (km)	Apogee height (km)	Orbital eccentricity	Argument of perigee (deg)
T	Cosmos 1355	1982-38A	1982 Apr 29.42 manoeuvrable	Cylinder?	-	1982 Apr 29.7	65.06	93.31	6812	425	443	0.001	278
D	Cosmos 1355 rocket	1982-38B	1982 Apr 29.42 <0.6 day 1982 Apr 29	Cylinder 1500?	8 long? 2.5 dia?	1982 Apr 29.6	65.03	89.40	6624	112	379	0.020	51
T	Cosmos 1356	1982-39A	1982 May 5.34 60 years	Cylinder + 2 vanes? 2500?	5 long? 1.5 dia?	1982 May 7.2	81.19	97.79	7030	632	671	0.003	48
	Cosmos 1356 rocket	1982-39B	1982 May 5.34 60 years	Cylinder 1440	3.8 long 2.6 dia	1982 May 6.4	81.21	97.92	7036	618	698	0.006	148
T	Cosmos 1357	1982-40A	1982 May 6.75 7000 years	Spheroid 40?	1.0 long? 0.8 dia?	1982 May 8.0	74.01	114.71	7820	1403	1480	0.005	105
T	Cosmos 1358	1982-40B	1982 May 6.75 8000 years	Spheroid 40?	1.0 long? 0.8 dia?	1982 May 8.0	74.01	114.90	7829	1418	1483	0.004	117
T	Cosmos 1359	1982-40C	1982 May 6.75 9000 years	Spheroid 40?	1.0 long? 0.8 dia?	1982 May 8.0	74.01	115.07	7836	1434	1482	0.003	121
T	Cosmos 1360	1982-40D	1982 May 6.75 9000 years	Spheroid 40?	1.0 long? 0.8 dia?	1982 May 8.0	74.00	115.24	7844	1448	1484	0.002	133

1982-40 continued on page 689

	Name		Launch date, lifetime and descent date	Shape and weight (kg)	Size (m)	Date of orbital determination	Orbital inclination (deg)	Nodal period (min)	Semi major axis (km)	Perigee height (km)	Apogee height (km)	Orbital eccentricity	Argument of perigee (deg)
T	Cosmos 1361	1982-40E	1982 May 6.75 10000 years	Spheroid 40?	1.0 long? 0.8 dia?	1982 May 8.0	74.01	115.42	7852	1463	1485	0.001	155
T	Cosmos 1362	1982-40F	1982 May 6.75 10000 years	Spheroid 40?	1.0 long? 0.8 dia?	1982 May 8.0	74.01	115.62	7861	1468	1498	0.002	195
T	Cosmos 1363	1982-40G	1982 May 6.75 10000 years	Spheroid 40?	1.0 long? 0.8 dia?	1982 May 7.5	74.01	115.81	7870	1480	1503	0.001	246
T	Cosmos 1364	1982-40H	1982 May 6.75 10000 years	Spheroid 40?	1.0 long? 0.8 dia?	1982 May 7.1	74.02	116.04	7880	1478	1526	0.003	240
	Cosmos 1357 rocket	1982-40J	1982 May 6.75 20000 years	Cylinder 2200?	7.4 long 2.4 dia	1982 May 8.0	74.05	117.81	7960	1481	1683	0.013	263
D	[Titan 3D]	1982-41A	1982 May 11.78 208 days 1982 Dec 5	Cylinder 13300? full	15 long 3.0 dia	1982 May 13.2	96.41	88.91	6598	177	262	0.006	142
D	Titan 3D rocket	1982-41B	1982 May 11.78 2 days 1982 May 13	Cylinder 1900	6 long 3.0 dia	1982 May 12.7	96.38	88.15	6559	155	207	0.004	140
T	Capsule	1982-41C	1982 May 11.78 70 years	Octagon? 60?	0.3 long? 0.9 dia?	1982 May 18.2	95.99	98.87	7082	701	707	0.0004	159

	Name		Launch date, lifetime and descent date	Shape and weight (kg)	Size (m)	Date of orbital determination	Orbital inclination (deg)	Nodal period (min)	Semi major axis (km)	Perigee height (km)	Apogee height (km)	Orbital eccentricity	Argument of perigee (deg)
D 2M R	Soyuz - Transport 5*	1982-42A	1982 May 13.42 106.21 days 1982 Aug 27.63	Sphere-cylinder+2 wings 6850	7.5 long 2.3 to 2.72 dia	1982 May 13.8	51.59	90.40	6675	269	325	0.004	178
D	Soyuz T-5 rocket	1982-42B	1982 May 13.42 2 days 1982 May 15	Cylinder 2500?	7.5 long 2.6 dia	1982 May 13.9	51.60	88.47	6580	186	217	0.002	122
D	Soyuz T-5† orbital module	1982-42C	1982 May 13.42 147 days 1982 Oct 7	Spheroid 1260?	2.3 dia	1982 Aug 28.2	51.62	90.47	6679	290	311	0.002	53
	Cosmos 1365 ††	1982-43A	1982 May 14.81 600 years	Cone-cylinder	6 long? 2.5 dia?	1982 May 15.2 1982 Nov 1.0	65.00 65.07	89.65 103.70	6636 7309	252 885	264 977	0.001 0.006	271 -
D	Cosmos 1365 rocket	1982-43B	1982 May 14.81 138 days 1982 Sep 29	Cylinder 1500?	8 long? 2.5 dia?	1982 Sep 28.1	65.00	89.13	6610	219	245	0.002	269
D	Cosmos 1365 platform	1982-43C	1982 May 14.81 158 days 1982 Oct 19	Irregular	-	1982 Sep 28.7	65.00	89.48	6628	239	260	0.002	263
2d	Fragments**	1982-43D-F											
T	Cosmos 1366	1982-44A	1982 May 17.99 > million years	-	-	1982 May 18.6	1.50	1436.83	42181	35803	35803	0	28

* Soyuz T-5 docked with Salyut 7 (1st airlock) on 1982 May 14.48. The Soyuz T-5 cosmonauts remained aboard Salyut, and the Soyuz T-5 spacecraft returned with the Soyuz T-7 crew (see page 703).

† Designated 1982-33V in the USA. Part of 1982-42A until 1982 Aug 27.

†† Cosmos 1365 jettisoned rocket and platform, before moving into 104 min orbit on 27 Sep 1982.

** Fragment 1982-43D is probably the uranium fuel core ejected from nuclear reactor of 1982-43A when in 104 min orbit.

1982-44 continued on page 691

	Name		Launch date, lifetime and descent date	Shape and weight (kg)	Size (m)	Date of orbital determination	Orbital inclination (deg)	Nodal period (min)	Semi major axis (km)	Perigee height (km)	Apogee height (km)	Orbital eccentricity	Argument of perigee (deg)
D	Cosmos 1366 launcher	1982-44B	1982 May 17.99 1 day 1982 May 18	Irregular	-	1982 May 18.0	51.62	88.21	6567	182	196	0.001	245
D	Cosmos 1366 launcher rocket	1982-44C	1982 May 17.99 2 days 1982 May 19	Cylinder 4000?	12 long? 4 dia	1982 May 18.0	51.62	88.46	6580	195	208	0.001	106
	Cosmos 1366 rocket	1982-44D	1982 May 17.99 8 months	Cylinder 1900?	3.9 long? 3.9 dia	1982 May 27.3	47.34	634.62	24464	350	35822	0.725	2
	Cosmos 1366 apogee motor	1982-44F	1982 May 17.99 > million years	Cylinder 440?	2.0 long? 2.0 dia?		Synchronous orbit similar to 1982-44A						
	Fragment	1982-44E											
T	Cosmos 1367	1982-45A	1982 May 20.55 100 years?	Windmill + 6 vanes 1250?	4.2 long? 1.6 dia	1982 May 21.1	62.86	707.44	26301	581	39264	0.735	318
D	Cosmos 1367 launcher rocket	1982-45B	1982 May 20.55 28 days 1982 Jun 17	Cylinder 2500?	7.5 long 2.6 dia	1982 May 21.7	62.82	92.18	6761	209	556	0.026	122
D	Cosmos 1367 launcher	1982-45C	1982 May 20.55 15 days 1982 Jun 4	Irregular	-	1982 May 20.7	62.86	92.48	6776	177	618	0.033	122
	Cosmos 1367 rocket	1982-45D	1982 May 20.55 100 years?	Cylinder 440	2.0 long 2.0 dia	1982 May 25.0	62.85	704.17	26218	583	39096	0.735	318

	Name		Launch date, lifetime and descent date	Shape and weight (kg)	Size (m)	Date of orbital determination	Orbital inclination (deg)	Nodal period (min)	Semi major axis (km)	Perigee height (km)	Apogee height (km)	Orbital eccentricity	Argument of perigee (deg)
D R	Cosmos 1368	1982-46A	1982 May 21.53 13.91 days 1982 Jun 3.44	Sphere-cylinder 6300?	6.5 long? 2.4 dia	1982 May 22.0	70.37	90.04	6654	211	341	0.010	60
D	Cosmos 1368 rocket	1982-46B	1982 May 21.53 8 days 1982 May 29	Cylinder 2500?	7.5 long 2.6 dia	1982 May 22.2	70.38	89.86	6646	209	326	0.009	54
D	Cosmos 1368 engine	1982-46E	1982 May 21.53 19 days 1982 Jun 9	Cone 600? full	1.5 long? 2 dia?	1982 Jun 3.6	70.39	89.18	6612	227	241	0.001	84
D	Fragments	1982-46C,D											
D	Progress 13*	1982-47A	1982 May 23.25 14 days 1982 Jun 6	Sphere-cylinder 7020	7.9 long 2.3 dia	1982 May 24.4	51.62	90.84	6697	290	347	0.004	203
D	Progress 13 rocket	1982-47B	1982 May 23.25 2 days 1982 May 25	Cylinder 2500?	7.5 long 2.6 dia	1982 May 23.9	51.62	88.58	6585	175	239	0.005	84
D	Fragment	1982-47C											
D R	Cosmos 1369	1982-48A	1982 May 25.37 13.87 days 1982 Jun 8.24	Sphere-cylinder 6300?	6.5 long? 2.4 dia	1982 May 26.2	82.31	89.98	6651	269	276	0.001	320
D	Cosmos 1369 rocket	1982-48B	1982 May 25.37 5 days 1982 May 30	Cylinder 2500?	7.5 long 2.6 dia	1982 May 26.7	82.32	89.10	6607	209	248	0.003	69
D	Cosmos 1369 engine	1982-48D	1982 May 25.37 33 days 1982 Jun 27	Cone 600? full	1.5 long? 2 dia?	1982 Jun 9.2	82.32	89.83	6643	252	278	0.002	280
D	Fragments	1982-48C,E											

* Progress 13 docked with Soyuz T5 - Salyut 7 (2nd airlock) on 1982 May 25.33 and undocked on 1982 Jun 4.27.

	Name		Launch date, lifetime and descent date	Shape and weight (kg)	Size (m)	Date of orbital determination	Orbital inclination (deg)	Nodal period (min)	Semi major axis (km)	Perigee height (km)	Apogee height (km)	Orbital eccentricity	Argument of perigee (deg)
D R	Cosmos 1370	1982-49A	1982 May 28.38 44 days 1982 Jul 11	Sphere-cylinder 6700?	7 long? 2.4 dia	1982 May 28.9	64.85	89.22	6614	197	275	0.006	60
D	Cosmos 1370 rocket	1982-49B	1982 May 28.38 3 days 1982 May 31	Cylinder 2500?	7.5 long 2.6 dia	1982 May 29.1	64.88	88.95	6601	192	254	0.005	56
T	Molniya 1BF	1982-50A	1982 May 28.92 11 years	Windmill + 6 vanes 1000?	3.4 long 1.6 dia	1982 May 31.0	62.83	736.13	27007	627	40631	0.741	288
D	Molniya 1BF launcher rocket	1982-50B	1982 May 28.92 46 days 1982 Jul 13	Cylinder 2500?	7.5 long 2.6 dia	1982 May 29.8	62.84	93.00	6801	213	632	0.031	120
D	Molniya 1BF launcher	1982-50C	1982 May 28.92 31 days 1982 Jun 28	Irregular	-	1982 May 30.1	62.81	92.78	6790	197	626	0.032	114
	Molniya 1BF rocket	1982-50E	1982 May 28.92 11 years	Cylinder 440	2.0 long 2.0 dia	1982 Jun 1.0	62.78	732.25	26909	614	40447	0.740	288
D	Fragment	1982-50D											
T	Cosmos 1371	1982-51A	1982 Jun 1.19 120 years	Cylinder + paddles? 750?	2 long? 1 dia?	1982 Jun 1.3	74.05	100.91	7179	790	812	0.002	24
	Cosmos 1371 rocket	1982-51B	1982 Jun 1.19 100 years	Cylinder 2200?	7.4 long 2.4 dia	1982 Jun 1.3	74.06	100.83	7175	783	811	0.002	43

	Name		Launch date, lifetime and descent date	Shape and weight (kg)	Size (m)	Date of orbital determination	Orbital inclination (deg)	Nodal period (min)	Semi major axis (km)	Perigee height (km)	Apogee height (km)	Orbital eccentricity	Argument of perigee (deg)
	Cosmos 1372*	1982-52A	1982 Jun 1.58 600 years	Cone-cylinder	6 long? 2.5 dia?	1982 Jun 1.6 1982 Aug 14.6	64.99 64.90	89.65 103.95	6636 7323	246 908	270 981	0.002 0.005	262 249
D	Cosmos 1372 rocket	1982-52C	1982 Jun 1.58 73 days 1982 Aug 13	Cylinder 1500?	8 long? 2.5 dia?	1982 Aug 12.4	64.99	88.97	6602	213	235	0.002	287
D	Cosmos 1372 platform	1982-52B	1982 Jun 1.58 101 days 1982 Sep 10	Irregular	-	1982 Aug 12.3	64.98	89.53	6630	242	262	0.002	271
	Fragment†	1982-52D											
D R	Cosmos 1373	1982-53A	1982 Jun 2.54 13.86 days 1982 Jun 16.40	Sphere-cylinder 6300?	6.5 long? 2.4 dia	1982 Jun 2.8	70.38	90.08	6657	210	347	0.010	62
D	Cosmos 1373 rocket	1982-53B	1982 Jun 2.54 9 days 1982 Jun 11	Cylinder 2500?	7.5 long 2.6 dia	1982 Jun 4.7	70.38	89.75	6640	204	320	0.009	50
D	Cosmos 1373 engine	1982-53E	1982 Jun 2.54 155 days 1982 Nov 4	Cone 600? full	1.5 long? 2 dia?	1982 Jun 26.8	70.38	91.89	6747	356	382	0.002	230
D	Fragments	1982-53C,D,F											
D R?	Cosmos 1374**	1982-54A	1982 Jun 3.90 0.07 day? 1982 Jun 3.97?	Delta wing? 1000?	2.5 long? 2.1 span?	Orbit similar to 1982-54B							
D	Cosmos 1374 rocket	1982-54B	1982 Jun 3.90 2 days 1982 Jun 5	Cylinder 2200?	7.4 long? 2.4 dia?	1982 Jun 4.2	50.66	88.32	6573	167	222	0.004	327
D	Fragment	1982-54C											

* Cosmos 1372 jettisoned rocket and platform before moving into 104 min orbit about 1982 Aug 11

† Probably the uranium fuel core ejected from nuclear reactor of 1982-52A when in 104 min orbit.

** Launched from Kapustin Yar, water recovery 109 min later about 17°S, 98°E (near Cocos Islands)

	Name		Launch date, lifetime and descent date	Shape and weight (kg)	Size (m)	Date of orbital determination	Orbital inclination (deg)	Nodal period (min)	Semi major axis (km)	Perigee height (km)	Apogee height (km)	Orbital eccentricity	Argument of perigee (deg)
T?	Cosmos 1375	1982-55A	1982 Jun 6.71 1200 years	Cylinder?	4 long? 2 dia?	1982 Jun 7.2	65.84	105.02	7374	981	1011	0.002	291
	Cosmos 1375 rocket	1982-55B	1982 Jun 6.71 600 years	Cylinder 2200?	7.4 long 2.4 dia	1982 Jun 7.1	65.84	104.93	7370	981	1002	0.001	290
	Fragment	1982-55C											
D R	Cosmos 1376	1982-56A	1982 Jun 8.33 13.86 days 1982 Jun 22.19	Sphere-cylinder 6300?	6.5 long? 2.4 dia	1982 Jun 10.4	82.34	89.88	6646	261	274	0.001	309
D	Cosmos 1376 rocket	1982-56B	1982 Jun 8.33 4 days 1982 Jun 12	Cylinder 2500?	7.5 long 2.6 dia	1982 Jun 8.7	82.34	89.07	6605	215	239	0.002	67
D	Cosmos 1376 engine	1982-56C	1982 Jun 8.33 36 days 1982 Jul 14	Cone 600? full	1.5 long? 2 dia?	1982 Jun 23.6	82.35	89.91	6647	257	281	0.002	302
D	Fragments	1982-56D,E											
D R	Cosmos 1377	1982-57A	1982 Jun 8.50 44 days 1982 Jul 22	Sphere-cylinder 6700?	7 long? 2.4 dia	1982 Jun 10.4	64.90	89.86	6646	173	363	0.014	56
D	Cosmos 1377 rocket	1982-57B	1982 Jun 8.50 4 days 1982 Jun 12	Cylinder 2500?	7.5 long 2.6 dia	1982 Jun 8.8	64.91	89.52	6629	171	331	0.012	57

	Name		Launch date, lifetime and descent date	Shape and weight (kg)	Size (m)	Date of orbital determination	Orbital inclination (deg)	Nodal period (min)	Semi major axis (km)	Perigee height (km)	Apogee height (km)	Orbital eccentricity	Argument of perigee (deg)
T	Westar 5	1982-58A	1982 Jun 9.02 > million years	Cylinder 1100 full 585 empty	2.74 to 6.84 long 2.16 dia	1982 Jun 11.7	0.34	1418.4	41819	34970	35912	0.011	198
	Westar 5 third stage (PAM)	1982-58B	1982 Jun 9.02 10 years	Cylinder 2154 full	2.29 long 1.5 dia	1982 Jun 23.6	27.44	625.06	24265	199	35574	0.729	188
T	Cosmos 1378*	1982-59A	1982 Jun 10.74 60 years	-	-	1982 Jun 12.0	82.51	97.72	7027	634	663	0.002	296
	Cosmos 1378 rocket	1982-59B	1982 Jun 10.74 60 years	Cylinder 2200?	7.4 long 2.4 dia	1982 Jun 12.9	82.52	97.74	7028	636	663	0.002	296
D	Cosmos 1379**	1982-60A	1982 Jun 18.46 0.16 day? 1982 Jun 18.62?	Cylinder?	4 long? 2 dia?	1982 Jun 18.4 1982 Jun 18.5	65.09 65.84	91.37 100.43	6723 7158	144 539	546 1021	0.030 0.034	48 250
D	Cosmos 1379 rocket	1982-60B	1982 Jun 18.46 < 1 day 1982 Jun 18	Cylinder 1500?	8 long? 2.5 dia?	1982 Jun 18.9	65.11	88.71	6589	130	292	0.012	52
D	Cosmos 1380†	1982-61A	1982 Jun 18.50 9 days 1982 Jun 27	Cylinder 700?	1.3 long? 1.9 dia?	1982 Jun 20.6	82.90	92.63	6780	145	659	0.038	73
D	Cosmos 1380 rocket	1982-61B	1982 Jun 18.50 7 days 1982 Jun 25	Cylinder 2200?	7.4 long 2.4 dia	1982 Jun 20.6	82.90	92.43	6771	141	644	0.037	72

* Oceanographic satellite.

** Probably passed close to Cosmos 1375 on 1982 Jun 18.60, then de-orbited and destroyed on re-entry.

† Intended navigation satellite; final-stage rocket failed to re-fire at apogee to circularize orbit.

	Name		Launch date, lifetime and descent date	Shape and weight (kg)	Size (m)	Date of orbital determination	Orbital inclination (deg)	Nodal period (min)	Semi major axis (km)	Perigee height (km)	Apogee height (km)	Orbital eccentricity	Argument of perigee (deg)
D R	Cosmos 1381	1982-62A	1982 Jun 18.54 12.87 days 1982 Jul 1.41	Sphere-cylinder 6300?	6.5 long? 2.4 dia	1982 Jun 19.0	70.36	90.34	6669	208	374	0.012	64
D	Cosmos 1381 rocket	1982-62B	1982 Jun 18.54 10 days 1982 Jun 28	Cylinder 2500?	7.5 long 2.6 dia	1982 Jun 20.7	70.37	90.01	6653	205	345	0.010	53
	Cosmos 1381 engine	1982-62F	1982 Jun 18.54 8 months	Cone 600? full	1.5 long? 2 dia?	1982 Jul 29.0	70.38	92.41	6773	361	429	0.005	152
D	Fragments	1982-62C-E											
D 3M R	Soyuz-Transport 6*	1982-63A	1982 Jun 24.69 7.91 days 1982 Jul 2.60	Sphere-cylinder +2 wings 6850	7.5 long 2.3 to 2.72 dia	1982 Jun 25.2	51.63	89.58	6635	248	265	0.001	2
D	Soyuz T-6 rocket	1982-63B	1982 Jun 24.69 2 days 1982 Jun 26	Cylinder 2500?	7.5 long 2.6 dia	1982 Jun 25.0	51.61	88.50	6581	190	216	0.002	115
D	Soyuz T-6** orbital module	1982-63C	1982 Jun 24.69 48 days 1982 Aug 11	Spheroid 1260?	2.3 dia	1982 Jul 5.6	51.63	90.44	6678	287	313	0.002	51
T	Cosmos 1382	1982-64A	1982 Jun 25.10 100 years?	Windmill + 6 vanes 1250?	4.2 long? 1.6 dia	1982 Jun 26.1	62.82	711.13	26392	592	39436	0.736	316
D	Cosmos 1382 launcher rocket	1982-64B	1982 Jun 25.10 43 days 1982 Aug 7	Cylinder 2500?	7.5 long 2.6 dia	1982 Jun 25.5	62.80	92.70	6786	218	597	0.028	118

* Soyuz T6 docked with Soyuz T5-Salyut 7 (2nd airlock) on 1982 Jun 25.75, with 2 Russian and first French cosmonaut. Undocked 1982 Jul 2.5.

** Designated 1982-33L in the USA. Part of 1982-63A until about 1982 Jul 2.

1982-64 continued on page 698

	Name		Launch date, lifetime and descent date	Shape and weight (kg)	Size (m)	Date of orbital determination	Orbital inclination (deg)	Nodal period (min)	Semi major axis (km)	Perigee height (km)	Apogee height (km)	Orbital eccentricity	Argument of perigee (deg)
D	Cosmos 1382 launcher	1982-64C	1982 Jun 25.10 20 days 1982 Jul 15	Irregular	-	1982 Jun 25.5	62.83	92.44	6774	185	606	0.031	119
	Cosmos 1382 rocket	1982-64D	1982 Jun 25.10 100 years?	Cylinder 440	2.0 long 2.0 dia	1982 Jun 28.1	62.78	708.50	26326	603	39293	0.735	316
D 2M R	STS 4* (Columbia F4)	1982-65A	1982 Jun 27.62 7.1 days 1982 Jul 4.7	Cylinder + delta wing 68800 empty	37.1 long 17.4 high 23.8 span	1982 Jun 28.6	28.52	90.33	6677	295	302	0.001	248
T	Cosmos 1383 (Cospas 1**)	1982-66A	1982 Jun 29.91 1200 years	Cylinder 700?	1.3 long? 1.9 dia?	1982 Jun 30.4	82.93	105.35	7388	991	1029	0.003	251
	Cosmos 1383 rocket	1982-66B	1982 Jun 29.91 600 years	Cylinder 2200?	7.4 long 2.4 dia	1982 Jun 30.3	82.93	105.25	7383	988	1022	0.002	238
D R	Cosmos 1384	1982-67A	1982 Jun 30.63 30 days 1982 Jul 30	Sphere- cylinder 6700?	7 long? 2.4 dia	1982 Jun 30.7	67.15	89.76	6641	170	355	0.014	70
D	Cosmos 1384 rocket	1982-67B	1982 Jun 30.63 5 days 1982 Jul 5	Cylinder 2500?	7.5 long 2.6 dia	1982 Jun 30.7	67.14	89.70	6638	168	351	0.014	69
D R	Cosmos 1385	1982-68A	1982 Jul 6.33 13.9 days 1982 Jul 20.2	Sphere- cylinder 6300?	6.5 long? 2.4 dia	1982 Jul 6.9	82.33	88.75	6590	186	237	0.004	81

* Launched from Cape Canaveral; landed at Edwards AFB, California. Cargo bay carried CIRRIS 1 (not separated). CIRRIS is Cryogenic infra-red radiation imaging system.

1982-68 continued on page 699

** Cosmos satellite for program of air and sea rescue.

	Name		Launch date, lifetime and descent date	Shape and weight (kg)	Size (m)	Date of orbital determination	Orbital inclination (deg)	Nodal period (min)	Semi major axis (km)	Perigee height (km)	Apogee height (km)	Orbital eccentricity	Argument of perigee (deg)
D	Cosmos 1385 rocket	1982-68B	1982 Jul 6.33 2 days 1982 Jul 8	Cylinder 2500?	7.5 long 2.6 dia	1982 Jul 6.6	82.32	88.61	6583	184	225	0.003	79
D	Cosmos 1385 engine	1982-68E	1982 Jul 6.33 104 days 1982 Oct 18	Cone 600? full	1.5 long? 2 dia?	1982 Jul 21.0	82.33	91.50	6727	306	391	0.006	187
D	Fragments	1982-68C,D,F,G											
T	Cosmos 1386	1982-69A	1982 Jul 7.41 1200 years	Cylinder 700?	1.3 long? 1.9 dia?	1982 Jul 9.2	82.96	104.78	7361	955	1011	0.004	290
	Cosmos 1386 rocket	1982-69B	1982 Jul 7.41 600 years	Cylinder 2200?	7.4 long 2.4 dia	1982 Jul 9.6	82.96	104.66	7356	955	1000	0.003	288
D	Progress 14*	1982-70A	1982 Jul 10.42 33.7 days 1982 Aug 13.1	Sphere cylinder 7020	7.9 long 2.3 dia	1982 Jul 12.5	51.64	90.72	6691	301	325	0.002	142
D	Progress 14 rocket	1982-70B	1982 Jul 10.42 2 days 1982 Jul 12	Cylinder 2500?	7.5 long 2.6 dia	1982 Jul 10.7	51.62	88.56	6584	183	229	0.004	95
D R	Cosmos 1387	1982-71A	1982 Jul 13.34 12.9 days 1982 Jul 26.2	Sphere-cylinder 5900?	5.9 long 2.4 dia	1982 Jul 13.8	82.34	89.08	6606	212	243	0.002	108
D	Cosmos 1387 rocket	1982-71B	1982 Jul 13.34 4 days 1982 Jul 17	Cylinder 2500?	7.5 long 2.6 dia	1982 Jul 13.6	82.34	88.96	6600	211	232	0.002	104

* Progress 14 docked with Soyuz T5 - Salyut 7 (2nd airlock) on 1982 Jul 12.49 and undocked 1982 Aug 11.

1982-71 continued on page 700

Year of launch 1982 continued

	Name		Launch date, lifetime and descent date	Shape and weight (kg)	Size (m)	Date of orbital determination	Orbital inclination (deg)	Nodal period (min)	Semi major axis (km)	Perigee height (km)	Apogee height (km)	Orbital eccentricity	Argument of perigee (deg)
D	Capsule	1982-71C	1982 Jul 13.34 18 days 1982 Jul 31	Ellipsoid 200?	0.9 long 1.9 dia	1982 Jul 26.4	82.33	88.68	6586	200	216	0.001	72
D	Fragment	1982-71D											
T	Landsat 4*	1982-72A	1982 Jul 16.75 80 years	Irregular + panels 1938	-	1982 Jul 19.3	98.26	98.63	7070	683	700	0.001	271
	Landsat 4 second stage	1982-72B	1982 Jul 16.75 2 years	Cylinder + annulus 350?	6.4 long 1.5 and 2.44 dia	1982 Jul 19.2	101.77	95.20	6906	359	696	0.024	293
T	Cosmos 1388	1982-73A	1982 Jul 21.27 7000 years	Spheroid 40?	1.0 long? 0.8 dia?	1982 Jul 22.2	74.02	114.58	7814	1395	1476	0.005	93
T	Cosmos 1389	1982-73B	1982 Jul 21.27 8000 years	Spheroid 40?	1.0 long? 0.8 dia?	1982 Jul 22.2	74.02	114.77	7823	1412	1477	0.004	103
T	Cosmos 1390	1982-73C	1982 Jul 21.27 8000 years	Spheroid 40?	1.0 long? 0.8 dia?	1982 Jul 22.2	74.02	114.95	7831	1429	1477	0.003	101
T	Cosmos 1391	1982-73D	1982 Jul 21.27 9000 years	Spheroid 40?	1.0 long? 0.8 dia?	1982 Jul 22.2	74.02	115.13	7839	1445	1477	0.002	106
T	Cosmos 1392	1982-73E	1982 Jul 21.27 10000 years	Spheroid 40?	1.0 long? 0.8 dia?	1982 Jul 22.2	74.02	115.32	7848	1462	1477	0.001	108

* Carries manoeuvring engine to assist possible retrieval by Space Shuttle.

1982-73 continued on page 701

Year of launch 1982 continued

	Name	Launch date, lifetime and descent date	Shape and weight (kg)	Size (m)	Date of orbital determination	Orbital inclination (deg)	Nodal period (min)	Semi major axis (km)	Perigee height (km)	Apogee height (km)	Orbital eccentricity	Argument of perigee (deg)
T	Cosmos 1393 1982-73F	1982 Jul 21.27 10000 years	Spheroid 40?	1.0 long? 0.8 dia?	1982 Jul 22.2	74.02	115.52	7857	1472	1485	0.001	197
T	Cosmos 1394 1982-73G	1982 Jul 21.27 10000 years	Spheroid 40?	1.0 long? 0.8 dia?	1982 Jul 22.2	74.01	115.71	7865	1476	1498	0.001	259
T	Cosmos 1395 1982-73H	1982 Jul 21.27 10000 years	Spheroid 40?	1.0 long? 0.8 dia?	1982 Jul 22.2	74.01	115.92	7875	1475	1518	0.003	254
	Cosmos 1388 1982-73J rocket	1982 Jul 21.27 20000 years	Cylinder 2200?	7.4 long 2.4 dia	1982 Jul 22.2	74.01	117.99	7968	1472	1708	0.015	274
T	Molniya 1BG 1982-74A	1982 Jul 21.41 $10\frac{3}{4}$ years	Windmill + 6 vanes 1000?	3.4 long 1.6 dia	1982 Jul 22.9	62.93	701.19	26145	617	38917	0.732	288
D	Molniya 1BG 1982-74B launcher	1982 Jul 21.41 44 days 1982 Sep 3	Irregular	-	1982 Jul 22.6	62.97	93.54	6827	207	691	0.035	65
D	Molniya 1 BG 1982-74C launcher rocket	1982 Jul 21.41 61 days 1982 Sep 20	Cylinder 2500?	7.5 long 2.6 dia	1982 Jul 22.2	62.96	93.42	6821	235	651	0.030	65
	Molniya 1BG 1982-74D rocket	1982 Jul 21.41 $10\frac{3}{4}$ years	Cylinder 440	2.0 long 2.0 dia	1982 Jul 26.8	62.98	698.72	26083	601	38809	0.732	288

	Name		Launch date, lifetime and descent date	Shape and weight (kg)	Size (m)	Date of orbital determination	Orbital inclination (deg)	Nodal period (min)	Semi major axis (km)	Perigee height (km)	Apogee height (km)	Orbital eccentricity	Argument of perigee (deg)
D R	Cosmos 1396	1982-75A	1982 Jul 27.52 13.76 days 1982 Aug 10.28	Sphere-cylinder 6300?	6.5 long? 2.4 dia	1982 Jul 27.6	72.86	89.47	6626	198	298	0.008	90
D	Cosmos 1396 rocket	1982-75B	1982 Jul 27.52 5 days 1982 Aug 1	Cylinder 2500?	7.5 long 2.6 dia	1982 Jul 27.7	72.84	89.38	6622	195	292	0.007	88
D	Cosmos 1396 engine	1982-75C	1982 Jul 27.52 23 days 1982 Aug 19	Cone 600? full	1.5 long? 2 dia?	1982 Aug 10.6	72.85	89.16	6611	222	243	0.002	104
D	Fragment	1982-75D											
T	Cosmos 1397	1982-76A	1982 Jul 29.82 1 year	Octagonal ellipsoid? 550?	1.8 long? 1.5 dia?	1982 Jul 30.2	50.69	93.37	6821	345	540	0.014	147
	Cosmos 1397 rocket	1982-76B	1982 Jul 29.82 9 months	Cylinder 2200?	7.4 long 2.4 dia	1982 Jul 30.6	50.69	93.28	6817	343	534	0.014	147
D R	Cosmos 1398	1982-77A	1982 Aug 3.48 9.89 days 1982 Aug 13.37	Sphere-cylinder 5900?	5.9 long 2.4 dia	1982 Aug 5.3	82.35	89.02	6603	216	234	0.001	71
D	Cosmos 1398 rocket	1982-77B	1982 Aug 3.48 4 days 1982 Aug 7	Cylinder 2500?	7.5 long 2.6 dia	1982 Aug 3.8	82.37	88.92	6598	212	228	0.001	35
D	Capsule	1982-77C	1982 Aug 3.48 10 days 1982 Aug 13	Ellipsoid 200?	0.9 long 1.9 dia	1982 Aug 13.5	82.35	88.45	6575	193	200	0.001	10
D	Fragment	1982-77D											

	Name		Launch date, lifetime and descent date	Shape and weight (kg)	Size (m)	Date of orbital determination	Orbital inclination (deg)	Nodal period (min)	Semi major axis (km)	Perigee height (km)	Apogee height (km)	Orbital eccentricity	Argument of perigee (deg)
D R	Cosmos 1399	1982-78A	1982 Aug 4.48 43 days 1982 Sep 16	Sphere-cylinder 6700?	7 long? 2.4 dia	1982 Aug 6.0	64.90	89.65	6636	171	344	0.013	61
D	Cosmos 1399 rocket	1982-78B	1982 Aug 4.48 4 days 1982 Aug 8	Cylinder 2500?	7.5 long 2.6 dia	1982 Aug 4.7	64.92	89.68	6637	171	347	0.013	60
D	Fragment	1982-78C											
T	Cosmos 1400	1982-79A	1982 Aug 5.29 60 years	Cylinder + 2 vanes? 2500?	5 long? 1.5 dia?	1982 Aug 7.6	81.16	97.57	7020	630	653	0.002	24
	Cosmos 1400 rocket	1982-79B	1982 Aug 5.29 60 years	Cylinder 1440	3.8 long 2.6 dia	1982 Aug 6.4	81.18	97.71	7026	608	688	0.006	164
D 3M R	Soyuz - Transport 7*	1982-80A	1982 Aug 19.72 113.07 days 1982 Dec 10.79	Sphere-cylinder+2 wings 6850	7.5 long 2.3 to 2.72 dia	1982 Aug 21.3	51.63	90.34	6672	289	299	0.001	297
D	Soyuz T-7 rocket	1982-80B	1982 Aug 19.72 2 days 1982 Aug 21	Cylinder 2500?	7.5 long 2.6 dia	1982 Aug 20.0	51.62	88.43	6578	194	205	0.001	91
	Soyuz T-7** orbital module	1982-80C	1982 Aug 19.72 7 months	Spheroid 1260?	2.3 dia	1982 Dec 11.66	51.62	91.51	6730	349	355	0.0005	117

* Soyuz T-7 docked with Soyuz T5 - Salyut 7 (2nd airlock) on 1982 Aug 20.77, with 1 female and 2 male cosmonauts. They returned to Earth in Soyuz T-5, undocking 1982 Aug 27 (see page 690). Soyuz T-7 (piloted by Soyuz T-5 crew) then undocked from 2nd airlock and re-attached to 1st airlock on 1982 Aug 29. Soyuz T-7 landed with Soyuz T-5 crew, who achieved a manned duration record of 211.37 days.

** Designated 1982-33AF in the USA. Part of 1982-80A until Dec 10.

	Name		Launch date, lifetime and descent date	Shape and weight (kg)	Size (m)	Date of orbital determination	Orbital inclination (deg)	Nodal period (min)	Semi major axis (km)	Perigee height (km)	Apogee height (km)	Orbital eccentricity	Argument of perigee (deg)
D R	Cosmos 1401	1982-81A	1982 Aug 20.41 13.9 days 1982 Sep 3.3	Sphere-cylinder 6300?	6.5 long? 2.4 dia	1982 Aug 21.4	82.34	89.89	6646	261	274	0.001	328
D	Cosmos 1401 rocket	1982-81B	1982 Aug 20.41 5 days 1982 Aug 25	Cylinder 2500?	7.5 long 2.6 dia	1982 Aug 21.2	82.34	89.12	6608	211	249	0.003	107
D	Cosmos 1401 engine	1982-81E	1982 Aug 20.41 33 days 1982 Sep 22	Cone 600? full	1.5 long? 2 dia?	1982 Sep 3.4	82.34	89.96	6649	259	283	0.002	305
D	Fragments	1982-81C,D											
T	Telesat 5* (Anik D1)	1982-82A	1982 Aug 26.97 >million years	Cylinder + panel 1238 full 660 empty	2.82 to 6.7 long 2.16 dia	1982 Aug 28.5 1982 Sep 3.3	24.10 0.16	639.54 1436.95	24588 42185	175 35795	36245 35819	0.734 0.0003	179 287
D	Telesat 5 second stage	1982-82B	1982 Aug 26.97 23 days 1982 Sep 18	Cylinder + annulus 350?	6.4 long 1.52 and 2.44 dia	1982 Aug 28.9	25.96	91.29	6723	217	472	0.019	79
	Telesat 5 third stage (PAM)	1982-82C	1982 Aug 26.97 16 months	Cylinder 2154 full	2.29 long 1.5 dia	1982 Aug 30.8	24.36	638.76	24568	175	36205	0.733	180
T	Molniya 3U	1982-83A	1982 Aug 27.01 19½ years	Windmill + 6 vanes 1500?	4.2 long? 1.6 dia	1982 Sep 1.2	62.87	736.60	27018	457	40823	0.747	280
D	Molniya 3U launcher rocket	1982-83B	1982 Aug 27.01 16 days 1982 Sep 12	Cylinder 2500?	7.5 long 2.6 dia	1982 Aug 28.8	62.87	90.89	6697	208	430	0.017	130

* Canadian communications satellite launched by NASA.

1982-83 continued on page 705

	Name		Launch date, lifetime and descent date	Shape and weight (kg)	Size (m)	Date of orbital determination	Orbital inclination (deg)	Nodal period (min)	Semi major axis (km)	Perigee height (km)	Apogee height (km)	Orbital eccentricity	Argument of perigee (deg)
D	Molniya 3U launcher	1982-83C	1982 Aug 27.01 14 days 1982 Sep 10	Irregular	-	1982 Aug 28.8	62.85	91.10	6708	194	465	0.020	120
	Molniya 3U rocket	1982-83E	1982 Aug 27.01 19½ years	Cylinder 440	2.0 long 2.0 dia	1982 Aug 31.1	62.87	733.20	26935	449	40665	0.747	280
D	Fragment	1982-83D											
D	Cosmos 1402* platform + reactor	1982-84A	1982 Aug 30.42 146.51 days 1983 Jan 23.93	Cylinder 3000?	6 long? 2.5 dia?	1982 Sep 1.2 1983 Jan 3.6	65.01 65.00	89.65 89.44	6636 6626	251 241	265 255	0.001 0.001	273 280
D	Cosmos 1402 rocket	1982-84B	1982 Aug 30.42 122 days 1982 Dec 30	Cylinder 1500?	8 long? 2.5 dia?	1982 Dec 28.4	65.00	89.50	6628	243	257	0.001	271
D	Cosmos 1402 fuel core	1982-84C	1982 Aug 30.42 161.04 days 1983 Feb 7.46	Cylinder 1000?	1.5 long? 1 dia?	1982 Dec 28.4	65.00	89.42	6625	240	254	0.001	57
D R	Cosmos 1403	1982-85A	1982 Sep 1.38 13.9 days 1982 Sep 15.3	Sphere-cylinder 6300?	6.5 long? 2.4 dia	1982 Sep 3.2	70.38	92.26	6763	354	416	0.005	212
D	Cosmos 1403 rocket	1982-85B	1982 Sep 1.38 8 days 1982 Sep 9	Cylinder 2500?	7.5 long 2.6 dia	1982 Sep 3.6	70.38	89.85	6645	205	328	0.009	54
	Cosmos 1403 engine	1982-85D	1982 Sep 1.38 7 months	Cone 600? full	1.5 long? 2 dia?	1982 Sep 15.3	70.38	92.23	6762	353	414	0.005	193
D	Fragments	1982-85C,E-G											

* Cosmos 1402 jettisoned rocket on 1982 Dec 28, but platform failed to separate from reactor and fuel core, so reactor and fuel core were not raised to high orbit. Fuel core was separated from platform + reactor and was designated 1982-84C. The satellite-rocket-platform combination was designated 1982-84A before 1982 Dec 28 and the first orbit for 1982-84A is for that combination.

	Name		Launch date, lifetime and descent date	Shape and weight (kg)	Size (m)	Date of orbital determination	Orbital inclination (deg)	Nodal period (min)	Semi major axis (km)	Perigee height (km)	Apogee height (km)	Orbital eccentricity	Argument of perigee (deg)
D R	Cosmos 1404	1982-86A	1982 Sep 1.49 13.7 days 1982 Sep 15.2	Sphere-cylinder 6300?	6.5 long? 2.4 dia	1982 Sep 2.7	72.85	92.31	6765	358	416	0.004	213
D	Cosmos 1404 rocket	1982-86B	1982 Sep 1.49 9 days 1982 Sep 10	Cylinder 2500?	7.5 long 2.6 dia	1982 Sep 2.7	72.85	89.93	6648	196	344	0.011	76
	Cosmos 1404 engine	1982-86G	1982 Sep 1.49 7 months	Cone 600? full	1.5 long? 2 dia?	1982 Sep 16.0	72.85	92.28	6764	356	415	0.004	182
D	Fragments	1982-86C-F,H											
T	Kiku 4 (ETS 3)*	1982-87A	1982 Sep 3.22 2000 years	Cylinder + paddles?	-	1982 Sep 6.9	44.62	107.10	7475	965	1228	0.018	219
	Kiku 4 rocket	1982-87B	1982 Sep 3.22 1000 years	-	-	1982 Sep 6.2	44.61	107.00	7470	965	1218	0.017	218
	Fragment	1982-87C											
T	Cosmos 1405	1982-88A	1982 Sep 4.74 manoeuvrable	Cylinder?	-	1982 Sep 7.3	65.02	93.30	6815	429	445	0.001	272
D	Cosmos 1405 rocket	1982-88B	1982 Sep 4.74 1 day 1982 Sep 5	Cylinder 1500?	8 long? 2.5 dia?	1982 Sep 4.9	65.02	89.41	6624	112	380	0.020	52
D R	Cosmos 1406	1982-89A	1982 Sep 8.43 12.9 days 1982 Sep 21.3	Sphere-cylinder 5900?	5.9 long? 2.4 dia	1982 Sep 11.3	82.31	88.94	6599	211	230	0.001	111

* Japanese Engineering Test Satellite.

1982-89 continued on page 707

	Name		Launch date, lifetime and descent date	Shape and weight (kg)	Size (m)	Date of orbital determination	Orbital inclination (deg)	Nodal period (min)	Semi major axis (km)	Perigee height (km)	Apogee height (km)	Orbital eccentricity	Argument of perigee (deg)
D	Cosmos 1406 rocket	1982-89B	1982 Sep 8.43 3 days 1982 Sep 11	Cylinder 2500?	7.5 long 2.6 dia	1982 Sep 8.7	82.30	88.86	6595	210	223	0.001	106
D	Capsule?	1982-89C	1982 Sep 8.43 17 days 1982 Sep 25	Ellipsoid? 200?	0.9 long? 1.9 dia?	1982 Sep 22.3	82.30	88.64	6584	200	211	0.001	314
D	Fragment	1982-89D											
D r	China 10*	1982-90A	1982 Sep 9.31 12 days 1982 Sep 21	Cylinder? 3600, then 1200?	-	1982 Sep 11.3	62.98	90.08	6658	174	385	0.016	148
D	China 10 rocket	1982-90B	1982 Sep 9.31 <1 day 1982 Sep 9	Cylinder	8 long 3 dia	1982 Sep 9.5	62.93	89.65	6636	169	347	0.013	150
D	Fragment	1982-90C											
D R	Cosmos 1407	1982-91A	1982 Sep 15.65 31 days 1982 Oct 16	Sphere-cylinder 6700?	7 long? 2.4 dia	1982 Sep 17.4	67.15	89.63	6634	173	339	0.013	67
D	Cosmos 1407 rocket	1982-91B	1982 Sep 15.65 4 days 1982 Sep 19	Cylinder 2500?	7.5 long 2.6 dia	1982 Sep 16.0	67.15	89.51	6628	172	328	0.012	67
T	Cosmos 1408	1982-92A	1982 Sep 16.21 60 years	-	-	1982 Sep 18.3	82.57	97.78	7030	635	668	0.002	289
	Cosmos 1408 rocket	1982-92B	1982 Sep 16.21 60 years	Cylinder 2200?	7.4 long 2.4 dia	1982 Sep 17.3	82.57	97.75	7028	633	667	0.002	299

* A capsule returned to Earth on 1982 Sep 14.

	Name		Launch date, lifetime and descent date	Shape and weight (kg)	Size (m)	Date of orbital determination	Orbital inclination (deg)	Nodal period (min)	Semi major axis (km)	Perigee height (km)	Apogee height (km)	Orbital eccentricity	Argument of perigee (deg)
T	Statsionar - Ekran 9	1982-93A	1982 Sep 16.78 > million years	Cylinder + plate	-	1982 Sep 19.7	0.45	1425.53	41958	35257	35902	0.008	201
D	Ekran 9 launcher rocket	1982-93B	1982 Sep 16.78 2 days 1982 Sep 18	Cylinder 4000?	12 long? 4 dia	1982 Sep 17.0	51.62	88.24	6569	186	195	0.001	199
D	Ekran 9 launcher	1982-93C	1982 Sep 16.78 1 day 1982 Sep 17	Irregular	-	1982 Sep 17.1	51.61	88.12	6563	183	186	0	106
	Ekran 9 rocket	1982-93D	1982 Sep 16.78 11 months	Cylinder 1900?	3.9 long? 3.9 dia	1982 Sep 25.1	47.46	629.53	24339	317	35604	0.725	2
	Ekran 9 apogee motor	1982-93F	1982 Sep 16.78 > million years	Cylinder 440?	2.0 long? 2.0 dia?		Synchronous orbit similar to 1982-93A						
	Fragment	1982-93E											
D	Progress 15*	1982-94A	1982 Sep 18.21 28.5 days 1982 Oct 16.7	Sphere-cylinder 7020	7.9 long 2.3 dia	1982 Sep 20.2	51.63	90.74	6692	301	326	0.002	329
D	Progress 15 rocket	1982-94B	1982 Sep 18.21 2 days 1982 Sep 20	Cylinder 2500?	7.5 long 2.6 dia	1982 Sep 18.5	51.64	88.60	6586	185	231	0.003	84

* Progress 15 docked with Soyuz T7-Salyut 7 (2nd airlock) on 1982 Sep 20.26.

	Name		Launch date, lifetime and descent date	Shape and weight (kg)	Size (m)	Date of orbital determination	Orbital inclination (deg)	Nodal period (min)	Semi major axis (km)	Perigee height (km)	Apogee height (km)	Orbital eccentricity	Argument of perigee (deg)
T	Cosmos 1409	1982-95A	1982 Sep 22.27 100 years?	Windmill + 6 vanes 1250?	4.2 long? 1.6 dia	1982 Sep 26.3	63.08	716.69	26530	613	39690	0.736	318
D	Cosmos 1409 launcher rocket	1982-95B	1982 Sep 22.27 25 days 1982 Oct 17	Cylinder 2500?	7.5 long 2.6 dia	1982 Sep 22.4	62.86	92.16	6760	216	547	0.024	120
D	Cosmos 1409 launcher	1982-95C	1982 Sep 22.27 21 days 1982 Oct 13	Irregular	-	1982 Sep 22.4	62.85	92.55	6779	199	603	0.030	120
	Cosmos 1409 rocket	1982-95D	1982 Sep 22.27 100 years?	Cylinder 440	2.0 long 2.0 dia	1982 Sep 23.3	63.02	707.43	26301	611	39234	0.734	318
T	Cosmos 1410	1982-96A	1982 Sep 24.38 10000 years	-	-	1982 Sep 26.2	82.61	115.99	7877	1495	1503	0.001	272
	Cosmos 1410 rocket	1982-96B	1982 Sep 24.38 10000 years	Cylinder 2200?	7.4 long 2.4 dia	1982 Sep 25.7	82.62	115.94	7875	1493	1501	0.001	303
T	Intelsat 5 F-5 [Atlas Centaur]	1982-97A	1982 Sep 28.97 > million years	Box + dishes + 2 vanes 1972 full	15.7 span 6.4 wide	1982 Sep 29.0 1982 Nov 1.0	24.35 1.59	634.11 1435.8	24450 42159	167 35767	35976 35795	0.732 0.0003	175 -
	Intelsat 5 F-5 rocket	1982-97B	1982 Sep 28.97 1½ years	Cylinder 1815	8.6 long 3.0 dia	1982 Sep 28.9 1982 Sep 29.0	28.34 24.33	89.56 620.22	6631 24019	148 185	358 35241	0.016 0.728	143 175

	Name		Launch date, lifetime and descent date	Shape and weight (kg)	Size (m)	Date of orbital determination	Orbital inclination (deg)	Nodal period (min)	Semi major axis (km)	Perigee height (km)	Apogee height (km)	Orbital eccentricity	Argument of perigee (deg)
D R	Cosmos 1411	1982-98A	1982 Sep 30.50 13.8 days 1982 Oct 14.3	Sphere-cylinder 6300?	6.5 long? 2.4 dia	1982 Sep 30.7	72.86	90.08	6656	197	358	0.012	86
D	Cosmos 1411 rocket	1982-98B	1982 Sep 30.50 6 days 1982 Oct 6	Cylinder 2500?	7.5 long 2.6 dia	1982 Oct 1.2	72.84	89.87	6645	194	340	0.011	80
D	Cosmos 1411 engine	1982-98C	1982 Sep 30.50 23 days 1982 Oct 23	Cone 600? full	1.5 long? 2 dia?	1982 Oct 18.5	72.83	88.81	6592	205	223	0.001	66
D	Fragments	1982-98D,E											
	Cosmos 1412*	1982-99A	1982 Oct 2.00 600 years	Cone-cylinder	6 long? 2.5 dia?	1982 Oct 4.5 1982 Nov 10.9	65.00 64.81	89.66 103.95	6637 7324	251 909	266 983	0.001 0.005	271 12
D	Cosmos 1412 rocket	1982-99D	1982 Oct 2.00 43 days 1982 Nov 14	Cylinder 1500?	8 long? 2.5 dia?	1982 Nov 10.7	64.99	89.25	6616	231	245	0.001	225
D	Cosmos 1412 platform	1982-99B	1982 Oct 2.00 63 days 1982 Dec 4	Irregular	-	1982 Nov 13.9	65.00	89.44	6626	242	253	0.001	286
1d	Fragments**	1982-99C,E											
T	Cosmos 1413†	1982-100A	1982 Oct 12.63 1 million years	-	-	1982 Oct 18.2	64.83	673.32	25448	19069	19070	0	140
D	Cosmos 1413 launcher rocket	1982-100B	1982 Oct 12.63 2 days 1982 Oct 14	Cylinder 4000?	12 long? 4 dia	1982 Oct 13.1	51.65	88.16	6565	179	194	0.001	203

* Cosmos 1412 jettisoned rocket and platform before moving into 104 min orbit on Nov 10.

1982-100 continued on page 711

** Fragment 1982-99E is probably the uranium fuel core ejected from nuclear reactor of 1982-99A when in 104 min orbit.

† See footnote on page 711.

	Name		Launch date, lifetime and descent date	Shape and weight (kg)	Size (m)	Date of orbital determination	Orbital inclination (deg)	Nodal period (min)	Semi major axis (km)	Perigee height (km)	Apogee height (km)	Orbital eccentricity	Argument of perigee (deg)
D	Cosmos 1413 launcher	1982-100C	1982 Oct 12.63 1 day 1982 Oct 13	Irregular	-	1982 Oct 12.9	51.64	88.24	6569	183	198	0.001	194
	Cosmos 1413 rocket	1982-100F	1982 Oct 12.63 70 years	Cylinder 1900?	3.9 long? 3.9 dia	1982 Oct 18.6	51.97	339.43	16127	408	19090	0.579	0
T	Cosmos 1414†	1982-100D	1982 Oct 12.63 1 million years	-	-	1982 Oct 18.2	64.82	673.44	25451	19065	19080	0.0003	238
T	Cosmos 1415†	1982-100E	1982 Oct 12.63 1 million years	-	-	1982 Oct 18.2	64.81	673.50	25452	19069	19079	0.0002	171
	Cosmos 1413 apogee motor?	1982-100H	1982 Oct 12.63 1 million years	Cylinder 440?	2.0 long? 2.0 dia?	1982 Oct 18.2	64.81	672.89	25437	19051	19066	0.0003	340
	Fragment	1982-100G											
D R	Cosmos 1416	1982-101A	1982 Oct 14.39 13.9 days 1982 Oct 28.3	Sphere-cylinder 6300?	6.5 long? 2.4 dia	1982 Oct 18.2	70.36	89.61	6633	231	278	0.004	62
D	Cosmos 1416 rocket	1982-101B	1982 Oct 14.39 9 days 1982 Oct 23	Cylinder 2500?	7.5 long 2.6 dia	1982 Oct 18.6	70.36	89.52	6628	195	305	0.008	55
D	Cosmos 1416 engine	1982-101D	1982 Oct 14.39 25 days 1982 Nov 8	Cone 600? full	1.5 long? 2 dia?	1982 Oct 28.7	70.35	89.38	6621	222	264	0.003	26
D	Fragment	1982-101C											

† Triple GLONASS payload (GLObal NAvigation Satellite System)

	Name		Launch date, lifetime and descent date	Shape and weight (kg)	Size (m)	Date of orbital determination	Orbital inclination (deg)	Nodal period (min)	Semi major axis (km)	Perigee height (km)	Apogee height (km)	Orbital eccentricity	Argument of perigee (deg)
T	Cosmos 1417	1982-102A	1982 Oct 19.25 1200 years	Cylinder 700?	1.3 long? 1.9 dia?	1982 Oct 21.6	82.97	104.86	7365	962	1012	0.003	282
	Cosmos 1417 rocket	1982-102B	1982 Oct 19.25 600 years	Cylinder 2200?	7.4 long 2.4 dia	1982 Oct 21.2	82.96	104.75	7360	962	1001	0.003	279
T	Gorizont 6	1982-103A	1982 Oct 20.69 > million years	-	-	1982 Oct 21.7	0.84	1434.08	42127	35685	35813	0.002	294
D	Gorizont 6 launcher rocket	1982-103B	1982 Oct 20.69 2 days 1982 Oct 22	Cylinder 4000?	12 long? 4 dia	1982 Oct 21.2	51.62	88.15	6564	182	190	0.001	197
D	Gorizont 6 launcher	1982-103C	1982 Oct 20.69 1 day 1982 Oct 21	Irregular	-	1982 Oct 21.2	51.60	87.34	6524	143	148	0	86
	Gorizont 6 rocket	1982-103D	1982 Oct 20.69 11 months	Cylinder 1900?	3.9 long? 3.9 dia	1982 Oct 21.2	47.46	633.82	24445	321	35812	0.726	0
	Gorizont 6 apogee motor	1982-103E	1982 Oct 20.69 > million years	Cylinder 440?	2.0 long? 2.0 dia?	1982 Oct 25.6	0.83	1434.08	42127	35685	35813	0.002	294
	Fragment	1982-103F											
T	Cosmos 1418	1982-104A	1982 Oct 21.59 1 year	Octagonal ellipsoid? 550?	1.8 long? 1.5 dia?	1982 Oct 23.0	50.67	92.33	6770	371	413	0.003	13

1982-104 continued on page 713

	Name		Launch date, lifetime and descent date	Shape and weight (kg)	Size (m)	Date of orbital determination	Orbital inclination (deg)	Nodal period (min)	Semi major axis (km)	Perigee height (km)	Apogee height (km)	Orbital eccentricity	Argument of perigee (deg)
	Cosmos 1418 rocket	1982-104B	1982 Oct 21.59 5 months	Cylinder 2200?	7.4 long 2.4 dia	1982 Oct 23.0	50.67	92.22	6765	362	411	0.004	17
T	RCA Satcom 5	1982-105A	1982 Oct 28.06 > million years	Box + vanes 589	1.42 × 1.62 × 1.75	1982 Oct 29.6 1982 Oct 31.2	24.84 0.17	635.01 1450.6	24484 42448	184 35994	36027 36146	0.732 0.002	179 317
	RCA Satcom 5 second stage	1982-105B	1982 Oct 28.06 5 years	Cylinder + annulus 350?	6.4 long 1.52 and 2.44 dia	1982 Oct 28.9	26.62	132.14	8600	236	4208	0.231	185
	RCA Satcom 5 third stage	1982-105C	1982 Oct 28.06 7 years	Sphere -cone 66	1.32 long 0.94 dia	1982 Oct 28.1	24.93	633.75	24440	172	35952	0.732	178
T	DSCS 15 (Type 2)	1982-106A	1982 Oct 30.17? > million years	Cylinder + 2 dishes 590	1.83 long 2.74 dia	1982 Oct 31.9	2.47	1437.63	42197	35772	35865	0.001	36
T	DSCS 16 (Type 3)	1982-106B	1982 Oct 30.17? > million years	Box + 2 vanes 1043	-	1982 Oct 31.9	2.48	1439.82	42239	35845	35877	0	101
D	IUS* lower section	1982-106C	1982 Oct 30.17? 2 months 1982 Dec?	Cylinder	3.6 long? 2.9 dia	1982 Oct 31.5	28.01	630.29	24362	130	35837	0.733	1
	IUS upper section	1982-106D	1982 Oct 30.17? > million years	Cylinder	1.4 long? 2.9 dia	1982 Nov 9.0	2.47	1448.98	42420	35863	36221	0.004	183

* First use of Inertial Upper Stage on new Titan 34D booster.

	Name		Launch date, lifetime and descent date	Shape and weight (kg)	Size (m)	Date of orbital determination	Orbital inclination (deg)	Nodal period (min)	Semi major axis (km)	Perigee height (km)	Apogee height (km)	Orbital eccentricity	Argument of perigee (deg)
D	Progress 16*	1982-107A	1982 Oct 31.47 44.2 days 1982 Dec 14.7	Sphere-cylinder 7020	7.9 long 2.3 dia	1982 Nov 1.6	51.62	90.94	6702	290	358	0.005	193
D	Progress 16 rocket	1982-107B	1982 Oct 31.47 2 days 1982 Nov 2	Cylinder 2500?	7.5 long 2.6 dia	1982 Oct 31.8	51.62	88.60	6586	183	233	0.004	96
D R	Cosmos 1419	1982-108A	1982 Nov 2.40 13.9 days 1982 Nov 16.3	Sphere-cylinder 6300?	6.5 long? 2.4 dia	1982 Nov 4.4	70.34	89.63	6634	230	282	0.004	78
D	Cosmos 1419 rocket	1982-108B	1982 Nov 2.40 4 days 1982 Nov 6	Cylinder 2500?	7.5 long 2.6 dia	1982 Nov 2.5	70.34	89.18	6611	209	257	0.004	54
D	Cosmos 1419 engine	1982-108C	1982 Nov 2.40 19 days 1982 Nov 21	Cone 600? full	1.5 long? 2 dia?	1982 Nov 16.8	70.34	88.99	6602	212	235	0.002	113
D	Fragments	1982-108D-F											
T	Cosmos 1420	1982-109A	1982 Nov 11.26 120 years	Cylinder + paddles? 750?	2 long? 1 dia?	1982 Nov 13.3	73.99	100.80	7174	780	811	0.002	340
	Cosmos 1420 rocket	1982-109B	1982 Nov 11.26 100 years	Cylinder 2200?	7.4 long 2.4 dia	1982 Nov 12.2	74.00	100.70	7169	774	808	0.002	359
D 4M R	STS 5** (Columbia F5)	1982-110A	1982 Nov 11.51 5.10 days 1982 Nov 16.61	Cylinder + delta wing 68800 empty	37.1 long 17.4 high 23.8 span	1982 Nov 13.3	28.47	90.48	6684	294	317	0.002	240

* Progress 16 docked with Soyuz T7 - Salyut 7 (2nd airlock) on 1982 Nov 2.56. Undocked 1982 Dec 13.65 and de-orbited over the Pacific Ocean.

** Launched from Cape Canaveral; landed at Edwards AFB, California.

1982-110 continued on page 715

	Name		Launch date, lifetime and descent date	Shape and weight (kg)	Size (m)	Date of orbital determination	Orbital inclination (deg)	Nodal period (min)	Semi major axis (km)	Perigee height (km)	Apogee height (km)	Orbital eccentricity	Argument of perigee (deg)
T	SBS 3*	1982-110B	1982 Nov 11.51 >million years	Cylinder + panel 1117 full 571 empty	2.82 long 2.16 dia	1982 Nov 13.9	0.60	1451.28	42463	34834	37336	0.029	160
T	Telesat 6** (Anik C3)	1982-110C	1982 Nov 11.51 >million years	Cylinder + panel 1238 full 660 empty	2.82 to 6.7 long 2.16 dia	1982 Nov 13.2 1982 Nov 19.2	23.30 0.06	661.54 1476.57	25161 42956	274 35863	37292 37293	0.736 0.017	358 27
	SBS 3 rocket (PAM)	1982-110D	1982 Nov 11.51 20 years	Cylinder 2154 (incl. fuel)	2.29 long 1.5 dia	1982 Nov 17.4	23.69	663.10	25200	318	37326	0.734	3
	Telesat 6 rocket (PAM)	1982-110E	1982 Nov 11.51 20 years	Cylinder 2154 (incl. fuel)	2.29 long 1.5 dia	1982 Nov 23.5	23.48	662.54	25186	331	37285	0.734	5
T	[Titan 3D]	1982-111A	1982 Nov 17.89 manoeuvrable	Cylinder 13300? full	15 long 3.0 dia	1982 Nov 20.4	96.97	92.59	6778	280	520	0.018	150
D	Titan 3D rocket	1982-111B	1982 Nov 17.89 40 days 1982 Dec 27	Cylinder 1900	6 long 3.0 dia	1982 Nov 20.3	96.97	92.00	6749	237	505	0.020	153
D R	Cosmos 1421	1982-112A	1982 Nov 18.40 13.9 days 1982 Dec 2.3	Sphere-cylinder 6300?	6.5 long? 2.4 dia	1982 Nov 20.3	70.34	89.64	6635	231	282	0.004	81
D	Cosmos 1421 rocket	1982-112B	1982 Nov 18.40 4 days 1982 Nov 22	Cylinder 2500?	7.5 long 2.6 dia	1982 Nov 18.5	70.35	89.14	6609	209	253	0.003	59

* Ejected from Shuttle on 1982 Nov 11.85; entered synchronous orbit 1982 Nov 13.94.
** Ejected from Shuttle on 1982 Nov 12.91; entered synchronous orbit 1982 Nov 16.

1982-112 continued on page 716

	Name		Launch date, lifetime and descent date	Shape and weight (kg)	Size (m)	Date of orbital determination	Orbital inclination (deg)	Nodal period (min)	Semi major axis (km)	Perigee height (km)	Apogee height (km)	Orbital eccentricity	Argument of perigee (deg)
D	Cosmos 1421 engine	1982-112F	1982 Nov 18.40 26 days 1982 Dec 14	Cone 600? full	1.5 long? 2 dia?	1982 Dec 4.4	70.34	89.29	6617	215	262	0.003	66
D	Fragments	1982-112C-E											
T	Statsionar-Raduga 11	1982-113A	1982 Nov 26.60 > million years	-	-	1982 Nov 26.8	1.24	1468.59	42800	36068	36776	0.008	306
D	Raduga 11 launcher rocket	1982-113B	1982 Nov 26.60 1 day 1982 Nov 27	Cylinder 4000?	12 long? 4 dia	1982 Nov 26.8	51.60	88.17	6565	182	192	0.001	105
D	Raduga 11 launcher	1982-113C	1982 Nov 26.60 2 days 1982 Nov 28	Irregular	-	1982 Nov 26.8	51.61	88.23	6568	186	194	0.001	219
	Raduga 11 rocket	1982-113D	1982 Nov 26.60 9 months	Cylinder 1900?	3.9 long? 3.9 dia	1982 Dec 1.2	47.32	649.92	24856	280	36676	0.732	1
	Raduga 11 apogee motor	1982-113F	1982 Nov 26.60 > million years	Cylinder 440?	2.0 long? 2.0 dia?		Synchronous orbit similar to 1982-113A						
	Fragment	1982-113E											
D R	Cosmos 1422	1982-114A	1982 Dec 3.50 13.8 days 1982 Dec 17.3	Sphere-cylinder 6300?	6.5 long? 2.4 dia	1982 Dec 5.6	72.85	89.68	6636	228	288	0.005	76
D	Cosmos 1422 rocket	1982-114B	1982 Dec 3.50 4 days 1982 Dec 7	Cylinder 2500?	7.5 long 2.6 dia	1982 Dec 3.6	72.85	89.26	6615	196	278	0.006	91

1982-114 continued on page 717

	Name		Launch date, lifetime and descent date	Shape and weight (kg)	Size (m)	Date of orbital determination	Orbital inclination (deg)	Nodal period (min)	Semi major axis (km)	Perigee height (km)	Apogee height (km)	Orbital eccentricity	Argument of perigee (deg)
D	Cosmos 1422 engine	1982-114D	1982 Dec 3.50 26 days 1982 Dec 29	Cone 600? full	1.5 long? 2 dia?	1982 Dec 18.3	72.85	89.30	6617	215	263	0.004	45
D	Fragments	1982-114C,E,F											
	Cosmos 1423*	1982-115A	1982 Dec 8.58 5 years	Windmill + cylinder 6050 full?	6.2 long? 1.6 and 2.0 dia	1982 Dec 12.4	62.83	93.77	6838	405	515	0.008	357
D	Cosmos 1423 launcher rocket	1982-115B	1982 Dec 8.58 19 days 1982 Dec 27	Cylinder 2500?	7.5 long 2.6 dia	1982 Dec 9.2	62.94	91.08	6707	232	425	0.014	55
D	Cosmos 1423 launcher	1982-115D	1982 Dec 8.58 14 days 1982 Dec 22	Irregular	-	1982 Dec 10.3	62.89	95.46	6918	471	609	0.010	22
7d	Fragments	1982-115C,E-AC											
T	Meteor 2-09	1982-116A	1982 Dec 14.94 500 years	Cylinder + 2 vanes 2750?	5 long? 1.5 dia?	1982 Dec 15.7	81.25	101.99	7230	812	892	0.006	274
	Meteor 2-09 rocket	1982-116B	1982 Dec 14.94 400 years	Cylinder 1440	3.8 long 2.6 dia	1982 Dec 16.4	81.26	102.04	7233	804	905	0.007	230
	Fragment	1982-116C											
T	Cosmos 1424	1982-117A	1982 Dec 16.42 43 days	Sphere- cylinder 6700?	7 long? 2.4 dia	1982 Dec 16.7	64.90	89.70	6638	171	349	0.013	64
D	Cosmos 1424 rocket	1982-117B	1982 Dec 16.42 3 days 1982 Dec 19	Cylinder 2500?	7.5 long 2.6 dia	1982 Dec 16.7	64.91	89.57	6632	171	336	0.012	63
T	DMSP 2-01 [Atlas Burner 2]?	1982-118A	1982 Dec 21.11 80 years	Irregular 751	6.40 long? 1.68 dia?	1982 Dec 22.6	98.72	101.36	7200	816	827	0.001	277

1982-118 continued on page 718

* Possible Molniya or early-warning Cosmos, where final-stage rocket disintegrated during firing.

	Name	Launch date, lifetime and descent date	Shape and weight (kg)	Size (m)	Date of orbital determination	Orbital inclination (deg)	Nodal period (min)	Semi major axis (km)	Perigee height (km)	Apogee height (km)	Orbital eccentricity	Argument of perigee (deg)
	Burner 2 rocket? 1982-118B	1982 Dec 21.11 60 years	Sphere - cone 66	1.32 long 0.94 dia	1982 Dec 23.3	98.63	101.60	7212	797	870	0.005	114
	Fragment 1982-118C											
T	Cosmos 1425 1982-119A	1982 Dec 23.38 13.9 days	Sphere-cylinder 6300?	6.5 long? 2.4 dia	1982 Dec 24.6	69.97	92.20	6760	348	416	0.005	214
	Cosmos 1425 rocket 1982-119B	1982 Dec 23.38 14 days	Cylinder 2500?	7.5 long 2.6 dia	1982 Dec 23.9	69.97	90.23	6664	231	340	0.008	59
	Cosmos 1425 engine 1982-119E	1982 Dec 23.38 1 year	Cone 600? full	1.5 long? 2 dia?	1983 Jan 6.6	69.97	92.15	6759	347	415	0.005	193
2d	Fragments 1982-119C,D,F,G											
T	Cosmos 1426 1982-120A	1982 Dec 28.50 manoeuvrable	Sphere-cylinder 6700?	7 long? 2.4 dia	1983 Jan 3.5	50.53	90.08	6661	209	356	0.011	103
	Cosmos 1426 rocket 1982-120B	1982 Dec 28.50 10 days	Cylinder 2500?	7.5 long 2.6 dia	1983 Jan 3.6	50.52	89.17	6615	190	284	0.007	113
	Fragments 1982-120C,D											
T	Cosmos 1427 1982-121A	1982 Dec 29.50 6 years	Cylinder?	4 long? 2 dia?	1983 Jan 3.3	65.84	94.03	6850	445	499	0.004	335
	Cosmos 1427 rocket 1982-121B	1982 Dec 29.50 4 years	Cylinder 2200?	7.4 long 2.4 dia	1982 Dec 29.7	65.85	93.93	6847	441	496	0.004	347

B

C

* These Cosmos satellites ejected a capsule

* These Cosmos satellites ejected a capsule

* These Cosmos satellites ejected a capsule

* These Cosmos satellites ejected a capsule

* These Cosmos satellites ejected a capsule

* These Cosmos satellites ejected a capsule

* These Cosmos satellites ejected a capsule

* These Cosmos satellites ejected a capsule

* These Cosmos satellites ejected a capsule

* These Cosmos satellites ejected a capsule

* These Cosmos satellites ejected a capsule

F

G

H

I

P

R

* Subgroups: (C) Calibration and diagnostic; (I) Interceptor; (T) Target.

* Subgroups: (C) Calibration and diagnostic; (I) Interceptor; (T) Target

*only [Thor Agena D] satellites with inital periods of about 95 minutes are indexed.